D1628059

Clinical Biochemistry

TRANSFUSION & TRANSPLANTATION SCIENCE

BIOMEDICAL SCIENCE PRACTICE

CYTOPATHOLOGY

CLINICAL BIOCHEMISTRY

IMMUNOLOGY

HAEMATOLOGY

MEDICAL MICROBIOLOGY

BIOLOGY OF DISEASE

HISTOPATHOLOGY

fundamentals OF **biomedical science**

Clinical Biochemistry

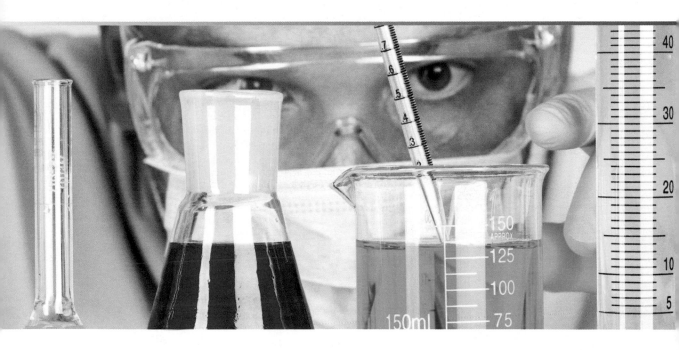

Edited by

Dr Nessar Ahmed
PhD, CSci, FIBMS
Reader in Clinical Biochemistry
School of Healthcare Science
Manchester Metropolitan University

OXFORD
UNIVERSITY PRESS

OXFORD
UNIVERSITY PRESS

Great Clarendon Street, Oxford OX2 6DP

Oxford University Press is a department of the University of Oxford.
It furthers the University's objective of excellence in research, scholarship,
and education by publishing worldwide in

Oxford New York

Auckland Cape Town Dar es Salaam Hong Kong Karachi
Kuala Lumpur Madrid Melbourne Mexico City Nairobi
New Delhi Shanghai Taipei Toronto

With offices in

Argentina Austria Brazil Chile Czech Republic France Greece
Guatemala Hungary Italy Japan Poland Portugal Singapore
South Korea Switzerland Thailand Turkey Ukraine Vietnam

Oxford is a registered trade mark of Oxford University Press
in the UK and in certain other countries

Published in the United States
by Oxford University Press Inc., New York

British Library Cataloguing in Publication Data

Data available

Library of Congress Cataloging in Publication Data

Data available

Typeset by MPS Limited, a Macmillan Company
Printed in Italy on acid-free paper by
L.E.G.O. S.p.A. – Lavis TN

ISBN 978-0-19-953393-0

3 5 7 9 10 8 6 4 2

*This book is dedicated to the one I love the most,
my daughter Neha Ahmed.*

Dr Nessar Ahmed
24 July 2010

Contents

Preface

Modern medicine is dependent on laboratory investigation of disease in order to confirm diagnosis, monitor treatment, and for screening and prognostic purposes. *Clinical Biochemistry* has been written with this in mind and is concerned with the biochemical basis of disease processes and their laboratory investigation.

This textbook should meet the needs of students studying modules in clinical biochemistry on BSc and MSc programmes in biomedical science or biochemistry. The topics covered are suitable for graduates entering the hospital clinical biochemistry service as trainee biomedical or clinical scientists. The book will be of value to practising biomedical scientists preparing for professional diploma examinations that lead to Fellowship of the Institute of Biomedical Science (FIBMS). It will also provide useful initial reading for clinical scientists and medical graduates studying towards postgraduate examinations in chemical pathology such as that leading to Fellowship of the Royal College of Pathologists (FRCPath).

I would like to acknowledge my colleague Dr Chris Smith for his friendship and guidance over the years. I am also grateful to all the contributors to this book for their co-operation and advice, in particular Dr Roy Sherwood for proofreading all the chapters. Mick Hoult deserves a special mention for his help with the illustrations. The preparation of *Clinical Biochemistry* has involved three years of hard work whilst trying to juggle my main academic activities of research, teaching, and programme management. However, despite this challenge, it has been a rewarding experience and well worth the effort as this book is dedicated to my daughter Neha.

Dr Nessar Ahmed, PhD, CSci, FIBMS
Reader in Clinical Biochemistry
Manchester Metropolitan University

An introduction to the Fundamentals of Biomedical Science series

Biomedical Scientists form the foundation of modern healthcare, from cancer screening to diagnosing HIV, from blood transfusion for surgery to food poisoning and infection control. Without Biomedical Scientists, the diagnosis of disease, the evaluation of the effectiveness of treatment, and research into the causes and cures of disease would not be possible.

However, the path to becoming a Biomedical Scientist is a challenging one: trainees must not only assimilate knowledge from a range of disciplines, but must understand—and demonstrate—how to apply this knowledge in a practical, hands-on environment.

The *Fundamentals of Biomedical Science* series is written to reflect the challenges of biomedical science education and training today. It blends essential basic science with insights into laboratory practice to show how an understanding of the biology of disease is coupled to the analytical approaches that lead to diagnosis.

The series provides coverage of the full range of disciplines to which a Biomedical Scientist may be exposed – from microbiology to cytopathology to transfusion science. Alongside volumes exploring specific biomedical themes and related laboratory diagnosis, an overarching Biomedical Science Practice volume provides a grounding in the general professional and experimental skills with which every Biomedical Scientist should be equipped.

Produced in collaboration with the Institute of Biomedical Science, the series

- Understands the complex roles of Biomedical Scientists in the modern practice of medicine.

- Understands the development needs of employers and the Profession.

- Places the theoretical aspects of biomedical science in their practical context.

Learning from this series

The *Fundamentals of Biomedical Science* series draws on a range of learning features to help readers master both biomedical science theory, and biomedical science practice.

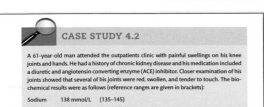

Case studies illustrate how the biomedical science theory and practice presented throughout the series relates to situations and experiences that are likely to be encountered routinely in the biomedical science laboratory. Answers to questions posed in each case study are available at: www.oxfordtextbooks.co.uk/orc/ahmed

Additional information to augment the main text appears in **boxes**.

> **BOX 5.1 Pseudohyponatraemia**
>
> Pseudohyponatraemia is a value for the measurement of sodium that is below the reference range (135 to 145 mmol/L) that has an artefactual cause. It is associated with increased amounts of lipid or protein in the specimen of plasma and with analytical techniques in which the sample is diluted prior to sodium analysis. These techniques include the use of indirect ISEs and, historically, flame photometers. Comparing sodium results from direct and indirect reading electrodes in specimens with increased lipid

Further features are used to help consolidate and extend students' understanding of the subject

Key points reinforce the key concepts that the reader should master from having read the material presented, while **Summary** points act as an end-of-chapter checklists for readers to verify that they have remembered correctly the principal themes and ideas presented within each chapter.

> using plasma for analysis is one of speed. Serum specimens require approximately 30 minutes for clot formation prior to centrifugation.
>
> **Key Points**
>
> Haemolysis is the destruction of red blood cells and can cause leakage or release of cellular contents into the serum, rendering them unsuitable for analysis of certain analytes. Haemolysis of blood samples often occurs due to their unsatisfactory collection from subjects or their storage.

Self-check questions throughout each chapter and extended questions at the end of each chapter provide the reader with a ready means of checking that they have understood the material they have just encountered. Answers to these questions are provided in the book's Online Resource Centre; visit www.oxfordtextbooks.co.uk/orc/ahmed

> acid, is excreted in urine and plasma uric acid concentrations are decreased. Puricase (PEG uricase) is a formulation containing the polymer, polyethylene glycol covalently attached to a genetically engineered (recombinant) form of urate oxidase. It has been shown to dramatically reduce the concentrations of uric acid in plasma with no adverse effects and is currently in phase III clinical trials. Present studies have shown puricase to be beneficial to patients with hyperuricaemia for whom conventional therapy is contraindicated or has been less effective. The medications used in treating gout together with their common side effects are listed in Table 4.4.
>
> **SELF-CHECK 4.5**
>
> What is the main treatment for acute gout?

Cross references help the reader to see biomedical science as a unified discipline, making connections between topics presented within each volume, and across all volumes in the series.

> tical error observed during routine laboratory analysis. It arises
> erythrocytes, which contain a higher concentration of potas-
> they circulate. Thus, when there has been a degree of trauma
> cimens have been handled carelessly between the time of col-
> example by over vigorous mixing, then lysis of the erythrocytes
> ad to a falsely high value for the observed potassium plasma
> then the effect is evident in the pink colouration imparted to
> emoglobin, which is also released from the lysed erythrocytes.
> ify haemolysis, but the process can also be automated on gen-
> rs with a channel set up to measure haemolysis at a suitable
>
> **Cross reference**
> Chapter 1 Biochemical investigations and quality control and Chapter 2 Automation

Online learning materials

online resource centre

The *Fundamentals of Biomedical Science* series doesn't end with the printed books. Each title in the series is supported by an Online Resource Centre, which features additional materials for students, trainees, and lecturers.

www.oxfordtextbooks.co.uk/orc/fbs

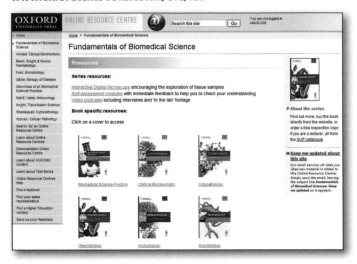

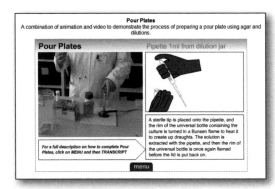

Guides to key experimental skills and methods

Multimedia walk-throughs of key experimental skills—including both animations and video—to help you master the essential skills that are the foundation of Biomedical Science practice.

Biomedical science in practice

Interviews with practising Biomedical Scientists working in a range of disciplines, to give you valuable insights into the reality of work in a Biomedical Science laboratory.

Digital Microscope

A library of microscopic images for you to investigate using this powerful online microscope, to help you gain a deeper appreciation of cell and tissue morphology.

The Digital Microscope is used under licence from the Open University.

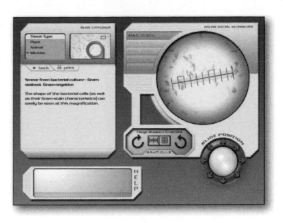

'Check your understanding' learning modules

A mix of interactive tasks and questions, which address a variety of topics explored throughout the series. Complete these modules to help you check that you have fully mastered all the key concepts and key ideas that are central to becoming a proficient Biomedical Scientist.

We extend our grateful thanks to colleagues in the School of Health Science at London Metropolitan University for their invaluable help in developing these online learning materials.

Answers to self-check, case study, and end-of-chapter questions

Answers to questions posed in the book are provided to aid self-assessment.

Lecturer support materials

The Online Resource Centre for each title in the series also features figures from the book in electronic format, for registered adopters to download for use in lecture presentations, and other educational resources.

To register as an adopter visit **www.oxfordtextbooks.co.uk/orc/ahmed** and follow the on-screen instructions.

Any comments?

We welcome comments and feedback about any aspect of this series. Just visit **www.oxfortextbooks.co.uk/orc/feedback/** and share your views.

Contributors

Dr Joanne Adaway
Department of Biochemistry, The Christie Hospital,
Manchester

Dr Nessar Ahmed
School of Healthcare Science, Manchester
Metropolitan University

Dr Gwendolen Ayers
Department of Clinical Biochemistry, Manchester Royal
Infirmary

Dr Farhad Behzad
School of Translational Medicine, University of
Manchester

Dr Nigel Brown
Institute of Liver Studies, King's College Hospital, London

Dr Gordon Brydon
Department of Clinical Biochemistry, Western General
Hospital, Edinburgh

David Cameron
Department of Biochemistry, Glasgow Royal Infirmary

Dr Paul Collinson
Department of Chemical Pathology, St George's Hospital,
London

Professor Robert Flanagan
Toxicology Unit, King's College Hospital, London

Dr Mike France
Department of Clinical Biochemistry, Manchester Royal
Infirmary

Dr John Honour
Department of Clinical Biochemistry, University College
Hospital, London

Dr Tim James
Department of Clinical Biochemistry, John Radcliffe
Hospital, Oxford

Dr Ian Laing
Department of Clinical Biochemistry, Manchester Royal
Infirmary

Dr Edmund Lamb
Department of Clinical Biochemistry, Kent and
Canterbury Hospital

Amy Lloyd
Oriel College, University of Oxford

Dr Gerald Maguire
Department of Clinical Biochemistry and Immunology,
Addenbrooke's Hospital, Cambridge

Dr Joanne Marsden
Department of Clinical Biochemistry, King's College
Hospital, London

Dr Garry McDowell
Faculty of Health, Edge Hill University, Ormskirk

Mary Anne Preece
Department of Clinical Chemistry, Birmingham Children's
Hospital

Walter Reid
Department of Clinical Biochemistry, Wythenshawe
Hospital, Manchester

Dr Roy Sherwood
Department of Clinical Biochemistry, King's College
Hospital, London

Dr William Simpson
Department of Clinical Biochemistry, Aberdeen Royal
Infirmary

Dr Julie Thornton
Department of Biomedical Sciences, University of Bradford

Dr David Tierney
Department of Clinical Biochemistry, Wythenshawe
Hospital, Manchester

Dr Pat Twomey
Department of Clinical Biochemistry,
The Ipswich Hospital

Dr Robin Whelpton
School of Biological and Chemical Sciences,
Queen Mary University of London

Gilbert Wieringa
Department of Biochemistry, The Christie Hospital, Manchester

Dr Allen Yates
Department of Clinical Biochemistry, Manchester Royal Infirmary

Online materials developed by

Sheelagh Heugh,
Faculty of Human Sciences,
London Metropolitan University

Dr Ken Hudson,
Faculty of Human Sciences,
London Metropolitan University

William Armour,
Faculty of Human Sciences,
London Metropolitan University

George Worthington,
Faculty of Human Sciences,
London Metropolitan University

Abbreviations

A	adenine
AAS	atomic absorption spectroscopy
AAT	α_1-antitrypsin
ABC	ATP binding cassette
ABCA1	ATP binding cassette type A1
ABP	androgen binding protein
ABV	alcohol by volume
ACA	automated clinical analyser
ACE	angiotensin converting enzyme
ACS	acute coronary syndrome
ACTH	adrenocorticotrophic hormone
ADH	antidiuretic hormone
ADP	adenosine diphosphate
ADPKD	autosomal dominant polycystic kidney disease
AFP	α-fetoprotein
AGE	advanced glycation endproduct
AIDS	acquired immunodeficiency syndrome
AIIRA	angiotensin II receptor antagonist
AIN	acute interstitial nephritis
AIP	acute intermittent porphyria
AKI	acute kidney injury
ALA	aminolevulinic acid
ALAD	ALA dehydratase deficiency
ALB	albumin
ALD	adrenoleukodystrophy
ALF	acute liver failure
ALP	alkaline phosphatase
ALS	acid labile subunit
ALT	alanine aminotransferases
AMH	anti-Mullerian hormone
AMI	acute myocardial infarction
AMP	adenosine monophosphate
ANP	atrial natriuretic peptide
Apo	apolipoprotein
APR	acute phase response
APRT	amidophosphoribosyl transferase (also called glutamine phosphoribosyl amidotransferase)
ARB	angiotensin receptor blocker
ARF	acute renal failure
ARH	autosomal recessive hypercholesterolaemia
ARPKD	autosomal recessive polycystic kidney disease
ARR	aldosterone to renin ratio
ASBT	apical sodium dependent bile acid transporter
ASP	acylation stimulatory protein
ASRM	American Society for Reproductive Medicine
AST	aspartate aminotransferase
ATN	acute tubular necrosis
A_{Tot}	total of non-volatile weak acids
ATP	adenosine triphosphate
AUC	area under the curve
AVP	arginine vasopressin
B_{48}	apoprotein B_{48}
BAPEN	British Association for Parenteral and Enteral Nutrition
BA	bile acid
BCG	bromocresol green
BCP	bromocresol purple
BEC	benzethonium chloride
BIL	bilirubin
BJP	Bence-Jones protein
BMI	body mass index
BNF	British National Formulary
BNP	B-type natriuretic peptide
BPH	benign prostatic hypertrophy
BSA	body surface area
BT-PABA	N benzoyl-L-tyrosyl para aminobenzoic acid
CA	carbohydrate antigen
CAH	congenital adrenal hyperplasia
cAMP	cyclic adenosine monophosphate
CARDS	Collaborative Atorvastatin in Diabetes Study
CBB	Coomassie brilliant blue

CBG	cortisol binding globulin	**DDAVP**	desmopressin
CBS	cystathionine-β-synthase	**DDT**	dichlorodiphenyltrichloroethane
CD36	cluster of differentiation 36	**DGAT**	diacylglycerol acyl transferase
CDC	Centers for Disease Control	**1,25-DHCC**	1,25-dihydroxycholecalciferol
CDG	congenital disorders of glycosylation	**DHEA**	dehydroepiandrosterone
CE	cholesterol ester	**DHEAS**	dehydroepiandrosterone sulphate
CEA	carcinoembryonic antigen	**DHT**	5α-dihydrotestosterone
CEDIA	cloned enzyme donor immunoassay	**DI**	diabetes insipidus
CEP	congenital erythropoietic porphyria	**DKA**	diabetic ketoacidosis
CETP	cholesterol ester transfer protein	**DLIS**	digoxin-like immunoreactive substances
CF	cystic fibrosis	**DNA**	deoxyribonucleic acid
CFTR	cystic fibrosis transmembrane conductance regulator	**DOPA**	dihydroxyphenylalanine
CHI	congenital hyperinsulinism	**DPP IV**	dipeptidyl peptidase IV
CK	creatine kinase	**DTPA**	diethylenetriaminepentaacetic acid
CKD	chronic kidney disease	**EA**	enzyme acceptor
CK-PZ	cholecystokinin-pancreozymin	**EBV**	Epstein-Barr virus
CML	chronic myeloid leukaemia	**ECD**	extracellular domain
CMV	cytomegalovirus	**ECF**	extracellular fluid
CNS	central nervous system	**ECG**	electrocardiogram
COAD	chronic obstructive airways disease	**ED**	enzyme donor
COHb	carboxyhaemoglobin	**EDIC**	Epidemiology of Diabetes Intervention and Complications study
COS	controlled ovarian stimulation	**EDTA**	ethylenediaminetetraacetic acid
COX	cycloxygenase	**EGF**	epidermal growth factor
CPA	Clinical Pathology Accreditation (UK) Ltd	**eGFR**	estimated glomerular filtration rate
CPR	chlorophenol red	**EHBA**	extrahepatic biliary atresia
CPS	carbamyl phosphate synthase	**EHSRE**	European Society for Human Reproduction and Embryology
CRH	corticotrophin releasing hormone		
CRP	C-reactive protein	**ELISA**	enzyme-linked immunosorbent assay
CSF	cerebrospinal fluid	**EMIT**	enzyme-multiplied immunoassay technique
CTx	C-telopeptide of type 1 collagen	**EMU**	early morning urine
cTnI	cardiac troponin I	**EP**	electrophoresis
cTnT	cardiac troponin T	**EPO**	erythropoietin
CV	coefficient of variation	**EPP**	erythropoietic porphyria
CVA	cerebrovascular accident	**EQA**	external quality assurance
CVID	common variable immunodeficiency	**ERCP**	endoscopic retrograde cholangiopancreatography
CYP450	cytochrome P450		
CZE	capillary zone electrophoresis	**ERF**	established renal failure
DAG	diacylglycerol	**ESR**	erythrocyte sedimentation rate
DCCT	Diabetes Control and Complications Trial	**ESRD**	end-stage renal disease

FA	fatty acid		**GLUT 2**	glucose transporter (type 2)
FAI	free androgen index		**GLUT 4**	glucose transporter (type 4)
FAL	fumaryl acetoacetate lyase		**GMP**	guanosine monophosphate
FAOD	fatty acid oxidation disorder		**GnRH**	gonadotrophin releasing hormone
FC	free cholesterol		**GP**	general practitioner
FCH	familial combined hyperlipidaemia		**GSD**	glycogen storage disease
FDA	Federal Drug Administration		**GSH**	glutathione
FDB	familial defective apoprotein B		**GSH-PX**	glutathione peroxidase
FH	familial hypercholesterolaemia		**GTP**	guanosine triphosphate
FHBL	familial hypobetalipoproteinaemia		**Hb**	haemoglobin
FID	flame ionization detector		**HbA$_1$c**	haemoglobin A$_1$c
FISH	fluorescence *in situ* hybridization		**HBD**	hydroxybutyrate dehydrogenase
FN	false negative		**HCC**	hepatocellular carcinoma
FOG	faecal osmotic gap		**25-HCC**	25-hydroxycholecalciferol
FP	false positive		**hCG**	human chorionic gonadotrophin
FPIA	fluorescence polarization immunoassay		**7-HCO**	7α-hydroxycholestenone
Free T4	free thyroxine		**HCP**	hereditary coproporphyria
FSH	follicle stimulating hormone		**HDL**	high density lipoprotein
G	guanine		**HDL-C**	high density lipoprotein cholesterol
G6P	glucose-6-phosphatase		**HEP**	hepatoerythropoietic porphyria
G6PD	glucose-6-phosphate dehydrogenase		**HFI**	hereditary fructose intolerance
GAD	glutamic acid decarboxylase		**HGPRT**	hypoxanthine-guanine phosphoribosyltransferase
GALT	gut associated lymphoid tissue		**HHS**	hyperosmolar hyperglycaemic syndrome
GALT	galactose-1-phosphate uridyl transferase		**5-HIAA**	5-hydroxyindole acetic acid
GC	gas chromatography		**HIS**	hospital information system
GC-MS	gas chromatography-mass spectrometry		**HIV**	human immunodeficiency virus
GCS	Glasgow Coma Score		**HLA**	human leukocyte antigen
GDM	gestational diabetes mellitus		**HMG CoA**	hydroxymethylglutaryl coenzyme A
GFR	glomerular filtration rate		**HNF**	hepatocyte nuclear factor
GGT	gamma-glutamyl transferase		**HoloTC**	holotranscobalamin
GH	growth hormone		**HOMA**	homeostasis model assessment
GHB	gamma-hydroxybutyrate		**HONK**	hyperosmolar non-ketotic state
GHRH	growth hormone releasing hormone		**HPA**	hypothalamic-pituitary-adrenal
GIP	glucose-dependent insulinotrophic peptide		**HPLC**	high performance liquid chromatography
GIT	gastrointestinal tract		**HPRT**	hypoxanthine guanine phosphoribosyl transferase
GLC	gas liquid chromatography			
GLP-1	glucagon-like peptide 1		**HRT**	hormone replacement therapy
GLP1-R	glucagon-like peptide 1 receptor		**HSL**	hormone sensitive lipase

5-HT	5-hydroxytryptamine
i.m.	intramuscular
i.v.	intravenous
IBD	irritable bowel disease
IBMS	Institute of Biomedical Science
IBS	irritable bowel syndrome
ICF	intracellular fluid
ICP-AES	inductively coupled plasma atomic emission spectroscopy
ICPMS	induction-coupled plasma mass spectrometry
ICSI	intracytoplasmic sperm injection
IDL	intermediate density lipoprotein
ID-MS	isotope dilution mass spectrometry
IEF	isoelectric focusing
IF	intrinsic factor
IFCC	International Federation of Clinical Chemistry
IFG	impaired fasting glycaemia
Ig	immunoglobulin
IGF-1	insulin-like growth factor-1
IGF-2	insulin-like growth factor-2
IGFBP3	insulin-like growth factor binding protein 3
IGT	impaired glucose tolerance
IL-6	interleukin 6
IMD	inherited metabolic disorder
IMP	inosine 5′ monophosphate
INR	international normalized ratio
IQC	internal quality control
IRMA	immunoradiometric assay
IRT	immunoreactive trypsin
ISE	ion-selective electrode
ITU	intensive therapy unit
IVF	*in vitro* fertilization
JBS2	Joint British Societies guidelines 2
JGA	juxtaglomerular apparatus
K_{ATP}	ATP sensitive potassium channel
kPa	kilopascals
LAGB	laparascopic adjustable gastric banding
LC	liquid chromatography
LCAT	lecithin cholesterol acyl transferase
LC-MS	liquid chromatography-mass spectrometry
LC-MS/MS	liquid chromatography-mass spectrometry/mass spectrometry
LD	lactate dehydrogenase
LDL	low density lipoprotein
LDLR	low density lipoprotein receptor
LFT	liver function test
LH	luteinizing hormone
LHRH	luteinizing hormone releasing hormone
LIS	laboratory information system
Lp(a)	lipoprotein(a)
LPH	lipotrophin
Lp-X	lipoprotein-X
LRP	low density lipoprotein receptor-related protein
LSD	lysergic acid diethylamide
LSDs	lysosomal storage disorders
MCAD	medium chain acyl CoA dehydrogenase
MCADD	medium chain acyl Co A dehydrogenase deficiency
MDMA	methylenedioxymethamphetamine
MDRD	modification of diet in renal disease
MEN	multiple endocrine neoplasia
MFOs	mixed function oxidases
MGUS	monoclonal gammopathy of undetermined significance
MHC	major histocompatability complex
MI	myocardial infarction
MMA	methylmalonic aciduria
MODY	maturity onset diabetes of the young
MPA	mycophenolic acid
MR	magnetic resonance
MRS	magnetic resonance spectroscopy
MS	mass spectrometry
MSH	melanocyte stimulating hormone
MS-MS	tandem mass spectrometry
mtDNA	mitochondrial DNA
MTTP	microsomal triacylglycerol transfer protein
MUST	malnutrition universal screening tool
NAC	N-acetylcysteine
NADH	reduced nicotinamide adenine dinucleotide

NAPQI	N-acetyl-*para*-benzoquinoneimine		PHHI	persistent hyperinsulinaemic hypoglycaemia of infancy
NCEP	National Cholesterol Education Programme		*pI*	isoelectric point
NEFA	non-esterified fatty acid		PICP	procollagen IC-terminal propetide
NEP	neural endopeptidase		PID	patient identification demographics
NETs	neuroendocrine tumours		PID	primary immunodeficiencies
NGSP	National Glycohaemoglobin Standardization Programme		PINP	procollagen IN-terminal propetide
NHS	National Health Service		PKD	polycystic kidney disease
NICE	National Institute for Health and Clinical Excellence		PKU	phenylketonuria
			PL	phospholipid
NICTH	non-islet cell tumour hypoglycaemia		PLA2	phospholipase A2
NPC1L1	Niemann-Pick C1 like 1		PLTP	phospholipid transfer protein
NPIS	National Poisons Information Service		PNP	purine nucleoside phosphorylase
NPV	negative predictive value		p.o.	by mouth (from Latin peros)
NSAID	non-steroidal anti-inflammatory drug		PO_2	partial pressure of oxygen
NSAIDs	non-steroidal anti-inflammatory drugs		POCT	point of care testing
NSF	National Service Framework		POMC	pro-opiomelanocortin
NSTEMI	non ST elevation myocardial infarction		PPAR	peroxisome proliferation-activated receptor
NTBC	(2-(2-nitro-4-trifluoromethylbenzoyl)-1,3-cyclohexanedione		PPi	pyrophosphate
			PPV	positive predictive value
NTproBNP	N-terminal proBNP		PRA	plasma renin activity
NTx	N-telopeptide of type 1 collagen		PRC	plasma renin concentration
OGTT	oral glucose tolerance test		PRG	progesterone
17-OHP	17α-hydroxyprogesterone		PRM	pyrogallal red-molybdate
OTC	ornithine transcarbamylase		PRO	prolactin
PA	propionic aciduria		PRPP	5-phosphoribosyl-1-pyrophosphate
PAC	plasma aldosterone concentration		PRPPS	phosphoribosylpyrophosphate synthetase
PBC	primary biliary cirrhosis		PSA	prostate specific antigen
PBG	porphobilinogen		PSAD	prostate specific antigen density
PC	pyruvate carboxylase		PSC	primary sclerosing cholangitis
PCI	percutaneous cardiac intervention		PT	prothrombin time
PCO_2	partial pressure of carbon dioxide		PTH	parathyroid hormone
PCOS	polycystic ovary syndrome		PTHrp	parathyroid hormone related peptide
PCR	polymerase chain reaction		QC	quality control
PCSK9	proprotein convertase subtilisin kexin type 9		QIS	Quality Improvement Scotland
PCT	porphyria cutanea tardia		RAA	renin-angiotensin aldosterone system
PEG	polyethylene glycol		RAGE	receptor for advanced glycation endproduct
PEPCK	phosphoenol pyruvate carboxy kinase		RDA	recommended dietary allowance
PGP	P-glycoprotein		REE	resting energy expenditure

R_f	retention factor		TG	thyroglobulin
RIA	radioimmunoassay		TGF-β	transforming growth factor-β
RID	radial immunodiffusion		THG	Tamm-Horsfall glycoprotein
RIQAS	Randox International Quality Assessment Scheme		TLA	total laboratory automation
RNA	ribonucleic acid		TLC	thin layer chromatography
RRT	renal replacement therapy		TN	true negative
RTA	renal tubular acidosis		TNFα	tumour necrosis factor α
s.c.	subcutaneous		TP	total protein
SA	sialic acid		TP	true positive
SCAD	short chain acyl CoA dehydrogenase		TPMT	thiopurine methyltransferase
SCC	side chain cleavage		TPN	total parenteral nutrition
SD	standard deviation		TRH	thyrotrophin releasing hormone
SeHCAT	selenium labelled homotaurocholic acid test		TRL	triacylglycerol-rich lipoprotein
SF-1	steroidogenic factor-1		TSH	thyroid stimulating hormone
SHBG	sex hormone binding globulin		TTG	tissue transglutaminase
SIADH	syndrome of inappropriate antidiuretic hormone		TTGA	tissue transglutaminase antibodies
SID	strong ion difference		U&E	urea and electrolytes
SLE	systemic lupus erythematosus		UBT	urea breath test
SLO	Smith-Lemli-Opitz		UDPGT	uridine diphosphate glucuronosyltransferase
SMA	sequential multiple analyser		UKNEQAS	United Kingdom National External Quality Assessment Service
SMAC	sequential multiple analyser with computer			
SOP	standard operating procedure		UKPDS	United Kingdom Prospective Diabetes Study
SR	scavenger receptor			
SRA	scavenger receptor type A		UTI	urinary tract infection
SRB1	scavenger receptor type B1		UV	ultraviolet
StAR	steroidogenic acute regulatory protein		VLCAD	very long chain acyl CoA dehydrogenase
STEMI	ST elevation myocardial infarction		VLCFA	very long chain fatty acids
SUDI	sudden unexplained death in infancy		VLDL	very low density lipoprotein
SUR1	sulphonylurea receptor type 1		VMA	vanillylmandelic acid
T1	mono-iodothyronine		VP	variegate porphyria
T2	di-iodothyronine		VRAC	volume-regulated anion channel
T3	tri-iodothyronine		VSCC	voltage-sensitive calcium channel
T4	thyroxine		WBR	whole body retention
TAG	triacylglycerol		WEQAS	Welsh External Quality Assessment Service
TBG	thyroxine binding globulin			
TC2	transcobalamin II		WHO	World Health Organization
TCA	tricarboxylic acid		WHR	waist hip ratio
TCA	trichloroacetic acid		WM	Waldenstrom's macroglobulinaemia
TCA	tricarboxylic acid (cycle)		XO	xanthine oxidase
TDM	therapeutic drug monitoring		ZS	Zellweger syndrome

Reference ranges

The typical adult reference ranges for common analytes measured in hospital laboratories are listed alphabetically. Hospital laboratories prepare their own reference ranges and so the values given here may differ somewhat from those in your hospital laboratory. The values for most analytes are expressed in concentrations except for enzymes where activities are given.

Analyte	Concentration/activity
Albumin	35–50 g/L
Alkaline phosphatase	95–320 IU/L
Ammonia	<100 µmol/L
Amylase	<300 IU/L
Alanine aminotransferase	5–42 IU/L
Aspartate aminotransferase	10–50 IU/L
Bicarbonate (hydrogen carbonate)	24–29 mmol/L
Bilirubin total	<17 µmol/L
direct	<5.2 µmol/L
Calcium (ionized)	1.20–1.37 mmol/L
Calcium (total)	2.20–2.60 mmol/L
Chloride	95–108 mmol/L
Cortisol time (9 am)	140–690 nmol/L
time (12 am)	80–350 nmol/L
C-reactive protein	<10 mg/L
Creatine kinase	55–170 IU/L
Creatinine	71–133 µmol/L
Follicle stimulating hormone	
follicular	4.0–13.0 IU/L
mid-cycle	5.0–22.0 IU/L
luteal	1.8–7.8 IU/L
post-menopausal	26.0–135 IU/L
male	0.7–11.0 IU/L
Gamma-glutamyl transferase	5–55 IU/L
Glucose (fasting)	3.5–5.5 mmol/L
Glucose (random)	<10 mmol/L
Glucose (urine)	negative
Growth hormone	<10 mU/L

Haemoglobin	male	13.0–18.0 g/dL
	female	11.5–16.5 g/dL
Insulin (fasting)		2–10 mU/L
Iron		10–40 µmol/L
Lactate		0.7–1.9 mmol/L
Lactate dehydrogenase		240–480 IU/L
Lipids	cholesterol (HDL)	>1.2 mmol/L
	cholesterol (total)	<5.0 mmol/L
	cholesterol (LDL)	<3.0 mmol/L
	triacylglycerols	<1.7 mmol/L
Luteinizing hormone		
	follicular	3–13 IU/L
	mid-cycle	14–96 IU/L
	luteal	1–11 IU/L
	post-menopausal	8–59 IU/L
	male	0.8–8.0 IU/L
Magnesium		0.8–1.2 mmol/L
Osmolality		275–295 mOsm/kg
Parathyroid hormone		1–6 pmol/L
PCO_2		4.7–6.0 KPa
PO_2		12–14.6 KPa
pH		7.35–7.45
Phosphate		0.80–1.4 mmol/L
Phenylalanine		<100 µmol/L
Potassium		3.8–5.0 mmol/L
Prolactin	male	86–324 mU/L
	female	103–497 mU/L
Protein (total)		60–80 g/L
Protein (urine)		<0.25 g/L
Sodium		135–145 mmol/L
Thyroid	free T4	10–25 pmol/L
	free T3	4.0–6.5 pmol/L
	total T4	60–160 nmol/L
	total T3	1.2–2.3 nmol/L
	TSH	0.2–3.5 mU/L
Testosterone	male	10–31 nmol/L
	female	0.5–2.5 nmol/L
Urate		0.1–0.42 mmol/L
Urea		3.3–6.7 mmol/L

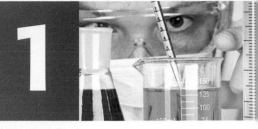

1

Biochemical investigations and quality control

David Cameron

Learning objectives

After studying this chapter you should be able to:

- Relate a number of key contributions made by eminent scientists to the development of modern clinical biochemistry laboratories

- Describe the path a typical laboratory sample follows after being obtained from the patient to the sending of the test results to the requesting clinician

- Define some of the terms used in clinical biochemistry laboratories

- Outline a number of tests performed by clinical biochemistry laboratories

- Discuss technical and clinical validation processes applied to a test result before its release from the laboratory

- Outline the quality assurance procedures used in clinical biochemistry laboratories to ensure that the standard of results meets the needs of the clinicians and their patients

Introduction

The analysis and testing of body fluids and tissues are fundamental to patient care and contribute significantly towards the **diagnosis**, **treatment**, **monitoring**, and **prognosis** of most disease processes. Diagnosis is the considered decision reached by a clinician after examining and investigating a patient's condition. Treatment is the application of medicines and therapies to try and cure the diagnosed condition, while monitoring is the use of diagnostic testing to assess any changes in the disease. A prognosis is a prediction of the clinical outcome of a disease.

In the United Kingdom, clinical testing is performed mainly by National Health Service (NHS) hospital laboratories, although it is also performed by numerous private laboratories. The role of the clinical laboratory is to provide timely, reliable, reproducible, relevant results, and expert, interpretative advice. Often, a particular disease or clinical condition can only be diagnosed on the basis of the results obtained from a number of clinical laboratory tests. Every disease produces a particular combination of test results that form a pattern characteristic of that illness. Although these patterns are indicative of particular clinical conditions, the situation is complicated in that the test results may overlap several laboratory disciplines. Thus, the clinical team must assess all clinical laboratory results and, indeed, many more considerations when they decide on the best course of action for their patient. Thus, modern medicine relies on the input from multiprofessional, multidisciplinary teams to choose the best pathway for patient care. Clinical tests are now available in general practitioners' surgeries, pharmacies, high street stores, and patients' homes, all of which pose different challenges in ensuring that the same quality of results is maintained. The hospital laboratory also has a role to play here in providing advice, training, and monitoring performance. The Clinical Pathology Accreditation (UK) Ltd (CPA) is the organization that accredits most clinical laboratories. At all times, the welfare and wellbeing of the patient is of paramount importance.

Three main groups of professional staff are responsible for providing the clinical biochemistry laboratory service: namely biomedical scientists, clinical scientists, and medical staff. Each laboratory has a lead consultant, who is medically qualified and acts as head of service, although this role can also be the responsibility of a clinical scientist. Biomedical scientists must have an honours degree in biomedical science from an Institute of Biomedical Sciences (IBMS) approved program. They are responsible for all technical aspects of the service, from the receipt of samples to the validation of the results of clinical tests. Clinical scientists usually have a specialist remit. Their entry qualifications reflect the needs of the service but they frequently have a PhD in a relevant subject, such as vitamin analysis or endocrinology. The traditional role of clinical scientists is to develop their specialist subject areas and provide advice on which clinical tests are required for a particular specimen and in interpreting the results of these tests. Given their high qualifications, the quality of advice they provide is correspondingly reliable.

The main disciplines involved in clinical testing are clinical biochemistry, haematology/blood transfusion, histopathology, and medical microbiology. However, continuous advances in laboratory medicine have also seen other disciplines, for example immunology, virology, cytology, and genetics, grow alongside, but also often as part of the four major biomedical disciplines. Clinical biochemistry is arguably the most diverse of the clinical laboratory disciplines. It provides such a wide array of test results that the clinical teams require frequent advice in interpreting results and which subsequent appropriate **cascade tests** to use, as outlined in Figure 1.1. In this chapter, you will learn about the roles of the clinical biochemistry laboratory in the diagnosis, treatment, monitoring, and prognosis of disease states.

Cascade tests are those generated in response to results from a preceding test to further aid diagnosis.

SELF-CHECK 1.1

What is the role of the clinical biochemistry laboratory service in the provision of patient care?

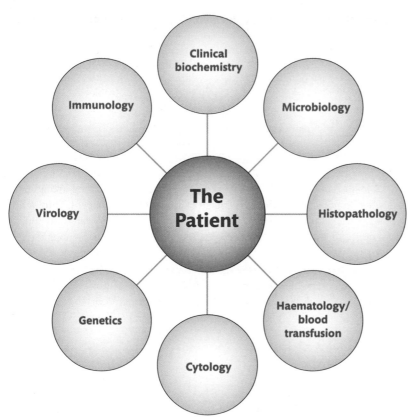

FIGURE 1.1
The patient is at the centre of all laboratory procedures. Laboratory testing is designed to support and improve patient care.

1.1 **Historical background**

Clinical biochemistry is a relatively young discipline in pathology, although knowledge of chemicals in body fluids dates back to ancient Egyptian times. It began to come alive in the mid-nineteenth century and grew through the early part of the twentieth century, when many eminent clinicians and scientists turned their interests towards the chemical components of body fluids and began to develop equipment and applications for measuring them. Since 1945, laboratory medicine and, in particular, clinical biochemistry has developed at an exponential rate. Evermore sophisticated analytical systems have become possible with developments in computing and miniaturization of components. There have been far too many contributors to mention them all but the individuals described in Box 1.1 have all made significant contributions to the development of the modern clinical biochemistry laboratory.

BOX 1.1 *Key contributors to clinical biochemistry*

Henry Bence Jones (1813–1873) is forever linked with the disease multiple myeloma and the presence of the so-called Bence-Jones protein in the urine of these patients. He promoted the chemical and microscopical examination of urine as a standard procedure in the diagnosis of disease. His publications on chemical analysis of urine and his application of chemistry to the human situation were well in advance of their time.

Otto Folin (1867–1934) was interested in colour reactions, which became the basis of quantitative clinical chemistry, and so established practical colorimetry and the analysis of body fluids. With Hsien Wu, he developed the classical Folin-Wu method to assay glucose in protein-free filtrates of blood.

Donald D. Van Slyke (1883–1971) introduced techniques for measuring the concentrations of amino acids and investigated the conversion of protein into urea by the liver. He made major advances to the understanding of acid-base and electrolyte problems and determining the concentration of urea in blood. These studies pointed the way to fluid and electrolyte therapy.

Arnold Beckman (1900–2004) founded Beckman Instruments, Inc. in 1935. He invented the pH meter, which was its first commercial product, and earned him a place in the National Inventors Hall of Fame in 1987. In total, Beckman registered 14 patents and his company, now merged with Coulter Electronics as Beckman-Coulter, is one of the largest laboratory diagnostics producers in the world.

Samuel Natelson (1909–2001) is often regarded as the father of paediatric clinical biochemistry. His studies, following World War II, on electrolyte and acid-base status in premature babies revolutionized neonatal care and has saved countless lives.

Leonard Skeggs (1918–2002) had a particular interest in the actions of renin and angiotensin during his early years; however, at heart he was an inventor. In the early 1950s he formulated the idea of continuous flow analysis and looked for a company to develop his ideas. He had almost given up when the company Technicon bought his idea in 1954 and went on to develop the automated continuous flow analyser. These instruments revolutionized clinical biochemistry, taking analyses out of individual test tubes and onto high throughput platform systems. In subsequent years, Skeggs was heavily involved in the design of the Technicon SMAC system, which became the main workhorse for most laboratories throughout the 1970s and early 1980s.

Rosalyn Yalow (1921–) and Solomon Berson (1918–1972) developed the technique of radioimmunoassay (RIA). They published the first analytical RIA method for determining the concentration of insulin, in 1960. This led the way for further developments in analysing corticotrophin, growth hormone, gastrin, and parathyroid hormone and significant discoveries regarding their physiological activities. Both received numerous prestigious awards for their work, culminating in the 1977 Nobel Prize to Yalow but, unfortunately, this recognition came too late for Berson who had died in 1972.

Georges Kohler (1946–1995), Cesar Milstein (1927–2002), and Niels Kaj Jerne (1911–1994) were responsible for developing procedures for producing monoclonal antibodies. Their discovery has had far-reaching effects not only in diagnostic testing and physiological understanding of disease processes but also in treatment of cancers. They were awarded the 1984 Nobel Prize in Physiology for their work.

Modern clinical biochemistry laboratories are an exciting mixture of highly automated high throughput systems, cutting-edge equipment, and advanced computer technology, which are controlled and monitored by highly skilled staff. Millions of tests are performed per year. Lord Carter's Independent Review of Pathology Service of 2006 found that 56% of all laboratory tests in the UK are performed by clinical biochemistry laboratories, which spend 23% of the overall pathology budget in England and Wales.

Pathology network describes a group of laboratories working together to share best practice and provide cost-effective care to the patient.

Pathology networks are used throughout the United Kingdom. These share resources and expertise across sites to maximize patient care and allow new ways of working as clinical biochemistry advances during the twenty-first century.

1.2 Modern clinical biochemistry laboratories

Modern clinical biochemistry laboratories are designed to provide results 24 hours per day and 7 days a week. Over 70% of patients require an analytical test to aid their diagnosis. Many of these results will be normal, helping the clinician to rule out potential diseases, point towards further investigations, and reassure the patient. However, unless a sample for analysis is delivered in the manner prescribed for that test, it is not possible to produce meaningful results. Therefore it is essential that the referring clinician adheres to the laboratory protocol for sample collection.

Samples and analytes

Clinical samples or specimens are materials collected from patients and used in the investigation of their diseases. Analytes are substances that are tested for in the sample, using appropriate analytical methods or chemical tests.

The sample must be collected into a suitable container before it can be analysed. If an error is made at this stage, it will invalidate the whole process so it is essential that care is taken here. Samples collected for clinical analysis include:

- blood
- urine
- other body fluids
- faecal material
- solid tissue samples

The majority of blood samples nowadays are drawn by trained phlebotomists and nurses under the instructions of medical staff. Urine for analysis is collected either as a spot sample (random, early morning, mid-stream) or a timed collection, which is usually over 24 hours. Other body fluids sampled include cerebrospinal fluid (CSF), gastric fluid, ascitic fluid, sweat, and amniotic fluid. Cerebrospinal fluid is the serum-like clear fluid that surrounds the brain, spinal cord, and subarachnoid space. It cushions these organs against injury and shock. Gastric fluid occurs in the stomach and varies in content as food is digested and in disease conditions. Ascitic fluid is the serous liquid that collects in the peritoneal cavity when there is an imbalance of plasma flow into and out of the lymphatic system. It often occurs in cirrhotic liver disease and malignancies. Sweat samples are infrequently required and are normally only analysed in cases of cystic fibrosis. Amniotic fluid is the watery liquid within the amniotic sac that cushions and protects the developing foetus in the womb. Its collection may require a clinical procedure for collection and will normally be the responsibility of the attending clinician. Saliva, because it is easy to collect, has become more popular and increasing numbers of suitable tests are being developed. However, there is still only a very limited repertoire of clinical tests, for example for measuring cortisol.

Faecal analysis is an unpleasant but necessary task for many gastrointestinal conditions. Tests are generally performed on a small random sample collected in a securely sealed pot. Other solid tissues include nail and hair clippings for drug and metal analyses.

Samples from neonates and children require special care as their veins and arteries are smaller than those of adults. Also, their total blood volume is progressively less, depending on their age and size. Very premature babies may weigh less than one kilogramme, but modern neonatal medicine has significantly improved their survival prospects.

Every laboratory should treat all samples as potentially infective. Known **high risk samples**, for example from patients carrying hepatitis B or C and HIV viruses, or any other category III pathogens must be labelled with an appropriate warning, for instance 'Dangerous Specimen'. They should be dealt with according to Health and Safety regulations for transport to and within the receiving laboratory. If a patient is suspected of carrying a category IV pathogen, such as Lassa fever, yellow fever, or Marburg virus, samples may only be handled in a specialized containment facility and they should not be sent to the local laboratory.

Cross reference

Chapter 7 Samples and sample collection in the *Biomedical Science Practice* textbook

Cross reference

Chapter 4 Health and safety in the *Biomedical Science Practice* textbook

An **analyte** is the substance whose nature and/or concentration is determined by a clinical test or analysis. Many analytes are investigated in clinical biochemistry; you will encounter a great number of them in the other chapters of this book. Specific requirements must be observed when analysing specimens, for example the volume collected, whether the patients require a specific diet or should fast, whether it is taken at a random time or a specific time of day, how the sample is preserved, transported, and treated on arrival at the laboratory, for instance does it need rapid centrifugation and separation? Each sample should have the name, date and time of sampling, and a unique patient identifier affixed to it.

The most frequent requests are for analyses of blood samples. These are usually centrifuged prior to analysis, as most biochemical estimations are carried out using plasma or serum. Following centrifugation, a visual assessment of the sample is necessary to decide whether there is sufficient volume for the analyses requested. In the absence of chemistry analysers that can measure serum indices, **haemolysis**, **icterus**, and **lipaemia** (see Section 1.4) are also assessed visually.

Blood tubes include a preservative and are generally colour coded in a consistent manner by manufacturers, depending on the anticoagulant used (Figure 1.2). Table 1.1 describes tubes commonly used in clinical biochemistry laboratories. Most blood tubes are moulded from a strong plastic, like polyethylene terephthalate, making them virtually unbreakable and safer to use than glass. Many also come with an inert separator gel barrier to aid separation of plasma or serum from the cells, which has increased the use of *primary* samples for clinical analyses. Primary samples are ones presented to the analyser in the same container in which they were originally collected from the patient.

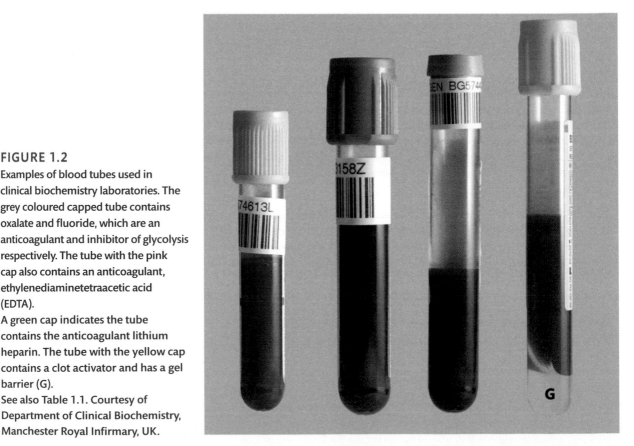

FIGURE 1.2
Examples of blood tubes used in clinical biochemistry laboratories. The grey coloured capped tube contains oxalate and fluoride, which are an anticoagulant and inhibitor of glycolysis respectively. The tube with the pink cap also contains an anticoagulant, ethylenediaminetetraacetic acid (EDTA).
A green cap indicates the tube contains the anticoagulant lithium heparin. The tube with the yellow cap contains a clot activator and has a gel barrier (G).
See also Table 1.1. Courtesy of Department of Clinical Biochemistry, Manchester Royal Infirmary, UK.

TABLE 1.1 Commonly used blood tubes and preservatives.

Tube type	Cap colour	Description
Plain serum See Figure 1.2	red	Coated with a clot activator Serum obtained by centrifugation which prevents exchange of analytes with clot Used for most routine chemistries
Serum separator See Figure 1.2	yellow	Coated with a clot activator Gel barrier allows for primary sampling and storage Used for most routine chemistries
Lithium heparin (plasma) See Figure 1.2	green	Available with and without a gel barrier The anticoagulant heparin allows the sample to be centrifuged immediately, so whole blood or plasma can be tested
Fluoride oxalate See Figure 1.2	grey	The anticoagulant oxalate binds calcium and the fluoride inhibits glycolysis, thus maintaining blood glucose for several days Osmotic effects of oxalate may cause haemolysis Used for blood glucose measurements
Ethylenediaminetetraacetic acid (EDTA) See Figure 1.2	pink	EDTA chelates calcium preserving cell morphology so that: • blood films can be used for haematological investigations • analytes requiring whole blood, for example haemoglobin A1c and erythrocyte enzyme activities

Separator gels are incorporated into blood collection tubes during their manufacture. The gel reacts minimally with clinical materials, and has a density such that during centrifugation it forms a barrier between the supernatant, which is the plasma or serum, and the pellet consisting of the clot and/or cellular material. Aliquots of plasma or serum can then be easily removed from the tube for analysis.

Whatever, the nature of the sample, it must be accompanied by a **request form**, which must be completed in full to allow the sample to be positively identified and linked to any previous reports. Request forms come in many shapes and sizes, although they are commonly in an A5 format. Figure 1.3 shows a typical example.

Each form should state the patient's name, address, hospital number, date of birth, time and date of sampling, the requests, and any other information which might help interpretation. This could include, for example, the patient's sex, last menstrual period, ethnic origin, drug monitoring information, such as the time of last treatment, and symptoms. Registering the tests requested on the form may require them to be hand written or there may be a list of tick boxes. Currently hardware and software applications are available for optical mark reading, allowing the test request form to be rapidly scanned, automatically allowing the analyser to perform the required test(s). This system is used widely in continental Europe (Figure 1.4). However, the United Kingdom has been relatively slow to use this technology: true text recognition software is less well-established and found in only a handful of UK laboratories.

> A **request form** is a document that accompanies a specimen, providing patient identification, sample details, tests requested, clinical details, and source of the request.

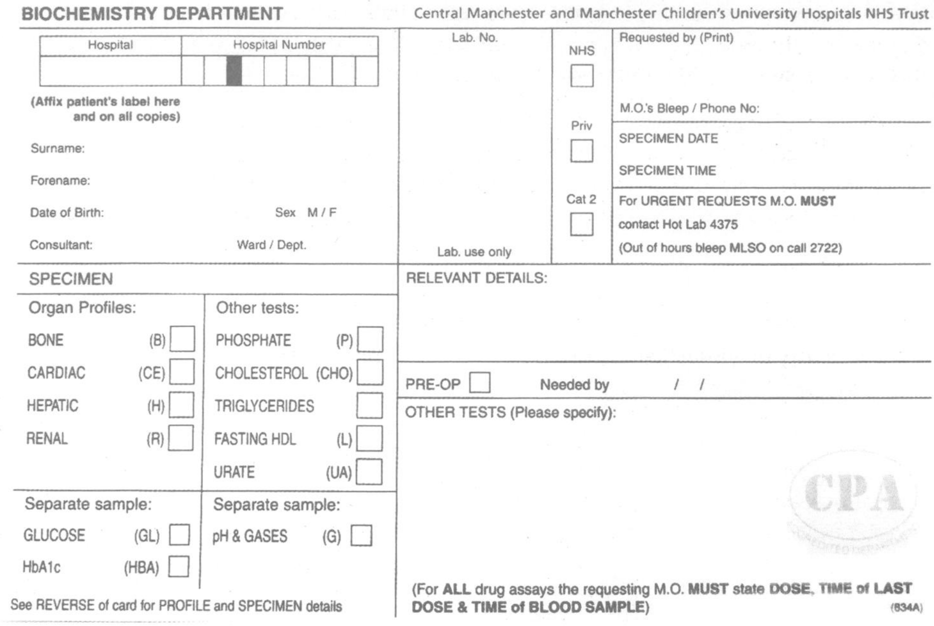

BIOCHEMISTRY DEPARTMENT   Central Manchester and Manchester Children's University Hospitals NHS Trust

Hospital	Hospital Number

(Affix patient's label here and on all copies)

Surname:

Forename:

Date of Birth: Sex M / F

Consultant: Ward / Dept.

Lab. No.

NHS ☐
Priv ☐
Cat 2 ☐

Lab. use only

Requested by (Print)

M.O.'s Bleep / Phone No:

SPECIMEN DATE

SPECIMEN TIME

For URGENT REQUESTS M.O. **MUST**
contact Hot Lab 4375
(Out of hours bleep MLSO on call 2722)

SPECIMEN

RELEVANT DETAILS:

Organ Profiles:

BONE   (B) ☐
CARDIAC   (CE) ☐
HEPATIC   (H) ☐
RENAL   (R) ☐

Other tests:

PHOSPHATE   (P) ☐
CHOLESTEROL (CHO) ☐
TRIGLYCERIDES ☐
FASTING HDL   (L) ☐
URATE   (UA) ☐

PRE-OP ☐   Needed by   / /

OTHER TESTS (Please specify):

Separate sample:

GLUCOSE   (GL) ☐
HbA1c   (HBA) ☐

Separate sample:

pH & GASES (G) ☐

See REVERSE of card for PROFILE and SPECIMEN details

(For **ALL** drug assays the requesting M.O. **MUST** state **DOSE, TIME** of **LAST DOSE & TIME of BLOOD SAMPLE**)   (834A)

CPA

FIGURE 1.3

An example of a biochemistry request form. Courtesy of Department of Clinical Biochemistry, Manchester Royal Infirmary, UK.

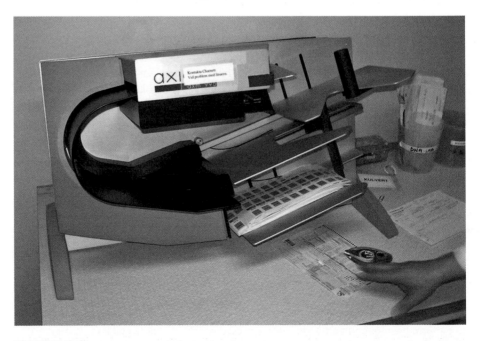

FIGURE 1.4

A sophisticated form scanner and optical mark reader.

The use of electronic requests is becoming more widespread, particularly within the hospital environment. In these cases, patient identification data and requests are entered into a **hospital information system** (HIS), which produces a unique label for the sample.

Depending on the system in operation, this label may contain all the information in a barcode format that is necessary to identify the sample, and so a paper request form is not required. Once the sample is sent to and arrives at the laboratory, its details are transferred electronically from the HIS to the **laboratory information system** (LIS), where they can be matched together.

SELF-CHECK 1.2

Why is it necessary to ensure that request forms and blood tubes match?

Whatever the type of sample, it must be delivered to the laboratory for analysis. Samples from outside the hospital must arrive packaged safely and securely, but will be in many different formats depending on their source. Those samples arriving from within the hospital may be delivered by a porter, in protective pods through a network of plastic pipes running throughout the hospital, called a **pneumatic tube system**, or even by hand by the person taking the sample.

Sample reception

The first point of contact for a clinical sample in most laboratories is the sample reception area. In some newer laboratories, which have been built to modern standards, this may function as a common reception for all the disciplines. Sample reception is where they are identified, sorted, and prepared for analysis. Mistakes made at reception will carry through the rest of the process and so this key area of any laboratory must be well organized.

When a sample reaches reception, the package must be opened and the samples checked against the request forms. This ensures the correct patient is booked into the system and the sample is of the type appropriate for the chosen analyses (see Figure 1.5). Staff must match the details of the patient on the request form with those on the sample itself. For example, patient's surname and forename, date of birth, hospital number or other unique identifier (the United Kingdom does not yet have a standardized health identifier for each individual) must all tally.

Samples and forms that do not correspond with each other at this stage cannot be processed. Electronic requesting, of course, reduces the problem of identification as a label or tag is added to the sample at source. This allows rapid identification in the laboratory and linkage to the LIS. Sample reception is a busy area of the laboratory and the staff working here, who are most often pre-registration grades, have a responsible role to perform in a very repetitive environment.

Once the sample has been correctly identified, a unique laboratory identifier is normally allocated to it in the form of a barcode number. This identifies each individual sample as it is transported through the various analytical areas. The barcode is also linked to the **patient identification demographics** (PID) in the LIS, so that when the analytical results become available they are transferred to the correct location.

Blood samples are centrifuged prior to measuring their serum indices. Haemolysis, icterus, and lipemia in the serum (or in plasma) can all affect their determination as described in

A **hospital information system** is the central patient database for a hospital and its patients. It is used to generate requests for diagnostic procedures, and is also a repository for the results from diagnostic tests.

A **laboratory information system** is also a database that contains all the laboratory diagnostic information. It is interfaced to the laboratory analysers and to the Hospital Information System.

Patient identification demographics is information which is unique to that individual and is stored on the HIS system. Details held include full name, date of birth, sex, address, and hospital number.

FIGURE 1.5
The sorting of clinical specimens in a busy but well-organized reception area.

Section 1.4. Following centrifugation, aliquots of the serum for specialist testing or offline storage may be taken. Again, it is necessary to ensure that this aliquot is correctly identified for future use. All samples must be appropriately stored until diagnostic testing has been completed, the results have been relayed to the clinician, and sufficient time has been allowed for any follow-up test on the same sample. The length of storage varies between laboratories because those with more space will be able to store samples for longer periods. Routine samples must be stored for at least three, and preferably seven, days before disposal. Samples for specialist testing must be stored according to local requirements, which may be months or even years in some cases.

Neonatal and paediatric samples require special attention because the volume of blood available is much less than that from an adult. It is desirable to maximize the use of the sample and special paediatric test profiles may be agreed in advance by the staff concerned with the infant.

The core laboratory

The **core laboratory** is the area where more than 90% by volume of tests are performed. Most hospital laboratories analyse 1,000 to 2,000 samples with a typical profile of 10 or more analytes per sample each day. The major pieces of laboratory equipment used to do this are high throughput chemistry and immunoassay analysers, which are the flagship instruments of most major diagnostic companies. The analysers can be linked together in some way to form an analytical **workcell**, as shown in Figure 1.6, which is a logical and productive grouping of analysers, tools, and personnel to maximize the efficiency and effectiveness of the analytical process. However, depending on the workload, these instruments can also function as standalone platforms for chemical or immunoassay analyses (see Figure 1.7).

Cross reference

Chapter 2 Automation

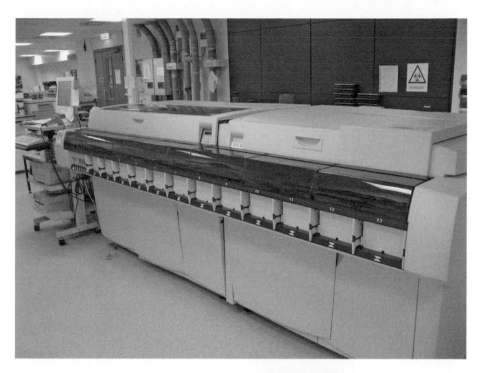

FIGURE 1.6
Chemistry and immunoassay platforms linked together by a common sample handler shown in the front of the picture to form a small workcell.

FIGURE 1.7
An example of a standalone immunoassay analyser.

FIGURE 1.8
Total laboratory automation system with multiple chemistry and immunoassay analysers linked together by a common track.

Ultimately the busiest laboratories require a **total laboratory automation** (TLA) system (Figure 1.8), where most common laboratory processes, including sorting, centrifugation, analysis, and storage of samples is performed automatically. The barcoded samples can be tracked through the whole process and are identified with the patient through his or her unique number. This type of system minimizes human intervention, as computer software instructed robotics perform all these processes within the TLA system.

Not all analysers are used overnight since the volume of work usually decreases significantly and staffing is also reduced accordingly. However, core laboratories operate 24 hours a day, 7 days a week and are the area of the laboratory where routine profiles are analysed (see Table 1.2), including, for example:

- urea and electrolytes
- liver function tests
- bone profile
- thyroid function tests
- a variety of other tests

TABLE 1.2 Common serum/plasma profile requests.

Profile name	Abbreviation	Tests					
Urea and electrolytes	U&E	Sodium	Potassium	Chloride	Bicarbonate	Urea	Creatinine
Liver function tests	LFT	AST	ALT	ALP	GGT	BIL	TP
Bone profile	BONE	Calcium	Phosphate	ALP	ALB		
Thyroid function tests	TFT	TSH	Free T4				

AST, aspartate aminotransferase; ALT, alanine aminotransferase; ALP, alkaline phosphatase; GGT, γ-glutamyl transferase; BIL, bilirubin; TP, total protein; ALB, serum albumin; TSH, thyroid stimulating hormone; Free T4, free thyroxine.

A urea and electrolytes (U&E) profile may or may not include chloride and hydrogen carbonate (commonly known as bicarbonate, HCO_3^-), depending on local circumstances. It is used to quickly assess renal function, dehydration, and, if hydrogen carbonate is included, acid-base status, before carrying out more specialized tests. A U&E profile is the most commonly requested in clinical biochemistry and is often accompanied by other requests, for example liver function tests (LFTs), C-reactive protein (CRP), glucose, amylase, uric acid, and magnesium. You can read more about U&E analyses in Chapter 5.

Liver function tests (LFTs), as their name suggests, are used to assess the various functions of the liver or the anatomical area which is affected. They include measuring the activities of aspartate aminotransferase (AST) and alanine aminotransferase (ALT). These enzyme activities are found in the hepatocytes and only released into the blood in significant quantities if liver damage occurs. The concentrations of albumin (ALB) and total protein (TP) in the plasma are both indicators of the liver's synthetic functions since, with the exception of immunoglobulins, blood proteins are formed in the liver. The presence of bilirubin (BIL) in the blood increases if the biliary system is impaired and is an indicator of impaired liver detoxification functions. The activities of alkaline phosphatase (ALP) and γ-glutamyl transferase (GGT) are also associated with biliary system impairment.

The bone profile assesses the amounts of calcium, phosphate, and ALP activity. The profile is often used as a general screen prior to more complex investigations.

Calcium is an essential mineral component of bone but also has many other uses in the body. About 50% of calcium in the blood is bound to albumin, whose concentration must also be measured to correct for this. Increased concentrations of calcium are found in patients with fractures who are immobile and in the condition sarcoidosis, a chronic disease of unknown cause that is characterized by the presence of nodules in the lungs, liver, and lymph and salivary glands. Decreased concentrations of calcium can be due to many factors, including a reduced dietary intake, acute pancreatitis, magnesium deficiency, bone disease, and chronic kidney disease.

Measurements of the concentration of phosphate in clinical samples of plasma or serum help diagnose conditions and diseases that affect the digestive system and interfere with its absorption as well as that of calcium. Phosphate concentrations are also carefully monitored in patients with renal disease to ensure that the patient is maintaining appropriate and constant levels. Alkaline phosphatase activity is also associated with bone and can be increased in Paget's disease or bony metastatic cancers.

Thyroid function tests measure thyroid stimulating hormone (TSH) released by the pituitary gland, which controls the synthesis of thyroxine. Patients with an underactive thyroid gland will have excessively high concentrations of TSH in their plasma. An overactive thyroid gland suppresses the concentration of TSH. The symptoms of an underactive thyroid gland include tiredness, lethargy, weight gain, and cardiac arrhythmias , whilst those associated with over activity include weight loss, irritability, insomnia, and cardiac arrhythmias . These are common symptoms and thyroid disease is relatively easy to treat so it is sensible to eliminate this as a cause at an early stage. Free thyroxine (free T4) can be measured at the same time as TSH or as a reflex test in response to an abnormal TSH result. Measuring free T4 has largely superseded the determination of total thyroxine (T4) in the UK.

Neonates are also screened at birth for congenital thyroid disease using blood spot samples (whole blood from a heel stab spotted on filter paper), which are sent to regional centres for analysis.

In addition to the common profiles described above, many other assays can be performed using core analysers. For example, immunoassay analysers will normally measure troponin I as a marker of cardiac damage, reproductive hormones such as luteinizing hormone (LH),

Cross reference
Chapter 5 Fluid and electrolyte disorders

Cross reference
Chapter 8 Liver function tests

Cross reference
Chapter 10 Disorders of calcium, phosphate, and magnesium homeostasis

Cross reference
Chapter 12 Thyroid disease

Cross reference
Chapter 15 Reproductive endocrinology and Chapter 19 Cancer biochemistry and tumour markers

follicle stimulating hormone (FSH), oestradiol, prolactin (PRO), progesterone (PRG), testosterone, and sex hormone binding globulin (SHBG). Tumour markers, for example prostate specific antigen (PSA), human chorionic gonadotrophin (hCG), α-fetoprotein (AFP), and carcinoembryonic antigen (CEA), and other analytes like cortisol may all be routinely measured.

Chemistry analysers are capable of measuring a variety of enzyme activities in addition to those listed above. These include those of creatine kinase (CK) and lactate dehydrogenase (LD). They can also determine the amounts of various other analytes including iron and transferrin, therapeutic drugs (digoxin, theophylline, phenytoin, carbamazepine), immunoglobulins, cholesterol, HDL-cholesterol, triacylglycerol, and the chemical analyses of biological fluids like urine and CSF in addition to plasma and serum. The list of analytes that can be studied is almost endless but ultimately each laboratory must decide on which repertoire to make available based on providing a cost-effective service to the hospital and its patients.

Cross reference
Chapter 2 Automation

SELF-CHECK 1.3

Why is U&E (Table 1.2) the most common test group requested in clinical biochemistry?

Emergency laboratory

The emergency laboratory is likely to have an analyser which is the same as that used in the core laboratory and provides a similar 24 hours, 7 days a week rapid turnaround service. Assays specific to the emergency analyser will include those for serum paracetamol, salicylate, and possibly ammonia. The emergency laboratory will also have a blood gas analyser to assess the acid-base status of patients and an osmometer to measure the osmolality of serum in unconscious patients who have ingested an unknown substance.

Each hospital has its own requirements for turnaround time of emergencies: the laboratory must be able to meet this. It is not unreasonable to expect results to be available 30 minutes after the laboratory receives the sample.

Specialist laboratory

Outside of the core and emergency areas, most laboratories also perform other tests depending on the volumes of work requested locally. This is likely to require specialist equipment and expertise. Table 1.3 gives examples of some of these specialized techniques.

Most laboratories will provide an electrophoresis service. However, for low volume specialist tests, which require expensive equipment and specialist expertise in interpreting and reporting the results, it is often more cost-effective to send some samples to a central laboratory that provides a regional service. Thus, for example, the screening and confirmation of drugs of abuse, screening for Down's syndrome and of neonatal samples, specialist immunoassay tests such as those for gastrin, insulin, androstenedione, and 17-hydroxyprogesterone, measuring trace elements such as lead, aluminium, cadmium, zinc, copper, manganese, and vitamins, A, B, C, D, and E, is available on a regional basis. There are also many assays that are only performed in a small number of centres throughout the UK, which provide a supraregional assay service.

SELF-CHECK 1.4

Explain why some specialist tests are sent to regional centres for analysis?

TABLE 1.3 Specialist analytical techniques and some of their common uses.

Technique	Abbreviation	Common uses
High performance liquid chromatography	HPLC	Serum, plasma or whole blood vitamins A, E, B_1, B_6 Urine or serum catecholamines
Gas chromatography mass spectrometry	GC-MS	Serum and urine drug analysis and confirmation Urinary steroids Urea breath test
Liquid chromatography mass spectrometry	LC-MS	Anti-fungal drugs such as itraconazole and ketoconazole used in post-transplant patients Drug analysis
Induction coupled plasma mass spectrometry	ICPMS	Urine and serum trace elements such as zinc, copper, lead, manganese, aluminium, and selenium
Tandem mass spectrometry	Tandem MS	Serum vitamin D and metabolites Steroid hormones
Electrophoresis	EP	Serum and urine protein fractions

1.3 **Point of care testing**

Technology has advanced to the stage where it is possible to perform biochemical tests alongside the patient's bedside. This is called point of care testing (POCT). Point of care testing offers a number of advantages over conventional laboratory tests. It eliminates transport of the sample to the laboratory, the tests are minimally invasive, usually requiring only a spot of whole blood or urine for analysis and the results are immediately available. Thus, the use of POCT has grown significantly.

There are four strands to POCT, which differ in who performs the test and where the test is performed. Thus tests may be performed:

- outside the laboratory but by laboratory staff
- outside the laboratory but by nursing and medical staff
- outside the laboratory by the patient
- in pharmacies, supermarkets, internet vendors

Tests carried out by laboratory staff but outside the clinical laboratory include those for haemoglobin A1c (HbA1c) at diabetic clinics and for therapeutic drugs in the epilepsy clinic. The analysers used are relatively sophisticated and so require trained staff to operate them but results are available in real time during the patient's consultation. The major advantage of this approach is that it allows the patient's therapy to be adjusted without the need for a recall and inconvenience of a further appointment. Tests performed outside the laboratory but by nursing and medical staff include monitoring the concentration of glucose in blood samples of patients in hospital wards and blood gases in patients in intensive therapy units (ITUs) and accident and emergency units. Performing such tests is relatively simple, but training of the relevant personnel by laboratory staff is necessary. Once trained, their performance must be monitored and this is also the responsibility of the clinical laboratory. Again, this type of POCT improves patient care because results are available in real time and can have an immediate impact on the clinical management of the patient.

Cross reference
Chapter 18 Point of care testing in the *Biomedical Science Practice* textbook

Tests can also be performed outside the laboratory but by the patient. This usually occurs at their home and as part of a disease monitoring program, for example, measuring the concentrations of glucose in blood and urine in diabetic patients. Used properly, this type of glucose monitoring can significantly improve the lifestyle of diabetic patients. Thus, blood glucose monitoring by diabetic patients in their own home has considerably improved the control of their blood glucose concentrations and reduced the number of visits they make to their general practitioner or hospital clinic. In this case, the training of the patient is usually carried out by diabetic nurse specialists but the clinical biochemistry laboratory is becoming increasingly involved in this area because of current POCT regulations. Laboratories will have a role in monitoring performance, which is a challenging task, given the tests are performed in the community.

Finally, pharmacies, supermarkets, and internet vendors all offer tests for a whole variety of clinical analytes, including glucose, cholesterol, and even PSA. Many of these tests are carried out using simple instruments. The key questions to be addressed with this form of POCT are: Who trains the test providers? How do these vendors deal with abnormal results? How are their performances monitored? There are a number of schemes to evaluate the effectiveness of this type of testing and also to address the regulatory issues which arise from them.

1.4 Evaluation of a clinical method

All methods used to measure particular analytes undergo evaluation prior to their routine use in clinical biochemical laboratories. This evaluation must consider the technical or analytical performance of the test and assess its clinical usefulness in investigating the disease in question. These two aspects of evaluation are considered in the next two subsections.

Technical or analytical validation

The results of all clinical tests must pass stringent checks to ensure that the value obtained is valid before it leaves the laboratory. These checks are determined by the individual laboratory and may vary in nature depending on the population being served. For example, a paediatric laboratory is likely to have specific criteria that differ from those operated by a hospital laboratory that deals with samples from adults. However, many of the criteria applied will be common or similar in all clinical laboratories and will include:

- limits of linearity
- analytical sensitivity and specificity
- accuracy and precision

Limits of linearity

All assays have a finite linearity, that is a limited range of values between which results can be regarded as accurate. Values above and below these defined levels are inaccurate. The values of the linearity limits depend upon the specific method. If the results exceed the upper linearity limit, then it may be necessary to perform the test on a diluted sample to obtain an accurate value. For values below the lower linearity limit, the result would normally be reported as less than some accepted value. In some circumstances, a re-analysis using a larger volume of sample would have to be performed if an absolute value is required. Table 1.4 shows some examples of **action limits**. These are test results set locally for each analyte, test results outside this range mean a course of action *must* be taken.

TABLE 1.4 Examples of action limits and the required response.

Analyte	Action limit	Response	Reason
Serum (potassium)	>6.5 mmol/L (high)	Check result and phone physician	Clinical implications for patient
Serum (potassium)	<2.5 mmol/L (low)	Check result and phone physician	Clinical implications for patient
Serum (calcium)	>3.00 mmol/L	Check result and phone physician	Clinical implications for patient
Serum (TSH)	<0.05 mU/L (low)	Analyse free T4	Aid to differential diagnosis

Analytical specificity and sensitivity

The **analytical specificity** of an assay is its ability to measure only the analyte in question. Analytical specificity is usually, but not exclusively, associated with immunoassays, which involve antigen-antibody interactions. Each of the antibodies to a particular antigen will react with it in a slightly different way and, indeed, may cross-react with other molecules that closely resemble it in structure. For example, the antibodies used in an assay to measure TSH in serum may show significant cross-reactivity with hCG. Thus, the results for pregnant women, who have high concentrations of hCG in their sera will be compromised. Hence, laboratory staff require a detailed knowledge of the components used in each assay and this enhances their ability to report results with confidence. Naturally, the goal of assay designers is to produce ones that are completely specific but this is a challenging task.

Cross reference

Chapter 15 Immunological techniques in the *Biomedical Science Practice* textbook

The **analytical sensitivity** of an assay is the smallest amount or concentration of an analyte it can detect. This is the limit of detection for the assay and numerical results below this value cannot be reported with the required degree of confidence.

SELF-CHECK 1.5

What is the difference between analytical sensitivity and specificity?

Accuracy and precision

Accuracy describes the ability of a test to produce the true value of an analyte in a sample. When the same analyte is measured 20 times in succession there will be a distribution of values reflecting the errors in the measuring process. The true value then becomes the average or mean ($\bar{x}$) of all the measurements. The mean can be calculated using the expression:

$$\bar{x} = \Sigma x / n$$

where Σx is the sum of the test results of n individual tests.

Precision describes the ability of a test to reproduce the same result consistently in the same specimen. The spread of the results or their distribution reflects the variability of the method used. Variability is usually expressed as the standard deviation (SD), which can be calculated using the formula:

$$SD = \sqrt{\frac{\Sigma [x-\bar{x}]^2}{n}}$$

A method with an SD of zero is perfectly precise. However, no assay exhibits a perfect performance. Thus, when examining laboratory data it is useful to express both the mean and the

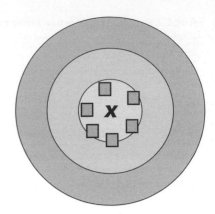

FIGURE 1.9
Results that are both accurate and precise.
See text for details.

SD, because both the accuracy and precision vary independently of each other. This means several different permutations are possible, any of which may have validity depending on the analyte being measured and the clinical requirements. Look at the target in Figure 1.9. Each concentric ring represents one standard deviation of the true mean value, which is shown at the centre of the target as X. The individual results are shown as squares and are both accurate and precise: accurate because their mean value is close to the true value and precise because they are within 1 SD of the mean.

If an assay is 100% accurate, the mean will equal the true value. In practice, the values are usually distributed evenly around the mean giving a normal or Gaussian distribution, which is illustrated in Figure 1.10. You can see that 66% of the population are between −1 and +1 SDs, 95% of the population lie between −2 and +2 SDs, and 99% of the population lie between −3 and +3 SDs of the mean. These values are confidence intervals and for any given method a range of −2 to +2 SDs is regarded as acceptable precision.

The **coefficient of variation** (CV) is another measure of precision, which describes the distribution of measurements for a repeated series of tests on the same sample. It is expressed as a percentage and can be calculated using the equation:

$$\%CV = \frac{SD}{\bar{x}} \times 100$$

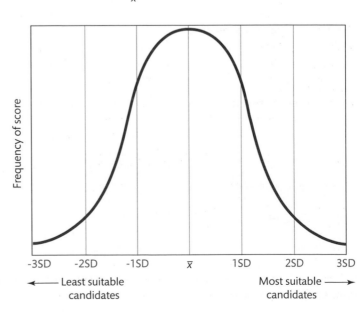

FIGURE 1.10
Normal distribution curve showing confidence
intervals expressed as standard deviations from
the mean.

where CV, $\bar{x}$, and SD have their usual meanings of coefficient of variation, mean, and standard deviation respectively. The value of the CV of an assay can be helpful when assessing its performance. The lower the CV, the more precise the method.

SELF-CHECK 1.6

Calculate the CV of an assay that gives results for the $\bar{x}$ and an SD of 5.0 and 0.5 mmol/L respectively.

Documentation and procedures

All analytical methods for testing clinical samples are carefully documented and any changes they display in performance recorded. The following should be provided for each analytical method and aid in result validation:

- standard operating procedures
- training log books
- analyser maintenance log books
- reagent log books
- calibration log books
- serum indices
- quality control

All the tests performed in a clinical laboratory must have full **standard operating procedures** (SOP), which is a document describing in detail, exactly how a measurement is performed to give a valid result. Procedures are reviewed annually as a minimum and updated as required.

Training log books ('logs') are records that are kept to confirm an operator has been fully trained in the use of an instrument and/or in performing a test. Thus, his or her competence is formally assessed and documented. **Analyser maintenance logs** are records that confirm an analyser has been maintained to the required standard. These standards are usually set by the manufacturer. **Reagent logs** are records which confirm that the reagents in use for a particular assay are in date and have been stored at the correct temperature. **Calibration logs** are records which confirm that the analyte to be measured has been calibrated according to manufacturer's instructions within the correct timescale.

Serum indices can be measured by most high volume chemistry analysers and are used to assess the effect of haemolysis, icterus, and lipaemia in serum or plasma. Haemolysis is the lysis of erythrocytes, which releases their contents into the plasma causing it to become pink or red coloured because of the presence of haemoglobin. It can be caused by poor venepuncture technique, over-vigorous mixing of the sample, overlong storage before centrifugation or an inappropriate anticoagulant in the tube. Assays that are particularly affected by haemolysis include those for potassium, lactate dehydrogenase, AST, ALP, phosphate, and magnesium. However, the interference also depends on the methods being used, as haemoglobin has an absorbance peak at 415 nm which will affect any method measuring at or near that wavelength. Icterus, more commonly known as jaundice, refers to the presence of high concentrations of the yellow pigment bilirubin, which appears in serum and in the other tissues most frequently in liver disease. Lipaemia is the presence of lipids and in particular triacylglycerols in the serum that give it a cloudy appearance.

High concentrations of bilirubin in serum or icterus occur in acute liver disease. Bilirubin exhibits a broad absorbance peak between 450 and 460 nm. Many assays of enzyme activities depend on measuring absorbances in this region of the spectrum and all of these can be affected. The higher the concentration of bilirubin present, the more likely it is to interfere with the assay. More significantly, non-enzymatic creatinine assays, which are based on the Jaffé reaction, are also measured at these wavelengths leading to possible spurious results.

Lipaemia is the cloudy appearance of serum samples usually caused by the presence of lipids and in particular triacylglycerols. It interferes in the analysis of most analytes and, in extreme cases, the initial absorbance may be so high that a result cannot be determined. The effect of lipaemia is difficult to eliminate. Most algorithms that have been applied to compensate for its effects have been based on the addition of intralipid, a synthetic product used to supplement nutrition in very ill patients, to normal serum.

These effects can interfere with clinical tests and lead to false low or high results being reported for particular analytes. A skilled biomedical scientist can make a visual assessment of the sample colour and equate that to any interference; however, this is time consuming and subjective. Measuring indices analytically is a reliable and consistent technique and easily applicable to large numbers of samples. The interference to the value of the index is calculated by measuring the absorbance of the diluted sample of plasma or serum at a series of wavelengths. The varying degrees of absorbance can then be used to estimate the amount of interference by applying a series of relatively simple rules. If the interference is more than 10% of the measured value, it is regarded as significant. If the interference exceeds this limit, then the results of the test can be automatically suppressed and an error code substituted for the analytical result.

SELF-CHECK 1.7

What is haemolysis and why does it cause interference in some biochemical assays?

Clinical evaluation and validation

Clinical biochemistry laboratory tests undergo clinical evaluations to assess their usefulness in diagnosing diseases. The evaluation involves determining their clinical sensitivities and specificities. You should not confuse these *clinical* terms with *analytical* sensitivity and specificity.

Clinical sensitivity and specificity

The **clinical sensitivity** of any assay describes its ability to detect only *patients* with a particular disease process. An assay with 100% sensitivity will always detect all patients with that disease. However, if sensitivity is only 50% then only one in two patients will be detected. The **clinical specificity** of an assay describes its ability to detect only that *disease process*. An assay with 100% specificity will detect patients with only that disease and not patients suffering from other conditions. In reality there are very few assays that can be claimed to have 100% clinical specificity and 100% clinical sensitivity. This knowledge is once again essential when interpreting results.

In numerical terms, clinical sensitivity is defined as the number of samples that give a true positive (TP) result, divided by the number of expected positives, which is the number of TP plus the values of false negative (FN) results, expressed as a percentage, that is:

$$\text{Clinical sensitivity} = TP \times 100 \, / \, TP + FN$$

Clinical specificity can also be expressed as a percent value. It can be calculated by dividing the number of measured true negative (TN) results by the number of expected negatives, which is the number of TN, plus the number of false positives (FP), which is again, multiplied by 100:

$$\text{Clinical specificity} = TN \times 100 / TN + FP$$

An assay may be excellent at detecting all of those who have the disease under investigation but if it also has a high rate of FPs, this will lead to many patients being erroneously classified as having the disease. This, of course, results in anxiety to the patients and further investigations by the clinical laboratory, both of which are unnecessary. In a similar manner, an assay which has a high rate of FNs will not detect all of those who have the disease.

SELF-CHECK 1.8

What is the difference between clinical specificity and clinical sensitivity?

Positive and negative predictive values

Generally, the majority of a population will not have the condition that is under investigation. This has implications for any test that is being used to predict the presence or absence of a specific disease. Calculating positive predictive values (PPV) and negative predictive values (NPV) using the expressions

$$PPV = TP / (TP + FP)$$

$$NPV = TN / (FN + TN)$$

can be used to predict the ability of a test to accurately detect disease in a defined population.

CASE STUDY 1.1

Ovarian cancer has an incidence of approximately 0.2% in the normal female population. A new tumour marker assay, called Q, for use in detecting ovarian cancer has been developed. A controlled clinical trial involving 100,000 females to evaluate its effectiveness is given in the table below.

	Disease present	
	Yes	No
Positive test	196 (TP)	6986 (FP)
Negative test	4 (FN)	92814 (TN)

(a) What is the positive predictive value of the assay Q?

(b) What is the negative predictive value of the assay Q?

(c) Comment on your findings.

Clinical validation

Clinical validation follows technical validation and examines the probability that an analytically correct result is really possible for that patient. Most clinical validations are performed using specific computer programs and the vast majority of results will be autovalidated. Some of the common criteria applied include:

- delta and range checks
- reference ranges

Delta checks examine any value obtained for an analyte against the previous result. Any changes found are compared to preset allowable differences that are determined locally. A **range check** looks at the value for an analyte to determine whether the result is physiologically possible. A **reference range** is defined as the range of values for a clinical test within which 95% of the healthy population fall. A reference range can only apply to a single, clearly defined population because age and sex related differences, variations between hospitalized and non-hospitalized patients, differences due to ethnic origin, the posture, standing or supine, of the patient when the sample was taken, the time of sampling, and even the time of year can all affect the range of possible values. For example, an analyte like 25-hydroxycholecalciferol (a form of vitamin D) is much higher in the summer months than in winter for the United Kingdom population. Reference ranges are also method dependent; this is particularly relevant for some assays, such as those for tumour markers where the clinician may be monitoring the effects of treatment or tracking a series of results to detect an early recurrence of a cancer.

Previous reports

If a patient has laboratory records from previous tests, it is helpful to view these when reporting the data of a new test. Any new and significant changes, which have not been flagged by other criteria, can easily be spotted.

CASE STUDY 1.2

A male patient diagnosed with a malignant tumour of the liver had it removed and was given a course of chemotherapy at the regional cancer centre. Originally his α-fetoprotein (AFP), a tumour marker, activity was 7500 KU/L, which is grossly elevated but after treatment this gradually declined to only 5 KU/L. His doctor removed a blood sample for a routine follow-up test after three months: its AFP result was 15 KU/L. Suspecting a recurrence of the tumour, the patient was rapidly referred back to the oncologists at the regional centre. Here, a re-check showed his AFP to be 5 KU/L. Detailed inquiries showed that the regional cancer centre laboratory uses a different instrument for analysis to the local hospital to which the doctor sent the sample. Both results are normal even though the values are significantly different because different methods have been used.

(a) How can the laboratories confirm that the results are not clinically significant?

(b) How can the laboratories avoid this happening again?

CASE STUDY 1.3

A sample from a 65-year-old male arrived in the laboratory at 9 am from a well-man clinic. The potassium is 8.5 mmol/L which is dangerously high. The sample is repeated and the same result is obtained. All other laboratory checks have been carried out and the result is analytically valid. The result is telephoned to the GP who reveals that the sample was taken at 4 pm the previous day by the nurse.

(a) What is the most likely cause of the high result?

(b) What would your recommended course of action be?

Clinical details

At this stage, it is necessary to check that clinical details pertinent to the patient, such as a preliminary diagnosis, current drug therapy or symptoms are present on the request form (we introduced request forms in Section 1.2). All this information can be used to complete the diagnostic picture. Unfortunately, this part of the request form is most frequently incomplete or even absent. When it is included, it can often be of great help when trying to interpret the results of a clinical test.

Communication

When all avenues to explain a difficult to interpret result have been explored, a phone call to the relevant clinician to ask appropriate questions will usually clarify the situation for both parties.

Each analyte must also have relevant quality controls associated with any test result. These quality controls must be validated before the result can be accepted. We will discuss quality control in more detail in the next section.

1.5 Quality control

Quality control is a set of procedures that maintain the validity of the clinical biochemical results. The excellence of the quality of clinical tests is one of the mainstays of any diagnostic service. The clinician needs to know that he or she can confidently rely on the results of any diagnostic test leaving the laboratory. Thus, they can be used to support diagnoses and in the monitoring of patients in care. This section will provide you with an overview of the quality processes used in the clinical laboratory to ensure that results are delivered with the required degree of confidence.

Internal quality control

Ideally, every test would be 100% **accurate** and have absolute **precision**. In practice, this ideal state can never be achieved because there are too many variables associated with the process of analysis, including the workings of the laboratory staff, the instrument used, the steps involved in the analytical method, the reagents used in the method and in calibrating the instrument. However, it is necessary to know the accuracy and precision of any test so that its results can be correctly interpreted when making a diagnosis or monitoring the treatment of a patient.

Systematic and random errors

A **systematic error** occurs when an assay has a constant bias against the true value. You can see an illustration of the effects of a systematic error in Figure 1.11 (a). Note that the results are precise: they are grouped closely together and are within 1 SD of the mean value. However, the mean of these results is different to the true mean. Furthermore, they differ from the true value by a constant amount in the same direction and therefore they were not obtained from an accurate method but one that shows systematic error. **Random errors** cause test values to vary either side of the mean and produce scattered results. You can see this illustrated in Figure 1.11 (b). Here the results are again accurate because the mean of all the values is equal to the true mean ($\bar{x}$). However, the method is not precise because the individual values are scattered up to 3 SDs apart. Random errors usually occur because of the inability of individuals to replicate a process or part of it consistently.

SELF-CHECK 1.9

What is the difference between the terms accuracy and precision for a biochemical test?

Selection of quality control materials

Quality control materials are the substances used to assess the quality control of clinical methods. They are usually serum containing a known quantity of the analyte in question. The main aim when choosing a material to use for internal quality assurance procedures is to have one that reacts in the same way as the serum or urine of a patient. Most major quality control manufacturers will provide, whenever possible, material based on a human matrix. Bovine and equine materials are also widely available and tend to be less expensive but may differ from human tissues for some analytes, particularly hormones. The material must be fully tested to be free from blood-borne viruses such as HIV and hepatitis B. The composition of all materials, human or non-human in origin, is adjusted to provide the required range of analytes and provide test results that could never occur using the serum or urine of any one

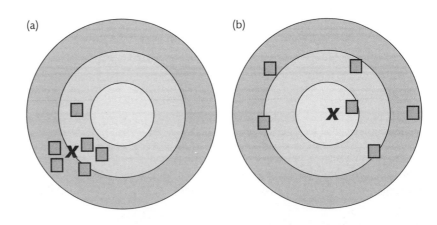

FIGURE 1.11
(a) Results that are precise but not accurate and show a systematic error. (b) Results that are accurate because the mean is close to the true mean but are imprecise because the values are scattered up to 3 standard deviations apart.

individual. Thus, it is possible to have most of the core analytes present in one vial, but there will be many assays that require specially manufactured material, which must be stored under stringent conditions.

Lyophilized or freeze-dried material is stable and can usually be stored safely between 4 to 8°C. The material must be reconstituted before use and the pipetting is a potential source of error. After reconstitution, the components will have a shortened period of stability, which may be only a few hours to several weeks if stored properly. Liquid-stable material is usually stored frozen until required and must be properly thawed before use. Unlike lyophilized material it does not require reconstitution; however, after thawing the components will have a similar shortened period of stability dependent on how they are stored. In both cases, good stability of the opened material optimizes its use.

Most clinical laboratories buy commercially manufactured material because of the difficulties involved in preparing stable in-house controls. A long expiry date for the material is desirable to minimize the disruption caused when a new material requires evaluation. Maintaining the same batch for one to two years should be possible and stocks need to be purchased in advance from the manufacturer. Unfortunately, for some specialized assays it is not financially viable for manufacturers to produce them commercially. Hence, they will need to be prepared, checked, and documented in-house by the clinical biochemistry laboratory.

Whatever, the QC material, it should be treated in exactly the same way as would a patient sample. Manufacturers will usually offer three target concentrations of material to cover the likely clinical range for each assay. However, there are no hard and fast rules governing usage, which will depend on the individual assay performance associated with local requirements. For qualitative assays, positive and negative QCs should be used with the negative control set just below the cut-off value for the assay.

Frequency of use

Most major manufacturers recommend that instruments should be tested at least once every eight hours using three levels of QC material. This is a good rule of thumb for assays on stable modern analytical platforms where maintenance, reagents, and calibrations are performed first of all and then QC analysed prior to patient samples. Providing there are no further changes made throughout the day, there is no real need to run more QC material. However, if any part of the assay is changed, for example a new reagent added, sample pipettes or lamp changed, then QC at all levels must be repeated to confirm that the assay results remain within control values. Older systems suffered from the inherent fault of the test result drifting from the true value. Consequently, instruments must be tested with QC material much more frequently than every eight hours. Results from tests on clinical samples can only be released if those obtained using QC before and after a batch of clinical samples are within the appropriate values. This situation is still the case for many manual assays, which involve multiple steps and, indeed, it is necessary to run QC interspersed throughout the samples to ensure the integrity of results.

Presentation of data

It is relatively quick and easy to assess whether a number is within or outside a specified range of say +/− 2 SDs of the mean. However, one value is only an assessment of an assay performance at that particular moment. The greater the number of estimations made, the more accurate the assessment becomes and, in general, an absolute minimum of ten data points are required to calculate any meaningful statistics. The greater the number of data points,

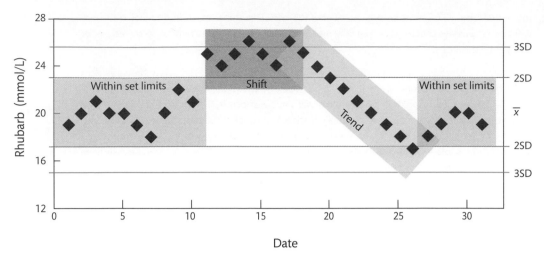

FIGURE 1.12
Levy-Jennings chart showing cumulative quality control data. See text for details.

the more difficult it becomes to interpret, However, graphically presented cumulative data is easier to interpret and a number of different methods have been developed to do this.

Levy-Jennings or Shewart chart

The Levy-Jennings or Shewart chart is the most commonly used graph to illustrate cumulative performance. If you examine Figure 1.12 you can see that concentration is plotted against time and that the values of the mean and −2 to +2, and −3 to +3 SDs are also indicated.

An assay is in control when results are distributed either side of, but close to the mean. Values within +/− 2 SD are generally regarded as acceptable. The presentation of QC results in this way makes it possible to see developing trends, such as sudden shifts in precision or accuracy, relatively easily. Any change from this pattern, such as a series of results that are evenly distributed but extend to +/− 3 SDs would result from a decrease in precision, and therefore indicate a deterioration in the performance of the test that requires investigation.

When six or more results show a consecutive move in the same direction, whether up or down from the mean, it is called a **trend**. Trends are suggestive of a developing problem, such as a deterioration in the quality of the reagents used in the test or in the QC material, or may indicate the analytical instrument requires maintenance or has laundry problems. If six or more results occur on one side of the mean, rather than scattered about it as in a trend, then this is called a **shift**. Shifts can be positive or negative depending on whether they occur above or below the mean respectively. In both cases, they indicate a change in accuracy of the assay. This may be caused by one of a number of factors, including changing a calibrator or a reagent by reconstituting them incorrectly, using materials from a different batch. Changes in the temperature at which the analysis is performed or simply due to a fault in the analytical instrument, such as a wrongly calibrated pipettor or faulty lamp in the spectrophotometer can also be causes.

Youden plot

The Youden plot is a graph of one QC against a second as you can see in Figure 1.13, where the coloured squares represent the values of standard deviations: yellow squares equal 1, green

CASE STUDY 1.4

When a Levy-Jennings graph for TSH was checked, the mean of results was found to have shifted by +1.5 SD over the four days since the last check. All other assays were found to be performing as expected. On investigation, it was discovered that the batch of reagent had been changed. A phone call to the manufacturer revealed that there were no other reports of problems with this batch. A more detailed investigation revealed that each batch of reagent has a specific calibrator, which can only be used for that batch.

(a) What type of error does this represent?

(b What is the most likely cause of the problem?

(c) How would you resolve the problem?

equals 2 and blue equals 3 SD respectively. The results of the two controls are plotted on the chart as usual x, y data points. A horizontal median line is drawn parallel to the x axis and a second median line is drawn parallel to the y axis. Their intersection is called the **Manhattan median**. One or two 450 lines then are drawn through the Manhattan median. The Youden plot is a useful graph in that it gives a clear indication of whether any errors are systematic or random in nature. Data points that lie near the 450 reference line and are within the 1 to 2 SD squares are acceptable results. However, points that lie near the 450 reference lines but are outside the 2 SD square are the result of systematic errors. Points that are distant from either 450 reference lines arise from random errors. Examples of data arising from both types of errors can be seen on Figure 1.13.

Use of rules

It is common practice to use rules as aids in interpreting QC analyses. These can be locally devised to suit particular assays or the requirements of patients. The commonest ones used are those devised by Westgard, which can be applied to QC results in a consistent and effective manner. The Method box shows a typical set of these rules, although there are many variations to them that can be applied to QC measurements and which can be tailored towards specific assay requirements maintaining the quality of results at the required level.

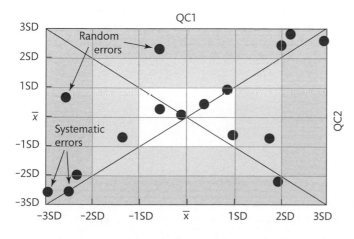

FIGURE 1.13

Example of a Youden plot. See text for details.

METHOD *The use of two or three QC materials*

Two QC materials

(1) When any QC measurement is greater than 3 SDs from the mean, the run should be rejected and corrective action is required.

(2) When any QC measurement is more than 2 SDs of the mean, a warning must be triggered and the run should be carefully monitored, although at this point the assay need not be rejected.

(3) When two consecutive QC measurements differ from the mean by +2 SD or −2 SD, the run should be rejected and corrective action taken.

(4) When one QC measurement exceeds +2SD and the next exceeds −2SD then the run should be rejected and corrective action taken.

Three QC materials

(1) When two out of three QC results differ from the mean by +/− 2 SDs, then the run should be rejected and corrective action taken.

(2) When six consecutive QC results occur to one side of the mean, the run should be rejected and corrective action taken.

(3) When seven consecutive results show an increasing trend the run should be rejected and corrective action taken.

Quality control software packages normally contain the Westgard rules so that anomalous results are automatically brought to the attention of whoever is operating the instrument and the results are held back. Instrument software, if appropriately written can also highlight situations where the QC is outside acceptable ranges usually using a colour-coded 'traffic light' type of system, where results within +/− 2 SDs are indicated in green, those between 2 and 3 SDs in yellow and results beyond 3 SDs are flagged in red.

Reviewing poor performances

Once a QC problem has been identified, its root cause must be found. A good analyst will systematically investigate a QC problem, eliminating potential causes one at a time until the basic cause of the problem is discovered. For instance, it would be easy for an operator to simply adjust a calibrator to compensate for a shift. However, this would not solve the problem but simply disguise it until the next problem appeared, which may be a shift back to the original values.

Quality control must be reviewed every time a control is run but it is good practice to review cumulative data on a regular basis. Every clinical laboratory has a QC officer whose responsibility is to ensure that this audit process takes place and timely corrective action is carried out when necessary.

External quality control

The purpose of an external quality control is to assess the performance of a laboratory, compare it to that of other laboratories that use the same analyser and method, and against other methods/analysers measuring the same analyte. The information gained from participation in such a scheme gives assurance that every laboratory achieves the same result as those using the same methodology and also indicates whether the method applied in one laboratory has a bias or reproducibility problem.

Selection of scheme

Laboratories in the UK must participate in external QC programs for all analytes for which one is available. The three main programs in the UK, are the United Kingdom National External Quality Assessment Service (UKNEQAS), Welsh External Quality Assessment Service (WEQAS), and Randox International Quality Assessment Scheme (RIQAS). None of the schemes offers a QC service for all the analytes measured in UK laboratories and so most are members of two or all three. Each program provides valuable information, although in different ways, to participators, but ultimately all serve the same overall purpose of improving the quality of clinical laboratory provision.

Presentation of data

External QC programs regularly, normally on a two to four-week cycle, send out samples for analysis and then issue reports on performances based on the returns they receive. The reports usually contain estimates of bias and reproducibility of the tests for the analytes measured and a score for a test that reflects its overall performance during a six-month rolling window. This score can be used to compare against the average results achieved for an analytical method and also against the overall results for the analyte, which may involve numerous different methods.

Reviewing poor performance

When the test for an analyte is deemed to have persistently poor performance in a QC program, the laboratory receives a formal letter highlighting this and is asked to explain how the problem will be resolved. The scheme organizers compare the laboratory's performance with that of other users of the same method and, if all experience the same poor performance, then the manufacturer will be approached as the error is specific to that method.

SELF-CHECK 1.10

What is the value of analysing external QC materials for any given analyte?

1.6 Audit

In 1993 the NHS defined **audit** as 'a quality improvement process that seeks to improve patient care through review of care against explicit criteria and the implementation of change'. Thus, both the internal and external QC programs used to assure the analytical validity of test results are types of audit. However, there is a potential for error in the pre- and post-analytical phases, which cannot be fully assessed by these programs. Thus, the process of audit has been introduced and accepted in clinical laboratories to evaluate both the pre- and post-analytical phases. Indeed, audit is regarded by CPA, which was mentioned in the Introduction, as essential to assure the continued validity of laboratory processes. The overall responsibility for ensuring audits are carried out effectively lies with the quality manager of the organization.

Audits are snapshots of a process as it would be impractical to evaluate the whole organization continuously. However a carefully constructed audit calendar gives a good view of the overall quality process. There are three main types of audit processes used in clinical laboratories:

- horizontal audit
- vertical audit
- examination audit

Horizontal audits are designed to examine one element of a process applied to more than one item. An example of this would be to look at the disposal of clinical waste throughout a department or how the department monitors the storing of reagents. Any anomalies identified by the audit must be addressed and the best practice to rectify the situation adopted. In contrast, a vertical audit is one designed to examine more than one element of a process but on a single item. A typical example would be to look at the analysis of a single sample for glucose and follow it through the reception process, PID entry, centrifugation, analysis, reporting, storage, and discard. Any problems discovered by the vertical audit, may require a horizontal audit of one or more elements of the process. An examination audit examines how a *person* performs a task to ensure that the SOP is correct and being accurately followed and to ensure that the person understands what he or she is doing and why they are following that particular procedure. The person in question, is observed carrying out an analysis. This type of audit is particularly useful because careful questioning of a person with experience of a particular SOP can lead to useful suggestions as to how it can be improved.

All three types of audits, vertical, horizontal, and examination, are designed to be used together to ensure that all aspects of the laboratory are working to agreed standards.

CASE STUDY 1.5

A vertical audit of a test for analysing concentrations of progesterone found that the reagents are stored in a refrigerator although the protocol stated they should be stored between 2 to 8°C. The temperature of the refrigerator was not monitored. A horizontal audit to measure the temperatures of refrigerators throughout the department showed none of them maintained temperatures between 2 to 8°C. The quality manager agreed with the departmental management team that the current storage of the reagents for this test was an unsafe practice and a critical non-compliance with CPA (Section 1.1) standards.

(a) What are the implications of storing temperature critical reagents in this way?

(b) What action is required to resolve this non-compliance?

(c) How would you determine whether the problem has been rectified?

SUMMARY

- Clinical biochemistry is a relatively young pathology discipline, with much of its development taking place in the last hundred years.

- Biochemical clinical tests are used to support the diagnosis, treatment, monitoring, and prognosis of the patient. Most clinical laboratories provide a similar core service, with the main analysers performing an extensive repertoire of chemical and immunoassay tests.

- The equipment used to analyse clinical samples is often complex and large laboratories will produce many thousands of results daily. However, advances made in automation, robotics, and computing have made it possible to cope with ever increasing workloads even when the numbers of staff are reduced.

- Specialist tests are often performed on clinical samples at regional centres where the required expertise is available and to reduce duplication of expensive equipment. In contrast, point of care testing is increasing and although presenting numerous benefits, it presents a new challenge to the clinical laboratory.

- Teamwork involving several groups of staff including medical, clinical, and biomedical scientists, laboratory support staff, and clerical workers is essential in obtaining accurate test results. However, a patient's sample is subject to many processes and checks in the clinical biochemistry laboratory before a result is made available to medical staff. Thus, the careful use of flags, colour codes, rules and graphics, and constant vigilance by the operator all contribute towards providing the patient with an accurate, reliable, and consistently reproducible result.

- Quality control is essential for all assays and when utilized properly gives confidence to the laboratory staff and to users of the service with regard to precision and accuracy of all the tests performed. Early detection of poor assay performance and proper corrective action minimizes the risk to patients of receiving an incorrect result.

- When a test result is queried, being able to quote the performance of the assay at the time of measurement allows clinical laboratories to confidently defend the result, which also gives confidence in the results to clinicians dealing with patients.

- The aims of an audit are to confirm that patients receive the best possible service in aiding diagnoses and in the monitoring and treatment in the healthcare environment. Thus, audits are major contributors to the continuous improvement program, which all good laboratories implement. They provide a transparent evidence base from which sound conclusions can be drawn and action plans constructed, ensuring that best practices are adopted within the clinical laboratory.

FURTHER READING

- **Beckett GJ, Ashby P, and Walker SW (2005)** *Lecture Notes on Clinical Biochemistry*, 7th edition. Oxford: Blackwell Publishing.

  This text explains the physiological rationale for any test in the context of a disease using a system-based approach. It encourages students to develop a critical approach to diagnostic investigations by appreciating the value and limitations of clinical biochemistry tests in diagnosis and patient management.

- Gaw A, Cowan RA, O'Reilly DJ, Stewart MJ, and Shepherd J (2004) *Clinical Biochemistry: an illustrated Colour Text*, 3rd edition. London: Mosby.

  This user-friendly book has minimal text but is full of excellent illustrations and uses case histories and problem-solving methods.

- Luxton, R. (2008) *Clinical Biochemistry*, 2nd edition. London: Scion Publishing.

  Useful reading to any undergraduate with an interest in the subject. Covers basic concepts and facts in a friendly manner and focuses on the areas of body function required to maintain health.

- Plebani M (2010) The detection and prevention of errors in laboratory medicine. *Annals of Clinical Biochemistry* 47, 101–10.

  A comprehensive review on types of errors in clinical biochemistry laboratories and their prevention.

- The Carter Review of NHS Pathology Services 2006. http://www.thecarterreview.com/publications.html

  Useful website to browse.

- Vermeer HJ, Thomassen E, and de Jongea, N (2005) Automated Processing of serum indices used for interference detection by the laboratory information system. *Clinical Chemistry* 51, 244–7.

- Westgard JO (2010) *Basic QC Practices*, 3rd edition. Madison, Wisconsin: Westgard Quality Corporation.

- Westgard JO, Barry PL, Hunt MR, and Groth T (1981) A multi-rule Shewhart chart for quality control in clinical chemistry. *Clinical Chemistry* 27, 493–501.

  Concise yet easy to understand descriptions of the steps used in planning, implementing, and operating quality control procedures.

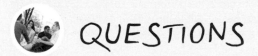

QUESTIONS

1.1 Henry Bence Jones is regarded as the father of paediatric clinical biochemistry:

  (a) True

  (b) False

1.2 Yalow and Berson developed which of the following?

  (a) Monoclonal antibodies

  (b) The Technicon Autoanalyser

  (c) Radioimmunoassay techniques

  (d) The pH meter

  (e) All of the above

1.3 Which of the following parameters are important when a blood sample is received by the clinical laboratory?

  (a) Sample and form have the same patient details attached

(b) Time of sampling

(c) Relevant clinical details

(d) Test(s) requested and return address

(e) All of the above

1.4 HDL is the abbreviation of which of the following?

(a) Human deficiency lymphocyte

(b) High dependency laboratory

(c) Heavy deuterium lamp

(d) High density lipoprotein

(e) All of the above

1.5 In the laboratory, which of the following does the abbreviation 'SOP' represent?

(a) Serum organic phosphate

(b) Standard operating procedure

(c) Shortened operating procedure

(d) Serum oligofactory protease

(e) All of the above

1.6 Clinical laboratories participate in external quality control schemes to allow which of the following?

(a) To ensure the day to day reproducibility of results

(b) To be able to spot random errors

(c) To compare performance in your laboratory against other laboratories

(d) To be able to calculate the SD of the method

(e) All of the above

1.7 What is the difference between the technical and clinical validation of a result generated in the laboratory?

1.8 Calculate the mean, SD and CV for the data shown below, which are a range of results obtained for the concentrations of oestradiol in pmol/L obtained using a new analytical method.

240, 260, 230, 280, 240, 220, 240, 230, 260, 250.

1.9 The new tumour marker assay (Q) for ovarian cancer introduced in Case study 1.1, was subjected to a controlled clinical trial to evaluate its effectiveness. A cohort of a thousand women with confirmed disease were used as positive controls and a cohort of a thousand women assessed as disease free acted as negative controls.

	Trial results positive	negative
1,000 Positive controls	980 True (TP)	20 False (FN)
1,000 Negative controls	70 False (FP)	930 True (TN)

(a) What is the clinical sensitivity of assay Q?

(b) What is the clinical specificity of assay Q?

(c) Comment on your findings.

1.10 A new method is introduced to the laboratory to measure the concentration of rhubarb in serum. A QC sample was repeated ten times and gave results of 10, 8, 10, 9, 9, 10, 11, 10, 8, 9. The method currently in use shows a mean of 10.0 and a SD of 0.667.

Calculate (a) the mean result (b) the SD, and (c) the CV for the new method? (d) How do the old and new methods compare?

Answers to self-check questions, case study questions, and end-of-chapter questions are available in the Online Resource Centre accompanying this book.

 Go to www.oxfordtextbooks.co.uk/orc/ahmed/

Automation

Tim James

Learning objectives

After studying this chapter you should be able to:

- Identify which tests can be performed on the two main types of analysers found in clinical biochemistry laboratories, the general chemistry analyser, and the immunochemistry analyser
- Explain the key automated steps within the analyser
- Describe the key mechanical components of the analysers and describe how they contribute to the analytical process
- Discuss the pre- and post-analytical stages of the laboratory process that can be automated and integrated with clinical laboratory analysers

Introduction

The workload of clinical laboratories worldwide has grown at a consistent rate of 5–10% per annum since the 1970s. During the same period, a technological revolution has developed analytical systems that allow hospital laboratories to process thousands of specimens and tests daily. Clinical biochemistry laboratories undertake a broad range of tests for a variety of different clinical conditions, as described throughout this textbook. A significant proportion of these tests utilize relatively uncomplicated reactions that may be automated with a common testing pattern, as shown in Figure 2.1. In its simplest format, a small volume of the sample is removed from the specimen and dispensed into a cuvette where reagent(s) is added to produce the reaction that is monitored. Each test is based on a different method but the component steps to produce the result are common. Also, it is possible to measure many different tests on the same instrument. A descriptive term applied to such instruments is **general chemistry analyser**, although historically other terms such as autoanalyser have also been used. Automated analysers are found in virtually every clinical biochemistry laboratory and provide the core high volume tests.

A major outcome of increased automation is the more rapid production of test results. This is measured as test **turnaround time**, the time taken to produce the result. The total turnaround

Cross reference
Chapter 17 Laboratory automation in the *Biomedical Science Practice* textbook

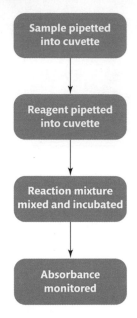

FIGURE 2.1
Automated chemistry analysers are highly complex instruments, but the basic sequence of events is simple: probes pick up and dispense sample and reagent(s) into a cuvette, a reaction is initiated and monitored. In most cases, the change monitored is in the absorbance.

time is the time taken from sample collection to the clinician receiving the result. The within laboratory turnaround time is the time between sample receipt in the laboratory to the production of results. Automation has speeded up the production of results, improved turnaround time, and consequently more timely clinical decisions.

The general chemistry analyser, while providing an analytical option for a significant proportion of the tests demanded of routine laboratories, has limitations for other testing areas. This is particularly true where there is a low concentration of the analyte in the test material. For such tests, a more sensitive method is required and a family of analysers called **immunochemistry analysers** or immunoassay analysers are employed. These are used to perform a wide range of tests, for example those for:

- hormones, such as thyroxine and cortisol
- drugs, such as phenytoin and theophylline
- vitamins, such as vitamin B12 and folate

Ideally, all tests would be performed on a single analyser, and consolidating as many assays as possible within the same analytical unit using similar technology has been an aim of both instrument manufacturers and laboratories. However, most clinical laboratories will have at least one general chemistry and one immunochemistry analyser; the majority will have more than one of each. Consequently, there has been an increased use of tracks and sample management systems to convey specimens between different processing and analytical units that automate the specimen handling functions. Such tasks have traditionally been undertaken manually. This has reduced the amount of time laboratory personnel spend undertaking simple repetitive tasks and reflects the aim of minimizing human intervention in the total analytical process. The term **total laboratory automation** has been used to describe the most advanced of these systems that enable procedures traditionally undertaken as discrete functions before and after analysis to be integrated in a single automated process. The extent of laboratory automation depends on the size and complexity of the laboratory service provided.

A considerable number of different analysers are available. However, the best design features within automation systems are maintained and copied between manufacturers. It is therefore possible to see common features within analytical systems manufactured by different suppliers. However, the variation of analysers from different manufacturers is such that there are also unique aspects in all instruments. Therefore, the descriptions given are general and not specific to any one instrument.

This chapter describes the basic characteristics and features of the two types of systems, that is the general chemistry analyser and the immunochemistry analyser, together with the automation elements that connect them together. Aspects of automation in more specialized areas of the biochemistry laboratory are also briefly discussed.

2.1 Historical context

The development of automation in clinical biochemistry has been steady and progressive since the early 1960s. The first automation in clinical laboratories employed the concept of **continuous flow analysis**, developed by Leonard Skeggs and commercialized by the Technicon Corporation. A defined volume of sample was introduced at regular intervals into a continuous flow of reagent within tubing where the chemical reaction could take place. The reaction most often resulted in a change in absorbance and this was monitored, in most cases, within a flow cell detector. Air bubbles were used to limit the sample to discrete portions of the moving reagent stream and coils ensured the sample and reagents were mixed in a standardized manner. Dialysis membranes reduced interference by providing protein-free samples. Reaction sequences could be modified with a range of additional devices. For example, by passing the tubing through longer mixing coils to provide extended incubation, or using heating blocks to achieve higher reaction temperatures.

A steady stream of applications of the 'autoanalyser' using different methods appeared from 1957, when Skeggs first presented the concept, through to the early 1970s. Automated methods included those for measuring calcium, creatinine, cholesterol, glucose, urea, and enzymes such as alkaline phosphatase, activities. The first methods were single test applications with a simple graphical output on a chart recorder and required a manual measurement of peak heights. Notable milestones occurred when the system adopted multiple channels and computerization as shown in Box 2.1. The application of continuous flow analysis has essentially disappeared from the clinical biochemistry laboratory. However, a derived technique, termed flow injection analysis, is still applied in other areas of analytical chemistry such as environmental and pharmaceutical laboratories.

In the late 1960s and early 1970s a number of companies developed automation concepts to compete with Technicon's autoanalyser, which utilized similar chemical methods but different application technology. The newer instruments made greater use of pipetting devices

BOX 2.1 Timeline of automation developments

The automation of methods to analyse clinical samples has a relatively long history. The following brief account illustrates some of the major developments.

1957	Skeggs introduced continuous flow analysis, which was developed commercially by Technicon into the autoanalyser
1960–1970	Growth in application of biochemical methods to the autoanalyser system
1969	Sequential Multiple Analyser (SMA)
1970	Dupont produce the Automated Clinical Analyser (ACA)
1974	Sequential Multiple Analyser with Computer (SMAC)
1978	Kodak produce the Kodak-Ektachem an analyser employing dry chemistry slide technology
1980–1990	Convergence and standardization of chemistry analysers
1990–1995	Automation of non-isotopic immunoassay methods
1995–2000	Development of combined general chemistry and automated immunoassay within a single instrument
2000–2005	Widespread utilization of total laboratory automation and tracking to link multiple analysers together

that allowed liquid transfers, such that both sample and reagents could be aspirated and dispensed from one location, either the sample cup or reagent container, to another, usually the reaction vessel. The newer analysers included the Automated Clinical Analyser (ACA), developed by Dupont, and the centrifugal analyser, and a dry chemistry slide technology analyser from Kodak.

The technical nomenclature used for these newer analysers highlighted their differences from and advantages over continuous flow systems. **Discrete analysers** are those capable of handling each sample in an individual vessel physically separated from the others. **Discretionary analysers** can perform one or more of the assays from those available from the analyser's repertoire, as determined by the operator. **Random access analysers** are those able to undertake an analysis independent of the position of the specimen in the sample queue. Early automated analysers were distinctly different from one another and each had individual advantages and disadvantages. However, the subsequent evolutionary developments have resulted in a range of instruments that utilize similar biochemical methods and similar application technologies even though they are produced by different manufacturers. These are described in more detail in the subsequent sections. General changes of note have been a growth in test and sample capacities, improved reproducibility, and greater robustness. The underlying methodological principles of many of the methods employed on chemistry systems have remained similar, although greater standardization has improved comparability.

2.2 General automated chemistry systems

General chemistry analysers automate the series of events necessary to initiate and monitor a defined reaction, which allows the quantitative, or occasionally qualitative, determination of the analyte of interest in clinical specimens. The analytical technique most commonly used in these systems is spectrophotometry, where a reaction is monitored by changes in optical characteristics in the ultraviolet or visible region of the spectrum. The other common technique used in general chemistry analysers is potentiometry, where an ion-selective electrode (ISE) is used to measure analytes such as sodium and potassium. Sodium and potassium are commonly requested investigations but they cannot be measured reliably using spectrophotometric methods and a separate ISE unit is required within the analyser. An example of a general chemistry analyser is shown in Figure 2.2.

Cross reference

Chapter 9 Electrochemistry in the *Biomedical Science Practice* textbook

General chemistry analysers may be small bench top devices with the capacity to undertake a few hundred tests per hour or may be large floor standing, high throughput instruments with a capacity to undertake several thousand tests hourly. The former are used in laboratories where relatively low specimen numbers are processed. Alternatively, they may be placed into rooms adjacent to wards where there is a need for rapid turnaround of results, or near emergency admissions or an outpatient clinic, where a decision on treatment may be dependent on a rapid result. Such smaller instruments can be considered an extension of the traditional single test **point of care testing** devices. Conversely, larger analysers are found in central laboratories where they undertake the core of the clinical biochemistry service. You can see the typical repertoire of assays these instruments perform in Table 2.1, together with the typical workload of medium to large UK laboratories.

Cross reference

Chapter 18 Point of care testing in the *Biomedical Science Practice* textbook

(a)

(b)

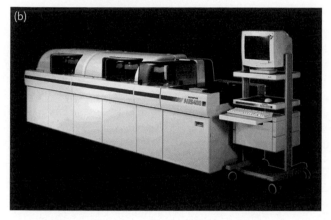

FIGURE 2.2

Examples of general chemistry analysers. (a) The ABX Pentra 400, a simple bench top system. (b) The Olympus AU5400, a high capacity analyser. See also Figures 1.6 and 1.8.

TABLE 2.1 Assays commonly undertaken on the main general chemistry analyser.

Tests for:	* Typical annual activity in medium to large laboratories in the UK
Sodium, potassium, chloride, bicarbonate, urea, creatinine, calcium, phosphate, albumin, total bilirubin, alanine aminotransferase, and alkaline phosphatase activities, C-reactive protein, glucose, total cholesterol, HDL , triacylglycerols	50,000 to 500,000
Uric acid, magnesium, total protein, creatinine kinase, iron transferrin, and aspartate aminotransferase, lactate dehydrogenase, γ-glutamyl transferase, amylase and lipase activities	10,000 to 50,000
Conjugated bilirubin, lactate, paracetamol, salicylate, ethanol, zinc, ammonia, bile acids, and angiotensin converting enzyme	500 to 10,000

* The workload associated with individual tests will depend on many factors, particularly local definitions of the key biochemistry profiles for organ function (renal function tests, liver function tests, etc.) and the patient population of the hospital. For example, ammonia is most often requested in neonates and children, therefore the absence of a paediatric ward or unit will lower demand for this test from the laboratory.

These tests may be requested individually or as a part of a profile. The exact definition of a test profile is determined by local arrangements but example profiles that are often used are:

- Renal profile: this usually includes the tests for creatinine, urea, sodium, and potassium.
- Liver profile: usually includes tests for alanine aminotransferase (ALT) and alkaline phosphatase (ALP) activities, albumin, and bilirubin.
- Lipid profile: this includes tests for total cholesterol, high density lipoprotein (HDL), and triacylglycerols.

The laboratory may also have arrangements with service users to provide more extensive profiles for patients with particular conditions or for specific clinical areas such as intensive care, where many biochemical investigations may be required daily to monitor the patient's metabolic status. It is, therefore, usual for each sample to require multiple tests. Most UK clinical biochemistry laboratories undertake an average of between six and ten tests per sample. This value is usually expressed as the test: sample or **test: request ratio**. An aim of automated testing is to perform all the assays required on one instrument, thus removing the need to move the sample between analysers.

The daily operation of analysers requires consistent maintenance schedules that are both planned and recorded. It is usual to have daily maintenance procedures, which may include cleaning key components such as sample probes, weekly maintenance, and monthly procedures, which may be less frequently needed housekeeping tasks, and six-monthly preventative maintenance when more fundamental instrument components are changed. A specialist engineer from the instrument manufacturer usually undertakes the preventative maintenance; however, the daily, weekly, and monthly maintenance should be within the capacities of competent laboratory staff.

Key Points

The capacity of an analyser is defined as the number of tests or samples that can be undertaken in a defined time period, often expressed per hour. Many analyser names incorporate a number in their title to indicate the number of tests it is capable of undertaking and some manufacturers have a family of analysers utilizing the same name but differing in capacity as indicated by the number. This number is usually the theoretical maximum capacity and may be reduced by a number of operational factors including the test: sample ratio, the actual tests required, and the sequence in which they are presented to the analyser.

User interface

All instruments use dedicated software that enables them to be operated. These user interfaces are designed for ease of use and have clear screen layouts that group similar functions together, often using icons. The access to the software can be organized hierarchically such that routine operation is restricted to the regular user, whereas the fundamental system setup is restricted to senior staff. Commonly grouped functions encountered within the system include routine operation, system configuration, chemistry parameters (see below), quality control, calibration, and maintenance. When the operator selects one of these functions, further sub-options are presented that can be selected as required. For example, if a maintenance procedure is required, the user will select the maintenance icon from which a series of options covering a range of maintenance protocols will be presented. All instruments have comprehensive manuals that allow users to understand the system and many incorporate 'help' modules such that when users encounter problems or error codes are presented, they can access electronic information that shows how to correct the problem. The help screens can also aid training.

A fundamental requirement for all analysers is to define what are generally called chemistry or test parameters for the instrument. These are the key features of each individual test, such as the volume of specimen and the detection wavelength. You can see a number of test parameters together with an associated common test in Table 2.2. Chemistry parameters are defined generally by the manufacturer and are only occasionally accessed because once they have been optimized and established, the parameters should remain constant to ensure reproducibility.

Specimen presentation

The analysis of blood specimens, in the form of serum or plasma, constitutes the majority of routine laboratory work undertaken on automated general chemistry analysers. However, it is possible to analyse a number of other body fluids if the appropriate method applications and validations have been undertaken. Therefore, laboratories also use the general chemistry

TABLE 2.2 Selected parameters as commonly defined for chemistry analysers, with an illustrative example to show how a calcium method may be set up. Each analyser will use instrument appropriate volumes.

Parameter	Parameter description	Values used in determining [calcium] (example on a typical automated system)
Sample volume	Volume of liquid specimen pipetted into the cuvette for the reaction	$2.4\ \mu L$
Reagent 1 volume	Volume of first reagent pipetted into the cuvette	$240\ \mu L$
Reagent 2 volume	Volume of second reagent (if required) pipetted into the cuvette	$0\ \mu L^*$
Primary wavelength	Wavelength in the UV-visible spectrum monitored to detect amount of product formed in test	660 nm
Secondary wavelength	Wavelength in the UV-visible spectrum monitored to correct for background reactions and absorbance	700 nm
Read points	Times at which the progress of the reaction is monitored, expressed either in a set time (seconds) or at defined intervals (for example every 3 seconds)	Reading at 90 seconds after reaction initiation
Type of reaction	Options are usually end point, where an absolute change is monitored or rate, where the continuous rate of change is monitored	End point

* Method uses a single reagent so no defined volume is required.

Cross reference

Chapter 7 Samples and sample collection in the *Biomedical Science Practice* textbook

analyser to test urine, cerebrospinal fluid (CSF), aspirates, and effusions using the same or slightly modified methods as those used with serum and/or plasma.

Specimen tubes may be presented to the analyser in racks, rotors, segmented rotors, or through an automated tracking system. When an instrument is connected to a tracking system, the option to present the sample directly on the analyser is usually retained and this gives the operator flexibility in specimen presentation.

Specimens are generally presented to analysers in tubes of consistent diameter, usually 13 or 16 mm, and height, usually 75 or 100 mm. The capacity of these tubes ranges between 4 and 10 mL. The majority of analytical systems can accommodate all of the common tube sizes. For small volume samples, for example paediatric specimens, presentation is often in a small cup, that may sit within an outer carrier tube of the dimensions listed above.

The term, **primary sample** is used for a blood specimen that is presented to the analyser in the same tube originally used when the sample was collected from the patient. In some instances, it may be necessary to transfer all or part of the specimen into a secondary container, for example if the specimen volume is small, as is the case with paediatric specimens. Aliquoting may also be necessary as part of the standard automated process on a particular instrument or because the specimen requires dividing for analysis by several analytical platforms. The use of primary samples was greatly enhanced with the introduction of separator gels, as shown in Figure 1.2, that are incorporated into the blood collection tubes during manufacture. These gels have minimal reactivity in the presence of clinical materials and a characteristic density so that during centrifugation they form a distinct barrier between the serum/plasma and the clot/cellular constituents of the blood specimen. The serum/plasma represents the lighter component of blood and following centrifugation will be in the upper part of the specimen tube, from where it can be directly sampled by the analyser.

Cross reference

Chapter 12 Centrifugation in the *Biomedical Science Practice* textbook

A key part of specimen presentation to any analytical platform is sample identification. Most laboratories use a barcode on each tube to identify the unique laboratory number allocated to the specimen. The quality of the barcode and its correct positioning on the specimen tube are critical for efficient processing. If the quality or alignment are compromised, the instrument will be unable to identify the specimen using the integrated barcode reader. Consequently, it will be unable to undertake testing. A number of common barcode formats are available and while most analysers can read and interpret a number of the common formats, the type employed in each laboratory usually requires defining in the analyser system configuration. The barcode is read at an early stage during the analytical cycle of the instrument. The analyser then interrogates its own database or queries a linked database to identify which tests are required on the specimen.

An alternative mode of operation is needed in the rare instance where barcodes are not used to identify specimens. This generally requires the programming of an association between the specimen and a unique position on the rack or rotor. This process is generally less efficient than using an interfaced, barcode driven operation and is more susceptible to specimen identification errors.

Quality control and calibration materials must also be presented to the analyser using a separate dedicated rack or rotor designed specifically for these materials. Alternatively, they may be presented in a tube that has a barcode that identifies them as a calibrator or quality control. As calibrators and quality control materials are expensive, a minimum volume is presented to the analyser in a small sample cup similar to that used for paediatric specimens. In some analytical systems, it is possible to hold quality control and calibration samples within a distinct compartment on the instrument. The use of such compartments depends on how stable the test analyte is within the quality control/calibrator material and how resistant the specimen aliquot is to evaporation. A range of design features, for example temperature control and/or covers for the compartment, are used to extend the time over which quality control or calibration aliquots may be used on an instrument.

The analysis of specimens requiring urgent or priority processing is often identified with a 'Stat' facility. This may be a dedicated rotor or rack position. Any sample occupying this position takes priority over samples that may be waiting to be analysed and so is processed as soon as it is loaded.

Cross reference
Chapter 1 Biochemical investigations and quality control

Specimen pipetting

All analytical systems require a mechanism to transfer a defined volume of the specimen from the sample tube into the reaction vessel of the instrument. These mechanisms use one of two types of sample probe:

- a disposable tip for each individual specimen
- a fixed probe that is only changed when it is damaged or has been used for a fixed number of sample cycles

Modern probes are precise, accurate components and on general chemistry systems typically transfer specimen volumes of between 1 to 20 µL, with increment intervals of 0.1 µL for each test. The uptake and transfer of specimens is most often achieved with a high precision syringe mechanism driven by a stepper motor. A commonly quoted figure for analysers is the **specimen dead volume**. Most sampling mechanisms cannot sample the complete contents of a sample cup and, consequently, a minimum or 'dead volume' of specimen in excess of that required to perform the analytical tests is required in the bottom of the cup. The dead volume varies with both the type of instrument and the dimensions of the sample cup. Typical dead volumes for general chemistry analysers are between 40 to 150 µL. A high dead volume requirement can be a problem when analysing low volume paediatric specimens.

Specimen probes incorporate a number of design features to identify sampling problems, which include:

- liquid level sensors
- probe crash detectors
- clot detectors

The amount of specimen volume in each sample tube is variable and it is therefore necessary for the probe to detect the liquid height using a liquid level sensor. If the probe is not lowered beneath the liquid level it will sample air, if lowered too far it may sample gel or hit the bottom of the tube. The liquid is generally detected by the probe as a change in electrostatic capacitance or conduction, which enables the probe to be lowered to a specified depth beneath the surface and an appropriate volume withdrawn. Physical barriers, for example a cap that has been inadvertently left on, or an empty cup or tube, are capable of significantly damaging probes. Therefore, modern sample and reagent probe arms are spring loaded, which provides a degree of tolerance such that if these problems are detected, the downward movement of the probe is interrupted. This prevents the probe from being damaged and continued operation is possible. The quality of specimens may also be compromised if the sample contains sediment, micro-clots or, indeed, large fibrin clots. Although these problems are not commonly seen they can occur in samples that have been stored for a long time or have not been mixed thoroughly at the time of collection, rendering the anticoagulant in the tube ineffective. All of these problems can affect the accuracy of the volume transferred and could potentially block the probe. However, the pressure changes associated with sampling a clean, uniformly viscose specimen are predictable so that deviations from them may be sensed and reported to the operator as a sampling error message. The sample can then be inspected for integrity problems. In some cases, these problems can be rectified by removal of the clot or re-centrifugation of the sample.

High volume general chemistry analysers require the specimen probe to move rapidly between a number of defined positions within a cycle time that may be as short as two to three seconds. Most instruments are able to track the probe's position within the cycle by detectors that are activated when the probe is aligned at a particular position. Failure to activate the detector indicates a problem with the robotic cycle. The ability to sample the specimen without removing the cap from the specimen tube is a feature of some general chemistry systems. Cap piercing technology has health and safety advantages in terms of reducing the exposure of staff to clinical specimens.

A significant problem, common to all analysers, is the possibility that material may be transferred by a probe from one sample to the next. This problem is called **carry over**. Carry over can be minimized by flushing both the internal and external surfaces of probes with de-ionized water and/or wash solutions during the sampling cycle. Carry over is not a problem with single sample disposable probes.

The integrity of plasma and serum samples can also be compromised by haemolysis, icterus, and lipaemia, which were described in Chapter 1. The presence of these factors will affect analytical procedures in different ways and are referenced by most manufacturers in the method descriptions. Traditionally, the specimen was visually inspected to assess these factors but increasingly this is automated on the general chemistry analysers by assessing the spectral characteristics of the sample, usually diluted in saline at specific wavelengths and calculating and reporting indices. The automated assessment of these factors has become more critical following the introduction of the larger automation systems as centrifugation and analyser presentation have become an integrated process without an opportunity for staff to assess the specimen visually. Equally the spectrophotometer unit on a general chemistry analyser is able to assess these factors more reproducibly than visual assessment, which may be subjective.

Cross reference

Chapter 1 Biochemical investigations and quality control

All of the sensing systems described are reported as alarms to the user and prevent misreporting of compromised results.

SELF-CHECK 2.1

The validity of test results produced by an automated analytical system are directly related to the quality of the specimen. What checks of specimen integrity are undertaken by the probe mechanism used to transfer specimens?

Reaction vessel: cuvette

In many analysers, the test reaction proceeds and is monitored in a cuvette. Cuvettes need to be made of materials that are inert to the reagents used and be optically clear at the wavelengths monitored for detecting analytes. Both disposable and re-usable washable cuvettes are utilized in automated chemistry systems. Disposable cuvettes are constructed from optically clear plastic and are consumables that require loading onto the instrument as part of a daily and ongoing maintenance. Following their use, they are a waste component that needs to be monitored and periodically removed. Often an automatic monitoring of the waste accumulated in the cuvette waste container is activated at critical thresholds to prevent it being over filled. Re-usable cuvettes are made from a range of optically pure and washable materials such as quartz or glass silica, and some plastics. Frequently they are made in the form of a ring. Following each reaction cycle, the liquid contents are aspirated to waste and the cuvette is washed. The washing or 'laundry' system is a series of probes that deliver cleaning agents and de-ionized water in a defined sequence to produce a clean, dry cuvette ready for the next sample and reaction cycle. Prior to re-use, each cuvette can be subjected to an optical check using de-ionized water to ensure clarity. Failure to meet minimal absorbance characteristics is suggestive of incomplete or inadequate cleaning and on some analysers it will not be re-used. The laundry system usually requires a significant supply of de-ionized water and consequently will require a water purification system to be connected.

The reaction temperature is maintained at a constant value, usually 37°C, by incubating the cuvette in a heated environment. This may be a water or oil bath, or may simply be warmed air. Temperature control is critical to ensure reproducible results since even a variation of 0.1°C can contribute to variation in results. All of the reactions performed in analysers are temperature dependent, but the analyses of enzyme activities, in which rates of reactions are monitored, are most temperature sensitive.

An alternative to the standard wet chemistry systems that use liquid reagents is the use of slide technology in which the reagents are incorporated into small individual slides onto which specimens may be dispensed. These 'dry' chemistry systems monitor similar chemical reactions to the analyses performed by traditional analysers, except the reaction takes place within the layers of reagents. The reaction is monitored by reflectance spectrophotometry from the surface of the slide. Slide-technology based systems are notable for their high degree of reproducibility.

HEALTH AND SAFETY

All of the analysers and robotic units employ probes and other rapidly moving components. Covers and lids are used to protect operators from these moving parts and these should be kept in place to prevent injury. To access these areas the unit should be placed into a STOP or PAUSE mode. Following manufacturers' safety warnings for automated units is essential.

Cross reference
Chapter 4 Health and safety in the *Biomedical Science Practice* textbook

Reaction sequence and reagents

Most analysers have a standard format of reagent addition in which one or two reagents are added to the specimen at fixed time points and the reaction proceeds for a period of about ten minutes. This is satisfactory for, and applicable to, most standard chemistry tests. A limited number of analysers have greater flexibility and it is possible to design a reaction sequence using more than two reagents, which may be added at a number of variable time points.

In the standard instrument design, pipetting the specimen into the cuvette is generally considered the start point of the reaction. Addition of reagent to the cuvette may occur at one of two time points: one immediately after specimen pipetting; and the second approximately five minutes later. Some methods only use a single reagent and this would usually be added at the first time point to allow a reaction to be monitored for the full ten minutes of the reaction cycle. Absorbance measurements to monitor the reaction may be taken over the entire period, from the initial mixing of specimen and reagent through to the end of the ten-minute reaction cycle. This is normally achieved by repeatedly moving the cuvette past a static spectrophotometer unit and recording the absorbance at a number of fixed wavelengths ranging from 340 to 800 nm. The number and setting of these wavelengths is instrument dependent but most utilize similar points as the methods employed are standard procedures. For example, 340 nm is the peak absorbance of NADH, a commonly used reactant with a characteristically high **molar absorptivity**. Therefore, all general clinical analysers have the ability to measure absorbances at this wavelength. Other wavelengths commonly used are 405, 540, and 700 nm. Instruments may have anywhere between 8 and 20 wavelength settings, any of which can be assigned to a particular test. Commonly two wavelengths are employed for each test. One, generally termed the primary wavelength, is where the major change in absorbance occurs. The second wavelength may be used to correct for background absorbances associated with the sample rather than the test reaction and so may minimize the effect of interfering factors such as haemolysis. Many analysers allow the user to review the full data of a reaction as a graph of absorbance against time. This can be useful when developing methods or to 'troubleshoot' problems with the analyser, and to review problems with specific specimens.

The absorbance changes that are recorded result from one of two general types of tests: those where the reaction reaches an end point and those where the rate of the reaction is determined. The former are often called *end point* and the latter as *kinetic reactions*. In an end point reaction, the absorbance ceases to change and reaches a constant value once the reaction finishes at some point within the ten-minute cycle. Examples of end point reactions are the tests used in calcium and glucose analyses. In contrast, in kinetic, or rate reactions the rate of change of the absorbance is continuous during the time period monitored. This type of monitoring is commonly associated with enzyme analyses, for example determining the activities of alkaline phosphatase and lactate dehydrogenase.

Instrument manufacturers provide pre-packaged reagents for the common tests listed in Table 2.1 and the analyser parameters are preset in the instrument memory. Instruments that restrict the user to the supplier's pre-packaged reagents only and do not allow new methods to be incorporated are called *closed* systems. However, most general chemistry analysers are *open* systems and allow the user to use their own reagents or those of other suppliers and apply user defined methods. The test methods developed by manufacturers have been subjected to comprehensive optimization to fulfil regulatory authority requirements prior to release: most laboratories use these in preference to developing their own. A major advantage of using pre-prepared reagents is that they are labelled with barcodes to identify the reagent. The barcodes allow the automatic auditing of the reagents with recording of the lot number and expiry date and allow fast and simple reagent loading.

Reagent replenishment is an integral part of daily maintenance and monitoring of reagent levels is part of the routine operation. All reagents have predictable 'on-board' stability and on most systems users are alerted when an expiry date and time is approaching. Many analysers will sense the liquid height each time the reagent probe samples the reagent and will calculate how much reagent remains in its container. This is often expressed as the number of remaining tests the analyser can undertake with the remaining reagent(s) and allows the user to decide when to replenish it (them).

The reagent compartments in the analyser are designed to hold multiple reagents to enable the full range of tests to be undertaken. These compartments are usually temperature controlled. Typically, the larger capacity instruments can hold in excess of 50 reagents at one time. Most systems can hold more than one cartridge of each test reagent to provide flexibility in terms of test capacity. Loading reagents into the reagent compartment(s) varies in complexity. The simplest systems have standard shaped reagent cartridges that may be loaded into a single reagent entry point and the cartridge is then automatically allocated to its position within the reagent compartment. Alternatively, there may be set positions for specific reagents and the operator must be careful in placing a reagent into its designated location.

The carry over of reagents, that is the retention of traces of reagent by their probes or within the cuvette system, may also occur for the same reasons as described above for specimen carry over. Manufacturers undertake extensive studies to identify potential sources of contamination and design processes within the analytical cycle to minimize any effects. This may involve general washing or may be a specific process such as having certain tests undertaken in a particular sequence to minimize a known problem. This can be necessary when a particular method is pH sensitive and so may be susceptible to carry over from a highly acidic or alkaline reagent. A more specific example is the test reagent for triacylglycerols, which contains a high lipase activity compared to that observed in plasma. Thus, if this test is performed following a triacylglycerol analysis, there is the potential to produce an artefactual high result.

Mixing of the reaction solution is required following the addition of each reagent. A number of different types of mixer probes are used to ensure complete mixing and consistent reaction kinetics. Generally, the mixers are lowered into the reaction mixture and agitate it in some way, often by rotation or vibration. A failure in mixing results in poor precision and is notable by a slowed, inconsistent reaction profile. Mixers are an additional source of carry over and also require cleaning between cycles.

Key Points

The operation of a large automated analyser that undertakes multiple tests can be better undertaken if the operator is aware of the rate of consumption of each of the reagents. It is not unusual to have 60 to 100 different reagents to monitor and, in an average day, these will be used at a variable rate dependent on the demand for each test. Consequently, it is necessary to ensure that reagents required for a frequently requested test, such as that to determine urea, are replenished to a much higher stock level than an infrequently requested test, such as one for ammonia.

Ion-selective electrode modules

Two of the commonest analytes requiring analysis in clinical biochemistry laboratories are sodium and potassium. Despite a number of attempts at monitoring using spectrophotometric methods, none have been successful. Their analysis on earlier automated analysers utilized integrated flame photometers but this approach has been completely replaced by incorporation of ISEs. All high throughput general chemistry analysers now have integrated ISE modules for measuring sodium, potassium, and in many cases chloride.

The ISE unit may be an integrated single module containing all the electrodes, or it may have separate individually changeable electrodes for each of the analytes. The electrodes have predictable lifetimes in terms of the number of test cycles and require periodic replacement as part of the maintenance program. The ISE module or unit has a separate fluidics cycle that is generally slower than the cycle for spectrophotometric tests. A significant consequence of this design feature is that it often compromises the manufacturer's quoted test throughput (see earlier Key point).

Cross reference

Chapter 9 Electrochemistry in the *Biomedical Science Practice* textbook

Data output

The usual output from automated instruments are the results of the clinical test on the specimen sample. These may be printed out and/or transmitted from the analyser to the laboratory

Cross reference
Chapter 1 Biochemical investigations and quality control

information management system, as described in Chapter 1. The arrangements for data review vary enormously between different departments but there are usually steps to both analytically and clinically check and validate the test result. The configuration of the software controlling the analyser is usually versatile, such that many rules may be incorporated. For example, if a result is higher or lower than a certain threshold it may force certain interventions.

Other outputs of data from instruments are method calibration and quality control data, and reagent lot and batch characteristics. Again, these data may be supplied in a printed form or transmitted for electronic archiving. An essential regulatory requirement is to keep such records for review.

Key Points

All automated instruments, no matter how reliable, will occasionally have a performance problem due to a component failure or deteriorating performance. Correction of the problem may require replacement, re-alignment or cleaning of the component in question. The instrument operators can 'trouble shoot' more effectively if they know which testing procedure is appropriate for each component.

SELF-CHECK 2.2

What is the difference between a closed and an open automated analyser?

2.3 Automated immunochemistry analysers

Immunochemistry analysers are automated instruments that analyse samples using analytical immunoassays. Immunoassays are analytical techniques that employ antibodies as reagents; they can be applied to a broad range of tests in clinical laboratories. In brief, immunoassays use an antibody raised against the analyte of interest to bind it within a reaction sequence that can be monitored. Unlike general chemistry analysers that often have similar reaction sequences irrespective of the manufacturer, the design of assays for an immunochemistry analyser is essentially instrument dependent. You can see an example of an immunoassay analyser in Figure 2.3.

FIGURE 2.3
The Siemens Centaur XP analyser; an example of an immunoassay analyser. See also Figures 1.6, 1.7, and 1.8.

Originally, immunoassay was in a manual format with the reaction being followed by attaching, often termed **labelling**, a radioactive element to either the antigen or antibody, a technique generally termed radioimmunoassay (RIA). Commonly used isotopes were Iodine-125 (^{125}I) and tritium (^3H). Radioisotopes have associated health and safety problems in terms of reagent handling and disposal, hence RIAs have limited potential for automation. Consequently, alternative labels that could be used to monitor the immunoassay reaction were developed, including a range of enzymes, for example alkaline phosphatase and horseradish peroxidase, and chemiluminescent and fluorescent molecules such as acridinium esters. These are safer than radioactive labels and are considerably more amenable to automation. As a result, RIAs have been largely replaced by other antibody-based techniques. The application of immuno-assays is considerable and includes all of the biomedical science disciplines, and the common testing areas. Table 2.3 list a number of these examples.

There is considerable variation to the basic immunoassay technique performed by automated instruments, but there are basically two types of immunoassay reaction: competitive and sandwich. The competitive assay uses a limited concentration of antibody to produce a reac-tion in which the test antigen and a labelled antigen compete for a limited number of sites on a complementary antibody. The competitive technique is usually used for analysing mol-ecules with a lower M_r, for example thyroxine, digoxin, and steroids, which have a character-istic inverse and non-linear relationship between the concentration of analyte and the signal produced. Sandwich immunoassays generally use two antibodies to test for an antigen, one of which is labelled. The reaction signal detected increases in intensity as the concentration of the test analyte increases and, in contrast to competitive assays, these assays are used to test for larger molecules, for example peptides and proteins.

Cross reference
Chapter 15 Immunological techniques in the *Biomedical Science Practice* textbook

The simplest automated systems for immunochemistry may be simple sample and reagent pipetting units that enable automation of the technique enzyme-linked immunosorbent assay (ELISA). Automated ELISA systems have reasonable flexibility and can be applied to most ELISA-based methods, irrespective of the manufacturer and are therefore open platforms. However, most immunochemistry analysers are closed systems, which can only be used for methods developed and supplied by the instrument manufacturer. The first automated immunochemistry analysers had the ability to perform a single method on a series of samples in a batch mode. These are still available and may be used for less common investigations or in a point of care setting. However, most biochemistry laboratories use the more complex random access analytical system that applies a patented immunochemistry technology within a closed system. The commonly found immunochemistry systems provide a wide repertoire of the common tests listed in Table 2.3, and have a throughput of up to several hundred tests per hour.

Immunochemistry analyser design incorporates many components and technical features devel-oped for general chemistry analysers. For example, the presentation of specimens uses similar racks and rotors to those found on general chemistry analysers. A notable difference between the two types of systems is that, in general, the sample capacity of racks and rotors for immuno-chemistry systems is less, with the racks typically holding only five or six specimens. The lower capacity racks reflect the relatively lower workload for immunochemistry instruments compared to general chemistry analysers.

Immunochemistry systems analyse analytes in samples that are generally at lower concen-trations than those measured with general chemistry analysers. Therefore, larger volumes of sample are needed to produce an adequate signal for measurement. Typical sample volumes are 20 to 200 µL. Carry over is also considered a more significant issue, as the concentra-tion range between specimens may be considerably larger. For example, the difference in the concentration of thyroid stimulating hormone (TSH) in a sample of serum from a patient with an increased (greater than 100 mU/L) compared to suppressed value (less than 0.01 mU/L)

TABLE 2.3 Assays commonly performed on immunochemistry analysers.

Grouping	Tests for:
Thyroid function	Thyroid stimulating hormone
	Thyroxine
	Triiodothyronine
	Thyroid peroxidase antibodies
Tumour markers	Prostate specific antigen
	α-fetoprotein
	Carcinoembryonic antigen
	β human chorionic gonadotrophin
	Ca 125, Ca 19-9
Haematinics	Vitamin B_{12}
	Serum folate
	Erythrocyte folate
	Erythropoietin
Steroids	Cortisol
	Testosterone
	Oestradiol
	Progesterone
	Androstendione
	Dehydroepiandrosterone sulphate
Peptide hormones	Follicle stimulating hormone
	Leuteinzing hormone
	Adrenocorticotrophic hormone
	Growth hormone
	Insulin-like growth factor-1
Bone metabolism markers	Parathyroid hormone
	Vitamin D
Therapeutic drugs	Carbamazepine, cyclosporine, digoxin, gentamycin, phenobarbitone, phenytoin, sodium valproate, tacrolimus, theophylline, vancomycin
Drugs of abuse	Amphetamines, barbiturates, benzodiazepines, cannibinoids, methadone, opiates
Infectious disease	Hepatitis A markers
	Hepatitis B markers
	Hepatitis C markers
	HIV
	Chlamydia
	Clostridium difficile
	Rubella
	Syphilis
Immune function	Allergens
	Complement C3 and C4
	Immunoglobulins

of TSH may be a factor in excess of 10,000. Consequently, the transfer of small volumes may have a significant effect on the interpretation of the test result. Disposable pipette tips or more thorough or complex washing routines between specimens are used to minimize specimen carry over in immunochemistry instruments.

SELF-CHECK 2.3

Two specimens were presented to an immunochemistry analyser for prolactin estimations. The tests required a specimen volume of 100 microlitres. The first and second specimens have true prolactin concentrations of 15,000 and 300 IU/L respectively. If the analyser developed a carry over problem in which 1 μL is carried over from one specimen to the next, what will be the apparent value for the concentration of prolactin in the second specimen?

The reaction sequence for immunochemistry systems is instrument dependent and shows considerable differences between available analysers. However, most of the methods have typical reaction times of 20–40 minutes, but may be longer. In general, longer incubation times are required when analytical sensitivity needs to be increased. For example, if the concentration of the test analyte is comparatively low it may only result in a small change in signal, which the monitoring system may not be able to detect. Extending the reaction time may produce a larger signal and therefore increase the signal to a detectable level. Many immunoassays are homogenous, a term that indicates the bound and unbound antibody are not separated in the immunoassay. Homogenous immunoassays can use conventional spectrophotometer units and some methods can be applied to the general chemistry systems described in Section 2.2. However, the majority of immunochemistry analysers utilize heterogeneous immunoassay where the bound and unbound antibody must be separated in the assay. Thus, a distinct additional automated step is required within the system. Examples include magnetic separation systems or other immobilized capture systems. Each system is distinct in this respect. Automated heterogenous immunoassay instruments are generally more sensitive than those utilizing homogenous immunoassays.

A distinct range of immunoassay instruments utilize immuno-nephelometric techniques, which are often used to test for specific proteins.

2.4 Integration of processes and robotics

The basic analyser types described in Sections 2.2 and 2.3 have been expanded in a number of ways to achieve:

- increased instrument capacity by linking two or more analyser units with a single sample presentation mechanism
- increased assay diversity, by combining general chemistry and immunoassay units within an apparently single instrument

These types of combinations have significantly increased the diversity of analyser types and help provide analytical solutions suited to any clinical laboratory setting. The features of these integrated analysers mirror those described earlier.

The majority of specimens processed in clinical biochemistry require the sample to be prepared in some way prior to presenting to the analysers described in Sections 2.2 and 2.3. The commonest preparatory process is centrifugation of blood to obtain serum or plasma. This can be described as a **pre-analytical step**, that is, it is part of sample preparation before the specific test is applied. Pre-analytical steps also include removing sample container caps and specimen sorting and aliquoting. The subsequent **analytical step** is the process undertaken by the analysers to produce a test result from a chemical or immunochemical method. Following analysis, the sample must be recapped and archived to enable rapid access should the specimen require further investigations. These processes are termed **post-analytical steps**.

Traditionally, all pre- and post-analytical processes were undertaken manually. Automation of these steps therefore reduces manual input and staff time, and speeds up the total laboratory process. It also provides health and safety benefits as laboratory staff are less likely to develop musculoskeletal problems such as repetitive strain injuries, which may develop from repeatedly undertaking the same manual task. It also reduces exposure of staff to clinical specimens, all of which represent an infection risk.

Key Points

An essential requirement of all of the automation units, as for the analysers themselves, is for specimens to be identified by a unique barcode number. Through this barcode the tube can be identified and the necessary specimen processing and testing undertaken. Therefore, good quality, correctly aligned barcodes are of critical importance for routing specimens in the system.

The development of integrated automation systems is closely related to greater utilization of robotic devices to pick samples up from one position and transfer them to another. The automation that is associated with such robotics can be broadly divided into two: standalone automation units and tracking systems.

Standalone automation units

Standalone automation units can undertake pre- and post-analytical processes but are a physically separate system to the analysers (Figure 2.4). Tracking systems allow the pre- and post-analytical processes undertaken by units connected to a track, which also conveys the specimens to the analysers. Standalone automation units have also been termed 'islands of automation', reflecting the characteristic feature that they are not directly connected to an analyser. They are able to automate the pre-analytical steps of sorting, uncapping and allocation of samples to analyser racks. Following these processes, laboratory staff are required to remove

FIGURE 2.4
The OLA standalone automation unit. This automation system allows samples to be allocated to a range of both analyser and storage racks, and can improve laboratory efficiency.

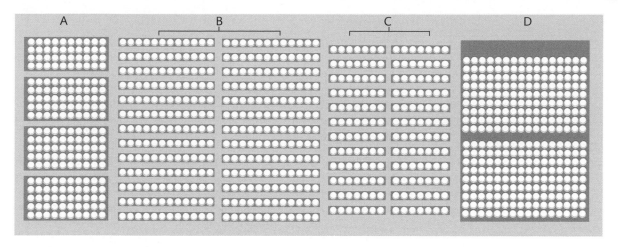

FIGURE 2.5

Simple layout of a standalone automation unit as observed from above. Samples are presented in racks at A, robotic arms will transfer the bar coded tubes into analyser racks at B and C. When all testing is complete, the tubes may be transferred into an archive rack at position D.

and present the racks to analysers for analysis. The major automation component of standalone systems is a robotic arm that moves the specimen, as required, between the different working areas. Figure 2.5 demonstrates the layout of an automation unit as seen from above. There is a series of working areas: some for sample racks, others for presenting samples to the unit, some for analyser racks, and a number for archive racks. When a sample is presented to the unit, it will read the barcode on the sample tube and identify whether there are tests to be undertaken or whether the tests are complete. If tests are required, it will allocate the sample to the analyser rack associated with this test. For example, if the sample must be analysed for glucose on a general chemistry analyser, it will present the sample to a rack position associated with that analyser. However, if it requires a test on an immunochemistry analyser the sample will be allocated to the immunochemistry analyser rack. Periodically the operator needs to remove full or partially full analyser racks from the automation unit and present them to the analyser. When they have been sampled and analyzed, they can be re-presented to the automation unit. At this point it will confirm electronically whether all the required tests have been completed; if this is the case, the sample will be allocated to an archive or storage rack. Alternatively, if a test or tests are still outstanding the sample will be allocated to another analyser rack as required. This cycle is repeated until all tests required on the sample are completed.

In addition to the allocation of specimen tubes to racks, the unit can also detect and remove the specimen cap and once all analyses are complete the tube can be resealed. Some of the units can prepare additional, labelled tubes for sample aliquots and in some cases may also include a centrifuge unit.

A notable feature of these systems is the ability to adjust the system to the requirements of different laboratories. In general, all of the systems have a specimen input area that can be configured to take a range of different racks, which may, if the system does not contain a centrifuge unit, be a centrifuge carrier or may be racks taken from analysers. Most laboratories have a range of analysers that have different rack, segment, or rotor characteristics. The automation units may be designed to accommodate a range of these and are therefore considered to be highly flexible.

Tracking systems

Many medium to large sized clinical laboratories reduce the number of manual interventions by linking the main analysers together using a tracking system. The specimen is carried by a conveyor belt system through a series of analytical steps that, in addition to connecting the standard analysers together, as described above, may include pre- and post-analytical processes such as centrifugation, archive, and retrieval from storage. You can see an outline of tracked systems in Figure 2.6 (a) and an example of such a system in Figure 2.6 (b).

Specimens are presented in a sample management or 'input' area, where a robotic arm lifts the specimen and places it onto the track. The specimen may be pre-centrifuged or the track may have an integrated centrifuge unit. If the latter is true, another robotic arm will remove the sample from the track into the centrifuge unit to spin the sample (see below). The majority of analysers currently available are incapable of piercing the caps of the specimen tube and it

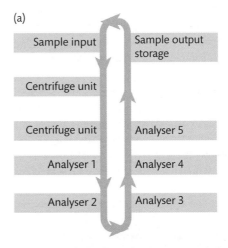

FIGURE 2.6

(a) Overview of a tracked system showing the connection between the various units. The detailed arrangement varies between different systems. (b) The Siemens Labcell is an example of a tracked automation system that can perform centrifugation, analyses, archiving, and storage of specimens.

is necessary for the caps to be removed. Therefore, most automated systems have the ability to remove and discard the specimen caps although this may or may not be integrated into the centrifuge unit.

Following centrifugation, the specimens are conveyed to one or more of the analysers for testing, usually by one of a number of the general chemistry and immunochemistry analysers as described in Sections 2.2 and 2.3 respectively. The tracking systems may also include haematology and coagulation analysers. The configuration of any one particular system needs to be carefully considered during the design and appraisal stages to provide optimal throughput and robustness and must reflect the laboratory's present and future testing needs. Increasingly, it is necessary to model workload patterns mathematically to ensure an appropriate balance of analytical systems to prevent any individual test unit becoming a rate limiting factor in the total automation cycle. To avoid any single points of failure, the design of automation systems usually incorporates adequate capacity and back-up options, often by duplicating some of the analysers.

Once a specimen has been fully analysed using all the tests requested, it may be archived for storage in a racking system. Generally, each rack has a unique identification and the specimen will have associated x, y coordinates to allow its subsequent recovery. The capacity of online storage units continues to grow and some systems also incorporate refrigerated units. The specimen integrity on storage is best achieved by recapping the sample primarily to reduce evaporation losses. Recapping is also possible on automated analysers and may use traditional caps or plugs, or foil, or a polymeric film. It is a relatively common occurrence for additional tests to be requested once the original test results have been considered. This may be part of a rules-based cascade system or may be based on clinical opinion during clinical review of the patient. If additional tests are required then systems that have retained the original specimen may re-present the specimen to the required analysers and the testing undertaken without manual intervention.

Key Points

The test turnaround time is dependent on multiple independent steps. When the transition from one step to another requires staff intervention, then it may result in variability and delays unless they are constantly available to undertake that step. A major benefit of tracking systems is a reduced and more consistent test turnaround times since most steps are integrated into a smooth process.

Centrifuge units

Centrifuge units may be found in both standalone automation units and as distinct units on a tracking system. A series of robotic transfers takes place within centrifuge units:

- Transfer of the sample tube onto a balance to weigh it. The weight will depend on the size and type of tube, for example glass tubes are heavier than plastic tubes, and on the volume of specimen present.

- Allocating the tube to a centrifuge rack. The allotted position will take into account the weight of the tube and contents to ensure the final load in each centrifuge rack is balanced. Failure to balance the load in a centrifuge can damage the centrifuge and compromise the operation.

Cross reference
Chapter 12 Centrifugation in the *Biomedical Science Practice* textbook

- The centrifuge racks are then transferred into the centrifuge. Blood specimens are centrifuged at 1,000 g or higher for 8–12 minutes.
- Following centrifugation, the samples are removed from the centrifuge and may be uncapped, before being transferred into a rack or onto a track for further processing.

Many laboratories have investigated the effect of changing the length of centrifugation. In general, there is a compromise between throughput of the centrifuge unit and the effectiveness in separating the blood components. On some automated systems, the centrifuge is the rate-limiting factor since its throughput, in terms of specimen numbers, is slower than that of the analysers. Consequently, the operation of the centrifuge system may be optimized by setting minimum waiting periods and offering priority spinning. The former enables the centrifuge to be used even if it is not fully loaded to maintain a steady stream of specimens to the analysers. The latter reduces the waiting period if the specimen is electronically tagged as requiring urgent or priority testing. Some tracking systems incorporate two or more centrifuge units to speed up the processing at this stage.

2.5 Automation in specialized areas of clinical biochemistry

Several more specialized pieces of clinical biochemistry equipment have incorporated elements of automation. A key area in this respect is the ability to load multiple specimens onto an autosampling system such that, once started, the system can be simply left to run. Analytical systems that incorporate automated injections of the specimen for subsequent analyses include most chromatography systems and those for elemental analysis such as atomic absorption and inductively coupled plasma spectrophotometers. The arrangements in each vary but the autosampler facilitates their continued operation without intervention. This is particularly helpful if individual specimens require a long period for analysis. For example, with some applications the run time for an individual specimen using high performance liquid chromatography (HPLC) may be in excess of an hour. Thus, overnight running using an automated specimen injection system has significant advantages.

An area of analysis yet to be seen in routine laboratories at the time of writing is microarray technology. This technique allows, for example, multiple immunoassay tests to be undertaken simultaneously. Also, polymerase chain reaction (PCR) analyses have seen a significant growth in many biomedical science disciplines and provision of greater automation for these techniques is in development.

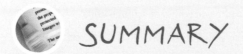

SUMMARY

- Clinical biochemistry laboratories have benefited significantly from developments in automation, instrumentation, and robotics.
- Two main types of instrument, the general chemistry analyser and the immunochemistry analyser predominate in clinical biochemistry laboratories, but a wide range of hybrid and combination analysers is available.

- Key steps in an automated analytical process are often common to many instruments, although there is often significant variation in the instrument design and operation.

- A thorough understanding of the individual components of an analyser and their relative mechanical and data interactions enables operators to troubleshoot problems more effectively.

- Consolidation of most of the biochemical test repertoire is now possible on a single system that comprises a set of integrated analysers.

- Total laboratory automation automates three phases of laboratory testing, namely pre-analytical, analytical, and post-analytical in a continuous process.

- A complex automation system requires extensive monitoring and careful management to achieve consistent and optimal efficiency.

- Ongoing maintenance and standard regimes of operation need defining and completing for optimal operation of automation.

- Automation has resulted in significant reduction in manual intervention in the total analytical process.

- Automation produces shorter and more consistent test turnaround times.

- For maximum utilization of automation, the total laboratory processes need to be reviewed regularly.

FURTHER READING

- Horowitz GL, Zaman Z, Blanckaert NJC, et al. (2005) Modular analytics: a new approach to automation in the clinical laboratory. *Journal of Automated Methods and Management in Chemistry* **27**, 8–25.

- Melanson SEF, Lindeman NI, and Jarolim P (2007) Selecting automation for the clinical chemistry laboratory. *Archives of Pathology and Laboratory Medicine* **131**, 1063–9.
  Describes all the practical issues associated with selection of an automated tracking system and compares the most widely available systems.

- Molloy RM, McConnell RI, Lamont JV, and Fitzgerald SP (2005) Automation of biochip array technology for quality results. *Clinical Chemistry and Laboratory Medicine* **43**, 1303–13.

- Ognibene A, Drake CJ, Jeng KY, et al. (2000) New modular chemiluminescence immunoassay analyser evaluated. *Clinical Chemistry and Laboratory Medicine* **38**, 251–60.

- Price CP (1998) Progress in immunoassay technology. *Clinical Chemistry and Laboratory Medicine* **36**, 341–7.

- Sarkkozi L, Simson E, and Ramanathan L (2000) The effects of total laboratory automation. *Clinical Chemistry* **46**, 751–6.

- Wheeler M (2007) Overview of robotics in the laboratory. *Annals of Clinical Biochemistry* **44**, 209–18.

- **Wild D (2005) The immunoassay Handbook, 3rd edition. Amsterdam: Elsevier Science.**

  This is a comprehensive text with individual chapters on each of the immunoassay technologies and separate chapters for many of the commonly available immunochemistry analysers.

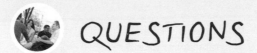 QUESTIONS

2.1. The usual centrifugal force and time of centrifugation used in automated centrifugation units for blood specimens is which of the following?

    (a) 1,000g for 8–12 minutes

    (b) 1,000g for 20–30 minutes

    (c) 5,000g for 8–12 minutes

    (d) 5,000g for 20–30 minutes

    (e) 5,000g for 30–60 minutes

2.2 Which of the following is (are) post analytical step(s)?

    (a) Centrifugation

    (b) Recapping

    (c) Decapping

    (d) Sample storage

    (e) Sample retrieval

2.3 List the differences between general chemistry analysers and immunochemistry analysers.

2.4 Describe the basic steps performed by automated analysers.

2.5 Which assays would be affected on an automated instrument if the sample probe were physically damaged?

2.6 Which analytes are measured using an ion-selective electrode module on a general chemistry analyser?

2.7 What materials may be used to recap samples on an automated recapping system?

2.8 Why is NADH a suitable reagent to use in biochemical analytical methods on an automated instrument?

2.9 When might a method reaction curve require reviewing?

Answers to self-check questions, case study questions, and end-of-chapter questions are available in the Online Resource Centre accompanying this book.

@ **Go to www.oxfordtextbooks.co.uk/orc/ahmed/**

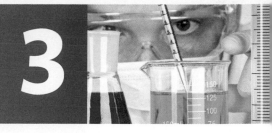

3

Kidney disease

Edmund Lamb

Learning objectives

After studying this chapter you should be able to:

- Distinguish between acute and chronic kidney diseases
- List the major causes of kidney disease
- Discuss the common clinical laboratory tests used to assess kidney function and typical laboratory values associated with health and the various stages of chronic kidney disease
- Describe different types of proteinuria and their clinical significance
- Critically appraise urinary total protein estimation as a marker of kidney disease
- Discuss the clinical utility and limitations of creatinine measurement as a marker of kidney function

Introduction

The renal system consists of the kidneys, ureters, bladder, and urethra, which are arranged as shown in Figure 3.1. The pair of kidneys is situated in the abdomen. Each is supplied with blood through renal arteries, which drain into renal veins.

The major physiological functions of the kidneys include homeostatic, excretion, biosynthetic, and catabolic roles. Their value in homeostasis includes maintaining the volume of blood plasma and the concentrations of its electrolytes, such as sodium and potassium, and stabilizing its pH all to within reference ranges. This is largely achieved by controlling the amounts of water and salts excreted, and the removal of excess H^+ and the regeneration of HCO_3^-. Kidneys also excrete waste products, such as urea, urate, and creatinine, in **urine**. Urine produced by the kidneys drains through the ureters to the bladder, where it is temporarily stored. Urine is expelled from the bladder to the exterior of the body through the urethra. The kidneys also synthesize the enzyme renin and the hormones **erythropoietin** (EPO) and calcitriol (also called 1α, 25-dihydroxycholecalciferol, 1,25-DHCC), which regulate blood pressure, stimulate the production of red blood cells by the bone marrow, and the absorption of calcium by the gastrointestinal tract (GIT) respectively. Kidneys also synthesize prostaglandins, which have a variety of physiological functions, and, finally, degrade hormones, such as insulin.

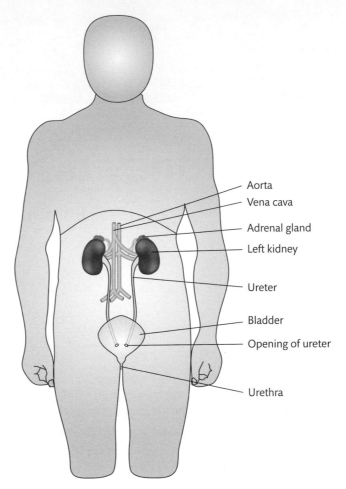

Aorta
Vena cava
Adrenal gland
Left kidney
Ureter
Bladder
Opening of ureter
Urethra

FIGURE 3.1
Schematic showing the renal system.

Given the variety of complex activities associated with the kidneys, it is not surprising that assessing renal function is a major component of the work of most clinical biochemistry laboratories. This chapter describes the basic anatomy and functions of the kidneys as a foundation to understanding their pathophysiology and the rationale underlying the diagnostic and management strategies involved in kidney disease. We will also outline the key analytical methods employed in the investigation of kidney disease.

3.1 **Renal anatomy**

The kidneys are a paired organ system found in the retroperitoneal space at a level between the lower part of the eleventh thoracic vertebra and the upper portion of the third lumbar vertebra. The right kidney is situated slightly lower than the left. Adult kidneys are about 12 cm long and weigh approximately 140 g. Each kidney receives blood through a single renal artery that branches off from the abdominal aorta. In adults, the kidneys receive approximately 25% of the **cardiac output**. Blood returns from each kidney in a renal vein, which drains into the inferior vena cava.

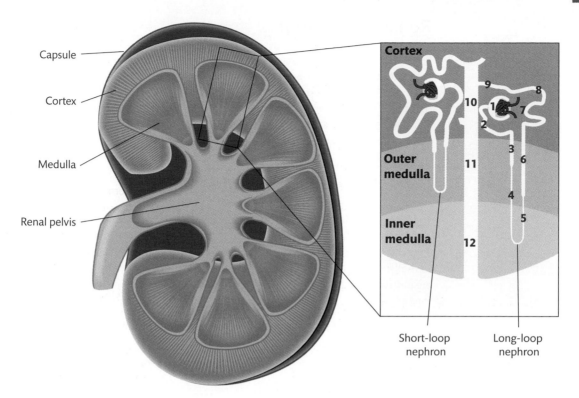

FIGURE 3.2

The organization of the kidney. The expanded view is a schematic of the kidney in section, showing the regions of the nephron and collecting duct system.

(1) Bowman's capsule containing glomerulus, (2) proximal convoluted tubule or pars convolute, (3) proximal straight tubule or pars recta, (4) descending thin limb of the loop of Henle, (5) ascending thin limb of the loop of Henle, (6) ascending thick limb of the loop of Henle or distal straight tubule, (7) macula densa, (8) distal convoluted tubule, (9) connecting tubule, (10) cortical collecting duct, (11) outer medullary collecting duct, (12) inner medullary collecting duct

Each kidney is composed of an outer fibrous capsule, a cortex, a middle medulla, and an inner pelvis region (Figure 3.2).

The tough capsule surrounding each kidney offers protection against trauma and prevents the entry of bacteria. The functional units of the kidney are the **nephrons**, each of which occupy the cortex and medulla. Nephrons consist of a **glomerulus** and renal tubule. The tubule contains several specialized regions: **Bowman's capsule**, proximal tubule, loop of Henle, distal tubule, and collecting duct, as seen in Figure 3.2. The Bowman's capsule is the double-walled, cup-like structure at the sealed end of the tubule, which encloses a cluster of capillaries forming the glomerulus. The capsule and glomerulus together constitute the renal corpuscle or **malpighian body**. The cortex consists of most of the glomeruli and the proximal and distal portions of the renal tubules, and surrounds the medulla. The medulla is subdivided into a number of conical areas known as the renal pyramids, the apex of which extends toward the renal pelvis, forming papillae. There are visible striations in the renal pyramids (the 'medullary rays'). These contain the straight tubular elements, which are the collecting ducts and the loops of Henle and their associated blood vessels, the vasa recta. Each kidney contains 0.4 to 1.2 million nephrons. Two populations of nephron occur: long- and short-loop nephrons. In humans, short-looped nephrons predominate.

Renal interstitium

The interstitium is the renal tissue found between nephrons. It constitutes 7–9% by volume of the cortex and a larger proportion of the medulla. The interstitium consists of a variety of cell types supported in an extracellular matrix of glycosaminoglycans. It provides structural support to the nephrons, helps stabilize the high osmotic gradient essential for urine production, and water retention, and is also a site of hormone production, for example erythropoietin and tissue factors such as prostaglandins.

Juxtaglomerular apparatus

The areas called singularly, the **juxtaglomerular apparatus** (JGA) are found in the cortex and are closely associated with the glomerulus and efferent arteriole of the nephron (Figure 3.3). They consist of a cluster of cells known as the macula densa. The cells of the macula densa are formed from the tubule. Adjacent arteriolar cells are filled with granules (containing renin) and are innervated with sympathetic nerve fibres. The JGA helps maintain systemic blood pressure through regulating the circulating intravascular blood volume and sodium concentration.

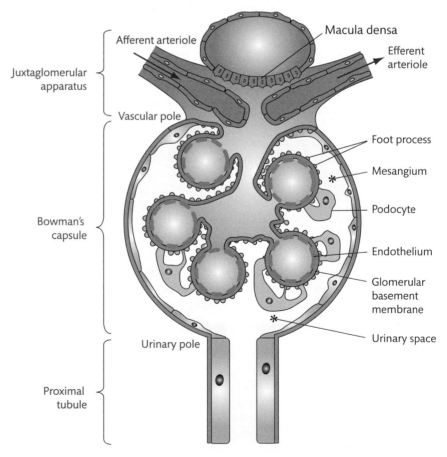

FIGURE 3.3
Section through the renal corpuscle and the juxtaglomerular apparatus.

Nephron

The Bowman's capsule forms the beginning of the tightly coiled proximal convoluted tubule, the pars convoluta. The tube becomes straighter in the renal medulla and this is called the proximal straight tubule or pars recta. Human proximal tubules are about 14 mm long. The pars recta drains into the descending thin limb of the loop of Henle, which, after passing through a hairpin loop, becomes first the ascending thin limb and then the ascending thick limb. The distal tubule includes the thick ascending limb of the loop of Henle, the macula densa, and the distal convoluted tubule. These form three morphologically and functionally distinct segments, seen in Figure 3.2. Other nephrons join to form the connecting tubule, cortical, outer medullary, and inner medullary-collecting ducts, which are formed from approximately six distal tubules. The collecting ducts ultimately drain into a renal calyx.

The renal artery divides into posterior and anterior elements that eventually subdivide into the afferent arterioles that supply the glomeruli. The glomerulus is often described as a knot of capillaries, which is enclosed within the Bowman's capsule. Each glomerulus is a highly specialized capillary bed consisting of approximately 40 glomerular loops about 200 μm in diameter and consists of several different types of cells. The renal corpuscle is shown in cross-section in Figure 3.3. The capillaries of the glomerulus rejoin to form the efferent arteriole, which then forms the capillary plexuses as well as the elongated vessels, the vasa recta, which surround the remaining parts of the nephron. The efferent arteriole then merges with renal venules to form the renal veins.

Cells of nephrons show regional specialization. For example, Bowman's capsules are lined with parietal epithelial cells. Within the central part of the glomerulus, between and within the capillary loops, are **mesangial cells**, which are suspended in a matrix (mesangium) that they synthesize. Mesangial cells have two major functions. First, they are phagocytic and participate in the clearance of macromolecules, for example circulating antigen-antibody complexes. Second, they provide structural support to the glomerular capillary loops and can contract in response to a variety of stimuli such as the hormones angiotensin II and antidiuretic hormone (ADH) or vasopressin, thus reducing the available surface area.

Capillary endothelial cells are found on the luminal side of the glomerular basement membrane. Between these are circular pores or **fenestra** with diameters of 70 to 100 nm, which are lined with a surface coating of negatively charged glycoproteins. The glomerular **basement membrane**, on which the **endothelial cells** sit, is approximately 300 nm thick and consists of three distinct layers: the lamina rara interna (lower layer in Figure 3.4(a)), the darker central lamina densa, and the lamina rara externa (upper layer in Figure 3.4 (a)). The lamina densa consists of a close network of fine, mainly type IV, collagen fibrils embedded in a gel-like matrix of glycoproteins and proteoglycans.

The narrow cavity on the opposite side of the basement membrane is Bowman's, or the urinary, space (Figures 3.3 and 3.4), where visceral epithelial cells called **podocytes** are attached onto the surface of the glomerular basement membrane. Podocytes can be visualized as being octopus-like in appearance with tentacles spreading out over the glomerular basement membrane. At right angles to each 'tentacle' are a large number of extensions or foot processes that embed into the basement membrane. The foot processes from adjacent podocytes are interdigitated to form filtration slits, which are covered by a loose, highly hydrated anionic glycoprotein coat. The resulting structure is relatively impermeable to molecules larger than M_r 60,000.

The epithelial cells lining the proximal convoluted tubule are cuboidal or columnar and have luminal brush borders consisting of millions of microvilli. This provides an extensive surface area for increasing the absorption of tubular fluid and filtered solutes.

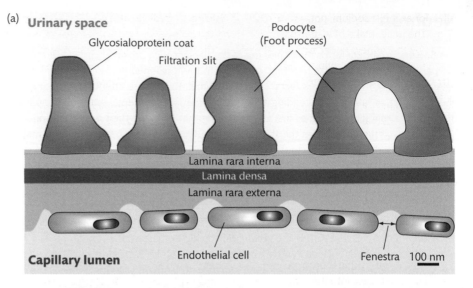

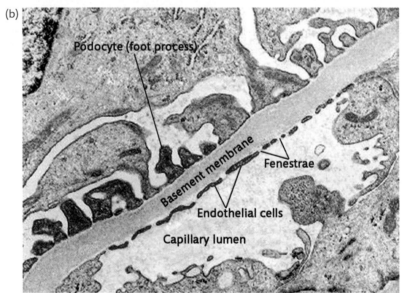

FIGURE 3.4

(a) Schematic showing the barrier between the capillary lumen and the urinary space. This is sometimes referred to as the glomerular filtration barrier. (b) Electron micrograph of the basement membrane. Courtesy of Dr JC Jeanette, Department of Pathology and Laboratory Medicine, University of North Carolina, USA.

SELF-CHECK 3.1

Describe the regions of a nephron.

3.2 Renal physiology

The excretory function of the kidney rids the body of the waste non-protein nitrogenous compounds urea, creatinine, and uric acid, as well as excess inorganic substances ingested in the diet, for example sodium, potassium, chloride, calcium, phosphate, magnesium, and sulphate ions. The daily intake of water may also exceed the requirements of the body and is eliminated by the kidneys under such circumstances. The kidneys produce urine by the processes of glomerular filtration followed by tubular reabsorption of solutes and water, which reduces the total urine volume excreted and modifies the composition of the urine.

The kidneys have a range of functions but their primary role is a homeostatic one to maintain a constant optimal composition of the blood and thus interstitial and intracellular fluids. The mechanisms of differential reabsorption and secretion, located in the tubule of a nephron, are the effectors of regulation.

Formation of urine

The basic stages in the formation of urine are shown in Figure 3.5. The first step is the filtering of plasma in the glomerulus to produce a filtrate consisting of water and solutes of small M_r within the tubule. Urine is produced from the filtrate by selectively reabsorbing metabolically useful materials back into the capillaries surrounding the tubule or by actively secreting other substances, for example toxins, into it from the capillary. The liquid, urine, is excreted.

The kidneys are central to the regulation of water, and all of the major anion and cation concentrations in the body; the maintenance of which are discussed in Chapters 5 (water, sodium, potassium, and chloride), 6 (H^+, HCO_3^-), and 10 (calcium and phosphate).

Cross reference

Chapter 5 Fluid and electrolyte disorders, Chapter 6 Acid-base disorders and Chapter 10 Disorders of calcium, phosphate, and magnesium homeostasis

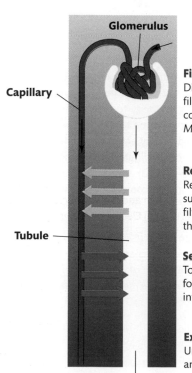

Glomerulus

Capillary

Tubule

Urine

Filtration:
Distal tubule collects a filtrate from the blood containing water and low M_r solutes

Reabsorption:
Reclaims valuable substances from the filtrate returning them to the body fluids

Secretion:
Toxins and excess ions, for example, are secreted into the distal tubule

Excretion:
Urine leaves the system and the body

FIGURE 3.5
Schematic showing an overview of the formation of urine.

The kidneys have other functions, but those concerned with excretion, and water and salt balance begin with filtration of plasma in the Bowman's capsule. Blood plasma filtered from the glomerular capillaries into the Bowman's space forms the initial glomerular filtrate. This requires water and solutes crossing the glomerular filtration barrier, which consists of the several cell types attached to the glomerular basement membrane and associated cells as shown in cross-section in Figures 3.3 and 3.4 and described in Section 3.1. The lamina densa of the basement membrane forms the main size discriminant barrier to the movement of proteins from the plasma into the tubular lumen. The other two layers of the membrane, the lamina rara externa and the lamina rara interna, are rich in negatively charged polyanionic glycoproteins such as heparan sulphate, and constitute the main charge discriminant barrier to the passage of circulating polyanions such as proteins.

A net filtration pressure of about 17 mm Hg in the glomerular capillary bed drives fluid across the glomerular basement membrane. Filtered fluid traverses the capillary wall via an extracellular route, that is, through endothelial fenestrae, basement membrane, and slit diaphragms (Figure 3.4). The net pressure difference across the glomerular filtration barrier must be sufficient not only to drive filtration but also to force the filtrate along the tubules. Maintaining renal blood flow is essential to kidney function and a complex array of intrarenal regulatory mechanisms ensure an adequate renal glomerular perfusion pressure across a wide range of systemic blood pressures. Modulators of vascular tone include nitric oxide, angiotensin II, prostaglandins, atrial natriuretic peptides and adenosine, and the renal sympathetic nervous system. In the absence of sufficient pressure, the lumina of the tubules will collapse. The composition of the filtrate is essentially the same as that of the plasma but with a notable reduction in molecules of M_r exceeding 15,000.

Each nephron produces about 100 µL of filtrate per 24 hours, which, assuming an average of one million nephrons per kidney, on a whole body basis equates to approximately 200 L/24 h of filtrate, or 140 mL/min. The rate of filtration is known as the **glomerular filtration rate** (GFR) and, in health, is therefore approximately 140 mL/min in adult humans. The use of GFR as a measure of kidney function is discussed in Section 3.3.

Of the 200 L of filtrate that passes through the glomeruli daily, approximately 99% is reabsorbed as it passes along the tubule during the production of urine. Different segments of the nephron have differing permeabilities to water, enabling the body to both retain water and produce urine of variable concentrations. Different regions of the tubule are specialized for certain functions. The proximal tubule is the most metabolically active part of the nephron and facilitates the reabsorption of 60 to 80% of the glomerular filtrate volume, including 70% of the filtered sodium and chloride, most of the potassium, HCO_3^{2-}, phosphate, and sulphate, and secretes 90% of the H^+ lost into the urine. Glucose is almost completely reabsorbed in the proximal tubule by a passive but Na-dependent process that is saturated at a blood glucose concentration of about 10 mmol/L. Thus, when the concentration of glucose in the blood exceeds this value, it will appear in the urine, a condition called glycosuria. Uric acid is reabsorbed in the proximal tubule by a passive Na-dependent mechanism but there is also an active secretory mechanism. Creatinine is also secreted at a rate of approximately 2.5 µmol/min. The loss of salt (sodium and chloride) from the proximal tube means the filtrate decreases in volume, its osmolarity remains similar to plasma at about 300 mOsm/L.

The concentration of the filtrate is greatly increased as it passes through the loop of Henle as shown in Figure 3.6. While the filtrate moves along the descending loop of Henle, water leaves the tubule by osmosis and the filtrate increases in concentration.

The ascending loop is also permeable to salts but is *impermeable* to water. Water, sodium, and a relatively small proportion of the urea also diffuse out of the collecting ducts into the

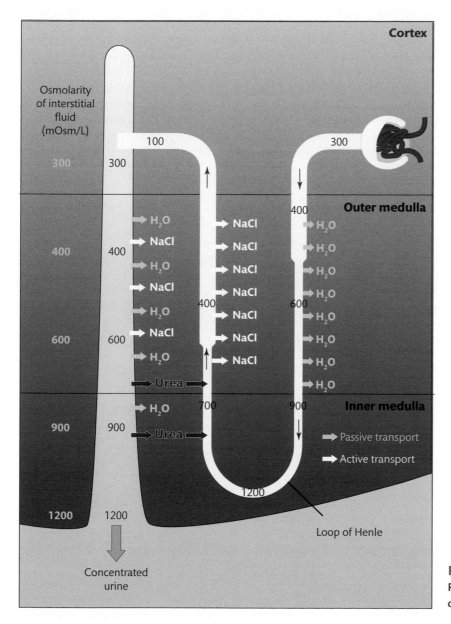

FIGURE 3.6
Role of the loop of Henle in concentrating urine.

interstitial fluid. Thus, the net effect is the formation of a gradient of osmolality in the interstitial regions surrounding the loop of Henle. Water leaves the tubule by osmosis in response to the concentration of solutes surrounding it. The highest concentration occurs in the region surrounding the hairpin turn of the loop of Henle; hence, its main function is to form concentrated urine that is hypertonic with respect to plasma. In the distal tubule, secretion is the prominent activity; organic ions, potassium, and H+ are transported from the blood in the efferent arteriole into the tubular fluid.

Urine formed by the kidneys drains into a renal calyx. Here urine collects before passing along the ureter and into the bladder, where it is stored before being discharged through the urethra. In health, urine is sterile, clear, has an amber colour, and characteristic odour. It has a slightly acidic pH, approximately 5.0 to 6.0, and a specific gravity of about 1.024. The osmolality of the urine produced varies from 50 to 1400 mOsmol/kg depending on water intake. In addition to dissolved compounds, urine contains a number of cellular fragments: complete cells, proteinaceous casts, and crystals, which are the so-called 'formed elements'. Urination or micturition is the discharge of urine. In healthy adults, adequate homeostasis can be maintained with a urine output of about 500 mL/24 h. Alterations in urinary output are described as **anuria** (less than 100 mL/24 h), **oliguria** (less than 400 mL/24 h), or **polyuria** (greater than 3 L/24 h or 50 mL/kg body weight per day). An overview of the reabsorption of substances from the glomerular filtrate by the renal tubules is shown in Figure 3.7.

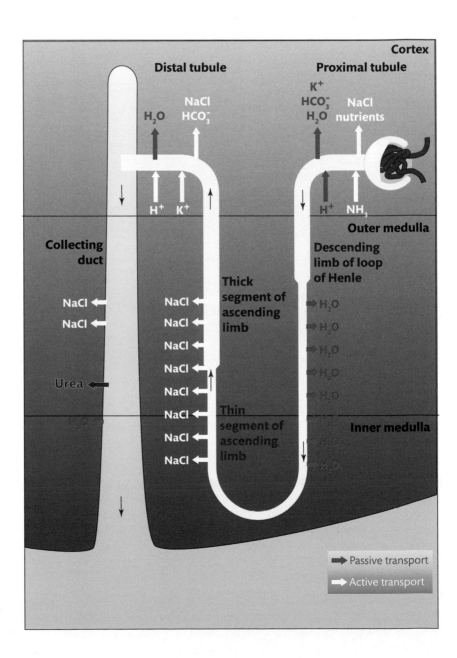

FIGURE 3.7
Overview of the reabsorption of material from the glomerular filtrate by renal tubules.

Renal protein handling

The combination of a specialized endothelium, epithelial cell barrier, and a basement membrane rich in negatively charged proteoglycans produces a fibre-matrix filter that restricts the passage of macromolecules (Figure 3.4). Molecules smaller than an M_r of approximately 5,000, that of inulin, are freely filtered whereas proteins larger than albumin, M_r 66,000 Da and molecular diameter 3.5 nm, are retained by the healthy glomerulus. In addition to size, molecular shape and charge also influence glomerular permeability.

The urinary concentration of proteins depends upon the filtered load and also on the efficiency of the proximal tubular reabsorptive process, which is saturable.

Following glomerular filtration, proteins are reabsorbed by receptor-mediated, low-affinity but high-capacity processes. The proteins are bound by two endocytic, multi-ligand receptors called megalin (M_r 600,000) and cubulin (M_r 460,000), which are located in apical clathrin-coated pits in intermicrovillar areas on the apical surface of tubular cell membranes. Following their binding, the proteins are internalized by **endocytosis** and transported by coated vesicles to endosomes where the ligands dissociate from the receptors. The ligands are transferred through endosomal compartments to lysosomes where they are hydrolyzed. The amino acids formed are released into the tubulo-interstitial space across the basolateral surface of the tubular epithelial cell. The receptors are returned to the apical plasma membrane through dense apical tubules. In health, the reabsorptive mechanism removes 99% of the filtered proteins, thus retaining most of their amino acid constituents for re-use.

The tubular secretion of proteins also contributes to urinary total protein concentration: in particular, the **Tamm-Horsfall glycoprotein** (THG), accounts for about 50% of normal urinary total protein excretion. The normal urinary total protein excretion is less than 150 mg daily.

Endocrine function

The endocrine functions of the kidneys may be regarded both as primary, because the kidneys are endocrine organs producing hormones, and as secondary, given the kidneys are sites of action for hormones produced or activated elsewhere. In addition, the kidneys degrade hormones such as insulin and aldosterone. In their primary role, the kidneys produce EPO, renin, prostaglandins, and 1,25-DHCC. Erythropoietin stimulates the synthesis of erythrocytes in the bone marrow. Prostaglandins and thromboxanes are synthesized from arachidonic acid by the cyclooxygenase enzyme system. This system is active in many parts of the kidney. Prostaglandins regulate the physiological action of other hormones affecting renal vascular tone, mesangial contractility, and the tubular processing of salt and water.

The proteolytic enzyme, renin is released from arteriolar cells in the JGA primarily in response to decreases in afferent arteriolar pressure and the intraluminal sodium delivery to the macula densa. The released renin hydrolyses the plasma protein angiotensinogen to angiotensin I. Angiotensin I is further metabolized in the lungs by angiotensin converting enzyme (ACE) to the potent vasoconstrictor and stimulator of aldosterone release angiotensin II. The vasoconstriction and release of aldosterone stimulates increases in the reabsorption of sodium by the distal tubular, intravascular volume, and blood pressure. The kidneys are primarily responsible for producing 1,25-dihydroxycholecalciferol from 25-hydroxycholecalciferol, a reaction catalysed by 25-hydroxycholecalciferol 1α-hydroxylase found in proximal tubular epithelial cells. The regulation of this system and its disturbance in kidney disease is considered in Chapter 10.

Cross reference

Chapter 10 Disorders of calcium, phosphate, and magnesium homeostasis

3.3 **Kidney function tests**

The evaluation of kidney function focuses on assessing the filtering capabilities of the kidney and examining the urine produced for the presence of markers of kidney disease.

Glomerular filtration rate

The GFR is the volume of plasma from which a given substance is completely removed from the plasma by glomerular filtration in a unit time. In essence, it is an estimation of the efficiency with which substances are cleared from the blood by glomerular filtration and so is a measure of nephron function.

The GFR is widely accepted as the best overall measure of kidney function, summarizing the complex functions of the kidney in a single numerical expression. A decrease in GFR precedes renal failure in all forms of progressive kidney disease. Measuring GFR is useful in identifying **chronic kidney disease** (CKD) in order to (1) target treatment to prevent progression and complications and (2) to monitor progression and predict the point at which renal replacement therapy (RRT) will be required. It is also used as a guide to the dosage of drugs that are excreted by the kidneys to avoid potential drug toxicity.

The GFR is determined by measuring the clearance of an exogenous or endogenous substance by the kidneys. A substance (S) is only suitable for estimating GFR if it occurs at a stable concentration in the plasma, is physiologically inert, freely filtered at the glomerulus and not secreted, reabsorbed, synthesized, or metabolized by the kidneys. In these conditions, the amount of S filtered at the glomerulus is equal to the amount excreted in the urine. The amount of S filtered at the glomerulus is equal to the GFR multiplied by the concentration of S in the plasma (PS), that is $GFR \times PS$. The amount of S excreted equals its concentration in the urine (US) multiplied by the urinary flow rate (V, volume excreted per unit time).

Since filtered S = excreted S:

$$GFR \times PS = US \times V$$

therefore

$$GFR = (US \times V) / PS$$

where GFR = clearance in units of mL of plasma cleared of a substance per minute, US = urinary concentration of the substance, V = volumetric flow rate of urine in mL/min and PS = plasma concentration of the substance.

The term $(US \times V) / PS$ is defined as the clearance of substance S and is an accurate estimate of GFR providing the criteria mentioned above are satisfied. The GFR varies with body size over a range from less than 10 in a neonate to 200 mL/min in a large adult. It is standard practice to normalize GFR to a measure of body size to allow comparisons of GFR between individuals. Conventionally, GFR is corrected to a body surface area (BSA) value of 1.73 m^2, which is the average BSA of adults aged 25 years.

$$GFR_{corrected} = GFR_{measured} \times (1.73/BSA\ m^2)$$

SELF-CHECK 3.2

A patient produced 1500 mL of urine in a 24-hour period. This patient had a urine creatinine concentration of 10 mmol/L and a plasma creatinine concentration of 100 μmol/L. Calculate the creatinine clearance in mL/min.

Measurement of GFR using exogenous substances

Generally, the most accurate measurements of GFR are obtained using infused exogenous molecules. This is recommended when an accurate knowledge of renal function is critical. A variety of exogenous radioisotopic, for example, ^{125}I-iothalamate, ^{51}Cr-ethylenediaminetetraacetic acid (EDTA), and ^{99}mTc-diethylenetriaminepentaacetic acid (DTPA), and nonradioisotopic, such as inulin or iohexol markers have been used to estimate clearance. These may be administered by constant infusion or as a single bolus. The determination of GFR may be based on the urinary or the plasma clearance of the marker. Following infusion, the GFR is calculated from knowledge of the amount of marker injected and the decrease in its concentration (activity) over time. In clinical practice in the UK, the majority of GFR measurements are undertaken in nuclear medicine departments using ^{51}Cr-EDTA.

Key Points

Inulin is a fructose polymer of plant origin, distinct from the peptide hormone insulin.

Measurement of GFR using endogenous substances

The overwhelming majority of GFR assessments in clinical practice are undertaken by clinical biochemistry laboratories using serum creatinine. As a GFR marker, creatinine can be expressed as its serum concentration or its renal clearance. The analytical measurement of creatinine is described in the Method box below.

Creatinine (M_r 113), derived from creatine and phosphocreatine breakdown in muscle, is freely filtered at the glomerulus. Its concentration is inversely related to GFR. As a GFR marker it is convenient and cheap to measure but its concentration in the plasma is affected by age, gender, exercise, muscle mass, certain drugs, for example cimetidine, trimethoprim, and nutritional status. Creatine in meat is converted to creatinine when it is cooked. Once eaten, it increases the creatinine concentration in plasma after ingestion. Further, a small, but significant and variable proportion of the creatinine, typically 7–10% of the total, in the urine is derived from tubular secretion. This proportion is increased in **renal insufficiency**, a partial failure of renal function. Additionally, in patients with advanced kidney disease, the extra-renal clearance of creatinine becomes significant due to its degradation from bacterial overgrowth in the small intestine. Thus, although an increase in serum creatinine concentration in response to falling GFR may be anticipated, its concentration may remain within the reference range and only increase when significant renal function has been lost. Hence, while an increased serum creatinine concentration is fairly specific for impaired kidney function, a normal serum creatinine does not necessarily equate to normal kidney function. Given these limitations, it is recommended that serum creatinine measurement is not solely used to assess kidney function.

SELF-CHECK 3.3

Describe the biological problems that limit the use of serum creatinine as a measure of GFR.

METHOD *Analytical measurement of creatinine*

Both chemical and enzymatic methods can be used to measure creatinine in body fluids. Most laboratories use adaptations of the same assay for measurements in both serum and urine. A method for the assay of serum creatinine was first described by Jaffe in 1886. This involves the reaction of creatinine with alkaline sodium picrate to form an orange-red coloured complex. There are various other approaches to creatinine measurement, including enzymatic, high performance liquid chromatography and isotope dilution-mass spectrometry (ID-MS) methods. However, the Jaffe assay in either end point (amount of coloured product formed at the end of the reaction), or, more commonly, kinetic (rate of coloured product formation) mode remains widely used. Measurement of creatinine using the **Jaffe reaction** suffers from three main problems: (1) non-specificity, (2) spectral interferences, and (3) non-standardized calibration. Many other compounds produce a Jaffe-like chromogen, including proteins, glucose, ascorbic acid, ketone bodies, pyruvate, guanidine, blood-substitute products, and cephalosporins. The degree of interference from these compounds varies both between patients and with the precise reaction conditions of the assay. Assays are also highly susceptible to spectral interference from bilirubin, haemoglobin, and lipaemia. Enzymatic assays are also not immune to these effects. Problems of both bias and imprecision with creatinine measurement remain and, in both cases, have their greatest impact in the near-normal range. There is significant between-laboratory variation in creatinine measurement, with much of the difference being due to calibration differences. Recently, however, a standardized reference material 967 suitable for calibration of serum creatinine assays has been prepared. It is also recommended that all methods should be traceable to (validated against) the reference method ID-MS. These two measures should address the major causes of differences between different creatinine assays.

SELF-CHECK 3.4

Describe the analytical problems that limit the use of serum creatinine as a measure of GFR.

Given that creatinine is endogenously produced and released into body fluids at a constant rate, its urinary clearance can be measured as an indicator of GFR. However, this requires a timed urine collection, which introduces its own inaccuracies, and is inconvenient and unpleasant. In adults, the intra-individual day-to-day coefficient of variation for repeated measures of creatinine clearance exceeds 25%. Additionally, due to tubular secretion of creatinine, creatinine clearance usually overestimates GFR in adults by 10–40% at clearances above 80 mL/min and by a much greater percentage at lower levels of GFR. Hence, at best, creatinine clearance can only provide a crude index of GFR.

Key Points

The relationship between creatinine clearance and reference measures of GFR is generally inferior to that shown by estimated GFR: the use of creatinine clearance to assess kidney function is declining in UK laboratories.

The mathematical relationship between serum creatinine and GFR can be improved by correcting for the confounding variables that make that relationship non-linear. Many equations have been derived that estimate GFR using serum creatinine corrected for some or all of gender, body size, ethnic origin, and age. These may produce a more reliable estimate of GFR than using serum creatinine alone. The Cockcroft and Gault and Modification of Diet in Renal Disease (MDRD) study equations have been widely used in adults. Two further equations (Schwartz and Counahan-Barratt) are recommended for use in children. It has recently been recommended by many healthcare organizations, including the UK Department of Health, that GFR should be estimated using the MDRD equation in addition to serum creatinine. The simplified version of this equation is now the most widely used approach to GFR estimation internationally (see the Method box below).

Since its publication a series of studies have confirmed that, for the detection of patients with stage 3 to 5 CKD, the MDRD equation provides a relatively accurate and clinically acceptable assessment of GFR. However, the equation demonstrates deteriorating performance (decreased accuracy and increased imprecision compared to a reference method) as GFR increases towards the physiological range. This is an important consideration when monitoring patients with stages 1 or 2 CKD, for example many patients with early diabetic nephropathy, or when assessing the suitability of potential kidney donors.

There are a number of clinical situations where an accurate knowledge of GFR is necessary, and where reliance on equation-based estimates of GFR should be avoided. These include cancer chemotherapy, or the use of any other drug that is excreted in urine and has a narrow therapeutic margin; the assessment of potential living related kidney donors; and the assessment of GFR in patients with muscle wasting disorders, including spina bifida and paraplegia. When accurate GFR information is required, reference methods such as those described above should be used.

Cystatin C is a relatively small protein of M_r 12,800, synthesized by all nucleated cells that has been proposed as a superior marker to creatinine of GFR. Its physiological role is that of a cysteine protease inhibitor. With regard to renal function, its most important attributes

METHOD *Estimation of GFR using the MDRD equation*

In the MDRD Study, GFR was determined in 1,628 predominantly middle-aged patients with known kidney disease using a reference ^{125}I-iothalamate method and expressed per 1.73 m² body surface area. Creatinine was measured in a central laboratory using a kinetic Jaffe assay. Stepwise multiple regression modelling on logarithm-transformed data was used to determine the set of variables that best predicted GFR. The data was later re-analysed to enable the equation to be used with ID-MS traceable creatinine assays. The equation is not easily remembered but can be programmed into laboratory computer systems to automatically produce a GFR estimate when serum creatinine is measured. There are also many calculators available online.

$$\text{GFR (mL/min/1.73 m}^2) = 175 \times [\text{serum creatinine (}\mu\text{mol/L)} \times 0.011312]$$

$$- 1.154 \times [\text{age}] - 0.203 \times [1.212 \text{ if black}] \times [0.742 \text{ if female}]$$

are its relatively small size and high pI (9.2), which enable it to be relatively freely filtered at the glomerulus. Serum concentrations of cystatin C appear to be unaffected by muscle mass, diet or gender. There are no known extrarenal routes of elimination, with clearance from the circulation occurring only by glomerular filtration. Cystatin C has mainly been measured using latex particle-enhanced turbidimetric or nephelometric immunoassays. Generally, the correlation between serum cystatin C and GFR is superior to that between creatinine and GFR. Further, cystatin C concentration increases sooner than that of creatinine as GFR declines: at approximately 60 mL/min/1.73 m^2 compared to about 40 mL/min/1.73 m^2 for serum creatinine. Cystatin C is therefore especially useful when trying to detect mild impairment of kidney function. However, due to its higher reagent costs compared to creatinine, cystatin C is not widely used in clinical practice.

Urinary examination: reagent strip ('dipstick') testing

Examination of urine or urinalysis, is invaluable in identifying urological and kidney disease. Examination of a midstream urine sample is often the first step in the assessment of a patient suspected of having, or confirmed to have, deteriorating kidney function. The appearance of urine itself can be helpful. A darkening from the normal pale straw colour indicates a more concentrated urine or the presence of another pigment. A variety of foods and drugs can also alter the colour of urine. For example, beetroot and levodopa may impart red and brown colorations respectively. Haemoglobin and myoglobin can give a pink-red-brown coloration, depending on their concentrations. Turbidity in a fresh sample of urine may indicate a urinary tract infection (UTI) but may also be due to fat particles in a patient with nephrotic syndrome. Excessive foaming of urine when shaken suggests proteinuria.

The microscopic examination of urine is also widely used. Nephrologists will often describe urine as being 'bland' or as having an 'active urinary sediment', which means that erythrocytes, white blood cells, and casts can be detected when the urine is examined using a microscope. These are signs of an active kidney inflammation.

Urine is also often chemically evaluated with the help of reagent strip ('dipstick') tests. Many tests of renal significance have been adapted for use by coating or impregnating pads of cellulose or strips of plastic with reagents for the test in question (Figure 3.8). Such dipstick tests detect analytes derived from the circulation that appear in the urine. These include glucose, ketones, and urobilinogen, in addition to changes in constituents that are more directly linked to pathology affecting the kidney or urinary tract, such as protein and haemoglobin. Urine samples for dipstick testing should be collected in sterile containers and dipstick testing performed using fresh urine. Dipsticks should only be used if they have been stored in a properly desiccated state, as they can deteriorate in a matter of hours. Other dipstick tests useful in assessing kidney function include those for leukocytes, nitrite, specific gravity, and pH.

Leukocyte esterase activity produced by neutrophils can be detected by reagent strip testing and may signal **pyuria** associated with a UTI. Pyuria is the presence of white blood cells in the urine. Significant pyuria is normally defined as a urinary white cell count greater than or equal to 10 leucocytes per mL. Nitrite is not normally present in urine but is found when bacteria reduce urinary nitrates (to nitrites). Many Gram-negative and some Gram-positive organisms are capable of this activity. The combination of nitrite and leukocyte esterase tests is valuable in identifying patients with a UTI, but negative results do not exclude this condition. Urinary specific gravity correlates with osmolality and gives an indication of urinary

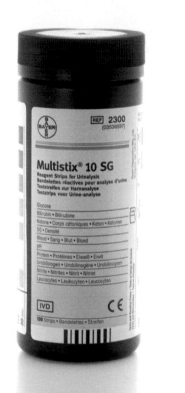

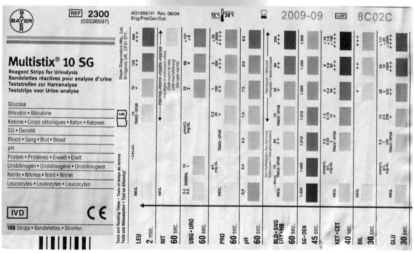

FIGURE 3.8
Renal dipsticks for testing urine.

concentration. Normal urinary specific gravity ranges from 1.003 to 1.030; higher values may indicate dehydration, glycosuria, or the syndrome of inappropriate antidiuretic hormone. Low values are associated with conditions where concentrated urine cannot be formed, for example diuretic use, diabetes insipidus, and adrenal insufficiency. In patients with intrinsic renal insufficiency, the specific gravity becomes fixed at 1.010; reflecting the specific gravity of the glomerular filtrate. Urinary pH is normally slightly acidic (5.5 to 6.5) but can range from 4.5 to 8.0. Measurement of urinary pH with reagent strips can be helpful in the assessment of patients with **renal tubular acidosis** (RTA) and those that form stones, although more accurate measurement using a pH electrode is preferred. Testing urine for total protein, albumin, and haemoglobin is considered below.

Proteinuria

Protein loss shows considerable biological variability and may be increased by upright posture, exercise, fever, UTI, and heart failure, as well as by kidney disease. Upright posture increases protein loss in both normal subjects and those with kidney disease. If **proteinuria** is postural, disappearing when the patient is recumbent and absent from early morning samples, the patient can be strongly reassured. This benign condition is called orthostatic proteinuria and the level of proteinuria will generally be less than 1,000 mg per day (approximately 100 mg/mmol creatinine). Any increase in the filtered load, due to glomerular damage, increased glomerular vascular permeability or increased circulating concentration of proteins of low M_r, or decrease in reabsorptive capacity, due to tubular damage, can result in increased urinary protein losses. Consequently, the appearance of significant amounts of protein in the urine generally suggests kidney disease. Commonly, proteinuria is classified as either tubular (low M_r proteinuria in isolation) or glomerular (higher M_r proteinuria) depending on the pattern of proteins observed. A third category, overflow proteinuria is also recognized in which filtration of excessive amounts of low M_r proteins exceeds the tubular capacity for reabsorption as shown in Table 3.1. Although a useful concept as an aid in understanding the pathophysiology of proteinuria, this classification has limited clinical application: in particular tubular and glomerular proteinuria commonly occur together.

TABLE 3.1 Characterization of proteinuria.

Type of proteinuria	Causes	Examples of proteins seen
Glomerular	Increased glomerular permeability, for example due to immune complex deposition, diabetic nephropathy	Progressively increasing excretion of high M_r proteins as permeability increases (for example albumin, IgG)
Tubular	Proximal tubular damage: decreased tubular reabsorptive capacity and/or release of intracellular components (for example due to nephrotoxic drugs or heavy metals, anoxia)	Predominantly lower M_r proteins (for example α_1-microglobulin, β_2-microglobulin, retinol binding protein) and enzymuria (for example N-acetyl-β-D-glucosaminidase, alkaline phosphatase, α-glutathione-S-transferase)
	Decreased nephron number due to progressive kidney disease: increased filtered load per nephron	As above
	Distal tubular damage	Tamm Horsfall glycoprotein π-glutathione-S-transferase
Overflow	Increased plasma concentration of relatively freely filtered protein (for example multiple myeloma, rhabdomyolysis)	Bence-Jones protein Myoglobin

> **Key Points**
>
> Quantitatively, albumin is the predominant protein in most proteinurias, although albumin is a relatively minor component of normal urinary protein loss.

Sample collection

The amount of urine produced and collected in a day, constitutes a 24-hour urine sample. This type of sample is used as the definitive means of demonstrating the presence of proteinuria, although it is widely accepted that collecting urine over such an extended period is a difficult procedure to control effectively. Indeed, more than 25% of samples are thought to represent incomplete collections. Overnight, first void in the morning ('early morning urine', EMU), second void in the morning, or random sample collections have also been used. However, the excretion of creatinine in the urine is fairly constant throughout the 24-hour period. Thus, measuring protein: creatinine ratios allows variations in urinary concentrations to be corrected. Measurements of spot or individual urine protein: creatinine ratios can be used to eliminate the need for 24-hour collections. An EMU sample is preferred since it correlates with 24-hour protein excretion and is required to exclude a diagnosis of orthostatic proteinuria. However, a random urine sample is acceptable if no EMU sample is available. If required, daily protein excretion (in mg/day) can be roughly estimated by multiplying the protein: creatinine ratio (measured in mg/mmol) by 10, since, although daily excretion of creatinine depends on muscle mass, an average value of 10 mmol creatinine excreted per 24 hours can be assumed. A suitable protocol for the further investigation of patients found to have proteinuria at screening is shown in Figure 3.9.

> **Key Points**
>
> Protein: creatinine ratios measured on a random urine specimen show excellent correlation with protein loss measured on a timed (for example 24-hour) collection and appear to have similar, or even superior, prognostic, predictive power.

SELF-CHECK 3.5

A patient's EMU sample has a protein: creatinine ratio of 125 mg/mmol. What is the approximate amount of protein excreted daily? Is this likely to represent a pathological protein loss?

Measurement of total protein

Urinary total protein concentration may be estimated using reagent strip test devices or measured by a variety of laboratory-based techniques. Standard urine dipsticks can be used to estimate protein concentrations. However, their accuracy varies with the concentration of the urine sample. Thus, these tests can only give a rough indication of the presence or absence of pathological proteinuria.

The dipstick test for total protein includes a cellulose test pad impregnated with tetrabromophenol blue and a citrate buffer, pH 3. The reaction is based on the 'protein error of indicators' phenomenon in which certain chemical indicators demonstrate one colour in the presence of protein

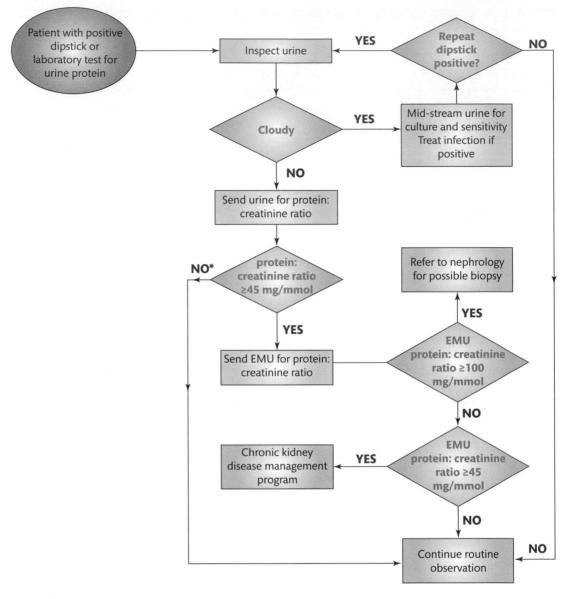

FIGURE 3.9

A typical protocol for investigating proteinuria in patients with a positive (+ or greater) dipstick or quantitative protein test. Patients with two or more positive (>45 mg total protein/mmol creatinine) quantitative tests one to two weeks apart should be diagnosed as having persistent proteinuria. Postural proteinuria should be excluded by the examination of an early morning urine (EMU).

*A serious primary renal pathology is unlikely in the absence of a systemic disease, such as diabetes or hypertension, or a borderline increase in total protein excretion (15–44 mg/mmol) without haematuria or a rise in serum creatinine concentration.

and another in its absence. Thus, tetrabromophenol blue is green in the presence of protein at pH 3 but yellow in its absence. The colour produced is observed after 60 seconds. The test has a lower detection limit of 150 to 300 mg/L, depending on the type and proportions of proteins present. The reagent is most sensitive to albumin and less so to globulins, **Bence-Jones protein**,

THG, and haemoglobin. Typically, if the colour of the tested strip matches that of the 'trace' block on the dipstick, it corresponds to a total protein concentration in the urine of approximately 150 mg/L. A colour matching the '+' block is equivalent to 300 mg/L. Significant or clinical proteinuria is deemed present when the colour change matches any block greater than that of the trace, that is greater than 300 mg/L. Assuming a typical urinary volume of 1.5 L daily, this is equivalent to a total protein excretion of 450 mg/day (approximately 45 mg/mmol creatinine). However, the specificity of protein dipsticks in detecting proteinuria is poor and misclassification errors are common. Positive dipstick tests should be confirmed in the laboratory by measuring the protein: creatinine or albumin: creatinine ratio on an EMU or random urine sample.

Key Points

Dipstick tests for urinary total protein have relatively poor sensitivity for proteinuria detection; they should never be used when screening for microalbuminuria in diabetics.

Laboratory methods used to measure total protein in urine include the Lowry method, turbidimetry after mixing with trichloroacetic or sulphosalicylic acid, or with benzethonium chloride (also called benzyldimethyl(2-[2-(ρ-1,1,3,3-tetramethylbutylphenoxy)ethoxy]ethylene) ammonium chloride) and dye binding using Coomassie brilliant blue or pyrogallol red molybdate. External quality assessment programs have highlighted significant variation between methods and, in particular, poor precision at low concentrations. Such concerns about the variable response of different proteins to these tests have led to many variants of the methods being published. The turbidimetric methods and the dye-binding methods do not give equal analytical specificity and sensitivity for all proteins; this may be of particular significance in the detection of immunoglobulin light chains, when an immunochemical method is more appropriate.

Measurement of specific proteins

Electrophoretic, including capillary electrophoresis, techniques can provide a semiquantitative or qualitative measurement of the range of proteins present in a sample of urine and estimations of a panel or pattern of proteins, when assays for the specific proteins are not available. Immunoassays are the preferred methods for the accurate, precise, and sensitive quantitation of individual proteins. A variety of analytical approaches are available but the commonest approach employs a light-scattering immunoassay, with turbidimetric or nephelometric detection. In addition to albumin, immunoassay techniques have also been used to measure low M_r proteins, most commonly α_1-microglobulin and retinol binding protein, in urine. Enzymatic assays have been used to measure tubular enzyme activities, for example those of N-acetyl-β-D-glucosaminidase and alkaline phosphatase, in urine.

Indications for urinary protein measurement.

It is mandatory to investigate for an increase in urinary protein loss in any patient suspected of kidney disease. In the international classification of CKD, stages 1 and 2 CKD require the presence of a marker of kidney damage other than a decreased GFR. The presence of proteinuria is the most frequently used alternative marker (Table 3.1). The presence of proteinuria is therefore necessary for both the identification of kidney damage and for guiding future treatment and surveillance.

Significance of proteinuria

Proteinuria greater than or equal to 450 mg per 24 hours is generally pathological, especially if there is concomitant **haematuria**. Proteinuria above 1,000 mg per 24 hours implies glomerular proteinuria. Glomerular proteinuria may be heavy and a mixed proteinuria with increases in both high and low M_r proteins may be observed. Numerous studies have demonstrated that proteinuria is a potent risk marker for both the progression of renal disease and cardiovascular risk. At all levels of GFR, the presence of proteinuria indicates a greatly increased risk of progressive disease. It is accepted that proteinuria is not just a consequence of, but contributes directly to, progression of kidney disease.

The accumulation of proteins in abnormal amounts in the tubular lumen may trigger an inflammatory reaction that contributes to interstitial structural damage and expansion. Glomerular filtration of an abnormal amount or types of proteins is thought to induce mesangial cell injury, leading to **glomerulosclerosis**. The use of ACE inhibitors and angiotensin receptor blockers (ARB), alone or in combination reduces protein loss by decreasing intraglomerular filtration pressure and, possibly, by stabilizing the glomerular epithelial cell slit diaphragm proteins (Figure 3.4). Consequently, the reduction of proteinuria is a therapeutic target.

Albuminuria

The negative charge of albumin hinders its filtration in the nephron resulting in its virtual exclusion from the urinary space. Nevertheless, approximately 1.3 g of albumin pass through the glomerular capillary walls daily. The majority of this is reclaimed by endocytosis (Figure 3.10) such that in health only the relatively small quantity of approximately 15 mg of intact albumin appears in the urine daily. During endocytosis the filtered proteins labelled as (•) in Figure 3.10 are reabsorbed following binding to endocytic multi-ligand receptors, such as megalin (M_r 600,000) and cubulin (M_r 460,000) located in intermicrovillar areas. They are then transported by coated vesicles to endosomes in which they dissociate from the receptors. The proteins are then transferred through endosomal compartments to lysosomes for degradation and further processing. The receptors return to the apical plasma membrane through dense apical tubules.

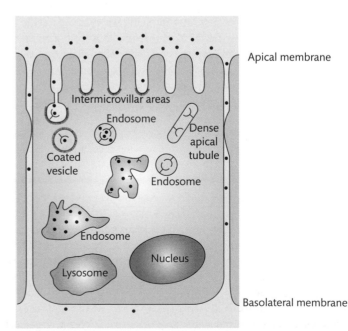

FIGURE 3.10
Reabsorption of filtered proteins by the proximal tubule. See text for details. Figure reproduced from Verroust PJ and Christensen EI (2002) Megalin and cubilin - the story of two multipurpose receptors unfolds. *Nephrology Dialysis Transplantation* **17**, 1867–1871.

Definition of microalbuminuria

Microalbuminuria is a term describing the loss of albumin in the urine in amounts that are abnormal but below the limit of detection of conventional urine dipsticks and can therefore only be detected using specific tests for albumin. The term is confusing in that it can mistakenly be taken to mean that there is abnormal excretion of *microalbumin* which is a small albumin molecule, whereas in fact the albumin excreted in this condition is exactly the same as in other conditions that cause proteinuria. In 'overt diabetic nephropathy' the amount of albumin present in the urine reaches concentrations that can be detected by conventional urine dipsticks, that is macroalbuminuria occurs.

Key Points

Microalbuminuria is a pathological increase in the rate of loss of albumin in the urine.

Establishing diagnosis of microalbuminuria

Urine albumin should be measured using a sensitive method in an EMU (preferred) or random mid-stream urine sample. Results are commonly expressed as albumin: creatinine ratios. The diagnosis of microalbuminuria requires the demonstration of increased albumin loss, as an increased albumin: creatinine ratio or increased rate of albumin loss, in at least two out of three urine samples collected in the absence of infection or of an acute metabolic crisis. Establishing the diagnosis has both prognostic and management implications in the care of patients with diabetes mellitus. The best possible metabolic control of diabetes should be achieved before investigating patients for microalbuminuria. Patients should not be screened during intercurrent illness.

Patients with diabetes mellitus who have persistent proteinuria, or microalbuminuria, need not be tested for it. All other patients with diabetes mellitus should undergo as a minimum, annual testing for microalbuminuria. Annual screening should commence five years after diagnosis in patients with type 1 diabetes mellitus and at diagnosis in patients with type 2 diabetes without proteinuria. Testing should continue on an annual basis up to the age of 75 years. There is currently no proven role in screening for microalbuminuria in people who do not have diabetes.

Significance of albuminuria

An albumin: creatinine ratio of less than 2.5 mg/mmol in males or less than 3.5 mg/mmol in females does not require further investigation until the patient's next annual review. Patients presenting with albumin: creatinine ratios above or equal to these cut-off values should have urine samples analysed for albumin estimation on two further occasions and ideally within one month. Patients demonstrating increased albumin: creatinine ratios in one or both of these samples have clinical microalbuminuria. It is, however, necessary to consider other possible causes for the increased loss of albumin, especially in the case of patients who have suffered from type 1 diabetes for less than five years. Possible reasons for the increase include non-diabetic renal disease, menstrual contamination, vaginal discharge, uncontrolled hypertension, UTI, uncontrolled diabetes, heart failure, intercurrent illness, and strenuous exercise. The recognition of microalbuminuria in patients with diabetes mellitus allows identification of diabetic nephropathy, and institution of treatment to reduce the risk of progressive kidney damage and cardiovascular mortality, at an earlier stage than would be possible with conventional protein dipstick testing (see above).

SELF-CHECK 3.6

How would you establish whether a patient has microalbuminuria?

Haematuria

Haematuria is the presence of haemoglobin or erythrocytes in the urine. Its occurrence may be due to glomerular, tubulointerstitial, or postrenal (bladder) disease; the latter two causes are the more common. Haematuria can present in a range of kidney diseases, including **glomerulonephritis** (especially IgA nephropathy), **polycystic kidney disease** (PKD), sickle cell disease, vasculitis and a number of infections. Urological diseases may also give rise to haematuria, including bladder, prostate, and pelvic or ureteric malignancies, kidney stones, trauma, bladder damage, and ureteric stricture.

Dipstick urinalysis is the test of choice for confirming frank or visible (macroscopic) haematuria and for detecting microscopic haematuria. In the dipstick test for haemoglobin the reagent pad is impregnated with buffered tetramethyl benzidine and an organic peroxide. Peroxidase activity of haemoglobin catalyses the reaction of cumene hydroperoxide and tetramethyl benzidine to produce a coloured product that ranges from orange through pale to dark green. Erythrocytes, free haemoglobin, and myoglobin are all detectable. Two reagent pads are employed for the low haemoglobin level; if intact red cells are present, the low-level pad will have a speckled appearance, with a solid colour indicating haemolysed red cells. The detection limit for free haemoglobin is 150 to 600 µg/L or 5 to 20 intact erythrocytes per µL. The test is equally sensitive to haemoglobin and myoglobin. Water must not be used as a negative control with this test due to the matrix requirements of the assay, and will give a false positive result.

3.4 **Kidney disease**

Kidney disease is classified on the basis of the level of GFR and the presence or absence of markers of kidney damage. Kidney disease may be either acute or chronic. These terms indicate the rate at which damage occurs or has occurred rather than its mechanism. The term *renal* has largely been replaced by the *kidney* when referring to chronic disease since it is more easily understood by patients and non-specialists. **End-stage renal disease** (ESRD) is a term describing the level of GFR and the occurrence of signs and symptoms of kidney failure that require treatment by RRT to maintain life. Typically, ESRD occurs at a GFR of less than 15 mL/min per 1.73 m^2. Renal replacement therapy options are dialysis and kidney transplantation. The term acute renal failure (ARF) has been replaced with **acute kidney injury** (AKI). However, ARF remains a widely used nomenclature and still appears in many textbooks.

Acute kidney injury

Acute kidney injury is diagnosed when the excretory function of the kidneys declines over hours or days.

Key Points

'Acute' in this context relates to the rapidity of onset of a disease rather than its severity. Similarly, 'chronic' in the context of disease implies a longer-term condition.

The biochemical clues to the development of AKI include rapidly rising serum creatinine concentration and severe and life-threatening metabolic derangements, particularly hyperkalaemia and metabolic acidosis. Acute kidney injury is a common condition complicating 5% of hospital admissions. It often inadvertently results, in part, from medical or surgical procedures. Intrinsic AKI can be caused by primary vascular, glomerular, or interstitial disorders but, in the majority of cases, the kidney lesion seen on histology is referred to as **acute tubular necrosis** (ATN). Acute tubular necrosis is caused by ischaemic or nephrotoxic injury to the kidney. Acute kidney injury develops rapidly, and therefore its sequelae are mainly a consequence of rapid electrolyte, acid-base, and fluid imbalances that are often difficult to control. The clinical assessment of AKI should consider whether the precipitant is prerenal (factors affecting the blood supply to the kidney), intrarenal (intrinsic kidney disease), or postrenal (diseases affecting the lower urinary tract). The commonest causes are listed in Table 3.2.

TABLE 3.2 Causes of acute kidney injury.

Type	Lesion	Cause
Prerenal AKI		Dehydration
		Haemorrhage
		Diarrhoea
		Post-operative fluid and blood losses
		Sepsis
		Acute cardiac failure
		Aortic dissection
Intrinsic renal disease	Glomerular disease	Rapidly progressive glomerulonephritis
		ANCA-associated vasculitides
		Goodpasture's disease
		Systemic lupus erythematosus
		Crescentic lesions complicating glomerular disease
		Cryoglobulinaemia
		Post-infectious glomerulonephritis
	Tubulointerstitial disease	Any prerenal cause leading to ATN (the most common cause of AKI)
		Drug nephrotoxicity (NSAIDs, ACE inhibitors, aminoglycoside antibiotics, amphotericin)
		Allergic TIN associated with antibiotics and NSAIDs
		Sarcoidosis
		Pyelonephritis
		Myeloma cast nephropathy
	Microvascular disease	Thrombotic microangiopathies (haemolytic-uraemic syndrome/thrombotic thrombocytopaenic purpura, pre-eclamptic toxaemia, scleroderma)
	Miscellaneous	Contrast nephropathy
		Poisoning (e.g. methanol)
		Rhabdomyolysis
		Urate nephropathy
		Hepatorenal syndrome
		Cholesterol embolism
		Renal vein thrombosis

(Continued)

TABLE 3.2 *(Continued)*

Type	Lesion	Cause
Postrenal AKI	Bladder outflow obstruction	Benign and malignant prostate disease
	Bilateral renal calculi or calculi within a single kidney	Invasive bladder carcinoma
	Retroperitoneal fibrosis	

ACE, angiotensin converting enzyme; ANCA, anti-neutrophil cytoplasmic antibody; ATN, acute tubular necrosis; NSAID, non-steroidal anti-inflammatory drug; TIN, tubulointerstitial nephritis.

CASE STUDY 3.1

A 24-year-old male presented with confusion, shortness of breath, and painful calves. It was reported by a friend that he had been lying on the floor for several hours. He was a known intravenous heroin and alcohol abuser.

On examination he appeared dehydrated and cold (temperature 35°C); his pulse was 75/minute and blood pressure 110/70 mmHg. Intravenous injection sites were apparent. His urine was dark coloured. His chest was clear. Arterial blood gases were done in the casualty department and a blood sample was sent to the pathology department and gave the following results (reference ranges are given in brackets):

Arterial blood

pH	7.276	(7.35–7.45)
PCO_2	4.82 KPa	(4.7–6.0)
PO_2	12.7 kPa	(12.0–14.6)
HCO_3^-	18.0 mmol/L	(24–29)

Serum

Sodium	138 mmol/L	(135–145)
Potassium	7.6 mmol/L	(3.8–5.0)
Creatinine	236 µmol/L	(71–133)
Calcium	1.66 mmol/L	(2.10–2.55)
Albumin	32 g/L	(35–50)
Phosphate	2.43 mmol/L	(0.87–1.45)
Creatine kinase	>140,000 U/L	(55–170)
C-reactive protein	73 mg/L	(<10)

The patient was considered to have AKI secondary to muscle breakdown (rhabdomyolysis) caused by prolonged limb compression (compartment syndrome). He was started on intravenous normal saline and bicarbonate (forced alkaline diuresis), calcium gluconate, calcium resonium, and dextrose, and began to feel better within 12 hours. His urine output was poor initially (250 mL output after 5 L infusion) and failed to improve, despite use of the diuretic frusemide. He continued to complain of painful thighs and his calves were swollen, particularly on the left leg. An ultrasound examination found no evidence of deep vein thrombosis. He underwent bilateral calf fasciotomies (i.e. surgical procedures in which the fascial compartment around the muscle is cut to reduce pressure) and was subsequently treated with haemofiltration and then haemodialysis. His creatine kinase gradually fell (4,600 U/L by day seven after admission). However, he remained on haemodialysis for nearly a month, by which point his own renal function had recovered.

(a) Why did the patient's urine appear dark?

(b) Why did he become hypocalcaemic and hyperphosphataemic?

Although the pathogenesis is uncertain, there is a well-recognized clinical pattern to AKI, with anuria or oliguria and abnormalities indicating tubular dysfunction. Emergency treatment is required to correct the biochemical abnormalities and this may include dialysis. If the patient survives, recovery will usually occur within days or weeks following removal of the initiating event. During the recovery phase, the role of the clinical laboratory in the monitoring of AKI is crucial to the assessment of electrolyte disturbance and fluid status. During recovery, there is an initial polyuric phase as glomerular function recovers before that of the tubules. Again, severe metabolic problems may persist, reflecting tubular damage such as potassium and phosphate wasting, and ongoing acidosis. The polyuric phase declines after a few days to weeks but requires careful monitoring to allow suitable fluid and electrolyte replacement. Uncomplicated AKI has a mortality rate of 5–10%, although AKI complicating non-renal organ system failure in intensive care patients is associated with mortality rates of 50–70%, despite advances in dialysis treatment.

Chronic kidney disease

Chronic kidney disease is defined as either kidney damage or GFR less than 60 mL/min/1.73 m^2 for at least three months. Kidney damage is defined as pathologic abnormalities or markers of damage, including abnormalities in blood or urine tests (for example proteinuria) or imaging studies (for example polycystic kidneys demonstrable on ultrasound Figure 3.11).

The internationally accepted classification of CKD is given in Table 3.3. It classifies kidney disease from stage 1, at the mild end of the spectrum, to stage 5, which is kidney failure or GFR less than 15 mL/min/1.73 m^2. This classification allows for consistency in reporting for epidemiological studies and also focuses treatment schedules for individual patients.

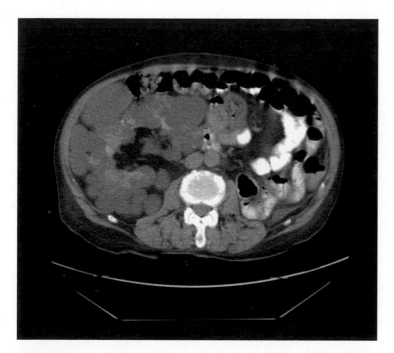

FIGURE 3.11
The detection of polycystic kidney disease using ultrasound scanning.
Courtesy of RadRounds (www.radrounds.com).

TABLE 3.3 The international classification of CKD indicating metabolic and management consequences. Prevalence data is based upon the NHANES 1999–2004 survey of a North American population.

Stage	Description	GFR, mL/min/ 1.73 m^2	Population prevalence /%	Metabolic consequences	Management
1*	Kidney damage with normal or increased GFR	>90	1.8	Hypertension more frequent than amongst patients without CKD	Diagnosis and treatment; treatment of comorbid conditions; slowing progression; CVD risk reduction
2*	Kidney damage with mildly decreased GFR	60–89	3.2	Hypertension frequent Concentration of PTH starts to rise (GFR 60–80)	Estimating progression
3**	Moderately decreased GFR	30–59	7.7	Hypertension frequent Decrease in calcium absorption (GFR <50) More markedly increased PTH concentration Reduced phosphate excretion Malnutrition (reduced spontaneous protein intake) Onset of left ventricular hypertrophy Onset of anaemia (erythropoietin deficiency)	Evaluating and treating complications
4	Severely reduced GFR	15–29	0.4	As above but more pronounced, plus: triglyceride concentrations start to rise, hyperphosphataemia Metabolic acidosis Tendency to hyperkalaemia Decreased libido	Preparation for RRT (creation of dialysis access, assessment for potential transplantation)
5	Kidney failure (established renal failure***)	<15	0.2	As above but more pronounced, plus: saline retention causing apparent heart failure Anorexia Vomiting Pruritus (itching without skin disease)	RRT, if uraemia present

CVD, cardiovascular disease; GFR, glomerular filtration rate; PTH, parathyroid hormone; RRT, renal replacement therapy.

* The diagnosis of stage 1 and 2 CKD requires the presence of kidney damage for ≥ 3 months manifest by pathological abnormalities of the kidney or abnormalities in the composition of urine, such as haematuria or proteinuria, or abnormalities in imaging tests either with (stage 2), or without (stage 1), decreased GFR.

** Stage 3 CKD is sometimes subdivided into stages 3A (45–59 mL/min/1.73 m^2) and 3B (30–44 mL/min/1.73 m^2), reflecting the more severe metabolic abnormalities and mortality risk that are associated with GFR values below 45 mL/min/1.73 m^2.

*** In the UK National Service Framework for Renal Disease, the term 'kidney failure' in the international classification system has been replaced by 'established renal failure' (ERF), defined as 'chronic kidney disease which has progressed so far that RRT is needed to maintain life'.

Key Points

Chronic kidney disease stages 1 and 2 cannot be diagnosed solely on the basis of GFR but require the presence of another marker of kidney disease, typically proteinuria or microalbuminuria, or the presence of abnormalities on imaging (for example polycystic kidneys). Chronic kidney disease stages 3–5 require only the demonstration of a persistent GFR less than 60 mL/min/1.73 m^2.

End-stage renal disease

In relation to the stages of kidney disease, kidney failure is present at a level of GFR less than 15 mL/min/1.73 m^2 (stage 5). At this level there are generally signs and symptoms of **uraemia** (azotaemia), or a need for RRT. The uraemic syndrome is the group of symptoms, physical signs, and abnormal findings that result from the failure of the kidneys to maintain adequate excretory, regulatory, and endocrine function. It is the terminal clinical manifestation of kidney failure. The classic signs of uraemia include progressive weakness and fatigue, loss of appetite followed by nausea and vomiting, muscle wasting, tremors, abnormal mental function, frequent but shallow respirations, and metabolic acidosis. The syndrome evolves to produce stupor, coma, and ultimately death unless support is provided by dialysis or successful kidney transplantation.

Many retained metabolites have been implicated in the systemic toxicity of the uraemic syndrome. The most characteristic laboratory findings are increased concentrations of nitrogenous compounds in plasma, including urea and creatinine, as a result of reduced GFR and decreased tubular function. Urea was the first metabolite identified as being increased in uraemia but it does not appear to be responsible for its systemic manifestations. Retention of nitrogenous compounds and of metabolic acids is followed by progressive hyperphosphataemia, hypocalcaemia, and potentially dangerous hyperkalaemia. Although most patients eventually exhibit acidaemia, respiratory compensation by elimination of carbon dioxide is important. In addition, inadequate synthesis of EPO is manifested by anaemia. Disordered regulation of blood pressure generally leads to hypertension. In addition to the consequences of reduced excretory, regulatory, and endocrine functions of the kidneys, the uraemic syndrome has several systemic manifestations including pericarditis, pleuritis, disordered platelet and granulocyte function, and encephalopathy.

Renal replacement therapy includes transplantation, haemodialysis, peritoneal dialysis, haemodiafiltration, and continuous haemofiltration. Dialysis removes toxic substances from the blood when the kidneys cannot satisfactorily remove them from the circulation. Extensive laboratory support is required by an RRT programme.

3.5 **Specific kidney disease**

In this section you will be introduced to the major causes of kidney disease, which will be described under the general headings:

- diabetic nephropathy
- hypertension and the kidney
- glomerular diseases

- tubulointerstitial disease
- renal tubular acidosis
- inherited renal disorders
- renal calculi
- monoclonal light chains and kidney disease
- toxic nephropathy

Diabetic nephropathy

Diabetes mellitus affects approximately 3% of the UK population. Chronic hyperglycaemia is the central feature of both type 1 and type 2 diabetes and is probably the major factor responsible for the long-term complications of diabetes, which include diabetic kidney disease (diabetic nephropathy). Diabetic nephropathy is promoted by a variety of other factors, including hypertension, hyperlipidaemia, and proteinuria. However, these factors need to occur against a background of genetic susceptibility; 30–40% of diabetic patients will develop nephropathy, irrespective of their glycaemic control. Diabetic nephropathy is a clinical diagnosis based on the finding of persistent increased concentrations of protein in the urine (proteinuria) in a patient with diabetes, but is often predated by a low but abnormally increased rate of albumin excretion. Patients with a urinary albumin excretion rate of 30 to 300 mg per 24 hours (approximately 3 to 30 mg/mmol creatinine) have microalbuminuria and are said to have 'incipient nephropathy'. Without intervention, most patients with incipient nephropathy progress to overt diabetic nephropathy ('macroalbuminuria'), which is characterized by a protein excretion exceeding 450 mg per 24 hours (equivalent to an albumin excretion rate of approximately 300 mg/24 hours or greater than 30 mg/mmol creatinine). As albuminuria worsens and blood pressure increases, there is a relentless decline in GFR: ESRD which develops in more than 75% of patients with overt nephropathy over 20 years. However, intervention can decrease the rate of decline. Diabetic nephropathy increases cardiovascular mortality and is the most common single cause of ESRD; approximately 20% of patients starting RRT programmes in the UK are doing so because of underlying diabetic kidney disease.

In type 1 diabetes it is unusual to develop microalbuminuria within the first five years of diagnosis but it can develop anytime thereafter, even after 40 years. The onset of type 2 diabetes is difficult to define and a higher proportion of these patients compared to type 1 diabetes are found to have microalbuminuria at diagnosis. Since microalbuminuria is the earliest sign of diabetic nephropathy and this can be ameliorated by treatment, routine screening of patients for microalbuminuria is recommended.

Major studies, including the Diabetes Control and Complications Trial (DCCT) and the UK Prospective Diabetes Study (UKPDS), have shown that intensive diabetes therapy (strict glycaemic control) can significantly reduce the risk of development of nephropathy. In addition, aggressive antihypertensive treatment can ameliorate the rate of fall in GFR: clinicians aim for tighter blood pressure control in diabetic than in non-diabetic patients. The use of ACE inhibitors or ARBs is particularly beneficial in this setting. These drugs appear to reduce albuminuria and the rate of decline in GFR to a greater extent than other antihypertensive agents for an equivalent reduction in blood pressure, presumably through intraglomerular effects on capillary blood pressure and protein filtration.

Hypertension and the kidney

Hypertension and CKD are inevitably intertwined: in many circumstances it is difficult to identify whether raised blood pressure is causing kidney disease or *vice versa*. Hypertension often develops as a consequence of CKD due to fluid and salt retention and activation of the sympathetic nervous and renin-angiotensin systems. Hypertension co-existing with CKD accelerates the development of ESRD and treatment of hypertension is critical in preventing the progression of ESRD. The majority of patients with CKD require a combination of three to four antihypertensive drugs to control their blood pressure. Treatment regimens should include diuretics and ACEs or ARBs to block the renin-angiotensin system.

Some patients with hypertension and CKD suffer from renal artery stenosis, which is normally caused by atheromatous disease. Atheromatous disease is progressive and may cause complete occlusion of the renal arteries. Treatment is possible through the surgical stenting of the diseased artery.

Patients that receive ACE inhibitors and ARBs may develop AKI and hyperkalaemia if they also have severe renal artery stenosis. In this setting, angiotensin II-dependent mechanisms are maintaining kidney function and this compensatory process clearly becomes abolished when ACE/ARBs are prescribed. Kidney function should be carefully monitored following the introduction of these drugs. A marked deterioration in function following their introduction should raise the suspicion of renovascular disease.

Patients with hypertension should have baseline tests performed to aid diagnosis of a specific cause and, equally importantly, to detect hypertensive end-organ damage. The presence of haematuria and proteinuria may identify those patients with underlying kidney disease and hypertensive **nephrosclerosis**. In patients with both hypertension and proteinuria there is an increased risk of death. Plasma sodium, potassium, and creatinine concentrations should be measured to detect hypokalaemia, which could suggest hyperaldosteronism, including Conn's syndrome, and to assess kidney function. Cholesterol measurement is also mandatory to assess overall cardiovascular risk.

Glomerular diseases

A number of distinct clinical syndromes result from glomerular injury. These include:

(1) IgA nephropathy

(2) rapidly progressive glomerulonephritis

(3) acute **nephritis**

(4) chronic glomerulopathies

(5) **nephrotic syndrome**

These may arise from specific glomerular diseases (for example IgA nephropathy, antiglomerular basement membrane disease), or follow infection. Alternatively, glomerular disease may be due to systemic diseases (for example systemic lupus erythematosus (SLE) and diabetic nephropathy can both result in nephrotic syndrome).

Glomerular disease presents clinically with abnormalities of the urine, including proteinuria and haematuria, hypertension, **oedema**, and, often, reduced renal excretory function. Establishing the diagnosis often requires histological examination of a kidney biopsy specimen.

Laboratory investigations performed to investigate glomerular disease and possible systemic disorders causing them include:

(1) urinary protein excretion

(2) serum creatinine concentration

(3) estimated GFR

(4) liver function tests

(5) glucose concentration

(6) urinary examination for Bence-Jones protein

(7) serum protein electrophoresis (if myeloma is suspected)

Serological testing is performed if either SLE or systemic vasculitis is suspected for the presence of autoantibodies to:

(1) antinuclear antigens

(2) double-stranded DNA

(3) extractable nuclear antigens

(4) antineutrophil cytoplasmic antibody

Antiglomerular basement membrane antibodies may be detected in rare cases of renal-limited antiglomerular basement membrane disease (**Goodpasture's disease**) and pulmonary-renal syndromes (Goodpasture's syndrome). Components of the complement system can be affected (for example reduced levels of C3 and C4) in several conditions, including SLE, infection, cryoglobulinaemia, and mesangiocapillary glomerulonephritis (also referred to as membranoproliferative glomerulonephritis). Blood samples are taken for bacteriological examination in suspected infection.

The nephrotic syndrome is characterized by gross changes in glomerular permeability. The diagnostic criteria for establishing nephrotic syndrome are the presence of proteinuria (total protein excretion greater than 3 g/24 hours or albumin excretion greater than 1.5 g/24 hours), hypoalbuminaemia, hypercholesterolaemia, and finally oedema.

Tubulointerstitial disease

A variety of chemical, bacterial, and immunological injuries to the kidney cause either generalized or localized changes that primarily affect the tubulointerstitium rather than the glomerulus. These disorders are characterized by alterations in tubular function that, in advanced cases, may cause secondary vascular and glomerular damage. Interstitial nephritis, including chronic **pyelonephritis**, is the primary tubulointerstitial disorder, accounting for 10% of new patients starting dialysis in the UK. There are both acute and chronic types of pyelonephritis, with the acute type most commonly associated with a UTI. This can then develop into chronic pyelonephritis, usually as a result of a renal tract abnormality such as abnormal urethral valves.

In addition to pyelonephritis, interstitial nephritis may present in acute and chronic forms usually as a result of drug toxicity or hypersensitivity. The most common drugs implicated are non-steroidal anti-inflammatory drugs (NSAIDs) and β-lactam antibiotics, although over one hundred different drugs have been implicated. Urinary findings may be normal or there may be low level proteinuria and eosinophils are often seen on light microscopy. Nephrotic syndrome may accompany an **acute interstitial nephritis** (AIN) associated with NSAIDs. Treatment of AIN is directed at removing any causative agent. Steroids are used to promote early resolution of clinical course.

Renal tubular acidosis

The renal tubular acidoses comprise a diverse group of both inherited and acquired disorders affecting either the proximal or distal tubules. They are characterized by a hyperchloraemic, normal anion gap metabolic acidosis with urinary bicarbonate or H^+ excretion inappropriate for the plasma pH. They result from failure to retain bicarbonate or an inability of the renal tubules to secrete H^+. Typically the GFR in RTA is normal, or only slightly reduced, and there is no retention of anions, such as phosphate and sulphate (as opposed to the acidosis of renal failure). The three categories of RTA are distal (dRTA, type I), proximal (pRTA, type II), and type IV, which is secondary to aldosterone deficiency or resistance. The finding of a hyperchloraemic metabolic acidosis in a patient without evidence of gastrointestinal bicarbonate losses and with no obvious pharmacological cause should prompt suspicion of an RTA. In addition to plasma electrolyte measurement, preliminary investigation should include determining the pH of a fresh EMU sample. A value greater than 5.5 in the presence of a systemic acidosis supports the diagnosis of RTA.

Inherited renal disorders

A comprehensive account of inherited kidney diseases is beyond the scope of this book. Several inherited disorders are described below to provide you with an introduction to the subject.

In polycystic kidney disease, there is enlargement of the kidneys due to cyst development. The autosomal dominant form of the disease (ADPKD) is the most common renal hereditary disease affecting between 1:400–1:1,000 live births and accounting for 10% of the ESRD population in the UK. Autosomal recessive polycystic kidney disease (ARPKD) is much less common (incidence 1:10,000 to 1:40,000 individuals). Figure 3.12 shows a markedly enlarged polycystic kidney from a patient with ADPKD.

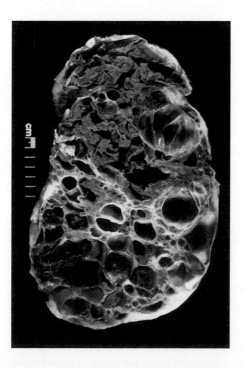

FIGURE 3.12
Polycystic kidney disease showing an enlarged polycystic kidney from a patient with ADPKD.

Autosomal dominant polycystic kidney disease can present at any age, but most commonly during adulthood. Presenting features include hypertension, haematuria, abdominal pain, and a UTI. The two major genes mutated in ADPKD have been identified, termed PKD1 and PKD2 and coding for polycystin-1 and -2 respectively. These proteins are involved in cell cycle regulation and intracellular calcium transport. Clinical manifestation within families is highly variable. Approximately 50% of ADPKD patients have reached ESRD by 50 years and 70% by 70 years of age.

Alport's syndrome (hereditary nephritis) is a disease that results from mutations in genes that code for collagen proteins found within the glomerular basement membrane. The primary abnormality involves the gene coding for the alpha 5 chain of type IV collagen (COL4A5). Alport's syndrome accounts for 1–2% of cases of ESRD and X-linked inheritance of the mutation occurs in the majority of cases (85%).

Inherited tubulopathies comprise a heterogeneous set of disorders often characterized by electrolyte disturbances. They include defects of the proximal tubule, for example Dent's disease, Lowe's syndrome, Wilson's disease, transport channel defects of the loop of Henle such as Bartter's syndrome, and distal tubular defects including Gitelman's syndrome and Liddle's syndrome. Features of the **Fanconi syndrome**, including glycosuria, hypophosphataemia, low M_r proteinuria, aminoaciduria, hypouricaemia, and type 2 RTA, accompany disorders of the proximal tubule. In addition to electrolyte disturbances, particularly of potassium, general reasons to suspect an inherited tubulopathy include a familial disease pattern, renal impairment, nephrocalcinosis, and stone formation, especially if these present at an early age. Although individually uncommon or rare, consideration of these conditions is a common reason for clinicians to seek laboratory support and advice. Discrimination from covert diuretic or laxative abuse and surreptitious vomiting, which can mimic these conditions biochemically, is an important differential.

Many inherited metabolic diseases affect the function of the kidney or are characterized by excessive urinary excretion of an intermediary metabolite(s) (for example phenylketonuria). Most are generalized disorders affecting other organs in the body and having renal sequelae as a systemic secondary consequence of the underlying disorder (for example due to vomiting). In some cases, kidney disease occurs as a result of accumulation of an intermediary metabolite, for example galactose-1-phosphate in galactosaemia. In others the disorder occurs due to failure of a transport mechanism. For example, cystinuria is an autosomal recessive, inherited disorder of tubular cystine reabsorption. Cystinuria should not be confused with cystinosis, which is a rare autosomal recessive disease due to defective export of cystine across the lysosomal membrane as a result of inactivating mutations in cystinosin, an integral lysosomal membrane protein. There is consequent cellular accumulation of cystine with crystallization that destroys tissues, which is shown in Figure 3.13. Classic cystinosis presents in the first year of life with failure to thrive, polyuria, polydipsia, hypophosphataemic rickets, and other features of Fanconi syndrome. Progressive renal damage generally results in ESRD by ten years of age. Diagnosis relies upon measurement of the cystine concentration of peripheral leucocytes or cultured fibroblasts. The renal effect often manifests as Fanconi syndrome as described above.

Renal calculi

Calculi ('stones') may occur in the renal pelvis, the ureter, or the bladder. Calcification can also occur scattered throughout the renal parenchyma (**nephrocalcinosis**). Approximately 5–10% of the population of the western world are thought to have formed at least one kidney stone

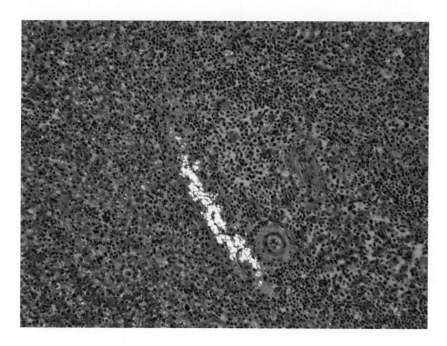

FIGURE 3.13
Post-mortem observation of crystals of cystine in the spleen from a patient who suffered from cystinosis. Courtesy of Department of Histopathology, Great Ormond Street Hospital, London, UK.

by the age of 70 years and the prevalence of kidney stones may be increasing. In both males and females, the average age of first stone formation is decreasing. For most stone types, there is a male preponderance. The passage of a stone is associated with severe pain called **renal colic**, which may last for 15 minutes to several hours and is commonly associated with nausea and vomiting.

The majority of kidney stones found in the western world are composed of one or more of the following substances:

(1) calcium oxalate with or without phosphate (frequency 67%)

(2) magnesium ammonium phosphate (12%)

(3) calcium phosphate (8%)

(4) urate (8%)

(5) cystine (1–2%)

(6) complex mixtures of the above (2–3%)

These poorly soluble substances crystallize within an organic matrix, the nature of which is not well understood.

SELF-CHECK 3.7

What are the biochemical features typical of a patient with Fanconi's syndrome?

Monoclonal light chains and kidney disease

Immunoglobulin molecules are formed from polypeptide heavy and light chains.

In normal individuals, the small quantity of circulating free polyclonal light chains (M_r ~22,000 Da) are filtered by the glomerulus and around 90% are reabsorbed in the proximal tubule.

Cross reference
Chapter 18 Specific protein markers

CASE STUDY 3.2

A male university student presented with left sided abdominal pain radiating to the groin. The pain had lasted ten days and was increasing. He felt nauseous and had been off food for two days. He reported that he had a UTI one year previously and his brother and aunt had both had renal stones in the past.

On examination his abdomen was tender, with pain localized to the left pelvic region. A urinary dipstick test was positive for blood (trace), protein, leukocytes, and ketones. Baseline laboratory investigations (serum sodium, potassium, creatinine, calcium, phosphate, full blood count) were all normal, but an X-ray demonstrated a 1.5 cm diameter stone at the junction of the left kidney and its ureter.

This was initially treated with stenting of the left ureter and the patient was discharged pending further investigation.

A full metabolic 'stone screen' demonstrated no abnormalities other than a positive cystine screening test. Urinary cystine excretion was 2008 µmol/24h (normal <420) and urinary excretion of lysine, ornithine, and arginine were also increased. Further management consisted of ensuring adequate fluid intake throughout the 24-hour period to ensure cystine was held below its urinary saturation point.

(a) What is the relevance of his family history?

(b) Why was excretion of lysine, ornithine, and arginine also increased?

When the concentration of filtered light chains is increased it leads to pathological alterations in the proximal tubule cells because fragments of light chains are resistant to degradation by tubular lysosomal proteases and accumulate in the cells. Light chains can also deposit in the kidney as casts, fibrils, and precipitates or crystals, giving rise to a spectrum of disease including:

(1) cast nephropathy

(2) amyloid

(3) light chain deposition disease

(4) Fanconi syndrome

However, not all patients with a large excess production of monoclonal light chains develop disease. Other promoters of kidney disease include dehydration, hypercalcaemia, contrast medium, and NSAIDs.

Multiple myeloma is a neoplastic proliferation of secretory B cells (plasma cells) that produce excess amounts of a monoclonal immunoglobulin (**paraprotein**), so-called M protein because of the characteristic peaks obtained from serum protein electrophoresis on agarose gel. In multiple myeloma, complete monoclonal immunoglobulins (usually IgG or IgA) are accompanied in the plasma by variable concentrations of free light chains that appear in the urine as Bence-Jones proteins. M proteins and light chains are identified in the blood and/or the urine in 98% of patients with myeloma using protein electrophoresis and immunofixation. Impairment of kidney function at presentation occurs in almost 50% of patients with myeloma for the variety of reasons given above.

Toxic nephropathy

The kidney is susceptible to insult from a variety of medicinal and environmental agents due to its large blood flow, concentration of filtered solutes during urine production, and the presence of a variety of xenobiotic transporters and metabolizing enzymes. Toxic nephropathy commonly occurs either as a result of decreased renal perfusion, due to precipitation within the tubule or to direct toxic effects at the proximal tubule level. In some cases the conjugation of environmental chemicals (for example mercury and cadmium) to glutathione and/or cysteine targets these chemicals to the kidney where inhibition of renal function occurs through a variety of mechanisms that are not completely understood. Renal injury induced by chloroform is dependent on its metabolism to toxic metabolites by the renal cytochrome P-450. Other compounds such as paraquat and diquat damage the kidney via the production of reactive oxygen species. Some drugs can cause kidney damage in the presence of normal renal function but a far greater variety cause problems in patients with kidney disease, predominantly due to accumulation because of decreased renal elimination. The British National Formulary has an appendix devoted to prescribing issues in the presence of renal impairment.

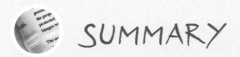

SUMMARY

- The kidneys perform a central role in the maintenance of homeostasis, affected by filtration, followed by differential reabsorption and secretion of solutes and water as the ultrafiltrate passes through the nephron.

- The kidneys also have a significant endocrine role, most notably in relation to maintenance of normal blood pressure, promoting the production of erythrocytes and synthesizing an active form of vitamin D.

- Kidney disease may be either acute or chronic. Acute kidney injury is associated with rapidly occurring metabolic derangement and carries a high mortality. With appropriate treatment, it is commonly reversible.

- Chronic kidney disease may occur as the result of a variety of specific kidney diseases, in particular diabetes mellitus, hypertension, glomerulonephritis, and polycystic kidney disease.

- Chronic kidney disease is classified into five stages based on GFR (measured in mL/min/1.73 m^2); greater than 90 (stage 1), 60–89 (stage 2), 30–59 (stage 3), 15–29 (stage 4), and less than 15 (stage 5 or established renal failure). Stages 1 and 2 require other evidence of kidney disease/damage in addition to GFR to establish the diagnosis.

- End-stage renal disease is the point at which renal function is no longer sufficient to support life; at some point most patients with stage 5 CKD will progress to the point where they will require dialysis or transplantation to prevent death.

- The general approach to kidney disease detection involves urinalysis, assessment of GFR and detection or measurement of proteinuria.

- Proteinuria is commonly detected using reagent strip devices. This is often followed by laboratory measurement of total protein using either turbidimetric or colorimetric approaches. There are many limitations to both reagent strip and laboratory measurement of total protein.

- In diabetic patients, urinary albumin excretion is measured using specific immunoassay approaches. Abnormally increased albumin excretion but at a level below that detectable by reagent strip devices is termed microalbuminuria. This finding has important prognostic and therapeutic implications for patients with diabetes.

■ Measurement of GFR is considered the best overall index of renal excretory function. A variety of approaches using either exogenous or endogenous markers are used. In clinical practice, measurement of creatinine and derivation of an estimate of GFR derived from it is the most widely used approach.

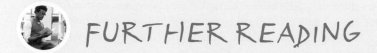

FURTHER READING

● **Brenner BM (2004)** *Brenner and Rector's the Kidney*, **7th edition. Philadelphia: Saunders.**

A two-volume 'bible' containing almost everything you might need to know about kidney disease and physiology.

● **Joint Specialty Committee on Renal Medicine of the Royal College of Physicians and the Renal Association, and the Royal College of General Practitioners (2006)** *Chronic Kidney Disease in Adults: UK Guidelines for Identification, Management and Referral.* **London: Royal College of Physicians.**

The basis of much of current UK practice in the identification and management of patients with CKD. The role of laboratory testing is clearly explained and there is a wealth of information in the supporting evidence.

● **Lamb E and Delaney M (2009)** *Kidney Disease and Laboratory Medicine*, **London: ACB Venture Publications.**

A review of kidney disease and its investigation in clinical biochemistry.

● **Levey AS, Coresh J, Greene T,** *et al.* **(2006) Using standardized serum creatinine values in the modification of diet in renal disease study equation for estimating glomerular filtration rate.** *Annals of Internal Medicine* **145**, 247–54.

The final in a series of articles explaining the derivation of the MDRD study equation, which is used to estimate GFR in most UK laboratories.

● **Myers GL, Miller WG, Coresh J,** *et al.* **(2006) Recommendations for improving serum creatinine measurement: a report from the laboratory working group of the National Kidney Disease Education Program.** *Clinical Chemistry* **52**, 5–18.

This article reviews many of the problems of serum creatinine measurement as a marker of GFR, together with recommendations for improvements.

● **Newman DJ (2002) Cystatin C.** *Annals of Clinical Biochemistry* **39**, 89–104.

An excellent review of the biology and clinical potential of cystatin C as a marker of kidney function.

● **Penney MD and Oleesky D (1999) Renal tubular acidosis.** *Annals of Clinical Biochemistry* **36**, 408–22.

Clear and accessible review of a confusing area.

● **Sayer JA and Pearce SHS (2001) Diagnosis and clinical biochemistry of inherited tubulopathies.** *Annals of Clinical Biochemistry* **38**, 459–70.

A clearly written account of the inherited tubulopathies. It also clearly explains, through excellent illustrations, facets of renal tubular electrolyte handling.

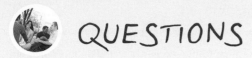

QUESTIONS

3.1 End-stage renal disease is typically characterized by:

(a) Hyperkalaemia

(b) Hypercalcaemia

(c) Hypophosphataemia

(d) Acidosis

(e) Anaemia

3.2 Arrange the following in decreasing order of frequency (i.e. commonest first) as a cause of kidney stones:

Urate

Magnesium ammonium phosphate

Calcium oxalate

Cystine

Calcium phosphate

3.3 Which of the following statements relating to kidney structure and function are correct?

(a) The majority of salt and water reclamation in the kidney takes place in the loop of Henle

(b) The glomerular blood supply is derived from the efferent arteriole

(c) Mesangial cells are responsible for the production of angiotensin II

(d) Approximately 99% of the glomerular filtrate is reabsorbed within the nephron

(e) Glycosuria occurs when the blood glucose concentration exceeds approximately 20 mmol/L

3.4 Which of the following statements relating to proteinuria are correct?

(a) The predominant protein in normal human urine is Tamm-Horsfall glycoprotein

(b) Megalin and cubulin are responsible for the tubular secretion of low molecular mass proteins into the urine

(c) Proteinuria may occur in patients with a UTI

(d) In health, the glomerular filtration barrier is relatively impermeable to molecules larger than 60 kDa

(e) An albumin:creatinine ratio greater than 2.5 mg/mmol in a male or 3.5 mg/mmol in a female is consistent with the presence of microalbuminuria

3.5 Which of the following statements relating to reagent strip (dipstick) tests are correct?

(a) The presence of nitrites and leucocytes in the urine is a strong indication of the presence of a UTI

(b) The normal pH of urine is approximately 7.0

(c) The reagent tetrabromphenol blue becomes yellow in the presence of protein

(d) Tetrabromphenol blue reacts strongly with Tamm-Horsfall glycoprotein but not with haemoglobin

(e) Urine of low specific gravity may be seen in patients with diabetes insipidus

3.6 Which of the following statements relating to the glomerular filtration rate (GFR) are correct?

(a) The GFR is the volume of urine from which a given substance is completely cleared by glomerular filtration per unit time

(b) The most accurate measurements of GFR are obtained using infused exogenous molecules such as inulin

(c) Creatinine is freely filtered at the glomerulus

(d) The creatinine clearance test typically overestimates GFR

(e) The MDRD equation estimates GFR using serum creatinine corrected for gender, weight, ethnic origin, and age

3.7 Which of the following statements relating to kidney diseases are correct?

(a) Acute kidney injury is a particularly severe form of irreversible kidney disease

(b) In the international classification of chronic kidney disease, five stages are recognized, with stage 5 representing end-stage renal failure

(c) Goodpasture's disease is characterized by a positive blood test for antiglomerular basement membrane antibodies

(d) Tubular proteinuria is a typical characteristic of nephrotic syndrome

(e) Autosomal dominant polycystic kidney disease is the most common renal hereditary disease

3.8 Describe the normal renal handling of proteins.

3.9 Discuss the classification and diagnosis of kidney disease.

3.10 Discuss common causes of kidney disease.

3.11 Describe the laboratory investigation of kidney disease, including any pitfalls.

Answers to self-check questions, case study questions, and end-of-chapter questions are available in the Online Resource Centre accompanying this book.

@ **Go to www.oxfordtextbooks.co.uk/orc/ahmed/**

Hyperuricaemia and gout

Joanne Marsden

Learning objectives

After studying this chapter you should be able to:

- Describe the structure of purine bases and nucleotides
- Outline purine metabolism and the excretion of uric acid
- Describe the consequences of hyperuricaemia in causing gout
- Discuss the laboratory investigation of gout
- List the drugs used in treating and managing gout
- Relate the mechanism of action of drugs used in the management of gout

Introduction

Purines are a group of aromatic, heterocyclic, nitrogenous bases. They consist of a five-membered imidazole ring fused to a six-membered pyrimidine ring forming a double-ring structure (Figure 4.1 (a)). All purines are derived from and are structurally related to the parent compound of that name, and include caffeine, xanthine, hypoxanthine, adenine, guanine, and **uric acid**. Different purines have differing amino and carboxyl side chains, as you can see in Figure 4.1 (b).

The purine bases, adenine (A) and guanine (G) can be thought of as part of a hierarchy of structures going from bases to **nucleosides**, **nucleotides**, and polynucleotides. Polynucleotides are more commonly called nucleic acids, deoxyribonucleic, and ribonucleic acids (DNA, RNA respectively).

Nucleosides are the chemical combination of a base and a sugar. The physiologically significant sugars are ribose and deoxyribose. The addition of one to three phosphate groups to a nucleoside forms a nucleotide. Table 4.1 summarizes the nomenclature associated with the purine

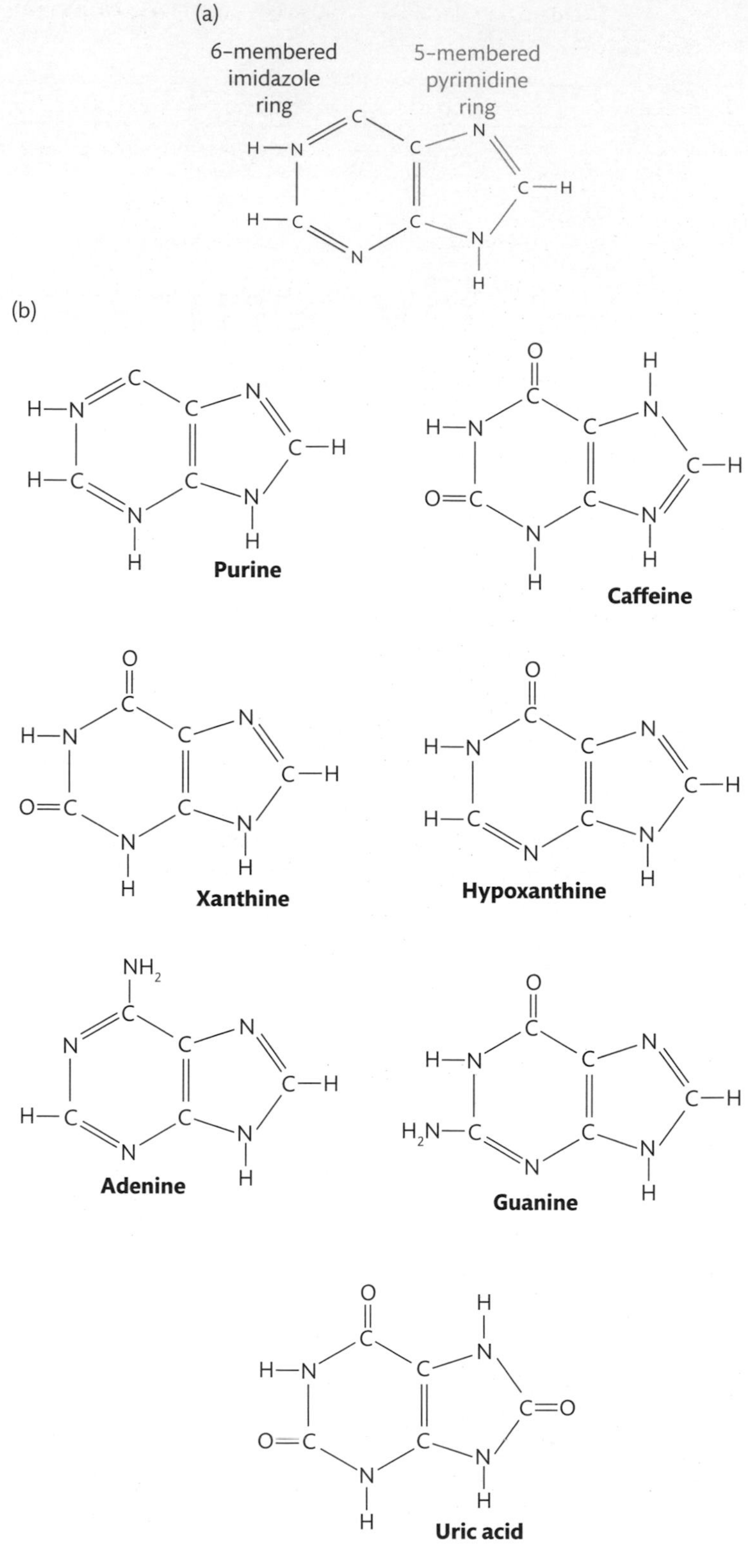

FIGURE 4.1

(a) The purine ring nucleus consists of a six-membered imidazole ring fused to a five-membered pyrimidine ring. (b) Examples of purine compounds including purine, caffeine, xanthine, hypoxanthine, adenine, guanine, and uric acid.

TABLE 4.1 Standard nomenclature of the major nucleic acid bases, nucleosides, and nucleotides.

Purine base	Ribonucleoside	Ribonucleotide*
Adenine (A)	Adenosine	Adenosine monophosphate (AMP)
Guanine (G)	Guanosine	Guanosine monophosphate (GMP)
Hypoxanthine	Inosine	Inosine monophosphate (IMP)
Xanthine (X)	Xanthosine	Xanthosine monophosphate (XMP)
Base	**Deoxyribonucleosides**	**Deoxyribonucleotides (as 5'-monophosphate)**
Adenine (A)	Deoxyadenosine	Deoxyadenosine 5'-monophosphate (dAMP)
Guanine (G)	Deoxyguanosine	Deoxyguanosine 5'-monophosphate (dGMP)

* Other nucleotides with two (di), three (tri), and, rarely four (tetra) phosphates are known. For example, adenosine 5'-triphosphate (ATP), deoxythymidine 5'-diphosphate (dTDP).

bases adenine and guanine and that of their nucleosides and nucleotides. You can see the structures of representative examples in Figure 4.2.

Nucleotides are the building blocks of nucleic acids, although many, for example ATP and GTP, are essential metabolites in their own right. Thus, they are involved in many metabolic activities. Although, the body obtains purine bases from a variety of sources, they are all degraded to uric acid.

Uric acid is the excretory end product of purine metabolism in humans. It is removed from the blood by filtration through the kidneys and excreted in urine. The excretion and poor solubility of uric acid ensure that it is normally present in only small concentrations in the plasma. However, some conditions can lead to decreased or increased concentrations, which are called **hypouricaemia** and **hyperuricaemia** respectively. Hyperuricaemia is the commoner of the two.

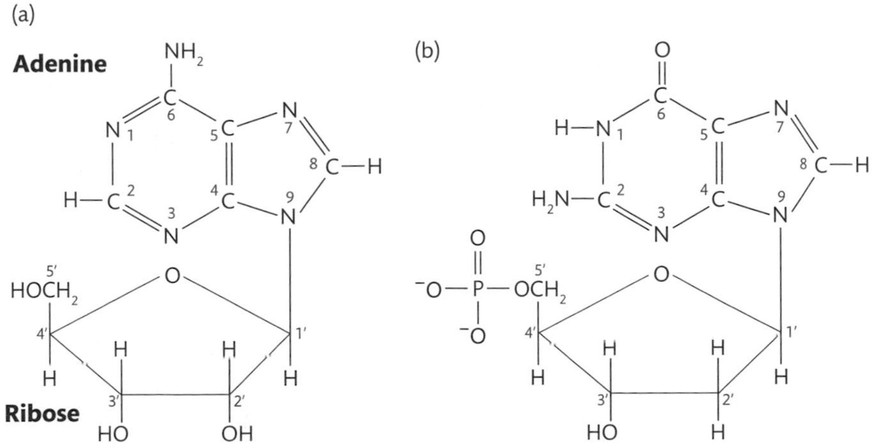

FIGURE 4.2
Structures of the ribonucleosides. (a) Adenosine. (b) The deoxyribonucleotide deoxyguanosine 5'-monophosphate (dGMP). See text for details.

A number of disorders associated with purine metabolism are predominantly the result of an abnormal catabolism. This either increases the amount of uric acid formed or decreases its excretion, which results in hyperuricaemia. The clinical outcome ranges from relatively mild, for example **gout**, to severe symptoms such as mental retardation and even death.

This chapter will outline the sources of uric acid and describe its metabolism, before discussing the diagnosis, clinical investigations, and management of conditions associated with hypouricaemia and hyperuricaemia.

4.1 **Sources of purines**

The three sources of purines in humans are:

- dietary intake
- *de novo* synthesis from small molecules present in the body
- salvage pathway

Dietary intake of purines

Dietary intake of purines accounts for about 30% of excreted uric acid. Red meat, game, shellfish, and yeast containing foodstuffs, such as Marmite and beer have high purine content and intake should be limited in patients with clinical gout.

SELF-CHECK 4.1

What is the end product of purine metabolism?

De novo synthesis of purines

The ***de novo* synthesis** of purine occurs mainly in the liver. A simplified outline of the synthesis is shown in Figure 4.3. The biochemical strategy is to combine several relatively small molecules together in a stepwise fashion and eventually form the complex double purine ring. Synthesis begins with the formation of 5-phosphoribosyl-1-pyrophosphate (PRPP) by condensation of pyrophosphate (PP_i), which is supplied from ATP, and ribose 5'-phosphate. This reaction is catalysed by phosphoribosylpyrophosphate synthetase (PRPPS). The addition of an amino group from the amino acid glutamine in a reaction catalysed by amidophosphoribosyl transferase (APRT, but also called glutamine phosphoribosyl amidotransferase) converts PRPP to 5-phosphoribosyl-1-amine. This step in the synthesis of purine nucleotides is often regarded as a rate limiting or controlling reaction. The presence of relatively high concentrations of purine nucleotides subject the enzyme APRT to negative feedback, which decreases the rate of synthesis of 5-phosphoribosyl-1-amine and therefore its subsequent conversion to purine nucleotides.

The 5-phosphoribosyl-1-amine is used to synthesize the first fully formed nucleotide, inosine 5' monophosphate (IMP) through a series of enzyme-catalysed steps. As you can see in Figure 4.3, once formed, IMP is a branch point in the biosynthesis of purines because it can be converted into adenosine monophosphate (AMP) *or* guanosine monophosphate (GMP). Each is formed through a different pathway that requires free energy from the hydrolysis of ATP *or* GTP respectively.

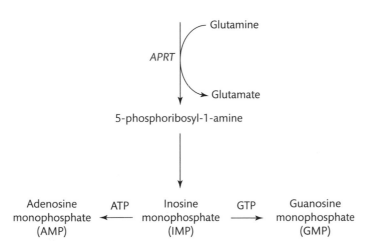

Ribose-5-phosphate + pyrophosphate

PRPPS

5-phosphoribosyl-1-pyrophosphate (PRPP)

Glutamine

APRT

Glutamate

5-phosphoribosyl-1-amine

Adenosine monophosphate (AMP)	ATP	Inosine monophosphate (IMP)	GTP	Guanosine monophosphate (GMP)

FIGURE 4.3
Outline of the *de novo* synthetic pathway for purines. See text for details.

Adenosine triphosphate and GTP can be phosphorylated to give other purine ribonucleotides. Deoxyribonucleotides are synthesized by reducing the corresponding diphosphate or triphosphate ribonucleotide, in reactions catalysed by ribonucleotide reductase, as you can see, for example, in the case of converting ADP to dADP, in Figure 4.4, Once formed, all the nucleotides are used in metabolic processes and for the synthesis of nucleic acids.

Salvage pathway for purine nucleotides

The synthesis of purine nucleotides is expensive in terms of the metabolic energy required. Strategies have evolved where free purine bases obtained in the diet or from the metabolic turnover of nucleotides can be reused or salvaged. This involves the activities of two phosphoribosyltransferases, adenine phosphoribosyltransferase and hypoxanthine-guanine phosphoribosyltransferase (HGPRT). The former salvages free adenine by converting it to AMP:

$$\text{Adenine} + \text{PRPP} \rightarrow \text{AMP} + \text{PP}_i$$

However, HGPRT converts guanine to GMP and can also catalyse the conversion of hypoxanthine to IMP, the precursor of AMP and GMP:

$$\text{Guanine} + \text{PRPP} \rightarrow \text{GMP} + \text{PP}_i$$

$$\text{Hypoxanthine} + \text{PRPP} \rightarrow \text{IMP} + \text{PP}_i$$

SELF-CHECK 4.2

Which enzyme catalyses the rate limiting step in purine synthesis?

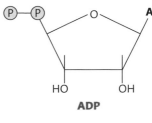

ADP

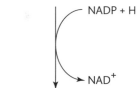

NADP + H

NAD$^+$

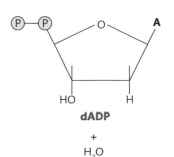

dADP

+

H$_2$O

FIGURE 4.4
Conversion of the ribonucleotide, ADP to the *deoxy*ribonucleotide, dADP.

4.2 Purine catabolism and excretion

Purine nucleotides are catabolized (degraded) by reactions that initially form their respective nucleosides: inosine, adenosine, guanosine, and xanthosine. An outline showing this can be seen in Figure 4.5.

Figure 4.6 shows that the commonly found nucleotides are eventually converted by a variety of enzyme catalysed reactions into xanthine. In turn, xanthine is oxidized to uric acid. Consequently, uric acid is the main end product of metabolism and the degradation of purines, whether they are derived from the diet, by *de novo* synthesis, or from the breakdown of nucleic acids in the body, leads to the formation of uric acid.

Uric acid is sparingly soluble but, despite this, the kidneys play the major role in removing it from the blood. This is possible because it can ionize in the presence of sodium, giving the salt, monosodium urate as shown in Figure 4.7. Monosodium urate is generally referred to as uric acid in clinical environments and we will adhere to this convention.

Approximately 98% of the uric acid filtered by the glomeruli is reabsorbed by the proximal tubules. Despite this, about two-thirds to three-quarters of the uric acid excreted is eliminated by the kidneys. The remaining uric acid is secreted into the gastrointestinal tract (GIT) where it is metabolized by gut bacteria in a process called **uricolysis** to form carbon dioxide and ammonia:

$$\text{Uric acid} + 4\ H_2O + \rightarrow 4\ NH_4^+ + 3\ CO_2 + \text{glyoxylic acid}$$

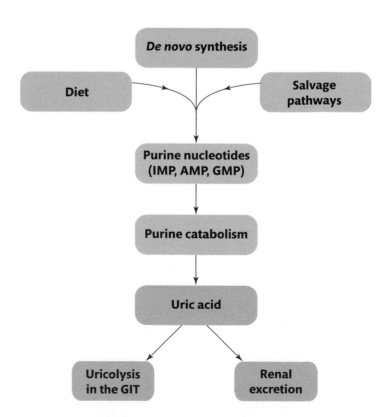

FIGURE 4.5
Overview of the breakdown of purines to uric acid and its excretion.

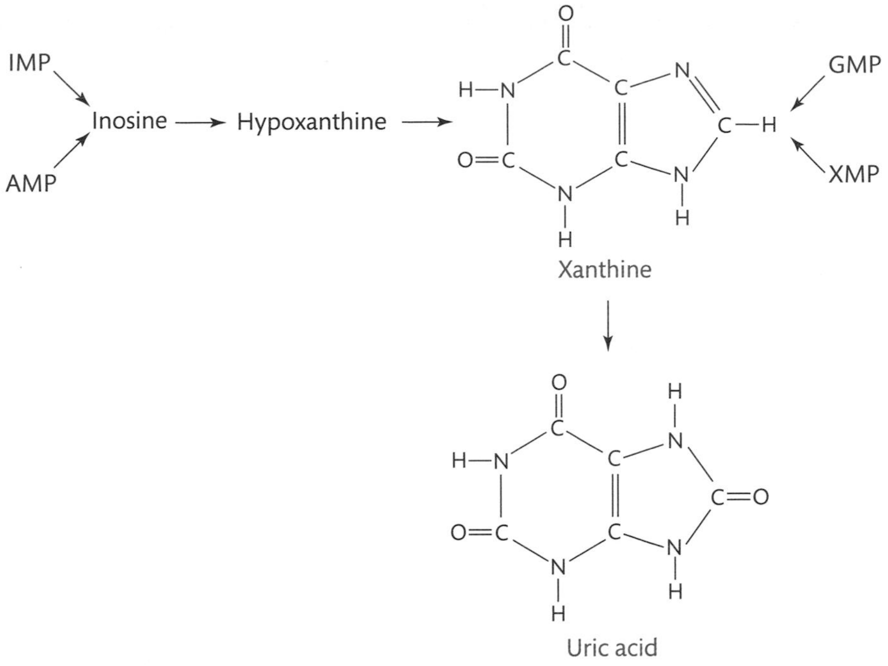

IMP, AMP → Inosine → Hypoxanthine → Xanthine ← GMP, XMP

Xanthine → Uric acid

FIGURE 4.6

The catabolism of common purine nucleotides. See text for details.

(a) monosodium urate structure

(b)

FIGURE 4.7

(a) The structure of monosodium urate. (b) Crystals of monosodium urate viewed using polarized light.

SELF-CHECK 4.3

Which organs are primarily responsible for the excretion of uric acid?

The concentrations of uric acid in blood are normally determined in samples of serum. Concentrations are generally higher in males than females, with a usual reference range for males of 0.1–0.42 and for females 0.1–0.36 mmol/L. Values below or above these limits are called hypouricaemia and hyperuricaemia respectively.

4.3 Hypouricaemia

Hypouricaemia is a measured serum concentration of uric acid below its reference ranges (Section 4.2). It is rare, being associated with relatively few clinical conditions. The primary metabolic cause is an autosomal recessive disorder in which there is a deficiency of xanthine oxidase activity in the liver. This results in increased excretions of xanthine and hypoxanthine and the formation of xanthine stones. This condition is called xanthinuria. Hypouricaemia can be caused by treatment with the drug allopurinol that inhibits xanthine oxidase (Section 4.7), thus reducing the synthesis of uric acid.

4.4 Hyperuricaemia

Hyperuricaemia is defined as an increase in the concentration of uric acid in samples of serum from patients to values above the reference ranges (Section 4.2). Given, the concentration of uric acid in plasma reflects an equilibrium between the amount ingested and produced, and the quantity excreted, then hyperuricaemia may be due to an increase in its formation or a reduction in its excretion, or a combination of both.

The causes of hyperuricaemia are divided into primary disorders due to inherited metabolic diseases involving purine metabolism, or secondary ones caused by a co-existing clinical condition. Both, of course, lead to an accumulation of uric acid in the body.

Primary causes of hyperuricaemia

More than 99% of causes of primary gout are idiopathic, that is there is no known cause, although contributory factors such as hormones, family history, and dietary causes are implicated. Some of the causes that increase the rate of formation of uric acid or decrease its excretion are shown in Table 4.2.

The X-linked recessive disorder, Lesch-Nyhan syndrome is characterized by an increase in the *de novo* synthesis of purines and is a major cause of primary hyperuricaemia. The increase in the amount of purines synthesized occurs because of an impaired feedback mechanism that severely reduces the activity of HGPRT and so impairs the salvage pathways for reusing purine nucleotides.

Cross reference

Chapter 20 Inherited metabolic disorders and newborn screening

Key Points

X-linked genetic diseases are generally passed from mother to son because the genetic lesion is found on the X chromosome. A recessive disorder is one in which a parent has a defective allele in their genome. Thus, although the parent is unaffected, he or she is a carrier of the disease.

TABLE 4.2 The main causes of hyperuricaemia resulting from overproduction and reduced excretion of uric acid.

Uric acid	Primary	Secondary
Overproduction of uric acid	Inherited metabolic disease (HGPRT or APRT deficiency) Idiopathic	Nutritional Excess intake Metabolic Increased ATP metabolism Cytotoxic drugs Drugs Alcohol Haematological Myeloproliferative disease Polycythaemia
Decreased excretion of uric acid	Idiopathic	Renal disease Chronic renal failure Hypertension Drugs Thiazide diuretics Salicylate Metabolic

Lesch-Nyhan syndrome is a rare condition and affects only 1 in 10,000 to 1 in 380,000 live births. It is characterized in infants by poor feeding, severe developmental delay, and self-mutilation. However, a *partial* deficiency in the activity of the HGPRT results in the less severe X-linked disease, Kelley-Seemiller syndrome. Kelley-Seemiller syndrome presents with a broader range of symptoms, including gout and kidney stones (**nephrolithiasis**). It is characterized according to the severity of neurological involvement.

An increased activity (superactivity) of PRPPS also results from an X-linked disorder. The mechanism of increased activity is thought to be due to resistance of the enzyme to negative feedback inhibition. The result is an increase in the synthesis of purine nucleotides. This disease is also characterized by hyperuricaemia and gout.

Von Gierke's disease or glycogen storage disease type 1 is caused by a deficiency in glucose-6-phosphatase (G6P) activity, which indirectly affects the *de novo* synthesis of purines. This recessive condition presents in infancy with symptoms of hypoglycaemia, failure to thrive, and hyperuricaemia (see Table 4.3). Glucose-6-phosphatase catalyses the dephosphorylation of glucose 6-phosphate, making glucose from stores of glycogen in liver cells available for release into the blood. A deficiency in G6P increases the amount of glucose 6-phosphate metabolized through the pentose phosphate pathway, which forms ribose 5-phosphate. Thus, the amount of ribose 5-phosphate available for purine synthesis (Section 4.1) is increased.

Secondary causes of hyperuricaemia

An increase in the concentration of uric acid in plasma may be secondary to an increase in the uric acid formed in the body. Thus, secondary hyperuricaemia may result from a number of factors such as a high dietary intake, increased metabolism of ATP, or an increase in the turnover of nucleic acids. Other causes include a reduced excretion of uric acid by the kidneys.

TABLE 4.3 Inherited disorders of purine metabolism.

Disorder	Enzyme affected	Presentation
Lesch-Nyhan syndrome	HGPRT (deficiency)	Hyperuricaemia Mental retardation
Kelley-Seemiller syndrome	HGPRT (partial deficiency)	Hyperuricaemia Nephrolithiasis
Gout	PRPPS (superactivity)	Hyperuricaemia
Von Gierke's disease	Glucose-6-phosphatase (deficiency)	Hyperuricaemia Hypoglycaemia
Xanthinuria	Xanthine oxidase (deficiency)	Hypouricaemia Xanthine stones

This may be due to renal disease or the stimulating effects of drugs, such as thiazide diuretics and low doses of salicylate, on its reabsorption from the kidney tubules, or to prolonged metabolic acidosis.

Irrespective of whether the hyperuricaemia is primary or secondary in origin, the build-up in the concentration of uric acid in the blood leads to the deposition of crystals of monosodium urate in the joints and tendons, a condition called gout.

4.5 Clinical features of gout

Gout is an inflammatory disease caused by accumulation of monosodium urate crystals in the joints particularly those of the big toe. It affects approximately 1% of the population; 7% of male patients are over 65 years in age. Typically, there is an increase in the concentration of uric acid in the plasma, as measured in serum samples from patients. The likelihood of developing gout and the onset of pain correlates with the amount of uric acid in the blood. Gout can be divided into four main phases:

- asymptomatic
- acute
- intercritical
- chronic tophaceaous gout

The progression of gout through these four stages is shown diagrammatically in Figure 4.8.

As the term suggests, a patient at the asymptomatic stage of gout does not experience pain or swelling of the joints although samples of serum have concentrations of uric acid above reference ranges. However, acute gout is characterized by a sudden onset of pain with swelling of the affected joint. The peak incidence of acute gout occurs between 30 to 50 years of age. In 60% of cases it affects the first metatarsal joint of the big toe or metacarpal of a finger. The joint is red coloured, swollen, warm, and very tender. Until the menopause, the incidence of gout in women is far less than that seen in men, but following menopause the numbers become greater although never equal to that seen in men. Intercritical gout occurs when these initial symptoms resolve within one to two weeks. During this phase the patient appears asymptomatic. In chronic tophaceous gout, the frequency of acute attacks of gout increases over time. Approximately 60% of patients have a second attack within one year and only 7% do

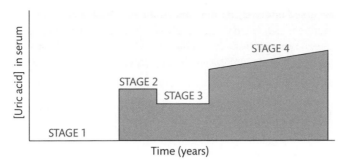

FIGURE 4.8
Schematic diagram showing the progression of gout from stage1 (asymptomatic hyperuricaemia), 2 (acute), 3 (intercritical), to 4 (chronic tophaceous). The time between stages 1 and 2 can last for many months until an acute attack occurs. The concentration of uric acid in serum samples generally increases and, if left untreated, gout may progress to stages 3 and 4 as shown.

not have recurrence within ten years. Patients often have symptoms that mimic **polyarthritis** with swellings around the joints that are swollen and painful. Tophi are chalky deposits of monosodium urate crystals that can be detected by radiological examination and are visible around the joints of the hands or feet. The duration of time between a first gouty attack and the development of tophi is highly variable, varying from 3 to 40 years. There is a correlation between the rate of tophi formation and the duration and increasing concentrations of uric acid in the plasma of the patient. Renal urate stones (**urolithiosis**) develop in patients who have prolonged increased levels of uric acid, They are produced in the kidney when the urinary uric acid concentration is increased. The most common urate stone is composed of ammonium urate.

4.6 **Diagnosis of gout**

A diagnosis of gout usually begins with a review of the family medical history and a physical examination. A swollen joint and red shiny skin above the affected area can indicate gout. However, a thorough medical examination is required to exclude other conditions such as **pseudogout** (Box 4.1), psoriatic arthritis, rheumatoid arthritis, or infection. Most patients with gout have concentrations of uric acid in samples of their serum above their reference ranges. Thus, this is a useful index to measure at the time the patient has symptoms.

BOX 4.1 *Pseudogout*

Pseudogout is a common inflammatory arthritis that results from an accumulation of calcium pyrophosphate crystals in synovial fluid surrounding the joint. It typically affects the knee joint and is often misdiagnosed as gout, whose clinical symptoms it resembles.

Pseudogout is not a disorder of purine metabolism and concentrations of uric acid in serum samples are within reference ranges. There is evidence of calcification of the cartilage of the knee joint seen by radiological examination.

However, an increased uric acid concentration is not universal in all patients with gout and, indeed, concentrations may fall in the acute phase. Thus, confirmation of an initial diagnosis of gout requires that **synovial fluid** be aspirated from the affected joint and examined using a microscope for the presence of monosodium urate crystals. Urine analysis and radiology may also be used.

Synovial fluid analysis

Synovial fluid examination is the most accurate method of diagnosing gout. The synovial membrane secretes synovial fluid into the cavities of synovial joints where it acts as a lubricant cushioning their movements. Samples of synovial fluid are collected by inserting a sterile needle into the joint and withdrawing the fluid into a universal container. The fluid is sent to the laboratory for microscopy examination. The sample is centrifuged to remove any sediment from the supernatant and a drop of the supernatant plated onto a microscope slide. In addition, some of the pellet formed from the centrifugation is also plated onto a separate slide. It is necessary to examine both supernatant and pellet as uric acid crystals may have precipitated in the fluid. A cover slip is placed on the slides, which are then examined using a microscope equipped with two polarizing filters on a rotating stage. This is necessary to detect the characteristic **birefringence** crystals, that is, they have two distinct indices of refraction, shown by monosodium urate. When observed in synovial fluid in this way, the monosodium urate appears as needle-shaped crystals packed in bundles or sheaves as seen in Figure 4.9.

In pseudogout (Box 4.1), crystals of calcium pyrophosphate can accumulate in the synovial fluid. However, these can be distinguished from crystals of monosodium urate because they show opposite and less marked colour changes when examined by polarizing microscopy.

Additional tests

The measurement of urinary uric acid excretion in samples of urine collected over 24 hours may have some value in diagnosis of gout. In patients identified as high uric acid excretors there is a tendency to treat the symptoms. Radiology, however, does not have a significant role in diagnosis because it is insufficiently sensitive or specific to detect the early stages of gout. However, X-ray investigation is useful in chronic tophaceous gout where it can distinguish gout from other inflammatory diseases, for example rheumatoid arthritis.

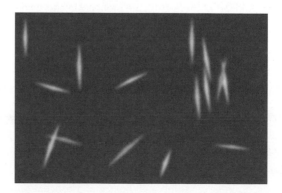

FIGURE 4.9
Synovial fluid examined using a polarizing microscope to show the presence of needle-shaped crystals of monosodium urate.

SELF-CHECK 4.4

What is the definitive diagnostic test for investigating gout.

4.7 **Treatment and management of gout**

Treatment is not usually warranted at the asymptomatic stage of gout. However, the physician can review the medical history of the patient, which involves discussing any current drug therapy and suggesting changes to the lifestyle, such as reducing the intake of purine-rich foods or giving dietary advice if the patient is overweight.

The treatment of acute gout is usually with **non-steroidal anti-inflammatory drugs** (NSAIDs) such as indomethacin or naproxen. In more than 90% of patients the attack is resolved within five to eight days from the start of treatment. There are side effects to this treatment, such as stomach ulcers and bleeding. The mode of action of NSAIDs is to block the action of enzymes called cyclooxygenases (COX-1 and COX-2) involved in the production of prostaglandins. There are alternative treatments recently available that will selectively block the COX-2 enzyme and reduce the side effects. Celecoxib is available and can be used to treat the acute phase of gout in patients at risk of stomach ulcers and bleeding.

Other treatments that have been used to treat acute gout are corticosteroids, for example prednisolone and adrenocorticotrophic hormone. Four- or five-day courses of these steroids are effective whether given orally, intramuscularly, or intravenously. They are particularly useful in treating patients who have a history of intolerance to NSAIDs. Colchicine, a tricyclic plant alkaloid that inhibits cell division, can be given orally to treat gout. The treatment is often effective and pain and swelling usually abates within twelve hours. However, a common side effect is diarrhoea and this may develop before an attack is resolved. It is advisable to rest the affected joint until the attack is resolved otherwise this will prolong the symptoms.

After treatment of the acute phase, risk factors associated with gout should be addressed to reduce the risk of further attacks. The use of drug medication, such as diuretics that can exacerbate the hyperuricaemia, should be discontinued where possible. Advice should be given regarding weight control, reducing intake of alcohol, and lifestyle changes such as limiting consumption of foods rich in purines. Following recurrent attacks, uric acid-lowering drugs (**uricosurics**) may be prescribed. The commonest of these drugs are probenecid and sulin pyrazole. Their action prevents the kidney from reabsorbing uric acid and so increases its urinary excretion. They can be beneficial when given in conjunction with the xanthine oxidase inhibitor, allopurinol, which inhibits the synthesis of uric acid from xanthine. Xanthine, being more water soluble than uric acid, is excreted in urine and plasma uric acid concentrations are decreased. Puricase (PEG-uricase) is a formulation containing the polymer, polyethylene glycol covalently attached to a genetically engineered (recombinant) form of urate oxidase. It has been shown to dramatically reduce the concentrations of uric acid in plasma with no adverse effects and is currently in phase III clinical trials. Present studies have shown puricase to be beneficial to patients with hyperuricaemia for whom conventional therapy is contraindicated or has been less effective. The medications used in treating gout together with their common side effects are listed in Table 4.4.

SELF-CHECK 4.5

What is the main treatment for acute gout?

TABLE 4.4 Medications used in treating gout shown at its different stages, together with some of their common side effects.

Drug	Stage of gout	Side effects
NSAIDs: indomethacin, Naproxen, celecoxib	Acute/intercritical	Can cause bleeding and ulceration
Colchicine	Acute/intercritical	Can cause diarrhoea
Corticosteroids	Acute	Depress immune system
Uricosurics: probenecid, sulin pyrazole	Chronic tophaceous gout	Skin rashes, gastrointestinal problems, kidney stones
Allopurinol	Chronic tophaceous gout	Diarrhoea, headache, rashes

CASE STUDY 4.1

An overweight 50-year-old man was awoken in the night with excruciating pain in the first metatarsal joint of his left foot. He was unable to put the limb on the ground and symptoms were temporarily relieved by an ice pack on his foot. On examination the joint was found to be red, shiny, and very tender. He has a family history of gout. Serum analysis showed the following (reference ranges are given in brackets):

Urate   0.56 mmol/L   (0.1–0.42)

(a) Comment on the clinical history and biochemical result.

(b) How should he be treated initially and then in the near future?

CASE STUDY 4.2

A 61-year-old man attended the outpatients clinic with painful swellings on his knee joints and hands. He had a history of chronic kidney disease and his medication included a diuretic and angiotensin converting enzyme (ACE) inhibitor. Closer examination of his joints showed that several of his joints were red, swollen, and tender to touch. The biochemical results were as follows (reference ranges are given in brackets):

Sodium	138 mmol/L	(135–145)
Potassium	4.4 mmol/L	(3.5–5.0)
Urea	16.0 mmol/L	(3.3–6.7)
Creatinine	150 µmol/L	(45–120)
*EGFR	35 mL/min	(60–89)
Urate	0.86 mmol/L	(0.10–0.42)

*EGFR (estimated glomerular filtration rate) is calculated from the serum creatinine, age, sex, and ethnicity of the patient.

(a) Comment on the results shown for this patient.

(b) What is the **differential diagnosis***?

(c) What additional tests would you ask to be performed?

*A differential diagnosis identifies which of the two or more diseases are responsible for the clinical features shown by the patient.

SUMMARY

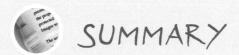

- Purine bases are double-ring structures, for example adenine, guanine, and uric acid.

- Sources of purines are foods such as red meat, shellfish, and cream sauces, their *de novo* synthesis and the salvage pathway.

- The purine synthetic pathway involves a series of enzyme catalysed reactions that result in the formation of uric acid.

- Hyperuricaemia is a concentration of uric acid in plasma above the reference range and may result from overproduction or decreased excretion of uric acid.

- Gout is an inflammatory disease diagnosed by hyperuricaemia and the presence of crystals of monosodium urate in the synovial fluid.

- There are four stages of gout: asymptomatic, acute, intercritical, and chronic tophaceous gout.

- Gout is treated with NSAIDs and analgesics.

- Few clinical conditions are associated with hypouricaemia, consequently it is rarely encountered.

FURTHER READING

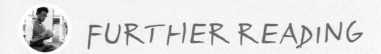

- **Luk AJ and Simkin PA (2005) Epidemiology of hyperuricaemia and gout.** *American Journal of Managed Care* **11**, 435–42.

  Gives a good review of risk factors associated with hyperuricaemia and problems in the treatment.

- **Mandell BF (2002) Hyperuricaemia and gout: a reign of complacency.** *Cleveland Clinic Journal of Medicine* **69**, 589–93.

  Presents a useful argument for individualizing treatment and management of gout—worth a read.

- **Martinon F and Glimcher LH (2006) Gout: new insights into an old disease.** *Journal of Clinical Investigation* **116**, 2073–6.

  Illustrates the mechanism behind the inflammatory response in gout.

- **Rott KT and Aguedelo CA (2003) Gout.** *Journal of the American Medical Association* **289**, 2857–60.

  Focuses on the diagnosis and treatment of gout.

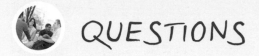

QUESTIONS

4.1   Which of the following are foods rich in purines?

    (a)   Red meat

    (b)   Apples

    (c)   Shellfish

    (d)   Chocolate

    (e)   None of the above

4.2   Which of the following are symptoms of gout?

    (a)   Red and swollen joint

    (b)   Arthritis

    (c)   Abdominal pain

    (d)   Fever

    (e)   None of the above

4.3   Which of the following increase the likelihood of developing gout?

    (a)   Drinking excess alcohol

    (b)   Family history of gout

    (c)   Choice of footwear

    (d)   Renal disease

    (e)   None of the above

4.4   List the main sources of purines in humans.

4.5   How is gout diagnosed?

4.6   List the four stages in the development of gout.

4.7   Describes the main causes of hyperuricaemia.

Answers to self-check questions, case study questions, and end-of-chapter questions are available in the Online Resource Centre accompanying this book.

Go to www.oxfordtextbooks.co.uk/orc/ahmed/

Fluid and electrolyte disorders

Tim James and Walter Reid

Learning objectives

After studying this chapter you should be able to:

- Define the term osmolality and describe how this differs from osmolarity
- Identify those plasma constituents that contribute to osmolality and use this information to calculate an osmolar gap
- Describe how the body responds to both increasing and decreasing osmolality
- Describe the mechanisms involved in controlling total body volume
- Describe the production sites and effects of antidiuretic hormone and aldosterone
- Explain the clinical need to measure plasma electrolytes and how the measured values may enable any abnormalities to be categorized
- Relate the clinical reasons why plasma electrolyte concentrations may be increased or decreased
- Discuss how specimen handling and storage conditions may cause artefactual errors

Introduction

Electrolytes are substances that will dissociate into ions in solution and so acquire the ability to conduct electricity. Examples of ionic compounds include acids, bases, and salts, such as sodium chloride. In a clinical setting, the term electrolyte is generally used in relation to the two most abundant cations present in biological fluids: sodium and potassium. Analyses to determine the concentrations of sodium and potassium in plasma are the most frequent tests undertaken in clinical laboratories. They are needed so often because disturbances to electrolyte balance are common and arise in many clinical settings. The consequences of severe disturbances to plasma electrolytes are life threatening and require correct identification and immediate treatment. It is, therefore, essential that clinical laboratories analyse and report the results for these tests with an appreciation of their significance and the potential to report erroneous results. For example, determining the concentration of potassium in plasma, although one of the most useful tests available to clinicians, is also one of the most vulnerable to variability and artefactual errors. The absolute volume and distribution of body fluids strongly influences plasma electrolytes; thus, an understanding of fluid balance is also essential to aid interpretation of electrolyte results.

This chapter approaches fluid and electrolyte balance from the viewpoint of the clinical biochemistry laboratory; it covers the major causes of each abnormality that may be encountered and provides guidance on the accurate reporting of these parameters.

5.1 Water homeostasis

The total water in an adult human male is approximately 60% of the total body weight. In a 70 kg man this has a volume of about 42 L. The total daily water losses should be balanced by an equivalent intake. In an average adult male the water losses in urine, sweat, faeces, and respiration; and intake in drinks and food, are both about 2 L.

Water in the body distributes between two major compartments. The larger *intra*cellular compartment (28 L in the average male) is the water found within cells. The *extra*cellular compartment (14 L), found outside the cells, has two main components: the plasma of the blood (3.5 L) and the interstitial fluid (10.5 L). Dissolved constituents in each compartment contribute to its effective **osmolality**. Osmolality is a measure of the number of osmotically active particles in solution. It is expressed as moles of solute per kg of solvent. In health, osmolality is maintained at about 285 to 295 mOsm/kg in both the extra- and intracellular compartments.

A free exchange of water between the compartments is ongoing, as water easily and rapidly crosses selectively permeable cell membranes. This is shown pictorially in Figure 5.1. We will discuss the osmolality of plasma and urine in Section 5.5.

Key Points

Osmolality is a measure of osmotically active particles in solution. In body compartments, water will move from a region of low osmolality to a region of high osmolality.

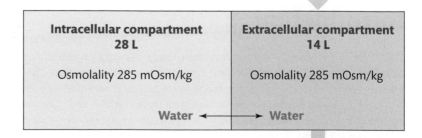

FIGURE 5.1
The relative distribution and relationship between water compartments in the body.

Movement of fluids

The osmolality of the extracellular compartment fluctuates with relative intake and output of fluid. If the osmolality of the extracellular compartment is greater than that in the intracellular, as might occur during dehydration, there will be an increased transfer of water from the cells into the extracellular compartment due to osmosis. Similarly, if the osmolality of the intracellular compartment is greater than that in the extracellular, as may happen during over-hydration, then the transfer of water into the cells will increase. The net transfer of water affects the volume of the cells. Changes that affect cellular volume will cause cells to swell when they take up water or shrink on losing it. This may adversely affect cellular functions. In healthy individuals, physiological mechanisms tightly control the extracellular osmolality to 285–295 mOsm/kg. Consequently, cellular volume and total body water remain in balance.

In contrast, the electrolyte compositions of the extracellular and intracellular fluids differ significantly. The major extracellular ions are sodium, chloride, and hydrogen carbonate (although this is often called by its ancient name, bicarbonate, in clinical circles). The predominant intracellular ions are potassium, magnesium, negatively charged proteins, and organic phosphates. The major mechanism for maintaining the gradient of sodium and potassium between the two compartments is the activity of a transmembrane enzyme frequently referred to as Na/K-ATPase or Na/K-pump. This activity transfers sodium from inside to the outside of the cell, in exchange for potassium in an energy dependent process that requires free energy from the hydrolysis of ATP:

$$\text{Na/K-ATPase}$$
$$3\,Na^+_{IN} + 2\,K^+_{OUT} + ATP + H_2O \rightarrow 3\,Na^+_{OUT} + 2\,K^+_{IN} + ADP + P_i$$

Regulation of blood volume

Water deprivation will lead to an increase in the osmolality of the plasma. Even relatively small increases of only 1–2% in plasma osmolality are detectable by **osmoreceptors** in the hypothalamus and lead to two physiological responses:

- stimulation of thirst

- production of antidiuretic hormone (ADH) or vasopressin

Increased drinking to slake the sensation of thirst obviously decreases plasma osmolality. Antidiuretic hormone is an octapeptide produced by the posterior pituitary gland that activates specific receptors within the kidneys to stimulate an increase in the permeability of the renal tubules to water. Thus, water retention, and therefore conservation, is increased. The action of ADH and satisfaction of thirst have the combined effect of increasing total body water and re-establishing plasma osmolality at around 285 mOsm/kg, as shown in Figure 5.2. At higher concentrations, ADH also stimulates the constriction of arterioles and so has a 'pressor' effect, hence the alternative name for ADH of vasopressin. The ADH of most mammals is the octapeptide, lysine vasopressin, but in humans is arginine vasopressin (AVP). The two forms differ by the single named amino acid residue. Antidiuretic hormone is rapidly metabolized by the liver and kidneys and has a half-life of only five minutes.

Plasma osmolality is also influenced by dietary intake and metabolic activities. For example, the diet contains constituents such as sodium chloride and glucose, which are osmotically active and therefore contribute to plasma osmolality (see Section 5.5 for additional details). Metabolic activities contribute to the osmolality in numerous ways, such as the removal or addition of water in condensation and hydrolytic reactions respectively.

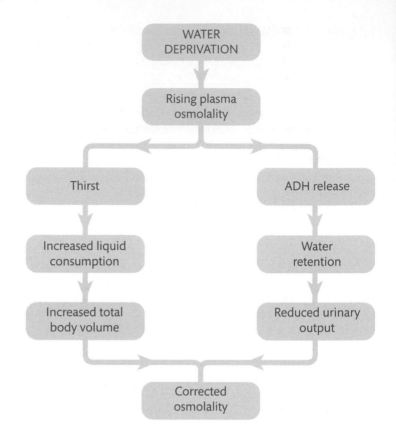

FIGURE 5.2

Flow diagram to show the major physiological responses following a rise in plasma osmolality.

Consequently, actions that maintain osmolality will change the total body volume of water and therefore blood pressure unless osmotic control is supplemented by mechanisms that control the volume of water. The primary control mechanism for body fluid volume is the renin-angiotensin-aldosterone (RAA) system. A low total body water or **hypovolaemia**, results in a lowering of blood pressure, which is detected by the juxtaglomerular apparatus (JGA) in the kidneys. The JGA then releases the specific proteolytic enzyme, renin, that catalyses the hydrolysis of circulating angiotensinogen to release an inactive decapeptide, angiotensin I. The subsequent activity of circulating angiotensin converting enzyme (ACE) on angiotensin I produces the active octapeptide, angiotensin II, which is the principal stimulant of aldosterone release.

Aldosterone is a steroid produced within the zona glomerulosa region of the adrenal cortex. Its release is also influenced by factors other than angiotensin II, notably blood concentrations of potassium above the reference range, a condition called **hyperkalaemia**. The main effect of aldosterone is on the distal renal tubule where it increases the exchange of sodium for potassium or hydrogen ions, increasing its retention. The net effect of increasing the concentration of sodium in the plasma is to increase blood osmolality and the concentration of ADH. Thus, water is retained, and the total body volume increased. Angiotensin II has additional effects, such as increasing myocardial contraction and vasoconstriction, stimulating the absorption of sodium and water by the gastrointestinal tract (GIT) and stimulating thirst.

These systems to control water and electrolyte balances are relatively simple and more complex mechanisms are also involved, including the natriuretic peptides and neuroendocrine control. Atrial natriuretic peptide (ANP) is released following atrial stretch and produces effects that essentially oppose those of the RAA system. Thus, ANP inhibits the production of renin and aldosterone, reduces sodium reabsorption by the kidneys, and acts as a vasodilator.

B-type natriuretic peptide (BNP) also stimulates similar effects. It and related peptides have received attention because of their possible use as cardiac markers, specifically of left ventricular failure. In essence, natriuretic peptides provide a mechanism to reduce cardiac demands by reducing blood volume and thereby blood pressure.

SELF-CHECK 5.1

What will be the immediate effect when there is a rise in plasma osmolality due to a rise in blood glucose as in diabetes mellitus?

5.2 Assessing fluid and electrolyte status

Disturbances of water and electrolytes are associated with a wide spectrum of clinical conditions. They are often multifactorial in origin. Consequently, testing serum or plasma for electrolytes is required for many patients' specimens in both the hospital and community settings. Patients within an acute medical or surgical setting require regular checks on their electrolyte status. In certain conditions, there may be a rapid disturbance to their plasma sodium and potassium concentration. Electrolyte analyses may be required to:

- Investigate a differential diagnosis to identify if an electrolyte abnormality is responsible for the clinical features shown by the patient.
- Detect an abnormality at the time of admission to hospital.
- Monitor correction of the abnormality as a marker of response to treatment.
- Identify a disturbance that arises during treatment.

Electrolyte analysis for primary care physicians, such as general practitioners, also forms a significant laboratory workload. The underlying reasons for requesting electrolyte tests in primary care situations are broadly similar to those in a hospital. A significant proportion of patients may be monitored to assess side effects of drug treatment, for example those used to reduce **hypertension**. Many of the pharmaceutical treatments for hypertension act upon the regulatory mechanisms of fluid and electrolyte homeostasis, which may cause side effects including abnormal concentrations of electrolytes.

Once identified, the electrolyte disturbance may be treated. Treatment is monitored by taking subsequent samples to check progression towards the correction of the imbalance. It is necessary for the laboratory to be alert to a number of problems that may subsequently arise, which include a worsening electrolyte abnormality, a rate of correction that is too fast or the possible over correction of the abnormality. Electrolyte concentrations can fluctuate quite markedly in a relatively short period. Therefore it is not uncommon for clinicians to monitor the plasma electrolytes of acutely unwell in-patients on a daily basis. In certain patient groups, for example those in neonatal or intensive care units, they may be monitored every eight hours.

Sample collection

Laboratory and clinical practice vary to the extent that three sample types are commonly used in the measurement of electrolytes in circulation. These are whole blood, plasma collected in lithium heparin tubes, and serum. These clinical materials produce similar but not identical results when subjected to electrolyte analyses. In practice, the differences do not prevent each of the major electrolyte abnormalities from being detected but the results are not directly interchangeable. Variations in results obtained using these test materials arise because of differences in the underlying analytical principles of detection, variation in calibration materials and strategies, and variability in the **specimen matrix**.

Electrolyte analysis on whole blood specimens is most frequently undertaken using point of care testing equipment in acute clinical settings, such as the emergency department or a medical admission ward. The benefits of this approach are the ready availability of results, which enables clinical management to be initiated immediately. There is ongoing interest in this analytical approach where a decision can be made in a clinical setting that is geographically remote from the laboratory. The disadvantages of this approach for electrolyte analyses include an inability to visually examine the specimen for evidence of **haemolysis** and its higher unit cost. However, although the unit expenditure may be higher for the result itself, the total healthcare costs may be reduced since an improved patient outcome may be possible.

Laboratory based analyses of electrolytes are predominantly performed on lithium heparin plasma or serum. A direct comparison of results from these two types of specimens shows that lithium heparin plasma and serum give generally comparable concentrations of sodium. However, for potassium there is consistent difference, with the use of serum giving a value about 0.2 mmol/L higher than that in lithium heparin plasma. This difference is due to the release of potassium from platelets during clotting. The difference is even greater in patients with a significantly increased platelet count. From a practical point of view, the advantage of using plasma for analysis is one of speed: serum specimens require approximately 30 minutes for clot formation prior to centrifugation.

Key Points

Haemolysis is the destruction of red blood cells and can cause leakage or release of cellular contents into the serum, rendering them unsuitable for analysis of certain analytes. Haemolysis of blood samples often occurs due to their unsatisfactory collection from subjects or their storage.

Artefactual changes to electrolyte values

Serum and lithium heparin plasma specimens are both generally accepted by laboratories for electrolyte determination. However, other specimen collection tubes contain anticoagulants that are unsuitable because they are salts of sodium or potassium. This includes tubes containing potassium ethylenediaminetetraacetic acid (EDTA), sodium citrate, or a mixture of sodium fluoride and potassium oxalate. All these anticoagulants lead to contamination of the plasma with sodium and/or potassium, if they are inadvertently used for electrolyte analyses. However, such contamination will be evident from the test results. You can see examples of this in Table 5.1. The magnitude of effect is significant in all of these cases, such that identification is relatively simple. However, errors are more difficult to detect when there has been only a partial contamination with an inappropriate anticoagulant. This can happen if samples are poured from one specimen collection tube into another type.

Electrolyte analyses and, indeed, other test parameters may also be compromised if patients have specimens taken from veins in the arm that has an intravenous drip. In these situations, the test results reflect the components of the fluid being given. For example,

TABLE 5.1 Artefactual electrolyte changes typically observed when specimens are collected into tubes containing inappropriate anticoagulants and preservatives. The changes observed will depend on the sample contents and the volume of blood collected.

Tube type	Lithium heparin	Sodium citrate	Potassium EDTA*	Sodium fluoride and potassium oxalate*
Sodium (mmol/L)	140	150–160	140	>180
Potassium (mmol/L)	4.0	–	>10.0	>10.0
Calcium (mmol/L)	2.30	?	<0.2	<0.3

* Both these collection tubes will adversely affect alkaline phosphatase activity and magnesium concentration, which will both be decreased, and osmolality, which will be increased.

a standard saline drip will contain sodium chloride at a concentration of approximately 150 mmol/L and a specimen taken from an adjacent vein will have a similar sodium concentration. In such cases, other blood parameters, notably the haematocrit (the relative volume of blood cells in the specimen), are reduced. The time between specimen collection and analysis can also affect the results of electrolyte analyses, particularly that of potassium as described in Section 5.4.

Measurement of electrolytes

Sodium and potassium ions are routinely analysed using **ion-selective electrodes** (ISEs). An ion-selective electrode measures the *activity* of the ion rather than its concentration. However, for dilute solutions, the concentration and activity are approximately the same and the results of the measurements for both sodium and potassium are generally given in mmol/L.

Cross reference
Chapter 2 Automation in this textbook and Chapter 9 Electrochemistry in the *Biomedical Science Practice* textbook

Two basic types of ISE are available, direct and indirect ISEs. Direct reading electrodes use undiluted specimen at the electrode surface, whereas indirect reading electrodes require the specimen to be diluted using a suitable buffer. Direct reading electrodes are most often found in the bench top analysers used in point of care instruments or extended repertoire blood gas instruments. In comparison, indirect reading ISEs are found in the high throughput laboratory based instruments described in Chapters 1 and 2. A fundamental operational difference between the two types of electrodes produces a phenomenon called the 'solvent exclusion effect'. Sodium ions are only distributed in the aqueous phase of the specimen and, if the non-aqueous fraction increases, for whatever reason, then a dilution effect will result. The degree of error will be proportional to the volume of the non-aqueous fraction. The two main components of this fraction are lipids and proteins. In most blood or plasma specimens, the non-aqueous phase is therefore a similar fraction of the sample. However, if the patient has an increased lipid or protein concentration in the plasma then the apparent sodium value will be lower it determined by an indirect ISE. The solvent exclusion effect changes the apparent concentration of any parameter that is found only in the aqueous phase of a specimen, although it is most pronounced for plasma sodium since the observed concentration range is comparatively narrow.

SELF-CHECK 5.2

What is the difference between direct and indirect ion-selective electrodes?

5.3 **Disturbances of plasma sodium**

The body of an average adult man contains approximately 3,700 mmoles of sodium. An outline of its distribution, intake, and losses is shown in Figure 5.3. Approximately 75% is found in the extracellular fluid (ECF) and the remaining 25% is located in bone and tissues. The intake of sodium in the diet is typically 100 to 200 mmol/day but the amounts in the body are maintained at a constant value by equal renal losses. About 5 mmol/day of sodium is also lost in the faeces and through the skin in sweat.

The concentrations of sodium in plasma is tightly controlled within the reference range 135 to 145 mmol/L. Values below this range are called **hyponatraemia**, those above, **hypernatraemia**. The majority of sodium analyses undertaken in a large clinical laboratory will be distributed closely around a value of 140 mmol/L, as seen in Figure 5.4.

Hyponatraemia

Hyponatraemia is the commonest electrolyte disorder and affects a significant proportion of both patients in hospitals and the elderly population in the community. Its clinical symptoms range from mild confusion, fatigue, muscle cramps, through to more serious consequences including brain oedema, seizures, respiratory arrest, and death. The concentration of sodium at which any of these events may occur varies depending upon the time over which the hyponatraemia has developed and concomitant pathologies. The clinical observations arise primarily from the cellular hypo-osmolality caused by the hyponatraemia.

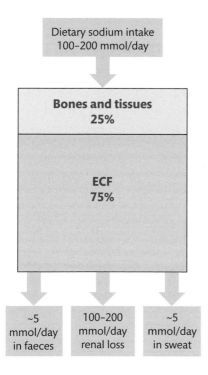

FIGURE 5.3

The intake, distribution, and excretion of sodium.

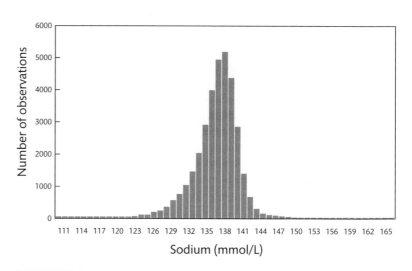

FIGURE 5.4

The distribution of 35,000 sodium measurements performed during a four-week period in the clinical laboratory of a large teaching hospital. The patient population tested was approximately 50% hospital patients and 50% patients referred from general practices. There is a greater frequency of hyponatraemia ([sodium] <135 mmol/L) compared to hypernatraemia ([sodium] >145 mmol/L); the majority of results are within the range 135–145 mmol/L.

TABLE 5.2 A classification of hyponatraemias and their causes.

Classification of condition	Causes	Comments
Hypervolaemic hyponatraemia	Congestive heart failure Liver failure Renal dysfunction (chronic renal failure and nephrotic syndrome)	Total body water and sodium are increased but the increase in water is greater than that of sodium
Euvolaemic hyponatraemia	Syndrome of inappropriate ADH Glucocorticoid deficiency Water overload Hypothyroidism Post-surgery	Total body water increases while total body sodium is normal or near normal
Hypovolaemic hyponatraemia	Gastrointestinal loss (vomiting and diarrhoea) Burns Diuretics Mineralocorticoid deficiency	Deficits of total body sodium and water but with a disproportionately greater loss of sodium

Hyponatraemia must be interpreted in relation to hydration status, that is, the amount of body fluid. It may be classified into three types relative to whether the total body fluid is high (hypervolaemia), normal (normo or euvolaemia), or low (hypovolaemia). A number of common and/or significant causes, based on this classification of hyponatraemia, are shown in Table 5.2.

Hyponatraemia also has artefactual causes, for example pseudohyponatraemia is mentioned in Box 5.1. The appropriate treatment of hyponatraemia depends on accurately identifying its cause. This includes establishing clinical history, medication, and undertaking an assessment of hydration status for the patient. Further laboratory investigations may include an assessment of renal function, endocrine tests, including those of the adrenal, thyroid, and pituitary glands, and determining the sodium and osmolality of urine specimens.

BOX 5.1 Pseudohyponatraemia

Pseudohyponatraemia is a value for the measurement of sodium that is below the reference range (135 to 145 mmol/L) that has an artefactual cause. It is associated with increased amounts of lipid or protein in the specimen of plasma and with analytical techniques in which the sample is diluted prior to sodium analysis. These techniques include the use of indirect ISEs and, historically, flame photometers. Comparing sodium results from direct and indirect reading electrodes in specimens with increased lipid (triacylglycerols greater than 10 mmol/L) and/or protein content (total protein greater than 90 g/L) to those with more normal concentrations, suggests that extreme pseudohyponatraemia can decrease results by 5–10 mmol/L. Pseudohyponatraemia is not a potential problem when the sodium concentration is determined by a direct reading ISE. Consequently, if a low plasma sodium result is thought to be due to this effect then a check using a direct reading ISE should be performed. Alternatively, osmolality can be used to exclude the effect (see Section 5.5).

Post-operative surgical patients are particularly vulnerable to hyponatraemia. This is more likely with increasing age. Post-surgical patients are subject to increased (ADH) concentrations exacerbated by stress, nausea, pain, or opiate analgesia. They may also be given fluids that increase total body water without increasing body sodium. Both these factors increase the relative amount of water compared to sodium. A common replacement fluid is isotonic dextrose solution, which provides both glucose and water to the patient. However, the metabolism of glucose will effectively leave a hypotonic solution. The dilutional hyponatraemia of post-surgical patients is similar to an increasingly reported problem that can occur following strenuous exercise, such as marathon running. Fluid replacement is often provided to the athletes in the form of an isotonic drink. However, the tonicity in the drink, as with the intravenous fluid described above, is in the form of glucose. Thus, the rapid depletion of the glucose by body cells is followed by a dilutional hyponatraemia. More recent isotonic replacement formulae contain a significant sodium chloride load.

Syndrome of inappropriate antidiuretic hormone

In some cases, an obvious cause for hyponatraemia is not apparent. The term syndrome of inappropriate ADH (SIADH) is used to describe these cases. It is common in the elderly and is used to describe the situation where, despite a low plasma osmolality, ADH production continues; thus the inappropriateness. Criteria for diagnosing SIADH include the following:

- A plasma osmolality below 270 mOsm/kg.
- An inappropriately high urinary concentration giving an osmolality greater than 100 mOsm/kg.
- Lack of conservation or renal leakage of sodium into the urine even while the patient is on a normal salt and water intake concurrent with euvolaemia.

In addition, diagnosis depends on the exclusion of the other causes of these features, which include thyroid, pituitary, adrenal diseases, renal insufficiency, or the use of diuretics. Measurement of antidiuretic hormone is restricted to a limited number of specialist laboratories and cannot be provided in a clinically useful timescale for most patients with hyponatraemia. Consequently, its use is restricted to unresolved and/or unexplained persistent electrolyte disorders.

There is significant debate about the best method to treat hyponatraemia, but simple initial measures include restricting fluid intake to 1 to 1.5 L per day. If hyponatraemia is corrected too rapidly then permanent cerebral damage can occur. Thus, the hyponatraemia must be corrected slowly over a period of days rather than hours. Potential new treatments include drugs that act on the ADH receptors of the kidney. In many cases, the hyponatraemia will resolve if the precipitating cause can be treated and resolved.

CASE STUDY 5.1

A 68-year-old lady with bowel cancer who had been admitted to hospital following a fall had a persistent low plasma sodium including the following plasma results (reference ranges are given in brackets):

Sodium	120 mmol/L	(135–145)
Potassium	3.9 mmol/L	(3.5–5.0)

Urea	3.4 mmol/L	(3.5–6.6)
Creatinine	53 µmol/L	(70–150)
Glucose	5.2 mmol/L	(3.5–5.5)
Osmolality	255 mOsm/kg	(285–305)

In this patient, the urine volume was low and relatively concentrated, the urine osmolality was 374 mOsm/kg and its sodium concentration was 75 mmol/L.

Comment on these results.

Hypernatraemia

Hypernatraemia is not as common as hyponatraemia. It is predominantly due to a deficit of water relative to sodium rather than any excess of total body sodium. A water deficit may occur following acute gastrointestinal loss, burns, exposure to hot dry conditions without adequate water intake, or it may arise because of lack of access to water that may occur following a stroke. However, the plasma sodium of a significantly dehydrated patient could be within the reference range due to water being redistributed between the intra-and extracellular compartments. Parameters that have been used to guide clinicians on the hydration status of a patient include the concentration of urea in the plasma, which may be increased, relative to plasma creatinine, and changes in the patient's weight. The latter can be particularly significant in neonates.

The early symptoms of hypernatraemia may be non-specific and include anorexia, restlessness, nausea, and vomiting. More serious symptoms that develop as the plasma sodium becomes progressively more concentrated include cerebral effects such as coma, bleeding, and subarachnoid haemorrhage. Correction of hypernatraemia, as with hyponatraemia, needs to be gradual to avoid cerebral damage. Neonates, children, and the elderly are vulnerable to hypernatraemia because they may have reduced abilities to form concentrated urine. This can lead to hypernatraemia following the ingestion or administration of excess salt. Health advice for new mothers in the UK includes the recommendation that babies should not be fed solid foods until they reach the age of six months. Adults who are unable to obtain water or ask for it are also vulnerable to this effect. It has also been suggested that some individuals are more susceptible to this condition than others.

Diabetes insipidus

A specific condition associated with hypernatraemia is diabetes insipidus (DI), which is characterized by the production of a high volume of dilute urine. The urine volume produced by a healthy adult is typically about 2 L per day but in patients with DI is usually much higher and may be as high as 6 L daily. There are two main types of DI: cranial and nephrogenic. Cranial DI arises from a failure to secrete ADH from the posterior pituitary gland. This may be congenital, or occur following a head injury, or it may be associated with a tumour. Nephrogenic DI occurs when the kidneys fail to respond appropriately to ADH. Patients may present with primary **polyuria**, an excessive urine output, and secondary **polydipsia**, which is excessive thirst, with the hypernatraemia only developing if urine losses exceed water intake. Indeed, patients with DI may not develop hypernatraemia if their water intake is adequate to match the urinary losses.

Cross reference

Chapter 11 Abnormal pituitary function

A specific test undertaken to diagnose DI is the water deprivation test, in which plasma electrolytes and osmolality, urine sodium and osmolality, and weight are all monitored, while the

CASE STUDY 5.2

A 2-year-old child, admitted to hospital following diarrhoea and vomiting, had the following results on analysis of plasma, 24 hours after admission (reference ranges are given in brackets):

Sodium	151 mmol/L	(135–145)
Potassium	3.7 mmol/L	(3.5–5.0)
Urea	4.9 mmol/L	(3.5–6.6)
Creatinine	65 µmol/L	(70–150)
Osmolality	314 mOsm/kg	(285–305)

The urine sodium concentration was 55 mmol/L and its osmolality was 314 mOsm/kg.

Comment on these results.

patient abstains from any intake of liquids. A normal response would be for the concentration of the patient's urine to increase with time. Diabetes insipidus is diagnosed if a high-volume, dilute urine continues to be formed despite an increasing plasma sodium concentration. The test is a dangerous procedure and requires careful assessment of the patient and the rapid production of laboratory results.

5.4 Disturbances of plasma potassium

The average adult man contains over 3500 mmoles of potassium, nearly all of which is found in the intracellular fluid (ICF). An outline of its distribution, intake and losses is shown in Figure 5.5. Approximately 55 mmoles is found in the ECF. The dietary intake of potassium is typically 30–100 mmol/day, About 5 mmoles is lost daily in faeces. The kidneys excrete 20–100 mmol of potassium per day, which maintains potassium balance.

Potassium is the predominant *intra*cellular cation and fulfils numerous roles, notably in neuro-muscular functions. In health, the concentrations of potassium in plasma are maintained in the range 3.5–5.0 mmol/L. Figure 5.6 illustrates a typical distribution of results from tests to measure plasma potassium concentrations, observed in a large teaching hospital laboratory. Most of the values obtained fall within this range.

A low concentration of plasma potassium below 3.5 mmol/L is termed **hypokalaemia**, while values greater than 5.0 mmol/L are called hyperkalaemia. The effect of too much, or too little, potassium is most apparent in skeletal muscle tissues and the heart. Hypokalaemia is associated with muscle weakness, fatigue, and in severe cases paralysis and cardiac arrhythmias. Hyperkalaemia produces an irregular heartbeat and may lead to cardiac arrest, which is life threatening. Both hypokalaemia and hyperkalaemia induce characteristic changes to the conduction of heartbeat that can be seen on an electrocardiogram (ECG). The clinical problems associated with abnormalities in plasma potassium are more dangerous if the rate of change in plasma potassium concentration is rapid. It is therefore essential for the clinical laboratory to communicate extreme abnormalities of plasma potassium to clinicians immediately the result is obtained; most laboratories will have a defined threshold at which the result will be given by telephone.

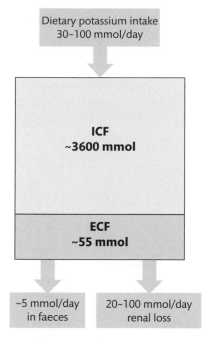

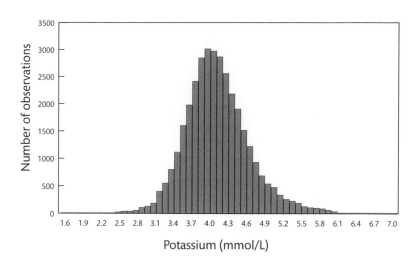

FIGURE 5.5
The intake, distribution, and excretion of potassium.

FIGURE 5.6
The distribution of 35,000 potassium measurements performed during a four-week period in a clinical laboratory of a large teaching hospital. The patient population tested was approximately 50% hospital patients and 50% patients referred from general practices. Results from samples that were haemolysed or had delayed centrifugation and caused artefactual increases in potassium have been excluded.

Potassium is one of the most clinically useful parameters used by clinicians and is affected by many commonly encountered conditions such as diabetes mellitus and renal disease. However, it is also one of the most sensitive to artefactual change, due to inappropriate specimen collection, handling, and storage. An artefactual increase of plasma potassium is probably the commonest pre-analytical error observed during routine laboratory analysis. It arises mainly through the fragility of erythrocytes, which contain a higher concentration of potassium than the plasma in which they circulate. Thus, when there has been a degree of trauma during the collection, or if specimens have been handled carelessly between the time of collection and centrifugation, for example by over vigorous mixing, then lysis of the erythrocytes (haemolysis) will occur and lead to a falsely high value for the observed potassium plasma concentration. If this occurs, then the effect is evident in the pink colouration imparted to the plasma or serum from haemoglobin, which is also released from the lysed erythrocytes. A visual examination can identify haemolysis, but the process can also be automated on general clinical chemistry analysers with a channel set up to measure haemolysis at a suitable wavelength.

Cross reference

Chapter 1 Biochemical investigations and quality control and Chapter 2 Automation

Sample transport effects

A significant logistical challenge to laboratories is to ensure that specimens requiring potassium analyses are transported within a short time and at a stable temperature. In a hospital, it is possible to transport the specimen to the on-site laboratory within an hour. However, the transport of specimens from general practices or from a hospital that does not possess its own laboratory make be several hours between specimen collection and sample centrifugation. This period gives rise to changes in the value of the measured potassium.

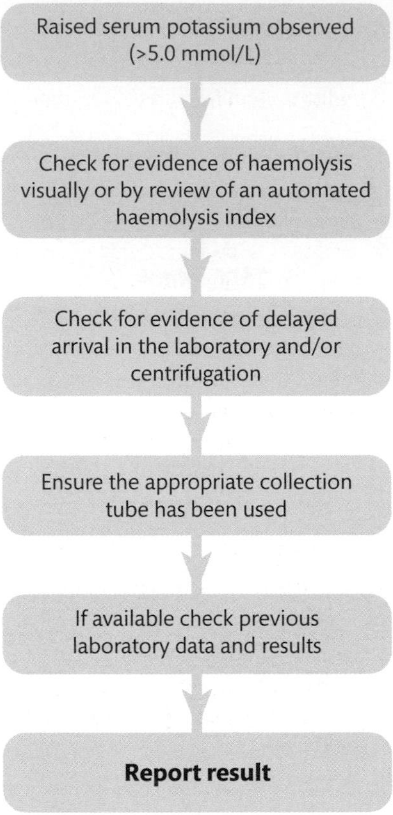

FIGURE 5.7

An increased potassium result is a relatively common observation when reviewing biochemistry data. However, since potassium may easily be artefactually increased by inappropriate sample collection and handling, many laboratories use a simple algorithm such as outlined in this flow diagram to minimize the reporting of misleading potassium results.

Raised serum potassium observed (>5.0 mmol/L)

Check for evidence of haemolysis visually or by review of an automated haemolysis index

Check for evidence of delayed arrival in the laboratory and/or centrifugation

Ensure the appropriate collection tube has been used

If available check previous laboratory data and results

Report result

Generally, the concentration of plasma potassium increases with longer storage times, a problem that is exacerbated if the specimens are stored in a refrigerator as the Na/K-ATPase is considerably less effective at reduced temperatures. However, an opposite effect can be observed when specimens are transported in vehicles during the summer months. The higher temperature appears to increase the effectiveness of the Na/K-ATPase and may lower the plasma potassium concentration. The magnitude of the effect can vary depending on the temperature and the length of time experienced but reports have suggested that the difference in potassium plasma concentrations between winter and summer months for some health centres averages about 0.5 mmol/L. A practical measure to reduce this problem is to centrifuge the specimen within the health centre, as the effect can only occur when the erythrocytes are in contact with plasma. Alternatively, the conditions of carriage should be changed by, for example, rescheduling transport to minimize the time between collection and centrifugation in the laboratory or by using air-conditioned vans.

The problem of artefactual elevation of potassium due to delayed centrifugation, sample haemolysis, or the use of an inappropriate specimen tube is such that most laboratories use an algorithm such as that presented in Figure 5.7 to prevent misinterpretation of the results.

Hypokalaemia

Hypokalaemia occurs in a variety of clinical situations. It may occur by any of three major causes:

- increased losses by the kidneys
- increased losses by the gastrointestinal tract
- redistribution from the extra- to the intracellular compartment

The reabsorption of potassium in the kidney occurs predominantly in the proximal tubule. Consequently, renal tubular failure that can occur during the recovery phase of oliguric renal failure and renal tubular acidosis, a group of conditions associated with an inability to excrete acid adequately or reabsorb bicarbonate by the tubules, can lead to hypokalaemia. Inappropriate loss of potassium by the kidney can also be a consequence of a range of drug treatments. For example, the commonest cause of hypokalaemia is a side effect of thiazide diuretics used to treat hypertension because they inhibit the reabsorption of sodium by the distal tubule of the nephron. The consequent exchange of sodium for potassium and hydrogen ions can cause both hypokalaemia and metabolic alkalosis. Chlorothiazide and bendroflumethiazide are two common examples of thiazide diuretics. A different group of diuretics, the loop diuretics, so-called because they inhibit sodium reabsorption in the ascending loop of Henle, can also cause hypokalaemia. An example of a loop diuretic is the drug frusemide.

The renal handling of potassium is primarily influenced by aldosterone, as described in Section 5.1, and hypokalaemia can arise when this hormone is present in excess. Conn's syndrome, or primary hyperaldosteronism, is a condition in which aldosterone is over-secreted, most commonly due to an adrenal adenoma. The high aldosterone concentration stimulates an inappropriately high excretion of potassium leading to hypokalaemia and the retention of sodium and fluid, which leads to hypertension that is resistant to treatment. Suspected Conn's syndrome may be diagnosed by measuring the plasma aldosterone concentration (PAC) and the plasma renin activity (PRA) with the result expressed as the PAC: PRA ratio. In Conn's syndrome, the plasma aldosterone concentration will be high due to over-secretion and renin will be low due to the suppression of its release by the hypertension and aldosterone feedback. The interpretation of the PAC: PRA requires care. Blood samples are analysed after the patient has been recumbent for at least eight hours and, again, after walking for 30 minutes. The value for aldosterone must not increase in the latter sample for diagnosis to be confirmed. A knowledge of the patient's drug history is also necessary as this test is sensitive to hypertensive treatments. A rare but notable cause of hypokalaemia is caused by a high consumption of liquorice, which contains glycyrrhizinic acid (Figure 5.8 (a)) and produces an effect in the kidneys similar to that of aldosterone Figure 5.8 (b).

Potassium ions may be lost from the GIT during prolonged diarrhoea and vomiting. Losses are also occasionally associated with misuse of laxatives. The movement of potassium from plasma, the extracellular compartment, into cells, the intracellular compartment, is associated with a variety of metabolic disturbances, which include alkalosis and treatment of diabetic ketoacidosis.

Cross reference
Chapter 13 Diabetes mellitus and hypoglycaemia

The treatment for hypokalaemia is dependent on its cause. If the underlying cause is a precipitating factor, such as diuretics, then a change in the treatment may be required. Treatment of true deficiency may require supplementation to increase the intake of potassium-rich foods, such as bananas, tomatoes, and orange juice, but the most effective supplement is a potassium salt preparation. Intravenous infusion of potassium salts can be dangerous and is restricted to only the most extreme circumstances. This treatment requires careful monitoring because of the dangerous effects of a *rapid* change in the concentration of potassium in the plasma.

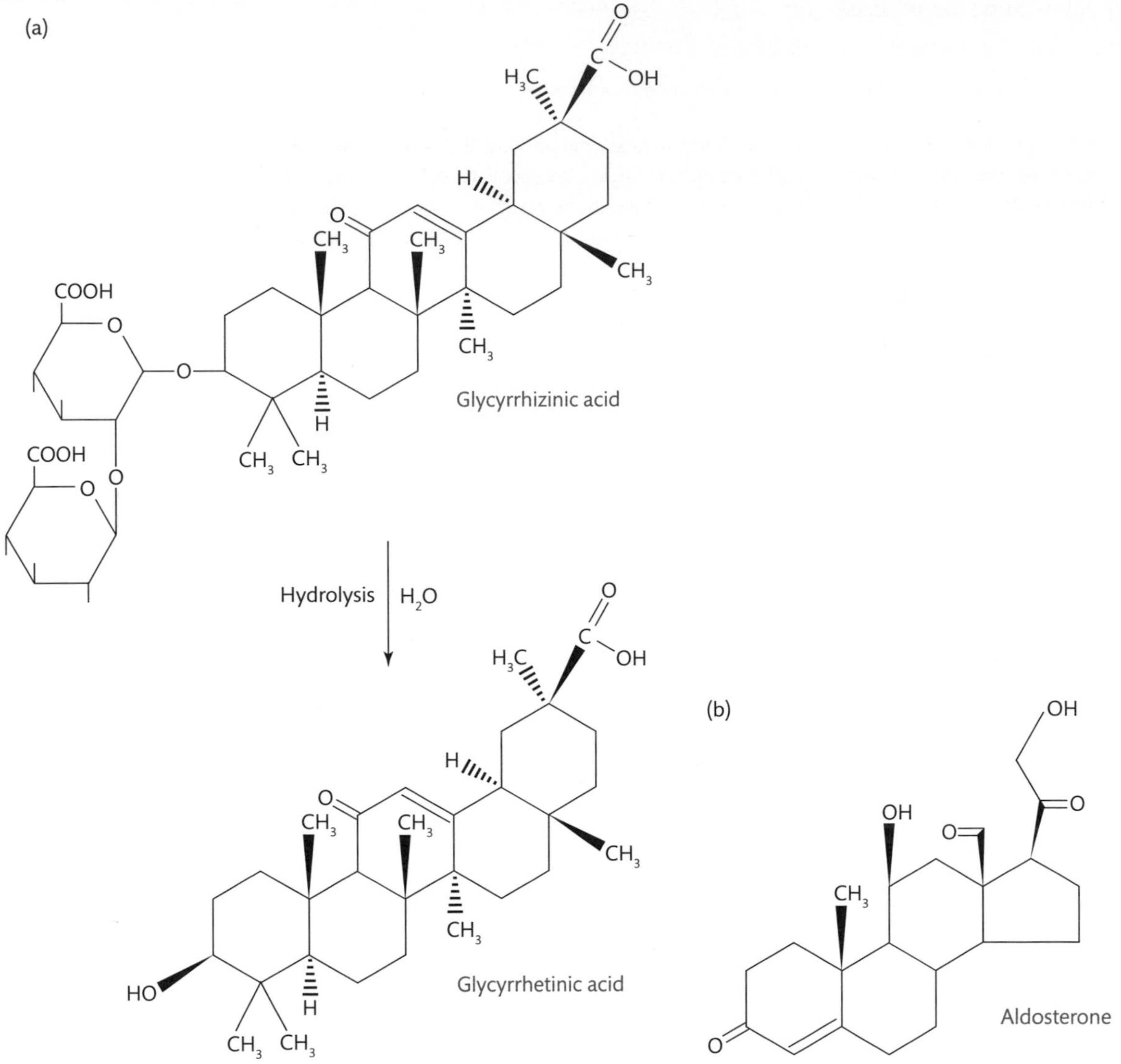

(a)

Glycyrrhizinic acid

Hydrolysis | H_2O

Glycyrrhetinic acid

(b)

Aldosterone

FIGURE 5.8
Excessive amounts of liquorice can mimic the effects of aldosterone because it contains relatively large concentrations of glycyrrhizinic acid. (a) Hydrolysis of glycyrrhizinic acid in the body releases glycyrrhetinic acid, which has a structural resemblance to aldosterone. (b) Aldosterone produces the physiological effect.

Key Points

Diuretics are drugs that increase production of urine and promote loss of water and salts from the kidneys. Diuretics are used to treat patients with oedema which is accumulation of fluids in body tissues. They are also used for the treatment of hypertension.

CASE STUDY 5.3

A 78-year-old lady receiving treatment for hypertension was investigated by her doctor by monitoring renal function tests (reference ranges are given in brackets):

Sodium	142 mmol/L	(135–145)
Potassium	2.2 mmol/L	(3.5–5.0)
Urea	8.9 mmol/L	(3.5–6.6)
Creatinine	142 µmol/L	(70–150)

Comment on these results.

CASE STUDY 5.4

A 30-year-old female patient with uncontrolled hypertension is suspected by an investigating endocrinologist of having Conn's syndrome. Results of routine biochemistry were (reference ranges are given in brackets):

Sodium	146 mmol/L	(135–145)
Potassium	2.1 mmol/L	(3.5–5.0)
Urea	7.2 mmol/L	(3.5–6.6)
Creatinine	146 µmol/L	(70–150)
Alkaline phosphatase	290 IU/L	(95–320)
Alanine aminotransferase	20 IU/L	(5–42)
Albumin	49 g/L	(35–50)
Bilirubin	8 µmol/L	(<17)
Calcium	2.19 mmol/L	(2.12–2.62)

(a) Are any of the electrolyte concentrations abnormal, and if so what condition is suggested?

(b) What further biochemistry investigations would you undertake? Explain your reasoning.

Hyperkalaemia

True clinical hyperkalaemia may arise due to its increased release from cells or because of impaired excretion. A rapid cellular release of potassium can follow significant tissue injury, for example a crush injury. Acute or end-stage chronic renal failure can both give rise to hyperkalaemia, which can be life threatening if not treated. Insulin can affect the movement of potassium into cells and, consequently, potassium abnormalities arise in diabetes mellitus. For example, diabetic ketoacidosis is associated with hyperkalaemia. Its subsequent treatment with insulin can lead to a rapid fall in the concentration of potassium in the plasma that requires careful monitoring. Hyperkalaemia is often associated with acidosis. Drugs that can induce hyperkalaemia include ACE inhibitors, angiotensin receptor blockers, spironolactone,

CASE STUDY 5.5

A sample was analysed for renal and liver function tests. The sample was from a 50-year-old man who had seen his GP to report problems with tiredness. The results observed were reviewed (reference ranges are given in brackets):

Sodium	200 mmol/L	(135–145)
Potassium	>10 mmol/L	(3.5–5.0)
Urea	6.2 mmol/L	(3.5–6.6)
Creatinine	87 µmol/L	(70–150)
Alkaline phosphatase	153 IU/L	(95–320)
Alanine aminotransferase	34 IU/L	(5–42)
Albumin	40 g/L	(35–50)
Bilirubin	12 µmol/L	(<17)

Explain these results.

β-adrenergic antagonists and non-steroidal anti-inflammatory drugs. The cardiac action of the drug digoxin is potentiated by potassium, and therapeutic monitoring of this drug should include an assessment of plasma potassium.

Treatment of hyperkalaemia is dependent on its severity. Severe disturbances require prompt correction to reduce the risk of cardiac arrest. Given that calcium and potassium have opposing

CASE STUDY 5.6

The results observed in a patient admitted to an emergency department with confusion and oliguria were reviewed (reference ranges are given in brackets):

Sodium	146 mmol/L	(135–145)
Potassium	7.2 mmol/L	(3.5–5.0)
Urea	18.5 mmol/L	(3.5–6.6)
Creatinine	256 µmol/L	(70–150)
Alkaline phosphatase	319 IU/L	(95–320)
Alanine aminotransferase	32 IU/L	(5–42)
Albumin	42 g/L	(35–50)
Bilirubin	13 µmol/L	(<17)
Calcium	2.47 mmol/L	(2.12–2.62)

The sample was not visibly haemolysed and was centrifuged and analysed in the laboratory within one hour of being collected.

(a) What is the electrolyte abnormality in this patient?

(b) What do the data suggest?

(c) What action should you take?

actions in the heart, the infusion of calcium salts may be undertaken. Alternatively, insulin may be employed to stimulate the movement of potassium into cells. However, this also requires glucose infusion to prevent any associated hypoglycaemia. Milder disturbances often require treatment of the underlying cause and include measures such as changing any medication that may be contributing to the hyperkalaemia.

SELF-CHECK 5.3

Why might vigorous shaking of a blood specimen collected into a lithium heparin preservative tube cause an apparent hyperkalaemia?

5.5 Plasma and urine osmolality

The analysis of osmolality of plasma and urine specimens is undertaken in a number of clinical situations including the investigation of hypo- and hypernatraemia. The osmolality of any solution affects its freezing point and this can be accurately determined using an **osmometer**, an instrument that is used for osmolality measurement (Figure 5.9). The higher the osmolality of a solution the more its freezing point is depressed below 0°C.

Osmolality results are generally expressed as mOsm kg^{-1} (or mOsm/kg). The similar parameter, osmolarity, is measured in moles per L of solvent (mol/L). If the solvent is water and the solution dilute, osmolality and osmolarity are practically equivalent. However, in plasma there is a small but significant difference between its osmolality and its osmolarity.

The main contribution to plasma osmolality arises from sodium, its counter anions, and two substances of relatively small M_n that are present at relatively high concentrations, namely urea and glucose. A number of formulae, of varying complexity, can be used to calculate the

FIGURE 5.9
Osmometer of the type routinely used to measure the osmolality of body fluids. Courtesy of Precision Systems Inc., MA, USA.

osmolality of plasma. However, in most situations the osmotic load of plasma can be calculated from its concentrations of sodium, potassium, glucose, and urea, using the expression:

$$\text{Osmolality} = 2\,[\text{Sodium}] + [\text{Potassium}] + [\text{Urea}] + [\text{Glucose}]$$

where all the concentrations are in mmol/L.

SELF-CHECK 5.4

A specimen of plasma from a patient yielded the following results:

Sodium	140 mmol/L	(135–45)
Potassium	4.0 mmol/L	(3.5–5.0)
Urea	6.0 mmol/L	(3.5–6.6)
Glucose	4.5 mmol/L	(3.2–5.5)

Calculate the osmolality of this plasma.

Osmolar gap

In practice, the formula to calculate plasma osmolality is useful in determining the **osmolar gap**, which is the difference between the measured and the calculated osmolality. Determining the osmolar gap is of value in that it may be used to predict the presence of significant quantities of osmotically active toxic substances, such as the poisons ethanol, methanol, and ethylene glycol (antifreeze). Large, unexplained osmolar gaps are clinically and physiologically significant and often need to be investigated using specific analyses to identify the osmotically active component.

The osmolar gap can also be of value in the management of patients who have been treated with an osmotically active substance, such as mannitol. Mannitol is an osmotic diuretic used in a number of clinical scenarios. These include patients with oliguric renal failure to ensure diuresis, and patients in neurology critical care settings in whom it is necessary to reduce cerebral oedema. Mannitol can also be used to enhance the renal excretion rate in certain cases of drug overdoses.

Clinical laboratories are also required to measure the osmolality of a significant number of urine specimens, usually in following up abnormalities in plasma sodium test results. Unlike plasma osmolality, which has very little variation in observed values, urine osmolality values do vary significantly throughout each day. These variations arise as the output of urine in health reflects water intake: a high intake results in dilute urine of a low osmolality (less than 100 mOsm/kg). Conversely, a low water intake should lead to concentrated urine with a high osmolality that may increase to above 800 mOsm/kg (reference range 285–295). Osmotically active constituents of urine include sodium, potassium, their counter ions particularly chloride, and generally a significant contribution from urea.

Key Points

The osmolar gap is the difference between measured and calculated osmolality of a plasma sample. Its measurement is of value in that it can indicate the presence of a foreign substance for example ethanol.

5.6 Chloride and bicarbonate

The cations sodium and potassium are counterbalanced by a number of anions, most notably chloride, which circulates in plasma usually at a concentration of 95 to 105 mmol/L and bicarbonate, which is usually observed to be between 22 to 30 mmol/L. Anions such as proteins, lactate, sulphate, and phosphate provide the remaining anion balance. In most clinical situations, anion measurement provides little clinical information beyond that of simply measuring sodium and potassium. However, in some metabolic disturbances it is of value to determine the **anion gap**, which is calculated as the difference between the total concentrations of sodium and potassium and those of chloride and bicarbonate.

Calculating the anion gap is most useful in diagnosing a cause for metabolic acidosis. Renal tubular acidosis involves a reduction in the concentration of bicarbonate that is balanced by an increase in that of chloride, thus maintaining the anion gap within the reference range (15–20 mmol/L). In metabolic acidosis caused by the production of endogenous acids, lactate or ketones, or the intake of endogenous substances, such as ethanol, methanol, ethane 1,2-diol, or salicylate (aspirin), which have been metabolized to acids, calculating the anion gap will show an increased value and so direct further investigations to eliminate each of these potential causes.

Analytically it is essential *not* to delay the analysis of specimens for bicarbonate since it is in equilibrium with carbon dioxide dissolved in the blood, referred to as PCO_2. Consequently, if the specimen is left in contact with air carbon dioxide will be released, shifting the equilibrium such that the measured bicarbonate value will be reduced.

CASE STUDY 5.7

A 49-year-old man with myeloma was referred for management by the clinical haematology team who requested and observed the following biochemistry results analysed on the main automated biochemistry analyser (reference ranges are given in brackets):

Sodium	131 mmol/L	(135–145)
Potassium	3.6 mmol/L	(3.5–5.0)
Urea	5.6 mmol/L	(3.5–6.6)
Creatinine	124 µmol/L	(70–150)
Alkaline phosphatase	135 IU/L	(95–320)
Alanine aminotransferase	42 IU/L	(5–42)
Albumin	30 g/L	(35–50)
Total protein	140 g/L	(60–80)
Bilirubin	15 µmol/L	(<17)
Calcium	3.02 mmol/L	(2.12–2.62)

(a) What is the electrolyte abnormality in this patient?

(b) The sample was subsequently analysed on a direct reading ion-selective electrode for sodium and a result of 144 mmol/L was produced. Explain the difference between the sodium results obtained and state which is the more clinically relevant result.

SUMMARY

- Accurate and reproducible plasma electrolyte analyses are core functions of the hospital laboratory service.

- The plasma electrolytes, sodium and potassium are the most frequently requested investigations in most clinical biochemistry laboratories.

- Ion-selective electrodes are the main analytical technique used for analysis of sodium and potassium.

- Circulating concentrations of sodium and potassium are under tight homeostatic regulation.

- Two distinct types of ion-selective electrode, direct and indirect, are used to measure electrolyte concentrations and these may produce different results when used for specimens in which the total protein and/or lipid contents are increased.

- Poor collection and handling of clinical samples can adversely affect plasma potassium test results, usually due to its release from damaged erythrocytes.

- Abnormalities of water and electrolyte balances are common and can give rise to a wide range of clinical conditions.

- Serious disturbances to plasma electrolyte concentrations can be life threatening and rapid communication of abnormal test results by laboratory staff to clinical teams is essential.

- Any follow-up tests to abnormally disturbed electrolyte concentrations require prior clinical evaluation and for sodium disturbances involve an assessment of the patient's hydration status.

- Persistent electrolyte disturbances may require additional laboratory testing to identify specific abnormalities of renal and endocrine functions.

FURTHER READING

- **Adler SM and Verbalis JG (2006)** Disorders of body water homeostasis in critical illness. *Endocrinology and Metabolism Clinics of North America* **35**, 873–94.

- **Antunes-Rodirguez J, De Castro M, Elias LLK, Valenca MM, and McCann SM (2004)** Neuroendocrine control of body fluid metabolism. *Physiological Reviews* **84**,169–208.

- **Goh KP (2004)** Management of hyponatraemia. *American Family Physician* **69**, 2387–94.

- **Gross P (2008)** Treatment of hyponatraemia. *Internal Medicine* **47**, 885–91.

- **Lien YH and Shapiro JI (2007)** Hyponatraemia: clinical diagnosis and management. *American Journal of Medicine* **120**, 653–8.

- Loh JA and Verbalis JG (2008) **Disorders of water and salt metabolism associated with pituitary disease.** *Endocrinology and Metabolism Clinics of North America* **37**, 213–34.

- Reynolds RM, Padfield PL, and Seckl JR (2006) **Disorders of sodium balance.** *British Medical Journal* **332**, 702–5.

- Young WF (2007) **Primary aldosteronism: renaissance of a syndrome.** *Clinical Endocrinology* **66**, 607–18.

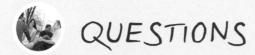

QUESTIONS

5.1 Which of the following may be associated with hyperkalaemia?

(a) Conn's syndrome

(b) Acute renal failure

(c) Diabetic ketoacidosis

(d) Treatment with thiazide diuretics

(e) Excessive laxative intake

5.2 Which of the following analytical techniques is/are commonly used to analyse electrolytes in clinical practice?

(a) Atomic absorption spectrophotometry

(b) High performance liquid chromatography

(c) Indirect ion-selective electrode

(d) Capillary electrophoresis

(e) Direct ion-selective electrode

5.3 Which of the following statements apply to antidiuretic hormone (ADH)?

(a) It has the alternative name of arginine vasopressin

(b) It activates receptors in the renal tubule to increase permeability to water leading to its conservation

(c) It is an octapeptide

(d) It is produced by the posterior pituitary

(e) It is associated with a 'pressor' effect

5.4 State whether the following statements are TRUE or FALSE

(a) Liquorice produces an effect similar to that of aldosterone and if ingested in excess can induce hypokalaemia

(b) Serum and lithium heparin plasma are both suitable specimens for the analysis of potassium, but the lithium heparin produces results that are greater in concentration than those for serum by approximately 0.2 mmol/L

(c) Aldosterone stimulates the reabsorptive exchange of sodium for calcium by the kidneys

5.5 What is meant by the term, pseudohyponatraemia?

5.6

(a) Calculate the osmolar gap for the sample analysed in Case study 5.1.

(b) Comment on your finding.

Answers to self-check questions, case study questions, and end-of-chapter questions are available in the Online Resource Centre accompanying this book.

 Go to www.oxfordtextbooks.co.uk/orc/ahmed/

Acid-base disorders

David Tierney

Learning objectives

After studying this chapter you should be able to:

- Define pH and relate it to hydrogen ion concentration
- Describe the significance of hydrogen ions and their physiological role
- Explain the different types and roles of buffer systems in acid-base homeostasis
- Discuss the roles of the renal and respiratory systems in acid-base balance
- Explain the clinical causes of acid-base disorders
- Apply the principles of hydrogen ion homeostasis to the interpretation of acid-base disorders
- Outline an alternative model of acid-base homeostasis

Introduction

The hydrogen ion (H^+) concentration in blood is maintained within narrow limits. Even small disturbances in the acid-base balance can severely affect physiological and biochemical processes. The body continuously produces enormous amounts of H^+ during metabolic processes. Complex homeostatic mechanisms have evolved to maintain acid-base balance, involving both the renal and respiratory systems. However, disturbances in acid-base balance are relatively common, occurring in a wide variety of disease conditions and can have potentially fatal consequences. The clinical laboratory plays a key role in the assessment of disorders of acid-base balance. This chapter will consider homeostasis of H^+ ions, the causes and consequences of acid-base disorders, and their laboratory investigation.

6.1 Hydrogen ion (H^+) and pH

The physiology and clinical significance of H^+ (or protons) contrasts with their relatively small concentration in the extracellular fluid (ECF), which is kept within tight limits (35–45 nmol/L). Table 6.1 shows the relatively low concentration of H^+ in comparison with other common

TABLE 6.1 The concentration of H^+ relative to other plasma cations.

Ion	Concentration in serum (mmol/L)
Na^+	140
K^+	4.5
Ca^{2+}	2.5
Mg^{2+}	0.8
H^+	0.00004

cations found in plasma. However, within nature H^+ are present over an enormous concentration range. Chemists are used to dealing with hydrogen concentrations varying by a factor of 10^{14}. In order to handle such a large, cumbersome range of concentrations, a Danish chemist, Soren Sorensen, devised the pH scale in 1909. The pH is calculated by taking the negative logarithm (to base 10) of the H^+ concentration:

$$pH = - \log [H^+] \text{ where } [H^+] \text{ is the hydrogen ion concentration.}$$

Sometimes the term H^+ activity is used instead of H^+ concentration; refer to Box 6.1 to differentiate between these terms.

TABLE 6.2 The relationship between pH and H^+ concentration.

To understand the definition of pH, see Box 6.2. The relationship between pH and H^+ concentration is emphasized in Table 6.2 and Figure 6.1. Examination of Figure 6.1 illustrates the logarithmic relationship over a scale to include the physiological reference range pH 7.35–7.45.

pH	H^+ (nmol/L)
0	1,000,000,000
1	100,000,000
2	10,000,000
3	1,000,000
4	100,000
5	10,000
6	1,000
7	100
8	10
9	1
10	0.1
11	0.01
12	0.001
13	0.0001
14	0.00001

SELF-CHECK 6.1

Which of the following three options involves the largest change in H^+ concentration: an increase in blood H^+ concentration of 80 nmol/L, a change in blood pH from 7.3 to 7.7, or a change in urine pH from 6.3 to 6.7?

Although both the pH scale and H^+ concentration are used in clinical laboratories, pH is the most commonly used notation. Both will be used in this text, as appropriate.

BOX 6.1 H^+ Activity or concentration?

More accurately, Sorensen defined pH as the *activity* of H^+, not the concentration. The 'p' is variously defined as the 'power' or 'potential' of H^+. In solutions containing other larger ions H^+ will become surrounded and shielded, reducing their activity and their ability to participate in chemical reactions. However, body fluids are relatively dilute and in effect the concentration and activity may be regarded as identical. Moreover, potentiometric measurement of blood pH measures activity and not H^+ concentration. Also, potentiometers can be calibrated against known H^+ concentrations, allowing compensation to be made within the calibration material for the effects of background ions.

BOX 6.2 Definition of pH

What does the negative logarithm to base 10 actually mean?

The log to power 10 is the power to which 10 must be raised to produce that number.

For example, $\log 100 = 10^2 = 2$, $\log 1,000 = 10^3 = 3$, $\log 10^7 = 7$, etc.

This means for every increase in pH of 1 there is a ten-fold increase in H^+ concentration.

The negative log gives an inverse ratio between pH and H^+ concentration. Therefore, the lower the H^+ concentration, the higher the pH value.

An example;

   if $[H^+] = 10^{-7}$ mol/L,

   then as $pH = -\log[H^+]$,

   therefore, $-\log[10^{-7}] = 7$, i.e. the $pH = 7$

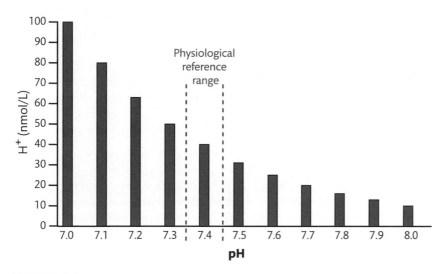

FIGURE 6.1

The relationship between pH and H^+ concentration. The logarithmic relationship means that for a pH change of 1, there is a ten-fold increase in H^+ concentration. The physiological reference range is highlighted to illustrate the narrow limits in which H^+ must be maintained in health.

Key Points

The pH scale is logarithmic, a change in pH of 1, for example a difference between a pH of 2 and a pH of 3, is equivalent to a ten-fold change in concentration. Similarly, a change of pH of 0.3 represents a doubling in concentration (consult Figure 6.1 for clarification). The pH scale is the inverse to the H^+ concentration; a low pH represents a high H^+ concentration. The pH scale is dimensionless, it has no units.

6.2 **Acids, bases, and buffers**

Acids and bases can be defined in several ways. The traditional approach used for clinical situations is the Bronsted-Lowry definition, which defines an **acid** as any compound that forms H^+ in water; they are proton donors. HA is an imaginary acid which dissociates in water to form H^+ and a conjugate **base** A^-. The acid is donating a proton to the solution.

$$HA \leftrightarrow H^+ + A^-$$

Acids are described as either strong or weak, depending on how easily the H^+ is able to dissociate from its conjugate base. A **strong acid** dissociates entirely to form H^+ and a base, with no recombination. **Weak acids** are only partially ionized in water and may recombine with the base until a chemical equilibrium is established. In the aqueous environment of the body H^+ combines with H_2O to form hydroxonium ions, H_3O^+, but for simplicity H^+ will only be used in this text. According to the Bronsted-Lowry definition a base is a compound which combines with H^+ in aqueous solution, they are proton acceptors. This is the reverse of the equation above with H^+ and base able to rejoin to form the original acid. A common cause of confusion is referring to an acid by the name of its anion, for example lactate and acetoacetate, instead of lactic and acetoacetic acids. At physiological pH these acids are almost fully dissociated, releasing H^+, resulting in the base and not the acid being the predominant form present. The Bronsted-Lowry definition of acids and bases is not the only definition; refer to Box 6.3 for an alternative definition.

A **buffer** is a weak acid in solution with its conjugate base. Consider the above equation again, as more H^+ are added they combine with the equivalent amount of conjugate base (A^-) to form more HA. Buffering replaces strong acids with weak acids, reducing the number of available H^+. Buffers have been called 'proton sponges', mopping up any excess H^+ that have been added to a solution and releasing them when H^+ are in short supply. Buffers play a major role in controlling acid-base balance in the body.

An experiment that reinforces how effective buffering is *in vivo* involved infusing the equivalent of 14 mmol of H^+, per L of body water into experimental dogs; the equivalent of 14,000,000 nmol/L of acid. However, the resulting measurable increase in plasma H^+ concentration was only 36 nmol/L, an increase of only 0.0026%. This experiment dramatically illustrates the huge buffering capacity that is available to absorb the infused H^+.

BOX 6.3 *Alternative to the Bronsted-Lowry definition of acids and bases*

There are several different definitions of acids and bases.

Bronsted, a Danish chemist and Lowry, an English chemist, worked independently to simultaneously introduce the definition by which they are known. Their definition differed from previous ones, such as that of Arrhenius, who defined acids and bases as compounds that produce H^+ and OH^- respectively in water. The Bronsted-Lowry theory has proven useful in clinical situations and for chemists in describing acid-base reactions in non-aqueous solvents.

SELF-CHECK 6.2

Which of the following compounds can act as an acid when dissolved in water: HCl, H_2CO_3, $H_2PO_4^-$, NH_4^+?

Henderson-Hasselbalch equation

We cannot avoid discussing buffer systems without mentioning the Henderson-Hasselbalch equation, the derivation of which is shown in Box 6.4. This equation relates pH to the components of any buffer pairs.

The Henderson-Hasselbalch equation helps explain the relative roles of the body's buffering compensatory mechanisms in regulating H^+ concentration during acid-base disturbances. The pH of a buffer system can be represented as:

$$pH \propto [Base] / [Acid]$$

That is, the pH depends on the ratio of concentration of the base to acid in the buffer pair. This is useful when discussing disturbances of acid-base balance.

The pK_a value expressed in the Henderson-Hasselbalch equation is significant as it indicates the relative strength of an acid, that is, the ease with which an acid is able to dissociate and release H^+. The lower the pK_a value the stronger the acid. At constant temperature and ionic strength, the pK_a value is a constant and in a similar way to pH the value of pK_a can vary over

BOX 6.4 Derivation of the Henderson-Hasselbalch equation

Consider again the equation showing the equilibrium between an acid (HA) and its products, A^- and H^+:

$$HA \rightleftharpoons H^+ + A^-$$

The relative concentrations at equilibrium can be summarized by the expression:

$$K_a = [H^+][A^-]/[HA]$$

where K_a is the dissociation constant of the acid, HA and $[H^+]$, $[A^-]$ and $[HA]$ are the respective molar concentrations.

Rearranging this expression gives:

$$[H^+] = K_a [HA]/[A^-]$$

Taking the negative logarithm of both sides gives:

$$-\log[H^+] = -\log K_a - \log([HA]/[A^-])$$

Remember the definition of pH is $-\log[H^+]$

Similarly, we can define $-\log K_a$ as pK_a, which may be regarded as the 'potential' of the acid dissociation constant.

Thus, $pH = pK_a - \log([HA]/[A^-])$

And this equation may be simply rearranged to give the usual form of the Henderson-Hasselbalch equation:

$$pH = pK_a + \log([A^-]/[HA])$$

Hence, the Henderson-Hasselbalch equation describes the relationship between the relative concentrations of an acid and its conjugate base to the pH of a solution.

several magnitudes of 10. The pK_a is also significant as the pH value at which the buffer acid and base are present in equal concentrations (that is, when $pH = pK_a$, then $[HA] = [A^-]$). Its value is also useful in buffering systems, as a buffer is most effective when the desired pH of the solution is equal to or +/−1.0 each side of the pK_a value. For example, if a buffer is required to maintain a solution at pH 8.4, the ideal buffer pair will have a pK_a value between pH 7.4–9.4, but as close to pH 8.4 as possible.

SELF-CHECK 6.3

The NH_3: NH_4^+ buffer pair has a pKa of 9.2. How effective would this buffer pair be in a urine solution at pH 6.2, and what information can you infer about the relative concentrations of NH_3 and NH_4^+ at pH 6.2?

Key Points

Acids donate H^+ in aqueous solution, bases accept H^+. Acids can be strong or weak. Buffers resist changes in pH; they are weak acids in solution with their conjugate bases. The Henderson-Hasselbalch equation describes the relationship between the relative concentrations of an acid and its conjugate base to the pH of a solution.

6.3 **Physiological role of H$^+$**

Why is the maintenance of H^+ concentration so important and why do relatively small changes in H^+ concentration have such a profound effect on our health? The H^+ concentration can affect the body's processes in several ways. Most proteins contain ionizable side chains. Changes in the H^+ concentration surrounding these groups can alter their charge, affecting their structure and therefore their biological function. These functions may be structural, transport, hormonal, enzymatic, or receptor. An extreme change in H^+ concentration may result in protein denaturation, resulting in total loss of biological activity. Despite the need for tight homeostatic control, the specific H^+ concentration required for the optimum activity of individual enzymes can vary enormously. For example, the optimum enzymatic activity of pepsin is between pH 1.5–3, suitable for the highly acidic environment of the stomach. In comparison, the enzyme alkaline phosphatase has an activity optimum at pH 10.

Variations in pH may adversely affect the biological activity of certain cations, such as calcium and magnesium. These cations are often found bound to proteins where they remain in a biologically inactive state. Increased H^+ concentration can result in the release of these cations from proteins, increasing the concentration of the biologically active form.

In addition, most metabolic processes occur within organelles and compartments within cells. Using a variety of techniques, the intracellular pH within cell compartments has been found to vary significantly, within skeletal cells the pH between compartments has been found to vary by 0.5. By influencing the ionization state of a range of both large and small molecules, H^+ are able to alter the ability of these molecules to pass across cell membranes, in effect they can compartmentalize the molecules within cells and between the intra- and extracellular compartments.

> ### Key Points
> Despite its small size and relatively low concentration H^+ can affect physiological processes by: altering protein structure and function, affecting the concentrations of biologically active cations, and by influencing the movement of molecules within cells.

6.4 Production of surplus H⁺

Most H^+ are produced by **aerobic** and **anaerobic** metabolism. A relatively small amount of acid may be ingested. All these acids are either metabolized or excreted. There are two main sources of acids, conventionally defined as **metabolic acids** or **respiratory acids**.

Metabolic acids are derived from three main sources:

- anaerobic metabolism of glucose to lactate and pyruvate
- anaerobic metabolism of fatty acids
- oxidation of sulphur-containing amino acids (cysteine and methionine) and cationic amino acids (arginine and lysine)

From these sources, our cells release 40–60 mmol of H^+ into the extracellular fluid per day (some references cite even higher amounts). A simple calculation shows that if all the acid produced by these sources was not neutralized and excreted, but was instead diluted into the 14 L of ECF present in an 'average' human body, then the concentration of H^+ in the blood would be approximately 90,000 times greater than normal, resulting in a plasma pH of less than 3.

Respiratory acids are derived from the 12–20 moles of CO_2 produced daily (the exact amount is dependent on the body's level of physical activity). Although CO_2 does not contain H^+, it reacts rapidly with water to form carbonic acid (H_2CO_3), which subsequently dissociates to produce bicarbonate (HCO_3^-) and H^+:

$$CO_2 + H_2O \leftrightarrow H_2CO_3 \leftrightarrow HCO_3^- + H^+$$

6.5 Hydrogen ion homeostasis

Despite the continuous production of H^+ by the normal process of metabolism, the body is able to maintain H^+ concentration within the reference range using three interrelated mechanisms; *buffering systems*, the *respiratory system*, and the *renal system*. It is beneficial to begin by summarizing the main features of these complex processes in advance:

- *Intracellular and extracellular buffering systems* provide an immediate, but limited, response to pH changes.

- The *respiratory system*, which can be activated almost immediately, controls PCO_2 by changing alveolar ventilation (PCO_2 is the partial pressure of CO_2 and refers to the amount of CO_2 dissolved in the blood).

- The *renal system* is the slowest to respond. However, its biological significance lies in the fact that it is the only system able to eliminate buffered H^+. Unlike the buffering and respiratory systems, H^+ is actually removed from the body simultaneously regulating plasma HCO_3^- concentration.

The buffer systems

Buffer systems are the first line of defence within blood and body tissues; they produce a rapid response and provide the body with huge reserves of **buffering capacity**. Buffering capacity is a term used to describe the ability of a buffer to resist changes in pH. It is proportional to the concentration of buffer present and the closeness of the pK_a of the buffer to the pH of the solution. In the body a buffer's effectiveness is also dependent on its availability within different cells, organs, and body fluids. Several different buffer systems exist in the body, including the bicarbonate-carbonic acid system, proteins, and phosphate.

The **bicarbonate-carbonic acid buffer system** is the major buffer system of the plasma, accounting for over 70% of the total plasma buffering capacity. It is also present at lower concentrations in erythrocytes. However, it is not an efficient system. Applying the Henderson-Hasselbalch equation to the bicarbonate-carbonic acid system:

$$pH = 6.1 + \log \left([HCO_3^-] / [H_2CO_3] \right)$$

where the pK_a for the bicarbonate-carbonic acid system is 6.1. In practice it is difficult to measure the concentration of H_2CO_3 in blood in a routine clinical laboratory. However, PCO_2 levels are in equilibrium with H_2CO_3 and are relatively simple to measure. Therefore the above can be expressed as:

$$pH \, \alpha \, [HCO_3^-] / PCO_2$$

As discussed previously an 'ideal' buffer should have a ratio [base]: [acid] of 1:1 and a pK_a close to the pH of the solution, which for plasma at physiological pH 7.4 would be pK_a 6.4–8.4. The bicarbonate: carbonic acid ratio in blood at physiological pH is 20:1 and the pK_a value is 6.1, both significantly different from the 'ideal'. However, two factors contribute in making the bicarbonate: carbonic acid pair effective in blood at physiological pH:

- It is present in high concentrations in blood.
- It is an *open system*, that is, it can be regulated by two mechanisms: by the excretion of CO_2 via the lungs and by the regulation of the rate of reclamation of HCO_3^- in the renal tubules. This unique property allows the body to produce and eliminate H_2CO_3 by two independent mechanisms:

$$CO_2 + H_2O \leftrightarrow H_2CO_3 \leftrightarrow HCO_3^- + H^+$$

To illustrate this relationship; any imbalance causing an accumulation of CO_2 will shift the equilibrium to the right, causing increased concentrations of HCO_3^- and H^+, both of which can be regulated by renal mechanisms. Similarly, an increase in the concentration of HCO_3^- and H^+ will cause a shift in equilibrium to the left, where CO_2 can be eliminated through the lungs.

Proteins, especially albumin, account for 95% of the non-bicarbonate buffering capacity of plasma. They act as buffers in both the intracellular and extracellular fluids. Albumin behaves as a weak acid, due to its high concentration of negatively charged amino acids (mainly carboxyl and aspartic acids). In addition, albumin contains 16 histidine residues whose imidazole groups, which have a pK_a value of 6.8, can react with H^+ and are the most important buffer groups of proteins at physiological pH. Histidine groups are also prevalent in haemoglobin within erythrocytes.

Phosphate occurs in both organic and inorganic forms within the body. At physiological pH 7.4 most phosphate within plasma and the initial part of the glomerular filtrate, exists as mono hydrogen phosphate (HPO_4^{2-}). This form of phosphate can accept H^+ to form dihydrogen phosphate ($H_2PO_4^-$). The relationship of this buffer pair is as follows:

$$pH = 6.8 + \log[HPO_4^{2-}]/[H_2PO_4^-]$$

As the pK_a of this buffer pair is 6.8, which is relatively close to the pH of glomerular filtrate, it is an effective buffer. Due to its relatively low concentration, phosphate tends to be a minor component of extracellular buffering, but it plays a significant contribution in buffering urine.

High concentrations of phosphate are prevalent in bone and the intracellular fluid. This is significant in some *acidotic* (low plasma pH) states when phosphate can be released from the bones and act as a buffer in the plasma. The relative distribution of the main buffer system between the body compartments is shown in Figure 6.2.

SELF-CHECK 6.4

The HCO_3^- : H_2CO_3 buffer system is the major buffer system in plasma. Explain how plasma does not become saturated with H_2CO_3 caused by the continuous production of large volumes of CO_2.

Urinary buffers

In comparison with plasma, urine pH may vary by a factor of a thousand-fold: pH 4.8–7.8. Urinary buffers are required in acid-base homeostasis as they provide the major mechanism by which H^+ ions are excreted from the body and are essential in the generation of HCO_3^-. Bicarbonate, ammonia, and phosphate are all involved in urine buffering mechanisms.

The ammonium ion (NH_4^+) is generated in renal tubular cells by the hydrolysis of the amino acid glutamine to glutamate, catalysed by the enzyme glutaminase. Further NH_4^+ is produced by the subsequent metabolism of glutamate to 2-oxoglutarate, which is further metabolized

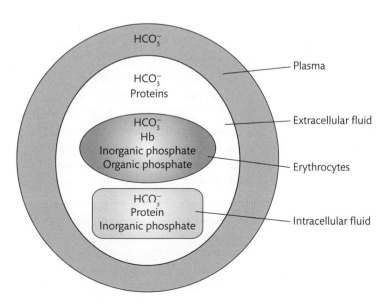

FIGURE 6.2
Schematic showing the relative distribution of the main buffer systems in the extracellular fluid, the intracellular fluid, erythrocytes, and plasma.

to glucose by the process of gluconeogenesis. Remember NH_4^+ is a weak acid and dissociates as follows;

$$NH_4^+ \leftrightarrow NH_3 + H^+$$

$$pH = 9.2 + \log [NH_3] / [NH_4^+]$$

If you look at the equation above, the pK_a of this buffer pair is 9.2. Therefore, at a physiological pH of 7.4 the equilibrium is in favour of the formation of NH_4^+, with a ratio of $[NH_3]$: $[NH_4^+]$ of 1:100. However, the cell membranes are relatively impermeable to NH_4^+, but they allow the diffusion of gaseous NH_3 into the tubular lumen. The continuous diffusion of NH_3 results in a steady dissociation of NH_4^+ within the tubular cells. The resulting H^+ is consumed in the conversion of 2-oxoglutarate to glucose. Within the acidic environment of the tubular lumen, the equilibrium is shifted in favour of the formation of NH_4^+. Due to its relative difficulty in passing back into the tubular cells, NH_4^+ is effectively trapped in the lumen and excreted in the urine. By this mechanism an excess of H^+ is eliminated from the body. The NH_4^+ accounts for over half the H^+ excretion derived from metabolic acids.

The phosphate buffer pair, described previously, is an effective urinary buffer, having a pK_a comparable with the pH of the glomerular filtrate and is present in relatively high concentrations in the tubular lumen.

Respiratory control of acid-base balance

The production of CO_2 by normal metabolism is the major source of H^+ *in vivo* and therefore the respiratory system plays a key role in the maintenance of acid-base balance. Haemoglobin has the property of binding with both H^+ and CO_2 with high affinity. The deoxygenated form of haemoglobin (Hb), as found in the body's tissues, has the strongest affinity for both CO_2 and H^+. In addition, the relatively high concentration of haemoglobin in the erythrocytes makes it an effective buffer in the blood, a significant feature in the respiratory control of acid-base homeostasis. Refer to Figure 6.3 (a) and (b) and consider the journey of a haemoglobin molecule from the tissues to the lungs:

In the tissues

- Erythrocytes are anaerobic, they produce little CO_2, therefore the CO_2 concentration in erythrocytes is low relative to the body tissues. The CO_2 produced in tissues diffuses freely down its concentration gradient into the ECF and subsequently into the erythrocytes.

- Some of this CO_2 (approximately 10%) remains dissolved as a gas, 20% of the CO_2 binds with proteins, mainly haemoglobin, to form carbamino compounds. However, most of the CO_2 combines with water in the aqueous environment of the erythrocytes, to form H_2CO_3, the reaction being catalysed by the enzyme carbonic anhydrase.

- Subsequent dissociation of H_2CO_3 results in the production of H^+ and HCO_3^-.

- As the HCO_3^- concentration rises within the erythrocyte, HCO_3^- passes into plasma along the concentration gradient, caused by the relative differences in concentration. In exchange, chloride diffuses in the opposite direction to maintain electrical neutrality (the **chloride shift**).

- The H^+ produced binds to the deoxygenated haemoglobin.

The net result is that the H^+ are buffered and HCO_3^- is produced and released into the plasma buffer pool.

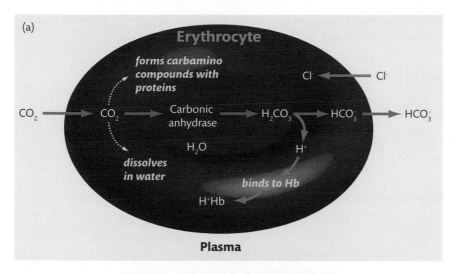

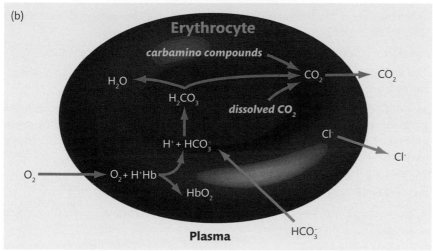

FIGURE 6.3
(a) The transport of CO_2 into the erythrocytes and the buffering action of haemoglobin in the peripheral tissue. (b) Uptake of oxygen and release of CO_2 in the lungs.

In the lungs

- On reaching the lungs, oxygen diffuses into the erythrocytes and forms oxyhaemoglobin, releasing the bound H^+. This is illustrated in Figure 6.3 (b).

- In a reverse process to that which occurred in the tissues, the released H^+ bind with HCO_3^- to form H_2CO_3, which subsequently dissociates to form CO_2 and H_2O. The dissociation of H_2CO_3 is catalysed by carbonic anhydrase, but this time in the reverse direction, reflecting its ability to catalyse the reaction in either direction, according to the relative concentration of the reactants and products.

- The released CO_2 enters the lungs where it is exhaled.

The amount of CO_2 eliminated from the lungs is determined by the rate of respiration, which is controlled by chemoreceptors in the medulla and in the carotid and aortic bodies. These receptors respond to changes in both the H^+ concentration and PCO_2 in plasma and cerebrospinal fluid (CSF). It should be noted that this mechanism relies on the efficient functioning of alveolar **ventilation**. Decreased ventilation efficiency may result in CO_2 retention within the blood and a corresponding increase in H^+ concentration. Even a relatively modest decrease in efficiency in the elimination of CO_2 may result in a significant decrease in pH.

Renal control of acid-base balance

The kidneys also play a major role in controlling acid-base homeostasis through their ability to recover filtered HCO_3^- and to generate HCO_3^-. It is during HCO_3^- generation that H^+ ions are excreted. The basic aspects of these mechanisms are identical, but for clarity are dealt with separately below.

Bicarbonate recovery

- HCO_3^- is freely filtered by the glomerulus into the tubular lumen, it cannot pass back as the lumen walls are impermeable to HCO_3^-. However, the body needs to reabsorb this filtered bicarbonate otherwise, the stock of HCO_3^- would be rapidly depleted and its buffering capacity severely compromised.

- Simultaneously, H^+ is actively secreted into the lumen, in exchange for sodium to maintain electrical neutrality, by a mechanism that requires metabolic energy.

- The H^+ combine with the filtered HCO_3^- in the tubular lumen to produce H_2CO_3.

- The H_2CO_3 formed subsequently dissociates to form CO_2 and H_2O, a reaction catalysed by carbonic anhydrase within the brush border of the proximal tubular cells.

- As the CO_2 concentration in the lumen increases, it is able to diffuse along its concentration gradient into the renal tubular cells, where it combines with H_2O (catalysed by carbonic anhydrase), resulting in the formation of HCO_3^- which diffuses back into the ECF to return to the buffer pool.

- The H^+ formed is actively secreted into the tubular lumen in exchange for sodium and the cycle continues again as illustrated in Figure 6.4.

- Approximately 90% of the filtered HCO_3^- is absorbed in the proximal tubule, the remainder is absorbed in the distal tubules and collecting ducts.

By this mechanism HCO_3^- is not lost.

It should be noted that evidence from *in vivo* and *in vitro* experiments with animal models indicates that there is active secretion of HCO_3^- into the cortical collecting tubules. The physiological role of this mechanism in humans is unclear, although experiments indicate that it may play a role in the compensation of metabolic alkalosis.

Bicarbonate generation

This mechanism is identical to the above bicarbonate recovery mechanism, the significant difference being that *there is a net loss of H^+* from the body during this process. Compare Figure 6.5 to Figure 6.4 to understand the difference.

This mechanism requires the presence of other filtered buffer bases, which, unlike HCO_3^-, are lost in the urine. These urinary buffers, predominantly phosphate and ammonia, have been previously described. Phosphate, as HPO_4^{2-}, is filtered by the glomerulus, and NH_3 diffuses into the tubular lumen, where they both react with H^+ to form $H_2PO_4^-$ and NH_4^+ respectively, which are subsequently excreted in the urine. For every H^+ excreted, HCO_3^- is also *generated* which is returned into the ECF.

By this mechanism HCO_3^- is generated.

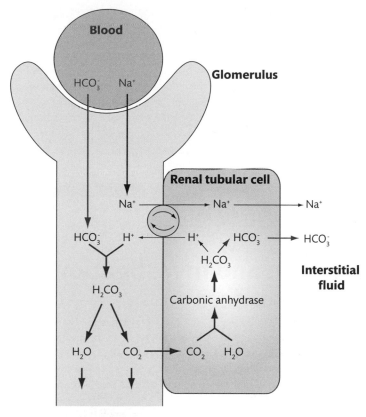

FIGURE 6.4
The reclamation of HCO_3^- by renal tubular cells. By this method most of the filtered HCO_3^- is reclaimed in the proximal tubule.

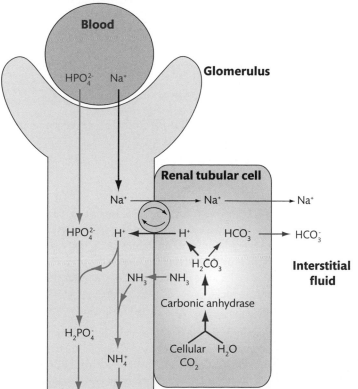

FIGURE 6.5
The generation of HCO_3^- by renal tubular cells. By this method H^+ is excreted into the tubular lumen with the simultaneous generation of HCO_3^-. The H^+ is buffered either by HPO_4^{2-} or NH_3. Ammonia is produced in renal tubular cells by the hydrolysis of the amino acid glutamine to glutamate, catalysed by the enzyme glutaminase. The NH_3 reacts with H^+ to form NH_4^+ that cannot re-enter the renal tubular cells and is lost in urine. This results in the excretion of H^+.

Key Points

Metabolic acids (that is H⁺) and respiratory acid (that is CO_2) are continuously produced in the body by cellular metabolism. To maintain acid-base balance the body has two systems to remove these acids: the renal system and the respiratory system. These systems are linked by the bicarbonate-carbonic acid buffer system. Excessive production of respiratory acid results in its conversion to H⁺ (and HCO_3^-) which is eliminated by the kidneys. Similarly, excess H⁺ is converted into respiratory acid, which is eliminated by the lungs.

SELF-CHECK 6.5

What is the role of carbonic anhydrase in acid-base balance?

6.6 Classification and investigation of acid-base disorders

Disturbances of acid-base metabolism may result in either excess H⁺ in the arterial blood— **acidaemia**, when the resulting pH is less than 7.35, or too few H⁺ —**alkalaemia**, with a pH greater than 7.45. The terms **acidosis** and **alkalosis** are clinical terms used to describe the underlying process causing the acidaemia or alkalaemia. It is possible that alkalosis and acidosis may co-exist, giving rise to a mixed acid-base disturbance, which may present with an arterial pH that is increased, decreased, or within the reference range. The causes of these disturbances may be due to **respiratory** disorders, with a primary change in PCO_2, or **metabolic**, also referred to as non-respiratory disorders, which initially cause changes in the concentration of HCO_3^-.

Compensation

Compensation is the physiological response to any acid-base disturbance, which tends to limit the change in pH caused by the primary disturbance.

Consider the equation:

$$\text{pH} \; \alpha \; [HCO_3^-] / PCO_2$$

Any imbalance in the equilibrium of this buffer pair causes an increase or decrease of either PCO_2 or HCO_3^- concentration, resulting in an initial compensatory response to restore the pH to normal. If the initial underlying disturbance is due to a respiratory cause, that is a change in PCO_2, then the compensatory mechanism is by a renal route, a *metabolic compensation*. This results in a change in HCO_3^- concentration *in the same direction* as PCO_2 caused by the primary disturbance. Similarly, an underlying metabolic disorder causing a change in HCO_3^- concentration results in a *respiratory compensation*, by changing PCO_2 in the same direction.

Consider an example: an acid-base disturbance has resulted in an acidosis, therefore the pH is lower than the reference range. If the underlying cause is a *metabolic* acidosis, there must have been a decrease in HCO_3^- concentration resulting in a decrease in pH. The compensatory response will be respiratory, by altering the PCO_2 *in the same direction* (that is decreasing) to bring the pH back to normal. Details as to how this is achieved will be discussed in the relevant sections.

Respiratory compensation tends to be relatively quick, within minutes or hours. However, the renal compensatory response may take 12–24 hours to be effective, possibly not reaching a peak until after 48–72 hours. *Complete compensation* is said to occur when the pH is returned within the normal reference range, or just outside it. However, the PCO_2 and/or HCO_3^- concentrations remain abnormal. Between the initial and complete compensation there is *partial compensation* when the H^+ concentration has not fully returned to normal.

Key Points

Compensation is the physiological process whereby the body attempts to restore the blood pH to normal reference values during an acid-base disturbance, by respiratory and renal mechanisms.

SELF-CHECK 6.6

An acid-base disorder has resulted in an acute increase in blood PCO_2 levels above the reference range. How would this affect the blood pH and what mechanism would you expect to compensate for this imbalance?

Reference ranges and units

Although there is some variation between laboratories, 'typical' reference range values used in acid-base analyses are shown in Table 6.3. It was previously mentioned that pH notation is used in preference to H^+ concentration. The units used to describe PCO_2 and PO_2 (the partial pressure of oxygen) are expressed in kilopascals (kPa) and in mmHg, as both units are still widely used.

Laboratory investigation of acid-base disorders

The clinical laboratory plays an essential role in the diagnosis and management of acid-base disorders. However, any laboratory assessment of acid-base balance should be considered in relation to the clinical history and physical examination of the patient. Acid-base status

TABLE 6.3 Typical reference range values for acid-base parameters and PO_2.

Parameter	Reference range
pH	7.35–7.45
[H^+]	35–45 nmol/L
PCO_2	4.7–6.0 kPa (35–45 mm Hg)
PO_2	12–14.6 kPa (90–110 mm Hg)
Serum [HCO_3^-]	24–29 mmol/L

can be fully characterized by the three parameters defined in the Henderson-Hasselbalch equation: pH, PCO_2, and plasma HCO_3^- concentration. If two of these parameters are known, the third can be calculated. Modern blood gas analysers, as shown in Figure 6.6, measure the basic parameters, pH and PCO_2, and are able to calculate the HCO_3^- concentration using the Henderson-Hasselbalch equation. Alternatively, HCO_3^- concentration may be directly measured as 'total bicarbonate' as part of a routine plasma analysis. In addition, two other parameters, **base excess** and **standard bicarbonate** are also calculated from the basic measured parameters and are used to determine *the metabolic component* of any acid-base disorder.

Standard bicarbonate is defined as the calculated concentration of bicarbonate in a blood sample corrected to a PCO_2 of 5.3kPa (40 mmHg). That is, it is the expected bicarbonate concentration if the PCO_2 is 'normal'. The base excess is defined as the number of mmoles of acid required to restore 1 L of blood, *in vitro*, maintained at a PCO_2 of 5.3 kPa (40 mmHg), back to pH 7.4. In a metabolic alkalosis, acid would have to be added, whereas a metabolic acidosis would require the removal of acid. The latter is called a **base deficit** and would be reported as a negative value. Calculated parameters remove the effects of PCO_2 on pH change. Abnormal standard bicarbonate or base excess values are used as an indication of the metabolic disturbances of the acid-base imbalance. A decreased standard bicarbonate value indicates a metabolic acidosis, whereas an increased value indicates metabolic alkalosis. In practice, however, clinical examination and the measurement of pH, HCO_3^-, and PCO_2 are usually sufficient for the interpretation and classification of acid-base disorders.

Key Points

Blood pH and PCO_2 are routinely measured in clinical laboratories. From these parameters HCO_3^- concentration can be calculated using the Henderson-Hasselbalch equation.

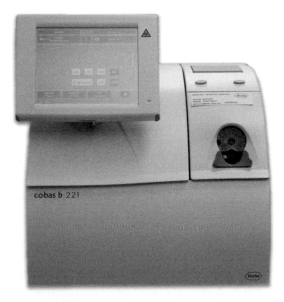

cobas b 221

FIGURE 6.6
Photograph of a blood gas analyser. Courtesy of Department of Clinical Biochemistry, Manchester Royal Infirmary, UK.

> ## Key Points
> Base excess and standard bicarbonate are calculated parameters that quantify the metabolic component of an acid-base disorder.

Specimen collection and transport

As with any patient's sample for laboratory analysis, a correct specimen must be taken and stored in an appropriate container and be delivered to the laboratory in a timely way. This is essential for samples taken for acid-base analyses. The sample must be:

- *Arterial* blood or *arterialized capillary* blood; the latter is not easy to obtain as it needs an arterial cannula *in situ*.
- *Heparinized*, to prevent clotting, but only using a small amount of heparin as heparin is acidic and an excess can dilute the sample and potentially cause haemolysis.
- *Well mixed*, but *free from bubbles*. Any air bubbles will increase the PO_2 and decrease the PCO_2 levels.
- *Chilled* or delivered to the laboratory without delay, as chilling reduces glycolysis and the production of lactate.

6.7 Metabolic acid-base disorders

Metabolic acid-base disorders are characterized by changes in the concentration of HCO_3^- in the ECF, which is a primary disturbance. This usually occurs due to build up or loss of H^+ in the body or direct gain or loss of HCO_3^- itself. These disorders can cause an acidosis or alkalosis.

Metabolic acidosis

The underlying problem in metabolic acidosis is a relative excess of H^+ which accumulates in the ECF, associated with a decrease in plasma HCO_3^- concentration. The decrease in plasma HCO_3^- may be due to a loss in HCO_3^- from the body, or caused by its reaction with increased levels of H^+, as part of the compensatory response. The decreased HCO_3^- concentration lowers the $[HCO_3^-]\,/\,PCO_2$ ratio, causing a decreased plasma pH.

> ## Key Points
> Regardless of their clinical cause, metabolic acidoses are always associated with a decrease in plasma HCO_3^- concentration.

Clinical causes

The causes of the relative H^+ excess may be broadly classified as due to an increased production or ingestion of H^+, a decreased excretion of H^+, or an increased loss of HCO_3^-.

Increased H^+ production may be due to an increased production of organic acids, for example as occurs in **lactic acidosis**. Lactate, an intermediate product formed during the anaerobic metabolism of carbohydrates, is catabolized in the liver to produce carbon dioxide and water. However, in circumstances causing increased production or decreased catabolism, lactate may accumulate. This is most commonly due to circulatory or respiratory insufficiency, resulting in poor tissue perfusion. Lactic acidosis may also occur in severe infections and some forms of leukaemia due to increased carbohydrate metabolism. The lactic acidosis associated with liver disease, some drugs, or as a result of inherited metabolic disorders is due to decreased hepatic catabolism.

Metabolic acidosis may also be caused by the accumulation of acetoacetate and β-hydroxy-butyrate (the so-called **ketone bodies**). This occurs in patients with diabetic ketoacidosis, due to a lack of insulin, or the lack of response of insulin receptor cells, the body is unable to metabolize glucose, resulting in the catabolism of lipids as an alternative energy source. **Ketoacidosis** is also associated with alcoholism and prolonged fasting. The ingestion of toxic acids is relatively rare and usually accidental. When ingested, ethylene glycol (antifreeze), results in the formation of the metabolites oxalate and glycolate. Ingestion of ethylene glycol may result in acute renal failure, due to the precipitation of calcium oxalate and hippurate in the urinary tract. Methanol is metabolized by a similar pathway to ethylene glycol, resulting in the formation of formaldehyde and formate. The optic nerve atrophy and neurological symptoms associated with methanol intoxication are attributed to the accumulation of formate. Metabolic acidosis associated with excessive ingestion of salicylate is caused by the production of anionic metabolites.

Decreased H^+ excretion may occur in any form of renal failure, resulting in metabolic acidosis. Renal failure due to glomerular dysfunction can decrease glomerular filtration, causing a reduction in the availability of sodium necessary for exchange with H^+ in the tubular lumen. Also, decreased glomerular filtration can reduce the availability of phosphate and other urinary buffer anions, which are essential for H^+ excretion. Hence, renal failure results in less H^+ being excreted and a consequent reduction in HCO_3^- generation. The **renal tubular acidoses** (RTA) are a diverse range of disorders causing metabolic acidosis due to abnormal H^+ secretion in the distal tubules (type 1), impaired HCO_3^- resorption in the proximal tubules (type 2), or aldosterone deficiency or impaired response to aldosterone in the distal tubule (type 4). Type 4 RTA occurs with **hypoaldosteronism**, as aldosterone is essential for the reabsorption of sodium in the distal tubule, a deficiency results in reduced sodium /H^+ exchange, so less H^+ are secreted.

Increased loss of HCO_3^- due to excessive loss of gastrointestinal fluids or increased renal excretion may result in a decrease in HCO_3^- buffering capacity, resulting in metabolic acidosis. Secretions into the intestinal tract contain relatively high concentrations of HCO_3^- (40–60 mmol/L), which is required to neutralize gastric contents as they enter the small intestine. Excessive loss of intestinal secretions, with a consequent loss of HCO_3^-, may result from several causes, including: prolonged diarrhoea, small bowel loss (such as in Crohn's disease), or cholera. Renal excretion of HCO_3^- may occur following the administration of acetazolamide, a carbonic anhydrase inhibitor, which slows the conversion of H_2CO_3 to CO_2 and water in the renal tubules. This leads to an increased urinary loss of H_2CO_3, with an associated decrease in the proximal reclamation of HCO_3^-. Type 2 RTA also results in the inability of the proximal tubule to reabsorb HCO_3^-. Refer to Table 6.4 for a summary of the causes of metabolic acidosis.

SELF-CHECK 6.7

Describe how chronic diarrhoea can affect plasma pH.

TABLE 6.4 Summary of the clinical causes of metabolic acidosis.

Primary cause	Mechanism
Increased H^+ production	
Lactic acidosis (sepsis, hypoxia, leukaemia, liver disease)	Increased [lactate] and [pyruvate] due to the anaerobic metabolism of glucose
Ketoacidosis (diabetes, starvation, alcoholism)	Increased acetoacetate and β-hydroxybutyrate due to anaerobic metabolism of lipids
Ingestion of H^+	
Ingestion of acid (rare)	
Ingestion of toxic substances (salicylate, methanol, ethylene glycol)	These are metabolized to toxic anionic products
Decreased H^+ excretion	
Renal failure—glomerular	Decreases available sodium (required for exchange with H^+) and phosphate (required to bind with, and excrete H^+)
Renal failure—tubular	Renal tubular acidoses, various mechanisms
Increased loss of HCO_3^-	
Gastrointestinal loss (prolonged diarrhoea, Crohn's disease)	Loss of fluids with high HCO_3^- concentration
Renal loss	
acetazolamide	Carbonic anhydrase inhibitor
RTA Type 2	Inability to reabsorb HCO_3^- in proximal tubule

Compensation

The compensatory response in metabolic acidosis is to try and restore the $[HCO_3^-]$ / PCO_2 ratio to normal values. This is initially achieved by reducing the PCO_2. The slower renal response also acts to increase the HCO_3^- concentration. Buffering, mainly the bicarbonate: carbonic acid buffer system in blood, provides the initial response. This is relatively rapid, the excess H^+ being removed by the reaction with HCO_3^- to form H_2CO_3. Stimulated by the effects of the increased H^+ concentration on the respiratory centre, the respiratory compensatory mechanism responds to decreased PCO_2 by causing deep gasping breaths, a form of **hyperventilation** called **Kussmaul respiration**. This response lowers the PCO_2 in the blood, increasing the $[HCO_3^-]$ / PCO_2 ratio. However, increasing ventilation has only a limited ability to lower PCO_2. After a delay, the renal compensatory mechanism responds by increasing HCO_3^- reabsorption and increasing the excretion of H^+. This is accomplished by increasing sodium/H^+ exchange and by increasing NH_3 production, the main mechanism for removing excess H^+.

Key Points

The respiratory compensatory response to metabolic acidosis is rapid, but has a limited effect. The slower renal response compensates by increasing H^+ excretion and HCO_3^- reclamation.

Clinical findings and management

Due to the compensatory need to lower PCO_2, increased ventilation may occur. Mild to moderate acidosis may result in an increased secretion of catecholamines, which can result in neuromuscular irritability, with possible arrhythmia and tachycardia. Due to the effects of catecholamines on the peripheral circulation the patient may present with pale, clammy skin.

More severe acidosis results in a decrease in blood pressure, potentially leading to loss of consciousness and coma. There is a movement of potassium from the intracellular to the extracellular compartments during acidosis, which can exacerbate the neuromuscular irritability, with the possibility of **cardiac arrest**.

The clinical assessment of the patient with metabolic acidosis is essential and results of acid-base analysis should be evaluated before initiating treatment. The initial approach is to treat the underlying cause of the disturbance, then assist the physiological compensatory mechanism. The use of intravenous alkali treatment (usually sodium bicarbonate) is controversial and appears to be reserved for patients with a severe, life-threatening acidosis.

Laboratory findings

Patients with metabolic acidosis have low HCO_3^- concentrations. The compensatory decrease in CO_2 results in decreased PCO_2 levels. Additional biochemistry tests may be undertaken to determine the specific underlying cause of the metabolic acidosis, for example blood glucose (for diabetes mellitus) and blood lactate concentrations for the diagnosis of lactic acidosis. Plasma urea and creatinine investigations may determine any degree of renal impairment. Refer to Table 6.5, which summarizes the laboratory findings in arterial blood in patients with metabolic acidosis.

The **anion gap** is a useful biochemical aid that may assist in the differential diagnosis of metabolic acidosis. The anion gap is the difference in concentrations (in mmol/L) between the two major plasma cations (sodium and potassium) minus the two major anions (chloride and bicarbonate). Chloride and bicarbonate make up 80% of plasma anions. The remaining 20% are from negatively charged proteins (mostly albumin), urate, lactate, sulphate, phosphate, and some organic anions. Chloride and bicarbonate concentrations are readily measured in most clinical laboratories; the remaining 20%, also referred to as the 'unmeasured anions', are less commonly measured. Sodium and potassium make up for over 90% of plasma cations. The remaining 10% is predominantly accounted for by calcium, magnesium, and some proteins, mainly γ-globulins. The anion gap is the difference between the total concentration

TABLE 6.5 Laboratory findings in metabolic acidosis. The findings indicate expected levels during the onset of the acidosis and in the fully compensated state.

	Acute	Compensated
pH	Decreased	Normal
PCO_2	Normal	Decreased
$[HCO_3^-]$	Decreased	Decreased

of measured cations minus the total concentration of measured anions, both measured in mmol/L:

$$\text{Anion gap} = [Na^+] + [K^+] - [Cl^-] - [HCO_3^-]$$

The anion gap reference range is *typically* cited as 10–18 mmol/L, although some laboratories cite different reference ranges. A possible explanation for these differences is the variety of analytical techniques and specific analysers used in the determination of the concentration of these contributing ions. In addition, some reference ranges exclude the measurement of potassium concentration in their calculations. An alternative classification of metabolic acidoses is based on differentiating disorders resulting in a metabolic acidosis with a *normal anion gap* and those with a *raised anion gap*. Box 6.5 illustrates some clinical conditions causing metabolic acidosis due to a raised anion gap and the associated 'unmeasured' anions. Metabolic acidosis with a normal anion gap may be associated with lost bicarbonate being replaced with chloride and is also referred to as *hyperchloraemic acidosis*.

Metabolic alkalosis

Metabolic alkalosis is characterized by an increased plasma pH due to a loss of H^+ or as a direct result of a gain in HCO_3^-. In both instances there is a relative increase in HCO_3^-, raising the $[HCO_3^-] / PCO_2$ ratio.

Key Points

Metabolic alkalosis is associated with increased HCO_3^- concentration.

BOX 6.5 *Use of anion gap in the differential diagnosis of metabolic acidosis*

This classification of the metabolic acidoses is based on a raised anion gap being caused by the consumption of HCO_3^- in buffering excess H^+, and unless the HCO_3^- is replaced by chloride, then the lost HCO_3^- will be replaced by unmeasured anions.

Metabolic acidoses with a *raised anion gap* may be caused by:

- Increased blood *lactate* concentration, resulting from lactic acidosis associated with shock, infection, hypoxia.

- Raised plasma *uric acid* concentration (for example due to renal failure).

- Ketoacidosis, resulting in increased acetoacetate and β-hydroxybutyrate production (diabetes mellitus, alcoholism).

- The ingestion of drugs or toxins (for example salicylate, methanol, ethylene glycol) that may result in *anionic metabolites*.

Metabolic acidoses with an apparently *normal anion gap* are associated with disorders where any loss of HCO_3^- is replaced by chloride ions:

- Gastrointestinal loss of HCO_3^- (diarrhoea, pancreatic fistulae).

- Renal loss of HCO_3^- (RTA, renal failure).

- Drugs or toxins (for example acetazolamide).

CASE STUDY 6.1

A 39-year-old man has suffered from a severe bout of diarrhoea lasting for a period of three days. When examined by his doctor he appeared moderately dehydrated. An arterial blood gas was taken. The laboratory results are shown below (reference ranges are given in brackets):

pH	7.30	(7.35–7.45)
PCO_2	4.4 kPa	(4.7–6.0)
PO_2	12.4 kPa	(12–14.6)
HCO_3^-	16 mmol/L	(24–29)

(a) Is an acid-base disorder present, if so what type?

(b) Is there any compensation?

Clinical causes

Clinical causes of metabolic alkalosis include excessive loss of H^+, increased reabsorption of HCO_3^- and excessive intake of alkali. There are two major routes by which H^+ can be lost from the body: through the gastrointestinal tract or via the kidneys. When H^+ are lost by either route, there is an equivalent gain of HCO_3^- by the ECF. Gastric secretions are rich in hydrochloric acid. Consequently, gastric aspiration or prolonged vomiting associated with **pyloric stenosis** can result in the loss of H^+, with the equivalent loss of chloride. Electrical neutrality is maintained as more HCO_3^- is reabsorbed, but these conditions are associated with a loss of body fluids resulting in hypovolaemia. This volume depletion often maintains the alkalotic state as it diminishes the ability of the kidneys to excrete the excess HCO_3^-. It is significant that in conditions of raised HCO_3^- levels, above a plasma HCO_3^- threshold of approximately 24 mmol/L, the kidneys are able to rapidly excrete any excess HCO_3^-. In metabolic alkalosis the low pH often persists due to associated hypokalaemia and volume depletion, by a mechanism that appears to raise or circumvent the HCO_3^- renal threshold. Hypokalaemia not only increases the intracellular exchange of potassium and H^+, but also increases potassium exchange for H^+ in the collecting tubules. Volume depletion stimulates the renin-angiotensin-aldosterone system, resulting in increased sodium reabsorption with the corresponding secretion of H^+ into the tubular lumen.

Renal loss of H^+ also occurs when there is an excessive secretion of mineralocorticoids or glucocorticoids. This stimulates the reabsorption of sodium by the distal tubule with the associated increased loss of H^+ and potassium into the tubular lumen to maintain electrical neutrality. In such cases the alkalosis is maintained, as the resulting hypokalaemia may further exacerbate the alkalosis as potassium moves out of the cells into the ECF and is replaced by H^+, again to maintain electrical neutrality. Conditions that cause the excessive production of mineralocorticoids and glucocorticoids include primary and secondary hyperaldosteronism, Cushing's disease, and bilateral adrenal hyperplasia. Certain diuretic drugs, such as frusemide and thiazide, increase the loss of chloride in the urine with an associated increase in HCO_3^- reabsorption.

A more unusual aetiology of metabolic alkalosis is the ingestion of HCO_3^- or other alkaline antacids to treat indigestion. It requires relatively large amounts to be absorbed to cause alkalosis. Study Table 6.6 to review the different clinical causes of metabolic alkalosis.

TABLE 6.6 Summary of the clinical causes of metabolic alkalosis.

Primary cause	Mechanism
Excessive loss of H^+	
Gastrointestinal loss of H^+ (prolonged vomiting, gastric aspiration)	Loss of H^+ without loss of HCO_3^-
Renal loss of H^+ (Cushing's disease, bilateral adrenal hyperplasia)	Increased plasma steroids, increases tubular sodium reabsorption, in exchange for H^+
Excessive reabsorption of HCO_3^-	
Drugs (frusemide, thiazide)	Cause increased loss of chloride, with associated HCO_3^- reabsorption
Intake of alkali	
Ingestion of antacids (rare)	Requires relatively large amounts to have a significant clinical effect

Compensation

The initial compensatory response to the increase in plasma pH, caused by an increase in the $[HCO_3^-]$ / PCO_2 ratio, is to increase plasma PCO_2. This decreases the $[HCO_3^-]$ / PCO_2 ratio back to normal levels, although the actual levels of HCO_3^- and PCO_2 will both be increased. This compensatory response can be achieved by **hypoventilation**. The low concentration of H^+ will inhibit the respiratory centre, decreasing ventilation. However, this response is self-limiting, as the increase in PCO_2 due to hypoventilation will stimulate the respiratory system and ventilation will increase. The renal compensatory response to chronic metabolic alkalosis is to decrease the reclamation of HCO_3^-. This reduces the plasma HCO_3^- concentration which ultimately reduces the $[HCO_3^-]$ / PCO_2 ratio.

The initial respiratory response to metabolic alkalosis is to increase PCO_2, but this is self limiting. Renal compensation reduces plasma $[HCO_3^-]$.

Clinical findings and management

Symptoms and signs of mild alkalosis are usually non-specific and are associated with the underlying disorder. The increased plasma pH may result in increased binding of calcium to proteins, as a consequence the decrease in ionized calcium may cause muscle cramps and tetany. Any related hypokalaemia may result in muscle weakness. The hypocalcaemia and hypokalaemia may also decrease vascular tone, decreasing cerebral circulation, which may result in confusion and possible coma. Associated hypovolaemia may cause skin turgor and thirst.

Management of metabolic alkalosis usually involves correction of true volume depletion and correction of any associated potassium and chloride depletion. The primary cause of the metabolic alkalosis must then be managed.

Laboratory findings

Study Table 6.7 to review the laboratory findings associated with metabolic alkalosis. In addition, the investigation of plasma sodium, potassium, urea, creatinine, albumin, and calcium

TABLE 6.7 Laboratory findings in metabolic alkalosis. The findings indicate expected levels during the onset of the alkalosis and in the fully compensated state.

	Acute	Compensated
pH	Increased	Normal
PCO_2	Normal	Normal or slight increase
$[HCO_3^-]$	Increased	Marked increase

CASE STUDY 6.2

A 4-year-old child was admitted to hospital hyperventilating, after drinking an unknown substance from a bottle found in a garage. The laboratory results are shown below (reference ranges are given in brackets):

Sodium	134 mmol/L	(132–144)
Potassium	6.1 mmol/L	(3.5–5.0)
Chloride	94 mmol/L	(95–108)
HCO_3^-	10 mmol/L	(24–29)
Urea	4.2 mmol/L	(2.7–7.5)
pH	7.20	(7.35–7.45)
PCO_2	3.2 kPa	(4.7–6.0)
PO_2	13.2 kPa	(12–14.6)

(a) Is there an acid-base disorder present, if so what type?

(b) Is there any compensation?

(c) What do the other laboratory investigations indicate?

concentrations would give an indication of hydration status as well as the possibility of hypokalaemia or hypocalcaemia.

SELF-CHECK 6.8

Why is plasma albumin a suggested test in the evaluation of a patient with metabolic alkalosis?

6.8 Respiratory acid-base disorders

Respiratory acid-base disorders are characterized by changes in the arterial PCO_2, which in turn, are related to changes in movement of air into and out of the lungs or the ability of gases to diffuse across the alveolar membrane. This group of acid-base disorders can cause a respiratory acidosis or alkalosis.

Respiratory acidosis

Respiratory acidosis is characterized by an increased arterial PCO_2 (**hypercapnia**), which decreases the $[HCO_3^-] / PCO_2$ ratio. The underlying problem is due to CO_2 retention, as a result of hypoventilation.

Key Points

Respiratory acidosis is associated with a rise in PCO_2.

Clinical causes

Hypercapnia and respiratory acidosis may occur when ventilation is impaired and the production of CO_2 in the tissues exceeds the ability of the lungs to remove the excess CO_2. In health, the rate and depth of respiration is controlled by neural and chemical receptors. The primary sites for sensing increased PCO_2, changes in arterial PO_2, and increased H^+ concentration are the peripheral chemoreceptors including the carotid and aortic bodies. Respiratory acidosis may occur due to acute or chronic clinical conditions. These causes may be broadly classified into disorders affecting the respiratory control mechanism, pulmonary disease, and neuromuscular disorders. Acute disorders are usually caused by obstruction to the airways or due to acute exacerbation of asthmatic conditions or **chronic obstructive airways disease** (COAD). Table 6.8 lists some of the different clinical causes of respiratory acidosis.

TABLE 6.8 Summary of the clinical causes of respiratory acidosis.

Primary cause	Mechanism
Disorders affecting the respiratory control mechanism	
Stroke	
Trauma to the central nervous system	These disorders and factors directly depress the respiratory centre causing retention of PCO_2
Cerebral tumours	
Infections of the central nervous system (such as meningitis)	
Drugs (such as barbiturates)	
Pulmonary disease	These disorders decrease the lungs' ability to eliminate CO_2, resulting in hypercapnia
Chronic obstructive airways disease	
Severe pneumonia	
Pulmonary fibrosis	
Emphysema	
Acute respiratory distress syndrome	
Neuromuscular disorders	Poliomyelitis, a viral disease, and Guillain-Barré syndrome, an autoimmune disease, both cause chronic breathing problems, resulting in hypercapnia
Poliomyelitis	
Guillain-Barré syndrome	
Obstruction of the airway	
Choking, obstruction by a foreign body	
Severe exacerbation of asthma	

Compensation

The increased production of H^+ is initially buffered by HCO_3^-, haemoglobin, and other ECF buffering mechanisms, but these have a relatively limited capacity. The increased H^+ concentration stimulates the respiratory system to increase the rate and depth of respiration. This results in an increase in the $[HCO_3^-] / PCO_2$ ratio as the PCO_2 decreases. This assumes the primary disorder is not caused by impairment to the respiratory centre or continued obstruction to the airways.

Unless the cause of the acute acidosis is rapidly resolved or treated, then further compensatory mechanisms are activated. The HCO_3^- concentration may also be increased using the chronic, renal response mechanism in order to return the $[HCO_3^-] / PCO_2$ ratio to normal. However, to be effective this response requires:

- adequate renal function, with no glomerular dysfunction and an adequate supply of sodium to exchange for H^+
- functioning tubular cells to produce and reclaim HCO_3^- and to produce ammonia (for urinary buffering)
- an adequate supply of other urinary buffers to accept the H^+

These mechanisms result in a loss of buffered H^+ in the urine and an increased production of HCO_3^-.

Clinical findings and management

Moderate increases in CO_2 concentration result in various cardiovascular effects, the consequences of which may cause an increased cardiac output, arrhythmia, warm skin, and sweating. The resulting acidosis can cause depression of the central nervous system, especially if the pH falls below 7.0. This may result in headaches, lack of concentration, and confusion. If the condition is severe, seizures may occur and the patient may become comatose.

The primary aim in the management of respiratory acidosis is to restore ventilation, then resolve the underlying dysfunction or disease.

Laboratory findings

Study Table 6.9, which summarizes the arterial blood results indicative of respiratory acidosis. The acidaemia may cause decreased calcium binding to protein, resulting in increased plasma ionized calcium concentrations.

SELF-CHECK 6.9

Explain how a cerebral tumour can cause a *respiratory* acidosis.

TABLE 6.9 Laboratory findings in respiratory acidosis. The findings indicate expected levels during the onset of the acidosis and in the fully compensated state.

	Acute	Compensated
pH	Low	Normal
PCO_2	Increased	Increased
$[HCO_3^-]$	Normal or slight increase	Increased

Respiratory alkalosis

The primary abnormality in respiratory alkalosis is a decrease in PCO_2 (**hypocapnia**), most commonly due to hyperventilation. This causes an increase in the $[HCO_3^-] / PCO_2$ ratio with a consequent rise in the plasma pH.

Key Points

Respiratory alkalosis is associated with a decreased PCO_2.

Clinical causes

The clinical causes of respiratory alkalosis, summarized in Table 6.10, may be due to overstimulation of the respiratory system or to pulmonary disease. Stimulation of the respiratory system may be due to anxiety (resulting in hysterical overbreathing), pain, sepsis, hypoxia, stroke, meningitis, cerebral tumours, hepatic cirrhosis, or salicylate overdose. Pulmonary disorders can be caused by chronic bronchitis, pulmonary embolism, or pulmonary oedema. Excessive mechanical ventilation can also result in decreased PCO_2.

TABLE 6.10 Summary of the clinical causes of respiratory alkalosis.

Primary cause	Mechanism
Overstimulation of the respiratory system	
Anxiety	These disorders result in hyperventilation, causing the body to breathe more deeply and more frequently than necessary. This exaggerated respiratory response results in decreased blood PCO_2 levels.
Pain	
Sepsis	
Hypoxia	
Stroke	
Meningitis	
Cerebral tumours	
Hepatic cirrhosis	
Salicylate overdose	
Pulmonary disease	
Chronic bronchitis	Pulmonary diseases cause reduced pulmonary ventilation. Examples of the underlying mechanism may include inflammation of the air passages (for example bronchitis) or accumulation of fluid in the pulmonary tissues and air spaces (oedema).
Pulmonary embolism	
Pulmonary oedema	
	The mechanism causing hypocapnia due to pulmonary embolism is less well understood, but is probably a combination of the above.
Iatrogenic	
Excessive mechanical ventilation	This causes a 'forced' hyperventilation.

Compensation

The initial compensatory response is a release of H[+] from ECF and erythrocyte buffers. This lowers the HCO_3^- concentration and readjusts the $[HCO_3^-] / PCO_2$ ratio. If low PCO_2 levels persist, the kidneys respond by reducing H[+] excretion by decreasing the formation of ammonia and decreasing H[+]/sodium exchange. There is also a decreased reclamation of HCO_3^-. The renal response usually takes between two to five days to achieve optimal effectiveness. These compensatory responses result in a reduction of plasma HCO_3^- concentration.

Key Points

The renal response to respiratory alkalosis acts to decrease plasma HCO_3^- concentration, but is relatively slow to achieve optimal effectiveness.

Clinical findings and management

Respiratory alkalosis may give rise to cardiovascular and neurological symptoms. As a consequence of hypocapnia blood vessels in the brain may constrict. This cerebral vasoconstriction may result in dizziness, confusion, and loss of consciousness. Patients also complain about tingling and numbness of their extremities. Tetany may occur, more as a consequence of the direct neuromuscular excitability due to alkalosis, than to the relatively modest decrease in ionized calcium.

Management of respiratory alkalosis is directed to treat the underlying cause. In acute cases symptoms may be relieved by increasing inspired CO_2 through rebreathing (such as from a paper bag), although without appropriate supervision this practice may be dangerous.

Laboratory findings

Refer to Table 6.11 for the laboratory findings associated with respiratory alkalosis. Mild hypokalaemia may result in the exchange of intracellular H[+] with extracellular potassium, as H[+] is removed from the cells as part of the compensatory response.

6.9 **Mixed acid-base disorders**

Mixed acid-base disorders occur when two or more primary disorders occur simultaneously. If the primary conditions are antagonistic in their effects on pH, for example one is acidotic the other alkalotic, then it may appear that one disorder is imitating the compensatory

TABLE 6.11 Laboratory findings in respiratory alkalosis. The findings indicate expected levels during the onset of the alkalosis and in the fully compensated state.

	Acute	Compensated
pH	Increased	Normal
PCO_2	Decreased	Decreased
$[HCO_3^-]$	Low normal or decreased	Decreased

CASE STUDY 6.3

A 62-year-old woman was admitted to hospital following an acute asthmatic attack. A sample of arterial blood was taken from the patient and analysed. The laboratory results are shown below (reference ranges are given in brackets):

pH	7.59	(7.35–7.45)
PCO_2	2.6 kPa	(4.7–6.0)
PO_2	12.9 kPa	(12–14.6)
HCO_3^-	24 mmol/L	(24–29)

(a) Is there an acid-base disorder present, if so what type?

(b) Is there any compensation?

response of the other. In such circumstances the effects on pH may partially or completely cancel out. Alternatively, the primary conditions may be additive, for example a respiratory acidosis with a simultaneous metabolic acidosis have a combined effect on exacerbating the decrease in pH.

Mixed acid-base disorders are relatively common and numerous combinations are possible. A knowledge of the clinical assessment of the patient is essential in the interpretation and differentiation of mixed acid-base disorders, but laboratory results may give the first indication of such disorders, especially interpretation of the anion gap. In such cases there may be an inadequate or extreme compensatory response or the compensatory response by PCO_2 or HCO_3^- are in opposite directions. Full compensation of pH is only seen in simple acid-base disorders due to respiratory derangements. A normal pH with abnormal PCO_2 or HCO_3^- concentration would probably indicate a mixed acid-base disorder. Compensation formulae are available to predict expected HCO_3^- changes derived from changes in PCO_2.

CASE STUDY 6.4

A 73-year-old man was admitted to hospital unconscious and cyanosed. The results of a blood gas analysis are shown below (reference ranges are given in brackets):

pH	6.90	(7.35–7.45)
PCO_2	10.5 kPa	(4.7–6.0)
PO_2	5.1 kPa	(12–14.6)
HCO_3^-	16 mmol/L	(24–29)

(a) Is there an acid-base disorder present, if so what type?

(b) Is there any compensation?

(c) What are the possible causes?

Any significant deviations between the predicted and actual HCO_3^- concentrations may indicate a mixed acid-base disorder. Box 6.6 illustrates an example of a mixed acid-base disorder and refer to Table 6.12 for examples of typical clinical conditions resulting in mixed acid-base derangements.

Key Points

Two or more primary acid-base disorders may occur simultaneously. Their effects on pH may be antagonistic or complementary.

BOX 6.6 Case history of a mixed acid-base disorder

A 63-year-old man was admitted to hospital with pneumonia. He had been taking a thiazide diuretic for several months. The results of blood tests on admission are given below (reference ranges are given in brackets):

Potassium	2.2 mmol/L	(3.5–5.0)
HCO_3^-	31 mmol/L	(24–29)
pH	7.65	(7.35–7.45)
PCO_2	4.3 kPa	(4.7–6.0)
PO_2	10 kPa	(12–14.6)

The raised pH indicates an alkalosis. The reduced PCO_2 indicates a respiratory alkalosis. However, the HCO_3^- would be expected to be decreased as part of the compensatory response and not increased as these results indicate. A possible explanation is that there is a mixed alkalosis present: a metabolic alkalosis due to the thiazide diuretic therapy and a respiratory alkalosis due to the pneumonia.

This diagnosis is indicated as the PCO_2 and HCO_3^- move in opposite directions relative to their reference ranges. Hypokalaemia is a relatively common finding in metabolic alkalosis.

TABLE 6.12   Examples of typical clinical conditions found in types of mixed acid-base disorders.

Mixed acid-base dysfunction	Clinical causes
Respiratory acidosis	Chronic bronchitis
Metabolic acidosis	Renal impairment
Respiratory alkalosis Metabolic acidosis	Salicylate overdose
Respiratory acidosis	Chronic obstructive airways disease
Metabolic alkalosis	Thiazide diuresis
Respiratory alkalosis	Hyperventilation
Metabolic alkalosis	Prolonged nasogastric suction

6.10 Interpretation of acid-base data

The interpretation of conditions resulting from disturbances in acid-base balance may be ascertained by the laboratory measurement of the three basic parameters: pH, PCO_2, and HCO_3^- concentration, and with appropriate reference to the clinical findings. This should be done in a systematic manner. The following is a simple guideline to understanding the interpretation of acid-base data and relates to equilibrium between PCO_2 and $[HCO_3^-]$. The need for reference to clinical details cannot be overstressed.

The logical sequence is to look first at the pH, which would indicate an acidosis or alkalosis. Then consider the direction of change of the PCO_2 which should then be considered in relation to any changes in plasma $[HCO_3^-]$. The degree and nature of any compensatory response must then be assessed. In simple acid-base disorders the compensatory change is in the same direction as the primarily changed variable.

What is the pH status? The pH determines whether the primary disorder is an acidosis or an alkalosis. A normal pH indicates no acid-base imbalance *or* an acidosis or alkalosis that has been compensated.

What is the PCO_2 result? If both *the pH* and PCO_2 are low then the patient has a compensated metabolic acidosis, the low PCO_2 being the compensatory response. The plasma HCO_3^- concentration would be expected to be relatively low, sometimes as low as 10 mmol/L.

If the *pH is decreased* and the *PCO_2 is increased* then this would indicate a respiratory acidosis. In acute conditions, plasma HCO_3^- concentration may be normal or only slightly increased due to the slow renal response. In chronic respiratory acidosis conditions HCO_3^- concentration would be markedly raised, but the pH would be closer to normal.

If *the pH is low* and the *PCO_2 normal* then this would indicate a possible uncompensated metabolic acidosis, the plasma HCO_3^- concentration would be decreased. The respiratory response tends to occur relatively quickly, and acts to decrease PCO_2. An alternative explanation is that there is a mixed acid-base disturbance, with a simultaneous respiratory acidosis maintaining the PCO_2 within the normal range.

If the *pH is increased* and the *PCO_2 is increased* then the patient may have a metabolic alkalosis, with a raised plasma HCO_3^- concentration. The plasma PCO_2 would be expected to increase as part of the compensatory response. However, due to the limited respiratory response associated with metabolic alkalosis, a more likely diagnosis is a mixed acid-base dysfunction, with metabolic alkalosis and a respiratory acidosis.

If *the pH is increased* and the *PCO_2 is decreased* a respiratory alkalosis is indicated. In this condition the plasma HCO_3^- concentration would be expected to decrease to compensate.

A *normal pH* with an *increased PCO_2* would indicate that the patient may have a mixed acid-base disturbance of a respiratory acidosis and a metabolic alkalosis (the plasma HCO_3^- concentration is increased in both these conditions). Alternatively, the patient has a fully compensated respiratory acidosis.

A *normal pH* with a *decreased PCO_2* indicates that the patient may have a respiratory alkalosis and a co-existing metabolic acidosis (the plasma HCO_3^- concentration is decreased in both these conditions) or, less likely, the patient has a fully compensated respiratory alkalosis.

A simplified strategy for investigation of acid-base disorders is outlined in Figure 6.7.

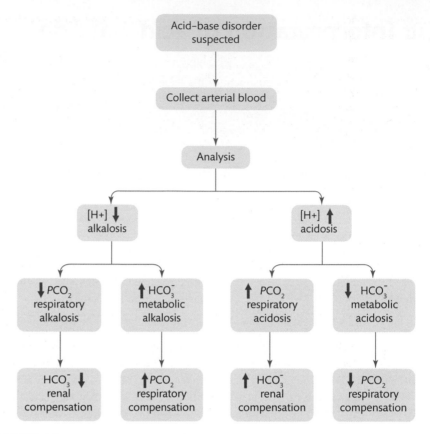

FIGURE 6.7
A schematic showing a simplified strategy for investigation of acid-base disorders.

An alternative approach to the interpretation of acid-base disorders

The approach to acid-base disturbances described so far is based on the 'traditional' model, using the Bronsted-Lowry definition of acids and bases and describing pH as primarily a function of the ratio of PCO_2 to plasma HCO_3^- concentration. However, acute care clinicians and anaesthesiologists have expressed concerns about the inability of the traditional model to describe complex acid-base disorders. In response, an alternative model was proposed by Stewart in 1981, radically redefining the basic concepts of acid-base balance. The fundamental principle of the Stewart approach is that plasma HCO_3^- concentration does not affect blood pH. Furthermore, Stewart defined an acid as any chemical species that shifts the dissociation constant of water to increase H^+ concentration and decrease OH^- concentration, comparable with the Arrhenius definition of acids and bases. In effect, the H^+ concentration is a 'dependent' variable and its addition or removal from a physiological system does not affect pH.

Stewart employed basic physicochemical principles and a mathematical approach to produce a series of six simultaneous equations linking the basic laws of mass action of water to various constituents in plasma. The resulting polynomial equation identified three main independent variables and concluded that plasma HCO_3^- concentration was a *dependent variable* and therefore *did not affect pH*. The independent variables are:

- **Strong ion difference** (SID). Plasma contains fully dissociated ions, 'strong ions'. The SID is the difference between the sum of these strong cations and anions, expressed as mmol/L. In practice only the major cations and anions are calculated, such as sodium, potassium, calcium, magnesium, chloride, and lactate. In health, the SID is approximately 40 mmol/L. Any derangement in acid-base balance causing an increase in the SID results in an increase in pH. Similarly, lowering the SID lowers pH.

- The total of non-volatile weak acids (A_{Tot}), which is defined as the total plasma concentration of all the non-volatile weak acids. In health A_{Tot} is mainly derived from the contribution made by phosphate and albumin.

- The third independent variable was identified as PCO_2

When applying Stewart's approach in a clinical context, respiratory disorders affecting H^+ concentration can still be explained in a similar way to the traditional approach, that is, due to changes in PCO_2, which is identified as one of the independent variables. However, Stewart explains metabolic disturbances as due to changes in either the concentration of the components of the SID or A_{Tot}, not due to changes in HCO_3^-. The SID may be affected by any disorder that significantly affects the concentration of any of the contributing strong ions, such as chloride or lactate. Alternatively, the SID may be affected by the hydration state of the body. Dehydration will increase the concentration of strong ions, increasing alkalinity. The major contributor to changes in A_{Tot}, are caused by changes in plasma albumin concentrations.

Key Points

Stewart's interpretation of acid-base disorders is based solely on the contribution of the SID, the total non-volatile weak acids (A_{Tot}) and PCO_2. It is independent of the plasma HCO_3^- concentration.

Stewart's model has been refined by other workers. Supporters claim that it improves the diagnostic and prognostic accuracy when evaluating the acid-base status of critically ill patients. This increased accuracy is attributed to the inclusion of the effects of albumin concentration and electrolyte balance, both of which may be significantly altered in acute care situations. This view is partially supported by a study of 255 paediatric intensive care patients. Analyses of arterial blood gases, plasma electrolyte, and albumin concentrations identified that 26% of these patients had significantly different clinical interpretations when compared to the traditional approach to interpretation. Critics of the Stewart model state that these differences are eliminated when the contributions of albumin concentration and the anion gap (which is comparable to SID) are included in the traditional interpretation. A further criticism is the relative complexity of the equations required by Stewart. Although calculators containing the appropriate equations are available, there is a requirement to measure several variables, with the potential risk of accumulating errors.

Both approaches purport to have the same purpose: to quantify the magnitude of an acid-base disturbance, to classify as respiratory or metabolic, and to identify and quantify the independent variables causing the disturbance. Stewart's main contribution to date is to emphasize the significance of other factors that affect pH. It is likely that both approaches will continue to be used according to the appropriate clinical situation.

Key Points

Stewart emphasized the effects of electrolyte and albumin concentrations on acid-base balance, especially in severely ill patients.

CASE STUDY 6.5

An 18-year-old woman was admitted to hospital in a comatose state with marked hyperventilation. A laboratory investigation revealed the following results (reference ranges are given in brackets):

Sodium	140 mmol/L	(132–144)
Potassium	3.0 mmol/L	(3.5–5.0)
Chloride	106 mmol/L	(95–108)
HCO_3^-	10 mmol/L	(24–29)
Urea	5.0 mmol/L	(2.7–7.5)
pH	7.53	(7.35–7.45)
PCO_2	1.6 kPa	(4.7–6.0)

Discuss these results.

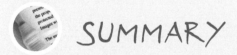

SUMMARY

- The processes of metabolism produces H^+ which are maintained in the blood at very low concentrations within a narrow reference range.

- The physiological control of H^+ concentration is maintained by three interrelated mechanisms: buffering systems, the respiratory system, and the renal system. All three mechanisms act to control the ratio of $[HCO_3^-]$: PCO_2.

- Intracellular and extracellular buffering systems, such as bicarbonate and haemoglobin, provide an immediate, but limited, response to pH changes.

- The respiratory system, which can be activated almost immediately, controls PCO_2 by changing alveolar ventilation.

- The renal system regulates $[HCO_3^-]$ and is the slowest to respond. However, its biological significance lies in the fact that it is the only system able to eliminate buffered H^+. Unlike the buffering and respiratory systems, H^+ is actually removed from the body, simultaneously regulating plasma HCO_3^- concentration.

- The physiological response to an acid-base disturbance, which limits the change in H^+ concentration, is referred to as compensation. Compensation alters the ratio of $[HCO_3^-]$: PCO_2 back to normal, although both $[HCO_3^-]$ and PCO_2 may be abnormal. The causes of these acid-base disorders may be respiratory, affecting PCO_2, or metabolic which initially affect the concentrations of HCO_3.

■ A decrease in the ratio $[HCO_3^-]$: PCO_2 causes acidosis. Acidosis may be caused by increased H^+ production or ingestion, excessive loss of HCO_3^-, or increased retention of CO_2.

■ Alkalosis is caused by an increase in the $[HCO_3^-]$: PCO_2 ratio, which may result from a variety of causes: loss of H^+ (including renal and gastrointestinal routes), retention of HCO_3^-, the ingestion of alkolytic agents, and disorders causing the reduction of PCO_2.

■ Mixed acid-base disorders occur when two or more primary disorders occur simultaneously. These can act in an additive or antagonistic way on their effect on pH

■ Acid-base data can be interpreted in a systematic manner, from laboratory results, by examining pH status, PCO_2 results, and the compensatory response by HCO_3^-.

■ The calculated parameters base excess and standard bicarbonate are derived from measurement of PCO_2 and HCO_3^-. They are used to quantify the metabolic component of acid-base disturbances.

■ Alternative models exist to explain acid-base homeostasis and disorders. These emphasize the importance of the contribution of plasma albumin concentration and 'strong ions' in the interpretation of acid-base data.

FURTHER READING

● **Fall PJ (2000)** A stepwise approach to acid base disorders. *Postgraduate Medicine* **107**, 249–58.

● **Gluck SL (1998)** Acid-base. *Lancet* **352**, 474–9.

● **Oh MS (1991)** A practical approach to acid-base disorders. *Western Journal of Medicine* **155**, 146–51.

● **Thomson WST, Adams JF, and Cowan RA (1997)** *Clinical Acid-Base Balance*. Oxford: Oxford University Press.

● **Williams A (1998)** Assessing and interpreting arterial blood gases and acid-base balance. *British Medical Journal* **317**, 1213–16.

QUESTIONS

6.1 Which of the following options are associated with metabolic acidosis? For each statement state whether TRUE or FALSE.

(a) A raised plasma HCO_3^-

(b) A decreased blood PCO_2

(c) It can be caused by hypoaldosteronism

(d) It is always associated with a normal anion gap

6.2 Which of the following are associated with metabolic alkalosis? For each statement state whether TRUE or FALSE.

(a) May be caused by prolonged vomiting

(b) Hypokalaemia may occur

(c) Can be associated with tetany

(d) Can be caused by hypoaldosteronism

6.3 Describe the physiological role of H^+ and the homeostatic mechanisms involved in maintaining H^+ concentration within narrow limits to maintain health.

6.4 What is the role of the laboratory in the investigation of acid-base disorders?

6.5 Calculate the anion gap from the following set of results and comment on the findings (reference ranges are given in brackets):

Sodium   134 mmol/L   (132–144)

Potassium   5.9 mmol/L   (3.5–5.0)

Chloride   94 mmol/L   (95–108)

Bicarbonate   14 mmol/L   (24–29)

Urea   5.3 mmol/L   (2.7–7.5)

The reference range for the anion gap is 10–18 mmol/L.

Answers to self-check questions, case study questions, and end-of-chapter questions are available in the Online Resource Centre accompanying this book.

 Go to www.oxfordtextbooks.co.uk/orc/ahmed/

Clinical enzymology and biomarkers

Paul Collinson and Amy Lloyd

Learning objectives

After studying this chapter you should be able to:

- Define the terms enzyme, enzyme activity, and biomarker
- Discuss how measurements of enzyme activities, and biomarkers can be used to detect disease
- Distinguish between a heart attack (acute myocardial infarction) and chest pains arising from a poor blood supply to the heart (angina)
- Discuss the use of clinical tests, such as the electrocardiogram, and biomarkers in diagnosis of heart attacks (acute myocardial infarctions)

Introduction

The basis of clinical enzymology is the measurement of enzyme activities in clinical samples that only occur when particular tissues in the patient have been damaged. Increases in activities above reference values are indicative of disease. The measurement of enzyme activities in the blood was the first step in what is now referred to as **biomarker** measurement. A biomarker is a biological molecule whose concentration in the blood changes in response to a specific disease. A range of biomarkers are available for use in clinical investigations. These include molecules of intermediary metabolism, cell signalling molecules, enzymes, or proteins whose concentrations change significantly in response to a disease state and so can be used to monitor the onset or progress of a disease, or predict the outcome in response to its treatment. There is now an overlap between traditional enzymology and determining the concentrations of tissue specific proteins. For this reason, it is simpler to consider enzyme measurements as part of the spectrum of biomarkers of disease. This chapter will discuss the clinical value of measuring enzyme activities and the concentrations of other biomarkers in the laboratory investigation and management of disease.

7.1 Enzymes and enzyme activities

An **enzyme** is a protein that functions as a biological catalyst for a specific chemical reaction. Enzymes have complex structures, therefore their concentrations cannot be measured directly using simple chemical methods. However, their concentrations can be determined using the chemical reactions they catalyse. Measuring enzyme activity uses natural compounds or synthesized molecules that resemble the natural chemical compound, called **substrates**, which the enzyme changes to a product. The activity of the enzyme is usually defined as the amount of product formed in a given time by a set amount of enzyme. Activity is proportional to the number of enzyme molecules present. The product formed in an enzyme-catalysed reaction may be coloured or, if not, may be linked (or coupled) to a series of reactions that produce a coloured compound. Figure 7.1 demonstrates examples of enzyme-catalysed reactions that produce coloured products.

Measuring enzyme activities

The amount of enzyme present in a clinical sample, such as blood, is proportional to its activity. Even a relatively small amount of an enzyme can produce comparatively large amounts of product in a short time. Thus, enzymes act as *amplifiers*. Hence, manipulating the quantities of the reagents in the reaction sequence can produce large amplification of the intensity of the colour. This increases the sensitivity of the test for the enzyme of interest. In addition, it reduces the time required to determine the activity, enabling faster generation of the results.

There are two ways to measure enzyme activity. The first is a **kinetic fixed time** or end point reaction, where the reaction is stopped after a set time and the final absorbance is taken as a measurement of the enzyme activity. The second method uses **kinetic continuous monitoring** or rate reaction, where the absorbance of the assay is determined at a series of defined time intervals, for example every 30 seconds. This method enables a graph of absorbance against time to be plotted and the rate determined as the slope of the graph.

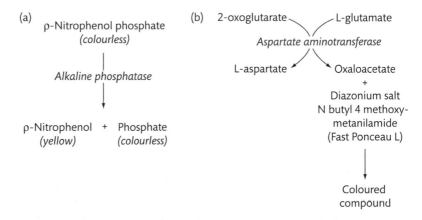

FIGURE 7.1

Reactions catalysed by enzymes. Note how in both cases colourless substrates give rise to a coloured product, directly in the case of alkaline phosphatase and indirectly with AST. The intensity of the colour is proportional to the amount of product formed and therefore the amount of enzyme present. (a) Alkaline phosphatase. (b) Aspartate aminotransferase.

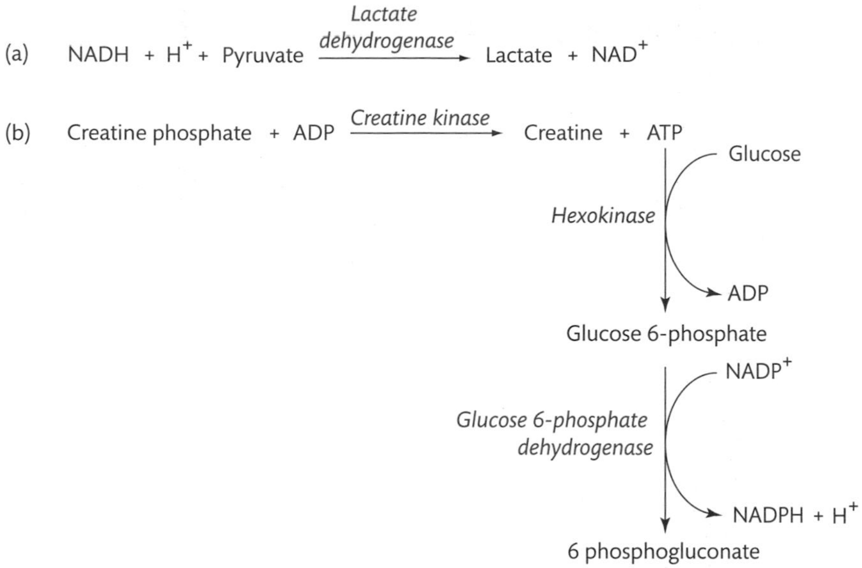

(a) $\quad NADH + H^+ + Pyruvate \xrightarrow{\text{Lactate dehydrogenase}} Lactate + NAD^+$

(b) $\quad Creatine\ phosphate + ADP \xrightarrow{\text{Creatine kinase}} Creatine + ATP$

Glucose

Hexokinase

ADP

Glucose 6-phosphate

$NADP^+$

Glucose 6-phosphate dehydrogenase

$NADPH + H^+$

6 phosphogluconate

FIGURE 7.2

Reactions catalysed by: (a) Lactate dehydrogenase. (b) Creatine kinase. The amount of NAD$^+$ produced or NADP$^+$ used can be used to determine the activities of the enzymes by measuring absorbance at 340 nm.

The development of sensitive spectrophotometers able to measure a wide range of wavelengths has allowed natural substrates and measurements in the ultraviolet region of the spectrum to be used in clinical tests. Assays using natural substrates and measurements in the ultraviolet region are the basis of most of the commonly used methods to determine enzyme activities in clinical biochemistry. Typically, they continuously measure the absorbances of NAD$^+$ or NADP$^+$ at 340 nm. Figure 7.2 shows the reactions used to monitor the activities of lactate dehydrogenase (LD) and creatine kinase (CK) in clinical laboratories by measuring changes in the absorbance of NAD$^+$ or NADP$^+$. Clinical investigations using these two enzymes that can be used to detect muscle damage are discussed in detail in Section 7.7.

Immunological methods

Given that enzymes are proteins, then the number of protein molecules in a specimen can be estimated using an immunoassay:

Many clinical tests for enzymes are now performed using immunoassays rather than by determining their catalytic activities. Indeed, for a variety of reasons, it is not practical to measure the activities of some enzymes, for example when they are present in such low amounts that a method more sensitive than the coupled chemical reactions must be used; immunoassays provide the necessary sensitivity. An extension of this idea is not to measure the enzyme itself, but to measure a protein or other molecule that is bound to it, since the value obtained is also a measurement of the amount of enzyme present.

Thus, although both chemical and immunological methods are used to determine enzyme activities, clinical enzymology has progressed from simple but relatively insensitive colorimetric methods to sensitive spectrophotometric techniques measuring into the ultraviolet range and very accurate immunoassays. In many cases, a coloured product is the result of the assay. Many of these methods have been automated.

Cross reference
Chapter 15 Immunological techniques in the *Biomedical Science Practice* textbook

Cross reference
Chapter 2 Automation

SELF-CHECK 7.1

What is an enzyme and how can it be measured?

7.2 Isoenzymes

Enzymes are found in all cells and tissues. However, the amount of any particular one present varies according to the type of cell and the tissue in which the cells are organized. One tissue may contain higher concentrations of an enzyme that is found in all cells of the body, while some will contain relatively large amounts of an enzyme that is present only in small quantities in others (and sometimes not be found in any others), such as CK in muscle. Enzymes can also occur in a number of different molecular forms, called **isoenzymes**, each of which can be tissue specific. Isoenzymes are made up of several subunits. Subunits are individual polypeptides encoded by different genes and therefore have different structures. Different combinations of subunits give rise to different isoenzymes. Examples of isoenzymes include those of CK, which has three different isoenzymes, and LD, where there are potentially five different isoenzymes as seen in Figure 7.3 (a) and (b).

Each type of isoenzyme has slightly different properties and may be present in varying amounts in multiple types of tissue as seen in Table 7.1 and Table 7.2.

Some tissues will respond to a particular disease by increasing the amount of an enzyme normally present within the cells. The amount of one particular type of tissue may be large compared with other tissues (muscle is a good example), so an enzyme found in that tissue will be a good marker of damage to that tissue as it simply swamps the amounts that might be found elsewhere.

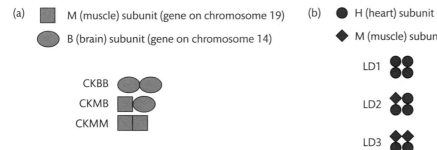

FIGURE 7.3
Cartoons showing the subunit arrangements of the isoenzymes of: (a) Creatine kinase (CK). Creatine kinase can form three different dimeric (two subunits) isoenzymes from its two different subunits. (b) Lactate dehydrogenase (LD). Lactate dehydrogenase is a tetramer (4 subunits) and its two different subunits can give rise to five isoenzymes.

TABLE 7.1 **The percentage distribution of the isoenzymes of creatine kinase (see Figure 7.3 (a)).**

Isoenzyme	Brain	GIT	Cardiac muscle	Skeletal muscle
CKBB (CK1)	100	95–96	1	0
CKMB (CK2)	0	1–2	22	3
CKMM (CK3)	0	3	77	97

TABLE 7.2 **The percentage distribution of the isoenzymes of lactate dehydrogenase (see Figure 7.3 (b)).**

Isoenzyme	Cardiac muscle	Kidney	Erythrocytes	Skeletal muscle	Liver
LD1	60	28	40	3	0.2
LD2	30	34	30	4	0.8
LD3	5	21	15	8	1
LD4	3	11	10	9	4
LD5	2	6	5	76	94

7.3 **Biomarkers**

A biomarker is a biological molecule whose concentration in the blood changes in response to a specific disease. Biomarkers are normally considered in relation to a specific type of tissue damage. In this chapter, we will explore the uses of biomarkers that indicate muscle damage, in particular damage to cardiac muscle, which can lead to heart disease. These types of biomarkers are usually referred to as cardiac biomarkers.

A number of different types of molecules have been exploited as biomarkers. These include:

- enzymes and their associated coenzymes or cofactors
- structural tissue proteins
- intermediates of metabolic pathways
- messenger molecules

Irrespective of its nature, an ideal biomarker will have a number of properties. It will not be present in the blood of normal, healthy individuals but be specific to a particular type of tissue. It will be structurally and functionally stable and easy to assay. Any changes in the concentration in clinical samples must be directly related to disease. Changes in measured concentration should result in different patient management strategies.

Biomarkers and tissue damage

A change in the concentration of a biomarker can be a signal that a tissue is undergoing injury. The tissue may find itself under abnormal stress due to injury from disease and respond by increasing the amount of the biomarker produced as a corrective response. Increased amounts of the biomarker may then leak into the bloodstream, causing its concentration there to increase to measurable levels. Alternatively, the biomarker may function as a biological messenger. Thus, when the tissue becomes stressed, its concentrations are increased to send a message to other parts of the body. Ideally, the body should act to correct the problem that is causing stress to the tissue. Correction of an abnormal physiological state is a normal homeostatic response. However, in disease it may not be possible to correct the problem and an increasing amount of the messenger may be produced, the equivalent to shouting louder and louder for help.

The concentration of a biomarker in the blood can also increase when a tissue is directly damaged and cellular contents leak into the bloodstream. If the damage to the cell is reversible

(a)

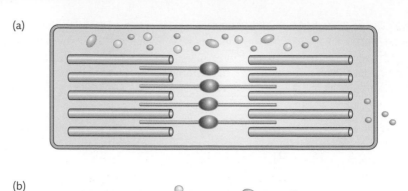

(b)

FIGURE 7.4
Schematics showing the degeneration of a healthy cardiac muscle cell to necrotic death.
(a) Shows the normal state, with background levels of cellular leakage and degradation and the normal levels of production of messenger molecules.
(b) Represents reversible damage to the cell. There is an increase in the amounts of some enzymes that may leak from the cell, the release of some small cytosolic molecules, and changes in the production of messenger molecules. (c) Cell death, with the loss of structural and other cytosolic proteins such as messenger molecules.

(c)

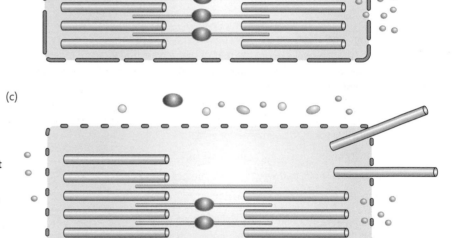

then the leakage may be slight and will only affect small molecules that can pass through the plasma membrane. If the damage to the cell is permanent and the cell dies, often referred to as **cellular necrosis**, then all of the cellular contents can leak into the bloodstream. Figure 7.4 shows a cartoon of a heart muscle cell (a cardiac myocyte) going through these stages. Notice the increased degradation of the cell membrane. It is always possible that any, or all, of these scenarios may occur simultaneously to different cells within a tissue.

7.4 **Cardiac disease**

Heart or cardiac disease kills more people in the developed world than any other single type of medical condition. In the UK alone, it accounts for one in five deaths in men and one in six deaths in women. Although deaths from heart disease are lower in the developing world, they are increasing with the moves towards a more western diet, rich in saturated fats, and an increase in cigarette smoking.

Heart disease occurs due to blockage of the blood vessels supplying oxygen to the heart muscle. In any tissue where the need for oxygen from the blood exceeds the rate at which it can be supplied, a deficiency of oxygen or **ischaemia** occurs. However, ischaemia is reversible provided it is not unduly prolonged. When the ischaemia is prolonged, irreversible cell damage and cell death can occur; this is known as **infarction**, and is followed by cellular breakdown and necrosis. Thus, infarction is irreversible.

In the heart, ischaemia of cardiac muscle will occur if the artery supplying blood to an area of cardiac muscle becomes partially or totally blocked. The reason that a partial blockage can also cause ischaemia is that some areas of cardiac muscle are on the borders of the area supplied with blood by the arteries, areas called watersheds. In a normal state they receive just enough blood to survive, but should one of the arteries become even partially blocked then the already barely adequate supply to this muscle slips below the minimum level; thus the muscle becomes inadequately supplied with oxygen and slides into a state of ischaemia. The ischaemia will be exacerbated if the need for oxygen increases at the same time, for example if heavy exercise is being undertaken.

SELF-CHECK 7.2

What is the key difference between myocardial ischaemia and myocardial infarction?

A number of terms are used to describe patients with a real or suspected heart attack, which is a **myocardial infarction** (MI). Myocardial infarction and acute myocardial infarction (AMI) are associated with cardiac pain and the death of cardiac tissue (myocardial cell necrosis). In some patients, such as diabetics, an infarction can occur without pain. It is then referred to as a *silent* MI.

Angina is often confused with MI but the term simply means heart pain. The difference between angina and MI is that damage to cardiac muscle does *not* occur in the former. Angina occurs in two principal forms: stable and unstable anginas. In stable angina, the cardiac pain occurs predictably and gradually and can be controlled by actions as simple as physically resting or by using appropriate drugs. Cardiac pain or breathlessness that is associated with exercise and relieved by rest is also stable angina. In contrast, in unstable angina the cardiac pain comes on unpredictably and is not relieved, or only partially, by rest or by drugs.

SELF-CHECK 7.3

Is angina a myocardial ischaemia or MI?

7.5 Cardiac disease, electrocardiogram, and biomarkers

When a patient first comes to hospital with suspected AMI, the diagnosis is undecided; it may be unstable angina, or AMI, or possibly not cardiac pain at all. This is why the initial diagnosis is of a suspected **acute coronary syndrome** (ACS). The crucial difference between angina and an AMI is whether or not myocardial damage has occurred. Myocardial damage can be detected by measuring the release of cardiac biomarkers into the blood. However, a diagnosis of MI is not made solely on the basis of measurements of cardiac biomarkers. The clinical findings such as the symptoms and examination must be suggestive of cardiac disease. The other crucial test that is always carried out before a blood test is an **electrocardiogram** (ECG) recording on the patient. The heart functions as a pump driving the blood around the body.

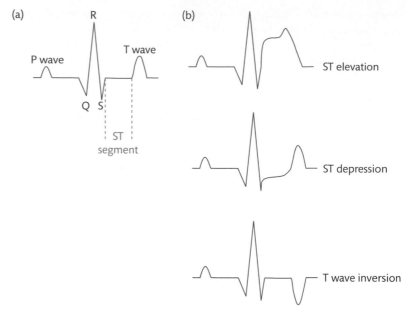

FIGURE 7.5
Electrocardiogram (ECG) recordings of: (a) Normal heart rhythm. (b) Various types of abnormal traces.

An electric signal is sent to all the muscle fibres in a well-defined sequence, which coordinates the contractions of the individual heart muscle fibres. The cardiac muscles in the atria of the heart contract and then as they relax, the muscular walls of the ventricles begin their contraction. The ECG is a record of this sequence of electrical events. Figure 7.5 (a) shows an example of a typical ('normal') ECG trace.

When the cardiac muscle becomes ischaemic, it can no longer conduct the electrical signal properly, in effect the area of ischaemia causes a short circuit. This can have an immediate and catastrophic effect, destabilizing the entire electrical signal and resulting in all the cardiac muscle fibres contracting separately and independently. Unless treated immediately, this ventricular fibrillation results in the death of the patient. If ventricular fibrillation does not occur, the change in the electrical signal can be detected by changes to the pattern of the ECG. Figure 7.5 (b) demonstrates these 'abnormal' ECG readings; observe the way that they differ from the typical ECG trace shown in Figure 7.5 (a). The changes tend to affect the T waves and/or the ST segments. For example, the elevation in the ST segments is typical of an AMI.

A combination of ECG findings and changes in cardiac biomarkers is used to classify AMI into two types. ST elevation myocardial infarction (STEMI) is indicated by a combination of ST elevation on the ECG reading and changes in cardiac biomarkers. Non-ST elevation myocardial infarction (NSTEMI) is classified by changes in cardiac biomarkers but the ECG readings do not show an ST elevation and sometimes no change at all, or only an ST depression or changes in the T wave (Figure 7.5 (b)).

7.6 Development of heart disease

Our understanding of how heart disease develops and progresses has undergone a radical change since the late 1980s. It is now accepted that the first change is the development of an atheromatous plaque in the wall of an artery (Figure 7.6 (a) and (b)). Plaques consist

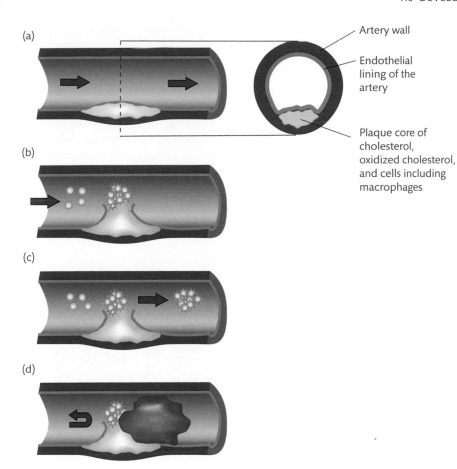

Artery wall

Endothelial lining of the artery

Plaque core of cholesterol, oxidized cholesterol, and cells including macrophages

FIGURE 7.6
Cartoons showing the stages in the development of a plaque. (a) Shows the formation of a plaque, its progression, and endothelial dysfunction with a schematic cross-section showing subendothelial plaque development. (b) and (c) Indicate the destabilization of the plaque. Tearing or erosion exposes the lipid core, causing the plaque to rupture and an intra-plaque thrombosis occurs. This is followed by activation and aggregation of blood platelets (white thrombus). An embolism may occur. An ECG would show ST depression, T wave inversion, or may be normal. (d) A thrombotic occlusion or red thrombus may occur. An ECG would show an ST elevation.

of cholesterol, oxidized lipids, together with inflammatory cells such as neutrophils and macrophages, which accumulate under the endothelial lining of the artery and extend into the arterial wall. The majority of atheromatous plaques do not cause problems; however, if the plaque reduces the diameter of the artery by more than 70% it can significantly reduce the blood flow, increasing the risk of ischaemia.

Heart disease begins with the development of a tear in or the erosion of the edge or the shoulder region of the plaque, which allows blood to enter (intraplaque haemorrhage) and mix with the plaque contents. The blood clots form an intraplaque **thrombus**. The next stage is the formation of a white thrombus, as platelets amass and aggregate over the tear in the plaque in an attempt to repair the damage. Platelet aggregation eventually reaches a stage where clumps break off into the blood and are swept downstream where they can lodge in and block small blood vessels, a process called downstream **embolization**. This pathological sequence is not inevitable and plaques may repair themselves even after they reach the stage of platelet aggregation.

It is necessary to know about the different stages of plaque evolution because this information will help in understanding why certain treatments are administered at specific stages in the process. At the platelet aggregation phase (white thrombus) anti-platelet and **anti-thrombotic** drugs are administered to limit platelet aggregation and prevent clot formation. At the stage of complete blockage, the treatment is to remove the clot from the blood vessel and restore flow in the artery. This is best performed mechanically by cardiac catheterization and **angioplasty** (per-cutaneous cardiac intervention or PCI). Alternatively, drugs to dissolve the clot (**thrombolytic**

therapy) may be given. Percutaneous cardiac intervention is the treatment of choice because it is more effective than thrombolysis and poses less risk to the patient. Both of these treatments work best if they are done as soon as possible. However, before treatment can take place an ECG is recorded on a patient when they first come to hospital. This particular test is performed because the ST elevation which can be indicated by an ECG is characteristic of AMI. The presumptive diagnosis for patients whose ECG indicates an ST elevation, and whose medical history corroborates this, is STEMI and they are treated accordingly. Measurement of cardiac biomarkers is used only to confirm the presumptive diagnosis. In contrast, when NSTEMI is suspected, the measurement of cardiac biomarkers is essential as a guide to diagnosis and treatment.

Heart attacks occur unexpectedly in most patients. It has been found that 70% of people admitted to hospital with AMI have no previous history suggestive of heart disease.

Patients who come to hospital with suspected STEMI, where the ECG shows the characteristic changes indicative of a MI, such as ST segment elevation, do not need biomarker measurement on arrival to make a diagnosis.

SELF-CHECK 7.4

What tests should be done on a patient who comes to the casualty department with chest pain?

7.7 Cardiac biomarkers in clinical practice

The first cardiac biomarkers were assays for enzyme activities in patients with suspected ACS, which includes stable angina, unstable angina, and MI. However, the use of cardiac biomarker testing has evolved to include immunoassays for cardiac enzymes and a number of non-enzymic cardiac specific proteins.

Muscle enzymes as cardiac biomarkers

A number of different enzymes are found in cardiac muscle tissue. The premise for assaying the activities of these enzymes was that muscle damage from trauma is clinically apparent and so any increase in their activities in blood samples, in the absence of obvious injury, must be due to damage to cardiac muscle tissue. There are two objections to this assumption. First, if there is muscle damage, the activities of enzymes released from skeletal muscle following its damage will swamp that due to loss from the cardiac muscle. The second is that there is a background level of enzyme activity in blood due to their leakage from skeletal muscle and other tissues. Thus, in practice it is difficult to assume that any increase in muscle enzyme activities can be used solely as a specific indicator of *cardiac* muscle damage.

The activities of three enzymes have been used to assess cardiac and skeletal muscle damage, namely aspartate aminotransferase (AST, Figure 7.1), LD, and CK (Figure 7.2).

Aspartate aminotransferase and lactate dehydrogenase

Aspartate aminotransferase activity in the blood starts to increase about 12 hours after a MI and reaches a peak at 36 hours. It usually reverts to normal levels 60 hours after an AMI. This was the first enzyme used as a cardiac biomarker. However, since 2000, AST has ceased to be recommended as a test for AMI, being superseded by other, more effective tests.

Lactate dehydrogenase is a ubiquitous enzyme in body tissue. Its main value as a cardiac biomarker is that its activity in the blood begins to rise later than AST, at 36 hours from the start of an infarction, and it is detectable for up to 72 hours. This means it can be used to detect an AMI in patients whose symptoms occurred more than 24 hours previously. As LD activity is so widespread in the body, an increase in LD levels is not sufficiently specific to diagnose a MI. However, this lack of specificity can be overcome by determining the pattern of isoenzymes of LD. An increase in LD1 or the ratio of LD1: LD2 activities was considered diagnostic of MI. This is called the LD1/LD2 flip. Measurement of LD activity using hydroxybutyrate as substrate provides a convenient way of measuring LD1 activity. This activity is often referred to as hydroxybutyrate dehydrogenase (HBD). Again, since 2000 it has been recommended that LD measurements should not be used in the clinical investigation of suspected AMIs.

Creatine kinase

Creatine kinase transfers a phosphate from creatine phosphate to adenosine diphosphate (ADP) to form adenosine triphosphate (ATP) as shown earlier in Figure 7.2. When muscle tissue contracts it consumes ATP; however, in some conditions muscles can potentially run out of ATP. Should this happen, the muscles would then stop working. However, muscles have stores of creatine phosphate that in the short term can be used to phosphorylate the ADP to ATP. The released creatine is subsequently broken down to creatinine, which, as discussed in Chapter 3, is used to test for renal function.

The level of CK activity in the blood is due to leakage from muscle tissues. This means that its activity in the blood will be affected by muscle mass and muscle composition. This is the reason that the reference value for CK activities in the blood of men is higher than that of women, who on average have less muscle mass. Slightly different types of muscle fibres are found in the muscles of different ethnic groups, which explains, for example, why CK levels are higher in Afro-Caribbeans than Caucasians.

Creatine kinase is found in all the muscles of the body. Its activity in the blood begins to increase about 4–6 hours following an AMI and peaks at 24 hours. However, there is much more skeletal muscle than cardiac muscle in the body. Consider someone involved in a road traffic accident who has a crushed chest and thus a damaged heart. There is also skeletal muscle damage, so the increase in CK from the damaged muscles may mask the increase resulting from the damaged heart. However, CK activity can be due to one of three isoenzymes, which were illustrated in Figure 7.3:

- CK1 or CKMM, found mostly in skeletal muscle
- CK2 or CKMB, found predominately in cardiac muscle
- CK3 or CKBB, found in smooth muscle

Assaying the individual isoenzymes can help in distinguishing between the CK activities from cardiac muscle and skeletal muscle damage. Thus, measuring CKMB activity is a more sensitive method of detecting an AMI, especially if there has been damage to skeletal muscle. Creatine kinase MB activity starts to increase slightly earlier than the overall CK levels and peaks 21 hours after a MI. To better distinguish between cardiac and skeletal muscle damage the amount of CKMB can be expressed as the ratio CKMB/total CK. The theory is that if the ratio increases it is more likely to be cardiac than skeletal muscle damage.

The usefulness of CKMB in detecting AMIs has led to the development of immunoassays for its measurement. However, there is a drawback to the use of CKMB in detecting AMI.

Although cardiac muscle contains more of the MB than other isoenzymes of CK it also contains significant amounts of CKMM. Similarly, although skeletal muscle also contains mostly CKMM, it does have significant amounts of CKMB. Crucially, this means that an increase in CKMB is not absolutely specific for cardiac damage, although for many years it was the best available test.

In addition to the isoenzymes of CK, there are also *isoforms* of CKMM and CKMB. Isoforms of a protein differ from one another due to post-translational modifications. Post-translational modifications include chemical modifications, such as glycosylation and phosphorylation, and, as is the case here, the removal of certain amino acid residues. Isoforms of CKMM and CKMB are formed by a deaminase in the bloodstream that removes the carboxy terminal amino acid residue from the M subunit of CK molecules. This means that there are potentially two isoforms of CKMB, called MB1 and MB2, and three isoforms of CKMM, called MM1, MM2, and MM3. There was initial interest in using the ratio of MB1 to MB2 as a very early test for AMI although this has been replaced by assays for cardiac troponin. However, measuring isoforms of CKMM is potentially a useful way of determining whether there has been recent skeletal muscle damage. The common form of CKMM is MM3, so an increased amount of MM1 would suggest that there has been recent skeletal muscle damage. This test is not yet used in routine clinical practice.

Muscle proteins as cardiac biomarkers

The muscle proteins, myoglobin and troponin, are used as cardiac biomarkers.

Myoglobin is a haem-containing protein of M_r 18,000 that strongly binds oxygen, allowing it to act as a local store of oxygen in the muscles. It is found in the cytoplasm of the cell and is released within 2–3 hours of an AMI reaching a peak concentration in the blood after 12 hours. As changes in myoglobin levels occur before those of any other cardiac biomarker there was, and is, interest in using measurement of myoglobin as a test for AMIs. However, like CK and CKMM, myoglobin is found in skeletal as well as cardiac muscle and thus possesses similar drawbacks.

Muscle tissue consists of bundles of muscle fibres. Each fibre is a single cell or syncitium containing many nuclei. This arrangement is a consequence of the fusion of many embryonic cells. The fibres contain typical organelles but are often given muscle-specific names, thus the plasma membrane is called the sarcolemma, the endoplasmic reticulum, sarcoplasmic reticulum, and the cytosol, sarcosol. Most of the sarcosol is occupied by longitudinally arranged bundles of contractile myofibrils. Myofibrils consist largely of two types of interdigitating protein filaments: one has a diameter of 12–16 nm and is referred to as thick filament, the other type have diameters of 7–8 nm and are called thin filaments. Contraction of the muscle fibre occurs in a sliding-type mechanism, with the two filaments moving towards each other.

The major proteins of the thick and thin filaments are myosin and actin respectively. Myosin is an adenosine triphosphatase (ATPase). When an ATP molecule binds to myosin, it can be enzymatically hydrolysed to ADP and phosphate. The free energy of this reaction is used to form links or cross bridges with the thin filament, which move the two filaments relatively closer together in a contractile event. The cross bridge is broken when a new ATP binds allowing the contractile cycle to repeat.

Calcium ions and regulatory proteins play key roles in muscle contraction and relaxation. Figure 7.7 shows a complex of molecules called the troponin-tropomyosin complex attached to the actin.

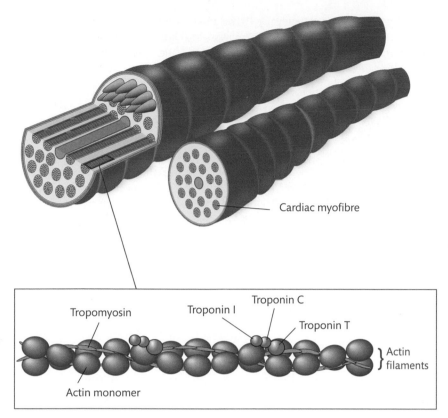

Cardiac myofibre

Tropomyosin

Troponin I

Troponin C

Troponin T

Actin filaments

Actin monomer

FIGURE 7.7
Cartoon showing the organization of the troponin-tropomyosin complex in a cardiac muscle fibre. The insert shows the coiled coil conformation of tropomyosin wrapping around an actin filament. The trimeric troponin complexes are bound to the tropomyosin as indicated. Its three subunits have separate functions: T is responsible for binding the complex to tropomyosin; I inhibits muscle contraction; and C binds calcium. See text for details.

Tropomyosin consists of two long α helices that coil around each other giving the molecule an extended conformation called a coiled coil. Each tropomyosin molecule is able to bind around seven actin molecules of the thin filament. Troponin molecules consist of three different subunits called T, I, and C respectively. Troponin T binds to tropomyosin, while troponin I inhibits the interactions between actin and myosin essential for contraction. Troponin C is able to reversibly bind calcium. In an absence of calcium, the binding of the troponin-tropomyosin complex to actin inhibits contraction. However, when a nerve impulse arrives at the motor end plate of a muscle fibre, it causes an electrical depolarization in the sarcolemma, which leads to the release of calcium from the sarcoplasmic reticulum into the sarcosol. The increased concentration of calcium allows it to bind to troponin C and causes a change in the conformation of the troponin molecule that reduces the affinity of the troponin-tropomyosin complex for actin. Thus, the inhibition is relieved and muscle contraction occurs. As the calcium is pumped back into the sarcoplasmic reticulum, the whole process is reversed and contraction ceases.

The amino acid sequences, and therefore structures, of both troponin T and troponin I in cardiac and skeletal muscle tissues are different. These differences have allowed immunoassays that are specific for cardiac troponin T (cTnT) and for cardiac troponin I (cTnI) to be developed. Assays for cTnT and cTnI have become the biochemical gold standard for detecting myocardial necrosis for three reasons. First, measurements of cTnT and cTnI are specific: cardiac damage is the only cause of any increase in cardiac troponin in the blood. Therefore, immunoassays for troponin have none of the drawbacks associated with measurements of CK activities and myoglobin concentrations. Second, cardiac troponin measurements are highly sensitive. The background level (reference value) of cardiac troponin in the circulation is extremely low.

Indeed, some of the methods currently used to measure the amounts of cardiac troponin are unable to detect any in the blood of healthy individuals. Third, and perhaps the most important reason, is that assaying for cardiac troponin detects patients with an AMI who are not detectable by measurements of CK and CKMB.

Approximately one-third of patients with chest pain, where a MI was excluded by serial measurement of CK and CKMB activities and therefore they were thought to be suffering from unstable angina, were found to have an increased cardiac troponin level in their blood samples. When these patients were monitored over time, it was found they had a greater risk of premature death or readmission with a MI. Numerous studies have confirmed these findings. A number of the standard treatments for AMI are only effective when given to individuals who have detectable troponin in the blood. An indication of the importance in diagnosis of assaying for cardiac troponin is that the definition of MI has been changed to include measurement of cardiac troponin as the test of choice. Acute myocardial infarctions are considered to have occurred when there are clinical suspicions of heart disease, preferably with changes on the ECG but accompanied by change in cardiac troponin in the blood, either an increase, decrease, or both.

Key Points

The definition of myocardial infarction is the detection of the rise and/or fall of cardiac biomarkers (preferably troponin) with at least one value above the 99th centile of the upper reference limit, together with evidence of ischaemia with at least one of the following: symptoms of ischaemia, ECG changes of new ischaemia (new ST-T changes), development pathological Q-waves in the ECG, or imaging evidence of new loss of viable myocardium or new regional wall motion abnormality.

SELF-CHECK 7.5

Which biochemical tests could be used to detect MI and what tests should not be used?

A range of medical conditions have been reported where the troponin concentrations in the blood have increased but the patient does not have an AMI. Thus, although an increased circulatory cardiac troponin concentration is absolutely specific for damage to cardiac muscle, an AMI is not the only cause of cardiac muscle damage. These conditions vary from the unexpected, such as patients with chronic kidney disease, to the exotic, for example puffer-fish poisoning. In all cases, the cause is injury to the cardiac muscle by a mixture of direct damage to the muscle itself and damage to the blood supply of the tissue.

7.8 Biomarkers of muscle damage for the diagnosis of AMI

Although assays for cTnT and cTnI are now the tests of choice for diagnosing AMI, they are not sufficient on their own and must be combined with a clinical assessment of the patient and the use of ECGs. When a patient comes to hospital with suspected ACS the initial assessment begins by performing an ECG and measuring the biomarkers cTnT or cTnI. If the ECG shows ST segment elevation, then the diagnosis is presumed to be STEMI and the patient will receive a thrombolytic drug or be taken to the cardiac catheterization laboratory. As you may remember from Section 7.5, assaying a cardiac biomarker is not required to make an initial diagnosis or to guide the treatment of these patients. However, cTnT or cTnI should be measured 10–12 hours after the patient first entered the hospital to confirm the diagnosis.

Unfortunately, in the majority of cases the diagnosis is less clear cut. An ECG may show only minor changes and can in fact be entirely normal. If this is the case, then the diagnosis depends on the cardiac troponin measurements, given that a change in troponin levels is necessary in order to diagnose AMI. Therefore, a blood sample and assays for cTnT or cTnI should be performed 6–8 hours after the patient first presented to hospital. If the diagnosis is still unclear and the patient's ECG changes or further symptoms develop that suggest an ACS, then the troponin measurement should be repeated six hours later. A normol troponin level in the blood 12–24 hours from first arrival at hospital excludes an AMI with a high degree of certainty.

Biomarkers for skeletal muscle damage

Damage to skeletal muscle can occur from a wide variety of causes although the most common is trauma. Skeletal muscle damage is easy to detect due to the release of CK into the blood but this can be misleading as vigorous exertion can also cause quite large increases in CK activity in the blood without any ill effects (apart from the aching you feel in the muscles afterwards). Increased CK activity due to damage to skeletal muscle is common following surgery, falls, or road traffic accidents. A more serious form of widespread muscle damage is **rhabdomyolysis**, which is the rapid lysis of skeletal muscle tissues due to injury to muscle tissue, accompanied by the excretion of myoglobin in the urine.

Although rhabdomyolysis may occur in association with traumatic damage, it is more commonly seen in association with other medical conditions or with drug treatment. Extensive rhabdomyolysis results in the widespread release of muscle constituents, especially myoglobin into the blood, which can result in damage to the kidneys and consequent acute renal failure. Kidney damage tends only to occur when extremely high CK activities are observed in the blood but there is not a very good correlation between the amounts of CK released and the probability of kidney damage. Activities of CK up to ten times the upper reference limit are not considered to be of clinical significance. However, values more than 50 times the upper reference limit are often seen in patients with rhabdomyolysis.

When estimating the degree of damage to skeletal muscle and determining whether or not any coincident cardiac damage has occurred, the combined measurement of CK and a cardiac troponin is immensely useful. However, measurement of myoglobin in the blood or in the urine does not assist in predicting whether or not rhabdomyolysis will be associated with acute kidney damage. In this circumstance, the simultaneous measurement of a marker of renal function is much more useful.

Cross reference
Chapter 3 Kidney disease

CASE STUDY 7.1

A 56-year-old male was admitted to the hospital with a six-hour history of central crushing chest pain which went into his left arm and neck. An ECG showed elevation of the ST segments in the chest leads.

He had a troponin value on admission of less than 0.01 µg/L and one measured six hours later of 1.5 µg/L (reference range <0.05 µg/L).

(a) Do you think this man was having a MI?

(b) Should you wait for the second troponin result taken six hours after admission to hospital before starting treatment?

(c) What is the role of the second troponin measurement in this patient?

CASE STUDY 7.2

A 65-year-old woman with a previous history of coronary artery bypass graft surgery was admitted to hospital with some mild central chest pain and breathlessness. Her ECG showed some T wave inversion.

Her cardiac troponin measured on admission was 0.1 µg/L and six hours later was 0.4 µg/L (reference range <0.05 µg/L).

(a) Could you make a confident diagnosis of MI when she was first admitted to hospital?

(b) Would you think this woman was having a MI after you received the results of the second blood test?

CASE STUDY 7.3

An 18-year-old man was involved in a fight outside a club and was stabbed in the left side of his chest. He was admitted to the hospital, where he collapsed with a poor pulse. There was evidence of a puncture wound to the heart. A blood sample with a request for a troponin assay was sent to the laboratory. The results reported a value of 1.5 µg/L (reference range <0.05 µg/L).

(a) Do you think this man has had an MI?

(b) Why is the troponin elevated in this man?

7.9 Natriuretic peptides

Natriuretic peptides are a series of ring-shaped molecules that promote an increased loss of sodium and water by the kidneys. Four natriuretic peptides occur naturally: atrial natriuretic peptide (ANP), B type natriuretic peptide (BNP), C type natriuretic peptide, and D type natriuretic peptide. They are all hormones which function in the homeostasis of sodium and water retention. Figure 7.8 indicates that they have many amino acid residues in common, but are slightly different and have varying properties.

Routine clinical practice concerns itself only with BNP as this is the only one measured routinely in clinical laboratories. It is the dominant natriuretic peptide produced by both the atria and the ventricles of the heart. As there is much more cardiac muscle in the ventricles than in the atria, it is the ventricles, which produce most of the BNP formed in the heart. B type natriuretic peptide is an unusual hormone in that it is produced constantly by the cardiac muscle cells. It does not appear to be stored within the cardiac muscle in any great amounts before its secretion. It is initially made as a prohormone called proBNP. It was believed that proBNP is hydrolysed into two portions: an N terminal portion called N-terminal proBNP (NTproBNP) and BNP. The NTproBNP is the more stable of the two and has a longer half-life in the bloodstream. However, recent studies have detected the presence of proBNP in the blood and so have challenged this mechanism.

The rate at which proBNP is manufactured by cardiac tissue is determined by the degree of tension on the cardiac muscle. Hence, when the cardiac muscle is stretched, more proBNP is made. This means that measuring the amounts of BNP or NTproBNP is extremely useful as a way of biochemically detecting an increase in the strain on the ventricle walls. An increase in BNP or NTproBNP will occur when the heart is stretched due to volume overload. The most common cause of this is **heart failure**, when the heart cannot pump blood at a rate sufficient to meet the needs of the body.

Heart failure is relatively common and usually occurs as a result of damage to the heart by a MI. Despite being common, heart failure is difficult to diagnose simply by taking a history and examining the patient. It has been estimated that as few as one-third of patients sent to hospital with a diagnosis of heart failure actually have that condition. A biochemical test for heart failure is therefore useful both for general practitioners and for doctors in an accident and emergency department. It has been found that measurement of BNP or NTproBNP can be reliably used to confirm or exclude a diagnosis of heart failure by general practitioners and in the emergency department.

7.10 **Possible future biomarkers in cardiac disease**

An ideal test for cardiac disease would be one that would predict that rupture of a vulnerable plaque was going to occur, or one that could detect ischaemia of cardiac muscle before it results in cell death. There have been a large number of different tests proposed for the early detection of both of these processes. To date, none of them has fulfilled the criteria discussed above. Some of the tests that have been examined may finally achieve this status but, to date, there have been many false starts.

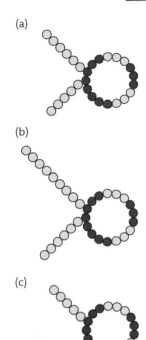

FIGURE 7.8
Schematics showing the structures of the natriuretic peptides. Each circle represents an amino acid residue. Residues common to all three are shown in red. (a) Atrial natriuretic peptide. (b) B type natriuretic peptide. (c) C type natriuretic peptide.

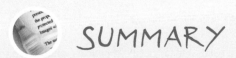

SUMMARY

- Enzymes are biological catalysts whose concentrations can be estimated using the reaction catalysed or measured by an immunoassay.

- Enzyme assays for diagnosing diseases depend on an increase in the activity or amount of enzyme present when a particular tissue is affected by illness and leakage or release of the enzyme into the blood.

- Enzymes are examples of biomarkers. Biomarkers are biological molecules whose concentrations alter when there is disease. They may be an intracellular metabolic intermediate, a messenger molecule, an enzyme, or a structural protein.

- Cardiac disease occurs when the blood vessels supplying the heart become blocked. This is usually due to sudden rupture of an atheromatous plaque, which will produce ischaemia and then infarction of cardiac tissue.

- The principal biomarkers of cardiac damage are the cardiac troponins. The combination of a change (appearance) of cardiac troponin in the blood, plus changes in the electrocardiogram and clinical features, can be used to diagnose and classify acute heart disease into stable angina, unstable angina, and myocardial infarction. Myocardial infarctions are divided into ST elevation myocardial infarction (STEMI) and non-ST elevation myocardial infarction (NSTEMI).

- Natriuretic peptides are biomarkers of cardiac failure and can be used to diagnose both acute and chronic heart failure.

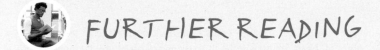

FURTHER READING

- **Collinson PO, Boa FG, and Gaze DC (2001) Measurement of cardiac troponins.** *Annals of Clinical Biochemistry* **38**, 423–49.

  A comprehensive review of cardiac troponin biochemistry and measurement methods.

- **Januzzi JL (2010) Cardiac Biomarkers in Clinical Practice. Sudbury, Massachusetts: Jones and Bartlett.**

  A comprehensive textbook on all aspects of cardiac biomarkers.

- **Januzzi J and Richards AM (2008) Review of natriuretic peptides.** *American Journal of Cardiology* **101**, S1–96.

  An illuminating review of natriuretic peptides.

- **Thygesen K, Alpert JS, and White HD (2007) Universal definition of myocardial infarction.** *Journal of American College of Cardiology* **50**, 2173–95.

QUESTIONS

7.1 A biomarker is:

    (a) A biological molecule which changes in concentration in the blood when there is disease

    (b) A biological molecule which acts as a catalyst for a specific chemical reaction

    (c) A biological molecule which decreases in concentration following a heart attack

    (d) A biological molecule which increases in concentration following a heart attack

    (e) A biological molecule which undergoes isomerization in the presence of disease

7.2 What is an enzyme and how can it be measured?

7.3 Is the following statement TRUE or FALSE?

Isoenzymes are composed of individual polypeptides which have undergone post-translational modification.

7.4 The difference between myocardial ischaemia and MI is:

(a) Myocardial ischaemia is always fatal

(b) Myocardial ischaemia only occurs when a blood vessel is completely blocked

(c) Myocardial ischaemia is reversible, but MI is not

(d) Myocardial infarction does not cause death, but myocardial ischaemia does

(e) Myocardial ischaemia will always occur when there is plaque rupture

7.5 Which tests can be used in the detection of MI?

7.6 Is the following statement TRUE or FALSE?

An elevated troponin level always means that the patient has had a MI (heart attack).

Answers to self-check questions, case study questions, and end-of-chapter questions are available in the Online Resource Centre accompanying this book.

 Go to www.oxfordtextbooks.co.uk/orc/ahmed/

8

Liver function tests

Roy Sherwood

Learning objectives

After studying this chapter you should be able to:

- Describe the functions of the liver
- Indicate the types of liver diseases that can occur
- Discuss the uses and limitations of the current liver function tests
- Explain the potential value of some new tests of liver function

Introduction

Tests for liver disease or dysfunction comprise a significant part of the routine workload of clinical biochemistry laboratories worldwide. Such tests may be carried out to establish whether an individual has liver disease or to determine the progression of existing disease. To understand and interpret the results of tests of liver disease it is necessary to have an understanding of the functions of the liver and also the limitations of the tests. This chapter will describe the liver and its functions in health and then consider its major diseases of interest in clinical practice. Liver function tests will be considered in some detail, in particular their use in investigation of disease, together with their advantages and limitations.

8.1 Liver

The liver is the largest solid organ in the body. In adults, it weighs 1.0–1.5 kg and its size is maintained in relatively constant proportion to body weight. It is comprised of a large right lobe and a smaller left lobe, and is located in the upper right quadrant of the abdomen under the lower edge of the diaphragm. A diagram of the liver is shown in Figure 8.1.

It is the only organ that has the capability to regenerate completely after it has been damaged or following a partial **hepatectomy**. Following liver transplantation in a child, the transplanted organ will grow with the child to the size appropriate for the child's size and weight.

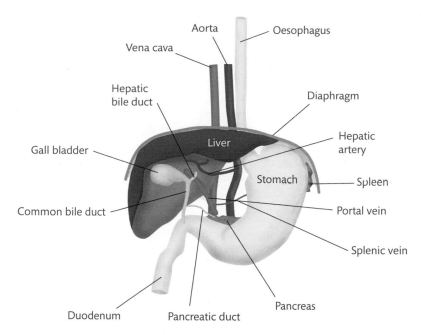

FIGURE 8.1
Anatomy of the liver (adapted from McFarlane I, Bomford A, and Sherwood R (2000) *Liver Disease and Laboratory Medicine.* **London: ACB Venture Publications, with permission).**

Parenchymal cells, or **hepatocytes**, comprise around 80% of the cells in the liver and perform the main metabolic functions associated with metabolism and **detoxification**. However, other important functions are performed by the non-parenchymal cells: endothelial lining cells, Ito cells, which store fat, Kupffer cells, which are modified macrophages, and pit cells, which are related to natural killer cells and are thought to play a role in defence against infection. Figure 8.2 shows the hepatocytes in the liver, bile canaliculi, and their association with the bile duct, branches of the hepatic artery and portal vein.

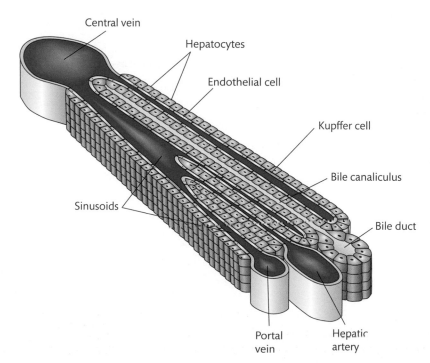

FIGURE 8.2
Schematic illustrating section through the liver showing organization of hepatocytes in relation to bile canaliculi, bile duct, hepatic artery, and portal vein.

Functions of the liver

The liver carries out numerous complex functions as shown in Table 8.1. The liver maintains the body's energy supply by its involvement in carbohydrate and fat metabolism.

Other than the immunoglobulins, most proteins in the circulation are synthesized wholly or partly in the liver. Except for **albumin**, transcobalamin II, and C-reactive protein (CRP) all are glycoproteins with carbohydrate side chains being added in the liver or in other organs. The liver is the site of the primary metabolism and detoxification of both endogenous and exogenous compounds, for example waste products of metabolism and drugs respectively. Many essential compounds such as vitamins and iron are stored in the liver, from where they are released into the circulation when required. The liver is also a site of hormone metabolism especially the 25-hydroxylation of vitamin D.

The maintenance of a stable glucose concentration in the circulation involves not just the hormones insulin and glucagon from the pancreas, but also the processes of **glycogenesis**, **glycogenolysis**, and **gluconeogenesis**, which take place in hepatocytes. Indeed, removal of the liver would lead to death from **hypoglycaemia** long before any toxic metabolites could accumulate. In situations where carbohydrates are readily available as is the case immediately following a meal, the liver synthesizes glycogen from the carbohydrates by glycogenesis and then breaks it down to form glucose when required, as in the fasting state by glycogenolysis. The liver can make glucose from other sources by gluconeogenesis using lactate, pyruvate, glycerol, and alanine. Glucose can also be used for fatty acid synthesis via the glycolytic and citric acid cycles.

The production and secretion of **bile** by the liver is essential to normal digestion and particularly to the absorption and metabolism of fats in the diet. Bile is a complex mixture containing

TABLE 8.1 Functions of the liver.

Liver function	Pathway/mechanism
Carbohydrate metabolism	Gluconeogenesis Glycogen synthesis and metabolism
Fat metabolism	Fatty acid synthesis Cholesterol synthesis and excretion Lipoprotein synthesis Bile acid synthesis
Protein metabolism	Synthesis of plasma proteins Urea synthesis and nitrogen removal
Hormone metabolism	Metabolism of steroid hormones 25-hydroxylation of vitamin D Metabolism of polypeptide hormones
Metabolism and excretion of drugs/toxins	
Storage	Glycogen Vitamins Iron
Metabolism and excretion of bilirubin	

cholesterol, cholesteryl esters, proteins, and bile salts (mainly taurocholate and glycocholate). Bile salts are required for the absorption from the small intestine of the free fatty acids and monoglycerides, formed in the gastrointestinal tract (GIT) by the action of lipase secreted from the exocrine cells of the pancreas. Mitochondrial β-oxidation of short-chain fatty acids occurs in the liver to form acetyl-CoA. The liver also synthesizes fatty acids, triglycerides, cholesterol, phospholipids, and lipoproteins.

Through **ureagenesis**, the liver plays a significant role in nitrogen and hydrogen ion homeostasis. A 70 kg human being needs to excrete about 10–20 g nitrogen a day from amino acids that are surplus to requirements and generated by protein catabolism. Nitrogen is converted to the water-soluble compound **urea** in the liver, which can then be excreted via the kidneys in the urine. Amino acids are first transaminated to glutamate then deaminated to ammonia, which can then enter the urea cycle.

When exogenous compounds, termed **xenobiotics**, reach the liver, it renders them water-soluble to permit excretion. The first phase of detoxification involves the creation of a polar group which is then available for conjugation, for example hydroxylation, which is mediated by **mixed function oxidases** (cytochrome P450 monooxygenases). Phase II is conjugation with glucuronic acid, acetyl or methyl radicals, sulphates, glycine, or taurine.

Bilirubin is a yellow-orange coloured pigment derived from haem (Fe-protoporphyrin IX). About 500 mmol (275 mg) of bilirubin is produced each day by the **reticuloendothelial system**, predominantly in the spleen and bone marrow. Erythrocytes are destroyed in the spleen, releasing the haemoglobin which is degraded further into haem and a globin chain. The haem is converted to protoporphyrin following release of the iron, which is either stored in the liver or bone marrow, or utilized during erythropoiesis in the bone marrow. The protoporphyrin is converted to bilirubin, which can leave the reticuloendothelial cells and bind to albumin. Figure 8.3 illustrates the formation of bilirubin from haem.

Figure 8.4 shows the formation of bilirubin following breakdown of haemoglobin, its subsequent conjugation in the liver, and excretion.

The bilirubin produced binds tightly, but reversibly, to albumin in a molar ratio of 1:1 at normal concentrations, but additional binding sites with lower affinity are found in **hyperbilirubinaemia**. Unconjugated bilirubin is insoluble in water at physiological pH and therefore binding to albumin prevents **extrahepatic** uptake and enables it to be transported to the liver from the site of production. This is particularly relevant in the neonate, where bilirubin deposited in the brain results in irreversible brain damage, a process called **kernicterus**. In the liver, the albumin-bilirubin complex dissociates and bilirubin enters the hepatocyte by a carrier-mediated process, where it binds to cytosolic proteins, mainly glutathione S-transferase B. Bilirubin is conjugated in the endoplasmic reticulum with glucuronic acid by the action of uridine diphosphogluconate glucuronosyl transferase to form mono- and di-glucuronides that are water-soluble. The bilirubin conjugates are excreted in the bile, probably via an energy-dependent, carrier-mediated process to overcome a significant concentration gradient. Bile pigments reaching the colon undergo bacterial degradation to urobilinogen and stercobilinogen. An extrahepatic circulation also exists for these compounds, a small fraction reaches the systemic circulation and is subsequently excreted in the urine. The remainder is excreted in the faeces. Some urobilinogens oxidize spontaneously to stercobilin, an orange-brown pigment responsible for the colour of stools.

SELF-CHECK 8.1

What is bile and what is its precise role?

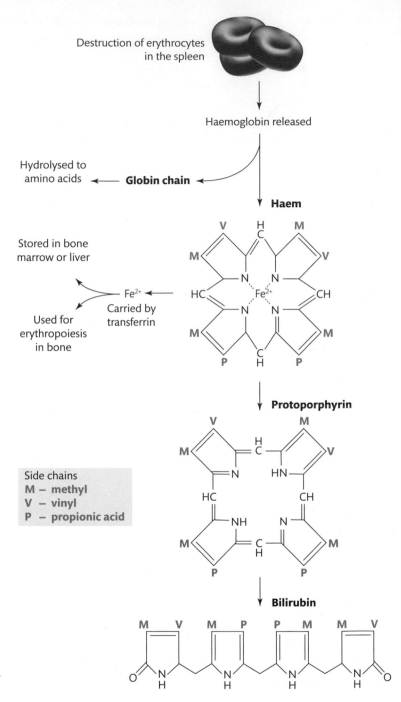

FIGURE 8.3
Schematic showing formation of bilirubin.
See text for details.

Key Points

Kernicterus is a neurological condition caused by deposition of bilirubin in the brain. It occurs in newborn children who have very high levels of blood bilirubin.

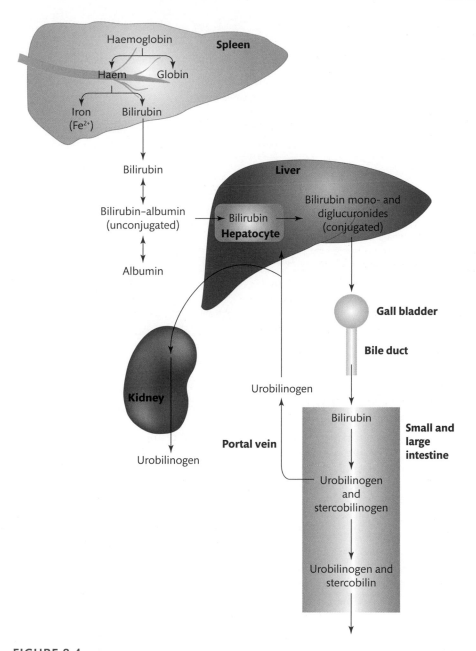

FIGURE 8.4
Formation of bilirubin, its subsequent conjugation in the liver, and excretion via the kidneys and the gut. See text for details.

8.2 **Diseases of the liver**

Diseases of the liver are classified according to their aetiologies, as shown in Table 8.2.

For clinical purposes, a subclassification defining the stage of the disease process, whether acute, subacute, or chronic and the pathological state of the liver assessed clinically, histologically, or radiologically is usually included.

TABLE 8.2 Diseases of the liver.

Viral	Biliary tract obstruction	Parasitic
Hepatitis viruses A to E	Gallstones	Schistosomes
Epstein-Barr virus	Sclerosing cholangitis	Liver flukes
Cytomegalovirus	Biliary atresia	Toxocara species
Arboviruses	Strictures	Tapeworms
Arenaviruses	Tumours	Leptospira species
Metabolic	**Vascular**	**Protozoal**
Haemochromatosis	Budd-Chiari syndrome	Amoebiasis
Wilson's disease	Portal vein thrombosis	Kala-azar
Hereditary hyperbilirubinaemias	Veno-occlusive disease	Malaria
Autoimmune	**Toxic/drug induced**	**Bacterial**
Autoimmune hepatitis	Alcohol	Tuberculosis
Primary biliary cirrhosis	Drugs	Pyogenic liver abcesses
Neoplastic	**Miscellaneous**	**Cryptogenic**
Primary	Polycystic liver disease	(cause unknown)
Secondary	Congenital hepatic fibrosis	

Cholestasis and jaundice

Cholestasis is the term used to describe the consequences of failure to produce and/or excrete bile. The failure may be related to hepatocyte damage or a reduction in the flow of bile out of the liver due to obstruction between the hepatic ducts and the sphincter of Oddi where the bile duct enters the duodenum. Cholestasis results in the accumulation of bilirubin in the blood, leading to **jaundice**. Three types of jaundice are usually recognized: hepatocellular (intrahepatic cholestasis), obstructive (extrahepatic cholestasis), and haemolytic. As the jaundice develops, darkening of the urine may occur as excretion of water

CASE STUDY 8.1

A 53-year-old woman presented to her doctor complaining of pruritis (itching), tiredness, and easy bruising. There was no previous medical history of note. On examination her liver and spleen were enlarged and she had spider naevi. The results of her biochemical tests for the liver were as follows (reference ranges are given in brackets):

Total protein	71 g/L	(60–80)
Albumin	30 g/L	(35–80)
Bilirubin	27 µmol/L	(0–20)
Alkaline phosphatase	856 IU/L	(30–130)
Aspartate aminotransferase	54 IU/L	(5–50)
Gamma-glutamyl transferase	643 IU/L	(5–55)

What is the differential diagnosis?

soluble bilirubin and its metabolites replaces biliary excretion. The lack of bile excretion into the duodenum, and therefore of bile pigments which are responsible for the characteristic colour of faeces, results in the stools becoming lighter in colour. Jaundice may also be due to excessive haemolysis, as occurs in the many different forms of haemolytic anaemia, in which case the excess bilirubin in the blood is predominantly unconjugated and does not appear in the urine.

Acute hepatitis and its sequelae

Acute inflammation of the liver, associated with hepatocyte damage, is most often caused by viruses or toxins, including drugs and alcohol, although there are many other potential causes, including autoimmune **hepatitis**, where the body produces antibodies against components of hepatocytes. The signs and symptoms shown depend on the severity of the process and the individual's response to the causative agent. Acute hepatitis may resolve, progress to chronic liver disease, or continue on to **acute liver failure** (ALF). In ALF, **hepatic encephalopathy** (swelling of the brain) develops, usually in association with severe coagulopathy, or prolongation of the time it takes for blood to clot, resulting from the failure of damaged hepatocytes to synthesize clotting factors.

Chronic liver disease

The distinction between acute and chronic liver disease is a temporal one. The initial onset of the disease is often difficult to define, but persistence of clinical or biochemical signs and/ or symptoms for more than six months is consistent with chronic liver disease. For example, **primary biliary cirrhosis** (PBC) and **haemochromatosis** have no recognizable acute phases and develop slowly with few, if any, symptoms over many months or years. In chronic hepatitis C, the initial viral infection may only produce mild non-specific symptoms and the development of chronic disease may not become apparent for many years.

Cirrhosis

The liver has a remarkable capacity to regenerate after it has been damaged. Following a single, short-lived, insult such as acute viral infection or paracetamol overdose it can often recover completely, with virtually normal architecture. When the cause of the damage persists, however, as, for example, in chronic viral infections, the capacity of the regenerative processes to keep pace with the liver cell death may be exceeded. Continuing hepatocyte breakdown can result in the formation of fibrous scar tissue. As the fibrous tissue deposition becomes more extensive, bridges of scar tissue form between adjacent portal tracts and/or between portal tracts and central veins, which is the hallmark of **cirrhosis**. The effects of cirrhosis on the liver are shown in Figure 8.5.

Cirrhosis is also the end result of the autoimmune diseases such as PBC and **primary sclerosing cholangitis** (PSC), where there is progressive destruction of the intrahepatic bile ducts. It can also develop when there is prolonged extrahepatic biliary obstruction. Even when frank cirrhosis has developed, there may be sufficient surviving parenchyma for the liver to continue to perform most of its normal functions, and patients may have no untoward symptoms. However, there may be rapid deterioration, often referred to as decompensation, with the development of complications and liver failure. Patients with long-standing cirrhosis are at increased risk of developing primary liver cancer or **hepatocellular carcinoma** (HCC).

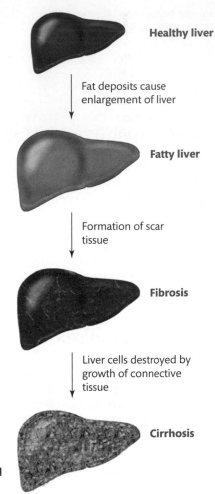

Healthy liver

Fat deposits cause enlargement of liver

Fatty liver

Formation of scar tissue

Fibrosis

Liver cells destroyed by growth of connective tissue

Cirrhosis

FIGURE 8.5
Schematic illustrating the changes in the physical appearance of a liver affected by cirrhosis.

Ascites

Ascites is the excessive accumulation of extracellular fluid in the peritoneal cavity. Most frequently, it occurs as a complication of advanced cirrhosis. Ascites can also develop during ALF in the absence of cirrhosis as well as in a number of non-hepatic conditions, particularly certain malignancies such as ovarian cancer. The precise mechanisms for its formation are not fully understood, but the primary event is probably reduced sodium excretion due to secondary **hyperaldosteronism**, compounded by **hypoalbuminaemia**. The urine may become virtually sodium free and, as long as sodium intake exceeds output, fluid will continue to accumulate.

Cross reference

Chapter 5 Fluid and electrolyte disorders

SELF-CHECK 8.2

What are the three types of jaundice?

8.3 Liver function tests

The liver has many diverse functions and it is, therefore, unlikely that any single biological marker would be able to detect all abnormalities associated with liver diseases. Many potential markers have been proposed over the years as tests of liver disease, but standard laboratory practice is for the initial assessment of liver disease to incorporate a profile of biochemical and haematological liver function tests (LFTs). This is something of a misnomer as few of the tests commonly incorporated in LFT panels are actually tests of liver function; most reflect the release of cellular contents following liver damage.

Why are liver function tests requested?

The first question that LFTs can help answer is 'does this patient have liver disease and if so how severe is it?' Negative results of LFTs are valuable as they indicate a low probability of significant disease. In the face of abnormal results for LFTs they provide valuable information in identifying the type of liver disease. The abnormalities in LFTs seen in disorders that are predominantly cholestatic (PBC, PSC) are quite different compared to those that are hepatitic (infective, autoimmune, toxins), which can guide further investigations. For example, in hepatitic disorders, viral serology and autoantibodies would typically be the second line of investigations, whereas in cholestatic conditions imaging techniques or **endoscopic retrograde cholangiopancreatography** (ERCP) would be more appropriate. Liver function tests can also provide information on the prognosis of the patient, particularly the likely survival times in chronic and acute disease, which has significant implications when liver transplantation is being considered. Monitoring patients with chronic liver disease with cirrhosis (for example hepatitis C, haemochromatosis, etc.) can detect progression to HCC at an early, potentially treatable stage. A typical panel of LFTs are shown in Table 8.3.

TABLE 8.3 Common liver function tests and the abnormalities they detect.

Test	Abnormality
Serum aminotransferases, for example AST, ALT	Parenchymal injury
Serum bilirubin	Cholestasis
Alkaline phosphatase, γ-glutamyl transferase	Biliary epithelial damage and biliary obstruction (alcohol)
Serum albumin	Synthetic function
Prothrombin time (INR)	Clotting (dynamic indicator)

BOX 8.1 Liver transplantation

Following trials in animals in the 1950s Thomas Starzl carried out the first human liver transplant in Colorado in 1963. Initially survival was measured in days but developments during the 1960s in the field of immunosuppressive agents, in particular cyclosporine developed from a fungus sample cultivated from soil, led to progressively longer survival times. Currently, the one-year survival rate is more than 90% and worldwide the five to ten-year survival rate is more than 80%. The 1980s was a period of significant growth in the field, with the establishment in the USA and UK of National Organ Transplant registries to identify donors and match them to recipients. Liver transplants in children with inherited metabolic liver disease were difficult initially until the introduction of reduced-size liver transplants (use of either the right or left lobe only) or split-liver transplants where the left lobe is given to one child and the right to another. A further development from this was living donor liver transplantation where a living person (usually a parent or close relative) donates a lobe or part of a lobe to a child. The laboratory has a major role in the assessment of patients prior to transplant and in monitoring following transplant for identification of rejection and optimization of the immunosuppressive therapy.

Aminotransferases

Aminotransferases, previously designated as 'transaminases', catalyse the transfer of an amino group from an α-amino acid to an α-oxo acid. The two most widely measured for clinical purposes are aspartate aminotransferase (AST) and alanine aminotransferase (ALT). They have no functional significance in blood, but are indicators of hepatic damage. Their activities are typically 3,000–7,000-fold higher in hepatocytes than in plasma. Although measurement of their activities is commonly used as a test of hepatocellular damage, these enzymes have a wide tissue distribution. Aspartate aminotransferase is found in liver, heart, skeletal muscle, kidney, brain, and red blood cells, and whilst ALT has a similar distribution, its concentrations are lower in extrahepatic tissues. Aspartate transaminase exists in both cytosolic and mitochondrial forms (half-life of cytosolic AST is 47 h whereas that for mitochondrial AST is 87 h), but ALT is in a cytosolic form only (half-life of 17 h). Consideration of the results of the aminotransferases as a ratio can be helpful as an AST/ALT ratio greater than two is suggestive of alcohol misuse because of release of mitochondrial AST due to damage from alcohol metabolites. The higher the relative increase in AST compared to alkaline phosphatase (ALP) the more likely it is that the patient has hepatitis. Although dependent on the methodology used for their measurement typical reference ranges for AST and ALT are 5–45 IU/L.

Intra-individual variation in the aminotransferases is low, with a day-to-day variation in AST of approximately 5–10% and for ALT 10–30%. There is no diurnal rhythm for AST, but ALT activity is up to 45% lower at night compared to a peak in the afternoon. Aspartate aminotransferase may be up to 15% higher in African-American subjects. There is a direct relationship between the measured serum activities of AST and ALT and body mass index (BMI); a BMI above 30 can increase both by 40–50%. Macro-AST and ALT have been reported where the enzyme is complexed to IgG molecules.

Increased AST/ALT activity due to liver disease

An increase in concentration of transaminases with a relatively smaller increase in other tests indicates hepatitic liver disease, for which there are many potential causes including:

- infective agents
- autoimmune disorders
- toxins

Infective agents such as hepatitis A/B/C, cytomegalovirus (CMV), Epstein-Barr virus (EBV), and others both viral and bacterial, all produce a significant increase in serum AST/ALT.

Alanine aminotransferase is better than AST for monitoring viral activity in chronic hepatitis B/C due to it being present only as a cytosolic form, which is released from damaged hepatocytes more easily than AST.

Autoimmune hepatitis can also produce an increase in AST and ALT and if suspected can be confirmed by detecting the presence of liver-kidney-muscle soluble liver antigen or liver cytosol autoantibodies. The range of toxins that can cause an increase in AST/ALT is wide. Alcoholic liver damage is usually chronic and cholestatic (cirrhotic), and alcoholic hepatitis is uncommon, but can occur in 'binge' drinkers, particularly in women. Many drugs have been reported to cause an increase in serum AST/ALT due to hepatocyte destruction and this is often idiosyncratic, especially for therapeutic drugs such as statins. A special case is the acute hepatocyte damage associated with overdosage of paracetamol and this is described in more detail below. Metals can be toxic to the liver and can cause increases in concentration of AST/ALT in the serum, in particular if associated with the inherited metabolic diseases of copper (Wilson's disease) and iron (haemochromatosis) where the metals accumulate in excess in the liver. Other poisons that can result in significant hepatocyte damage and release of AST/ALT include mushrooms (*Amanita phalloides*).

Non-hepatic disease and AST/ALT

The relative non-specificity of serum measurements of AST, and to a lesser extent ALT, for liver disease must always be taken into consideration when interpreting results. Skeletal muscle disease can produce a significant increase in AST with a smaller increase in ALT, which is proportional to the rise in creatine kinase (CK) and the extent of muscle damage. Causes of skeletal muscle disease include: hypothyroidism, rhabdomyolysis, drug misuse (especially cocaine or alcohol), surgery or trauma, fitting, polymyositis, dermatomyositis, and extreme exercise. Aspartate aminotransferase is present in cardiac myocytes so serum AST can be increased following **acute myocardial infarction** (AMI), cardiac surgery, or trauma. This can be confirmed by the finding of a positive cardiac troponin and a raised CK. Aspartate aminotransferase is increased in the presence of haemolytic anaemia more than ALT due to release from the breakdown of red blood cells. The extent of increase is dependent on the degree of haemolysis and is usually proportionally less than the increase in lactate dehydrogenase (LD). In appropriate ethnic populations inherited disorders that cause haemolytic anaemia must be considered, for example sickle cell disease (HbS and C), **hereditary spherocytosis** or glucose 6-phosphate dehydrogenase (G6PD) deficiency. Drug induced haemolytic anaemia although relatively uncommon must not be overlooked.

Paracetamol overdosage and the aminotransferases

Paracetamol is a major hepatic toxin if more than 10 g is ingested, due to formation of a reactive intermediate N-acetyl-*para*-benzoquinoneimine (NAPQI) when the normal metabolic pathway is exhausted. Unfortunately, hepatic symptoms are often delayed, with presentation occurring at 48–72 h after ingestion. Paracetamol overdosage is the commonest cause of ALF in the UK and for super-urgent transplant listing. If given in the first 36 hours, N-acetylcysteine (NAC) infusion is an effective treatment by replenishing hepatic glutathione to restore the normal sulphation and glucuronidation pathways of paracetamol metabolism and minimizing NAPQI production. In severe cases of paracetamol overdosage, multi-organ damage can occur, with pancreatic, renal, adrenal, and pituitary glands all suffering damage. A rise in serum AST/ALT indicates hepatocyte destruction and values can often be as high as 10,000 IU/L, with the magnitude of the rise being a rough guide to the extent of damage that has occurred. However, AST needs to be considered with bilirubin in assessing prognosis using serial results, as if the AST and bilirubin fall together this suggests hepatic recovery and good prognosis, whereas a falling AST with a rising bilirubin indicates critical loss of hepatocytes and poor prognosis without transplantation.

Key Points

N-acetyl cysteine can be rapidly metabolized to glutathione, which can protect against NAPQI toxicity in paracetamol poisoning.

Bilirubin

Bilirubin is a yellow-orange coloured pigment derived from haem. Bilirubin is produced when red blood cells are destroyed; therefore excessive haemolysis or cell destruction causes an increase in serum bilirubin (usually unconjugated). Measurement of serum/plasma bilirubin is one of the true 'liver function' tests. A normal serum bilirubin concentration is less than 20 μmol/L. Increased plasma bilirubin concentrations above about 50 μmol/L cause the yellow colouration of individuals, referred to as jaundice. Differentiation of conjugated from unconjugated hyperbilirubinaemia is a valuable tool in determining the cause of jaundice. Conjugated bilirubin reacts with diazotized sulphanillic acid to produce a coloured compound, azobilirubin, which is measured spectroscopically. Intramolecular hydrogen bonding in unconjugated bilirubin prevents it reacting with the sulphanillic acid unless the bonds are first disrupted by caffeine or alcohol. Hence, conjugated bilirubin has been termed 'direct reacting' and unconjugated 'indirect reacting'. In obstructive liver disease, biliary excretion of conjugated bilirubin is inadequate to meet the demand so conjugated hyperbilirubinaemia occurs, whereas in hepatic failure there is an insufficient hepatocyte mass for conjugation as well as failed excretion, and a mixed pattern emerges. Hyperbilirubinaemia of the newborn is unconjugated due to hepatocyte immaturity and should normalize within 2–3 days after birth as the liver matures.

The daily production of unconjugated bilirubin is 250–350 mg whilst clearance is approximately 400 mg/day in adults (5 mg/kg/day). The half-life of unconjugated bilirubin is less than five minutes, whereas δ-bilirubin (conjugated bilirubin-albumin complex) has a half-life of 17–20 days. The serum bilirubin concentration in an individual shows a day-to-day variation of 15–30%. Concentrations are 15–30% lower in African-Americans. Food ingestion does not alter the bilirubin concentration but there is a two-fold increase after 48 hours of fasting.

Key Points

Bilirubin that has been conjugated to glucuronide is referred to as conjugated bilirubin and is water-soluble. Unconjugated bilirubin is free and is not water soluble.

Non-hepatic causes of raised bilirubin

As for AST, in appropriate ethnic populations inherited disorders that cause haemolytic anaemias must be considered in unconjugated hyperbilirubinaemias, for example sickle cell disease (HbS and C), hereditary spherocytosis, or G6PD deficiency. In a sickle cell crisis the plasma bilirubin is typically 30–60 μmol/L. Intravascular haemolysis due to other causes includes toxins: for example in Wilson's disease free copper can cause red blood cell destruction and a normocytic anaemia with raised bilirubin; infections, for example malaria, can cause a hyperbilirubinaemia; and various others—PBC, autoimmune hepatitis, and alcoholic hepatitis (**Zieve's syndrome**).

Gilbert's syndrome

Bilirubin conjugation is catalysed by UDP-glucuronosyl transferase 1A in the liver. **Gilbert's syndrome** is a common autosomal dominant benign condition that results in a mild unconjugated hyperbilirubinaemia and has an estimated prevalence of 3–8% in the general population. The defect is a TA insertion in the TATA repeat box of the gene promoter region. Homozygosity reduces the capacity for gene expression. Normally this is of no consequence, but in situations of increased demand, for example illness, infections, and fasting, a marked rise in plasma bilirubin (40–80 µmol/L) can occur.

Bilirubin in the neonate

Unconjugated hyperbilirubinaemia in the first few days of life may be physiological, that is, due to immaturity or breast milk jaundice. However, hereditary unconjugated hyperbilirubinaemias may occur in certain pathological conditions, for example haemolytic disorders, hypoxia, galactosaemia, and fructosaemia. Conjugated hyperbilirubinaemia is always pathological, including: infections (bacterial or viral); **extrahepatic biliary atresia** (EHBA); endocrine, such as congenital hypothyroidism; vascular abnormalities; hereditary conjugated hyperbilirubinaemias; or associated with total parenteral nutrition.

Alkaline phosphatase

Alkaline phosphatases are a group of enzymes that hydrolyse phosphate esters in alkaline solutions. Four structural genes encoding ALP have been identified and sequenced. They consist of a tissue non-specific *ALP* gene on chromosome 1 and genes for the intestinal, placental, and germ cell ALP forms on chromosome 2. As its name suggests, the tissue non-specific gene is widely expressed, being present in osteoblasts, hepatocytes, and other cells. Tissue specific differences in the properties of the ALP from these various cells exist, originating

CASE STUDY 8.2

A 22-year-old biomedical scientist recovering from flu was noticed to be slightly jaundiced. He was concerned that he might have hepatitis and therefore had some blood tests (reference ranges are given in brackets):

Bilirubin	58 µmol/L	(0–20)
Alkaline phosphatase	89 IU/L	(30–130)
Aspartate aminotransferases	42 IU/L	(5–50)

His urine bilirubin was negative and both his haemoglobin and reticulocyte count were both normal.

What could be the possible cause of the jaundice?

from post-translational modifications in the carbohydrate side chains. In the circulation, the plasma ALP activity is mainly derived from liver and bone isoforms in approximately equal amounts with contributions from the intestinal form after eating and from placental ALP during pregnancy. The liver-derived ALP isoform is located on the outside of the bile canalicular membrane and increased plasma ALP activity reflects intra- or extrahepatic bile duct obstruction.

Plasma ALP activity varies with age in childhood due to growth spurts, with bone formation releasing the bone isoenzyme; age dependent reference ranges are therefore usually used up to 18 years of age. Adult ranges for ALP are heavily method dependent but a common range would be 30–120 IU/L. Serum ALP activities are 10–15% higher in African-Americans. Following food ingestion, increases of up to 30 U/L may be seen. In patients with blood groups B and O; these increases can persist for up to 12 hours; attributable to the intestinal isoenzyme. An unexplained high serum ALP result should always be confirmed with a repeat fasting sample. Increases up to two- or three-fold are normal in the third trimester of pregnancy due to the presence of the placental isoenzyme.

Key Points

Alkaline phosphatase concentrations in the plasma of children are higher than those of adults. This is believed to be due to increased release from the bones of growing children.

Raised plasma ALP

Any obstruction of the bile duct or ductules can cause an increase in plasma ALP activity. This may be intrahepatic as in PBC or cryptogenic cirrhosis, or may be because of obstruction of the common bile duct by gallstones, strictures (for example PSC), or biliary or pancreatic malignancy. Portal hypertension or reduced blood flow due to left heart failure also result in raised plasma ALP.

Disorders that cause an increase in plasma ALP not of liver origin include increased bone isoform from osteoblasts in **Paget's disease** or malignancy in the bone. Germ cell tumours can produce either the germ cell isoenzyme or in some cases the placental isoenzyme. A condition that can result in significantly increased plasma ALP is benign transient **hyperphosphatasaemia**. Originally described in infants, transient hyperphosphatasaemia can also occur in adults and during pregnancy. There is a marked rise in ALP, often to several thousand IU/L, which would usually indicate significant pathology. It is, however, a benign condition with a return to normal of the ALP in 6–8 weeks. Transient hyperphosphatasaemia is associated with concurrent infections in over 60% of cases, particularly GIT infections. There is a characteristic pattern on polyacrylamide gel electrophoresis with the normal pattern of isoenzymes being accompanied by variant forms which react with neuraminidase. It is believed to be due to changes in carbohydrate side-chains causing failure of recognition by receptors and reduced clearance, thus prolonging half-life.

Gamma-glutamyl transferase

Gamma-glutamyl transferase (GGT) is an enzyme responsible for the transfer of glutamyl groups from γ-glutamyl peptides to other peptides or amino acids. It is present throughout the liver and biliary tract and in smaller amounts in other organs, such as the heart, kidneys, lungs, pancreas, and seminal vesicles. Gamma-glutamyl transferase activity in the plasma is predominantly (95%) due to the liver isoenzyme. Plasma GGT is increased by enzyme inducing drugs, such as phenytoin and carbamazepine, as well as alcohol, because of increased cellular concentrations of GGT and release due to cell breakdown. There is a direct relationship between plasma GGT and the amount of alcohol consumed; measurement of plasma GGT has therefore been used as a marker of excess alcohol intake. A raised plasma GGT can persist for weeks after cessation of chronic alcohol misuse and often does not return to normal even if an individual abstains from alcohol entirely as there is underlying liver damage. As for the aminotransferases and ALP the reference range is method dependent but is typically 5–50 IU/L. Plasma GGT activity decreases immediately after food, then increases with increasing time after ingestion until it returns to its starting value. There is a relationship between plasma GGT activity and weight, with values being 50% higher in individuals with a BMI >30. This is believed to be due to fat deposition in the liver (**steatosis**) in obese subjects. Steatosis with a raised plasma GGT also occurs in diabetes mellitus, non-alcoholic steatohepatitis, and non-alcoholic fatty liver disease. Any liver disease that results in fibrosis and/or cirrhosis, such as alcoholic cirrhosis, PBC, PSC, haemochromatosis, α_1-antitrypsin deficiency, and Wilson's disease will cause a raised plasma GGT. Space occupying lesions, including malignancy (HCC or metastases secondary to malignancy elsewhere in the body), and granulomatous disease, for example sarcoidosis and TB are also associated with a raised plasma GGT.

Key Points

Serum GGT concentrations are often very high in individuals who suffer from alcoholic liver disease.

Plasma proteins

Plasma proteins, with the exception of the immunoglobulins, are produced in the liver. The protein produced in largest amounts (about 12 g a day) is albumin, which is a critical osmotic regulator and transporter of many hydrophobic compounds. The acute phase proteins, for example anti-proteases, are synthesized in the liver, as are the metal and vitamin transport proteins, for example transferrin, caeruloplasmin, and vitamin D binding protein.

Plasma albumin

Maintenance of plasma albumin concentrations can be achieved with only 10% of normal hepatocyte mass. Once the plasma albumin falls below 20 g/L, the risk of ascites and/or **oedema** increases and drug binding/transport decreases. A low plasma albumin usually suggests chronic liver disease as the half-life of albumin in the circulation is 21 days.

Other causes of a low plasma albumin are protein loss via the kidney in **nephrotic syndrome**, through the skin in burns, or the GIT in protein-losing enteropathy. A low protein intake in cases of malnutrition or patients on low protein diets will result in a reduced plasma albumin, as will failure to absorb adequate amounts of amino acids for protein synthesis due to **malabsorption** associated with GIT disease. It must be remembered that the dye bromocresol green used in the method most commonly used for the measurement of plasma albumin also reacts with α-globulins for example α_1-antitrypsin and α_2-macroglobulin so that in the total absence of albumin the method will still produce a result of up to 5 g/L. The bromocresol purple method is not subject to this interference but is more expensive and therefore less commonly used.

Plasma globulins

Increases in plasma globulins occur in alcoholic cirrhosis due to immunoglobulin production. Similar increases are often associated with autoimmune liver disease, for example autoimmune hepatitis, PBC, etc. Transferrin may be increased if there is concurrent iron deficiency, in an attempt to scavenge as much available iron as possible.

International normalized ratio

The **international normalized ratio** (INR) is the prothrombin time (PT) of a patient's blood sample divided by the PT time of a normal control sample analysed in the same batch (12–16 seconds). A normal result for this ratio is 0.8–1.1. The INR is a dynamic true 'liver function' test rising rapidly in ALF. The PT time measures the rate at which prothrombin is converted to thrombin in the presence of activated clotting factors, calcium and thromboplastin. Raised INR indicates deficiency of factors II, VII, IX, or X, or of vitamin K, which is involved in conversion of precursors. An INR >3.0 is a contraindication for liver biopsy.

Other secondary biochemical liver tests

Alpha-fetoprotein (AFP) measurements are used as a tumour marker for the detection and monitoring of primary hepatocellular malignancies, such as hepatoblastoma and HCC. Alpha-fetoprotein is produced by hepatoblasts and is therefore raised in the regenerating liver particularly in chronic viral hepatitis. Carbohydrate deficient transferrin is a test that has high specificity for the detection of excess alcohol intake as a cause of liver damage. The carbohydrate antigen CA19-9 is useful in monitoring the activity of the autoimmune disease PSC, which often progresses to a tumour of the bile ducts or **cholangiocarcinoma**. It is, however,

CASE STUDY 8.3

A 14-year-old boy presented to his doctor complaining of abdominal pain, nausea, and weakness. His previous medical history included recent travel to Africa to visit relatives. On examination he was jaundiced, had a fever, and an enlarged liver with tenderness. The results of his biochemical tests for the liver were as follows (reference ranges are given in brackets):

Total protein	71 g/L	(60–80)
Albumin	40 g/L	(35–80)
Bilirubin	233 µmol/L	(0–20)
Alkaline phosphatase	307 IU/L	(30–130)
Aspartate aminotransferase	1825 IU/L	(5–50)
Gamma-glutamyl transferase	282 IU/L	(5–55)

What is the differential diagnosis?

not specific and is elevated in many cholestatic liver diseases as well as in colorectal and pancreatic tumours. Measurement of serum ferritin can be useful in identifying haemochromatosis, but ferritin is a positive acute phase reactant so is raised in many illnesses as well as being released from damaged hepatocytes in acute hepatic failure. There are many specific diagnostic tests, for example copper/caeruloplasmin for Wilson's disease, α_1-antitrypsin that are beyond the scope of this chapter.

SELF-CHECK 8.3

What are the different forms of alkaline phosphatase?

8.4 The current liver function tests: pros and cons

The current profile of LFTs are inexpensive to run, are widely available, and interpretable. The major role of LFTs is to direct subsequent investigations, which may include imaging techniques, viral serology, and measurement of autoantibodies.

The biggest disadvantage is that they are over 30 years old and there are many newly recognized diseases for which they have no specific diagnostic value. They have little prognostic value, which is necessary in liver transplantation, or use for evaluating therapeutic success. They do not assess liver 'function' quantitatively or dynamically.

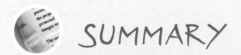 SUMMARY

- The liver has many metabolic functions, being involved in carbohydrate, fat, steroid, and protein synthesis; storage of essential vitamins and metals; and the detoxification of endogenous and exogenous substances.

- Liver function tests are requested to aid in diagnosis, to determine prognosis and to monitor response to treatment.

- A typical panel of LFTs includes the aminotransferases AST and ALT to detect hepatocyte damage, bilirubin to assess conjugation and to detect biliary obstruction, ALP and GGT to identify cholestasis, and albumin and INR to measure synthetic function.

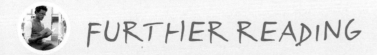 FURTHER READING

- **American Gastroenterological Association (2002) Medical position statement: evaluation of liver chemistry tests. *Gastroenterology* 123, 1364–6.**

  Guidelines on the use of biochemical tests of liver disease in clinical practice.

- Bogdanos DP, Invernizzi P, Mackay IR, and Vergani D (2008) **Autoimmune liver serology: current diagnostic and clinical challenges.** *World Journal of Gastroenterology* **14**, 3374–87.

  Review of autoantibodies used in the diagnosis of autoimmune liver diseases.

- Fontana RJ (2008) **Acute liver failure including acetaminophen overdose.** *Medical Clinics of North America* **9**, 761–94.

  Comprehensive review of causes and effects of acute liver failure, with particular reference to paracetamol overdose.

- McFarlane I, Bomford A, and Sherwood R (2000) *Liver Disease and Laboratory Medicine.* **London: ACB Venture Publications.**

  A textbook covering all aspects of laboratory involvement with liver diseases.

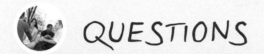

QUESTIONS

8.1 What is the role of the liver in carbohydrate metabolism?

8.2 Which of the following proteins is NOT made in the liver?

(a) Albumin

(b) Transferrin

(c) Immunoglobulins

(d) C-reactive protein

8.3 Assuming that no further hepatocyte damage occurs how long will it take an ALT of 120 IU/L to fall to within the reference range?

8.4 How does N-acetylcysteine help protect the liver following paracetamol overdosage?

8.5 In the measurement of bilirubin, is indirect bilirubin conjugated or unconjugated bilirubin?

Answers to self-check questions, case study questions, and end-of-chapter questions are available in the Online Resource Centre accompanying this book.

 Go to www.oxfordtextbooks.co.uk/orc/ahmed/

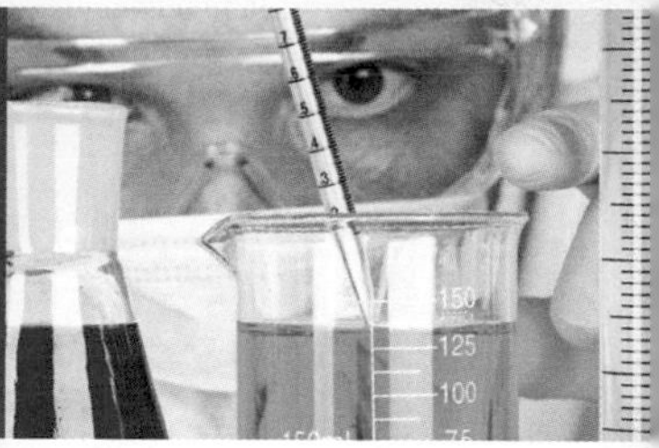

Abnormalities of lipid metabolism

Mike France

Learning objectives

After studying this chapter you should be able to:

- Define what is meant by the term lipid
- Outline a simple classification of the lipids and explain their nomenclature
- Illustrate the structures of common lipids
- Discuss the metabolism of the various types of lipids
- Classify and describe lipid disorders
- Discuss the roles of lipids in health and disease
- Reflect on the basis for deciding to treat hypercholesterolaemia with drugs
- Evaluate the key role of calibration in the accuracy of lipid assays
- State the advantages of using apoprotein rather than cholesterol measurements

Introduction

Lipids are any greasy materials that can be extracted from organisms. They do not usually dissolve easily in water and so give a permanent dark stain when rubbed on absorbent paper. However, they are normally more soluble in organic solvents. Lipids show an extraordinarily diverse array of different structures, which is a reflection of their significant roles in organisms. In cells, they form substantial parts of biological membranes, serve as metabolic fuels, are precursors of hormones, and act in cell signalling and control systems.

Given their diversity of structures, lipids can be classified in a number of ways. One of the simplest is to divide them on the basis of the products released on hydrolysis. This gives three major groups:

(1) simple lipids

(2) compound or complex lipids

(3) derived or polyprenyl lipids

The products obtained by hydrolysing a simple lipid are an alcohol and a variable number of fatty acids (FAs). Common examples of simple lipids are waxes and **triacylglycerols** (TAGs). Triacylglycerols are still often referred to by their old and misleading name of triglycerides. Examples of compound lipids include the **phospholipids** (PLs) and **glycolipids** found in biological membranes. On hydrolysis, the complex or compound lipids produce an alcohol and FAs but also other compounds such as phosphate groups, complex alcohols, or sugars. The third group, sometimes simply called *others*, are all derivatives of isoprene and cannot be hydrolysed. They are often referred to as polyprenyl compounds, although their old and again misleading name, of isoprenoid compounds is still used. Polyprenyl compounds include the fat-soluble vitamins A and D, steroids, for example cholesterol, and some of its hormone derivatives such as aldosterone, testosterone, and oestrogen. Fatty acids, TAGs, PLs, and cholesterol and its derivatives are all of clinical interest because of their roles in the development of deposits in arteries, called **atheromas** (Section 9.8), which are associated with heart disease, strokes, and peripheral vascular disease.

This chapter describes the structures and metabolism of lipids, their role in disease and emphasizes their role in the development of heart disease.

9.1 **Types of lipids**

Given the large numbers of different types of lipids, this subsection will concentrate on FAs and TAGs, PLs, and steroids (largely cholesterol), which can be regarded as representative types.

Fatty acids

Fatty acid molecules usually consist of an even numbered chain of carbon atoms that terminates with a carboxyl group (Figure 9.1). The normal chemical numbering would be to count or name from the carboxylic carbon atom, with the carboxylic carbon being designated C-1. The carbon adjacent to it is called the α carbon, that furthest is called the ω carbon. Fatty acids may be **saturated** or **unsaturated** depending on the presence of double bonds.

Fatty acids with several double bonds are called **polyunsaturated**. In natural FAs, the double bonds are in the *cis* configuration (Box 9.1), meaning the hydrogen atoms are attached to the same side of the double bond. *Trans* FAs, where they are on opposite sides, are harmful and are not part of normal metabolism.

In chemical nomenclature, the double bonds would be identified by the numbered positions from the carboxylic carbon. It is, however, becoming increasingly common to number their positions counting from the ω carbon. For example, the 20 carbon FA with five double bonds, called eicosapentaenoic acid, can be described numerically as $\triangle^3, \triangle^6, \triangle^9, \triangle^{12}, \triangle^{15}\text{-}C_{20:5}$, using the ω nomenclature. The 20 refers to the number of carbon atoms and the 5 to the number of double bonds. The $\triangle^3$ means, for example, there is a double bond between the third and fourth carbon atoms counting from the ω carbon and so on.

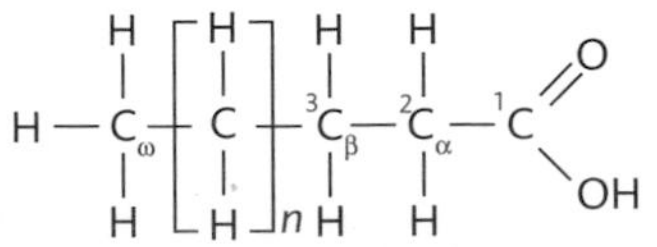

FIGURE 9.1
Structure of a fatty acid.

BOX 9.1 *cis fatty acids*

Unsaturated FAs produced industrially have double bonds in the *trans* configuration. A healthy, natural diet does not contain *trans* FAs. Polyunsaturated FAs can be extracted from fish oils and are used therapeutically as Omacor, which contains the ω-3 FAs, eicosapentaenoic, and docosahexaenoic acids. Their beneficial effects arise partially from their *cis* configurations, which impart a relatively rigid shape to their molecules.

SELF-CHECK 9.1

Arachidonic acid has 20 carbon atoms and four *cis* double bonds between carbons 5 to 6, 8 to 9, 11 to 12, and 14 to 15. Is it an ω-3 or an ω-6 fatty acid?

Free FAs are metabolic fuels. They link the metabolisms of carbohydrates and lipids, and their transport in the blood is interlinked with that of cholesterol. Fatty acids are stored in **adipose tissues** in the form of TAGs.

Triacylglycerols

Triacylglycerols consist of a three carbon glycerol backbone esterified with three FA residues. The three carbon atoms of the glycerol backbone are referred to as Sn 1, 2, and 3. Note in Figure 9.2 that the Sn2 carbon is usually esterified to an unsaturated FA.

Triacylglycerols are synthesized in **adipocytes** of white adipose tissue using glycerol 3-phosphate, derived mainly from glycolysis, to which FAs are added by fatty acid acyl transferases found in membranes of the smooth endoplasmic reticulum. The final step that converts a diacylglycerol (DAG) to TAG is catalysed by diacylglycerol acyl transferase (DGAT). The activity of DGAT is stimulated by acylation stimulatory protein (ASP). Triacylglycerols form a lipid droplet in the adipocyte, which is then coated with the protein **perilipin**.

Triacylglycerol (fat) stored in adipocytes is a significant store of energy in the body. Its deposition is increased when fuel molecules act on nuclear receptors called peroxisome proliferation activating receptors of the γ type (PPAR γ). Figure 9.3 shows that the lipolysis of TAGs, in the reaction catalysed by hormone sensitive **lipase** (HSL), is inhibited by the actions of insulin, but stimulated by the hormones of fasting such as catecholamines, glucagon, and growth hormone. Insulin acts to inhibit the phosphorylation of perilipin, which prevents HSL from acting

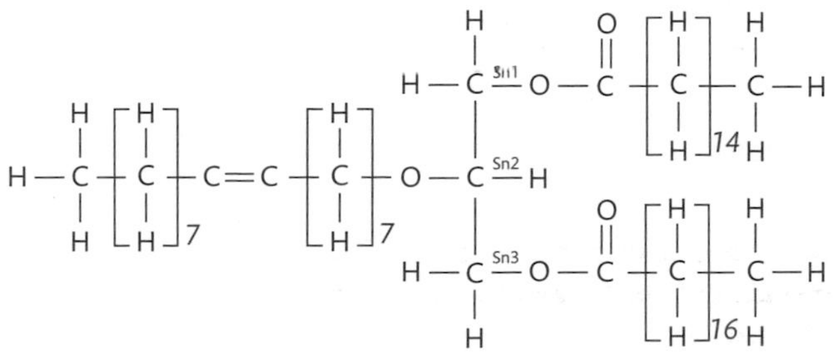

FIGURE 9.2
Structure of a typical triacylglycerol.

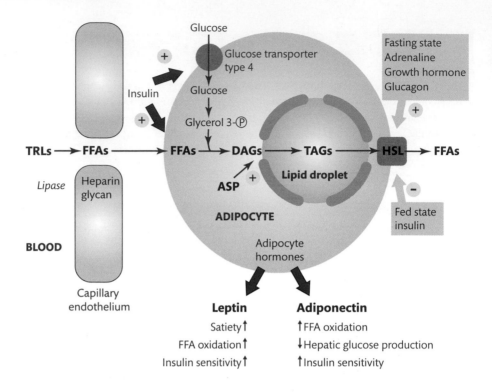

FIGURE 9.3
Metabolism of an adipocyte.
See main text for details.

TRL	Triacylglycerol rich lipoproteins
FFA	Free fatty acid
DAG	Diacylglycerol
TAG	Triacylglycerol
ASP	Acylation stimulating protein
HSL	Hormone sensitive lipase
+	Stimulation
−	Inhibition

on the lipid droplet. This system ensures that fat is stored after eating and free FAs are released when you have not eaten for a while.

Phospholipids

The structures of the PLs based on glycerol, show some similarity to those of TAGs but the Sn3 position is occupied by a phosphate ester. Figure 9.4 shows the structure of the PL, phosphatidylcholine, or lecithin.

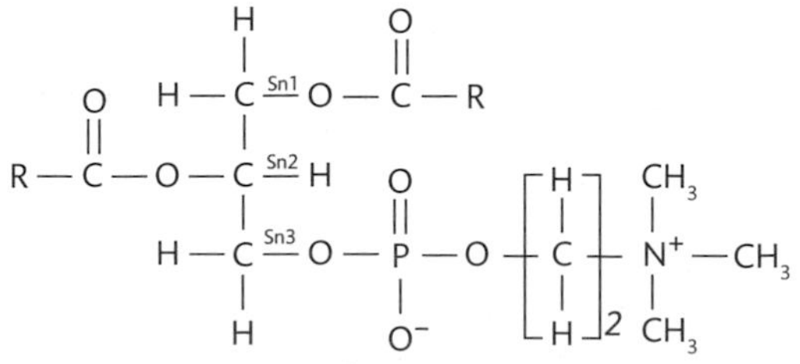

FIGURE 9.4
Structure of the phospholipids,
phosphatidylcholine (lecithin).

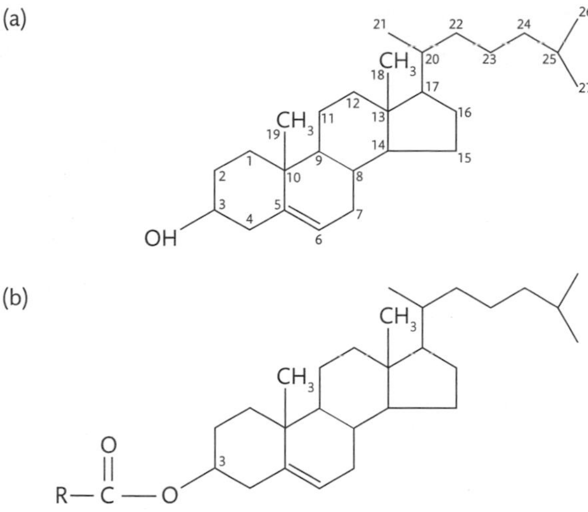

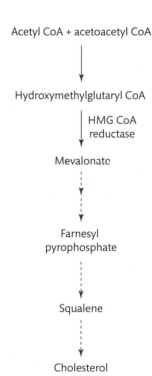

FIGURE 9.5
Structures of: (a) Cholesterol.
(b) Cholesterol esterified with
a fatty acid.

Phospholipids comprise about 40–50% of the lipids found in many biological membranes.

Cholesterol

Cholesterol is a sterol. Figure 9.5 (a) shows the structure of cholesterol and the convention for numbering its carbon atoms.

It can be esterified by attaching a FA to the hydroxyl group, C-3, which converts free cholesterol (FC) to a cholesterol ester (CE), as seen in Figure 9.5 (b). Cholesterol is the precursor of steroid hormones, D vitamins, bile acids/salts, and is a key component of biological membranes. Cellular cholesterol is derived by *de novo* synthesis using acetyl CoA or from the dietary intake.

Biosynthesis of cholesterol

The initial step in biosynthesis of cholesterol is the production of hydroxymethylglutaryl CoA (HMG CoA) from acetyl CoA and acetoacetyl CoA as shown in Figure 9.6.

The next step is the reduction of HMG CoA to form mevalonate in a reaction catalysed by HMG CoA reductase. This reaction is a regulatory step in cholesterol biosynthesis; the enzyme is the target for inhibition by the statin group of cholesterol lowering drugs (Section 9.11).

Absorption of dietary cholesterol

Dietary cholesterol is absorbed mainly in the duodenum and jejunum of the gastrointestinal tract (GIT). Absorption depends on solubilization with other fats and bile salts to form minute droplets called mixed **micelles**. Figure 9.7 shows that CEs are first hydrolysed to FC which is absorbed from micelles along with plant sterols by a sterol transporter called Niemann-Pick C1 like 1 protein (NPC1L1).

However, the plant sterols are subsequently returned to the GIT lumen by a sterol transporter, making the system specific to cholesterol. Cholesterol, together with TAGs, is eventually

Acetyl CoA + acetoacetyl CoA

↓

Hydroxymethylglutaryl CoA

HMG CoA
reductase

Mevalonate

↓

Farnesyl
pyrophosphate

↓

Squalene

↓

Cholesterol

FIGURE 9.6
Biosynthetic pathway of
cholesterol. The step catalysed
by the key regulatory enzyme
hydroxymethylglutaryl CoA
reductase is indicated.

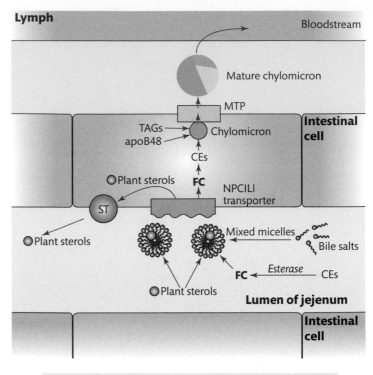

CE	Cholesterol ester
FC	Free cholesterol
NPC1L1	'Nieman-Pick C1 like 1' intestinal sterol transporter
MTP	Microsomal transport protein
TAGs	Triacylglycerols
ST	Sterol transporter

FIGURE 9.7
Outline of the absorption of cholesterol by the gastrointestinal tract (GIT). See main text for details.

delivered to the blood as CEs in mature **chylomicrons**. The same transporter system secretes plant sterol into **bile**. Defects in this sterol absorption system can lead to increased concentrations in the blood of plant sterols called **sitosterols** (see Section 9.5).

9.2 **Lipoproteins**

Given their water insolubility, lipids associate with proteins called **apoproteins** to form water soluble complexes or **lipoproteins** if they are to circulate. Apoproteins are classified into structurally related groups that reflect their functions. The major groups are called A, B, C and are referred to as apoA, apoB and so on; numbers are used to indicate their different subtypes. Typically, apoproteins are **amphipathic**, meaning that their molecules have water soluble and water insoluble parts. Figure 9.8 shows how this property allows **hydrophobic** lipids to be packaged in association with the hydrophobic portion of the apoprotein, while its **hydrophilic** areas are on the outside in contact with water. Thus, the complex is water soluble.

ApoB protein forms a fixed structural element of lipoprotein particles. There is a single molecule of it attached to each particle and it remains with it throughout its metabolism. Other apoproteins are present in multiple copies and are freely exchangeable between the different

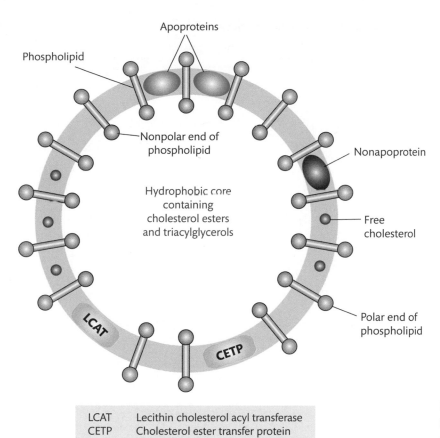

| LCAT | Lecithin cholesterol acyl transferase |
| CETP | Cholesterol ester transfer protein |

FIGURE 9.8
Structure of a typical lipoprotein particle.

types of lipoproteins. They have both structural and regulatory functions. As lipid is lost during metabolism lipoprotein particles become smaller and denser and shed transferable apoproteins, which alters their subsequent metabolism.

SELF-CHECK 9.2

(a) How many apoB protein molecules are there in a lipoprotein particle? (b) Is this information useful clinically?

Classification of lipoproteins and lipoprotein particles

Lipoprotein particles are mainly classified according to their density, giving:

- chylomicrons
- very low density lipoproteins
- intermediate density lipoproteins
- low density lipoproteins
- high density lipoproteins

TABLE 9.1 Properties and composition of lipoproteins.

	Diameter (nm)	Density (g/L)	Composition percentage				
			CEs	FC	TAGs	PLs	Protein
Chylomicrons	80–500	<950	1–3	1	86–94	3–8	1–2
VLDL	30–80	950–1006	12–14	6–8	55–65	12–18	8–15
IDL	25–30	1006–1019	20–35	7–11	25–40	15–22	12–19
LDL	19–25	1019–<1063	35–45	6–10	6–12	20–25	20–25
Lp(a)	0–25	1052–<1063	30–36	8–10	3–4	20–25	30–35
HDL$_2$	8–11	1063–<1125	15–20	4–6	3–8	30–40	35–40
HDL$_3$	6–9	1125–1210	10–18	1–4	3–6	25–35	45–55

However, each class of lipoprotein exists as a spectrum of similar particles. Table 9.1 shows the physical characteristics of the different lipoproteins.

Chylomicrons are assembled in the GIT from dietary and bile-derived lipids. Very low density lipoproteins (VLDLs) are formed in the liver and metabolized to give intermediate density lipoproteins (IDLs) and low density lipoproteins (LDLs) following increasing losses of TAGs. Chylomicrons, VLDL, and IDL are collectively known as triacylglycerol-rich lipoprotein (TRL) particles. High density lipoproteins (HDLs) are assembled from hepatic and tissue lipids, and apoA proteins. They function in the transport of cholesterol. Other lipoproteins, such as Lp(a) and **lipoprotein-X** (Lp-X), have unusual structures and, as yet, no clear physiological functions.

SELF-CHECK 9.3

When lipid is lost from lipoprotein particles, do they become more or less dense?

Lipid transfer proteins mediate the transfer and exchange of lipids between the different types of lipoprotein particles. The three main lipid transport proteins are cholesterol ester transfer protein (CETP), lecithin cholesterol acyl transferase (LCAT), and phospholipid transfer protein (PLTP).

Cholesterol ester transfer protein is synthesized by the liver, small intestine, adipose tissue, and macrophages. In the blood, CETP binds to lipid molecules and facilitates their association with lipoproteins. Its main action is to promote the transfer of CEs from HDL to VLDL particles. The resulting cholesterol depleted HDL is then capable of receiving further cholesterol. Lecithin cholesterol acyl transferase is made in the liver. Three-quarters of its plasma catalytic activity is associated with HDL and a quarter with LDL. Molecules of LCAT have separate binding sites for HDL and LDL particles. Figure 9.9 shows LCAT catalysing the transfer of the FA from the Sn2 position of phosphatidycholine (lecithin) to the C-3 hydroxyl group of cholesterol to give a CE and lysolecithin. This CE is non-polar and therefore sinks to the core of HDL particle, which becomes spherical, leaving its surface free to accumulate more cholesterol. Phospholipid transfer protein promotes the interchange of PL molecules between lipoprotein particles, particularly to HDL. It has a role with the cholesterol transporter ABCA1 in transporting PLs from

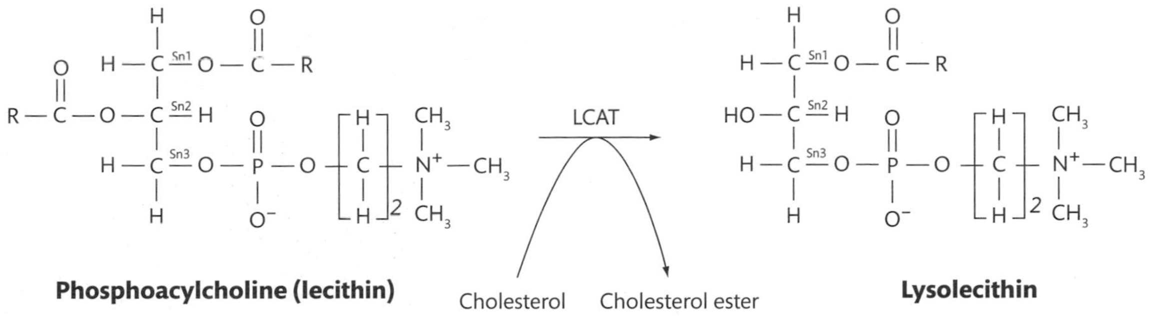

Phosphoacylcholine (lecithin)   Cholesterol   Cholesterol ester   **Lysolecithin**

FIGURE 9.9
The conversion of a phosphoacylcholine (lecithin) to lysolecithin in a reaction catalysed by lecithin cholesterol acyl transferase (LCAT).

tissue membranes to apoA1 proteins to form a precursor of HDL, which can accept more cholesterol from the tissues. Phospholipid transfer protein may have a role in cholesterol efflux from macrophages and **foam cells** (Section 9.8).

SELF-CHECK 9.4

What reactions do (a) LCAT and (b) CETP promote in lipoprotein metabolism?

Lipid transfer reactions are essential to the correct metabolism of lipoproteins. Several inherited **dyslipidaemias** are caused by mutations in the genes for individual apoproteins, lipid receptors, or transfer proteins. Inherited defects may also occur in regulatory genes whose products control the expression of individual genes or of clusters such as the apoC3, A1, A4, and A5, and the apoE, C1, C2, and C4 groups on chromosome 11 and 19 respectively.

Lipoprotein metabolism

Lipoprotein metabolism is mediated by receptors that recognize and bind apoproteins and transfer proteins. Lipoprotein receptors may be relatively specific or capable of interacting with a broad range of lipoproteins.

Chylomicrons

Chylomicrons are involved in the transport of dietary lipids. They are assembled by microsomal triacylglycerol transfer protein (MTP) from lipid, absorbed from the intestine, and a truncated form of the apoB, formed exclusively in the intestine, called $apoB_{48}$ protein. This protein shares a common N-terminal sequence with $apoB_{100}$. Indeed, $apoB_{48}$ constitutes 48% of the sequence for $apoB_{100}$ protein. It is the product of the same gene on chromosome 2 shortened by post-transcriptional editing of its mRNA. $ApoB_{48}$ protein functions as a structural component of chylomicrons but does not act as a **ligand** in interactions between chylomicrons and receptors. Rather, particles acquire surface apoC1, C2, C3, and E proteins, which control their metabolism. Mature chylomicrons are transported into lymph and then to blood where tissue lipases remove TAGs. During this process apoC particles are shed and acquire additional apoE proteins. Figure 9.10 shows how chylomicrons bind to multiple heparin sulphate glycans in

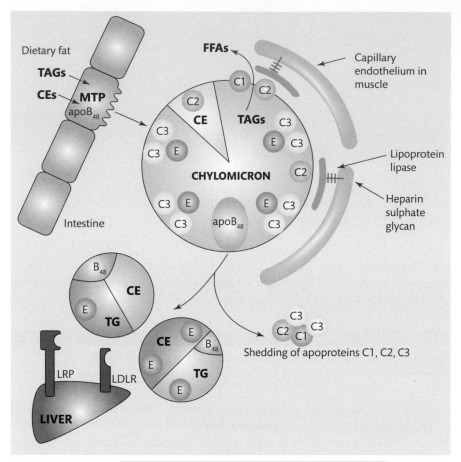

C1, C2, C3	Apoprotein C1, C2, C3
E	Apoprotein E
LRP	Receptor-like protein
LDLR	LDL receptor
CE	Cholesterol ester
MTP	Microsomal triacylglycerol transfer protein
FFA	Free fatty acid
B_{48}	Apoprotein B_{48}
TAG	Triacylglycerol

FIGURE 9.10
Metabolism of a chylomicron.
See main text for details.

the capillary **endothelium** so increasing their access to lipoprotein lipase, which catalyses the hydrolysis of TAGs to release FAs.

ApoC2 protein is an activator of lipoprotein lipase. The hepatic uptake of chylomicrons is inhibited by apoC1 and apoC3 proteins, which prevent the apoE protein binding to hepatic receptors. Thus, the delivery of TAGs to energy requiring tissues, such as muscle, is promoted. The progressive removal of TAGs produces chylomicron remnants, which may exchange additional TAGs with other lipoproteins. When they are finally taken up by the liver, cholesterol may constitute 50% of their lipid mass. ApoE protein is the most significant ligand for chylomicron remnant removal by hepatocytes; it interacts with both the low density lipoprotein receptor (LDLR) and another receptor, long-windedly called the LDL receptor-related protein (LRP). Both receptors are structurally similar but LRP has a longer extracellular domain; it has the major role in clearing chylomicron remnants.

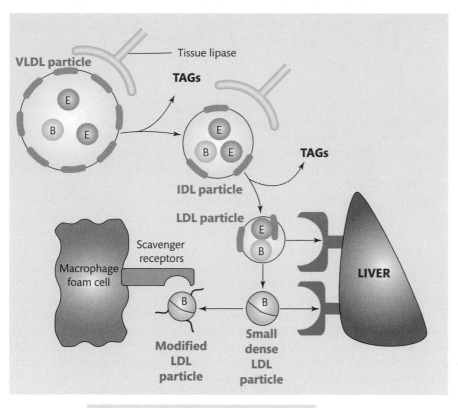

VLDL	Very low density lipoprotein
LDL	Low density lipoprotein
IDL	Intermediate density lipoprotein
E	Apoprotein E
B	Apoprotein B_{100}

FIGURE 9.11
The conversion of very low density lipoprotein (VLDL) to low density lipoprotein (LDL) particles. See main text for details.

Very low density lipoproteins

ApoB$_{100}$ is the main structural apoprotein in VLDL. Initially it is shielded from its receptors by associated lipid. However, as with chylomicrons, the removal of TAGs exposes it to receptors to which it can bind. The assembly of VLDL particles is also similar to that of chylomicrons. ApoB$_{100}$ protein associates with TAGs under the influence of a hepatic MTP before being secreted into the blood. The rate of VLDL synthesis is determined mainly by the delivery of FAs to the liver. Notice in Figure 9.11 that VLDL is catabolized in a similar way to chylomicrons with TAG removal following interactions with endothelial lipoprotein lipase and heparin sulphate glycans.

Triacylglycerol removal is a complex process controlled by apoC1, C2, C3, and apoA5 proteins. ApoC1 inhibits CETP, reducing the transfer of TAGs to other lipoprotein particles. ApoC3 protein inhibits lipoprotein lipase activity and that of other lipases, such as hepatic lipase, thus increasing TAG delivery to extrahepatic tissues. In contrast, ApoC2 protein *activates* lipoprotein lipase and promotes the removal of TAGs. Apoprotein A5 promotes the removal of VLDL and chylomicron remnants by facilitating binding of TRLs to lipoprotein lipase. ApoA5 protein is produced only in the liver and is remarkable in that its circulating concentrations in plasma

BOX 9.2 Discovery of the LDL receptor

The LDL receptor was first described by Goldstein and Brown, who received the Nobel Prize in Physiology or Medicine in 1985. Brown was interested in the enzymatic control of cholesterol biosynthesis and succeeded in partially purifying HMG CoA reductase. Goldstein worked as a clinician under Fredrickson, whose lipid classification is still widely used. Brown had a background in molecular biology, medical genetics, and tissue culture. The symbiotic partnership of Goldstein and Brown was forged during discussions as bridge partners. Together, they have made several fundamental discoveries about the molecular mechanisms causing hypercholesterolaemia (Section 9.5).

are low, at about one molecule of apoA5 to 24 molecules of VLDL. ApoA5 protein has a stabilizing role and is rapidly recycled. Very low density lipoprotein is converted to IDL by the loss of TAGs and the acquisition of cholesterol, with consequent reductions in particle sizes but increases in their densities. ApoC1 and C2 proteins are lost, as the TAG content and surface areas of VLDL particles are reduced and they acquire apoprotein E. Figure 9.11 demonstrates that as TAGs and apoC proteins are lost, apoE proteins become more accessible to hepatic receptors; hence apoprotein E mediates their hepatic removal.

Low density lipoporoteins

Most of the cholesterol in the blood is associated with LDL. There is a correlation between increasing concentrations of cholesterol in samples of serum and the risk of the patient developing **coronary heart disease**; most of this risk is accounted for by that of LDL. Concentrations above 2 mmol/L are associated with depositions of cholesterol in arteries called fatty streaks, which are a prerequisite for the damaging effect of other risk factors. Low density lipoproteins are subject to lipid exchanges with other lipoproteins, resulting in small dense LDL that are particularly damaging to arteries because their small sizes allow them to penetrate arterial walls.

The predominant apoprotein in LDL is $apoB_{100}$, a large amphipathic glycoprotein synthesized in the liver. The subscript 100 indicates the M_r of the full sized protein. Low density lipoproteins are the end product of the metabolism of VLDL. Low density lipoprotein is removed from blood by the hepatic LDLR (Box 9.2), which binds $apoB_{100}$ proteins. The LDLR also binds apoE protein and takes up apoE protein-containing triacylglycerol-rich lipoproteins that are the precursors of LDLs.

The LDL receptor is a glycoprotein, which spans the plasma membrane. The receptors cluster in a region of the membrane called a **clathrin** coated pit. The extracellular portion of the receptor binds LDL particles. On binding LDL, the receptor-LDL complex moves inwards and buds off to form a clathrin coated vesicle within the cytoplasm. Several vesicles then fuse to form a membranous sac called an **endosome**. In the acidic environment of the endosome, the LDL binding domain changes conformation and releases the LDL. The LDLR is then recycled back to the cell surface and is once again available to bind LDL. Mutations that result in malfunctions to this molecular machinery cause severe inherited forms of **hypercholesterolaemia** because of a reduced removal of LDL from blood. Figure 9.12 illustrates the key process of receptor mediated uptake of LDL.

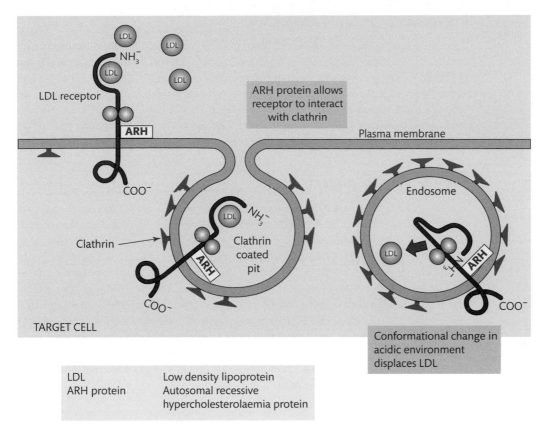

FIGURE 9.12
Receptor mediated endocytosis of low density lipoprotein (LDL) particles. See main text for details.

SELF-CHECK 9.5

(a) Does the LDL receptor recognize apoB and apoE proteins? (b) If so, what role does this play in the uptake of triacylglycerol-rich lipoproteins?

Low density lipoproteins are modified by **glycation**, **oxidation**, or other chemical reactions that can enhance their uptake by scavenger receptors (SR). A number of different receptors that scavenge for modified LDL have been identified: SRA, SRB, SRC, SRD, SRE, SRF, SRPSOX, and CD36 and the list is growing. Modified LDL is the main culprit in the formation of an atheroma (Section 9.8).

Small dense LDL

Cholesterol in LDL particles continues to exchange with TAGs in TRL particles due to binding of LDL to LCAT. Triacylglycerols in LDL are subsequently lost by the action of lipases leaving smaller, more dense cholesterol containing particles. Small dense LDL particles accumulate even with modest increases in VLDL. Small dense LDL remains in blood longer than normal LDL and is more prone to modification and uptake by scavenger receptors. Uptake into macrophage foam cells is a key step in the development of atheroma.

Intermediate density lipoproteins

Apoprotein E is mainly produced by the liver. It is the main ligand for IDL interactions with the LDL receptor when apoB proteins are shielded from receptors by TAGs. ApoE protein has binding sites complementary to a number of receptors. As well as IDL, apoprotein E acts as the ligand that mediates uptake of all classes of triacylglycerol-containing lipoprotein particles, including VLDL, chylomicrons, and chylomicron remnants. These receptor-ligand interactions mediate the transfer of TAGs to tissues. Intermediate density lipoprotein is converted to LDL by further loss of TAGS and the acquisition of cholesterol.

High density lipoproteins

The main structural protein in HDL is apoprotein A1, which is synthesized in the liver and intestine. Typically, there are four to seven molecules of apoA1 per HDL particle. ApoA1 protein acquires FC from tissues through the actions of an ATP-binding cassette (ABC) transporter called ABC type A1 (ABC A1) transporter to form small discoidal-shaped HDL particles. These mature into larger forms of HDL by lipid transfer reactions. Apoprotein A1 binds and activates LCAT, which esterifies FC acquired from the tissues. Esterified cholesterol is less water-soluble than the free form and sinks to the core of the particle. The particles become more spherical in shape and can accept more FC at their surfaces. Other ATP-dependent transporters, ABC G1 and ABC G5 add FC and PLs to HDL3 particles. These exchange their CEs for TAGs from other lipoproteins by interacting with CETP and PLTP transfer proteins to form HDL2. In reality, each category of particle represents a spectrum of particles. Figure 9.13 shows two pathways for the transfer of cholesterol from tissues to the liver.

Esterified cholesterol in HDL may be transferred to other lipoproteins and subsequently taken up by the liver or HDL may be directly taken up by the liver. These two pathways constitute a process called *reverse cholesterol transport*. About 1500 mg of FC per day is esterified and transported in this way, which is a protective mechanism against cholesterol accumulation, given that the dietary intake of cholesterol is about 500 to 1,000 mg daily. However, this view of reverse cholesterol transport may be too simple. The liver itself is a major source of HDL cholesterol. Peripheral and hepatic secretions of cholesterol are similar and form discoidal HDL particles that are capable of accepting cholesterol from peripheral tissues and transporting it to the liver.

High density lipoprotein particles may have a role in transport of cholesterol to cholesterol requiring tissues such as the adrenal glands. Particles of HDLs that contain the apoprotein A2 are relatively resistant to hepatic uptake and may be involved in this process. Two-thirds of circulating HDL particles contain the apoA2 protein, which is structurally distinct from apoA1. High density lipoprotein has an **antiatherogenic** effect, which may be related to its **antioxidant** properties or to the role of HDL in reverse cholesterol transport, particularly from macrophages. Table 9.2 shows the components of HDL and their protective mechanisms.

Particles of HDL interact with the two types of SRB receptors. Scavenger receptor type B1 (SRB1) is involved in the uptake of HDL by the liver as well as by cholesterol-requiring tissues such as those of the adrenal glands. Heparin sulphate proteoglycans bind with bulky HDL particles rich in TAGs and CEs giving access to hepatic lipase. Triacylglycerols are removed, which facilitates closer interactions with the SRB1 receptor. The SRB1-HDL complex is internalized but may be returned to the membrane surface following removal of cholesterol and can therefore transport more cholesterol to the liver. Alternatively, the interaction of HDL with SRB2 receptors promotes the excretion of cholesterol by facilitating the intracellular trafficking of HDL from the sinusoids across hepatocytes to the biliary canalicular membrane and into bile.

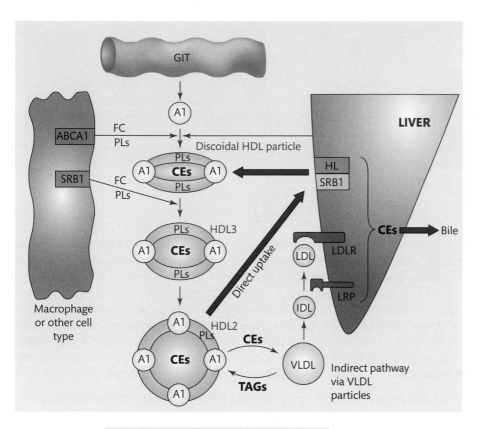

A1	Apoprotein A1
ABC	ATP-binding cassette (type)
CE	Cholesterol ester
FC	Free cholesterol
GIT	Gastrointestinal tract
HDL	High density lipoprotein
HL	Hepatic lipase
IDL	Intermediate density lipoprotein
LDL	Low density lipoprotein
LDLR	LDL receptor
LRP	LDL receptor-like protein
PL	Phospholipid
SRB1	Scavenger receptor type B1
TAG	Triacylglycerol
VLDL	Very low density lipoprotein

FIGURE 9.13
Reverse cholesterol transport.
See main text for details.

TABLE 9.2 Properties of some of the components of HDL particles.

Component	Function
Apoprotein A1	Removes LDL lipid hydroperoxides
Paraoxonase 1 and 3	Destruction of lipid hydroperoxides
Platelet activating factor acetyl hydrolase	Hydrolysis of oxidized PLs
LCAT	Removal of oxidized lipid
Apoprotein J	Anti-inflammatory
Caeruloplasmin	Anti-inflammatory

BOX 9.3 *Kringle domains*

The dominant structural features in Lp(a) are called kringle domains because they have a shape similar to the eponymous Danish doughnut. However, a kringle is also a doughnut-shaped metal ring sewn into the bottom, front, and back edges of sails for the purpose of securing the sail when it is shortened in high winds. I am sure that both the doughnut and the structural motif are named after these metal devices!

9.3 Lipoprotein (a)

Lipoprotein (a) (Lp(a)) or *lipoprotein little a* consists of one molecule of apoB protein attached to apoprotein (a) or apo(a). Lipoprotein (a) occurs as a number of isoforms due to variation of structural elements called **kringle** domains (Box 9.3). The diversity of the molecule is also increased because one of the kringle domains can be present in multiple copies, giving molecules with a wide range of M_r. The rate of secretion of apo(a) protein decreases with increasing M_r resulting in an inverse relationship between the M_r of an Lp(a) and its concentration. Apoprotein (a) is a protease. It is similar in structure to **plasminogen**, which degrades blood clots. Lipoprotein (a) inhibits the action of plasminogen and so tends to promote clotting. The extent to which Lp(a) causes clotting depends on the isoform composition.

When assessing the contribution of Lp(a) to cardiovascular risk, both the isoform composition and their individual concentrations need to be taken into account. This has not always been done but, in general, studies point to increasing risk both for cardiovascular and cerebrovascular disease with increasing Lp(a). The risk is increased particularly in the presence of increased LDL cholesterol, such as occurs in familial hypercholesterolaemia (FH) or with other **prothrombotic** risks such as tobacco smoking and increased **homocysteine** concentrations in the blood. Lipoprotein (a) is not part of routine evaluation of lipid disorders because its contribution to the overall risk is unclear and treatments are unavailable to reduce its circulating concentration.

SELF-CHECK 9.6

Is the concentration of Lp(a) routinely measured as part of an evaluation of cardiovascular risk?

9.4 Classification of lipid disorders

Classification of lipid disorders is partly based on knowledge of the underlying metabolic defect and partly on pragmatic considerations. Such classification is clinically useful, to identify appropriate treatments, and in research. A widely used classification is that of Frederickson, one of the pioneers of clinical lipidology, which has been adopted by the World Health Organization (WHO). This classification originally emphasized the observed lipoprotein pattern or phenotype, but now is expressed mainly in terms of lipid measurements. The Frederickson classification recognizes five types, numbered 1 to 5.

Type 1 is associated with marked **hyperchylomicronaemia**. It is caused by a deficiency of lipoprotein lipase or a deficiency of its activator, apoprotein C2. There are two subtypes of type 2. Type 2a implies hypercholesterolaemia without **hypertriglyceridaemia** and includes **monogenic** inherited disorders, such as FH, but is more often **polygenic** in origin. Type 2b implies combined hypercholesterolaemia with increased amounts of circulating TAGs. It is the pattern seen in familial combined dyslipidaemia, but there are several other causes, including secondary hyperlipidaemia that is associated with the overproduction of VLDL.

Type 3 lipid disorder is associated with variants of apoE protein, which leads to an accumulation of abnormal IDL particles. This disorder may be suspected when cholesterol and TAGs increase in equimolar amounts, reflecting the cholesterol enrichment of IDL. Type 4 is associated with hypertriglyceridaemia that implies increases in VLDL without chylomicrons.

Type 5 is associated with marked hypertriglyceridaemia and increased VLDL and chylomicrons. The primary cause may be increases in VLDL and IDLs that interfere with the receptor-mediated removal of chylomicrons. Chylomicrons can easily be identified by their ability to float to the surface of plasma as a creamy layer when it is stored at 4°C. However, chylomicrons are always present when the concentration of TAGs is greater than 10 mmol/L.

The Frederickson classification does not include the combined dyslipidaemia associated with the **metabolic syndrome** with abnormally dense LDL and does not include Lp(a) or HDL abnormalities, nor does it address the secondary hyperlipidaemias.

Secondary lipid disorders

Secondary lipid disorders are those arising as a consequence of a disease, drug treatment, or defective nutrition. Modifications to the last two may be the most effective treatment if they are the cause of the disorder.

Table 9.3 shows a number of the metabolic or drug-related factors that can disrupt lipid metabolism.

> The **metabolic syndrome** consists of a number of interrelated factors that increase the risk of developing cardiovascular disease. They include central obesity hypertriglyceridaemia, type 2 diabetes and insulin resistance, and are characterized by increased catabolism of apoA1 protein.

TABLE 9.3 Examples of metabolic effects and drugs that can be secondary causes of hyperlipidaemia.

Metabolic effects	Drugs
Hypothyroidism	β blockers
Excessive alcohol consumption	Corticosteroids
Obesity	Oestrogen replacement
High energy diet especially saturated fat	Androgen excess in women and replacement in men
Type 2 diabetes (less common in type 1)	Cyclosporine and related immunosuppressants
Metabolic syndrome	Antidopamine agents
Renal disease especially with proteinuria	HIV and HIV antiretroviral regimes
Lipodystrophy	Isotretinoin analogs (used to treat acne)
Anorexia nervosa	
Paraproteinaemia	
Kidney disease especially with gross proteinuria	
Cholestatic liver disease (LpX)	

The ingestion of excess dietary energy, whether in the form of fat (lipid) or carbohydrates, drives the production of VLDL. This is because of the excessive amounts of FAs delivered to the liver. Overproduction of VLDL may cause **mixed dyslipidaemia** because it is the precursor of LDL. A genetic or an acquired inability to store fat in adipose tissue, called **lipodystrophy**, has similar effects because more dietary-derived fat reaches the liver. An acquired *lack* of fat as seen, for instance in **anorexia nervosa,** is also associated with increased VLDL and LDLs.

Obesity appears to be associated with either replete or dysfunctional fat metabolism, which, again, has the same consequence of excessive FA accumulation by liver cells. **Central obesity** is more often associated with hypertriglyceridaemia than is general obesity because the central fat stores are more prone to release their FAs.

Alcohol consumption promotes an overproduction of FAs and provides a source of extra energy, thus increasing VLDLs. Paradoxically, alcohol can cause a marked increase in HDLs. A major mechanism for VLDL overproduction is **insulin resistance** because insulin normally suppresses the release of FAs from adipose stores. Resistance of the adipocytes to insulin allows the excessive release of FAs with a hepatic overproduction of VLDL. Insulin resistance is largely responsible for the type IV hyperlipidaemia seen in type 2 diabetic patients.

An accumulation of TRLs is commonly due to secondary causes such as in lipid disorders of chronic renal failure. Protein losing renal disease is associated with the hepatic overproduction of apoprotein B and VLDL, which correlates with reduced concentrations of blood albumin. Hypothyroidism is a particularly common cause of increased cholesterol and, occasionally, TAGs due to reduced LDLR and LRP-mediated lipoprotein uptake of LDL and TRLs. Hypothyroidism should always be considered a possible cause if this lipid profile is encountered in a patient. Paraproteins may interfere with the receptor-mediated uptake of lipoproteins with the accumulation of IDL. They also form complexes with lipoproteins and reduce the rate of their removal. Such complexes are prone to infiltrate the skin, causing flat lipid-containing patches called planar **xanthomas**.

Therapeutic drugs used in treating acquired immunodeficiency syndrome (AIDS) can to a greater or lesser degree damage fat cells and lead to insulin resistance because of impaired insulin signalling and reduced glucose uptake by cells relying on the glucose transporter type 4 (GLUT 4) transporter (Box 9.4).

Protease inhibitors particularly interfere with the LRP and lipoprotein lipase removal of TRLs. To a certain extent, however, these problems may be due to the human immunodeficiency virus (HIV) infection itself, given that they are occasionally seen in drug-naive HIV patients. Thiazide diuretics increase LDL by an unknown mechanism. Beta blockers raise TAGs, sometimes

BOX 9.4 Glucose transporters

Glucose is moved across cell membranes by a family of proteins called glucose transporters (GLUT). Different transporters, numbered 1–4, are found in different types of cells. Some are stimulated to transport glucose by insulin, others, however, are active without the hormone. The GLUT 1 transporter is involved in basal and non-insulin mediated glucose uptake by many types of cells. Glucose uptake by hepatocytes is mediated by GLUT 2. In normal metabolism, the brain derives all of its metabolic energy from glucose, following its entry by GLUT 3 in an insulin dependent manner. Lastly, GLUT 4 is responsible for the insulin dependent uptake in muscle and adipose tissues.

markedly, due to a reduced clearance of TRL particles. Oestrogen, as a component of oral contraceptives or post-menopausal hormonal replacement therapy, increases HDL and TAGs and, at least in post-menopausal women, reduces LDL. However, this seemingly beneficial change fails to protect against heart disease, perhaps because of the production of small dense LDL particles arising from the increases in TAGs. Progestogens counter the oestrogen-induced lipid changes by raising LDL and lowering HDL. Androgens cause an increase in LDLs with a reduction in VLDL and HDL. This shift towards an atherogenic profile is of concern in men receiving hormonal supplements but it also occurs in androgen excess syndromes in women, notably polycystic ovary syndrome.

Derivatives of retinoic acid used in the treatment of **psoriasis**, markedly increase VLDL formation, probably by promoting hepatic FA and apoprotein B production.

Finally, cholestatic liver disease is associated with increases in cholesterol due to accumulation of Lp-X, which consists of cholesterol and PLs but not apoprotein B. This is not atherogenic but it may interfere with hepatic chylomicron removal. Patients with cholestatic liver disease may also by chance have other forms of hyperlipidaemia; these can be distinguished by measuring apoprotein B concentrations in clinical samples (Section 9.9).

Cross reference
Chapter 15 Reproductive endocrinology

9.5 Hypercholesterolaemia

Hypercholesterolaemia is defined as high concentrations of cholesterol in blood that increase the risk of **cardiovascular disease**. It is recommended that total cholesterol concentration should be less than 5.0 mmol/L and that LDL cholesterol should be less than 3.0 mmol/L in healthy people. The underlying genetic causes of hypercholesterolaemia may be monogenic or polygenic in nature.

Monogenic hypercholesterolaemias

Monogenic hypercholesterolaemias include:

- familial hypercholesterolaemia
- familial defective apoprotein B_{100}
- autosomal recessive hypercholesterolaemia
- mutations in the gene encoding proprotein convertase subtilisin kexin type 9
- sitosterolaemia

Familial hypercholesterolaemia

A diagnosis of FH is complex, as can be inferred from the definition of the condition given in Table 9.4. It is caused by defective LDLR-mediated endocytosis. Over 900 different mutations of the LDLR gene are known, which affect all aspects of the process, such as ligand binding, internalization, ligand displacement, and receptor recycling. About 1 in 500 people worldwide are heterozygotes for mutations that lead to FH. The clinical manifestations of arterial disease vary somewhat in **heterozygous** FH depending on other lifestyle factors, but most patients show some signs of it by the age of 50. Women have some protection before the **menopause**. **Homozygous** disease, although partially dependent on the amount of functioning LDLR, is associated with disastrously premature **coronary artery** disease.

TABLE 9.4 Criteria used in the diagnosis of heterozygous familial hypercholesterolaemia.

Definition of familial hypercholesterolaemia

Definite familial hypercholesterolaemia

Total cholesterol >7.5 mmol/L in adults and >6.7 mmol/L in children under 16 years of age

or

LDL cholesterol >4.9 mmol/L in adults and 4.0 mmol/L in children

plus one of the following:

The presence of tendon xanthomas

A * first or ** second-degree relative with FH

Possible familial hypercholesterolaemia

Same cholesterol criteria as above

plus one of:

Family history of myocardial infarction before age 50 years in a **second-degree relative or before age 60 years in a *first-degree relative

Cholesterol >7.5 mmol/L in a first or second degree relative or >6.7 mmol/L in a sibling

* A first-degree relative has a 50% genetic link with parents, children, full siblings.
** A second-degree relative has a 25% genetic link with parents, grandparents, grandchildren, aunts and uncles, nieces and nephews, half-brothers and sisters.

Low density lipoprotein receptor-apoprotein E mediated uptake of IDL is also deficient in FH. This does not manifest as hypertriglyceridaemia unless other factors intervene because LRP (Section 9.2) function is still active.

A diagnosis of FH is based on finding increased LDL cholesterol, a family history of premature coronary heart disease and the presence of cholesterol deposits called xanthomas on tendons, particularly on the Achilles tendon (see Box 9.5). Table 9.4 demonstrates how a combination of increased plasma cholesterol, family history and clinical examination are used as criteria in diagnosis. However, in the absence of tendon xanthomas, the diagnosis often remains in doubt. A diagnosis based on analysing patient DNA is expensive because of the large number of different mutations that may underlie the condition. However, some populations such as Afrikaners, Christian Lebanese, Finns, and French Canadians show a founder effect, with FH being associated with a limited number of characteristic mutations. The Department of Health in the UK is evaluating a screening programme based on DNA analysis; such a programme is already well established in Holland.

Familial defective apoprotein B$_{100}$

In hyperlipidaemia due to familial defective apoprotein B (FDB), the domain of apoB protein that is recognized by the LDLR is defective. The phenotypic expression of the disease is milder than in FH, probably because the LDLR is intact and apoprotein E-mediated uptake of IDL is normal. Homozygous FDB is clinically similar to heterozygous FH. Surprisingly, FDB responds just as well to treatment with statins (Section 9.11) as heterozygous FH because of facilitation of apoE-mediated IDL uptake by the LDLR with less conversion to LDL.

Autosomal recessive hypercholesterolaemia

The cytosolic tail of the LDLR is anchored in clathrin coated pits by an adaptor called **autosomal** recessive hypercholesterolaemia (ARH) protein. This protein allows the receptor to interact with other molecules concerned with endocytosis and receptor recycling (Section 9.2).

BOX 9.5 Hypercholesterolaemia

A 43-year-old man had a health check and found that his blood cholesterol concentration was 10.6 mmol/L (desirable <5.0 mmol/L). He appeared otherwise healthy although he complained of painful Achilles tendons and had to cut the backs off his trainers. His mother was 70 years old and had not had any heart trouble, but her two brothers had heart attacks in their 50s. On examination he was found to have enlarged Achilles tendons because of the presence of tendon xanthomata due to cholesterol deposition. He also had finger tendon xanthomata and lipid deposits in both corneas. Fortunately, he had no evidence of arterial disease and presented a normal heart exercise stress test. Based on the clinical and laboratory findings he had heterozygous FH and was treated with a statin drug (Section 9.11). His blood cholesterol reduced to 4.6 mmol/L and he remained well after 12 years of treatment.

Indeed in this form of FH, binding of the ligand to the hepatic receptor is increased but there is a defect in recycling the receptor back to the pits. Different tissues use different types of adapter proteins, thus the defect is tissue specific. Clinical presentation is similar to FH, with cholesterol concentrations lying between those of heterozygous and homozygous FH. There is some response to the cholesterol-lowering drugs, statins (Section 9.11), presumably due to up-regulation of the LDLR but less than in heterozygous FH. The usual treatment is **plasmapheresis**.

Mutations in the gene encoding proprotein convertase subtilisin kexin type 9

The proprotein convertase subtilisin kexin type 9 (PCSK9) acts intracellularly on the LDLR, directing it towards catabolism rather than its recycling. The molecular mechanism by which mutations to the PCSK9 gene cause hypercholesterolaemia is not completely clear, although mutations result in a protein of increased activity that promotes the LDLR catabolism and, consequently, leads to hypercholesterolaemia.

Sitosterolaemia

Sitosterolaemia is an inherited disease caused by a defect in efflux of sterols from cells due to mutations in the gene for a membrane sterol transporter. When there is an excess of cholesterol or sterol in cells, two half transporters, ABCG5 and ABCG8, combine to form an active sterol transporter. This is the transporter we saw in Figure 9.7 that is responsible for transferring plant sterols from intestinal cells back to the GIT lumen. It is also responsible for moving them from biliary cells into bile. Consequently, defects in the system enhance absorption of plant sterols by the GIT but reduce biliary secretion of cholesterol and plant sterols into bile. The result is sitosterolaemia and xanthoma formation.

Most routine clinical laboratory methods (Section 9.9) cannot distinguish plant sterols from cholesterol and so their presence indicates an *apparent* hypercholesterolaemia. Their identification requires that the sterol content of blood be evaluated by more sophisticated techniques, such as gas chromatography. Also, apoB protein can be measured to estimate the concentration of LDL particles. Patients with sitosterolaemia respond poorly to treatment with cholesterol-lowering statins (Section 9.11). Therapy is therefore aimed at reducing the absorption of cholesterol and plant sterol by restricting dietary cholesterol and promoting their excretion by ingesting oral resins. Ezetimibe, which inhibits both plant sterol and cholesterol absorption, is predictably effective. The resins bind bile acids and prevent their recycling.

SELF-CHECK 9.7

Are plant sterols absorbed from the diet?

Polygenic or non-familial hypercholesterolaemia

The genes responsible for the commonest polygenic form of hypercholesterolaemia are unknown. However, nutritional studies have established a link between a diet high in saturated fats and an increased concentration of blood cholesterol. There is a graded relationship: countries with low saturated fat intake, like Japan, have lower levels of cholesterol compared to countries with a high one, such as Finland. Dietary *trans* fat and high levels of cholesterol intake also promote hypercholesterolaemia.

9.6 Hypocholesterolaemia

An abnormally low concentration of cholesterol in the blood is called **hypocholesterolaemia**. The questions of whether the concentration of blood cholesterol can be too low and whether this causes cancer sometimes arise. Cancer is often associated with low cholesterol but it is likely that the cancer is responsible for the low cholesterol and not the other way round. As for cholesterol lowering, a newborn infant has a blood cholesterol concentration of approximately 1.5 mmol/L. People on low fat diets may have cholesterol values of about 2.5 mmol/L and this is entirely beneficial. In addition, the LDLR appears to be saturated by concentrations of cholesterol of about 3 mmol/L. 'Natural' cholesterol levels are therefore likely to be lower than this.

A low blood cholesterol reflects a low value for LDL cholesterol and the condition, **hypobetalipoproteinaemia**. The causes of hypobetalipoproteinaemia include genetic defects that result in familial hypobetalipoproteinaemia (FHBL) and **abetalipoproteinaemia**, and a number of non-inherited forms arising from malignancy, malnutrition, intestinal malabsorption, and liver disease.

Familial hypobetalipoproteinaemia

Familial hypobetalipoproteinaemia heterozygotes have half the normal average cholesterol concentration and homozygotes have concentrations less than 1.3 mmol/L. Over 35 mutations that lead to the production of truncated forms of $apoB_{100}$ and result in FHBL have been described. Common truncated forms are $apoB_{37}$, $apoB_{46}$, and $apoB_{31}$; the subscript indicates the size of the truncated protein, as a percentage of the full-sized protein. Truncations shorter than $apoB_{25}$ are undetectable in plasma because of their rapid degradation.

Heterozygotes are generally asymptomatic and are at low risk of atheroma. In contrast, the very low amounts of apoB protein present in the homozygous state lead to deficiencies in assembling chylomicrons and the malabsorption of fats. Impairments in forming VLDL particles may predispose the patient to developing a fatty liver. However, most of the clinical manifestations reflect deficiencies of the fat-soluble vitamin E, which also occurs in abetalipoproteinaemia.

Abetalipoproteinaemia

Abetalipoproteinaemia or Bassen-Kornzweig syndrome is a rare autosomal recessive condition in which lipoproteins containing apolipoprotein B (chylomicrons, VLD, and LD lipoproteins) fail to be synthesized. This occurs as a result of mutations to the gene encoding MTP. The resulting defective MTP causes deficiencies in assembling chylomicrons and VLDL particles. Defects in forming chylomicrons leads to the malabsorption of fats and fat-soluble vitamins, especially vitamin E. Vitamin E is normally packaged in chylomicrons and transported to the liver. Here, it is incorporated into VLDL particles and subsequently into LDLs.

Abetalipoproteinaemia and a deficiency of vitamin E result in the progressive neurological defect, **retinitis pigmentosa**, and erythrocytes with abnormal membranes called **acanthocytes**. Dietary supplementation with vitamin E is effective in preventing neurological degeneration.

Low numbers of blood HDL particles

A low amount of HDL particles may have genetic origins or be secondary to other factors such as metabolic syndrome. Insulin resistance *decreases* endothelial lipoprotein lipase activity leading to HDL particles enriched in TAGs. In contrast, insulin resistance *increases* hepatic lipase activity altering the intravascular and hepatic remodelling of HDL and generating lipid poor apoprotein A1, which is rapidly catabolized. An increased catabolism of apoprotein A1 and low numbers of HDL particles also occur in malnutrition, intestinal malabsorption, and end-stage renal failure.

Cross reference
Chapter 3 Kidney disease

Severe inflammation lowers HDL due to the effects of cytokines on lipase activity. Hypertriglyceridaemia secondarily lowers HDL by increasing cholesterol exchange to TRL particles. Severe liver disease reduces the biosynthesis of apoprotein A1. Drugs, such as androgens, progestogens, high doses of thiazide diuretics, and β blockers lower HDLs by a variety of mechanisms.

Familial causes of low HDLs

Table 9.5 illustrates that low HDLs are caused by several inherited defects affecting apoprotein A1, ABCA1, or LCAT that are inherited in an autosomal dominant or co-dominant fashion.

Co-dominant inheritance means that the abnormal gene interferes with the function of the normal gene. This produces effects in heterozygotes similar to those seen in homozygotes. There are about 50 known mutations in the gene for apoA1 that affect its binding to LCAT. Poor binding results in a reduced transfer of cholesterol to apoprotein A1 and low amounts of HDL particles because poorly lipidated apoA1 is rapidly catabolized. Mutations that do not affect LCAT binding may produce a rapidly catabolized, structurally and functionally abnormal protein, which can aggregate into a material called amyloid that damages tissues. Other mutations have no apparent effect on HDL metabolism.

Familial **hypoalphalipoproteinaemia** is associated with structurally normal apoA1 protein with an unknown inherited factor causing increased catabolism. In familial apoA1 deficiency, virtually no apoA1 is synthesized because of an autosomal co-dominant inheritance. Tangier disease (Table 9.5) is associated with low amounts of HDL particles and reduced apoA1 protein levels due to mutations in the ABCA1 transporter gene resulting in poor transfer of FC from tissue to apoA1 lipoprotein particles. Poorly lipidated apoA protein is rapidly catabolized.

TABLE 9.5 Inherited causes of low concentrations of blood HDL particles and apoprotein A1.

Disease (mode of inheritance)	Metabolic defect	Ratio HDL/ other lipid	Clinical features	Premature atheroma
Structural mutations in apoprotein A1 (autosomal dominant)	Rapid catabolism of apoA1; some cause amyloid	0.4–0.9	Sometimes corneal opacities	No
Familial (autosomal dominant)	Normal apoA1 structure but rapid catabolism	0.4–0.9	None or corneal opacities if HDL very low	Yes
Familial apoprotein A1 deficiency (autosomal dominant)**	No synthesis of apoA1 protein	<0.1*	Planar xanthomas and corneal opacities	Yes
Tangier disease (autosomal dominant)**	Lack of lipidation of apoA1 due to mutation in ABC A1 transporter	<0.11	Cholesterol deposits in tonsils and other reticuloendothelial system; corneal opacities	Uncertain
Familial LCAT deficiency (autosomal recessive)	Rapid HDL catabolism. Reduced binding of LCAT to HDL and LDL	<0.3 Raised [TAG]	Corneal opacities, proteinuria, nephropathy and anaemia	No
Partial LCAT deficiency or fish eye disease (autosomal recessive)	Rapid HDL catabolism Reduced binding of LCAT to HDL only	<0.3 Increased [TAG]	Corneal opacities	Uncertain

* Cannot be measured accurately in homogeneous assays.

** Means that the abnormal gene interferes with the function of the normal gene.

Lecithin cholesterol acyl transferase deficiency causes low HDL by reducing esterification of FC, which limits the capacity of HDL to accept tissue cholesterol.

Corneal opacities are associated with inherited low HDL, presumably reflecting reduced cholesterol export from tissue. The association between low HDL due to gene defects and atheroma is variable. A virtual complete absence of apoprotein A1 in familial apoA1 deficiency is associated with the development of atheroma, presumably reflecting foam cell formation. Low HDL in familial hypoalphalipoproteinaemia also predisposes to cardiovascular disease, which may reflect reduced reverse cholesterol transport (Figure 9.13). On the other hand, Tangier disease is not associated with accelerated atheroma formation but with cholesterol accumulation in reticuloendothelial tissue, such as the tonsils, causing orange discolouration. This suggests an alternative mechanism for removal of cholesterol from macrophage foam cells in atheroma because ABCA1 is crucial for the initial transfer of cholesterol from cells as the first stage of reverse cholesterol transport formation. Clinical manifestations of LCAT deficiency depend on whether there is complete or partial deficiency. Partial deficiency is associated with reduced binding of LCAT to HDL but not to LDL. In complete or familial LCAT deficiency, binding of LCAT to both LDL and HDL are reduced. Partial LCAT deficiency carries the greater risk of heart disease because binding of LCAT to LDL allows remodelling to more atherogenic, small, dense LDL.

SELF-CHECK 9.8

What is the cause of Tangier disease?

TABLE 9.6 Primary and secondary causes of increased blood HDL particles.

Primary	Secondary
CETP deficiency	Regular excessive alcohol consumption
Hepatic lipase deficiency	Primary biliary cirrhosis
Familial hyperalphalipoproteinaemia possibly monogenic or polygenic	Oral oestrogen
	Sustained strenuous exercise
	Phenytoin

High numbers of blood HDL particles

Table 9.6 shows that HDL may be increased due to a reduced cholesterol exchange with TRL particles, an increased biosynthesis of apoA1 protein, increased efflux of cholesterol from tissues to apoA1, or a reduced uptake of HDLs by hepatocytes.

Prolonged and severe exercise, such as running, increases the removal of TRLs from muscle capillaries and so limits cholesterol exchange between HDL and TRL particles. Reduced CETP activity caused by mutation or overexpression of the gene for apoC1 protein also increases HDLs by restricting exchange with TRL particles. Increases in LCAT activity raise HDL by promoting cholesterol transfer from tissues. A deficiency of hepatic lipase activity also increases HDL by inhibiting uptake into hepatocytes.

Mechanisms such as those that increase the numbers of HDL particles will not necessarily stimulate reverse cholesterol transport (Figure 9.13). This shows that increased HDLs may not always be beneficial. Indeed, hyperalphalipoproteinaemia may be associated with corneal opacification, which also occurs when HDLs are present in only extremely low amounts, suggesting that the removal of cholesterol from tissue may be abnormal. Nevertheless, HDL concentrations greater than 1.7 mmol/L reduce the cardiovascular risk associated with increased LDLs. It is essential to evaluate the likely clinical significance of an HDL value above 3.0 mmol/L; its common causes are excessive alcohol consumption and oral oestrogen treatment.

SELF-CHECK 9.9

A blood sample from a woman patient was found to have a low HDL concentration of 0.3 mmol/L. What are its possible causes?

9.7 **Hypertriglyceridaemia**

Hypertriglyceridaemia is a concentration of blood TAGs above accepted desirable values. There is a positive correlation between concentrations of TAGs in blood samples and the risk of developing coronary heart disease. However, when HDL is also accounted for, this association disappears and concentrations of TAGs are strongly *inversely* related to HDL. The poorer predictive value of TAG measurements compared to those of HDLs may be because their concentrations in blood are more variable and so a single estimate will be less representative of the true mean value. This is not to say that TAG concentrations should be ignored: the atherogenic potential of TRLs is inversely linked to the size of the particle, which is determined by their TAG content. An accumulation of large chylomicrons caused, for instance, by a deficiency of lipoprotein lipase activity, is not atherogenic. Metabolic syndrome is associated with concentrations of blood TAGs above 1.7 mmol/L. Even this degree of mild hypertriglyceridaemia is often associated with accumulation of atherogenic small dense LDL particles due to an increase in the TAGs exchanged with LDLs. Triacylglycerol lowering treatment may be beneficial by preventing the accumulation of small, dense LDL and as an indirect way of increasing HDL particles. Hypertriglyceridaemia is associated with a number of inherited conditions and is also associated with acute **pancreatitis**.

Inherited hypertriglyceridaemias

Inherited hypertriglyceridaemias include:

- familial endogenous hypertriglyceridaemia
- familial combined hyperlipidaemia
- type 3 hyperlipidaemia
- apoprotein E polymorphisms

Familial endogenous hypertriglyceridaemia

A diagnosis of inherited hypertriglyceridaemia is rarely made in clinical practice but may be suggested by the family history of the patient. Candidate genes for its cause are those that encode hepatic lipase or apoC3 and apoA5 proteins. The production of these defective proteins may interfere with the interactions of TRL particles and their receptors, or mutation in any number of other genes involved in VLDL synthesis or catabolism. Secondary hypertriglyceridaemia is commoner than the inherited form, and may contribute to its presentation.

Familial combined hyperlipidaemia

Familial combined hyperlipidaemia (FCH) is common in families whose members have suffered a heart attack. It presents between the ages of 20 to 30 years of age, rather than early childhood as in FH. There may be increased LDL or TRL particles, or both. A combined increase in the blood concentrations of cholesterol and TAGs is referred to as combined hyperlipidaemia. Familial combined hyperlipidaemia presents many of the features of the metabolic syndrome. Obesity and insulin resistance will obviously contribute towards increased VLDL production by interacting with genetic factors. The rate of apoprotein B production is increased in FCH and hyperbetalipoproteinaemia is a feature. Family studies have linked **haplotypes** of the apoA1-C3-A4-A5 gene cluster to hypertriglyceridaemia.

SELF-CHECK 9.10

What is the underlying defect in FCH that causes increased concentrations of LDL and VLDL?

Type 3 hyperlipidaemia

Type 3 hyperlipidaemia is caused by an accumulation of IDLs. Clinically it presents as a mixed hyperlipidaemia, sometimes associated with tuberous xanthomas not attached to tendons, and palmar xanthomas that cause an orange discolouration of the palmar creases. Over 90% of type 3 hyperlipidaemia is associated with apoE2/2 homozygosity and a markedly reduced binding of IDL to cell surface receptors. The remaining 10% are mainly associated with mutations that affect the heparin binding site of apoprotein E, which have similar effects on IDL removal and a dominant mode of inheritance. However, 90% of E2/2 homozygotes do not develop type 3 hyperlipidaemia and may even show low concentrations of cholesterol and TAGs in their blood. Clearly, lifestyle factors are necessary for its expression. The E2/2 genotype without hyperlipidaemia does not confer excess cardiovascular risk. In contrast, ApoE4/4 has increased affinity for receptors and is associated with raised concentrations of blood LDLs and increased cardiovascular risk. This may be due to efficient LRP-mediated uptake of cholesterol from TRL particles causing reduced expression of the LDLR.

Apoprotein E polymorphisms

Three common isoforms of apoE, E2, E3, E4, are expressed, together with over 20 rare variants. These are the products of three alleles of the single gene locus called $\varepsilon2$, $\varepsilon3$, and $\varepsilon4$. The isoforms differ in amino acid residues at positions 112 and 158, which affect receptor binding. The combination, apoE3 cysteine^{112} and arginine^{158} is likely to be the original form. The residues in apoE2 are cysteine^{112} and cystine^{158}, and for apoE4 arginine^{112} and arginine^{158}. Apoprotein E2 has less than 1% affinity for LDLR compared with other polymorphisms. Overall, six phenotypes are possible, which occur in order of frequency E3/3, E4/3, E3/2, E4/4, E4/2, and E2/2. Only E2/2 gives the propensity to develop type 3 hyperlipidaemia. The apoE protein is important in neuronal lipid metabolism and the E4 isoform increases the risk of developing Alzheimer's disease and of abnormal protein metabolism resulting in amyloid deposition. Many clinical laboratories restrict reporting of apoE genotype to E2/2 to avoid inadvertently screening for Alzheimer's disease.

Triacylglycerols and acute pancreatitis

Acute pancreatitis causes mild hypertriglyceridaemia as part of the general inflammatory stress response. However, chylomicronaemia, which is associated with blood concentrations of TAGs greater than 10 and usually above 20 mmol/L, causes acute pancreatitis. The mechanism by which this occurs may reflect the production of pro-oxidant FA metabolites. Reduced lipoprotein lipase activity is a recognized cause of recurrent acute pancreatitis. Increases in VLDLs, for instance due to type 2 diabetes, hypothyroidism, ethanol abuse, or inherited hypertriglyceridaemia may also precipitate chylomicronaemia and pancreatitis by interfering with the removal of chylomicrons.

SELF-CHECK 9.11

A serum sample from a patient had a milky appearance and produced a very high result for its triacylglycerol content of 23 mmol/L. (a) What lipoprotein is causing this? (b) What is the clinical significance of this test result?

9.8 Lipids, atheroma, and heart disease

Atheromas are complex lipid-rich lumps that occur on the inner lining of arteries. The maturation of atheromas proceeds by the acquisition of lipid material and by cell proliferation and damage as part of inflammatory and oxidative processes.

Cross reference
Chapter 7 Clinical enzymology and biomarkers

The first step in atheroma formation is the appearance of fatty streaks, which are common in young adults and even children. The fat is purely intracellular and largely reflects CEs within macrophages called foam cells. Endothelial dysfunction may be the initiating event in forming an atheroma. Adhesion of platelets and monocytes to the area of damage produces an inflammatory response, causing smooth muscle proliferation and lipid uptake. Inflammatory processes break down foam cells and release lipids, which accumulate in the atheroma core.

Phospholipase A2 (PLA2), which circulates in plasma and is associated with lipoproteins, especially small dense LDL particles, is also found in atheromas. Phospholipase A2 activity catalyses the hydrolytic removal of the FA residue from the Sn2 carbon of the glycerol producing an FA and lysophospholipid (Figure 9.14), which can damage the surfaces of LDL particles and promote their uptake through scavenger receptors on the surfaces of macrophages.

Cholesterol efflux from macrophages constitutes only a small fraction of total tissue cholesterol turnover and reverse cholesterol transport (Figure 9.13), but it is vital in preventing the formation of an atheroma. Some apoE protein is produced by macrophages where it also has a role, along with SRB1 and cholesterol transporters, in cholesterol efflux and preventing the formation of foam cells. Figure 9.15 shows a mature atheroma consisting of a large lipid core within the intima and media of the arterial wall but with the luminal surface covered with a fibrous cap containing smooth muscle cells within a connective tissue matrix.

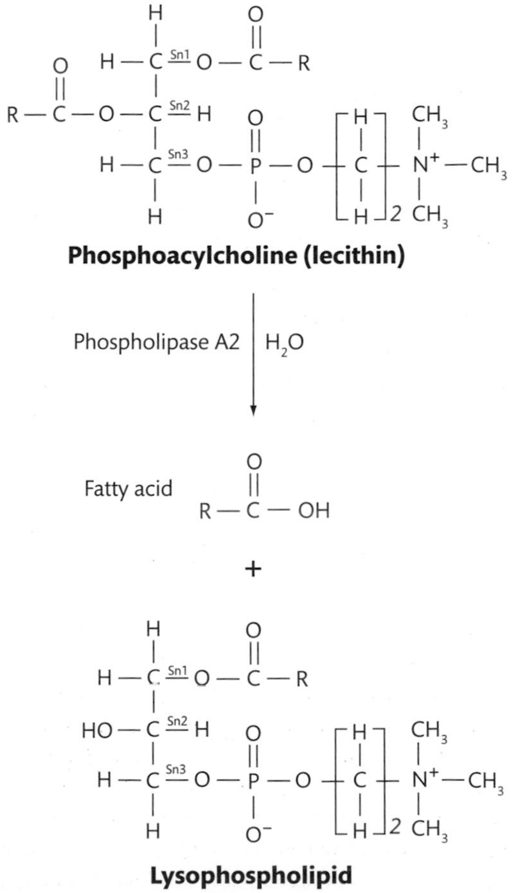

Phosphoacylcholine (lecithin)

Phospholipase A2 H_2O

Fatty acid $R-C(=O)-OH$

+

Lysophospholipid

FIGURE 9.14

The hydrolytic conversion of a phosphoacylcholine (lecithin) to a fatty acid and lysophospholipid in a reaction catalysed by phospholipase A2.

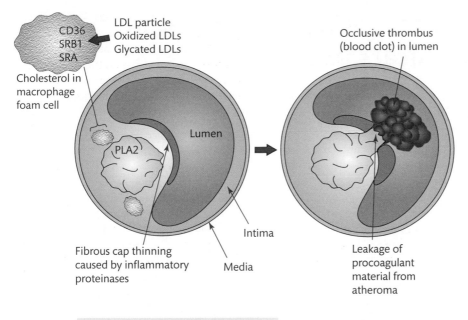

LDL	Low density lipoprotein
PLA2	Phospholipase type A2
SRB1	Scavenger receptor type B1
CD36	Cluster of differentiation 36
SRA	Scavenger receptor class A

FIGURE 9.15
Outline of the formation and rupture of an atheroma. See main text for details. See also Figure 7.6.

The atheroma may not block the lumen of an artery to the extent that it restricts the flow of blood, and so it may not cause symptoms associated with a lack of oxygen (**ischaemia**). Rather, the rupture of the fibrous cap, and consequent thrombosis and obstruction of blood flow, is the crucial event which precipitates a myocardial infarction. In most myocardial infarctions, the culprit atheroma blocks less than 50% of the arterial diameter prior to thrombosis. Atheromatous plaques prone to rupture are called unstable plaques and are characterized by a jelly-like lipid core.

Atherogenesis is a disease of the *arterial wall* rather than just an obstruction of the lumen. Deposition of lipid causes thickening of the inner lining or intimal media of arteries. Intimal medial thickness, measured by intraluminal ultrasound techniques, is an excellent index of atheromatous disease and is increasingly used as a proxy measure of **ischaemic heart disease** in clinical trials. One of the benefits of cholesterol lowering treatments (Section 9.11) may be to remove the cholesterol from the central lipid core, making the plaque more stable. This is clearly distinct from treatments to lower the concentration of blood cholesterol to prevent the initial accumulation of lipid. Intensive cholesterol lowering treatment has been shown to cause a regression of atheromatous plaques.

SELF-CHECK 9.12

Do native LDL particles accumulate in arteries and cause atheroma?

The future of lipidology will certainly be concerned with developing ways for more accurately measuring lipids in clinical samples, assessing cardiovascular risk and treatments using safer, more effective, and better tolerated lipid-lowering therapies.

9.9 Measurement of lipids in clinical samples

The Joint British Societies Guidelines 2 (JBS2) in the UK and National Cholesterol Education Program (NCEP) of the USA both recommend that concentrations of cholesterol and other lipids in the blood should not exceed specific values.

In developed countries, the bulk of the population has greater than ideal blood cholesterol concentration and so *desirable* levels, rather than reference ranges are used. The expert panel of JBS2 has adopted values for both males and females of total cholesterol less than 4 mmol/L and TAGS less than 1.7 mmol/L. For HDL cholesterol, values greater than 1.1 and 1.2 mmol/L have been recommended for males and females respectively. Statistical normal ranges are useful for evaluating the significance of abnormalities but because they are gender and age related they are not presented here. However, if concentrations in the samples of patients are in excess of the values given, then they are used to assess the need for treatment.

Numerical values set by such bodies based on clinical studies should be equivalent to results produced in clinical laboratories. Consistency is achieved by being able to trace the results obtained with commercially supplied chemical analysers back to a common reference method and the establishment of calibration protocols that align commercial systems with that reference method. This is known as an accuracy base.

Calibration of cholesterol assays

The Centers for Disease Control (CDC) in the USA have established reference methods for determining the concentrations of cholesterol and HDL. Satisfactory field methods have been calibrated relative to this and disseminated to clinical laboratories by a cholesterol reference method laboratory network to allow them to calibrate commercial methods using automated systems. Calibration is easier for total cholesterol methods measured after removal of lipoproteins from samples, but more difficult for homogeneous assays where separation is avoided for purposes of automation. Standardization of homogeneous assays for LDL cholesterol and HDL cholesterol is complicated because both analytes are present as a range of particles, with differing physical characteristics and cholesterol content.

Total allowable error

The determination of the concentration of an analyte may be different from its true value because of poor accuracy, causing a bias between methods, or to poor precision of the methods. Multiple measurements may correct for poor precision but it is not usual practice to make several measurements. The practice of estimating total allowable error incorporates acceptable bias and precision.

$$\text{Total allowable error} = \text{bias} + 1.96 \, (\text{coefficient of variation})$$

The accepted value for total allowable error is 12%. As an example, a bias of 4% and a CV of 4% meet the criterion.

Routine lipid analyses

Chemical autoanalysers are routinely used to determine the concentrations of cholesterol and triacylglycerol. The methods are based on the sequences of reaction given in Figure 9.16, which shows how enzymatic reagents are used.

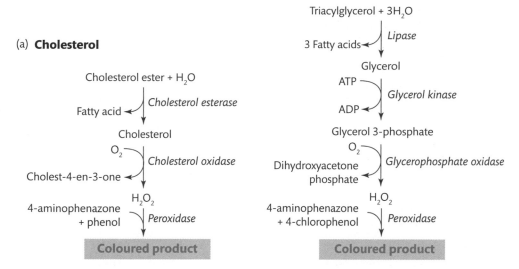

FIGURE 9.16
Typical reaction sequences for measuring the concentrations of:
(a) Cholesterol. (b) Triacylglycerols on an autoanalyser.

These reactions are carried out in a single reaction vessel and measure *total* cholesterol, unless various steps are taken to shield lipoproteins from the reagents. Triacylglycerol determinations are based on the amount of glycerol released by lipases; any contribution by free glycerol is usually ignored.

The concentrations of lipids in lipoprotein fractions in clinical samples can be estimated using one of several methods. These include:

- HDL cholesterol precipitation methods
- homogeneous HDL methods
- LDL cholesterol methods
- homogeneous LDL cholesterol methods
- measuring apoproteins

HDL cholesterol precipitation methods

Early methods for measurement of HDL in clinical laboratories involved precipitating apoB-containing lipoproteins using polyanionic compounds and divalent ions, such as manganese-heparin, magnesium-dextran sulphate, magnesium-phosphotungstate, and polyethylene glycol. Following the removal of the precipitate by centrifugation, the concentration of non apoB-associated cholesterol in the supernatant was measured. Such centrifugation methods, however, are susceptible to interference from VLDLs and lipoproteins with disproportionately high lipid contents, such as IDL, and are not amenable to complete automation.

Homogeneous HDL methods

Homogeneous HDL methods mask non-HDL particles from the cholesterol measuring reagents but retain them in solution so that analyses can be made in a single reaction vessel. This has been achieved by a variety of innovative technologies. A common method involves production of soluble complexes of chylomicrons, VLDL, and LDL using α cyclodextrin and magnesium chloride. The reagent enzymes, cholesterol oxidase and cholesterol esterase are attached to polyethylene glycol to make them inaccessible to the complexes, but remain accessible to HDL particles. Homogeneous methods are relatively robust and are not interfered with by concentrations of TAGs up to about 10 mmol/L. They may, however, be susceptible to interference by unusual lipoproteins, as occurs in some forms of inherited dyslipidaemia, renal disease, and diabetes that are characterized by an accumulation of IDL.

LDL cholesterol methods

Most routine clinical biochemistry laboratories calculate the concentrations of LDL cholesterol using the formula derived by Friedewald, Levy, and Fredrickson in 1972. The formula was designed for epidemiological studies to describe average population relationships between lipids but is now used for individual estimates of LDL cholesterol:

$$\text{LDL cholesterol} = \text{total cholesterol} - \text{HDL cholesterol} - \frac{\text{TAG}}{2.2}$$

The formula is based on two main assumptions:

(1) most circulating TAGs are in VLDL

(2) there is a fairly consistent relationship between the concentrations of VLDL, TAG, and VLDL cholesterol

On average, the molar concentration of VLDL cholesterol equals that of TAGs divided by 2.2. Given that the concentration of HDL cholesterol can be measured directly, then that of LDL cholesterol can be calculated using the above equation.

Lipoproteins with unusual lipid content including postprandial TRL particles give misleading results. Ideally patients should undergo a 12-hour fast to provide a standardized sample (Box 9.6). However, because the *calculated* LDL cholesterol relies on three separate measurements, which is subject to the summation of measurement errors, the calculated LDL cholesterol does not meet the CDC total allowable error criterion of 12%. Thus, there is considerable

BOX 9.6 Should only fasting lipids be measured?

The concentrations of circulating cholesterol and HDL cholesterol are only a little affected by meals. Triacylglycerol concentrations, however, may be markedly changed. The concentrations of TAGs are required to estimate LDL cholesterol using the Friedewald formula. This would not be the case if LDL cholesterol concentrations were measured directly. Increasingly, LDL cholesterol is used in clinical decision making, so patient fasting is required. For pragmatic reasons, lipids may not be measured in samples from fasting patients. Treatment decisions are therefore based on *total* cholesterol estimations. When it is necessary to estimate TAG concentrations, particularly when they are increased, standardized sampling conditions are required. This usually means fasting, but a light fat-free breakfast may suffice, for instance in the case of diabetic patients.

interest in the development of homogeneous methods for determining LDL cholesterol concentrations.

Homogeneous LDL cholesterol methods

Automated homogeneous assays to measure the concentrations of LDL cholesterol have been designed but are not in common use. All existing commercial methods meet the CDC criteria for total allowable error for normolipidaemic samples. However, they are prone to interference from abnormal lipoproteins, particularly IDLs. The task facing the designer of a homogeneous LDL cholesterol assay is to mask non-LDL cholesterol and measure LDL cholesterol without precipitating lipoproteins, since this would necessitate a separation step. This is achieved by employing detergents and reagents that differentially block off and solubilize lipoproteins to allow the concentration of LDL cholesterol to be measured directly.

Measuring apoproteins

Atherogenic small dense LDL and small TRL particles are underestimated by cholesterol measurements because they are both depleted in cholesterol. Only one molecule of apoB protein occurs in each apoB-containing lipoprotein; its number is a measure of the amount or concentration of such particles. However, one caveat is that Lp(a) may constitute 20% of circulating apoprotein B. Non-HDL cholesterol (total cholesterol minus HDL cholesterol), which reflects LDL cholesterol and VLDL cholesterol, is advocated as a measure of apoB containing-particles in the NCEP guidelines from the USA. The advantages of this type of estimation are that it does not require a calculation to give the concentration of LDL cholesterol and it is not affected by the differing cholesterol contents of the various TRL particles. The Quebec Cardiovascular Study found considerable disagreement between LDL cholesterol, non-HDL cholesterol, and apoB protein concentrations. However, apoB estimations better identified patients with features of metabolic syndrome.

The lack of an accuracy base is a barrier to the use of measuring apoB protein in routine practice. Although the WHO has advocated for standardized reference methods there has not, as yet, been a CDC calibration initiative. The bias between methods makes it impossible to set clear decision limits as currently exist for cholesterol measurements. There is, however, a clear role for apoB protein measurements in situations where determining cholesterol concentrations is misleading, such as in primary **biliary cirrhosis** when Lp-X may be present, and sitosterolaemia when non-cholesterol sterols interfere in the standard cholesterol assays. Whether measuring the concentrations of apoB proteins offers an improved assessment of cardiovascular risk because they better reflect LDL particle numbers in situations where small, dense LDL accumulate, is controversial and is not common practice at the moment.

High density lipoprotein particles contain variable numbers of apoA1 molecules and so its measurement less closely reflects particle number. Nevertheless, it is possible that determining the apoB/apoA1 protein ratio may have advantages over using the LDL/HDL ratio when assessing cardiovascular risk.

Key Points

Lipoprotein-X is an abnormal lipoprotein found in serum of patients with obstructive jaundice, and is an indicator of cholestasis. In patients with familial LCAT deficiency, there is an inverse relationship between plasma Lp-X levels and LCAT activity.

TABLE 9.7 Conditions that justify cholesterol lowering treatment.

Coronary heart disease or other atherosclerotic disease such as stroke or peripheral vascular disease

Familial hypercholesterolaemia or other inherited lipid disorders

Renal disease

Type 1 or type 2 diabetes

9.10 Assessment of cardiovascular risk

Most people who have heart attacks have 'normal' values for their blood cholesterol concentration. The average cholesterol concentration in patients who survive a heart attack is only slightly higher than seen in a control population. Cholesterol as a risk factor for heart attacks applies to the whole population: it is not possible to identify a high cholesterol group with high risk of heart attack and a low cholesterol group with low risk within the general population. It is more profitable to combine multiple risk factors with cholesterol to identify those individuals at high risk. Algorithms for calculating risk have been generated from studies that have followed a large group of patients over time having made baseline assessments of risk factors. One such, the Framingham study, named from the small town in the USA, and the risk factors for heart disease and stroke (cardiovascular disease) identified by the study have been incorporated into an equation that uses age, systolic blood pressure, the total cholesterol/HDL cholesterol ratio, and smoking to assess the overall risk of developing cardiovascular disease over ten years. The same information has also been represented in the form of a chart such as is available in the British National Formulary. Separate charts are produced for men and women. The charts show three bands of risk: less than 10%, 10–20%, and greater than 20%, or the risk for an individual can be calculated using the equation.

Other risk factors, such as a family history of heart disease, hypertriglyceridaemia (greater than 1.7 mmol/L), impaired fasting glucose or glucose tolerance, and premature menopause are not taken account of in estimation of risk but a clinical weighting is given for these factors. Table 9.7 shows that some conditions carry sufficient risk to justify cholesterol lowering treatment without considering its level or other factors.

The prediction of risk has not been validated in other ethnic groups such as Asians or Afro-Caribbeans but a weighting may be given for ethnicity.

Intervention with drug treatment is generally reserved for those at higher risk even though it has been demonstrated to be of benefit for those at lower risk because risk benefit analysis and consideration of costs have restricted more widespread intervention.

9.11 Management of hyperlipidaemias

The first line of any treatment for hyperlipidaemia is to amend lifestyle factors as recommended in the JBS2, which are summarized as:

- Do not smoke.
- Maintain an ideal body weight and avoid central obesity.
- Eat a diet containing less than 30% fat, of which saturated fats constitute less than 10% of total fat by substituting monounsaturated fats for them.

BOX 9.7 *Discovery of statins*

The potential therapeutic uses of statins were first recognized by a Japanese biochemist. In the 1970s, Endo extracted 6,000 substances from fungi and screened them for their ability to inhibit cholesterol synthesis. He eventually succeeded in extracting a potent inhibitor called compactin. This had previously been identified by Beechham's pharmaceutical company, although they failed to recognize its potential because of its inability to lower blood cholesterol concentrations in rats. However, rats have much lower values than humans, and Endo was able to show potent cholesterol lowering activity in dogs, which metabolize cholesterol in a manner similar to humans. This was the first step towards the development of modern statin drugs.

- Ingest less than 300 mg of cholesterol daily (roughly the equivalent of one egg).
- Eat five portions of fresh fruit or vegetables daily.
- Eat fatty fish twice a week.
- Restrict alcohol intake to less than 21 units (1 unit equals 10 g) for men and 14 units for women weekly.
- Restrict salt (NaCl) to less than 100 mmoles or 6 g per day.
- Take regular aerobic exercise such as fast walking for half an hour or swimming on most days.

Drug treatments for hypercholesterolaemia

The target concentrations for cholesterol lowering therapy recommended in JBS2 are: a total cholesterol value of less than 4.0 mmol/L and an LDL cholesterol concentration of less than 2.0 mmol/L in all high risk people.

Statins (Box 9.7) are the most effective cholesterol lowering drugs and are used as first-line treatment to clinically reduce the concentration of cholesterol in blood. It is generally accepted that increased concentrations of cholesterol in the blood cause arterial disease and its clinical manifestations.

Figure 9.17 is a re-drawing of Figure 9.6, which shows the biosynthetic pathway of cholesterol, but indicates that statins are highly active inhibitors of HMG CoA reductase, the enzyme that regulates the biosynthesis of cholesterol in hepatocytes. Hence, treatment with statins reduces its biosynthesis by the liver. This leads to a depletion of hepatic intracellular cholesterol stores and an increase in its uptake from the plasma, so reducing the concentration of cholesterol in the blood of the patient. Statins in common use in the UK in order of potency are rosuvastatin (Figure 9.18), atorvastatin, simvastatin, pravastatin, and fluvastatin.

They all work in a similar manner but differ in their pharmacokinetic and pharmacodynamic properties.

The clinical use of statins can delay the progression of cardiovascular disease and strokes in patients with known ischaemic heart disease or those at risk of it, and so are taken by patients suffering from such disease. Statin therapy is also of value to patients with an increased likelihood of developing cardiovascular disease, for example diabetic patients.

Cross reference
Chapter 21 Therapeutic drug monitoring

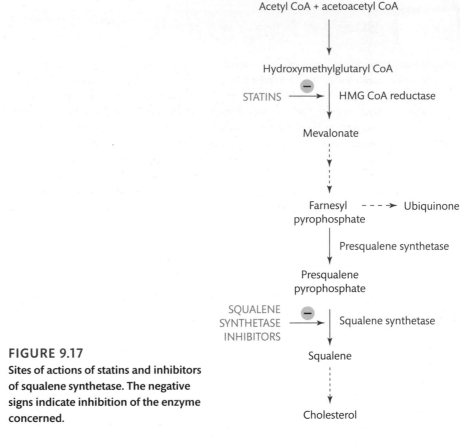

FIGURE 9.17
Sites of actions of statins and inhibitors of squalene synthetase. The negative signs indicate inhibition of the enzyme concerned.

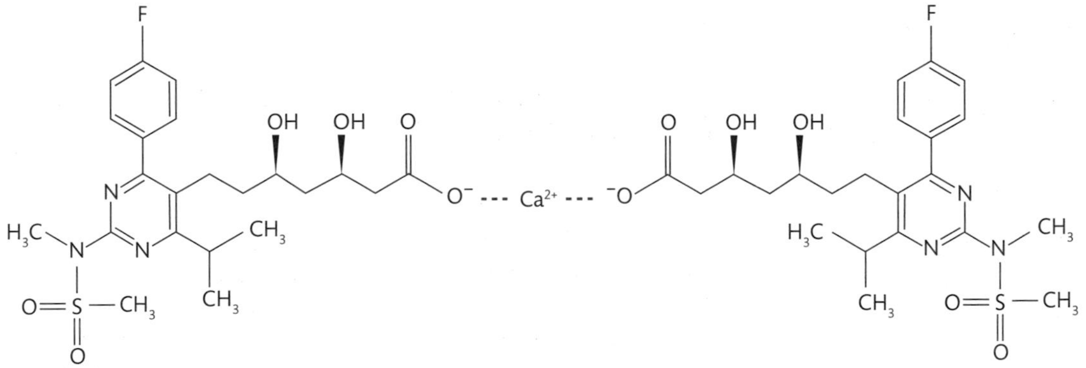

FIGURE 9.18
Structure of calcium rosuvastatin.

If you re-examine Figure 9.17, you can see that another of the biological products formed from farnesyl pyrophosphate is ubiquinone. The clinical use of statins may also inhibit its formation. Ubiquinone is a component of the electron transport chain and reduced amounts of it can interfere with energy metabolism in muscle tissues. A number of other unwanted side effects are also known. Thus, drugs that inhibit squalene biosynthesis but do not affect the biosynthesis of farnesyl pyrophosphate may offer advantages and are under development. Any cholesterol lowering effect of such inhibition may augment that of statins.

BOX 9.8 Functional foods

Cholesterol lowering functional foods contain plant sterols and stanols, which are 5 α reduced forms of sterols. Sterols and stanols are structurally similar to cholesterol and serve similar functions in plants to those of cholesterol in animals. They differ from cholesterol only in minor side chain modifications. Their similar structural and physical properties mean they can displace cholesterol from mixed micelles and so limit the absorption of cholesterol. This is not a pharmacologically potent effect. A daily intake of about 3 g of plant sterol is recommended, justifying its description as food.

Statins alone can reduce cholesterol by 30–40%. When combined with the cholesterol absorption inhibitor, exetimibe, a further 15% reduction can be obtained. The third-line agent is a bile acid sequestrant, which interferes with the enterohepatic circulation of bile salts and promotes the loss of cholesterol into bile.

In combined hyperlipidaemia, where increased VLDL production may be contributing to increased LDLs, a fibrate may be preferred as the third line. Fibrates are PPAR α (Section 9.1) agonists, which encourage hepatic triacylglycerol-rich lipoprotein uptake and FA oxidation. Polyunsaturated long-chain FAs may also be helpful in this situation by reducing the biosynthesis of VLDLs and promoting peroxisomal FA oxidation. Niacin, which acts mainly by inhibiting peripheral lipolysis and reducing VLDL output, may also be useful; it is the only drug currently available that has the effect of significantly increasing HDLs. In heterozygous FH (Section 9.5) the functioning gene is induced to produce LDLRs but these strategies are largely ineffective in homozygous disease because there are no functioning genes to induce. Low density lipoproteins can be removed using affinity absorbents (plasmapheresis) but this can only be performed intermittently and, therefore, success is limited. Uses of cholesterol lowering functional foods are outlined in Box 9.8.

Side effects of statins

Statins have some significant side effects. Very rarely a muscle breakdown called **rhabdomyolysis** occurs; usually when statins are combined with fibrates that interfere with statin disposal. However, complaints from patients of minor muscle aches and pains are common. Some of these may be related to drug interaction or possibly to some mild background muscle metabolic problem made worse by ubiquinone depletion. Supplements of ubiquinone improve statin toleration in some individuals but there have been no clinical trials to study this. Muscle problems can occur even if this is not indicated by increases in the activity of the muscle enzyme, creatine kinase, in blood samples. Thus, increases in activity of CK do not predict the development of rhabdomyolysis. Measuring the value of CK activity in routine monitoring has, therefore, been questioned. Minor changes in liver enzyme functions are common with statins but serious liver problems attributable to statins are extremely rare. There is a consensus that statin treatment is safe in patients with chronic liver disease provided the disease is stable. A transaminase activity staying less than three times the upper limit of normal is the usual criterion for stability.

Treatment of hypertriglyceridaemia

The cornerstone of treatment to reduce chylomicronaemia is to limit fat intake to 20–25 g/day. Secondary causes of hypertriglyceridaemia, such as diabetes or even impaired glucose metabolism, hypothyroidism, and use of drugs such as β blockers, should be dealt with.

CASE STUDY 9.1

A 55-year-old woman consulted her GP complaining of tiredness. She did not appear to be hypothyroid but the GP measured her thyroid function tests. The patient also took the opportunity to have a general health check. She was a little overweight, had normal blood pressure and she did not smoke or drink. Her blood glucose concentration was normal. Her fasting lipids were assessed and gave the following data (reference ranges and desirable values for lipids are given in brackets):

TSH	16 mU/L	(0.2–5.2)
Free T4	18 pmol/L	(9–24)
Cholesterol	7.2 mmol/L	(<4.0)
HDL	1.2 mmol/L	(>1.2)
TAG	1.5 mmol/L	(<1.7)

Explain these results.

Drug treatment using fibrates, very long chain polyunsaturated FAs, or statins is relatively ineffective in reducing chylomicronaemia. Chylomicronaemia may cause acute pancreatitis and antioxidant supplements have been shown to reduce the frequency of attacks in uncontrolled case studies. More drastic ways of restricting fat intake, such as pancreatic lipase inhibition or even bariatric surgery may be contemplated if attacks of acute pancreatitis are difficult to control.

In milder forms of hypertriglyceridaemia (blood concentrations less than 10 mmol/L) when the risk of pancreatitis is low, the need for drug treatment is based on an assessment of global

CASE STUDY 9.2

A 36-year-old woman was reviewed at the lipid outpatient clinic. She had suffered several attacks of acute pancreatitis in previous years. She was moderately obese, had type 2 diabetes and required insulin treatment. She had been instructed to follow a diet very low in fat, of less than 25 g fat daily. She was prone to occasional binges of alcohol and had been advised to stop drinking. Her blood lipid concentrations were determined (desirable values are given in brackets):

Cholesterol	9.2 mmol/L	(<4.0)
TAG	36.2 mmol/L	(<1.7)

Her plasma was very milky in appearance and so HDL cholesterol could not be measured.

(a) Comment appropriately on these results.

On her next review, the lipid results were cholesterol 3.3, HDL 0.7, and TAGs 3.0 mmol/L.

(b) Explain these results.

CASE STUDY 9.3

A 55-year-old woman developed type 2 diabetes. She was a little overweight at 103.4 kg, with a body mass index of 33 Kg/m². She did not smoke, drank 8 to 12 units of alcohol a week, and had a normal blood pressure. Routine blood testing showed the following (desirable values are given in brackets):

Cholesterol	7.2 mmol/L	(<4.0)
HDL	1.0 mmol/L	(>1.2)
TAG	16.6 mmol/L	(<1.7)

Her apoprotein E polymorphism was E2E2.

She was advised to eat a very low fat diet of less than 25 g daily, and prescribed small doses of statins. When she returned to the clinic her lipid profile was:

Cholesterol	2.2 mmol/L
HDL	1.0 mmol/L
TAG	2.1 mmol/L

(a) From which of the dyslipidaemias does this patient suffer? Comment appropriately on your answer.

(b) Why were statins prescribed?

cardiovascular risk. Weight loss and reversal of any secondary causes of dyslipidaemia must be addressed. Regular exercise is helpful. The initial priority for drug treatment based on assessment of cardiovascular risk is to lower the LDL cholesterol using statins. Combination therapy with fibrates, niacin, and long-chain polyunsaturated FAs may be contemplated if the response is inadequate. Drugs that increase the sensitivity to insulin, such as metformin and thioglitazones, may be useful. Aids to weight loss, such as intestinal lipase inhibitors and centrally acting appetite suppressors may help by restricting fat intake.

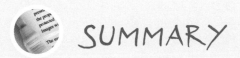

SUMMARY

■ Different types of lipids occur in the body and include fatty acids, triacylglycerols, phospholipids, and cholesterol.

■ Lipids are insoluble in water and associate with apoproteins in the blood to give lipoproteins, and this is the form in which they are transported in the circulation.

■ The lipoprotein particles include chylomicrons, which largely transport dietary triacylglycerols from GIT to peripheral tissues; VLDL particles that transport triacylglycerols from the liver to peripheral tissues; LDL particles, which transport cholesterol in the blood and are taken up by the liver and, finally, HDL particles, which transport cholesterol from peripheral tissues to the liver for excretion and have cardioprotective properties.

- Lipid disorders can either be genetic in origin or secondary to other diseases, drug treatment, or defective nutrition.

- Hypercholesterolaemia is a key risk factor for cardiovascular disease and the underlying causes can be either familial or monogenic, or alternatively non-familial and polygenic in nature.

- Hypertriglyceridaemia is also a risk factor for cardiovascular disease and is caused by a number of inherited conditions or is part of the metabolic syndrome. Severe hypertriglyceridaemia causes acute pancreatitis.

- Deposition of lipids in arterial walls and the subsequent formation of an atheroma are key features of atherogenesis and coronary heart disease.

- In addition to cholesterol, there are other factors that increase the risk of cardiovascular disease. These can be used to detect high risk patients.

- Management of hyperlipidaemia involves using a combination of lifestyle changes aimed at reducing risk factors and the use of lipid lowering drugs such as statins.

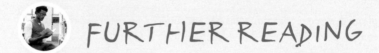

FURTHER READING

- Brunzell JD (2007) **Hypertriglyceridaemia.** *New England Journal of Medicine* **35**, 1009–17.

- Frank PG and Marcel YL (2000) **Apolipoprotein A1: structure-function relationships.** *Journal of Lipid Research* **41**, 853–72.

- JBS2 (2005) **Joint British Societies guidelines on prevention of cardiovascular disease in clinical practice.** *Heart* **91**, 1–52.

- Lewis GF and Rader DJ (2005) **New insights into the regulation of HDL metabolism and reverse cholesterol transport.** *Circulation Research* **96**, 1221–32.

- Marks D, Thorogood M, Neil HA, and Humphries SE (2003) **A review on the diagnosis, natural history and treatment of familial hypercholesterolaemia.** *Atherosclerosis* **168**, 1–14.

- Moller DE and Kaufman KD (2005) **Metabolic syndrome: a clinical and molecular perspective.** *Annual Review of Medicine* **56**, 45–62.

- Nauck M, Warnick GR, and Rifai N (2002) **Methods for measurement of LDL-cholesterol: a critical assessment of direct measurement by homogeneous assays versus calculation.** *Clinical Chemistry* **48**, 236–54.

- Packard CJ (2006) **Small dense low-density lipoprotein and its role as an independent predictor of cardiovascular risk.** *Current Opinions in Lipidology* **17**, 412–17.

- Rader DJ, Cohen J, and Hobbs HH (2003) **Monogenic hypercholesterolaemia: new insights into pathogenesis and treatment.** *Journal of Clinical Investigation* **111**, 1795–1803.

- Steinberg D (2005) **An interpretive history of the cholesterol controversy: part II: the early evidence linking hypercholesterolaemia to coronary disease in humans.** *Journal of Lipid Research* **46**, 179–90.

- Trigatti BL, Rigotti A, and Braun A (2000) **Cellular and physiological roles of SR-B1, a lipoprotein receptor which mediates selective lipid uptake.** *Biochimica et Biophysica Acta* **1529**, 276–86.

- Wang M and Briggs MR (2004) **HDL: the metabolism, function and therapeutic importance.** *Chemical Reviews* **104**, 119–37.

- Warnick GR, Nauck M, and Rifai N (2001) **Evolution of methods for measurement of HDL-cholesterol: from ultracentrifugation to homogeneous assays.** *Clinical Chemistry* **47**, 1579–96.

- Whitfield AJ, Barret PHR, Van Bockxmeer FM, and Burnett JR (2004) **Lipid disorders and mutations in the apoB gene.** *Clinical Chemistry* **50**, 1725–32.

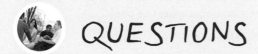

QUESTIONS

9.1 Which one of the following is correct?

   (a) Chylomicrons contain $apoB_{100}$

   (b) The LDL receptor only binds with apoB

   (c) Familial hypercholesterolaemia is caused by abnormal apoB

   (d) Each molecule of VLDL, IDL, and LDL contains one molecule of apoB

   (e) Apo B moves between different lipoproteins as they are metabolized

9.2 State whether the following are TRUE or FALSE.

   (a) Hyperthyroidism causes raised cholesterol

   (b) Familial hypobetalipoproteinaemia is caused by mutations causing truncations in apoB

   (c) ApoE 4/4 polymorphism is linked to Alzheimer's disease

   (d) Statins prevent cholesterol absorption from the diet

   (e) Total cholesterol is the most important risk factor for heart attacks

9.3 Outline the processes that transfer cholesterol from general body tissues to the liver.

9.4 Describe the role(s) of apoproteins in the metabolism of triacylglycerol-rich lipoproteins.

9.5 **(a)** What metabolic processes lead to the production of LDL particles?

   (b) What is the metabolic fate of LDL particles?

   (c) What makes LDL particles become more atherogenic?

9.6 What are the difficulties that would have to be faced in order to screen populations for familial hypercholesterolaemia?

9.7 **(a)** Why is calibration necessary when determining cholesterol concentrations in clinical samples?

   (b) How is consistency between laboratories achieved?

9.8 **(a)** How frequent is homozygous familial hypercholesterolaemia in the population?

(b) Is statin treatment useful in this condition?

9.9 Is measurement of blood cholesterol useful in assessing whether someone will develop coronary artery disease?

9.10 A male patient had the following lipid pattern (desirable values are given in brackets):

Cholesterol   7.3 mmol/L (<4.0)

HDL   0.9 mmol/L (>1.2)

TAG   9.6 mmol/L (<1.7)

Further testing at the lipid clinic showed the patient was homozygous for apoprotein E2.

(a) What is the Frederickson type of this disorder?

(b) Why are both TAGs and cholesterol concentrations increased?

(c) What other factors may have contributed to this dyslipidaemia?

Answers to self-check questions, case study questions, and end-of-chapter questions are available in the Online Resource Centre accompanying this book.

 Go to www.oxfordtextbooks.co.uk/orc/ahmed/

10

Disorders of calcium, phosphate, and magnesium homeostasis

Nessar Ahmed and Farhad Behzad

Learning objectives

After studying this chapter you should be able to:

- Describe the roles, distribution and metabolic regulation of calcium, phosphate, and magnesium ions in the human body
- Discuss the causes and consequences of disorders of calcium, phosphate, and magnesium homeostasis
- Describe the uses of laboratory tests in investigating disorders of calcium, phosphate, and magnesium homeostasis
- Appreciate the principles underlying management of disorders of calcium, phosphate, and magnesium homeostasis
- Discuss the metabolic disorders affecting bone and their laboratory investigations

Introduction

Calcium, phosphate, and magnesium ions are inorganic minerals and key components of bone. Bone contains 99% of the body's calcium, 85% of its phosphate, and 55% of its magnesium. In addition to its mechanical role in locomotion and protection of organs, bone also acts as a reserve for these minerals, in particular calcium and phosphate. Indeed, the concentration of calcium in the plasma is maintained within its reference range by absorption of dietary calcium and also by exchange with calcium between blood and bone. Changes in plasma calcium concentrations above or below its reference range can result in clinical disorders that have characteristic clinical features. Similarly, homeostatic control of phosphate and magnesium concentrations is essential to prevent disorders associated with changes in their concentrations in the plasma.

This chapter will describe the distribution, function, and metabolic regulations of calcium, phosphate, and magnesium in health. The disorders causing an alteration in their plasma concentrations will then be discussed, together with their clinical presentation and laboratory investigations and management. Finally, the clinical value of bone markers in **metabolic bone disease** will be discussed.

10.1 Distribution, function, and regulation of calcium

Calcium is the most abundant mineral in the body. An average adult has approximately 25,000 mmoles which is equivalent to 1 kg of body calcium. The distribution of calcium in the various body compartments, its exchange between bone and extracellular fluid (ECF), and dietary intake and losses are shown in Figure 10.1.

Nearly all body calcium (99%) is present in the bone in the form of crystals similar to those of hydroxyapatite ($Ca_{10}[PO_4]_6[OH]_2$). The total calcium content of ECF is approximately 23 mmoles of which 9.0 mmoles is present in the plasma. About 100 mmoles of calcium in the bone is exchangeable with the ECF, and indeed, a total of 500 mmoles is exchanged daily.

The normal dietary intake of calcium is approximately 25 mmoles per day. Gastrointestinal secretions also contain calcium and some of this is reabsorbed together with dietary calcium. Around 20 mmoles of calcium are lost in the faeces each day. Approximately 240 mmoles of calcium are filtered by the kidneys every day, with the bulk of this filtered calcium being

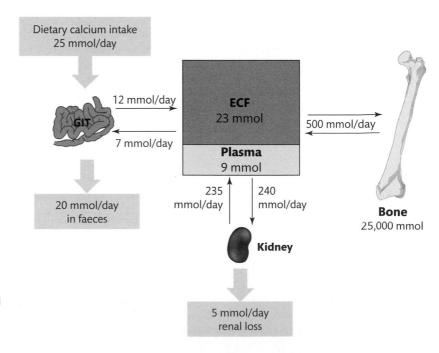

FIGURE 10.1

Calcium distribution in the body and its exchange between different body compartments.

reabsorbed in the tubules and the remainder, some 5 mmoles per day, being lost in the urine. A minute amount of calcium is lost via skin.

Calcium has a structural function in the body in that it is incorporated in the bones and teeth. It is also required for release of neurotransmitters and the initiation of muscle contraction. Calcium is crucial as an intracellular signal and acts as an intracellular second messenger for hormones following their interaction with target cells. Calcium also acts as a co-enzyme for coagulation factors.

Calcium is present in the plasma in three forms:

- free ionized calcium (Ca^{2+}): ~47%
- protein bound calcium (mainly with albumin): ~46%
- complexed calcium (with citrate, phosphate): ~7%

The reference range for calcium in the serum or plasma is typically 2.20–2.60 mmol/L for total calcium and 1.20–1.37 mmol/L for ionized calcium (Ca^{2+}). However, only the ionized form of calcium is physiologically active and it is this fraction which is regulated by homeostatic mechanisms.

SELF-CHECK 10.1

What proportion of plasma calcium is biologically active?

Plasma calcium concentrations

The most common methods for measuring calcium determine its total concentration. However, measurements of total calcium are affected by changes in plasma albumin concentrations which can give rise to misleading results. Changes in the concentrations of plasma albumin, for whatever reason, can affect that of plasma total calcium because approximately half of the calcium in blood is bound to albumin. For example, when plasma albumin concentrations increase, total calcium measurements are also increased. Similarly, when plasma albumin concentrations decline, then total calcium concentrations also decline. However, the concentrations of free unbound, ionized calcium are *not* affected by changes in albumin so there is no physiological or pathological change. Two formulae are used to modify measured calcium concentrations to give a *corrected* calcium value. For albumin concentrations less than 40 g/L, the corrected calcium can be determined using:

corrected [calcium] = measured total [calcium] + 0.02(40 − [albumin])

For albumin concentrations greater than 45 g/L, the corrected calcium is determined using:

corrected [calcium] = measured total [calcium] − 0.02([albumin] − 45)

The alternative is to measure the ionized calcium directly and, indeed, this can be done using an ion-selective electrode.

Key Points

Changes in albumin concentrations affect total calcium but not free ionized calcium concentrations.

Changes in acid-base balance can influence concentrations of ionized calcium but have no effect on plasma total calcium concentrations. Ionized calcium ions compete with hydrogen ions (H^+) for negative binding sites on albumin and changes in the ionized fraction may occur in acute acid-base disorders.

In **alkalosis**, the concentration of H^+ in the plasma is low and so more calcium ions bind to albumin as H^+ dissociate from albumin, thus increasing protein bound calcium. Thus, the concentration of plasma ionized calcium may decline below its reference range and produce the associated symptoms of low plasma calcium (see Section 10.2).

However, in **acidosis**, the concentration of H^+ in the plasma increases and more of these will bind to the negative sites on the albumin displacing the calcium ions; therefore the effect is to decrease protein bound calcium but increase free ionized calcium in the plasma. When the plasma ionized calcium concentration rises above its reference range, then symptoms typical of high plasma calcium can be seen in affected patients (see Section 10.2).

Cross reference
Chapter 6 Acid-base disorders

The change in ionized plasma calcium affects release of hormones which will correct plasma ionized calcium concentrations and return them within their reference range.

The analytical measurement of calcium is described in Box 10.1.

SELF-CHECK 10.2

Calculate the corrected calcium for a serum specimen with an albumin concentration of 56 g/L and a total calcium of 2.1 mmol/L.

Regulation of plasma calcium

In the blood, the concentration of ionized calcium is maintained within its reference range largely by the action of two hormones that work together and include parathyroid hormone

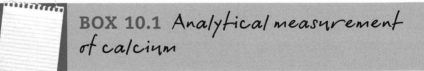

BOX 10.1 *Analytical measurement of calcium*

Serum is the preferred sample for determination of calcium although heparinized plasma is also acceptable. However, blood samples for calcium determination should not be collected in tubes containing EDTA as this will chelate the calcium. Measurements of calcium can determine the total or free ionized fraction. The latter measurement has the advantage that it reflects the biologically active form of calcium and although more useful clinically, is more difficult to perform analytically. Total calcium measurements are made in either serum or urine. Most of these methods rely on reaction of calcium with σ-cresolphthalein in alkaline conditions to give a red-coloured complex whose absorbance is measured at 570–580 nm. In many procedures, the sample is pre-treated with acid to release protein-bound and complexed calcium ions. Atomic absorption spectroscopy is used as a reference method for determination of total serum calcium. In this method, the samples are pre-treated with lanthanum-HCl to reduce interference from protein, phosphate, and other anions.

Free or ionized calcium measurements are now provided by many hospital laboratories and are made using ion-selective electrodes.

(PTH) and calcitriol, which is also known as 1,25-dihydroxycholecalciferol (1,25-DHCC) or 1,25-dihydroxyvitamin D_3.

Parathyroid hormone

Parathyroid hormone is released from the parathyroid glands which are four or more small glands found behind the thyroid gland in the neck. You can see the anatomical structure of the parathyroid glands in Figure 10.2.

When the concentration of plasma calcium falls below a certain level, the change is detected by calcium-sensing receptors in the parathyroid glands that are stimulated to release PTH into the bloodstream.

Parathyroid hormone is a polypeptide hormone consisting of 84 amino acid residues. It is synthesized as a large precursor preproPTH (115 amino acid residues). In the rough endoplasmic reticulum of parathyroid cells, 25 residues are cleaved to produce proPTH and then further processing in the Golgi apparatus results in cleavage of six residues to give PTH. Figure 10.3 shows how PTH is produced from preproPTH.

The sequences of amino acids removed are thought to be necessary in the intracellular transport of the hormone. The biological activity of PTH is situated in the amino terminal 1 to 34 amino acid residues. Parathyroid hormone is stored in granules until its release and has a half-life of 5–10 minutes in the blood before being rapidly metabolized in the liver and kidneys.

Following metabolism, fragments of PTH are produced and can be detected in the plasma, together with intact PTH. These fragments include an amino terminal fragment which is biologically active and exists for 5–10 minutes, a carboxyl terminal fragment with a half-life of 2–3 hours, and other smaller fragments. Immunoassays for PTH had the limitation in that they detected some of these fragments which were not biologically active, thus making

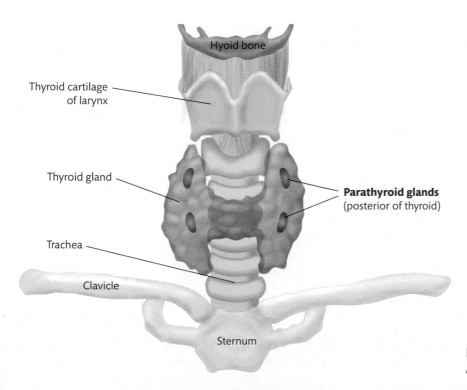

FIGURE 10.2
Anatomy of the parathyroid glands.

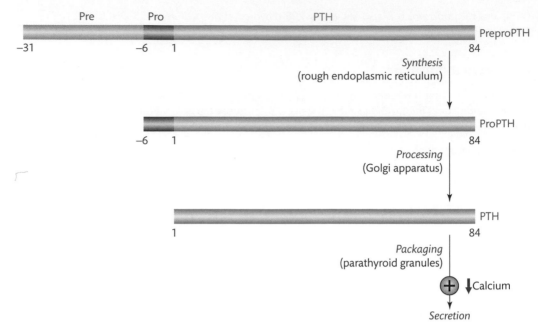

FIGURE 10.3
A schematic illustrating conversion of preproPTH to PTH in parathyroid cells.

interpretation of the results difficult. However, more recent immunometric assays measure only the intact PTH and provide more reliable results.

After its release into the bloodstream, PTH acts on the bone and causes bone **resorption**, a process that releases bone calcium and increases the plasma calcium concentration. Another effect of PTH is to stimulate the synthesis of calcitriol in the kidneys. Calcitriol circulates in the blood and acts on the gastrointestinal tract (GIT) where it stimulates absorption of calcium. Parathyroid hormone also increases calcium reabsorption in the kidneys, which again helps to increase plasma calcium concentrations. Parathyroid hormone decreases bicarbonate reabsorption by the kidneys so that more bicarbonate is lost in the urine. This produces more acidic conditions in the blood, which in turn helps to raise the plasma ionized calcium concentration. The regulation of plasma calcium by release of PTH is outlined in Figure 10.4.

The action of PTH on the kidneys also reduces reabsorption of phosphate, thus promoting its loss in the urine. Phosphate regulation is discussed in Section 10.4.

Calcitriol

Calcitriol is a hormone derived from cholecalciferol (vitamin D). It is formed in the skin by the action of ultraviolet (UV) light on 7-dehydrocholesterol, a derivative of cholesterol. In the liver, cholecalciferol is hydroxylated in a reaction catalysed by 25-hydroxylase to form 25-hydroxycholecalciferol (25-HCC). Further hydroxylation of 25-HCC occurs by the activity of 1-hydroxylase in the kidneys to form 1,25-dihydroxycholecalciferol (1,25-DHCC) or calcitriol. The synthesis of calcitriol is outlined in Figure 10.5.

When plasma calcium concentration is low, release of PTH occurs and can stimulate activity of the enzyme 1-hydroxylase in the kidneys, increasing formation of calcitriol. When plasma calcium is normal or high then the enzyme 24-hydroxylase becomes active and converts

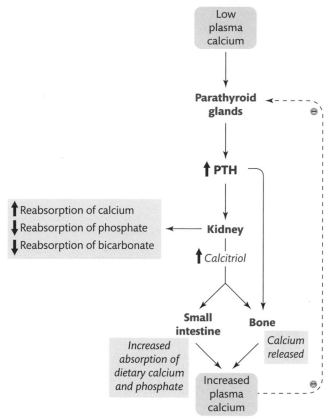

FIGURE 10.4

Regulation of plasma calcium concentrations by the combined action of PTH and calcitriol. Parathyroid hormone and calcitriol also influence plasma phosphate concentrations. See text for details.

25-HCC to 24,25-DHCC, which has very weak activity with regard to calcium absorption from the gut.

Calcitriol can stimulate increased absorption of calcium from the GIT. It does this by interacting with a calcitriol receptor located in intestinal epithelial cells. This interaction results in increased synthesis of a calcium binding protein which enables these cells to actively transport more calcium from the GIT across the mucosa and into the blood, thus increasing plasma calcium concentration. To maintain electroneutrality, the transport of calcium ions by the intestinal epithelial cells is accompanied by inorganic phosphate; thus calcitriol also stimulates the intestinal absorption of phosphate. Calcitriol also stimulates an increase in the renal tubular reabsorption of calcium therefore reducing losses of calcium in the urine and helping to increase plasma calcium concentrations.

Calcitriol also stimulates the release of calcium from the bone by acting on **osteoclasts** in the bone, causing bone resorption which increases plasma calcium concentrations. Calcitriol also inhibits the release of calcitonin, a hormone that reduces plasma calcium by inhibiting release of calcium from bone.

Key Points

It is the plasma free ionized calcium concentrations that are biologically active and influence the negative feedback effect on release of PTH.

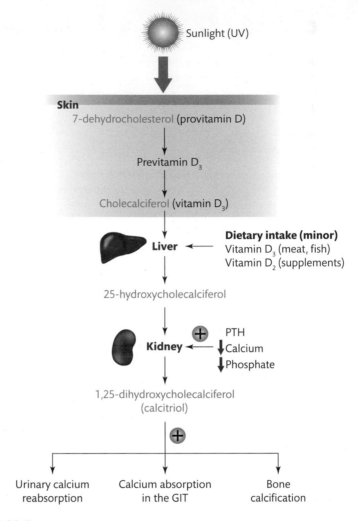

FIGURE 10.5
Synthesis of calcitriol following action of sunlight on 7-dehydrocholesterol in the skin. See text for details.

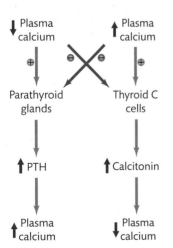

FIGURE 10.6
The opposite effects of PTH and calcitonin on plasma calcium concentrations.

Calcitonin

The hormone, calcitonin consists of 32 amino acid residues and is secreted by the parafollicular or C cells of the thyroid gland. The functions of calcitonin oppose those of PTH. Figure 10.6 outlines how calcitonin opposes the effects of PTH.

For example, calcitonin reduces calcium absorption by the GIT and reduces bone resorption and renal tubular reabsorption of calcium in the kidneys. The effects of all three are to reduce the concentration of calcium in the plasma. However, the precise physiological role of calcitonin is unclear.

SELF-CHECK 10.3

What effect does the release of (a) PTH and (b) calcitriol have on plasma calcium concentrations?

10.2 **Disorders of calcium homeostasis**

Disorders of calcium homeostasis occur when the free ionized or corrected calcium concentration falls outside its reference range. **Hypercalcaemia** describes high plasma calcium concentrations above the reference range, whereas **hypocalcaemia** is characterized by calcium concentrations below its reference range.

Abnormalities of calcium homeostasis are often detected by accident, when blood is tested for another reason, or when calcium is monitored in diseases associated with abnormal calcium homeostasis, or because of the presence of clinical features of hypo- or hypercalcaemia.

Key Points

Disorders of calcium homeostasis can give rise to an increase or decrease of plasma calcium concentrations, referred to as hyper- or hypocalcaemia respectively.

Hypercalcaemia

Hypercalcaemia is often an asymptomatic laboratory finding but care must be taken as hypercalcaemia is often indicative of other diseases. In general, hypercalcaemia arises due to increased intestinal absorption, decreased renal excretion, or an excessive loss of calcium from the bones.

- Increased GIT absorption:
  - excess vitamin D intake
  - tuberculosis or sarcoidosis
  - acromegaly
  - idiopathic hypercalcaemia of infancy
- Decreased renal excretion:
  - thiazide diuretics
  - milk-alkali syndrome
- Increased bone loss:
  - malignancy
  - primary hyperparathyroidism
  - Paget's disease
  - hyperthyroidism
  - lithium therapy
  - familial hypocalciuric hypercalcaemia

An excess intake of vitamin D may lead to hypercalcaemia. This may be iatrogenic, for example when treating **hypoparathyroidism**, or can occur following accidental ingestion.

Hypercalcaemia may occur in patients with graulomatous diseases such as tuberculosis or sarcoidosis. A **granuloma** is a mass of granular tissue produced in response to chronic inflammation. The granulomas contain macrophages in which there is increased conversion of 25-HCC to 1,25-DHCC that, in turn, increases plasma calcium concentration that is hypercalcaemia.

Hypercalcaemia is occasionally seen in **acromegaly** probably due to stimulation of 1-hydroxylase activity in the kidneys (Section 10.1) by excess growth hormone. Idiopathic hypercalcaemia of infancy is a condition associated with hypercalcaemia due to increased sensitivity to vitamin D by bone and gut, but the precise mechanism by which this occurs is unknown.

The use of thiazide **diuretics** can reduce renal calcium excretion and can lead to hypercalcaemia, although this rarely occurs in people with otherwise normal calcium metabolism. It is more likely to occur in patients with increased bone resorption or **hyperparathyroidism**.

The **milk-alkali syndrome** was a significant cause of hypercalcaemia, particularly in people who ingested large amounts of milk together with alkali antacids, such as bicarbonate, to relieve symptoms of peptic ulceration. This can cause an alkalosis and reduce renal calcium excretion giving rise to hypercalcaemia although the precise mechanism by which this occurs is still unknown. Milk-alkali syndrome is now rare, as antacids have been replaced as the treatment for peptic ulcers by drugs which inhibit the secretion of gastric acid.

The most common cause of acute hypercalcaemia is the presence of malignant tumours such as breast and certain types of lung cancers. These tumours can produce parathyroid hormone related peptide (PTHrp), which has an activity similar to that of PTH in that it can increase the concentration of plasma calcium. However, unlike PTH, the release of PTHrp is not controlled by the concentration of ionized calcium. Furthermore, cytokines and prostaglandins produced by tumours that metastasize to bones may increase bone resorption, which, in turn, releases calcium and contributes towards the raised plasma concentrations.

A common cause of chronic hypercalcaemia is hyperparathyroidism, which is an overactivity of the parathyroid glands. This can occur at any age and affect both men and women, but is common in post-menopausal women. It results in excessive production of PTH leading to hypercalcaemia. More details on hyperparathyroidism as a cause of hypercalcaemia are given in Box 10.2.

BOX 10.2 Hyperparathyroidism

Hyperparathyroidism due to a defect in the gland itself is referred to as primary hyperparathyroidism. This commonly occurs due to a parathyroid adenoma (a benign tumour) or parathyroid hyperplasia (an increased production/growth of normal cells). Parathyroid carcinomas (cancerous tumours) are rare. These patients have high concentrations of plasma PTH and calcium.

Secondary hyperparathyroidism can occur in patients who suffer from vitamin D deficiency or chronic kidney disease. Both vitamin D deficiency and kidney disease reduce the production of calcitriol causing hypocalcaemia. The parathyroid glands respond by producing PTH to correct the hypocalcaemia. Concentrations of plasma PTH in these patients are increased, whereas those of calcium are normal or low.

Autonomous PTH secretion occurs in end-stage renal failure. It is believed that PTH production escapes from the control of plasma ionized calcium as a result of the prolonged hypocalcaemia in severe renal failure. The hypercalcaemia usually becomes evident first when the patient has a renal transplant. The new kidney can synthesize calcitriol and together with the high PTH gives rise to a hypercalcaemia referred to as tertiary hyperparathyroidism.

Most patients with hyperparathyroidism have no signs or symptoms but present only with hypercalcaemia. When symptoms do occur, they are commonly associated with hypercalcaemia and include muscle weakness, bone pain, nausea, vomiting, constipation, and depression.

The management of primary, secondary, and tertiary hyperparathyroidism does differ. In general, treatment is first aimed at correction of the hypercalcaemia. If the patient is symptomatic, then surgical removal of the parathyroid gland or parathyroid tumour will alleviate the symptoms of most patients.

Individuals who suffer from hyperthyroidism often suffer from hypercalcaemia. Thyroid hormones have no direct effect on calcium homeostasis, but can increase bone turnover by increasing osteoclastic activity, giving rise to mild hypercalcaemia.

When patients are immobilized for significant periods of time, there is decreased bone formation but continual resorption of bone and release of calcium that is lost in the urine. Thus, hypercalcaemia may occur in such patients who also suffer from increased bone turnover, for example as in **Paget's disease**.

Patients on lithium therapy may develop mild hypercalcaemia. Lithium probably raises the set point for PTH inhibition by calcium. Hypercalcaemia due to lithium therapy usually resolves when use of the drug is discontinued.

Familial hypocalciuric hypercalcaemia is an autosomal dominant condition which develops from childhood and is characterized by chronic hypercalcaemia but is usually asymptomatic. Most cases are due to mutations in the *CASR* gene that code for calcium-sensing receptors in cells of the parathyroid glands and kidneys. Since these defective receptors cannot detect calcium concentrations, the parathyroid gland produces inappropriately high levels of PTH which causes the hypercalcaemia. In addition to hypercalcaemia, these patients present with **hypocalciuria**, that is low concentrations of urinary calcium. The reduced loss of calcium in the urine is due to the failure of kidney cells to recognize and excrete excessive calcium from the body.

Key Points

The common causes of hypercalcaemia include malignancy and hyperparathyroidism.

Clinical effects of hypercalcaemia

The signs and symptoms of hypercalcaemia are often minimal, particularly in mild hypercalcaemia. However, as the hypercalcaemia worsens, the symptoms become more obvious, particularly with acute increases in plasma calcium. In severe cases, hypercalcaemia can cause abnormal electrical impulses, **arrhythmias**, and severe cases may result in **cardiac arrest**. The sites of the body affected by hypercalcaemia and the symptoms produced are given in Table 10.1.

The most serious consequence of prolonged hypercalcaemia is renal damage. Hypercalcaemia may also depress neuromuscular excitability and this may present as constipation and abdominal pain. Hypercalcaemia also has an adverse effect on the central nervous system, causing depression, nausea, vomiting, and often dehydration in affected patients. Gastrin release can

TABLE 10.1 Clinical features of hypercalcaemia.

Site affected	Symptoms
Abdominal	Pain
	Nausea
	Vomiting
	Anorexia
	Constipation
Kidneys	Thirst
	Frequent urination
	Kidney stones
Muscular	Muscle atrophy
	Muscle weakness
	Muscular twitching
Bone	Bone pain
	Fractures
	Curvature of spinal column
Psychological	Coma
	Apathy
	Depression
	Irritability
	Memory loss

also be stimulated by hypercalcaemia leading to excessive gastric acid secretion in the stomach, which in turn can cause peptic ulceration.

Investigation of hypercalcaemia

A number of biochemical tests can assist in the diagnosis of hypercalcaemia and its causes. Tests used to investigate suspected hypercalcaemia include measurement of serum or plasma concentrations of ionized or total calcium. If total calcium is measured then plasma albumin concentrations need to be determined in case corrected calcium measurements are required. Measurements of plasma PTH are also valuable in diagnosis of primary hyperparathyroidism, one of the common causes of hypercalcaemia. A strategy for clinical investigation of hypercalcaemia is outlined in Figure 10.7.

If a patient is suspected of hypercalcaemia then usually the corrected calcium is determined and if greater than 2.6 mmol/L indicates hypercalcaemia. Patients who have very high concentrations of plasma calcium equal to, or in excess of, 3.5 mmol/L require urgent treatment to correct the hypercalcaemia. Measurement of plasma PTH will help to determine whether the cause is malignancy where there will be a low or undetectable PTH concentration or primary hyperparathyroidism which is characterized by a high plasma PTH concentration.

Management of hypercalcaemia

How hypercalcaemia is treated depends on both its severity and cause. Patients with high concentrations of plasma calcium, above 3.5 mmol/L, need to be treated urgently and may require dialysis or emergency **parathyroidectomy**. A common approach is to identify the underlying

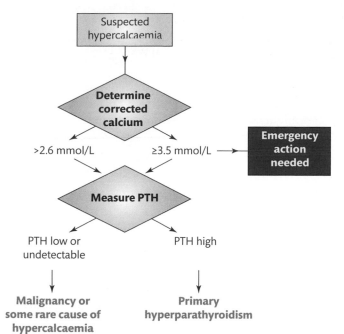

FIGURE 10.7

A possible strategy for investigation of suspected hypercalcaemia. See text for details.

cause of the hypercalcaemia and treat this wherever possible. For example, surgical removal of a parathyroid adenoma may be necessary in patients with hypercalcaemia due to primary hyperparathyroidism. Intravenous saline should be administered in dehydrated patients to restore the glomerular filtration rate, which not only improves hydration but also enhances calcium excretion. In some patients, drugs such as frusemide may be used to inhibit the renal reabsorption of calcium and promote calcium excretion. Other drugs, such as **bisphosphonates**, lower plasma calcium concentration by inhibiting bone resorption and may be of value in treating hypercalcaemia.

SELF-CHECK 10.4

Which drugs might be useful in the management of hypercalcaemia? Explain why.

CASE STUDY 10.1

A 52-year-old woman was being treated with vitamin D therapy for her hypocalcaemia, which she developed following parathyroidectomy for hyperparathyroidism. She was admitted to hospital suffering from nausea, vomiting, and muscle weakness. Biochemical analysis of her serum sample gave the following results (reference ranges are given in brackets):

Total calcium	3.70 mmol/L	(2.20–2.60)
Albumin	41 g/L	(35–45)
Phosphate	1.2 mmol/L	(0.8–1.4)
Alkaline phosphatase	82 IU/L	(30–130)

Explain these results.

CASE STUDY 10.2

A 58-year-old woman presented to her doctor suffering from muscle weakness, fatigue, thirst, and polyuria. She provided a urine sample for glucose estimation and a serum sample for biochemical analysis (reference ranges are given in brackets):

Urine
Glucose	−ve	(−ve)

Serum
Sodium	155 mmol/L	(135–145)
Potassium	3.6 mmol/L	(3.5–5.0)
Urea	7.0 mmol/L	(3.5–6.6)
Total calcium	3.3 mmol/L	(2.20–2.60)
Albumin	32 g/L	(35–45)

(a) What is the corrected calcium?

(b) What is the likely diagnosis?

(c) What further investigations might be useful?

(d) How should this patient be treated?

Hypocalcaemia

Hypocalcaemia arises from a decreased absorption of calcium from the GIT, increased renal excretion, decreased loss from bones, or it can be an **artefactual** result, arising from, for example, improper specimen collection. Some of the causes of hypocalcaemia are:

- Decreased GIT absorption:
  - vitamin D deficiency
- Increased renal loss:
  - renal failure
- Decreased bone loss
  - hypoparathyroidism
  - pseudohypoparathyroidism
  - hungry bone syndrome
  - rhabdomyolysis
  - phosphate administration
  - magnesium deficiency
  - acute pancreatitis
- Artefactual:
  - collection of blood in tube containing EDTA

Vitamin D deficiency can arise in an individual due to dietary deficiency of vitamin D or its malabsorption in the GIT. It can also occur because of inadequate synthesis of vitamin D in the skin due to lack of exposure to sunlight, for example as in certain communities who cover themselves up for religious and/or cultural reasons.

In chronic renal failure there may be a decrease in the reabsorption of calcium from the GIT, and decreased synthesis of 1,25-DHCC from 25-HCC may occur causing hypocalcaemia. Increased output of PTH due to hypocalcaemia may cause metabolic bone disease because of its effect on osteoclastic activity.

Hypocalcaemia may occur in patients suffering from hypoparathyroidism which is the reduced activity of the parathyroid glands and characterized by decreased output of PTH. Hypoparathyroidism can be divided into two types: congenital hypoparathyroidism, where there is congenital absence of the parathyroid glands, and acquired hypoparathyroidism, which can be idiopathic or autoimmune in nature, or arise following surgery of the parathyroid glands, referred to as parathyroidectomy.

Pseudohypoparathyroidism is a condition where excessive secretion of PTH occurs because the target tissues fail to respond to this hormone (resistance) and hypocalcaemia persists. This condition is more common in males than females. Patients present with skeletal abnormalities, including a short stature, mental retardation, **cataracts**, and testicular atrophy.

Hungry bone syndrome can cause hypocalcaemia following surgical treatment of hyperparathyroidism in patients who have had prolonged secondary or tertiary hyperparathyroidism. In this condition, calcium from the plasma is rapidly deposited in the bone causing hypocalcaemia.

During **rhabdomyolysis**, calcium may be moved from the plasma when large amounts of intracellular phosphate are released and precipitate this calcium in the bones and extraskeletal tissues. Similarly, phosphate administration may cause hypocalcaemia.

Magnesium deficiency can also cause hypocalcaemia. Magnesium is required for both secretion and action of PTH and a deficiency produces hypocalcaemia. During acute pancreatitis, calcium can precipitate in the abdomen, causing hypocalcaemia and there is often decreased PTH secretion and increased calcitonin release which may contribute towards the hypocalcaemia.

Artefactual indications of non-existent hypocalcaemia are common and arise when blood is accidentally collected into tubes, for example those containing the anticoagulant ethylenediaminetetraacetic acid (EDTA). The calcium is removed from solution by the EDTA and thus subsequent determination of its concentrations give a low value which is indicative of hypocalcaemia.

SELF-CHECK 10.5

Why is it not a good idea to use tubes containing EDTA as an anticoagulant for collection of blood for calcium determination?

Clinical effects of hypocalcaemia

The clinical effects of hypocalcaemia are listed in Table 10.2.

The major effect of hypocalcaemia is on neuromuscular function. Indeed, the earliest symptom of hypocalcaemia is usually paraesthesiae or 'pins and needles' sensation affecting the hands, feet, and perioral regions of the body. In some patients with hypocalcaemia, fatigue and anxiety may occur, along with muscle cramps which can be painful and sometimes may progress to **tetany** affecting, face, hand, and feet muscles. A common symptom of hypocalcaemia is **petechiae**. These are small red spots on the skin that arise due to broken capillaries. Sometimes, petechiae may develop into a rash.

TABLE 10.2 Clinical features of hypocalcaemia.

Site affected	Symptom
Eyes	Cataracts
Skin	Petechiae
Neuromuscular	Paraesthesiae
	Tetany
	Convulsions
	Cardiac arrhythmias
	Bronchial spasms
	Laryngeal spasms
Psychological	Behavioural disturbances
	Depression
	Memory loss
	Hallucinations
	Anxiety

In some extreme cases, patients may have bronchial or laryngeal spasms in addition to life-threatening complications such as cardiac arryhthmias.

Hypocalcaemia may also cause depression, memory loss, and hallucinations. In chronic hypocalcaemia due to hypoparathyroidism, the high phosphate concentration leads to a precipitation of calcium phosphate in the eye lens and this underlies the development of cataract.

A feature to look out for in hypocalcaemia is the **Trousseau's sign**, which is a spasm of the hand and forearm that occurs when the upper arm is compressed, for example by a blood pressure cuff. Similarly, tapping of the cheekbone can cause a spasm of the face muscles, which is referred to as the **Chvostek's sign**. These are shown in Figure 10.8.

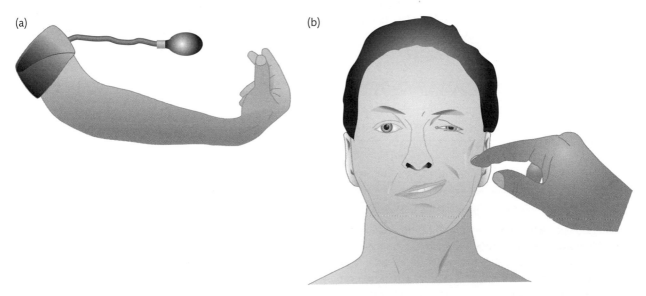

(a) (b)

FIGURE 10.8
Clinical features of hypercalcaemia including: (a) Trousseau's sign. (b) Chvostek's sign.

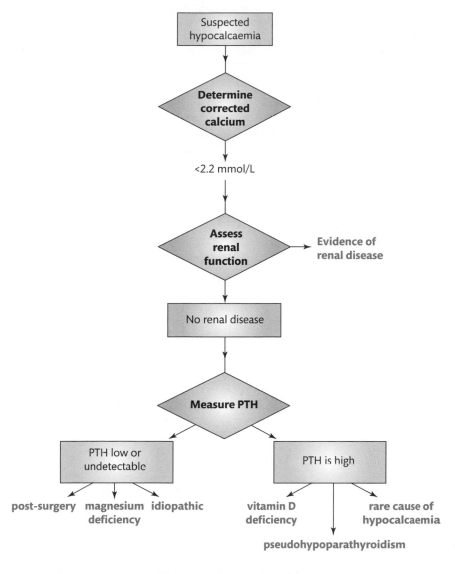

FIGURE 10.9
A possible strategy for investigation of suspected hypocalcaemia. See text for details.

Key Points

Tetany is a neurological symptom characterized by muscular contractions, cramps and seizures, all a consequence of hypocalcaemia.

Investigation of hypocalcaemia

Biochemical tests, which may be useful when investigating a patient with suspected hypocalcaemia, include measurement of serum calcium, albumin, PTH, urea and creatinine, magnesium, and vitamin D concentrations. A possible strategy for investigation of hypocalcaemia is outlined in Figure 10.9.

If a patient is suspected of hypocalcaemia then usually the corrected calcium is determined and if less than 2.2 mmol/L confirms hypocalcaemia. These patients should be assessed for

CASE STUDY 10.3

A 36-year-old man who had previously suffered a viral infection had a blood sample taken for routine monitoring. The results for calcium and albumin were as follows (reference ranges are given in brackets):

Albumin	38 g/L	(35–45)
Total calcium	0.3 mmol/L	(2.20–2.60)

Suggest a possible explanation for these results.

their renal function as this is a common cause of hypocalcaemia. If they do not have renal disease then measurement of plasma PTH concentrations can reveal possible causes of the hypocalcaemia. Low or undetectable plasma PTH concentrations often occur in patients with magnesium deficiency, post-surgery, or those with idiopathic hypocalcaemia. Plasma PTH concentrations are high in patients with hypocalcaemia due to vitamin D deficiency, pseudohypoparathyroidism, and in other rare causes of hypocalcaemia.

Management of hypocalcaemia

The main approach to managing hypocalcaemia is to treat the underlying cause wherever possible. Mild cases of hypocalcaemia are often treated with oral calcium supplements. Patients with hypocalcaemia due to vitamin D deficiency may be placed on calcitriol or its precursors. Magnesium supplements may be prescribed for patients with hypocalcaemia due to magnesium deficiency.

10.3 Distribution, function, and regulation of phosphate

The total body content of phosphate in the average male is over 20,000 mmols. The majority of this phosphate, that is about 80–85%, is in the bone, 15% in the intracellular fluid (ICF), and 0.1% in the ECF. Most of the intracellular phosphate is attached covalently to lipids and proteins.

The daily intake of phosphate is around 40 mmols (derived mainly from dairy products and green vegetables) of which 14 mmols is lost in the faeces. The kidneys lose approximately 26 mmols of phosphate every day in the urine. You can see the distribution of body phosphate in Figure 10.10.

In biology, phosphate in solution is usually referred to as inorganic phosphate, to distinguish it from phosphates bound in various phosphate esters. Inorganic phosphate is usually denoted as P_i and at physiological pH consists of a mixture of HPO_4^{2-} and $H_2PO_4^-$ ions.

A crucial role of phosphate is that it combines with calcium to form hydroxyapatite which is the mineral component of bone and teeth.

Phosphate is required in the body to maintain cell wall integrity and is required in metabolic processes such as glycolysis and oxidative phosphorylation. It is a key component of

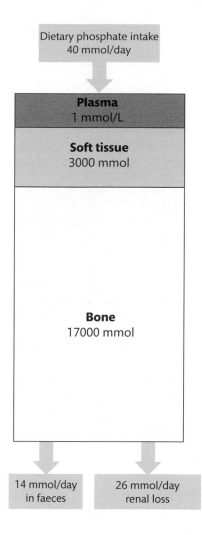

FIGURE 10.10
Distribution of phosphate in body compartments, its intake, and excretion from the body.

molecules such as adenosine tri- and diphosphates (ATP and ADP) respectively. Phosphate functions as a urinary buffer for excretion of H^+ ions in the kidneys. It is also needed for synthesis of 2,3-diphosphoglycerate, which regulates dissociation of oxygen from oxyhaemoglobin. Phosphate is required for phosphorylation and dephosphorylation reactions, that control the activity of many enzymes.

Inorganic phosphate in the plasma exists in three forms:

- free inorganic phosphate: ~80%
- protein-bound phosphate: ~15%
- complexed with calcium or magnesium: ~5%

The analytical measurement of phosphate is outlined in Box 10.3.

Regulation of plasma phosphate

Phosphate concentrations in the plasma are controlled by two hormones, PTH and calcitriol. Regulation of plasma phosphate by PTH and calcitriol has already been shown in Figure 10.4.

BOX 10.3 *Analytical measurement of phosphate*

A common method for measurement of phosphate relies on its reaction with ammonium molybdate under acid conditions to form a colourless phosphomolybdate complex that can be measured directly by its absorption of light in the UV region of the spectrum i.e. at 340 nm. Alternatively, this complex can be reacted further with reducing agents to give a molybdenum blue complex which is measured at 600–700 nm.

Serum samples or heparinized plasma should be used for determination of phosphate. Anticoagulants such as EDTA, citrate, and oxalate have the disadvantage in that they interfere in the formation of the phosphomolybdate complex.

Parathyroid hormone decreases phosphate reabsorption in the kidneys causing loss of phosphate in the urine and a fall in plasma phosphate concentrations. Calcitriol increases phosphate absorption in the gut and therefore raises the concentration of plasma phosphate.

SELF-CHECK 10.6

What effect does (a) PTH and (b) calcitriol have on plasma phosphate concentrations?

10.4 Disorders of phosphate homeostasis

Plasma phosphate concentrations are kept within their reference range of 0.8–1.4 mmol/L. The reference range for plasma phosphate is higher during infancy and childhood. Disorders of phosphate homeostasis occur when the concentration of phosphate falls outside its reference range. **Hyperphosphataemia** describes high concentrations of plasma phosphate above its reference range, whereas **hypophosphataemia** refers to plasma phosphate concentrations below their reference range.

Key Points

Disorders of phosphate homeostasis can cause an increase in its plasma concentrations either above or below its reference range referred to as hyper- and hypophosphataemia respectively.

Hyperphosphataemia

High concentrations of plasma phosphate can be due to increased intake, reduced loss, movement of phosphate from ICF to ECF, or can be artefactual. Some causes of hyperphosphataemia include:

- Increased intake:
  - oral
  - intravenous
  - vitamin D intoxication
- Reduced renal loss:
  - renal failure
  - hypoparathyroidism
  - acromegaly
  - pseudohypoparathyroidism
- Cellular release:
  - tissue destruction
  - intravascular haemolysis
  - catabolic states
  - diabetic ketoacidosis
- Artefactual
  - haemolysis
  - delayed separation of serum

Excessive dietary intake whether oral or intravenous is a rare cause of hyperphosphataemia and is more likely in patients suffering from renal failure. Excessive intake of vitamin D may increase calcitriol formation which may cause hyperphosphataemia because calcitriol increases phosphate absorption from the GIT.

Renal failure is one of the commonest causes of hyperphosphataemia. During renal failure, the glomerular filtration rate falls and therefore loss of phosphate in the urine declines and its plasma concentration increases, causing hyperphosphataemia. Hypoparathyroidism is characterized by low PTH secretion which causes reduced renal excretion of phosphate, giving rise to hyperphosphataemia. In acromegaly there is increased activity of 1-hydroxylase, which results in an increased synthesis of calcitriol. This in turn, increases absorption of dietary phosphate from the GIT. In addition, the excess growth hormone acts directly to reduce renal excretion of phosphate and both effects contribute towards the hyperphosphataemia. In pseudohypoparathyroidism, there is resistance to PTH action, causing decreased renal excretion of phosphate, producing hyperphosphataemia.

Increased release of phosphate from cells can cause a shift of phosphate from ICF to ECF causing hyperphosphataemia. For example, phosphate is released from blood cells into the plasma during intravascular haemolysis and this could give rise to hyperphosphataemia. During diabetic ketoacidosis, deficiency of insulin prevents uptake of phosphate by cells causing hyperphosphataemia. In catabolic states, that is any condition where there is increased turnover of cells (for example when treating malignancy with chemotherapy), it results in release of phosphate during cell destruction, causing hyperphosphataemia.

Artefactual causes of hyperphosphataemia include delayed separation of serum from blood before analysis or haemolysis of blood collected for phosphate determination. In both cases, measurement of phosphate will reveal high concentrations even though the patient has no phosphate disorder.

TABLE 10.3 Clinical features of hyperphosphataemia.

Site affected	Symptom/complication
Neuromuscular	Tetany
Soft tissues	Metastatic calcification
Muscle	Calcification
Kidneys	Renal osteodystrophy

Clinical effects of hyperphosphataemia

Hyperphosphataemia can affect calcium metabolism causing hypocalcaemia as the calcium is precipitated out as calcium phosphate. As a consequence these patients can suffer from tetany, particularly if the plasma phosphate rises rapidly. Hyperphosphataemia can also cause metastatic **calcification**, that is, deposition of calcium phosphate in soft tissues which often occurs in severe hyperphosphataemia, as in patients with renal failure.

Excess free phosphate can be taken up into vascular smooth muscle cells and activate a gene *cbfa-1* which promotes calcium deposition in the vascular cells causing calcification. Hyperphosphataemia-induced resistance to PTH is believed to contribute towards **renal osteodystrophy**. The various consequences of hyperphosphataemia are listed in Table 10.3.

Investigation of hyperphosphataemia

Biochemical tests useful for investigating hyperphosphataemia include measurement of serum phosphate, calcium, urea/creatinine, and urinary phosphate. A typical strategy that might be useful when investigating causes of hyperphosphataemia is outlined in Figure 10.11.

When hyperphosphataemia is detected, laboratory artefacts are considered and excluded first. Increased phosphate intake is then considered and if excluded, then renal function is assessed to exclude renal failure. Conditions which cause a shift in phosphate from ICF to ECF can be considered and, if excluded, then tests are done to assess whether the hyperphosphataemia is due to endocrine causes such as hypoparathyroidism. If these are excluded, then the hyperphosphataemia is probably due to a rare cause which will need investigation.

Management of hyperphosphataemia

Hyperphosphataemia is managed by identifying and treating the underlying cause wherever possible. However, oral intake of aluminium, calcium, and magnesium salts may be of help as they can bind to phosphate in the gut, reducing its absorption. Caution is required when using calcium salts as they can cause hypercalcaemia and promote vascular calcium phosphate deposition. Resin binders like sevelamer (Renagel) promote phosphorus excretion without affecting calcium concentrations. Sevelamer decreases vascular calcium deposition in patients with renal failure. Phosphate binders that contain aluminium should not be used in patients with renal failure as they can cause aluminium toxicity. Haemodialysis may be required in

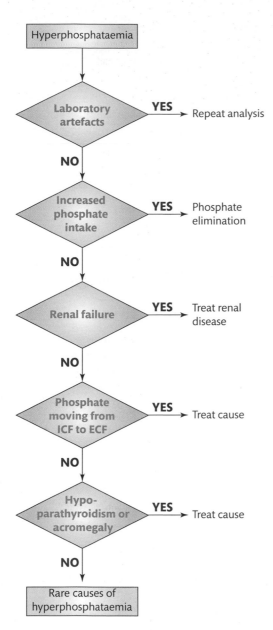

FIGURE 10.11

A possible strategy for clinical investigation of hyperphosphataemia. See text for details.

severe cases of hyperphosphataemia and particularly for patients suffering from renal failure. Infusion of insulin will promote cellular uptake of phosphate and should be used in patients with diabetic ketoacidosis. In cases of phosphate toxicity, **gastric lavage** and oral phosphate binders are utilized to prevent further absorption.

SELF-CHECK 10.7

Why does measurement of phosphate in a haemolysed blood sample often reveal hyperphosphataemia?

CASE STUDY 10.4

A 62-year-old woman was admitted to hospital to have a cataract removed. Her medical history included thyroidectomy some ten years previously as she had suffered from hyperthyroidism. Routine biochemical analysis of her serum gave the following results (reference ranges are given in brackets):

Total calcium	1.5 mmol/L	(2.20–2.60)
Albumin	42 g/L	(35–45)
Phosphate	2.61 mmol/L	(0.8–1.4)

(a) Explain these results.

(b) What might have caused the cataract?

Hypophosphataemia

Hypophosphataemia is less common than hyperphosphataemia but when it does occur causes more damage. It occurs because of a decreased intake of phosphate, increased renal loss, and increased uptake by cells. Some of the causes of hyperphosphataemia include:

- Decreased intake:
  - starvation
  - parenteral
  - vitamin D deficiency
  - phosphate binding agents
- Increased renal loss:
  - primary hyperparathyroidism
  - secondary hyperparathyroidism
  - diuretic therapy
  - renal tubular defects
- Cellular uptake:
  - diabetic ketoacidosis
  - alkalosis
- Multiple causes:
  - chronic alcoholism

Inadequate intake of phosphate is a rare cause of hypophosphataemia, but when it occurs it could be either dietary or is more likely to be seen in patients on parenteral nutrition. Vitamin D deficiency results in decreased synthesis of calcitriol and therefore decreased absorption of phosphate from the gut, giving rise to hypophosphataemia. Phosphate binding agents such as certain antacids that contain aluminium hydroxide can bind phosphate in the gut, preventing its absorption into the blood, causing hypophosphataemia. Increased renal loss of phosphate may cause hypophosphataemia. For example, in primary hyperparathyroidism, there is increased secretion of PTH, which causes excessive loss of phosphate via the kidneys, giving rise to hypophosphataemia. In addition, certain diuretics can cause increased loss of phosphate via the kidneys, resulting in hypophosphataemia. A number of congenital renal tubular defects cause hypophosphataemia because the mechanism for phosphate reabsorption is defective and so much of the phosphate is lost in the urine, for example as in the Fanconi syndrome and vitamin D resistant **rickets**.

Increased cellular uptake of phosphate may cause hypophosphataemia. For example in the recovery phase of diabetic ketoacidosis, patients are administered insulin which in turn promotes cellular uptake of phosphate giving rise to hypophosphataemia. The hypophosphataemia develops because of depletion of total body phosphate as a consequence of **osmotic diuresis**.

Cross reference
Chapter 13 Diabetes mellitus and hypoglycaemia

Respiratory alkalosis can cause hypophosphataemia by stimulating the enzyme phosphofructokinase and therefore use of phosphate to form glycolytic intermediates in glycolysis. Chronic alcohol intake is a rare cause of hypophosphataemia and the pathogenesis is complex and multifactorial. Reduced absorption, poor diet, vomiting, and diarrhoea probably play a role.

Clinical effects of hypophosphataemia

The following clinical effects may be seen in patients with hypophosphataemia. They may suffer paresthesia, **ataxia**, and coma.

Muscle weakness can occur, causing respiratory dysfunction in patients with severe hypophosphataemia. Often patients with hypophosphataemia have an increased susceptibility to infections (possibly due to defective phagocytosis). Hypophosphataemia may cause **osteomalacia** or softening of the bones. The major clinical effects of hypophosphataemia are summarized in Table 10.4.

Investigation of hypophosphataemia

Biochemical tests useful for investigating hypophosphataemia include the measurement of serum phosphate and calcium concentrations, and urinary phosphate measurements. A strategy for clinical investigation of hypophosphataemia is outlined in Figure 10.12.

When hypophosphataemia is detected in a blood sample from a patient, it is crucial to first exclude medication that may cause hypophosphataemia. Possible causes for movement of phosphate from ECF to ICF should be investigated and if excluded then measurement of urinary phosphate may reveal causes that are either due to inadequate phosphate intake or excessive renal losses.

TABLE 10.4 Clinical features of hypophosphataemia.

Site affected	Symptom
Neuromuscular	Paraesthesiae Ataxia
Muscular	Weakness Respiratory dysfunction
Immune system	Increased susceptibility to infections
Bone	Osteomalacia
Psychological	Coma

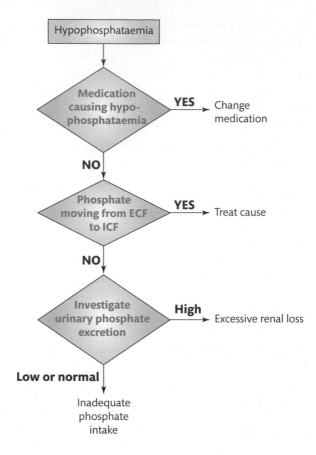

FIGURE 10.12

A possible strategy for clinical investigation of hypophosphataemia. See text for details.

Management of hypophosphataemia

A major aim of management is to treat the underlying cause wherever possible as this will resolve the hypophosphataemia. Oral or parenteral administration of phosphate may be required to correct a phosphate deficit in the body.

SELF-CHECK 10.8

Why does treatment of diabetic patients with ketoacidosis cause hypophosphataemia?

10.5 Distribution, function, and regulation of magnesium

Magnesium is the fourth most common cation in the body. The adult human body contains more than 1,000 mmoles of magnesium. About 750 mmoles of magnesium is found in the bone and around 450 mmoles is in the muscle and soft tissues. The ECF including plasma contains only 15 mmoles of magnesium, that is, about 1% of the total body magnesium. The normal dietary intake of magnesium is about 15 mmoles per day and ~30% of this is absorbed in the GIT, whereas the remaining 70% is lost in the faeces. The kidneys control magnesium homeostasis by losing 5-10 mmols of magnesium per day. You can see the distribution of body magnesium in Figure 10.13.

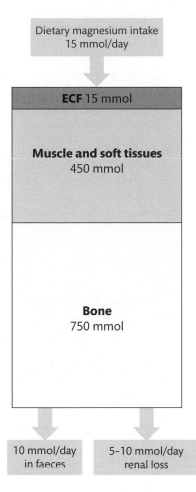

FIGURE 10.13
Distribution of magnesium in various body compartments, its intake, and excretion from the body.

Magnesium acts as a cofactor for about 300 enzymes, including many involved in energy metabolism, and protein and nucleic acid synthesis. It is also required to maintain the structures of ribosomes, nucleic acids, and certain proteins. Magnesium interacts with calcium and is required for normal cell permeability and neuromuscular function. Finally, magnesium is required for synthesis and secretion of PTH.

Magnesium exists in the plasma in three different forms:

- free ionized magnesium (Mg^{2+}): ~55%
- protein-bound magnesium (mainly albumin) ~32%
- complexed (with phosphate or citrate): ~13%

The analytical measurement of magnesium is outlined in Box 10.4.

Regulation of plasma magnesium

Regulation of the plasma magnesium concentration is achieved mainly by reabsorption of magnesium in the proximal tubules and loop of Henle in the kidneys. Increased dietary intake of magnesium results in greater loss of magnesium via the kidneys and vice versa, thus a balance is maintained. A number of factors can influence the rate of excretion of magnesium.

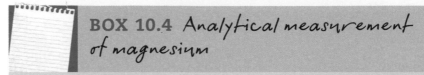

BOX 10.4 *Analytical measurement of magnesium*

Serum or heparinized plasma is used to measure magnesium. Anticoagulents such as EDTA, citrate, and oxalate form complexes with magnesium and should not be used.

Methods to measure magnesium either determine total or free magnesium. Most methods for total magnesium rely on its reaction with dyes such as formazan or methylthymol blue that change colour on reaction and can be measured by their absorbance. The reference method for measurement of total magnesium is atomic absorption spectroscopy where the samples are pre-treated with lanthanum-HCl to exclude interference from proteins and anions. Ion-selective electrodes are now commercially available to measure free magnesium.

For example, hypercalcaemia and hypophosphataemia both decrease renal reabsorption of magnesium, promoting its loss in the urine. However PTH has the effect of increasing renal reabsorption of magnesium.

The concentration of magnesium in the plasma reflects its dietary intake and the ability of the kidneys and GIT to retain it. Given that most magnesium is present inside cells, the relationship between total body deficit and plasma concentration is rather weak. However, in cases of severe magnesium deficiency, a reduction in its plasma concentration does occur.

10.6 **Disorders of magnesium homeostasis**

The typical reference range for magnesium in the plasma is 0.8–1.2 mmol/L. Disorders of magnesium homeostasis can increase plasma magnesium concentrations above the reference range, referred to as **hypermagnesaemia**, or below the reference range, as in **hypomagnesaemia**. It is worth emphasizing that plasma measurements of magnesium are not reliable indicators of the body status as only 1% of body magnesium is in the ECF.

Key Points

Disorders of magnesium homeostasis can cause an increase in its plasma concentrations either above or below its reference range, referred to as hyper- and hypomagnesaemia respectively.

Hypermagnesaemia

Hypermagnesaemia arises due to increased intake of magnesium, decreased excretion from the body, or due to release of magnesium from cells. Some of the causes of hypermagnesaemia are:

- Increased intake:
  - oral
  - parenteral
  - antacids
  - laxatives
- Decreased excretion:
  - renal failure
  - mineralocorticoid deficiency
  - hypothyroidism
  - hypocalciuric hypercalcaemia
- Cellular release:
  - cell necrosis
  - diabetic ketoacidosis
  - tissue hypoxia

Hypermagnesaemia that is due to increased intake of magnesium is rare and when it does occur is usually in combination with renal failure. Increased intake could be oral or parenteral, or due to increased intake of magnesium-containing antacids or laxatives. Acute or chronic renal failure is the most common cause of hypermagnesaemia seen in clinical practice. Mild hyper-magnesaemia may occur in patients who have mineralocorticoid deficiency, for example those suffering from **Addison's disease,** where there is a decline in the ECF volume. Hypothyroidism is a rare cause of hypercalcaemia probably because of reduced excretion of magnesium via the kidneys. Similarly hypocalciuric hypercalcaemia may cause hypermagnesaemia because it is associated with reduced excretion of calcium and probably magnesium via the kidneys.

Cell **necrosis,** for example during crush injuries, may cause release of intracellular magnesium ions (from muscle/soft tissues) into the plasma resulting in hypermagnesaemia. In diabetic ketoacidosis and **hypoxia** there is leakage of magnesium ions out of cells and into the blood, causing hypermagnesaemia.

Key Points

Renal failure causes reduced excretion of magnesium allowing its blood concentration to rise and is the most common cause of hypermagnesaemia.

Clinical effects of hypermagnesaemia

The clinical affects of hypermagnesaemia are largely due to its effects on neuromuscular func-tion and these are listed in Table 10.5.

The clinical effects of hypermagnesaemia include muscle weakness, respiratory paralysis, and, in very severe cases, it can cause cardiac arrest.

Investigation of hypermagnesaemia

The common causes of hypermagnesaemia are usually evident from clinical examination of patients. Severe hypermagnesaemia tends to occur in patients with chronic renal failure

TABLE 10.5 Clinical features of hypermagnesaemia.

Site affected	Symptom
Neuromuscular	Respiratory paralysis Weakness
Cardiovascular	Hypotension Arrhythmia Cardiac arrest
Abdominal	Nausea Vomiting

especially if they take laxatives or antacid preparations containing magnesium. A strategy for investigation of clinical hypermagnesaemia is outlined in Figure 10.14.

If hypermagnesaemia is detected, then it might be appropriate to consider whether the patient is taking medication that may cause hypermagnesaemia. Renal function should be

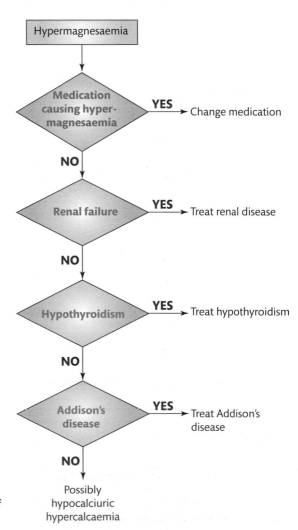

FIGURE 10.14

A possible strategy for the clinical investigation of hypermagnesaemia. See text for details.

assessed to exclude renal failure as a cause of hypermagnesaemia. The patient can then be assessed for hypothyroidism and mineralocorticoid deficiency, as in Addison's disease, as they can both cause hypermagnesaemia. If all of these are ruled out, then it is worth considering whether the patient suffers from hypocalciuric hypercalcaemia, which can be a rare cause of hypermagnesaemia.

Management of hypermagnesaemia

Hypermagnesaemia has a number of different causes, as mentioned above, and it might be appropriate to detect and treat the underlying cause wherever possible as this will usually correct the concentration of blood magnesium.

Another approach might be to increase renal excretion of magnesium in patients with normal renal function by intravenous administration of diuretics.

However, severe hypermagnesaemia due to renal failure may require dialysis.

Hypomagnesaemia

Hypomagnesaemia is not the same as magnesium deficiency. Indeed, hypomagnesaemia can occur in patients who do not have a magnesium deficiency and vice versa. Hypomagnesaemia is more common than hypermagnesaemia and arises due to either decreased intake of magnesium or increased losses of magnesium from the body via the renal and/or gastrointestinal routes. The causes of hypomagnesaemia include:

- Decreased intake:
  - starvation (especially protein-energy malnutrition)
  - malabsorption
  - parenteral
  - prolonged gastric suction
- Increased renal loss:
  - osmotic diuresis
  - diuretic therapy
  - alcoholism
  - hypercalcaemia
  - hypoparathyroidism
  - hyperaldosteronism
  - cis-platinum therapy
- Increased losses via GIT:
  - prolonged diarrhoea
  - laxative abuse
  - gut fistula

Decreased intake of magnesium may occur in patients suffering from starvation or gastrointestinal diseases that cause malabsorption of nutrients. Decreased intake of magnesium may also occur in patients receiving parenteral nutrition or those who have had prolonged gastric suction.

A significant proportion of patients with diabetes mellitus suffer from hypomagnesaemia due to loss of magnesium in osmotic diuresis.

Cross reference
Chapter 13 Diabetes mellitus and hypoglycaemia

TABLE 10.6 Clinical features of hypomagnesaemia.

Site affected	Symptom
Neuromuscular	Tetany Convulsions Cardiac arrhythmias
Muscular	Weakness Muscle cramps
Abdominal	Nausea Vomiting
Psychological	Ataxia Depression Psychosis Disorientation

Patients using loop and thiazide diuretics frequently present with hypomagnesaemia as do those using certain antibiotics. These antibiotics act on the loops of Henle in the kidneys to reduce absorption of magnesium, promoting its loss in the urine, causing hypomagnesaemia.

In alcoholism, hypomagnesaemia is believed to occur due to a number of reasons, including increased renal excretion, inadequate dietary intake, vomiting, and diarrhoea. Hypercalcaemia increases renal excretion of magnesium and can cause hypomagnesaemia. Hypoparathyroidism causes renal wasting and this may account for increased loss of body magnesium in the urine, whereas hyperaldosteronism increases renal flow promoting loss of magnesium in the urine. In patients on *cis*-platinum therapy (cytotoxic drug therapy) there is damage to the kidneys preventing renal reabsorption of magnesium, giving rise to hypomagnesaemia.

Increased losses of magnesium via the GIT occur in prolonged diarrhoea, laxative abuse, and gut fistulae, and this can cause hypomagnesaemia.

Clinical effects of hypomagnesaemia

The clinical effects of hypomagnesaemia are very similar to those seen in hypocalcaemia and arise largely due to the role of magnesium in neuromuscular function. The symptoms of hypomagnesaemia are listed in Table 10.6.

Key Points

The clinical features of hypomagnesaemia are very similar to hypocalcaemia.

Investigation of hypomagnesaemia

In the majority of cases, hypomagnesaemia is usually obvious following clinical examination of the patient. However, measurement of the following biochemical analytes may be useful when investigating hypomagnesaemia. Urinary magnesium measurements reflect the rate of magnesium excreted per day by the kidneys and decline when there is decreased intake of magnesium. If hypomagnesaemia occurs with high urinary magnesium measurements, then the losses of urinary magnesium are likely to be due to renal damage, for example when

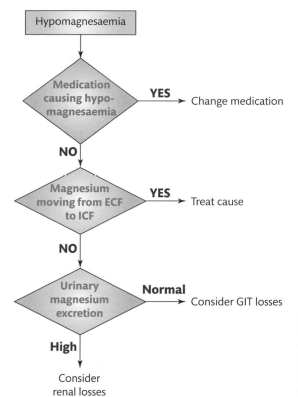

FIGURE 10.15
A possible strategy for the clinical investigation of hypomagnesaemia. See text for details.

using certain diuretics or in patients on *cis*-platinum therapy. Serum calcium measurements may reveal hypercalcaemia which is known to increase renal magnesium excretion causing hypomagnesaemia. Alternatively, hypocalcaemia may occur in patients with hypomagnesaemia as a consequence of hypoparathyroidism.

Serum potassium measurements may reveal hypokalaemia. This finding in patients with hypomagnesaemia may give an indication as to the cause of the hypomagnesaemia, for example hyperaldosteronism or diuretic therapy.

A possible strategy for investigation of hypomagnesaemia in a patient is outlined in Figure 10.15.

On finding hypomagnesaemia, it is necessary to review the medication the patient is receiving in order to exclude drugs capable of causing hypomagnesaemia. Any conditions that cause shifts of magnesium from ECF to ICF should then be considered and excluded. Urinary magnesium measurements over 1 mmol/day in patients with hypomagnesaemia reveal excessive urinary loss, whereas values below 1 mmol/day are indicative of magnesium losses via the GIT.

Management of hypomagnesaemia

Management of a patient with hypomagnesaemia should begin with identifying and treating the underlying cause wherever possible. Use of oral magnesium supplements may be adequate for mild cases of hypomagnesaemia. However, severe magnesium deficiency in a patient suffering from malabsorption may require intravenous infusions of magnesium.

SELF-CHECK 10.9

Why do patients with hypomagnesaemia often suffer from hypocalcaemia?

CASE STUDY 10.5

A 63-year-old woman presented at hospital with vomiting, diarrhoea, and muscular weakness. She was a non-smoker and did not suffer from diabetes. Biochemical tests were performed on her serum specimen and gave the following results (reference ranges are in brackets):

Sodium	138 mmol/L	(135–145)
Potassium	3.0 mmol/L	(3.5–5.0)
Urea	3.1 mmol/L	(3.5–6.6)
Total calcium	1.3 mmol/L	(2.20–2.60)
Albumin	41 g/L	(35–45)
Magnesium	0.3 mmol/L	(0.8–1.2)

(a) Explain these results.

(b) How should this patient be treated?

10.7 Bone metabolism

Bones are rigid parts of the skeleton. Their functions include movement, support, and protection of various body organs. Other functions of bone include production of red and white blood cells and storage of minerals, particularly calcium and phosphate. The adult human body contains 206 bones whereas 270 are found in infants.

Bone consists of **osteoid**, which is an organic matrix containing deposits of hydroxyapatite. Osteoid contains mainly type I collagen with some non-fibrous proteins. Hydroxyapatite is a mineral composed of calcium, phosphate, and water and gives the bone its hardness and weight. Bone is metabolically active and undergoes continuous remodelling that, in turn, comprises two processes:

- bone formation
- bone resorption

Bone formation is mediated by cells called **osteoblasts** which synthesize the bone matrix. Bone resorption is mediated by large multinucleated cells called osteoclasts, which can secrete acid and cause proteolytic digestion. Osteoblasts and osteoclasts work together when remodelling bone. Indeed, resorption of old bone first occurs by osteoclasts, and then osteoblasts appear at the site, filling the cavity with osteoid which becomes calcified. The process of bone remodelling is shown in Figure 10.16.

The two processes of bone formation and resorption are tightly linked, so that when a certain amount of bone is resorbed an equivalent amount is formed to replace it. The proliferation of osteoblasts is stimulated by some growth factors, for example transforming growth factor-β (TGF-β). In contrast, PTH and some cytokines arrest osteoblast proliferation and indirectly stimulate bone resorption. The activity of osteoclasts is inhibited by calcitonin. The process of bone formation and resorption is carefully balanced and any imbalance will result in disease.

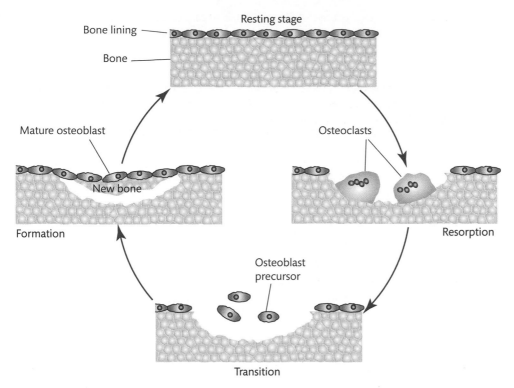

FIGURE 10.16
Remodelling of bone illustrating the role of osteoclasts in bone resorption and osteoblasts in bone formation.

There are two types of bone: **cortical** which is sometimes referred to as compact bone and **trabecular** that is often referred to as spongy bone. Most of the bones in the body are made from both these types, although there is a variation within the bones at different sites. The hard outer surface of bone is cortical and accounts for 80% of the bone mass in a healthy adult. The bone interior consists of the trabecular type, accounting for the remaining 20% of bone mass.

Key Points

Osteoblasts are cells involved in bone formation, whereas osteoclasts are involved in bone resorption or breakdown of bone to release minerals such as calcium.

10.8 **Markers of bone disease**

Normal bone health is dependent on the balance between bone formation and resorption; decoupling of this balance leads to metabolic bone disease. Assessment of bone health is best achieved by having a biochemical marker of bone formation and another of resorption. Markers that can indicate bone formation are serum alkaline phosphatase (ALP) activity and osteocalcin, whereas those for bone destruction are urinary hydroxyproline and urinary pyridinoline. Increased osteoclastic activity can be assessed by measuring urinary N- and

C-telopeptides of type I collagen (NTx and CTx), whereas collagen synthesis by osteoblasts is assessed by measuring serum procollagen I C-terminal propeptide (PICP) and procollagen I N-terminal propeptide (PINP).

Bone contributes 15 to 30% of the total plasma ALP activity; the remainder comes from the liver, GIT, and some other sources. In the bone, ALP activity is found in the membranes of osteoblasts and is released during osteoblast turnover and is therefore a good marker of bone formation. Children who have active growth have high concentrations of ALP in their plasma. There are several ELISAs for bone ALP available with varying degrees of cross-reactivity to the liver isoenzyme.

Osteocalcin is a non-collagenous protein synthesized by osteoblasts and released into the plasma during bone formation hence its use as a bone marker. Osteocalcin can be measured using ELISA. However, osteocalcin is much less stable than any of the other markers.

The imino acid hydroxyproline is a component of collagen which is released and excreted in the urine during bone resorption. Urinary hydroxyproline measurements can be significantly influenced by the ingestion of dietary gelatin. Pyridinoline and deoxypyridinoline are collagen crosslinks not influenced by diet and found in the bone and cartilage. They are released during bone resorption and their urinary concentrations can be used as markers for bone resorption. Urinary pyridinium crosslinks are usually estimated by chromatographic or radioimmunoassay techniques.

The discovery of cross-linked N-telopeptides of type I collagen (NTx) in urine has provided a specific marker of bone resorption. The NTx molecule is specific to bone because of the unique amino acid sequence and orientation of the cross-linking alpha-2 N-telopeptide. The production of NTx molecules is mediated by osteoclasts in the bone and the urinary concentrations are directly related to the extent of bone resorption. Studies have shown that increased bone resorption is the major cause of bone loss in the elderly (osteopenia) which in turn is a major cause of osteoporosis.

The C-telopeptide of type 1 collagen (CTx) is a peptide fragment from the carboxyl terminal end of the protein. It is useful for monitoring anti-resorptive therapies, such as bisphosphonates and hormone replacement therapy, in postmenopausal women and people with low bone mass or osteopenia. The concentration of CTx reflects resorption of mature collagen by osteoclasts. Any process increasing osteoclast activity results in an increase in CTx. Both NTx and CTx can be measured using immunoassays.

The soluble C-terminal and N-terminal propeptides of procollagen appear in the blood following cleavage from newly synthesized collagen. They include procollagen I C-terminal propeptide (PICP) and procollagen I N-terminal propeptide (PINP). They are not filtered by the kidneys so their plasma concentrations are not affected by renal function. Their concentrations reflect the rate of collagen synthesis by osteoblasts and will be raised in any disease that increases the activity of osteoblasts or activation. Measurements of PINP are useful in monitoring response to anti-resorption therapy. Antibodies have been raised to both propeptides and they can be measured using radioimmunoassay or ELISA techniques.

10.9 Metabolic bone disease

The term metabolic bone disease refers to abnormal bone function arising from a number of disorders that are often associated with abnormal calcium, vitamin D, phosphate, and magnesium homeostasis. They are often detected and monitored using bone markers and most are clinically reversible once the underlying defect has been rectified. The major metabolic bone diseases are **osteoporosis**, osteomalacia and rickets, and Paget's disease.

Osteoporosis

Osteoporosis is a bone disease that increases the risk of fractures and is characterized by reduced bone mineral density. Osteoporosis is the most prevalent metabolic bone disease and is particularly common in women after the menopause. In the UK, one in four women and one in twenty men after the age of 60 years will have osteoporosis. About one-third of hospital orthopaedic beds in the UK are occupied by patients with osteoporosis.

Osteoporosis is characterized by equal losses of osteoid and minerals, resulting in a decrease in bone mass. The rate of bone formation is usually normal but the rate of resorption increases and there is greater loss of trabecular than cortical bone. With normal ageing there is a progressive decline in osteoblastic activity compared to osteoclastic activity. Therefore a progressive decline in bone mass occurs. In women, oestrogen deficiency after the menopause leads to an increased rate of bone loss but the underlying aetiology of this primary osteoporosis is still unclear. However, osteoporosis is often secondary to endocrine diseases, malignancy, drug treatments, and miscellaneous causes of osteoporosis. Such cases are often referred to as secondary osteoporosis. Some causes of osteoporosis include:

- Ageing:
  - especially post-menopausal (primary osteoporosis)
- Endocrine:
  - thyrotoxicosis
  - Cushing's syndrome
  - diabetes mellitus
  - oestrogen deficiency (women)
  - androgen deficiency (men)
  - hyperparathyroidism
- Malignancy:
  - multiple myeloma
- Drugs
  - prolonged heparin treatment
  - alcoholism
- Miscellaneous
  - immobilization
  - malabsorption of calcium
  - weightlessness

Clinical effects of osteoporosis

Osteoporosis results in bones that are brittle and easily fractured. In its early stages, it is asymptomatic but as the disease progresses bone pain (for example severe backache), spontaneous fractures and collapse of vertebrae, and fractures of ribs and hips occur readily with minor trauma.

Key Points

Osteoporosis increases the risk of bone fractures in affected patients because it causes defective mineralization of bone.

Investigation of osteoporosis

The diagnosis of osteoporosis depends largely on clinical examination and radiological investigations. The diagnosis of osteoporosis is determined by the bone mineral density, which in turn is measured by dual-energy X-ray absorptiometry. The diagnosis of osteoporosis is established when the bone mineral density is less than or equal to 2.5 standard deviations below the reference range of a young adult population.

Biochemical tests are not of much use in the investigation of primary osteoporosis as concentrations of plasma calcium, phosphate, and ALP are within reference range, although the plasma ALP activity may rise after a fracture. Urinary hydroxyproline excretion may be increased in patients with osteoporosis if there is rapid bone loss, but is often within reference range. Urinary excretion of pyridinium crosslinks are increased in patients with osteoporosis but also in other conditions associated with increased bone resorption. Biochemical tests may, however, be of use in detecting the underlying causes of secondary osteoporosis. Serum concentrations of CTx may be twice the upper limit of normal in patients with osteoporosis.

Management of osteoporosis

Hormonal replacement therapy may be of value in preventing primary osteoporosis after the menopause. Patients at risk of developing osteoporosis should be counselled to avoid known risk factors, for example excessive alcohol intake. Adequate dietary intake of calcium, vitamin D, and regular exercise are necessary in individuals at risk of developing osteoporosis.

Treatment with bisphosphonates (drugs which suppress bone resorption) and oestrogens may be beneficial in individuals suffering from osteoporosis.

Osteomalacia and rickets

Osteomalacia refers to the softening of bones due to a defect in mineralization of the bone matrix or cartilage, or both. In children, the condition is known as rickets. Both osteomalacia and rickets are rare conditions that affect certain communities such as south Asians and the elderly in the UK. The causes of osteomalacia include:

- dietary deficiencies of calcium, vitamin D or phosphate
- insufficient exposure to sunlight
- malabsorption of vitamin D
- chronic renal failure
- renal tubular defects
- tumour-induced osteomalacia

A common cause is calcium or phosphate deficiency, or both (both are required for bone mineralization). Calcium deficiency is usually due to deficiency of vitamin D, therefore the high incidence of osteomalacia in Asian and elderly individuals. In these individuals, the vitamin D deficiency may be dietary or due to reduced exposure to sunlight. The addition of adequate vitamin D in most foods has virtually eliminated rickets in children in the developed world. Vitamin D deficiency may also occur in patients with malabsorption, for example those suffering with coeliac disease.

Cross reference
Chapter 17 Gastrointestinal disorders and malabsorption

Renal disease can result in reduced formation of calcitriol required for absorption of dietary calcium. Phosphate deficiency occurs in renal tubular defects (reduced reabsorption of phosphate) such as the Fanconi syndrome giving rise to vitamin D resistant rickets or osteomalacia. Tumour-induced osteomalacia is uncommon, but causes increased renal phosphate excretion resulting in hypophosphataemia and osteomalacia.

Clinical effects of osteomalacia and rickets

Patients suffering from osteomalacia can present with bone pain, muscular weakness, and increased susceptibility to fractures even after minor trauma. Children with rickets may suffer from bone pain and have bowed legs and cranial defects.

Investigation of osteomalacia and rickets

The biochemical findings typically include hypocalcaemia, hypophosphataemia, and elevated PTH concentrations. Concentrations of serum vitamin D and its metabolites are often low in osteomalacia or rickets due to vitamin D deficiency. The activities of serum ALP are usually raised. Serum concentrations of CTx may be twice the upper limit of normal in patients with osteomalacia.

Key Points

Diagnosis and treatment of rickets in children is crucial in order to prevent bowing of the long bones in the legs.

Management of osteomalacia and rickets

When managing patients with osteomalacia or rickets, a common approach is to identify the causal factors and then aim to treat the underlying cause wherever possible. Oral or parenteral administration of vitamin D may be required in patients with vitamin D deficiency. Phosphate supplements may be required to correct phosphate deficiency.

SELF-CHECK 10.10

What is the common cause of vitamin D deficiency in certain communities in the UK?

Paget's disease

Paget's disease of the bone, also known as osteitis deformans, is named after Sir James Paget, the British surgeon who described this condition in 1877. It is characterized by excessive osteoclastic bone resorption followed by formation of new bone of abnormal structure that is laid down in a disorganized manner resulting in deformed bones. Paget's disease is rarely diagnosed in those under 40 years of age. However, males are more commonly affected than females. Many individuals with Paget's disease are asymptomatic, therefore it is difficult to estimate the prevalence of this condition. It is believed to affect about 5% of the UK population over the age of 50 years.

The aetiology of Paget's disease is still unclear. However, certain genes, including the sequestrosome 1 gene on chromosome 5, are associated with Paget's disease. It has been suggested that viral infections trigger Paget's disease in people with an inherent tendency to develop the condition.

Clinical effects of Paget's disease

Paget's disease may be asymptomatic in some patients but clinical features can occur in other sufferers. These include bone pain and bone deformities, such as bowed tibia, **kyphosis** (hunching of the back), increasing skull size, and bone fractures.

> *Key Points*
>
> **Paget's disease affects the elderly and can cause serious bone deformation.**

Investigation of Paget's disease

The diagnosis of Paget's disease is usually made on the basis of clinical examination supported by radiological investigations. Paget's disease is diagnosed using X-rays and imaging of the affected bone.

Calcium and phosphate concentrations in serum samples tend to be within their reference ranges but hypercalcaemia may occur in immobilized patients. The abnormal bone turnover causes increased activities of serum ALP. There is increased urinary excretion of hydroxyproline. Serial measurements of serum ALP and urinary hydroxyproline may be used when monitoring treatment of Paget's disease. Concentrations of serum CTx are 2–4 times the upper limit of normal in patients with Paget's disease.

Management of Paget's disease

The aim of managing Paget's disease is to give relief from bone pain and to prevent disease progression. Analgesics may be used to give relief from the pain. In severe cases, bisphosphonates may be used to reduce osteoclastic activity. Patients with bone deformity may require supports such as heel lifts or specialized footwear. Corrective surgery may be required where joints are damaged, for fractures, or severely deformed bones, or where nerves are being compressed by enlarged bones.

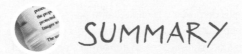

SUMMARY

- Calcium is largely found in bone and its plasma concentrations are regulated by PTH, calcitriol, and calcitonin.

- Disorders of calcium homeostasis present as either abnormally high plasma calcium or hypercalcaemia, or abnormally low plasma calcium or hypocalcaemia.

- Phosphate occurs largely in the bones and its plasma concentrations are controlled by PTH and calcitriol.

- Disorders of phosphate homeostasis cause hyperphosphataemia or hypophosphataemia.

- Magnesium occurs largely in the bones and to a lesser extent in the muscles and soft tissues.

- Disorders of magnesium homeostasis can cause either hypermagnesaemia or hypomagnesaemia.

- Biochemical tests can be used to diagnose disorders of calcium, phosphate, and magnesium homeostasis and to identify their causes.

- The major metabolic diseases affecting the bone are osteoporosis, osteomalacia, and Paget's disease.

- Markers of bone disease have a role in diagnosis and in the monitoring of treatment.

FURTHER READING

- Colina M, La Corte R, De Leonardis F, and Trotta F (2008) Paget's disease of bone: a review. *Rheumatology International* **28**, 1069-75.

  A comprehensive review of Paget's disease of bone.

- Cooper MS and Gittoes NJ (2008) Diagnosis and management of hypocalcaemia. *British Medical Journal* **336**, 1298-1302.

  A review of hypocalcaemia and particularly its diagnosis and management.

- Cremers S, Bilezikian JP, and Garnero P (2008) Bone markers—new aspects. *Clinical Laboratory* **54**, 461-71.

  Describes the value of current and new bone markers in the investigation of bone disease.

- Moe SM (2008) Disorders involving calcium, phosphorus and magnesium. *Primary Care* **35**, 215-37.

  A general review of disorders of calcium, phosphorus, and magnesium metabolism.

- Owens BB (2009) A review of primary hyperparathyroidism. *Journal of Infusion Nursing* **32**, 87-92.

  A user-friendly review of primary hyperparathyroidism.

- Peacock M (2010) Calcium metabolism in health and disease. *Clinical Journal of the American Society of Nephrology* **5**, 523-30.

  A comprehensive review of the metabolism of calcium in health and disease.

- Scharla S (2008) Diagnosis of disorders of vitamin D metabolism and osteomalacia. *Clinical Laboratory* **54**, 451-9.

  A comprehensive review of disorders of vitamin D metabolism and osteomalacia and their laboratory investigation.

- Shepard MM and Smith JW (2007) Hypercalcemia. *American Journal of Medical Sciences* **334**, 381-5.

  A review of the laboratory investigation of hypercalcaemia.

- Shoback D (2008) Clinical practice: hypoparathyroidism. *New England Journal of Medicine* **359**, 391-403.

  A comprehensive review of hypoparathyroidism.

- Tiosano D and Hochberg Z (2009) **Hypophosphatemia: the common denominator for all rickets.** *Journal of Bone and Mineral Metabolism* **27**, 392–401.

  Considers the role of hypophosphataemia in bone metabolism and particularly in rickets.

- Topf JM and Murray PT (2003) **Hypomagnesemia and hypermagnesemia.** *Reviews in Endocrine and Metabolic Disorders* **4**, 195–206.

  A discussion of hypo- and hypermagnesaemia.

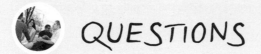

QUESTIONS

10.1 Which one of the following is the commonest cause of hypercalcaemia?

   (a) Excessive vitamin D intake

   (b) Milk-alkali syndrome

   (c) Malignancy

   (d) Familial hypocalciuric hypercalcaemia

   (e) Acromegaly

10.2 Which one of the following is NOT a cause of osteomalacia?

   (a) Malabsorption of vitamin D

   (b) Renal tubular defects

   (c) Tumour-induced osteomalacia

   (d) Inadequate exposure to sunlight

   (e) Excessive vitamin D intake

10.3 State whether each of the following statements is TRUE/FALSE.

   (a) Serum ALP activity is raised in patients with Paget's disease

   (b) Osteocalcin cannot be synthesized by osteoblasts

   (c) Growing children have higher concentrations of serum ALP than adults

   (d) Bone formation is mediated by osteoblasts

   (e) Urinary excretion of pyridinium crosslinks is always reduced in patients with osteoporosis

10.4 Outline the main steps in the biosynthesis of calcitriol.

10.5 Outline the major causes of hyperphosphataemia.

Answers to self-check questions, case study questions, and end-of-chapter questions are available in the Online Resource Centre accompanying this book.

 Go to www.oxfordtextbooks.co.uk/orc/ahmed/

11

Abnormal pituitary function

Garry McDowell

Learning objectives

After studying this chapter you should be able to:

- Describe the basic structure and function of the endocrine system
- Explain the mechanisms of hormone action and control of their secretion
- Describe the structure and function of the pituitary gland
- Explain the functions of anterior and posterior pituitary hormones
- Describe the conditions which lead to abnormal hormone production
- Describe and explain the investigation of suspected pituitary dysfunction

Introduction

The endocrine system controls body functions through the action of mediators known as hormones. A **hormone** is a molecule which is released by an endocrine gland directly into the bloodstream in one part of the body and exerts its effects in other parts. The blood supply circulates the hormone throughout the body. Hormones then exert their effect by binding to receptors on their target cells. The hormones and the glands which secrete them form the endocrine system and the study of diseases associated with the endocrine system is known as **endocrinology**. This chapter will review the general functions of hormones, their chemical nature, mechanism of action, and regulation, and will then focus on disorders of hormones released by the pituitary gland.

11.1 Endocrine system

The body has two types of glands, called exocrine and endocrine glands. Exocrine glands secrete their products into ducts that carry the secretions into body cavities, the lumen of the gut, or to the outer surface of the body. In contrast, endocrine glands release their products into the fluid that surrounds cells (interstitial fluid) rather than into ducts. From the interstitial fluid the hormones diffuse into blood capillaries and the blood carries them to the target cells throughout the body. The concentration of hormones in the blood is generally low. The endocrine glands include:

- pituitary
- thyroid and parathyroid
- adrenal
- pancreas
- ovaries
- testes

Chemical classification of hormones

Hormones can be classified into two broad chemical categories: lipid-soluble or water-soluble.

Lipid-soluble hormones

The lipid-soluble hormones include steroid and thyroid hormones. Steroid hormones are derived from cholesterol. Each steroid hormone is unique due to the presence of different chemical groups attached at different locations on the 'core' of the molecule. Steroid hormones can be divided into five groups depending on the receptors they bind to on target cells. These five groups include glucocorticoids, mineralocorticoids, oestrogens, androgens, and progestogens. Thyroid hormones are derived from attaching an iodine molecule to the amino acid tyrosine contained within the thyroglobulin molecule in the follicles of the thyroid gland.

Water-soluble hormones

The water-soluble hormones include amines, peptide hormones, and the eicosanoid hormones. The amine hormones are derivatives of amino acids. They are synthesized by decarboxylation (removing CO_2) from certain amino acids. They all contain an amino group ($-NH_3^+$) which is where their name derives from. The catecholamines, for example, are derived from the amino acid tyrosine. Peptide and protein hormones are derived from amino acids and include, for example, antidiuretic hormone (ADH), insulin, and growth hormone. In addition, a separate subgroup of peptide hormones exists and this is the glycoprotein hormones. These have the same basic structure as other protein hormones, being composed of amino acids, but they also have a carbohydrate molecule attached. Examples include the gonadotrophins (follicle stimulating hormone (FSH), luteinizing hormone (LH)), and thyroid stimulating hormone (TSH). The eicosanoid hormones are derived from arachidonic acid, for example prostaglandins and leukotrienes. The chemical structure of hormones is important when considering how they are transported in the body, which will be discussed in the next section.

Transport of hormones

The water-soluble hormones are soluble in the blood, circulate in their free form, and are not attached to any other molecule. On the other hand, lipid-soluble molecules are not soluble in the water component of blood and must be carried bound to a transport protein. For example, the steroid hormone cortisol which is produced by the adrenal glands is lipid-soluble. Cortisol is bound to a transport protein called cortisol binding globulin or transcortin. Likewise, thyroxine binding globulin (TBG) is a transport protein that binds thyroid hormones. All transport proteins are synthesized in the liver and increase the solubility of lipid-soluble hormones in the blood; they also decrease loss of small hormones by filtration from the kidney and act as a reserve of hormones present in the bloodstream.

Usually >99% of lipid-soluble hormones are bound to a transport protein in the blood, with only a small fraction (<1%) free in the circulation. Free hormone is not bound to a transport protein and is the biologically active fraction. The free or unbound hormone can diffuse out of the circulation, bind to cellular receptors, and exert its physiological effect.

Mechanism of hormone action

Hormones bring about their biological effects by binding to specific cell receptors. The receptors for lipid-soluble hormones are located inside their target cells, while the receptors for water-soluble hormones are located on the cell surface as part of the cell's plasma membrane.

Lipid-soluble hormones diffuse from the blood into the interstitial fluid and across the lipid bilayer of the cell membrane. Within the cell, the hormone binds to receptors in the cytoplasm or nucleus. The activated hormone-receptor complex then alters gene expression within the nucleus. Gene expression results in transcription forming messenger RNA (mRNA). The mRNA then leaves the nucleus and enters the cytoplasm where it undergoes translation producing a new protein. The new protein then exerts its effect by altering cellular function to bring about the biological effect of that hormone.

The mode of action of water-soluble hormones is different as they are not lipid-soluble so cannot diffuse across the lipid bilayer of the plasma membrane and bind to internal receptors. Instead they bind to receptors which are integral to the plasma membrane. The hormone binds to a membrane receptor which then brings about the production of a second messenger molecule within the cell and it is this second messenger that mediates the cellular response; in this case, the hormone itself is acting as a first messenger. Let us examine this in more detail. A water-soluble hormone diffuses from the circulation into the interstitial fluid and binds to a membrane receptor. The hormone-receptor complex in turn activates a G protein, which in turn activates the enzyme adenylate cyclase. The adenylate cyclase, on the cytosolic surface of the plasma membrane, catalyses the conversion of adenosine triphosphate (ATP) into cyclic adenosine monophosphate (cAMP). The concentration of cAMP within the cell therefore increases. Cyclic AMP acts as the second messenger and activates a protein kinase enzyme, which results in the phosphorylation (adding a phosphate group to) of other cellular proteins, for example enzymes. The effect of phosphorylation is that some of these proteins will become activated or deactivated and it is these phosphorylated proteins that produce the cellular response to the hormone. The final stage is for the cell to turn off the response to the stimulus. This is achieved by the action of an enzyme called phosphodiesterase, which inactivates cAMP. Many hormones bring about their physiological effect through an increase in intracellular cAMP, they include ADH, TSH, and adrenocorticotrophic hormone (ACTH).

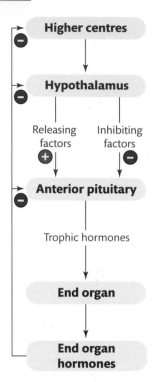

FIGURE 11.1
A schematic showing the hypothalamic-pituitary-end organ axis to illustrate positive and negative feedback.

Control of hormone secretion

The regulation of hormone production is by a feedback mechanism which can be positive or negative. A positive feedback mechanism tends to strengthen or reinforce a change in one of the body's control mechanisms. Negative feedback tends to weaken or diminish the response. Figure 11.1 illustrates this diagrammatically using the hypothalamus and pituitary as an example.

Notice how the hypothalamus can produce a releasing factor that acts on the pituitary, stimulating the release of a trophic hormone, which in turn stimulates an end organ to produce another hormone; this is an example of positive feedback. The end organ hormone then acts on the hypothalamus and pituitary to reduce the release of the hypothalamic releasing hormone and the pituitary trophic hormone—this is termed negative feedback. Notice how in Figure 11.1 negative feedback can occur at all levels including the direct inhibition by a hormone of its own secretion, but also those of the end hormone in the hypothalamic-pituitary axis. Knowledge of negative feedback is necessary when interpreting the results of biochemical data.

11.2 Structure of the hypothalamus and pituitary gland

The pituitary gland lies in the pituitary fossa beneath the hypothalamus, to which it is connected by the pituitary stalk or infundibulum. The normal adult pituitary has two lobes, the anterior and posterior lobes and measures 1-1.5 cm in diameter. Release of anterior pituitary hormones is stimulated by releasing factors or suppressed by inhibiting factors from the hypothalamus. Neuronal cell bodies located in the paraventricular nuclei synthesize specific hormones which pass along their respective axons and are released from the nerve terminals in the median eminence into the hypothalamic-hypophyseal portal system. Hypothalamic hormones are carried to the anterior pituitary via the portal system. A portal system is one in which blood flows from one capillary network into a portal vein and then into a second capillary network without passing through the heart. In the hypophyseal portal system, blood flows from the capillaries of the hypothalamus into portal veins that carry blood to the anterior pituitary, as shown in Figure 11.2.

The superior hypophyseal arteries which branch from the internal carotid arteries supply blood to the hypothalamus. At the junction of the median eminence and the infundibulum these arteries divide into a capillary network called the primary plexus of the hypophyseal portal system. From here the blood drains down into the hypophyseal portal veins that pass down outside the infundibulum. In the anterior pituitary, the portal veins divide again into the secondary plexus of the hypophyseal portal system as shown in Figure 11.2. The hypothalamic-hypophyseal portal system therefore has a crucial role in the control of the anterior pituitary. Thus, any disease process that interferes with the portal circulation, for example, non-functioning tumours of the hypothalamus or pituitary will result in pituitary dysfunction, even though the pituitary hormone secreting cells remain undamaged. Hormones secreted by the anterior pituitary pass into the anterior hypophyseal veins and from there into the general circulation.

The anterior lobe contains a number of different cell types which secrete a number of peptide and glycopeptide hormones. Table 11.1 shows the various cell types of the anterior pituitary and the hormones which they secrete.

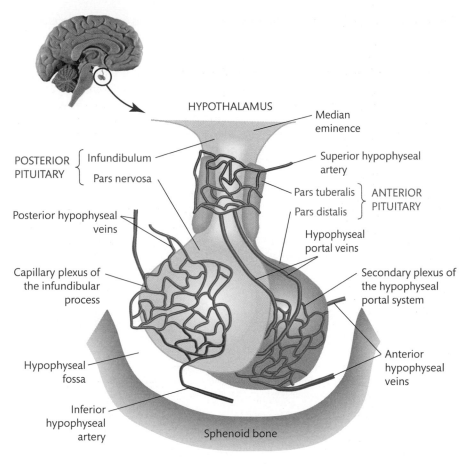

FIGURE 11.2
Hypothalamus and pituitary showing their blood supply.

TABLE 11.1 Pituitary cell types and the hormones which they secrete.

Anterior pituitary cell type	Hormone secreted
Corticotrophs	Adrenocorticotrophic hormone (ACTH)
Lactotrophs	Prolactin
Gonadotrophs	Luteinizing hormone (LH) and follicle stimulating hormone (FSH)
Thyrotrophs	Thyroid stimulating hormone (TSH)
Somatotrophs	Growth hormone (GH)

Secretion of anterior pituitary hormones

The secretion of anterior pituitary hormones is controlled by two mechanisms. The first is the releasing factors from the hypothalamus, which have a stimulatory effect on the anterior pituitary and two inhibitory factors that have a suppressive effect. Table 11.2 shows the

TABLE 11.2 Hypothalamic stimulatory and inhibitory hormones and their effects on the pituitary gland.

Hypothalamic stimulatory/inhibitory hormone	Pituitary hormone released	End organ	Effect
CRH	ACTH	Adrenal	Stimulates cortisol production
GHRH	GH	Liver	Stimulates IGF-1 production
Somatostatin	Blocks GHRH action	Pituitary	
Dopamine	Inhibitory effect on prolactin	Breast	Milk production
TRH	TSH	Thyroid	T4 production
GnRH	LH	Testis (male)	Testosterone production
	LH	Ovary (female)	Ovulation
	FSH	Testis (male)	Sperm production
	FSH	Ovary (female)	Oestradiol

CRH, corticotrophin releasing hormone; GHRH, growth hormone releasing hormone; TRH, thyrotrophin releasing hormone; GnRH, gonadotrophin releasing hormone; ACTH, adrenocorticotrophic hormone; GH, growth hormone; TSH, thyroid stimulating hormone; LH, luteinizing hormone; FSH, follicle stimulating hormone.

hypothalamic stimulatory and inhibitory hormones. The second mechanism is negative feedback, where the hormones released by the target glands, for example thyroid, decrease the secretion of the anterior pituitary hormone. This is described diagrammatically in Figure 11.1.

SELF-CHECK 11.1

What type of hormones are secreted by the following anterior pituitary cells: gonadotrophs and lactotrophs?

Secretion of posterior pituitary hormones

The posterior pituitary does not synthesize hormones but it does store and release two hormones. These are ADH and oxytocin and both are secreted directly from nerve endings, whose cell bodies are located within the supraoptic and paraventricular nuclei of the hypothalamus. Their axons form the hypothalamo-hypophyseal tract which begins in the hypothalamus and ends near the blood capillaries in the posterior pituitary. Oxytocin is synthesized in the paraventricular nucleus and ADH in the supraoptic nucleus. Following synthesis in the cell bodies, the hormones are packaged in secretory vesicles which are transported to the nerve terminals in the posterior pituitary where they are stored until release. Blood is supplied to the posterior pituitary by the inferior hypophyseal arteries which originate from the internal carotid arteries. In the posterior pituitary, the inferior hypophyseal arteries drain into the capillary plexus of the infundibular process into which oxytocin and ADH are secreted. Blood then flows into the posterior hypophyseal vein and the systemic circulation as shown in Figure 11.3.

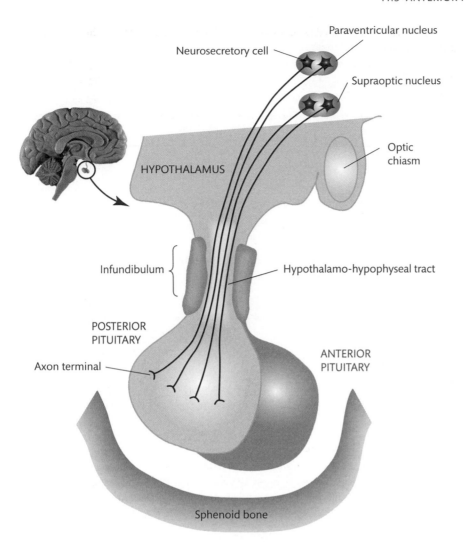

FIGURE 11.3
The structure of the posterior pituitary.

11.3 **Anterior pituitary hormones**

The hormones produced by the anterior pituitary together with the end organ on which they act are summarized in Table 11.2.

The anterior pituitary hormones can be classified into one of two groups: the glycoproteins (FSH, LH, and TSH) and single chain polypeptides (ACTH, growth hormone (GH)), and prolactin.

The pituitary glycoprotein hormones each comprise two subunits (α and β). The α subunit is common to all, while the β subunit is unique to each hormone and confers its biological activity.

Adrenocorticotrophic hormone

Release of cortisol from the adrenal glands is controlled by ACTH released from the anterior pituitary which in turn is controlled by corticotrophin releasing hormone (CRH) from the hypothalamus, as shown in Figure 11.4. Adrenocorticotrophic hormone, and therefore cortisol, is secreted in a circadian rhythm, with concentrations being low at midnight and rising

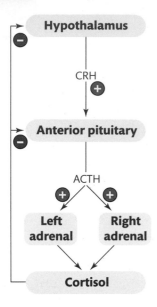

FIGURE 11.4
Schematic showing the hypothalamic-pituitary-adrenal axis. Positive control is from CRH and ACTH, whereas the negative feedback is from cortisol.

CASE STUDY 11.1

A 30-year-old woman complaining of tiredness was seen by her physician. A serum sample was collected at 2 pm. The following results were obtained (reference ranges are given in brackets):

Urea	5.2 mmol/L	(3.4–6.8)
Sodium	140 mmol/L	(136–146)
Potassium	4.2 mmol/L	(3.5–5.0)
TSH	1.3 mU/L	(0.5–5)
Free T4	15 pmol/L	(9–22)
Cortisol	220 nmol/L	(>550 at 9 am)

The physician has asked for advice regarding the interpretation of the results and any further investigations you would recommend.

What advice would you give to the physician?

in the final hours of sleep to peak shortly after awakening. The concentration then steadily declines throughout the day. Stress is also a major determinant of ACTH release and hence the concentration of cortisol, being increased by fear, illness, hypoglycaemia, and surgery. Note in Figure 11.4 how cortisol has a negative feedback effect on the pituitary and hypothalamus.

Growth hormone

Growth hormone is produced by somatotroph cells which are the most abundant cells in the anterior pituitary. Growth hormone secretion occurs in pulses, most frequently at night during the early hours of sleep. During the day, plasma levels are usually undetectable. Growth hormone action is mediated via the action of insulin like growth factor-1 (IGF-1), which is predominantly synthesized in the liver. Insulin-like growth factors have a number of actions. They can:

- stimulate cell growth and multiplication
- increase protein catabolism
- enhance lipolysis in adipose tissue
- decrease uptake of glucose by cells

Control of GH and IGF-1 release is via negative feedback to the pituitary and hypothalamus as shown in Figure 11.5.

Prolactin

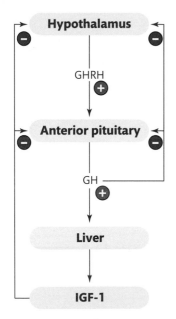

FIGURE 11.5
Schematic showing the hypothalamic-pituitary-GH axis. Positive and negative feedback mechanisms are shown.

Look again at Table 11.2. Notice how prolactin is unique in that its secretion is controlled by an inhibitory factor (dopamine). Concentration of prolactin in the plasma is increased during pregnancy and lactation. Prolactin is also increased in response to stress. The physiological role of prolactin in non-lactating women and in men remains unknown, although a role in immune regulation has been suggested. By itself prolactin has only a weak effect in stimulating lactation. Only after the breast has been primed by oestrogen, progesterone, glucocorticoids, GH, insulin, and thyroxine will prolactin bring about milk secretion.

SELF-CHECK 11.2

What is the most common cause of an increased serum prolactin concentration in a woman of childbearing age?

Thyroid stimulating hormone

Thyroid stimulating hormone secretion is under the control of thyrotrophin releasing hormone (TRH) from the hypothalamus. Thyroid stimulating hormone acts on the thyroid gland to stimulate the synthesis and secretion of thyroxine (T4) and tri-iodothyronine (T3). Both of these hormones then exert a negative feedback effect on the pituitary and hypothalamus to reduce TSH and TRH secretion respectively, as shown in Figure 11.6.

> *Key Points*
>
> **It is the free T4 and free T3 that are biologically active and exert a negative feedback effect on the anterior pituitary and hypothalamus.**

Gonadotrophins (follicle stimulating hormone and luteinizing hormone)

The gonadotrophins, FSH and LH, are both regulated by the hypothalamic gonadotrophin releasing hormone (GnRH). The major role of LH in the male is to stimulate testosterone secretion by Leydig cells of the testes, whilst in the female, LH stimulates ovulation at mid-cycle.

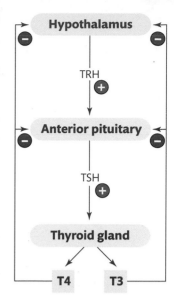

FIGURE 11.6
Schematic showing the hypothalamic-pituitary-thyroid axis. Positive control is from TRH and TSH, whereas the negative feedback is from free T4 and free T3.

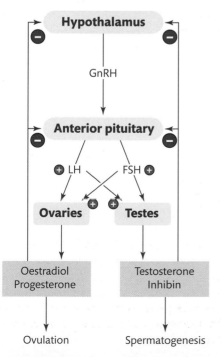

FIGURE 11.7
Schematic showing the hypothalamic-pituitary-gonadal axis. Positive control is from GnRH, FSH, and LH, whereas the negative feedback is from testosterone, inhibin, oestradiol, and progesterone.

In the male, FSH controls spermatogenesis, whereas in the female it controls ovarian follicular development and hence oestradiol secretion. In both sexes, FSH stimulates the secretion of inhibin, which is responsible for some of the negative feedback control as shown in Figure 11.7.

11.4 Posterior pituitary hormones

Antidiuretic hormone acts on the kidneys, decreasing urine output and resulting in the production of concentrated urine. Antidiuretic hormone secretion is controlled by the **osmolality** of the blood and the blood pressure.

The secretion of ADH is increased by:

- low blood pressure
- high blood osmolality
- stress

High blood osmolality stimulates **osmoreceptors** in the hypothalamus which cause an increased synthesis of ADH and its secretion from the posterior pituitary. Antidiuretic hormone acts on the kidneys and sweat glands to retain more water and decrease water loss by perspiration from the skin respectively. The action of ADH on arterioles causes them to constrict and increase blood pressure. Vasopressin is another name used for ADH because of its effect on blood pressure.

During delivery of a baby, oxytocin enhances contraction of smooth muscle cells in the wall of the uterus and after delivery it stimulates milk ejection from the mammary gland in response to mechanical stimulation from the suckling infant. The function of oxytocin in males and non-pregnant females is still not established.

11.5 Disorders of pituitary function

From a clinical perspective, disorders of pituitary function can be classified into two broad categories:

- hyperfunction states, where a pituitary hormone is produced in excess
- hypofunction states, where there is a deficiency of a pituitary hormone

We will now consider the most common disorders under each category.

11.6 Anterior pituitary hyperfunction

Tumours resulting in excess growth hormone, ACTH, and prolactin have been reported. Let us now consider those you will encounter most commonly in clinical practice.

Hyperprolactinaemia

Hyperprolactinaemia is the term given to an increased concentration of prolactin circulating in the blood and is the result of excessive prolactin production by the lactotrophs.

Prolactinomas are the most common benign adenomas of the pituitary. As we discussed earlier, prolactin is under negative control by hypothalamic dopamine and hyperprolactinaemia may be the result of stalk compression by any pituitary mass interrupting the portal blood flow and preventing hypothalamic dopamine reaching the lactotroph.

Clinical features of hyperprolactinaemia

The clinical presentation of hyperprolactinaemia varies according to the patient's age, sex, and duration of the hyperprolactinaemia. Pre-menopausal female patients usually present with amenorrhoea (an absent menstrual cycle), infertility, and galactorrhoea, or inappropriate release of breast milk. Men and post-menopausal women usually present with symptoms of a pituitary mass, for example headache and visual disturbance due to compression of the optic chiasm. Men usually have larger tumours at presentation as pre-menopausal women will notice amenorrhoea and galactorrhoea sooner. In addition, men can present with poor libido and impotence. Compression of normal pituitary tissue may result in deficiency of other pituitary hormones, for example GH, ACTH, LH, FSH, or TSH.

Investigation of hyperprolactinaemia

A single basal measurement of prolactin will often give a full picture regarding this hormone. However, care is required in the interpretation of what constitutes an increased concentration as prolactin can be increased by physiological as well as pathological causes. The concentrations of prolactin can also vary throughout the day so it is necessary to confirm high results over a few days. Prolactin concentrations are high during sleep and lowest between 9 am–12 pm. The circulating plasma concentration of prolactin can also give an indication of the likely cause:

- Prolactin >5,000 mU/L is almost always due to a prolactinoma and can be associated with large macroprolactinomas.
- Prolactin <5,000 mU/L may be due to a prolactinoma or a non-functioning pituitary tumour. It is unlikely for a macroprolactinoma to be associated with a prolactin concentration in this range.
- Prolactin <1,000 mU/L is often associated with polycystic ovarian syndrome and physiological causes.

By following a systematic approach, one can determine the cause of hyperprolactinaemia. We will now outline an approach for the investigation of an individual with hyperprolactinaemia. The first stage is to exclude physiological causes resulting in hyperprolactinaemia such as:

- pregnancy
- lactation
- stress, which can be physical or psychological
- neurogenic, for example breast manipulation

Key Points

When investigating a patient with hyperprolactinaemia rule out physiological causes, non-hypothalamic/pituitary causes, and drug effects.

Having excluded physiological causes, we need to exclude non-hypothalamic or non-pituitary causes, for example:

- Hypothyroidism
- Renal failure
- Drugs, for example:
  - dopamine receptor antagonists
  - metoclopramide
  - chlorpromazine
  - methyldopa
  - cimetidine

Then finally it is appropriate to consider hypothalamic or pituitary causes of hyperprolactinaemia:

- Hypothalamic disease, such as:
  - tumour
  - infiltrative disease
  - cranial irradiation
- Pituitary disease, such as:
  - prolactinoma
  - pituitary stalk lesion
  - Cushing's disease
  - acromegaly
  - infiltrative diseases

Breast manipulation can result in hyperprolactinaemia due to a reduction in hypothalamic dopamine. This is in effect similar to the effect of a suckling infant. Non-hypothalamic and non-pituitary causes can be excluded by a careful clinical and drug history, while hypothyroidism can be excluded by measuring serum levels of TSH and free T4, the results of both of which should be within the reference range. Hypothalamic disorders, such as a tumour, result in hyperprolactinaemia due to dopamine being unable to reach the anterior pituitary to inhibit prolactin release.

Prolactinomas are benign adenomas of the lactotrophs and are classified as either macroprolactinomas or microprolactinomas on the basis of their size. A microprolactinoma is less than 10 mm in diameter, while a macroprolactinoma is greater than 10 mm in diameter. Diagnosis is made on the basis of a clinical examination by a physician, the prolactin result and the result of imaging of the pituitary gland, either by a computed tomography scan or a magnetic resonance imaging scan.

Laboratory 'artefact'

When evaluating patients with hyperprolactinaemia, one has to consider laboratory artefacts which can significantly affect the results obtained and particularly when prolactin results do not correlate with clinical findings. These are the high dose hook effect and macroprolactin. The hook effect can be seen in samples with extremely high prolactin levels which cause antibody saturation in two-site assays leading to artificially low results. This can be eliminated by serial dilutions of samples.

CASE STUDY 11.2

A 25-year-old woman went to see her physician complaining of amenorrhoea. The physician requested FSH, LH, prolactin, and TSH to be measured on a blood sample collected at the surgery. The results reported by the laboratory are shown below (reference ranges are given in brackets):

FSH	<0.5 U/L	(2–8 in follicular phase)
LH	<0.1 U/L	(2–10 in follicular phase)
Prolactin	1026 mU/L	(<500)
TSH	1.28 mU/L	(0.3–4.1)

(a) What is the most likely explanation for these results?

(b) What further investigations would you initiate on this sample?

Macroprolactin is a complex of prolactin with usually an IgG antibody. Hyperprolactinaemia results from reduced clearance of this complex. It should be noted that not all assays detect macroprolactin to the same degree. In clinical practice, therefore, liaison between the laboratory and the clinician may be required in circumstances where there is a discrepancy between the clinical findings and laboratory results. Macroprolactin can be removed by polyethylene glycol precipitation and the native prolactin reassayed. Whether it is appropriate to check for macroprolactin in all patients with signs and symptoms of hyperprolactinaemia remains controversial, but it is reasonable to check for it in patients with hyperprolactinaemia and those with less typical symptoms.

SELF-CHECK 11.3

What are the common, non-pathological causes of hyperprolactinaemia?

Management of hyperprolactinaemia

The management of hyperprolactinaemia not due to secondary causes such as hypothyroidism is predominantly medical. Dopamine agonists can be used to suppress prolactin release and control symptoms. Monitoring of therapy is by serial prolactin measurements usually taken at each clinic visit or annually for patients on long-term therapy.

For patients with large tumours or those resistant to medical therapy, surgery remains a possibility and the reader is advised to consult specialist texts should they require further information.

Acromegaly

Acromegaly is usually caused by excess GH production from a pituitary adenoma in more than 99% of adults with this condition. Very rarely the cause of acromegaly may be tumours affecting other parts of the body, for example the lungs producing large amounts of GH or the growth hormone releasing hormone (GHRH). The latter can stimulate the anterior pituitary to produce GH. Acromegaly is rare, with an incidence of 3–4 cases per million per year.

Clinical features of acromegaly

The symptoms usually begin insidiously, that is appear slowly, and changes occur over a long period of time, meaning that diagnosis is usually delayed for up to 15–20 years.

The clinical features are related to the effects of GH and IGF-1. Patients often exhibit coarse facial features, including frontal bossing, a large protruding jaw, referred to as **prognathism**, enlarged tongue and nose, and enlarged hands and feet due to the growth of bone and soft tissue. They often complain of an increasing shoe size. Patients may also complain of headache, increased sweating, and visual and sleep problems. There is also enlargement of internal organs such as the liver, so affected individuals may have **hepatomegaly**. Excessive secretion of GH during childhood results in abnormal height (known as gigantism) unless treated. Figure 11.8 shows a patient who has had excessive secretion of GH throughout life and suffers from gigantism and acromegaly.

On biochemical testing patients usually have impaired glucose tolerance or frank diabetes mellitus. The major causes of death in patients with acromegaly are coronary heart disease and cerebrovascular disease.

Investigation of acromegaly

As we discussed earlier in this chapter, GH is secreted in a pulsatile fashion especially during sleep, with daytime concentrations often being undetectable. Increased release of GH also occurs as a consequence of stress. In patients with acromegaly, however, GH remains

FIGURE 11.8
Photograph of a patient with features of both acromegaly and gigantism standing next to the editor of this book. Courtesy of Dr N Ahmed, School of Healthcare Science, Manchester Metropolitan University, UK.

detectable throughout the day. Some authorities therefore advocate measuring a single post-prandial GH as a first-line test. If the GH is undetectable this excludes acromegaly, but if GH is detectable, this requires further investigation.

Key Points

A random serum GH concentration within the reference range does not exclude acromegaly.

An increased IGF-1 concentration is highly specific for acromegaly and correlates well with disease activity. It is worth noting that since the liver is the main source of IGF-1 a decline is seen in starvation and diseases associated with malnutrition such as hepatic failure, inflammatory bowel disease, and renal failure.

Serum IGF-1 concentrations are increased in acromegaly, compared with age and sex-matched controls, and vary minimally, leading some authorities to suggest that an undetectable GH and normal IGF-1 can be used to effectively exclude acromegaly.

The definitive test for acromegaly is the response of GH to the oral glucose tolerance test (OGTT). Briefly, the patient fasts from midnight the night before the test. Just before the start of the test, a blood sample is drawn from the patient for the measurement of GH. The patient is then given 75 g anhydrous glucose orally. Further blood samples are then collected at 30, 60, 90, 120, and 150 minutes for the analysis of GH. Normal healthy subjects will show suppression of GH to undetectable levels during the test, while patients with acromegaly will show no such suppression due to the autonomous nature of GH secretion in affected patients.

Management of acromegaly

Drugs used to treat patients with acromegaly include bromocriptine, which acts by reducing GH secretion from pituitary tumours but is effective in less than 50% of patients. An analogue of somatostatin called octeotride may be used and acts by inhibiting release of GH. Pegvisamant, which is a GH receptor antagonist, is a more recent form of medical therapy. Surgical removal of pituitary tumours may be attempted, depending on the size and location of the tumour. Acromegaly caused by non-pituitary tumours can be treated by their surgical removal and this usually causes a fall in serum GH concentrations.

It is also necessary to monitor patients during and after therapy. While there is no general agreement on the best way to do this, most endocrinologists will routinely monitor plasma GH throughout the day and IGF-1 either after a change in therapy or annually.

SELF-CHECK 11.4

Why does a glucose load suppress release of GH from the anterior pituitary?

Adrenocorticotrophic hormone dependent Cushing's syndrome

Cushing's syndrome may be ACTH independent as a result of an adrenal tumour, or ACTH dependent due to a pituitary tumour or ectopic ACTH production from, for example, a small

CASE STUDY 11.3

A 55-year-old man presented to his physician complaining of headaches and an increasing shoe size. On examination he had coarse facial features. The physician requested a random GH measurement and the following results were returned from the laboratory (reference ranges are given in brackets):

Glucose	15 mmol/L	(2.7–7.9)
Random GH	5 mU/L	(<10)

(a) What diagnosis is the physician considering?

(b) Do these results rule it in or out?

The physician also performed an oral glucose tolerance test and the results obtained are given below:

Time (minutes)	GH (mU/L)
0 (fasting)	6
30	7
60	10
90	8
120	8

(c) Comment on the results of the oral glucose tolerance test.

(d) What other investigations would be useful?

Cross reference

Chapter 14 Adrenal disease

cell carcinoma of the bronchus. Cushing's disease, however, occurs due to uncontrolled production of ACTH from a pituitary adenoma.

Investigation of ACTH dependent Cushing's syndrome

The investigation of a patient with Cushing's syndrome can be divided into two stages: diagnosis of hypercortisolism and localization of the source of excess cortisol.

Here we will confine ourselves to a discussion of the localization of excess ACTH production. Broadly speaking, investigation of this differential diagnosis involves measuring plasma ACTH, serum potassium, and serum cortisol following high dose dexamethasone suppression, and venous sampling. We will now consider each one in turn.

In Cushing's syndrome due to an adrenal tumour, plasma ACTH is undetectable because of the negative feedback of excess cortisol on the pituitary and hypothalamus, while patients with pituitary dependent or ectopic Cushing's syndrome will have detectable concentrations of ACTH. Patients with ectopic ACTH production from, for example, a small cell carcinoma often have very high levels of ACTH, although there can be a considerable overlap. Patients with ectopic ACTH production nearly always display hypokalaemic alkalosis due to excessive mineralocorticoid action. In such cases, serum potassium is often less than 2.0 mmol/L.

Cross reference

Chapter 5 Fluid and electrolyte disorders

A high dose dexamethasone suppression test can aid in the diagnosis of ACTH dependent Cushing's syndrome. Dexamethasone is a member of the glucocorticoid class of hormones and, like cortisol, will inhibit release of ACTH by negative feedback. Dexamethasone is administered at 2 mg every 6 hours for 48 hours and in the majority of patients with Cushing's disease it will suppress serum cortisol by at least 50%. There will be no suppression if the patient has ectopic ACTH secretion or an adrenal tumour. In this case, an ACTH concentration will be diagnostic, that is raised in ectopic ACTH secretion, and suppressed in patients with an adrenal tumour, due to negative feedback.

Bilateral inferior petrosal sinus sampling is a specialist invasive procedure performed to ascertain the source of ACTH secretion. Under radiographic (X-ray) control, catheters are passed from both femoral veins into the left and right inferior petrosal sinuses which drain blood from the pituitary. The most widely used technique is then to take samples for ACTH before and after CRH stimulation. Blood samples are collected from the left and right petrosal sinus and also from a peripheral vein. Patients with pituitary driven Cushing's disease will show a central: peripheral gradient of ACTH of ≥3:1 which is diagnostic of a pituitary source. Petrosal sinus sampling, however, remains a specialist test and as such is available in only a few regional centres.

Management of ACTH dependent Cushing's syndrome

The management of these patients is aimed at removing the ACTH secreting tumour by surgery. Certain drugs that can inhibit cortisol synthesis, for example metyrapone, have also been used but offer limited efficacy.

11.7 Anterior pituitary hypofunction

Having considered disorders resulting from hypersecretion of pituitary hormones let us now consider hypofunction. **Hypopituitarism** or deficiency of one or all of the pituitary hormones has an incidence of approximately 4 per 100,000 per year and can be due to a number of causes, for example:

- Tumours:
  - pituitary adenoma
  - secondary deposits, for example in breast
- Infarction
- Granuloma:
  - sarcoidosis
- Infection:
  - tuberculosis
  - abscess
- Iatrogenic:
  - surgery
  - irradiation
- Hypothalamic disease:
  - tumour
  - trauma
  - infection

TABLE 11.3 Clinical signs and symptoms of anterior pituitary hormone deficiency (adapted from Thorogood and Baldeweg, 2008).

Anterior pituitary hormone	Signs and symptoms
ACTH	Non-specific: lethargy, anorexia, weight loss, Addisonian crisis, shock, fever, confusion, hypoglycaemia, hyperkalaemia
FSH and LH	Decreased libido, impotence and infertility Amenorrhoea and menopausal symptoms Change in body composition
GH	Decreased quality of life, lethargy, depression, increased cardiovascular risk, change in body composition
TSH	Hypothyroidism, lethargy, depression, hypothermia, hypotension, weight gain

All these result in destruction of normal pituitary structure or function and can result in the deficiency of one or more of the pituitary hormones.

Clinical features of hypopituitarism

The clinical presentation will vary depending on the hormones affected and the duration of this deficiency. Table 11.3 shows the classical signs and symptoms in adults.

Key Points

The signs and symptoms of pituitary failure depend on the hormone affected and the extent and duration of hormone deficiency.

Investigation of hypopituitarism

The following baseline tests would be recommended in evaluating suspected hypopituitarism. Ideally these tests should be performed on a blood sample collected at 9 am:

- Serum cortisol: a sample collected at 9 am with a cortisol >550 nmol/L indicates an intact hypothalamic-pituitary-adrenal axis and effectively excludes cortisol deficiency. Although one should note that in some cases subjects may have a cortisol concentration >550 nmol/L, but be unable to respond appropriately to an increase in physiological stress, for example during trauma, severe illness, or surgery. In addition, oestrogen containing medications, for example the oral contraceptive pill, will increase the concentration of CBG and therefore the serum concentration of cortisol will be higher.

- Luteinizing hormone, FSH, and oestrogen or testosterone: a low FSH and LH together with a low oestrogen or testosterone would suggest a pituitary or hypothalamic cause.

- Thyroid stimulating hormone and free T4: a low TSH in the presence of a low free T4 would be consistent with secondary hypothyroidism.

Further evaluation requires dynamic function testing. The insulin tolerance test is considered the gold standard for the assessment of the pituitary-adrenal and pituitary-growth

BOX 11.1 Growth hormone deficiency in adults

The most frequent cause of adult growth hormone deficiency is a benign tumour of the pituitary gland, either due to the pressure effects of the tumour itself or as a consequence of treatment by surgery or radioiodine.

It is estimated that approximately 3 in every 10,000 adults in the UK are growth hormone deficient due to a number of reasons, including pituitary tumour, pituitary surgery, radiotherapy, or idiopathic causes, where no cause has been found. The symptoms include:

- Increase in adipose tissue around the waist
- Decrease in lean body mass
- Decrease in strength and stamina
- Decrease in bone density
- Increase in LDL cholesterol, with a decrease in HDL cholesterol
- Excessive tiredness
- Anxiety and depression
- Feelings of social isolation
- Reduced quality of life
- Increased sensitivity to heat or cold

Adults who have biochemically documented GH deficiency and who have the symptoms listed above may benefit from GH replacement. Recombinant GH is administered by daily subcutaneous injection. Fluid retention has been reported as a side effect although this is rare with carefully titrated therapy.

hormone axes. The patient is fasted from midnight the night before the test. A baseline sample is taken for the measurement of glucose, cortisol, and GH before insulin is administered intravenously. Further samples are then drawn at 30, 60, 90, and 120 minutes for the measurement of glucose, cortisol, and GH. For the result to be valid, the patient must achieve adequate hypoglycaemia and have a blood glucose of <2.2 mmol/L for significant stimulation of the pituitary gland. An adequate pituitary reserve is demonstrated by an increase in cortisol concentration of at least 200 nmol/L to a value >550 nmol/L. With cortisol deficiency there is a diminished response. A normal GH response is an increase in GH concentration to >20 mU/L with a diminished response in deficiency states. The combined pituitary function test, which tests all the anterior pituitary hormone reserves, is now rarely used in clinical practice.

Management of hypopituitarism

Glucocorticoid replacement in patients with hypopituitarism is usually initiated first. Thyroxine replacement can be initiated once glucocorticoid treatment has been started. Sex hormone treatment may be required to restore normal sexual function and can be given as testosterone replacement in males or as oestrogen replacement in females.

The benefit of growth hormone replacement in adults is difficult to assess. The National Institute for Health and Clinical Excellence (NICE) recommends that growth hormone replacement should only be used in adults with confirmed severe GH deficiency and reduced quality of life as assessed by the Adult Growth Hormone Deficiency Questionnaire. The consequences of a deficiency of GH secretion in adults is outlined in Box 11.1.

11.8 Posterior pituitary dysfunction

Damage to the posterior pituitary can occur due to head injuries, by neoplasms affecting the hypothalamus and pituitary, following surgical or radiation therapy, or it can be idiopathic, that is of unknown cause. A feature of posterior pituitary dysfunction is deficiency of ADH which can cause cranial **diabetes insipidus**.

Cranial diabetes insipidus

Diabetes insipidus results from an absolute or relative deficiency of ADH or vasopressin. Diabetes insipidus can be divided into two categories designated cranial or nephrogenic diabetes insipidus. Cranial diabetes insipidus is due to a deficiency of ADH from the posterior pituitary, while nephrogenic diabetes insipidus is caused by a failure of the kidneys to respond to ADH. Patients with cranial diabetes insipidus fail to concentrate their urine and as a result produce a large volume of dilute urine which can lead to severe dehydration if access to water is restricted.

Clinical features of cranial diabetes insipidus

Patients develop thirst and polyuria with urine volumes ranging from 3–20 L every 24 hours. The causes of cranial diabetes insipidus are listed below

- Familial
- Acquired:
  - idiopathic
  - trauma
    - head injury
  - tumour
  - granuloma
    - sarcoidosis
  - infection
    - meningitis

It should be noted that the familial causes are rare. However, at least 30% of cases are idiopathic.

Investigation of cranial diabetes insipidus

The simultaneous measurement of plasma and urine osmolality will give a good assessment of ADH secretion. Diabetes insipidus can be excluded by a normal plasma osmolality of 280–295 mOsm/kg and a urine osmolality >750 mOsm/kg. Diabetes insipidus is suggested by an elevated plasma osmolality of >300 mOsm/kg together with a urine osmolality <750 mOsm/kg.

The water deprivation test can be used to diagnose diabetes insipidus in difficult cases and can also be combined with a desmopressin (DDAVP) test to differentiate between cranial and nephrogenic diabetes insipidus. The water deprivation test is a dynamic test of posterior pituitary function, which is potentially dangerous and must be performed under close medical supervision. During the test, the patient is not allowed to drink for up to seven hours, during which serum and urine samples are collected for the measurement of osmolality. Normal subjects concentrate their urine to >700 mOsm/kg while maintaining a normal plasma osmolality of 295 mOsm/kg. In diabetes insipidus, the plasma becomes abnormally concentrated with an osmolality >295 mOsm/kg, whilst their urine remains dilute with an osmolality of <270 mOsm/kg. Desmopressin is a synthetic analogue of ADH and when given to patients with cranial diabetes insipidus, it mimics the action of ADH, causing the urine to become concentrated. In patients with nephrogenic diabetes insipidus, the kidneys are unresponsive to ADH, consequently DDAVP is ineffective and the subject does not concentrate their urine. Thus, this test can be used to discriminate between cranial and nephrogenic diabetes insipidus.

Management of cranial diabetes insipidus

Desmopressin reduces the amount of water lost in the kidneys and is used in the treatment of patients with cranial diabetes insipidus. It can be administered orally, nasally, or parenterally to treat patients affected by cranial diabetes insipidus.

CASE STUDY 11.4

A 30-year-old nurse was referred to the endocrinology department of a hospital with a three-month history of polyuria.

The endocrinologist performed a water deprivation test which gave the following results:

Time	Osmolality (mOsm/kg)	
	Plasma (285–295)	Urine
7 am	289	70
8 am	295	78
10 am	300	72
12 pm	305	75
2 pm	306	76

(a) Comment on the results.

(b) What is the differential diagnosis at this point in the test?

She was then administered desmopressin (DDAVP) by an intramuscular injection.

Intramuscular injection of DDAVP

	Osmolality (mOsm/kg)	
	Plasma (285–295)	Urine
4 pm	295	750
6 pm	296	860

(c) Comment on the results.

(d) What is the diagnosis?

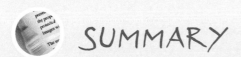

SUMMARY

- Hormones are chemical messengers produced by endocrine glands that circulate in the blood and act on target cells via receptors.

- Feedback mechanisms, both negative and positive, regulate release of hormones.

- The pituitary gland is influenced by release of peptides from the hypothalamus and also releases peptide hormones itself, which influence release of hormones from other endocrine glands located in the thyroid, adrenals, and gonads.

- Hyperfunction of the anterior pituitary may result in increased production of prolactin, GH, and ACTH causing hyperprolactinaemia, acromegaly, and ACTH-dependent Cushing's syndrome respectively.

- Hypofunction of the anterior pituitary can cause reduced output of a number of anterior pituitary hormones and result in hypopituitarism.

- Posterior pituitary dysfunction can result in low ADH secretion which presents clinically as cranial diabetes insipidus.

- Release of hormones from the pituitary gland can be investigated by measuring the concentration of single hormones in serum or by dynamic function testing.

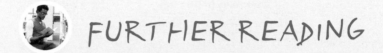

FURTHER READING

- **Ben-Shlomo A and Melmed S (2008) Acromegaly.** *Endocrinology and Metabolism Clinics of North America* **37**, 101–22.

  Detailed account of acromegaly and its investigation.

- **Sam S and Frohman LA (2008) Normal physiology of hypothalamic pituitary regulation.** *Endocrinology and Metabolism Clinics of North America* **37**, 1–22.

  Very detailed discussion of hypothalamic pituitary regulation.

- **Thorogood N and Baldeweg SE (2008) Pituitary disorders: an overview for the general physician.** *British Journal of Hospital Medicine* **69**, 198–204.

  Excellent, easy to read review on this topic.

- **Toogood AA and Stewart PM (2008) Hypopituitarism: clinical features, diagnosis and management.** *Endocrinology and Metabolism Clinics of North America* **37**, 235–61.

  Very comprehensive account of hypopituitarism.

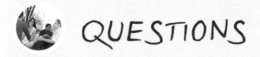

QUESTIONS

11.1 Which one of the following findings would be diagnostic of acromegaly in a patient who has undergone an OGTT?

(a) A decrease in plasma glucose concentration

(b) A rise in plasma glucose concentration

(c) A decrease in serum GH concentration

(d) No reduction in serum GH concentration

(e) An increase in serum TSH concentration

11.2 Are the following statements TRUE or FALSE?

    (a) Thyrotrophin releasing hormone from the hypothalamus can stimulate release of TSH from the anterior pituitary

    (b) Somatostatin can inhibit release of GH from the anterior pituitary

    (c) Cranial diabetes insipidus occurs due to a deficiency of ACTH released from the posterior pituitary

    (d) Dopamine agonists can be used to suppress prolactin release in patients with hyperprolactinaemia

    (e) The water deprivation test is used for diagnosis of Cushing's syndrome

11.3 Both ACTH and GH are secreted in a circadian and pulsatile manner. Discuss the clinical implication of this in the investigation of patients with pituitary dysfunction.

11.4 Some laboratories use TSH alone when screening for hypothyroidism. What are the potential limitations of this strategy in a hospital with a large endocrine department seeing patients with pituitary dysfunction?

11.5 Why do men usually present with larger macroprolactinomas than women?

Answers to self-check questions, case study questions, and end-of-chapter questions are available in the Online Resource Centre accompanying this book.

 Go to www.oxfordtextbooks.co.uk/orc/ahmed/

12

Thyroid disease

Garry McDowell

Learning objectives

After studying this chapter you should be able to:

- Describe the structure and function of the thyroid gland
- Explain the function of thyroid hormones
- Outline the action of thyroid hormones and control of their secretion from the thyroid gland
- Describe the conditions which lead to abnormal thyroid hormone production
- Discuss the investigation of suspected thyroid dysfunction

Introduction

The thyroid gland secretes thyroid hormones that are required for normal metabolism of body cells. Disorders of thyroid function can result in either inadequate or excess production of thyroid hormones causing altered cellular metabolism and development of associated clinical features.

This chapter will describe the nature and role of thyroid hormones, their regulation in the blood and the consequences of changes in their secretion. The value of laboratory investigations in diagnosis and monitoring of treatment will be discussed.

12.1 Structure of the thyroid gland

The thyroid gland is found below the larynx and is a butterfly shaped gland composed of a right and left lobe on either side of the trachea. Both lobes are joined by an isthmus in front of the trachea. The normal thyroid gland weighs approximately 30 g and is highly vascularized, receiving 80–120 mL of blood per minute, as shown in Figure 12.1.

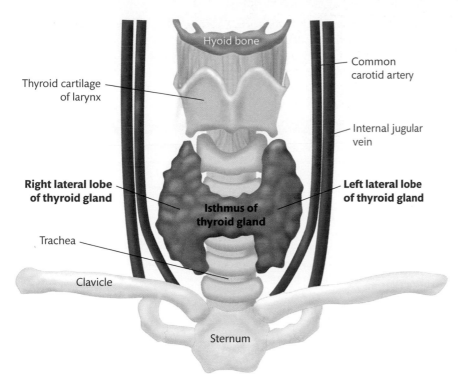

Hyoid bone

Common carotid artery

Thyroid cartilage of larynx

Internal jugular vein

Right lateral lobe of thyroid gland

Left lateral lobe of thyroid gland

Isthmus of thyroid gland

Trachea

Clavicle

Sternum

FIGURE 12.1
Anatomical location of the thyroid gland in the neck.

Microscopic examination of thyroid tissues shows small spherical sacs called thyroid follicles that make up most of the thyroid gland. The wall of each follicle is composed mainly of follicular cells, most of which extend to the lumen of the follicle. Figure 12.2 shows the structure of thyroid follicles.

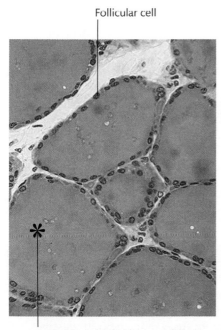

Follicular cell

Follicle containing thyroglobulin

FIGURE 12.2
Histological structure of the thyroid gland showing the follicles in which thyroid hormones are made. Courtesy of Dr A L Bell, University of New England College of Osteopathic Medicine, USA.

A basement membrane surrounds each follicle. Follicular cells produce two hormones: thyroxine (T4), which contains four iodine atoms and tri-iodothyronine (T3), which contains three iodine atoms. Together T4 and T3 are known as thyroid hormones. The parafollicular cells or C-cells lie in between the follicles and produce a hormone called calcitonin, which regulates calcium homeostasis.

SELF-CHECK 12.1

What are the two cell types in the thyroid gland and what hormones do they secrete?

12.2 Thyroid hormones

The thyroid hormones T4 and T3 are produced by the incorporation of iodine into tyrosyl residues in **thyroglobulin** in a series of steps which are described as:

- iodide trapping
- synthesis of thyroglobulin
- oxidation of iodide
- iodination of tyrosine
- coupling
- pinocytosis of colloid
- secretion of thyroid hormones
- transport of thyroid hormones in the blood.

We will now consider each step in a little more detail. Figure 12.3 shows the steps involved in the synthesis of thyroid hormones.

Follicular cells in the thyroid gland trap iodide ions by active transport from the blood into the cytosol. The synthesis of thyroglobulin also occurs in the follicular cells. Thyroglobulin is a large glycoprotein that is produced in the rough endoplasmic reticulum, modified by the attachment of a carbohydrate molecule in the Golgi apparatus and packaged into secretory vesicles. The vesicles then release thyroglobulin in a process known as **exocytosis** into the follicle. Thyroglobulin contains a large number of tyrosine residues that will ultimately become iodinated. In the diet, iodine is present in the form of iodide and this must be oxidized to iodine which can be used for iodination of tyrosine residues of thyroglobulin. As iodide becomes oxidized to iodine it passes across the cell membrane into the lumen of the follicle. As iodine molecules form they are incorporated into tyrosine residues of thyroglobulin. The binding of one atom of iodine to the tyrosine residues results in the formation of monoiodothyronine (T1), whilst the binding of two iodine atoms results in the formation of di-iodothyronine (T2). During the coupling step, two molecules of T2 join to form thyroxine (T4), while a coupling of T1 and T2 results in tri-iodothyronine (T3). Iodinated thyroglobulin incorporating T4 and T3 is stored in the colloid. Oxidation of iodide, iodination of tyrosine residues, and coupling reactions are all catalysed by the enzyme thyroid peroxidase. Then, under the control of thyroid stimulating hormone (TSH) which is produced by the anterior pituitary, droplets of colloid re-enter the follicular cells by a process known as **pinocytosis** and merge with lysosomes. The enzymes present in lysosomes catalyse the proteolytic digestion of thyroglobulin releasing T4 and T3, whose structures are shown in Figure 12.4.

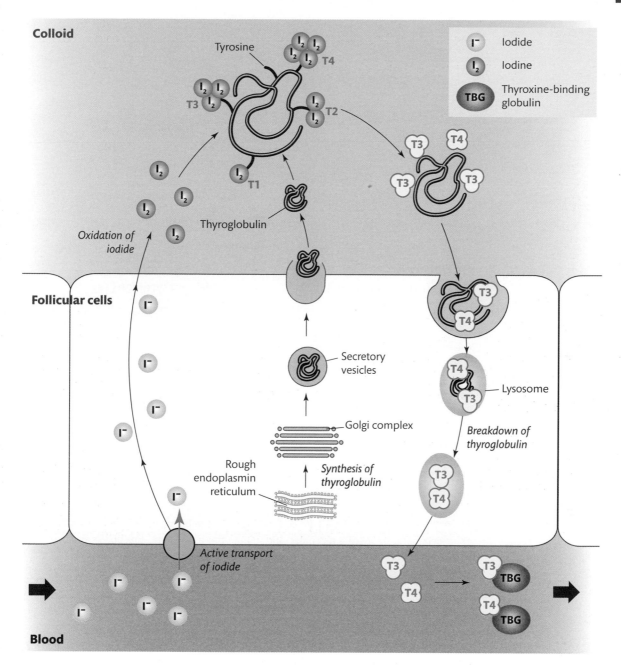

FIGURE 12.3
Synthesis of thyroid hormones T4 and T3.

Since T4 and T3 are lipid-soluble, they diffuse across the plasma membrane and enter the circulation. Due to their lipophilic nature, more than 99% of T4 and T3 are bound to the transport protein **thyroxine binding globulin** (TBG). Thyroxine is released from the thyroid gland in greater amounts than T3, although T3 is the more biologically active hormone. Thyroxine enters cells and is deiodinated (removal of one I atom) to form T3.

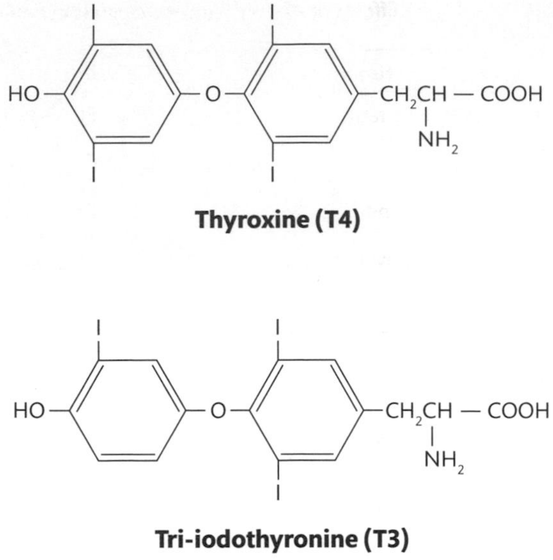

Thyroxine (T4)

FIGURE 12.4
Chemical structures of T4 and T3.   **Tri-iodothyronine (T3)**

The majority of thyroid hormones in plasma are bound to specific proteins in order to render them water-soluble, reduce renal loss, and to provide a large pool of hormones, whilst protecting the cells from the physiological effect of the hormone. The plasma binding proteins are TBG and to a lesser extent albumin and pre-albumin. The plasma concentrations and proportions of thyroid hormones which are bound are shown below:

	Concentration	T4 (%)	T3 (%)
TBG	20 mg/L	70–75	75–80
Pre-albumin	0.3 g/L	15–20	Trace
Albumin	40 g/L	10–15	10–15

The unbound or free T4 and T3 are considered to be the biologically active fraction that can enter cells, bind to specific receptors, and initiate the physiological response and cause the negative feedback regulation of thyroid hormone secretion.

The approximate reference ranges for serum concentrations of total and free thyroid hormones are:

	Total	Free
T4	60–160 nmol/L	10–25 pmol/L
T3	1.2–2.3 nmol/L	4.0–6.5 pmol/L

Thyroxine is the major hormone secreted by the thyroid gland, which is converted by specific de-iodinase enzymes, particularly in the liver and kidney, to form T3, the biologically active hormone. The peripheral deiodination of T4 provides approximately 80% of plasma T3, the remainder being derived from thyroid gland secretion.

SELF-CHECK 12.2

What are the steps involved in the synthesis of thyroid hormones?

TABLE 12.1 Effects of thyroid hormones on metabolic indices.

Increased by a rise in [thyroid hormone]	Increased by a decline in [thyroid hormone]
Basal metabolic rate	Plasma cholesterol
Plasma calcium	Creatine kinase
Sex hormone binding globulin	Creatinine
Angiotensin converting enzyme	Thyroxine binding globulin
Liver enzymes (gamma-glutamyl transferase)	

12.3 Function of thyroid hormones

Table 12.1 shows the effect of thyroid hormones on metabolism. They increase intracellular transcription and translation, bringing about changes in cell size, number, and differentiation. They also promote cellular differentiation and growth.

SELF-CHECK 12.3

What are the effects of thyroid hormones on metabolism?

12.4 Control of thyroid hormone secretion

Thyroid hormone production is under both positive and negative feedback control as shown in Figure 12.5.

Thyrotrophin releasing hormone (TRH) from the hypothalamus acts on the anterior pituitary causing release of TSH, which in turn acts on the thyroid gland and stimulates the synthesis and release of thyroid hormones. Briefly, a low blood concentration of free T4 or T3 stimulates the hypothalamus to secrete TRH, which enters the hypothalamic portal veins and flows to the anterior pituitary where it stimulates **thyrotrophs** to secrete TSH. The TSH then acts on the follicular cells to stimulate T4 and T3 production and their subsequent release. A rise in the concentration of unbound T4 and T3 in the blood inhibits further release of TRH and TSH from the hypothalamus and anterior pituitary respectively, via a negative feedback effect.

SELF-CHECK 12.4

What is the name given to the control mechanism where thyroxine controls its own release?

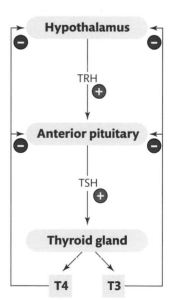

FIGURE 12.5
Regulation of thyroid hormone secretion.

12.5 **Disorders of thyroid function**

From a clinical perspective disorders of thyroid function can be classified into two broad categories: hyperfunction states where thyroid hormones are produced in excess, referred to as **hyperthyroidism**, and hypofunction states where there is a deficiency of thyroid hormones, referred to as **hypothyroidism**.

12.6 **Hyperthyroidism**

Hyperthyroidism has a significant short- and long-term morbidity and mortality. The prevalence of hyperthyroidism in women is ten times more common than in men. The annual incidence of hyperthyroidism is quoted as 0.8/1,000 women.

Causes of hyperthyroidism

The most common causes of hyperthyroidism are **Graves' disease** and toxic multi-nodular **goitre**. Less commonly, hyperthyroidism may occur in patients on thyroxine therapy or due to excess thyroid hormones being produced by ectopic thyroid tissue. Very rarely, hyperthyroidism may be a consequence of TSH secreting tumours.

Graves' disease is an autoimmune condition characterized by the presence of diffuse thyroid enlargement, eye abnormalities and thyroid dysfunction. The disease predominantly affects females, with a peak incidence in the third and fourth decades of life.

Hyperthyroidism can often arise in patients with a multi-nodular goitre and occurs in an older population than affected by Graves' disease. The age of onset is typically over 50 years, with females being affected more than males.

Drugs such as **amiodarone** can have a significant effect on thyroid function. Amiodarone is used in the treatment of cardiac **arrhythmias**, has a structure similar to that of thyroid hormones, and interferes with the peripheral conversion of T4 to T3. Consequently the concentrations of T4 may be increased while T3 is low. In practice, it is advisable to check thyroid function by assay of TSH and free T4 before commencing amiodarone treatment. Interpretation of thyroid function test results can be problematic during treatment and assessment of thyroid status during this time is best undertaken by careful clinical assessment.

Clinical features of hyperthyroidism

The clinical condition is often referred to as **thyrotoxicosis** and affected individuals present with characteristic features. The common symptoms and signs of hyperthyroidism are shown in Table 12.2.

On clinical examination of patients with Graves' disease, a large and diffuse goitre is usually present which is soft to the touch. A bruit is frequently heard over the thyroid and its blood vessels due to increased blood flow through the hyperactive gland. Patients with Graves' disease have characteristic eye signs, with a staring expression due to lid retraction, the white of the eye or the sclera being visible above and below the iris. In addition there is a tendency for

TABLE 12.2 Symptoms and signs of hyperthyroidism.

Symptoms	Signs
Increased irritability	Tachycardia
Increased sweating	Goitre
Heat intolerance	Warm extremities
Palpitations	Tremor
Lethargy	Arrhythmias
Loss of weight	Eye signs
Breathlessness	Proximal myopathy
Increased bowel frequency	Muscle weakness

the movement of the lid to lag behind that of the globe as the patient looks downwards from a position of maximum upward gaze, referred to as 'lid-lag'.

In patients with a toxic multi-nodular goitre, the cardiovascular features tend to predominate in this often older population. The goitre is classically nodular and may be large.

Investigation and diagnosis of hyperthyroidism

Measurement of TSH will in most cases of hyperthyroidism show suppression of TSH to a concentration below the lower limit of the reference range and in many cases to less than the limit of detection for the assay. The exception to this is a TSH secreting pituitary tumour in which case the concentration of TSH may be normal or at the top of the laboratory reference range. Thyroid stimulating hormone secreting pituitary tumours, however, are extremely rare. The concentration of free T4 is increased, often in association with a significant increase in free T3 concentration. In some cases free T3 alone may be increased, with a normal T4 and low or undetectable levels of TSH, and this is referred to as T3-toxicosis.

The diagnosis of Graves' disease is made by the finding of hyperthyroidism on biochemical testing, the presence of goitre, and extra-thyroidal signs such as eye signs. In other cases the presence of a thyroid stimulating antibody (TSH receptor antibody) and diffuse increased iodine uptake on thyroid scanning confirms the diagnosis.

The biochemical diagnosis of hyperthyroidism due to a toxic multi-nodular goitre is fairly straightforward with suppression of TSH concentration. Free T4 and T3 concentrations are increased although they may not be grossly abnormal, with values at or just above the reference range. Thyroid scintillation scanning shows patchy uptake of isotope with multiple hot and cold areas being seen throughout the gland.

A TSH secreting pituitary **adenoma** is a rare cause of hyperthyroidism. In these cases TSH is usually within the reference range, or inappropriately normal, or only slightly raised above it, often around 6 mU/L, with an increased free T4 and T3. In such cases imaging will often identify a pituitary lesion.

SELF-CHECK 12.5

What are the common clinical features of hyperthyroidism?

Management of hyperthyroidism

The treatment of hyperthyroidism including Graves' disease falls into three broad categories. These are anti-thyroid drugs, radioactive iodine, or subtotal **thyroidectomy**. Some of the symptoms such as **tachycardia** and tremor can be controlled with β-blocking drugs for the first few weeks of therapy. Radioactive iodine (^{131}I) can be used to treat hyperthyroidism and works by initially interfering with organification of iodine and then induces radiation damage to the thyroid. The major side effect of radioiodine treatment is that approximately 80% of subjects will develop hypothyroidism as a result. There is no evidence of an increase in the risk of malignancy following radioiodine therapy. Subtotal thyroidectomy is highly effective although surgical complications can occur in some patients. In elderly patients with a multi-nodular goitre, radioiodine is the treatment of choice, although anti-thyroid drugs can be used until radioiodine treatment becomes effective. Surgery may be required in patients who present with symptoms of hyperthyroidism and an enlarged thyroid gland compressing structures in the neck.

SELF-CHECK 12.6

What are the three broad categories of treatment for a patient with hyperthyroidism?

CASE STUDY 12.1

A 30-year-old housewife presented with weight loss, irritability, and had been feeling uncomfortable whilst on holiday in Spain. She was taking oral contraceptive pills and was not pregnant. On examination, her palms were sweaty, she had a fine tremor, and there was no enlargement of the thyroid gland. The following results were obtained for thyroid function tests (reference ranges are given in brackets):

TSH   <0.1 mU/L   (0.2–3.5)

Free T4   20 pmol/L   (9–23)

Free T3   22 pmol/L   (4.0–6.5)

(a) Comment on these results.

(b) What is the likely diagnosis?

CASE STUDY 12.2

A 40-year-old medical secretary attended for review of her treatment for Graves' disease. She was taking carbimazole which was started one month previously. The results for her thyroid function tests are given below (reference ranges are given in brackets):

TSH   <0.1 mU/L   (0.2–3.5)

Free T4   <5 pmol/L   (9–23)

Free T3   2.5 pmol/L   (4.0–6.5)

Comment on these results.

12.7 **Hypothyroidism**

Hypothyroidism is an insidious condition with significant morbidity and the subtle and non-specific signs are often associated with other conditions. Hypothyroidism is more common in elderly women and ten times more common in women than in men. The annual incidence of hypothyroidism is 3.5/1,000 women.

Causes of hypothyroidism

The causes of hypothyroidism can be primary where they affect the thyroid gland, or secondary where the anterior pituitary or hypothalamus is affected.

The most common cause of primary hypothyroidism is the autoimmune condition called **Hashimoto's thyroiditis** where autoantibodies cause progressive destruction of the individual's own thyroid gland.

Loss of functioning thyroid tissue occurs following thyroidectomy or radioiodine treatment and may lead to hypothyroidism. These patients will have an increased TSH concentration with a low free T4 concentration, provided they are not receiving any form of thyroid hormone replacement therapy. Drug treatment with compounds such as lithium and iodine can also result in hypothyroidism.

Other causes of hypothyroidism include congenital hypothyroidism, which occurs in newborn children with a defect in the development of the thyroid gland, resulting in either its absence or an undeveloped gland. Untreated children develop a condition referred to as **cretinism**. Children with cretinism present with growth failure, developmental delay, and are often deaf and mute. Box 12.1 gives further information about congenital hypothyroidism.

BOX 12.1 *Congenital hypothyroidism*

Congenital hypothyroidism is caused by a deficiency of thyroid hormones at birth, usually due to an absent thyroid gland or by an ectopic gland, which means the thyroid gland is not in the correct anatomical position in the neck.

Congenital hypothyroidism in the UK occurs in approximately 1:3500 births. Most babies with congenital hypothyroidism are diagnosed very early before symptoms develop by means of the neonatal screening program, where thyroid hormones are measured in a sample of blood collected on a special card from a heel prick. If signs and symptoms are present, they may include feeding difficulties, sleepiness, constipation, and jaundice (yellow colouration to the skin caused by excess bilirubin).

Children with congenital hypothyroidism are treated with thyroxine and placed on life-long therapy. The prognosis is generally good and experience from the UK national screening program has shown that almost all children with congenital hypothyroidism who are diagnosed and treated early will mature normally.

A small proportion who have been diagnosed late or who have severe hypothyroidism may develop difficulties later in life such as poor hearing, clumsiness, and learning difficulties.

Diseases or injuries affecting the hypothalamus or anterior pituitary can result in reduced production of TRH and TSH respectively, causing a decline in production of thyroid hormones from the thyroid gland. This is referred to as secondary hypothyroidism.

Clinical features of hypothyroidism

The clinical condition is often referred to as **myxoedema** and affected patients present with features associated with reduced cellular metabolism. The common symptoms and signs of hypothyroidism are shown in Table 12.3.

In parts of the world where there is iodine deficiency some patients may present with a goitre and the thyroid gland undergoes hyperplasia. However, goitres also arise due to other reasons, as given in Box 12.2.

Investigation and diagnosis of hypothyroidism

The routine biochemical assessment involves the measurement of TSH and free T4 concentration. As the concentration of thyroid hormones declines, the concentration of TSH increases. The concentration of T3 is preferentially maintained and so measurement of T3 is not recommended as this could be misleading. Thyroxine concentration correlates better with thyroid activity than that of T3 for diagnosis of hypothyroidism. A guideline for the interpretation of thyroid hormone results is shown in Table 12.4.

Individuals with hypothyroidism due to Hashimoto's thyroiditis will have an increased TSH concentration with low free T4 and the majority will have detectable thyroid antibodies. Thyroid peroxidase antibodies may also be detected. The patient may also present with a history of other autoimmune diseases such as diabetes, Addison's disease, and pernicious anaemia.

Patients with secondary hypothyroidism will have a low serum TSH concentration together with a low free T4. The distinguishing feature here is that the TSH concentration is inappropriately low.

SELF-CHECK 12.7

What are the common clinical signs of hypothyroidism?

TABLE 12.3 Symptoms and signs of hypothyroidism.

Symptoms	Signs
Lethargy	Periorbital and facial oedema
Dry coarse skin	Pale dry skin
Slow speech and mental function	Goitre
Cold intolerance	Cool peripheries
Pallor	Bradycardia
Hoarse voice	Median nerve compression
Constipation	Delayed relaxation of reflexes
Weight gain	

BOX 12.2 Goitre

A goitre is an enlarged thyroid gland and can mean that all the thyroid gland is swollen or enlarged, or one or more swellings or lumps develop in a part or parts of the thyroid. There are different types of goitre, such as:

- Diffuse smooth goitre

  This means that the entire thyroid gland is larger than normal. The thyroid feels smooth but large. There are a number of causes. For example:

  - Graves' disease, an autoimmune disease which causes the thyroid to swell and produce too much thyroxine
  - thyroiditis (inflammation of the thyroid), which can be due to various causes, for example viral infections
  - iodine deficiency, the thyroid gland requires iodine to make T4 and T3
  - some medicines can cause the thyroid to swell, for example lithium

- Nodular goitres

  - A thyroid nodule is a small lump which develops in the thyroid. There are two types:
  - a multinodular goitre; this means the thyroid gland has developed many lumps or 'nodules' and feels generally lumpy
  - single nodular goitre, for example a cyst, an adenoma, or a cancerous tumour

Symptoms of goitre

In many cases there are no symptoms apart from the appearance of a swelling in the neck. The size of a goitre can range from very small and barely noticeable, to very large.

Most goitres are painless. However, an inflamed thyroid (thyroiditis) can be painful. There may be symptoms of hypo- or hyperthyroidism.

A large goitre may press on the trachea or even the oesophagus. This may cause difficulty with breathing or swallowing.

Treatment of goitre

Treatment depends on the cause, the size of the goitre, and whether it is causing symptoms. For example, a small goitre that is not due to a cancerous nodule, when the thyroid is functioning normally, may not require treatment. An operation to remove some or the entire thyroid may be an option in some cases.

Management of hypothyroidism

Management of hypothyroidism involves the replacement of thyroid hormones, usually T4, although T3 may sometimes be used. Treatment should be commenced carefully with elderly patients, especially those with pre-existing ischaemic heart disease, being started on a low dose and titrating the dose slowly. Thyroxine replacement therapy is monitored by regular measurement of TSH and free T4. Adequate replacement is achieved when the TSH is within the lower part of the reference range with a normal free T4. It should be noted, however, that the concentrations of free T4 can vary post-dose, although this is not clinically significant.

TABLE 12.4 Guide to the interpretation of thyroid function tests.

	TSH low	TSH normal	TSH high
T4 low	Severe non-thyroidal illness Hypopituitarism	Sick euthyroid syndrome NSAIDs Some anticonvulsants TBG deficiency Hypopituitarism	Hypothyroidism
T4 normal	Thyrotoxicosis Sub-clinical thyrotoxicosis Treated thyrotoxicosis Over-treated hypothyroid	Euthyroid Adequate T4 replacement	Subclinical hypothyroidism Inadequate T4 replacement Recovery from non-thyroidal illness
T4 high	Thyrotoxicosis, T4 replacement	Sick euthyroid syndrome Erratic compliance with T4 replacement Increased TBG	Erratic compliance with T4 therapy.

SELF-CHECK 12.8

How do you treat a patient with hypothyroidism?

CASE STUDY 12.3

A 63-year-old man, who was previously fit and well, presented with a five-day history of shortness of breath associated with wheeze and dry cough. He denied symptoms of hyperthyroidism and his family, social, and past medical history were unremarkable. The electrocardiogram was consistent with atrial fibrillation and a fast ventricular response. The results are as follows (reference ranges are given in brackets):

TSH	6.4 mU/L	(0.4–4)
Free T3	12.5 pmol/L	(4–6.5)
Free T4	51 pmol/L	(10–30)
Testosterone	43.1 nmol/L	(10–31)
FSH	18.1 IU/L	(1–7)
LH	12.4 IU/L	(1–8)

GH, prolactin and IGF-1 normal.

(a) Comment on these results.

(b) What further investigations would you suggest?

(c) Can you provide an explanation for these results?

12.8 Laboratory tests to determine the cause of thyroid dysfunction

Thyroid peroxidase antibodies are present in about 95% of patients with autoimmune hypothyroidism secondary to Hashimoto's thyroiditis. They may also be found in a small number of healthy individuals but their appearance usually precedes the development of thyroid disorders.

Thyroglobulin antibodies are found in many patients with autoimmune thyroid disease; however, measurement of thyroglobulin antibodies has no additional value to measuring thyroid peroxidase antibodies alone.

Thyroid stimulating hormone receptor antibodies are measured in most routine laboratories using methods that quantify the inhibition of TSH binding to porcine or human TSH receptors. In most patients the measurement of TSH receptor antibodies is not essential for diagnostic purposes.

The response of plasma TSH to a standardized challenge of infused TRH has been used for many years to investigate patients with borderline hyperthyroidism. A marked TSH response to >2 times the baseline value excludes hyperthyroidism. With the development of new sensitive TSH assays it has been shown that a normal basal serum TSH predicts a normal TSH response to TRH stimulation, whilst a suppressed basal TSH predicts a failure to respond during TRH stimulation. The TRH test is now not routinely performed in clinical practice. The measurement of free T4 and T3 is outlined in Box 12.3.

BOX 12.3 Measurement of free T4 and T3

The measurement of TSH in a basal blood sample by a sensitive immunometric assay provides the single most sensitive, specific, and reliable test of thyroid status. As we have already discussed, the free hormones (free T4 and free T3) are widely held to be the biologically active fractions. Direct methods involve measurement of the free hormone in the presence of protein bound hormone. The analogue methods use tracer derivatives of T4 or T3 capable of binding to the antibody but not reacting with the binding proteins. The two-step assays involve the binding of the free hormone in the sample with solid phase antibody, removal of the sample and back titration of unoccupied binding sites on the antibody with labelled hormone. Interference in free T4 and T3 assays by, for example, abnormal binding proteins and *in vivo* antibodies that bind T4 and T3, can cause problems in the interpretation of thyroid hormone results.

The reference method for free T4 and free T3 measurement is equilibrium dialysis using undiluted serum, but this cannot be performed in large numbers on a routine basis.

12.9 Interpretation of thyroid function tests

Interpretation of thyroid function tests can be difficult; however, there are a few basic principles which can help. Table 12.4 shows the most common causes of changes in the hormone pairs TSH and free T4.

Pregnancy can have a significant effect on the result of thyroid hormone testing. In a normal pregnancy the concentration of TBG increases due to the action of oestrogen. Free thyroid hormone concentrations also increase due to the weak thyroid stimulating effect of high concentrations of human chorionic gonadotrophin (hCG) in early pregnancy. The concentration of TSH is increased compared to the non-pregnant state, but remains within the non-pregnant reference range. **Hyperemesis gravidarum** or a state of severe vomiting during the first trimester is frequently associated with very high concentrations of free T4 and free T3 making it difficult to differentiate from true thyrotoxicosis. It is thought that very high concentrations of hCG are also responsible for this condition.

Severe non-thyroidal illness can also affect the concentrations of thyroid hormones. Interpretation of results should take into account the patient's general clinical state and bear in mind that during the illness and recovery the thyroid axis will not be in a steady state. A general scheme for the interpretation of thyroid function tests is shown in Figure 12.6.

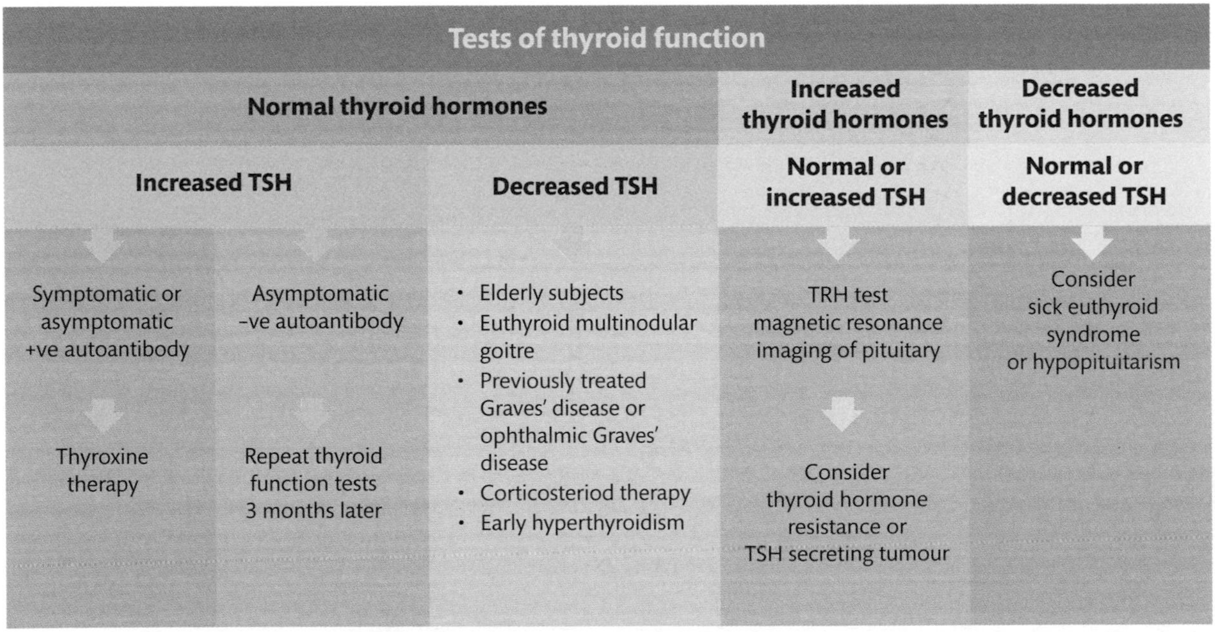

Tests of thyroid function				
Normal thyroid hormones			**Increased thyroid hormones**	**Decreased thyroid hormones**
Increased TSH		**Decreased TSH**	**Normal or increased TSH**	**Normal or decreased TSH**
Symptomatic or asymptomatic +ve autoantibody	Asymptomatic –ve autoantibody	• Elderly subjects • Euthyroid multinodular goitre • Previously treated Graves' disease or ophthalmic Graves' disease • Corticosteriod therapy • Early hyperthyroidism	TRH test magnetic resonance imaging of pituitary	Consider sick euthyroid syndrome or hypopituitarism
Thyroxine therapy	Repeat thyroid function tests 3 months later		Consider thyroid hormone resistance or TSH secreting tumour	

FIGURE 12.6

A flowchart for the interpretation of thyroid function tests.

SUMMARY

- The thyroid gland produces hormones called thyroxine (T4) and tri-iodothyronine (T3) which are required for normal cellular metabolism.

- Release of T4 and T3 is controlled by thyroid stimulating hormone (TSH) produced by the anterior pituitary, which in turn is controlled by release of thyrotrophin releasing hormone (TRH) from the hypothalamus.

- Disorders of thyroid function can result in either excess or reduced secretion of thyroid hormones.

- Hyperthyroidism occurs due to increased release of thyroid hormones and produces the clinical features of thyrotoxicosis.

- Hyperthyroidism can be treated with anti-thyroid medication, radioiodine, or surgery to remove all or part of the thyroid gland.

- Hypothyroidism occurs due to deficiency of thyroid hormones and produces the clinical features of myxoedema.

- Hypothyroidism can be treated by thyroid hormone replacement, usually with T4 alone.

- Thyroid dysfunction can be investigated by measuring the concentration of serum TSH, free T4, and free T3.

FURTHER READING

- **Association for Clinical Biochemistry, British Thyroid Association and British Thyroid Foundation (July 2006)** *UK Guidelines for the Use of Thyroid Function Tests*. The Association for Clinical Biochemistry, British Thyroid Association, and British Thyroid Foundation. Available from the Association for Clinical Biochemistry.

- **Carson M (2009)** Assessment and management of patients with hypothyroidism. *Nursing Standard* **23**, 48–56.

- **Cooper DS (2003)** Hyperthyroidism. *Lancet* **362**, 459–68.

- **Cooper DS (2005)** Anti-thyroid drugs. *New England Journal of Medicine* **352**, 905–17.

- **Dayan CM (2001)** Interpretation of thyroid function tests. *Lancet* **357**, 619–24.

- **Kharlip J and Cooper DS (2009)** Recent developments in hyperthyroidism. *Lancet* **373**, 1930–2.

- **Roberts CG and Ladenson PW (2004)** Hypothyroidism. *Lancet* **363**, 793–803.

- **Shivaraj G, Prakash BD, Sonal V, Shruthi K, Vinayak H, and Avinash M (2009)** Thyroid function tests: a review. *European Review for Medical and Pharmacological Sciences* **13**, 341–9.

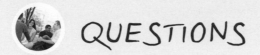

QUESTIONS

12.1 Which one of the following may cause hyperthyroidism?

(a) Graves' disease

(b) Hashimoto's thyroiditis

(c) Thyroidectomy

(d) Carbimazole

(e) Cushing's disease

12.2 The most common cause of primary hypothyroidism is:

(a) Graves' disease

(b) Hashimoto's thyroiditis

(c) Pituitary apoplexy

(d) Thyroid hormone replacement

(e) Cushing's disease

12.3 Patients with hypothyroidism may have a TSH result that is above the reference range.

(a) True

(b) False

12.4 Which of the following methods can be used to treat hyperthyroidism? (select all that apply)

(a) Carbimazole

(b) Thyroxine

(c) Surgery

(d) Insulin

(e) Radioiodine

12.5 Which of the following proteins is the main plasma binding protein for thyroid hormones?

(a) Transferrin

(b) Thyroxine binding pre-albumin

(c) Thyroxine binding globulin

(d) Caeruloplasmin

(e) Fibrinogen

12.6 Which of the following is a symptom of hypothyroidism?

(a) Heat intolerance

(b) Weight loss

(c) Cold intolerance

(d) Palpitations

(e) Increased bowel frequency

12.7 What is the most likely cause of the results below, obtained on a 25-year-old medical secretary who is on 100 µg of thyroxine per day (reference ranges are given in brackets)?

TSH: 7.7 mU/L (0.2–3.5)

Free T4: 25 pmol/L (9–23)

Answers to self-check questions, case study questions, and end-of-chapter questions are available in the Online Resource Centre accompanying this book.

 Go to www.oxfordtextbooks.co.uk/orc/ahmed/

13

Diabetes mellitus and hypoglycaemia

Allen Yates and Ian Laing

Learning objectives

After studying this chapter you should be able to:

- Describe the mechanism of glucose induced insulin secretion from the pancreatic β cell
- Describe the control of blood glucose concentration by insulin and by the counter-regulatory hormones glucagon, cortisol, adrenaline, and growth hormone
- Identify the target tissues of insulin action
- List the actions of the incretin hormones
- Define insulin resistance and the metabolic syndrome
- Describe the classical clinical features of and list the diagnostic criteria for diabetes mellitus
- Classify the different types of diabetes
- Identify and classify the medical emergencies of diabetic ketoacidosis, hyperosmolar hyperglycaemic syndrome, and hypoglycaemia
- List the long-term complications of diabetes
- Identify treatment strategies for diabetes

Introduction

Diabetes mellitus is caused by an absolute or functional deficiency of circulating insulin, resulting in an inability to transfer glucose from the bloodstream into the tissues where it is needed as fuel. Glucose builds up in the bloodstream (**hyperglycaemia**) but is absent in the tissues. The hyperglycaemia overwhelms the ability of the kidney to reabsorb the sugar as the blood is filtered to make urine. Excessive urine is made as the kidney loses the excess sugar. The body counteracts this by sending a signal to the brain to dilute the blood, which is translated into thirst, expressed by frequent fluid intake called **polydipsia**. As the body spills glucose into the urine, water is taken with it, increasing thirst and the frequency of urination (**polyuria**). Polydipsia and polyuria, along with weight loss (despite normal or increased food intake) and fatigue (essentially because ingested energy cannot get to the tissues where it is needed) are the classic symptoms of diabetes. One of the first to describe the disorder was the ancient Hindu surgeon and physician Susruta, around 600 BC, who described a condition 'brought on by a gluttonous overindulgence in rice, flour and sugar' in which the urine is 'like an elephant's in quantity'.

Key Points

Insulin deficiency can result from autoimmune attack and destruction of the insulin secreting tissue (type 1 diabetes), or from the gradual overstressing of the insulin secreting tissue, due to a diet too rich in carbohydrate and fat and a lack of exercise (type 2 diabetes). The World Health Organization (WHO) has defined it on the basis of laboratory measurements of glucose.

As expressed graphically in Figure 13.1 the WHO estimated that in 1995 the worldwide prevalence of diabetes was 30,000,000 people, in 2005 it was 217,000,000, and by the year 2030 it will be 366,000,000; a ten-fold increase in the world's diabetic population in just 30 years. This increase will be most prevalent in the developing world, in countries such as India and China. Worldwide someone dies from diabetes-related causes every ten seconds, during which time two other people will develop the condition. It ranks among the top three killer diseases along with coronary heart disease and cancer. The treatment of diabetes and its complications will have a significant impact on healthcare resources throughout the world for many years to come.

Figures for the UK are no less depressing. According to the WHO, there were 1.76 million diagnosed diabetics in the UK in 2000 and it is estimated there will be 2.67 million by 2030. The charity **Diabetes UK** estimates that there may be a further one million people with the condition who haven't been diagnosed. The UK national audit office calculated that from 1998 to 2008 the incidence of type 2 diabetes rose by 54%.

The already extensive economic burden diabetes puts on healthcare is set to rise further. In 2002 the first Wanless Report estimated the total annual cost of diabetes to the NHS to be £1.3 billion, with the total cost to the UK economy much higher. In 2004, Diabetes UK

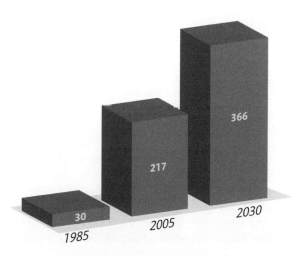

FIGURE 13.1
Global growth in diabetes (millions). See text for details.

estimated that diabetes accounted for around 5% of all NHS spending. Approximately one in every ten people treated in UK hospitals attend for treatment of diabetes and its complications. Consequently, regular attendance at diabetes and lipid clinics, and monitoring of cardiac and renal function generate a significant workload for the laboratory.

SELF-CHECK 13.1

What are the four classic symptoms of untreated diabetes?

13.1 The islets of Langerhans

The human pancreas contains small groups of easily recognizable, specialized, **endocrine** secretory cells surrounded by a sheath of collagen called the **islets of Langerhans** (Figure 13.2a), named after Paul Langerhans who first described them in 1869. The main cell type within the islet is the **beta (β) cell**, which secretes **insulin** and comprises over 80% of the islet mass (Figure 13.2b).

The **alpha (α) cell** secretes **glucagon**, the **delta (δ) cell** secretes **somatostatin**, and the **PP cell** secretes **pancreatic polypeptide**. Other cell types are also present in pancreatic islets, for example **ghrelin**-secreting cells, which are involved in **appetite** (eat) signalling. Somatostatin exerts an inhibitory paracrine effect on other islet endocrine cells, in addition to having several extra-islet actions. A definitive function for pancreatic polypeptide has yet to be uncovered.

13.2 Glucose-induced insulin secretion

Key Points

The β cells of the pancreatic islet act as glucose sensors; they secrete insulin in response to rising levels of glucose in the bloodstream and they reduce insulin output in response to falling glucose levels. This is known as stimulus-secretion coupling.

The mechanism of **glucose-induced insulin secretion** from the pancreatic β cell can be broken down into three main phases, namely transport and metabolism of glucose, metabolically

FIGURE 13.2
Section of human pancreas containing islets stained with: (a) Haemotoxylin/eosin. (b) Immunostained for insulin. Tissue section and photomicrograph are courtesy of Ms C Glennie and Dr G Howarth, Department of Histopathology, Manchester Royal Infirmary, UK.

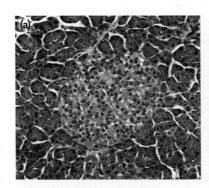
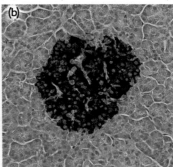

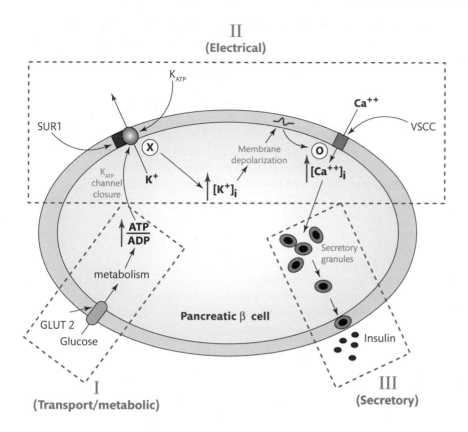

FIGURE 13.3
The three main phases of glucose-induced insulin secretion. See text for details.

generated changes in cellular ion flux, and finally the entry of calcium and the initiation of calcium-dependent insulin release. We can see these phases in Figure 13.3.

Glucose enters the cell via a membrane-bound **glucose transporter**, known as **GLUT 2** to initiate step 1 of the secretory mechanism. Once inside the cytosol, glucose is **phosphorylated** to **glucose 6-phosphate** by a **high K_m hexokinase** enzyme called **glucokinase** (or hexokinase IV). Unlike hexokinase I, II, and III, glucokinase has a high K_m for glucose ($K_m = 5.5$ mmol/L), in other words it has a much lower affinity for glucose and can thus 'sense' this hexose over its physiological range. We can see this in Figure 13.4.

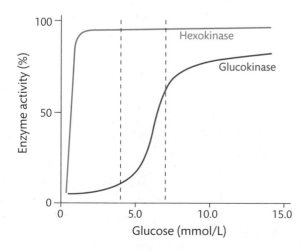

FIGURE 13.4
Glucokinase vs hexokinase activity. See text for details.

The expression and activity of glucokinase constitutes the rate-limiting step of stimulus secretion coupling and it has been called the β cell 'glucose-sensor'. Genetic mutations in the gene (**GK**) encoding this enzyme have been implicated in some types of diabetes as mentioned in Section 13.7.

Once phosphorylated, glucose enters the **glycolytic pathway** to produce **pyruvate**, which is further metabolized within the mitochondria. This in turn leads to an increase in generation of the secretory signals **NADH** and ATP.

These metabolically generated signals initiate step II of the process, which involves the closure of a membrane-bound potassium ion channel. Beta cells are excitable cells and under fasting conditions this channel is open and the membrane potential is maintained at around $-70mV$ by the relatively high intracellular to extracellular potassium gradient. The generation of ATP from glucose metabolism causes a rise in the cytosolic ATP/ADP ratio, which closes this nucleotide or ATP-sensitive potassium channel, the **K_{ATP} channel**. We can see this outlined in Figure 13.3. Closure of this channel depolarizes the membrane to around -40 to $-30mV$. This in turn opens a membrane potential-sensitive calcium channel in the plasma membrane, the **voltage-sensitive calcium channel** (VSCC). As we can see in Figure 13.3, opening of the VSCC facilitates the entry of extracellular calcium ions, allowing step III of the secretory process, calcium-induced, insulin secretion (Figures 13.3 and 13.5).

There are several operational ion channels in β cell membranes which could initiate depolarization in order to open calcium channels and generate calcium influx. Whilst the importance of the K_{ATP} channel cannot be understated, K_{ATP} channel-independent pathways for glucose-stimulated insulin release also exist. One alternative pathway is via the activation of the **volume-regulated anion channel** (VRAC), which is also operative in β cells. In terms of depolarization, the loss of anions, or negative charge (for example chloride ions via VRAC opening), or the build up of cations, or positive charge (potassium ions via K_{ATP}–closure) are slightly different routes to the same outcome.

The first response to the influx of calcium is the exocytotic release of insulin granules from stores close to the cell membrane (step III). This chemically primed pool of granules is called the **readily releasable pool**. Its size determines the magnitude of the first phase secretory response. For secretion to continue beyond this, granules must be mobilized from other stores within the β cell. Increasing the cytosolic Ca^{2+} concentration initiates first phase secretion. However, sustained insulin secretion can only be maintained if the cell is stimulated by metabolizable secretagogues such as glucose.

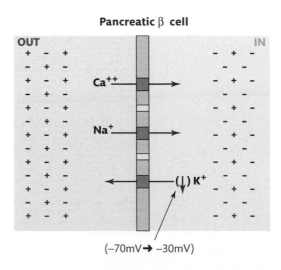

Pancreatic β cell

$(-70mV \rightarrow -30mV)$

FIGURE 13.5

Ionic movements across β cells. This shows the principal cationic fluxes in pancreatic beta cells, determined by normal ionic gradients across the plasma membrane. The negative resting membrane potential ($-70mV$) is a result of the relative greater outward K^+ current rather than the combined inward Na^+ and Ca^{2+} current. Depolarization (i.e. a shift to a more positive membrane potential) is achieved by reducing the K^+ current (via closure of K_{ATP} channels) upon stimulation of beta cells with glucose.

SELF-CHECK 13.2

What is the correct sequence of events in glucose-induced insulin secretion?

Other insulin secretagogues

Substrates other than glucose can also initiate insulin secretion. These include **leucine, ketoisocaproate**, and **methyl succinate**. Other agents, called **potentiators** of insulin secretion, have the ability to 'amplify' the effect of glucose on the β cell. Potentiators include some fatty acids, the amino acid **arginine**, and the **incretin hormones** (mentioned in Section 13.5). The sulphonylurea **tolbutamide**, a pharmaceutical used to treat type 2 diabetes, also has a direct stimulatory effect on β cells. It acts by binding to a **sulphonylurea receptor 1** (SUR1) found on the plasma membrane of the β cell. This receptor is intimately linked with the K_{ATP} channel (see Figure 13.3) such that when tolbutamide binds to its receptor, K_{ATP} channels close and the cell depolarizes. **Prandial glucose regulators** also stimulate insulin release via the K_{ATP} channel as described in Section 13.11.

Insulin processing

Insulin biosynthesis starts from the translation of a single chain 86 amino acid precursor, **preproinsulin**, from insulin mRNA, as we can see in Figure 13.6. As the molecule is inserted into the β cell endoplasmic reticulum the amino terminal signal peptide is cleaved to form **proinsulin**. In the endoplasmic reticulum the proinsulin is enzymically cleaved by several endopeptidase enzymes to give insulin and what was the connecting peptide, **c-peptide** (referred to as the C chain in Figure 13.6). Insulin consists of an aminoterminal B chain of 30 amino acids and a carboxyterminal A chain of 21 amino acids, which are connected by disulphide bridges occurring at cysteine residues in the protein. Insulin and c peptide are packaged in the Golgi apparatus into secretory granules. **Zinc** is also present in and released from the secretory granule. In a normal individual insulin and c-peptide are co-secreted in a molar ratio into the circulation. Both molecules are cleared from the bloodstream at different rates, resulting in differing insulin to c-peptide ratios in the blood. In patients with type 2 diabetes, incomplete processing of the proinsulin molecule in the secretory granule results in the release of various components of proinsulin (intact and split proinsulins) into the bloodstream. Detectable concentrations of proinsulins have been observed in these patients. The peptide hormone, **amylin**, is also co-secreted from the β cell. Amylin inhibits glucagon secretion, delays gastric emptying, and acts as a satiety signal to the brain.

13.3 **Glucagon secretion**

In normal metabolism the concentration of circulating glucose is regulated by a balance between the secretion of insulin and its opposing hormone, glucagon. Secreted from α cells of the pancreatic islet, glucagon is one of the **counter-regulatory hormones** in glucose homeostasis. Its secretion is influenced by a variety of different stimuli, including hormones, nutrients, and neurotransmitters. Glucose is a potent physiological regulator of α cell function. Insulin has been proposed as one of the main facilitators of glucose action on α cell activity, and α cells are known to express large numbers of insulin receptors. Alpha cells also have a K_{ATP} channel that is activated by zinc ions, which reduces glucagon secretion. Hyperglycaemia rapidly suppresses glucagon release, whereas low blood glucose (**hypoglycaemia**) rapidly facilitates glucagon secretion.

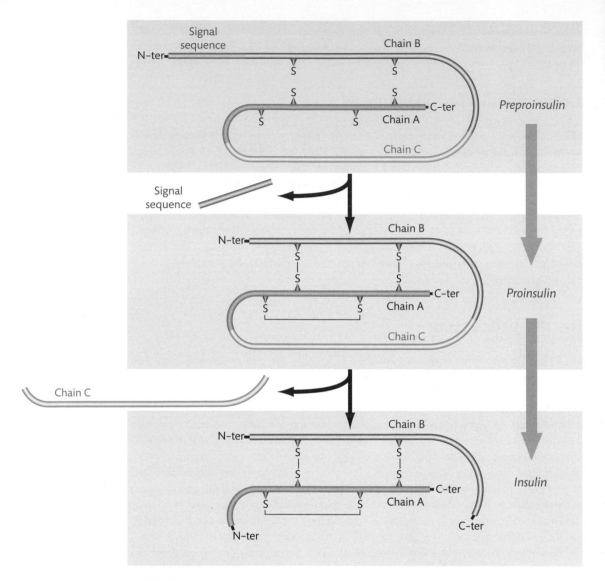

FIGURE 13.6
The steps involved in biosynthesis of insulin. See text for details. This figure is reproduced with kind permission from the Beta Cell Biology Consortium (www.betacell.org), funded by NIDDK U01DK072473.

Amino acids, such as arginine, **non-esterified fatty acids** (NEFAs), and **ketones**, suppress glucagon secretion, as do insulin, zinc, and somatostatin. Stress hormones, such as adrenaline and activation of the autonomic nervous system stimulate glucagon release. These effects are especially pronounced during periods of stress, for example during hypoglycaemia, hypoxia, and hypothermia, where readily available access to metabolic fuel can protect against potentially life-threatening situations. In healthy subjects, glucagon secretion is stimulated by a high protein meal but inhibited by those rich in carbohydrate, or by oral glucose, helping to maintain blood glucose within the normal physiological range.

In diabetes there is a relative glucagon hypersecretion at normal and increased glucose concentrations and impaired responses to hypoglycaemia, resulting in a deterioration in glucose-sensing by the α cell.

SELF-CHECK 13.3

Which of the following will suppress glucagon secretion: glucose, NEFA, catecholamines, or insulin?

13.4 Insulin, glucagon, and the counter-regulatory hormones

The main target organs for the glucose regulatory hormones are the liver, muscle, and adipose tissues.

Key Points

Insulin is an anabolic hormone; its main function is to clear the bloodstream of post-prandial glucose and transfer it into the tissues where it can be used as fuel.

If we look at Figure 13.7 we can see that binding of insulin to its receptor on target tissues first triggers autophosphorylation of the receptor and subsequent initiation of intracellular signalling cascades. Recruitment of glucose transporters (**GLUT 4 transporters**) from intracellular stores to the cell membrane, facilitating increased glucose uptake, is a key consequence of this activation. Insulin post-receptor binding also stimulates the conversion of the transported glucose into suitable storage products, namely an increase in glycogen synthesis and an increase in glycolysis and fatty acid synthesis.

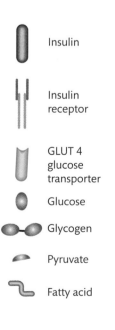

Insulin

Insulin receptor

GLUT 4 glucose transporter

Glucose

Glycogen

Pyruvate

Fatty acid

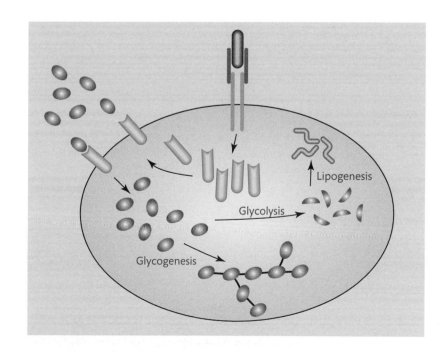

FIGURE 13.7
Insulin activation of target cell. See text for details.

TABLE 13.1 Metabolic effects of insulin.

Liver	Muscle	Adipocyte
Glycogenolysis $\downarrow\downarrow$	Glucose uptake $\uparrow\uparrow\uparrow$	Glucose uptake $\uparrow\uparrow\uparrow$
Gluconeogenesis $\downarrow\downarrow$	Ketone metabolism $\uparrow$	Lipolysis $\downarrow\downarrow\downarrow$
Ketogenesis $\downarrow$		

Key: $\uparrow$ = stimulatory, $\downarrow$ = inhibitory.

Reproduced with kind permission from Smith J and Nattrass M (2004) *Diabetes and Laboratory Medicine*. Marshall W and Horner J eds. London: ACB Venture Publications.

As shown in Table 13.1, insulin inhibits the breakdown of glycogen to glucose (**glycogenolysis**) and inhibits both the formation of glucose from non-carbohydrate sources (**gluconeogenesis**) in the liver and the breakdown of lipids to NEFA (**lipolysis**) in adipocytes.

Glucagon and the other counter-regulatory hormones, **cortisol**, **adrenaline**, and **growth hormone**, are **catabolic** hormones which when stimulated by low blood sugar oppose the actions of insulin, raising the concentration of glucose in the blood and preventing the brain being starved of fuel (Table 13.2). Glycogenolysis and gluconeogenesis in the liver are promoted and hepatic glucose uptake is greatly reduced. Lipolysis is stimulated in the adipocyte, as is the production of ketones (**ketogenesis**) in the liver.

Key Points

The net result of the balancing act between anabolic and catabolic hormones is an optimal blood glucose level, ensuring a continual supply of glucose to the brain with few, if any, interruptions.

TABLE 13.2 Metabolic effects of catabolic hormones.

	Catecholamines	Glucagon	Cortisol	Growth hormone
Liver				
Glycogenolysis	$\uparrow\uparrow\uparrow$	$\uparrow\uparrow$	None	None
Gluconeogenesis	$\uparrow$	$\uparrow\uparrow\uparrow$	$\uparrow\uparrow\uparrow$	$\uparrow$
Ketogenesis	$\uparrow$	$\uparrow\uparrow$	$\uparrow$	$\uparrow$
Muscle				
Glucose uptake	$\downarrow$	None	$\downarrow\downarrow$	$\downarrow$
Ketone metabolism	$\downarrow$	?	?	?
Adipocyte				
Glucose uptake	$\downarrow$	None	$\downarrow\downarrow$	$\downarrow$
Lipolysis	$\uparrow\uparrow\uparrow$	None	$\uparrow\uparrow$	$\uparrow$

Key: $\uparrow$ = stimulatory, $\downarrow$ = inhibitory, None = no effect, ? = uncertain.

Reproduced with kind permission from Smith J and Nattrass M (2004) *Diabetes and Laboratory Medicine*. Marshall W and Horner J eds. London: ACB Venture Publications.

CASE STUDY 13.1

A 22-year-old mother of two was found unconscious at 5 am by her sister and brought to the hospital in a coma. Her blood glucose was 0.9 mmol/L. She 'woke up' after intravenous glucose infusion, was admitted to a ward, and maintained on a glucose/saline drip. Investigations were arranged and the following evening she had a grand mal epileptic seizure, which was treated with intravenous diazepam. At this time her blood glucose was 1.1 mmol/L. She then inhaled her own vomit and was transferred to intensive care, where she died 48 hours later.

The results of her investigations arrived after she died and were:

Urine drug screen	negative
Serum cortisol on admission	<30 nmol/L (150–700)

A post-mortem revealed small, withered adrenal glands and lymphocytic infiltration of the pituitary gland.

(a) Why did this patient present with a hypoglycaemic coma?

(b) What is the diagnosis?

SELF-CHECK 13.4

What effect does insulin have on gluconeogenesis and glycogenolysis?

13.5 Incretin hormones

In 1964, teams headed by Elrick and McIntyre noted that when glucose was given orally it stimulated the release of three times more insulin than the same amount of glucose given intravenously. This was called the incretin effect, using a term that had been introduced by La Barre in 1932 to describe hormonal activity deriving from the gut that was able to increase the endocrine secretion from the pancreas. We now know that this is due to the release of the incretin hormones, **glucagon-like peptide 1** (GLP-1) and **glucose-dependent insulinotropic peptide** (GIP).

Key Points

The incretin hormones are released from the small intestine in response to oral glucose or a mixed meal. Their effect on the endocrine pancreas brings about the three-fold amplification of glucose-induced insulin release.

Glucose-dependent insulinotropic peptide is released from K cells in the duodenum and jejunum. The oral ingestion of food stimulates a twenty-fold increase in blood GIP levels. Glucose-dependent insulinotropic peptide binds to its receptor on β cells and triggers a cAMP-mediated rise in insulin secretion. L cells in the mucosa of the ileum and colon secrete

GLP-1, and its secretion rises three-fold after the ingestion of food. Glucagon-like peptide 1 is one of the most potent insulin secretagogues known and its role is to stimulate insulin release, especially the first phase of insulin release, in response to a meal. This effect is glucose dependent such that endogenous GLP-1 secretion alone does not cause hypoglycaemic episodes. Glucagon-like peptide 1 also acts by binding to its receptor, a **G-protein** coupled receptor, **glucagon-like peptide 1 receptor** (GLP1-R), found in the brain, lung, islets, stomach, hypothalamus, heart, intestine, and kidney. In the pancreatic β cell GLP-1 receptor binding leads to a cAMP-mediated rise in insulin secretion. The incretin hormones also promote insulin gene expression and insulin biosynthesis; they prolong β cell survival, reduce apoptosis and stimulate proliferation and differentiation of new β cells. Glucagon-like peptide 1 suppresses glucagon secretion (possibly via the β cell K_{ATP} channel) in a glucose-dependent manner, meaning that the counter-regulatory effect of hypoglycaemia on glucagon release is unaffected. Extra-pancreatic effects of GLP-1 include slowing down gastric emptying, delaying nutrient delivery to the small intestine, and reducing rapid post-prandial glucose excursions. Glucagon-like peptide 1 also activates satiety regulating areas in the brain, reducing food intake.

GLP-1 metabolism

Glucagon-like peptide 1 is rapidly metabolized, with a half-life in the circulation of less than two minutes. The hormone is catabolized by **dipeptidyl peptidase IV** (DPP-IV), a membrane-bound enzyme found in capillary endothelia of the kidneys and intestine. It catalyses the hydrolysis of GLP-1 from its active form (called **GLP-1 (7-36) amide**) to its inactive form (called **GLP-1 (9-36) amide**). The inactive GLP-1 (9-36) may act as a GLP-1 receptor blocker at the β cell, further reducing the activity of GLP-1 (7-36) amide. Glucose-dependent insulinotropic peptide is also metabolized by DPP-IV, with a half-life of around seven minutes in the circulation, and the inactive metabolite, GIP (3-42) amide, may also act as an antagonist at the GIP receptor. Another enzyme, **neutral endopeptidase** (NEP), can also break down GLP-1. This enzyme is found mainly in the kidney and is thought to be involved in the renal clearance of active GLP-1 from the circulation.

13.6 Impaired glucose and lipid handling

Glucose levels in the blood are normally regulated within very tight margins in normal healthy individuals. Breakdown in this regulation can be due to disease processes, such as diabetes, cancer, or a range of associated complaints outlined in Section 13.7, or by the introduction, sometimes inappropriately, of external substances which can change the blood glucose concentration. The resulting loss of glucose homeostasis can result in inappropriately high (hyperglycaemia) or inappropriately low (hypoglycaemia) blood glucose concentrations.

Hypoglycaemia can have several causes and is described in detail in Section 13.13 of this chapter but is generally defined as being a fasting blood glucose concentration of 2.5 mmol/L or below in a symptomatic patient.

The rest of this section is concerned with the development of hyperglycaemia, the reasons for it and the metabolic conditions which are a consequence of it.

Key Points

Metabolic hyperglycaemia arises from a combination of a reduction in the efficiency with which insulin can move glucose into tissues and by a reduction in the number of functioning β cells. This results in a surplus of glucose in the bloodstream.

Normal fasting glycaemia is quantified as a blood glucose concentration greater than 4.5 mmol/L and less than 5.2 mmol/L in a normal healthy adult after an overnight fast. As defined by the WHO, **impaired fasting glycaemia** (IFG) is defined in individuals with a fasting plasma glucose concentration higher than 6.0 mmol/L, but below 7.0 mmol/L, the diagnostic cut-off for the diagnosis of diabetes. **Impaired glucose tolerance** (IGT) is a state of impaired glucose regulation, diagnosed during **oral glucose tolerance testing** (i.e. after a 75 g oral glucose load, see Box 13.1), and is defined as a two-hour post-glucose load plasma glucose level of greater than 7.8 mmol/L and less than 11.1 mmol/L, with a non-diabetic (i.e. less than 7.0 mmol/L) fasting glucose level.

The WHO definitions for diabetes and intermediate hyperglycaemia are outlined in Box 13.2.

Blood glucose measurements for the diagnostic purposes outlined above and for the diagnosis of diabetes should only be done on approved *clinical laboratory* based analysers. Portable point of care glucose meters, such as those used by patients with diabetes are not as accurate and are more prone to operator error, variation in storage conditions, and age of equipment.

Individuals classified as IFG or IGT have worse glucose control than normal individuals, but not as severe as in diabetes. They carry a higher risk for diabetes and cardiovascular disease than in the normal state.

Insulin resistance and the metabolic syndrome

To understand how diabetes develops we need to understand the concepts of beta cell function and insulin resistance. Beta cell function is a quantitative measure of the ability of the endocrine pancreas to secrete insulin (itself a measure of total beta cell mass and beta cell

BOX 13.1 The oral glucose tolerance test

Patients are fasted for 8–14 hours (i.e. from the evening before) prior to the test, although water is allowed. The oral glucose tolerance test is normally scheduled for the morning (glucose tolerance exhibits diurnal variation). A baseline or zero time blood sample is drawn just before a drink containing 75 g of glucose is given, which should be consumed within five minutes. Blood is then drawn at timed intervals for the measurement of blood glucose and sometimes other analytes, for example insulin levels. The number of samples and the sampling interval, for example every 30 minutes, can vary according to the purpose of the test. For simple diabetes screening the most important sample is the two-hour sample, and this and the baseline sample may be the only bloods taken.

BOX 13.2 WHO definitions of diabetes and intermediate hyperglycaemia

Diabetes

Fasting plasma glucose	≥ 7.0 mmol/L (126 mg/dL) or
Two-hour plasma glucose*1	≥ 11.1 mmol/L (200 mg/dL)

Impaired glucose tolerance (IGT)

Fasting plasma glucose	≤ 7.0 mmol/L (126 mg/dL) and
Two-hour plasma glucose*1	≥ 7.8 and <11.1 mmol/L (140 mg/dL and <200 mg/dL)

Impaired fasting glycaemia (IFG)

Fasting plasma glucose	6.1 mmol/L to 6.9 mmol/L (110 mg/dL to 125 mg/dL) and, if measured
Two-hour plasma glucose*1,2	<7.8 mmol/L (140 mg/dL)

*1 Venous plasma glucose two hours after ingestion of 75 g oral glucose load.
*2 If two-hour plasma glucose is not measured, status is uncertain as diabetes or IGT cannot be excluded.

Data from *Definition and Diagnosis of Diabetes Mellitus and Intermediate Hyperglycaemia*. Report of a WHO/IDF consultation, 2006.

glucose sensitivity). Insulin sensitivity is a measure of insulin action at its target tissues, that is, how efficient insulin is at getting extracellular glucose into its target tissues. The reciprocal measure of insulin sensitivity is insulin resistance, so that as sensitivity falls, insulin resistance rises. Glucose homeostasis is thus a balance between insulin resistance and beta cell function, outlined in Figure 13.8. A person with significant insulin resistance will have normal glucose homeostasis if they have a large beta cell capacity, whereas someone with low beta cell function will have normal glucose homeostasis only if they have low insulin resistance.

Key Points

Diabetes results when the beta cell function is insufficient to overcome the insulin resistance. In type 1 diabetes beta cell function is destroyed. In type 2 diabetes, beta cell function cannot overcome the insulin resistance.

A simple way of assessing beta cell function and insulin resistance is to model fasting blood glucose and insulin values in an algorithm designed to take into account tissue glucose utilization and pancreatic beta cell function during steady-state (fasting) conditions. This model was developed in Oxford and is called **homeostasis model assessment** (HOMA). The two derived variables are HOMA-B, an estimate of pancreatic β cell function; and HOMA-IR, an estimate of tissue insulin resistance (this is the reciprocal of tissue insulin sensitivity, HOMA-S).

I. Normal

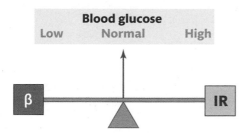

II. Developing insulin resistance

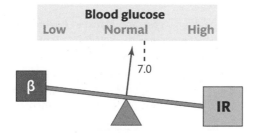

III. Compensatory increase in β cell output

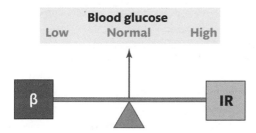

IV. β cell exhaustion/depletion and type 2 diabetes

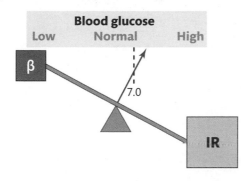

FIGURE 13.8

Development of insulin resistance (IR) in type 2 diabetes. See text for details.

Using this method a young, fit, and healthy individual will have a HOMA-B of 100% and a HOMA-IR of 1.0.

Figure 13.9 outlines the gradual decline in β cell function with age in an unselected healthy population. There is a slow but significant decrease in HOMA-B of about 1% per year. If we extrapolate this we find that half the population will have diabetes by an age of about 120.

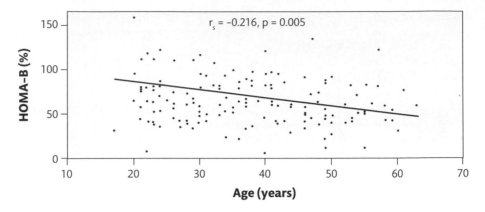

FIGURE 13.9
Declining β cell function (HOMA-B) with age in a healthy population. The r_s = Spearman rank correlation for HOMA-B versus age.

This decline is accompanied by a subtle rise in fasting glucose and **glycated haemoglobin**, (HbA$_{1C}$), in the blood. Newly presenting type 2 diabetes patients have a HOMA-IR of approximately 2 and a HOMA-B of approximately 40%. In order to maintain normal glucose homeostasis the patient with diabetes needs a HOMA-B of 250%. Young women with the insulin-resistant form of polycystic ovary syndrome are as insulin resistant as newly presenting middle-aged patients with type 2 diabetes, but they have high β cell activity and hence normal glucose homeostasis.

Key Points

Declining β cell function and increasing insulin resistance are equally important causative factors in the development of type 2 diabetes.

The clustering of several metabolic abnormalities, including obesity, insulin resistance, dyslipidaemia, and hypertension, all of which are highly predictive for the development of cardiovascular disease, and type 2 diabetes make up the **metabolic syndrome**. The key features of the metabolic syndrome are listed in the International Diabetes Federation consensus definition outlined in Box 13.3.

BOX 13.3 The International Diabetes Federation (IDF) consensus definition of the metabolic syndrome

According to the new IDF definition, for a person to be defined as having the metabolic syndrome they must have:

Central obesity (defined as waist circumference ≥94 cm for Europid men (≥90 cm for South Asian men) and ≥80 cm for Europid and South Asian women

plus any two of the following four factors:

- raised triacylglycerol level: ≥1.7 mmol/L (150 mg/dL) or specific treatment for this lipid abnormality
- reduced HDL cholesterol: <1.03 mmol/L (40 mg/dL) in males and <1.29 mmol/L (50 mg/dL) in females, or specific treatment for this lipid abnormality

- raised blood pressure: systolic BP ≥130 or diastolic BP ≥85 mm Hg, or treatment of previously diagnosed hypertension
- raised fasting plasma glucose (FPG) ≥5.6 mmol/L (100 mg/dL) or previously diagnosed type 2 diabetes
- If above 5.6 mmol/L or 100 mg/dL, OGTT is strongly recommended but is not necessary to define presence of the syndrome
- Reproduced with kind permission from the International Diabetes Federation, The IDF consensus worldwide definition of the metabolic syndrome, 2006

Insulin has many actions, not just on glucose homeostasis. Insulin signalling is involved in correct lipid and lipoprotein metabolism and in blood pressure regulation (see below).

Mechanisms of beta cell loss

Loss of β cell function is a key process in the development of type 2 diabetes. The maintenance of β cell mass is a balance between the development of new cells derived from stem cells and their death from apoptosis and necrosis. Evidence suggests that pancreatic islets can differentiate from pancreatic ductal cells in the adult as well as in the embryo. Autopsies of islets from patients with type 2 diabetes indicate that accelerated apoptosis may be the main mechanism of beta cell loss. Inflammatory cytokines (see below) can induce beta cell apoptosis.

Dysregulated NEFA metabolism may also promote beta cell apoptosis. In the short term, β cells can use NEFA as an energy substrate for the ATP generation which drives insulin release. However, with chronic exposure of islets to high NEFA levels this property is lost. Furthermore, glucose stimulated insulin release is reduced when NEFAs are chronically elevated. Chronic high glucose levels exhibit toxicity to beta cells, compromising insulin release and contributing to β cell loss. It is thought that a build up of glucose and/or NEFA in β cells causes a backlog in normal metabolic pathways such that glucose metabolism does not produce the metabolic signals for insulin secretion as efficiently as it should. In addition, excess glucose can be shunted down the glyoxalase pathway to produce toxic metabolites such as methylglyoxal, further contributing to beta cell dysfunction and apoptosis.

Transgenic studies with mice that have had the pancreatic beta cell insulin receptor deleted show a defect in glucose handling similar to that in type 2 diabetes, indicating that insulin itself has a role in maintaining beta cell function.

Incretins, insulin resistance, and type 2 diabetes

Incretin hormones are secreted in the intestine in response to ingested glucose and carbohydrate. They act on the pancreas to increase the release of insulin in response to glucose, and they also send signals to the brain which are involved in satiety signalling. The incretin response is almost completely lost in type 2 diabetes. This is mainly due to reduced beta cell function and mass (see above), but deficient secretion of incretins may also contribute to this pathophysiology. The release of GLP-1 is decreased in type 2 diabetes. Consequently, treatment with GLP-1 may be of some benefit to patients with type 2 diabetes. The infusion or subcutaneous injection of GLP-1 does improve blood glucose control, promotes weight loss, and improves insulin sensitivity and insulin secretion. The insulinotropic actions of GLP-1 are well preserved in type 2 diabetes, making it a good therapeutic target.

The release of glucagon in response to normal or high glucose levels can be suppressed by GLP-1. This does not happen during hypoglycaemia, making GLP-1 an efficient α cell sensitizer. In patients with type 2 diabetes GLP-1 directly inhibits glucagon release independently of any other effects on insulin release or gastric emptying. This suppression is at least as effective in diabetes as in health. In contrast, nutrient-stimulated GIP secretion is relatively normal in type 2 diabetic patients, but its insulinotropic action is significantly impaired. The resistance of β cells to GIP severely limits its potential as a therapeutic agent in the treatment of diabetes. However, recent research has identified GIP receptors outside the pancreas and gastrointestinal tract, most notably on adipocytes. Glucose-dependent insulinotropic peptide is secreted strongly in response to fat ingestion and may have a role in the translation of excessive amounts of dietary fat into adipocyte stores. This has opened the possibility of using GIP receptor antagonists for the treatment of obesity, insulin resistance, and diabetes.

SELF-CHECK 13.5

What effect does the incretin hormone GLP-1 have on appetite and on glucagon secretion?

Impaired lipoprotein regulation

The main lipoprotein abnormalities in type 2 diabetes are raised **triacylglycerols** and reduced **high density lipoprotein** (HDL) **cholesterol** in the bloodstream. Excess plasma triacylglycerol (**hypertriglyceridaemia**) is linked to an increased incidence of coronary artery disease. High density lipoprotein cholesterol is known to be cardioprotective, therefore low levels are associated with an increased cardiovascular risk.

Cross reference

Chapter 9 Abnormalities of lipid metabolism

During fasting triacylglycerols are secreted by the liver in **very low density lipoprotein particles** (VLDL), a process facilitated by **microsomal triacylglycerol transfer protein** (MTTP) in a rate-limiting step that is negatively regulated by insulin. Very low density lipoprotein transports cholesterol and triacylglycerol from the liver to peripheral tissues. Triacylglycerol is removed in target tissues as it circulates, and is stored, particularly by adipose tissue. The removal of triacylglycerols converts the VLDL to **low density lipoprotein** (LDL) particles. This contains most of the circulating cholesterol, which is then taken up by peripheral cells in a regulated manner by their surface LDL receptors. **Lipoprotein lipase**, the enzyme which hydrolyses lipids in lipoproteins, is upregulated by insulin. Thus insulin resistance causes triacylglycerols to rise by increased secretion of VLDL and delayed clearance.

High density lipoprotein cholesterol is involved in the transport of cholesterol from peripheral tissues to the liver. The main lipoprotein in HDL is **apolipoprotein** (apoA1). Nascent apoA1 is secreted by the liver and combines with other lipoproteins. It takes up cholesterol from peripheral tissue and other lipoproteins and the mature HDL is then taken up by a specific liver receptor. Insulin is involved in apoA1 production by the liver and so influences circulating HDL concentrations.

Non-esterified fatty acids are a source of energy for aerobic respiration. Circulating NEFAs are released from adipocytes following hydrolysis of triacylglycerol. The controlling enzyme for this process is called **hormone sensitive lipase** (HSL), which is negatively regulated by insulin. Normally, when blood glucose is low, insulin levels are also low and adipose tissue releases NEFAs. After a meal, NEFA release from adipose tissue is suppressed by the increased insulin release, so NEFA levels fall. As obesity develops, the total amount of adipose tissue increases and the ability of insulin to decrease circulating NEFAs is impaired. Obese individuals have high fasting and post-prandial NEFA levels. Non-esterified fatty acids are also taken up by the liver and incorporated into triacylglycerols, promoting VLDL secretion.

Non-esterified fatty acids are substrates for aerobic oxidation by metabolism to acetyl coenzyme A. This is further metabolized in the TCA cycle to yield ATP. High levels of NEFA inhibit insulin-mediated glucose oxidation in striated muscle, thus contributing to insulin resistance.

Key Points

In obese individuals, triacylglycerols can also be deposited in non-adipose tissues, particularly in the liver and striated muscle. This also inhibits insulin action and further contributes to the insulin resistance.

SELF-CHECK 13.6

What effect does hormone sensitive lipase have on serum NEFA concentration?

Hypertension, oxidative stress, and inflammation

Hypertension (high blood pressure) is a component of the metabolic syndrome and is common in type 2 diabetes. The vascular endothelium (the layer of cells lining the inside of the blood vessels) is important in circulatory homeostasis, where it controls the contraction of arterial smooth muscle. Blood pressure is maintained by a series of control mechanisms balancing contraction and relaxation of the smooth muscle in the walls of small arteries (arterioles). In both the metabolic syndrome and type 2 diabetes, the ability of the endothelium to induce arteriolar dilation is reduced. A contributing factor to this may be oxidative stress.

Increased arterial blood flow affects shear receptors on the endothelial wall, which induce the production of **nitric oxide** (NO). Nitric oxide stimulates the arterial smooth muscle to relax, thus reducing blood pressure. **Reactive oxygen species** such as oxygen ions, peroxides, and free radicals react with NO to produce **peroxynitrite** ($ONOO^-$), a highly reactive, toxic species, which consumes NO, reducing its availability. Peroxynitrite also nitrates a wide range of proteins, altering their functions. Endothelial function is further impaired by LDL oxidation.

Chronic low-grade inflammation is seen in both the metabolic syndrome and type 2 diabetes and is thought to increase cardiovascular risk. Levels of **C-reactive protein** (CRP), **amyloid A**, and other inflammatory markers are increased, as are levels of circulating inflammatory cytokines. As the adipose tissue mass increases, it is invaded by macrophages which secrete **interleukin 6** (IL-6), which in turn triggers the release of CRP from the liver. **Tumour necrosis factor** α (TNFα) is also produced by macrophages but most of its action is local within the adipose tissue itself.

Key Points

Both IL-6 and TNFα act on adipocytes and other cell types to inhibit insulin action. This adipocyte insulin resistance compromises the controlling action of insulin on triacylglycerol uptake and NEFA release, further contributing to dyslipidaemia.

13.7 Diagnosis, classification, and aetiology of diabetes

The WHO criteria for the diagnosis of diabetes have recently been reviewed in 2006 (see Box 13.2) and have been adopted for use in the UK. A person can be diagnosed as having diabetes if they exhibit clinical symptoms of the disease: thirst, polyuria, fatigue, weight loss, and have a fasting plasma venous glucose level, measured in an accredited laboratory, greater than 7.0 mmol/L, and/or a two-hour oral glucose tolerance test (as described in Section 13.6) plasma glucose level greater than or equal to 11.1 mmol/L, or if they have a random plasma glucose higher than 11.1 mmol/L. In the UK this must normally be confirmed on at least two different occasions before a definitive diagnosis is made.

Type 1 diabetes

Formerly known as insulin-dependent diabetes or child (youth) onset diabetes, type 1 diabetes is characterized by a complete lack of endogenous insulin due to the autoimmune destruction of the pancreatic β cells. Type 1 accounts for approximately 10% of all diabetes and most patients present with it before the age of 40, with a peak incidence at around 9–13 years of age. Disease onset is usually acute and the initiation of autoimmune destruction has been linked to exposure to certain infectious triggers (notably viral, for example coxsackie B, flu, rubella), or environmental triggers (for example nitrosamines used in smoking meat and fish) in genetically susceptible individuals. Type 1 diabetes has a strong association with components of the **major histocompatability complex** (MHC), notably the **human leukocyte antigens** (HLA), and 95% of type 1 diabetic patients express either **HLA DR3** or **HLA DR4** antigens. Pancreases from patients with type 1 diabetes show lymphocytic infiltration and almost complete destruction of β cells. Almost 85% of type 1 diabetes patients have circulating islet cell antibodies, most of which are directed against β cell **glutamic acid decarboxylase** (GAD). Treatment is achieved by lifelong injections of insulin.

Type 2 diabetes

Type 2 diabetes is a term used for diabetes in older people whose glucose homeostasis is abnormal but who do not have the dramatic presentation of the disease seen in type 1 diabetes. Type 2 diabetes is an advanced stage of a disease process starting in early adult life (and, more frequently, in childhood) which becomes manifest in middle age. Although not strictly defined in genetic terms there is nonetheless a genetic predisposition to the condition, such that if both parents are affected the lifetime risk of an individual for type 2 diabetes is increased to about 60%. In type I diabetes the concordance risk for identical twins is 40%. Thus both types 1 and 2 have genetic and environmental components.

Studies involving the Pima Indian tribe in Arizona highlight the importance of the environment. Members of the tribe adopt one of two lifestyles: one urban, the other agricultural. The urban dwellers are markedly obese, with a diabetes incidence of 50%, whereas the incidence of diabetes in the slimmer, more active, agricultural group is less than 10%.

Key Points

The environmental components associated with type 2 diabetes are chronic over-nutrition and lack of exercise.

Weight gain and related metabolic changes lead to the development of insulin resistance and eventually diabetes. In the early twentieth century, even in developed countries, most people had limited access to food. Nowadays in developed countries most people can eat what they want, not just what they can get. This means that type 2 diabetes is becoming more common in younger people. Modern diets contain less carbohydrate and more fat, particularly saturated fatty acids. This is illustrated by the fact that in Britain in World War 1 the average height of a male army recruit was five feet four inches, some six inches shorter than present day (early twenty-first century) standards. In 1900 the average age at menarche (the onset of menstruation), which is critically dependent on nutritional status, was 16 whereas it is now under 13.

Obesity can be evaluated as the **body mass index** (BMI). This is calculated as weight in kilograms divided by height in metres squared (kg/m^2). A normal BMI is 20–25 kg/m^2. People with a BMI in the range of 25–30 kg/m^2 are described as being overweight, whilst a BMI higher than 30 kg/m^2 is regarded as obese. A BMI higher than 40 kg/m^2 is regarded as morbidly obese. Diabetes risk increases with rising BMI. The risk curve starts to rise at a BMI of 22.5 kg/m^2, which is normal, and from 25–30 kg/m^2 the risk doubles. Subcutaneous fat such as that found on the thighs and buttocks seems less harmful than abdominal fat accumulating around the waist, which is accompanied by fat deposition within the abdomen (**omental** fat). Abdominal obesity is often known as **android** obesity, and tends to occur in middle-aged men who become more sedentary, especially in those who overindulge in 'fatty' foods and alcohol, easily observed in the UK as a 'beer belly'. This can be easily assessed in clinical practice by measuring waist circumference and comparing it to hip size (**waist hip ratio**, WHR). In women, who typically accumulate fat on the thighs and buttocks (termed **gynoid** obesity), this becomes more evident after the menopause due to redistribution of adipose tissue.

SELF-CHECK 13.7

What is the predominant abnormality in type 2 diabetes mellitus?

Diabetes in pregnancy

Diabetes during pregnancy, known as **gestational diabetes mellitus** (GDM), affects about 4–5% of pregnancies. It has varying severity with an onset, or at least is first detected, during pregnancy. In most women it presents during the second or third trimester and probably occurs because the body cannot produce enough insulin to meet the extra demands of pregnancy.

CASE STUDY 13.2

A 59-year-old man saw his doctor because he was feeling tired and lethargic. He used to be physically active and involved in several different sports. However, for the previous six years he had stopped playing sport and had since gained three stone in weight and had a body mass index of 32 kg/m^2 (normal 20–25). On clinical examination he had a raised blood pressure and laboratory tests gave the following results (reference ranges are given in brackets):

Blood glucose (fasting)	6.2 mmol/L	(3–6)
Blood glucose 2 hours after an OGTT	12.5 mmol/L	(<7.8)
Total cholesterol	5.2 mmol/L	(<5.0)
Fasting triacylglycerols	3.5 mmol/L	(0.8–2.2)
HDL cholesterol	1.0 mmol/L	(>1.2)

(a) What is the diagnosis?

(b) What is the cause of these results?

(c) How should he be treated?

In some women, however, GDM can be found during the first trimester of pregnancy. In some of these women, it is likely that the condition existed before the pregnancy. Pregnancy puts a stress on glucose homeostasis in all women, not just in those with diabetes, so there is currently some debate as to how high the blood sugar needs to be before making a diagnosis of GDM. Some clinicians prefer to use a lower level of glucose in a random sample, above 9.0 mmol/L rather than above 11.1 mmol/L, but keeping the fasting plasma glucose cut-off at 7.0 mmol/L or above. Gestational diabetes mellitus usually appears during the second trimester, by which time the baby's major organs are well developed so the risk to the baby from a GDM mother is less than to a baby from a mother with type 1 or 2 diabetes, as this would have been present from the beginning of the pregnancy.

Gestational diabetes usually resolves after the birth of the baby; however, women with GDM have a greater risk (30%) of developing type 2 diabetes than the general population (10%). Women from ethnic groups that show a higher rate for type 2 diabetes (south Asian, Afro-Caribbean) are more likely to develop type 2 diabetes if they have had GDM.

Monogenic forms of diabetes

Diabetes arising from single gene defects (**monogenic**) account for less than 1% of all cases. Diabetes diagnosed before the age of six months can arise from mutations in genes that encode the K_{ATP} channel or the SUR1 sulphonylurea receptor. Glycaemic control can normally be achieved by treatment with high-dose sulphonylureas rather than insulin. In young patients with stable, mild, fasting hyperglycaemia, glucokinase gene mutations should be considered. These patients might not need specific treatment. Familial, young-onset diabetes not typical of type 1 or type 2 diabetes can be due to mutations in the transcription factors **hepatocyte nuclear factor** 1-α (HNF-1α), hepatocyte nuclear factor 4-α (HNF-4α), and hepatocyte nuclear factor 1-β (HNF-1β). These relatively mild conditions have been termed **maturity onset diabetes in the young** (MODY). Hepatocyte nuclear factor 4-α mutations give rise to MODY1, which involves dysregulated function of several processes including GLUT 2 glucose transport and lipid metabolism. Hepatocyte nuclear factor 1-α mutations cause MODY3, the most common form of MODY in western and Asian countries, which results in impaired insulin secretion. Maturity onset diabetes in the young is caused by mutations in the HNF-1β gene which precipitate early-age diabetes associated with renal disease. Patients with these mutations can often be treated with low-dose sulphonylureas, occasionally with insulin. Mitochondrial DNA mutations can cause diabetes accompanied by deafness, which usually requires insulin treatment.

Secondary diabetes

It is well recognized that diabetes can result from the consequences of other primary disorders or conditions, which are outlined in Table 13.3. In some countries severe malnutrition in children, with chronic lack of protein, can cause severe insulin-requiring diabetes. General pancreatic disease such as pancreatitis, haemochromatosis, and pancreatic carcinoma can give rise to mild to severe diabetes. Total **pancreatectomy** can also give rise to insulin-requiring diabetes, although the requirement is usually small. Certain pancreatic tumours such as **glucagonoma** can induce increased glycogen breakdown, gluconeogenesis, and increased ketogenesis, whereas **somatostatinoma** inhibits insulin secretion and glucagon secretion, both resulting in mild to severe diabetes. Mild forms of diabetes can be induced by chronic exposure to a whole range of drugs. **Thiazide diuretics**, **beta blockers**, and β_2 **adrenergic agonists**, in addition to **immunosuppressants**, may all directly affect β cell function. **Corticosteroids** act as insulin

TABLE 13.3 Secondary causes of impaired glucose tolerance and diabetes.

Type	Cause	Effect	Severity
Malnutrition	Childhood malnutrition with or without chronic pancreatitis	Lack of protein? $\uparrow$ dietary cassava	Severe, insulin requiring
Pancreatic disease	Pancreatitis	Chronic damage to endocrine pancreas	Mild, progressing to more severe
	Pancreatic carcinoma	?	Mild to severe
	Total pancreatectomy	No endocrine pancreas	Treat with insulin but requirement usually small
	Haemochromatosis	Fibrosis due to iron overload in pancreas	Mild to severe
Drugs	Thiazide diuretics (heart failure)	Impaired insulin secretion via $\downarrow$ K$^+$	Mild
	Beta blockers (hypertension)	? unclear, ? direct action on β cells	Mild
	β_2 adrenergic agonists (asthma)	$\uparrow$ glucose, $\uparrow$ insulin, $\uparrow$ lactate, $\downarrow$ K$^+$	Mild
	Immunosuppressants (transplants)	? insulin resistance, ? insulin secretion	Mild
	Corticosteroids (inflammation, overtreatment of Addison's)	Insulin antagonist, $\uparrow$ hepatic gluconeogenesis, $\downarrow$ glucose uptake in muscle and adipose	Mild to severe
Endocrine disorders	Cushing's (corticosteroids)	See above. Severe when due to ectopic ACTH	Mild to severe
	Acromegaly (growth hormone)	$\uparrow$ hepatic glucose output, $\downarrow$ peripheral glucose uptake	Mild
	Phaeochromocytoma (adrenaline)	α-adrenergic $\downarrow$ insulin secretion, $\uparrow$ insulin resistance, $\uparrow$ hepatic glucose output	Mild
	Conn's syndrome (hyperaldosteronism)	Probably due to impaired insulin secretion via $\downarrow$ K$^+$	Mild
Pancreatic endocrine tumours	Glucagonoma	$\uparrow$ glycogen breakdown, gluconeogenesis, $\uparrow$ ketogenesis	Mild to very severe
	Somatostatinoma	$\downarrow$ insulin secretion, $\downarrow$ glucagon secretion	Mild to severe

antagonists, increasing hepatic gluconeogenesis and suppressing glucose uptake in muscle and adipose tissue. Diabetes can also be associated with other endocrine disorders, usually as a result of excess counter-regulatory hormones. **Cushing's disease**, due to excess cortisol, **acromegaly**, due to excess growth hormone, **phaeochromocytoma**, due to excess adrenaline, and **Conn's syndrome**, due to excess **aldosterone**, can all cause relatively mild forms of diabetes.

13.8 **Acute complications of diabetes**

In this section we discuss some of the acute complications that can arise from poor control of diabetes. These can be split into hyperglycaemic states, namely **diabetic ketoacidosis** (DKA) and **hyperosmolar hyperglycaemic syndrome** (HHS), and hypoglycaemic states or **acute hypoglycaemia**.

Diabetic ketoacidosis

Cross reference

Chapter 6 Acid-base disorders

The diagnosis of DKA is based on the clinical features of uncontrolled diabetes (i.e. hyperglycaemia, uncontrolled thirst, and urination) in the presence of high levels of ketones in an acidic bloodstream, referred to as acidosis, that is, arterial blood pH of less than 7.3.

In practice, patients with normal blood pH but who have a hydrogen carbonate (bicarbonate, HCO_3^-) level below 15 mmol/L are often classified as having ketoacidosis. Diabetic ketoacidosis is a severe complication which carries a significant mortality (about 6%) and occurs most frequently in type 1 diabetes, but is also seen in patients with type 2 diabetes. It can occur in patients with established diabetes or can be the presenting feature. Pneumonia or urinary tract infections are common precipitating factors and heart attack, trauma, and stroke also need to be considered. In about 25% of cases no precipitating cause can be identified.

Patients typically have a short (a few days) history of symptoms which can include polydipsia, polyuria, variable weight loss, weakness, and drowsiness. Severely affected individuals can present with dehydration, a racing pulse (tachycardia), low blood pressure, and can have deep and rapid breathing (called air hunger or Kussmaul breathing), a characteristic of any metabolic acidosis. Reduced consciousness is common but less than 10% present with coma.

Severe insulin deficiency prevents the down-regulation of glucagon secretion, exacerbating the hyperglycaemia. High glucagon and low insulin levels, as shown in Figure 13.10, promote lipolysis and NEFA release from adipose tissue. Non-esterified fatty acids are normally metabolized by beta oxidation to acetyl CoA in the liver and other tissues. Acetyl CoA is metabolized in the TCA cycle to produce ATP. This pathway is overwhelmed in DKA and the terminal metabolites of NEFA oxidation, acetoacetate, and beta hydroxybutyrate (ketone bodies) accumulate, causing an acidosis. Acetone, produced from acetoacetate, can be detected as a characteristic 'pear drops' smell on the breath of affected patients.

Hyperglycaemia and hyperketonaemia contribute to an osmotic diuresis, leading to loss of both sodium and potassium. Total fluid loss at presentation can be as high as 8 litres since the onset of symptoms and the sodium loss can be just as significant (400–700 mmol). Despite the significant depletion of total potassium, hyperkalaemia may be present due to the movement of potassium from the intracellular to the extracellular space (caused by the lack of insulin and the prevailing acidosis).

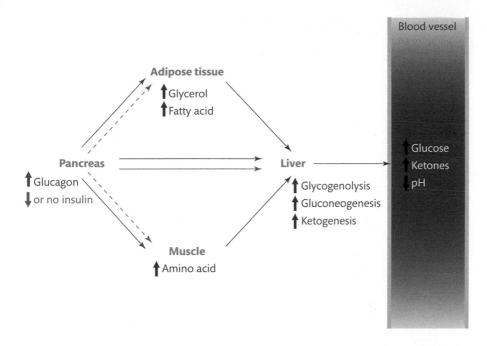

FIGURE 13.10
The development of ketoacidosis. See text for details.

Controlled replacement of fluid loss and low dose intravenous insulin restores glucose homeostasis and turns off ketone production. High blood potassium may initially fall to low levels which can be rectified with intravenous potassium. As the metabolic derangements are controlled a stable insulin regime can be re-established.

Hyperosmolar hyperglycaemic syndrome (HHS)

The main finding in hyperosmolar hyperglycaemic syndrome is a marked hyperglycaemia (sometimes above 50 mmol/L) without the presence of high levels of ketones in the bloodstream and the attendant acidosis that is seen in DKA. The **osmolality** of the blood is frequently high, between 320 and 340 mOsm/kg (compared to 280–295 mOsm/kg in a normal adult). This syndrome was formerly known as **hyperosmolar non-ketotic state** (HONK) or coma; however, it is now known that the syndrome can be associated with minor degrees of ketosis and acidosis, hence the new name. It occurs principally in type 2 diabetic patients with some endogenous insulin release, which reduces the ketogenesis and lipolysis encountered in DKA, but is insufficient to suppress hepatic glucose production and promote tissue glucose uptake.

Hyperosmolar hyperglycaemic syndrome develops over many days and is often associated with a high intake of sugary drinks and polyuria in the days prior to admission. Precipitating factors can include pneumonia, urinary tract infection, myocardial infarction, or stroke and some drugs (antipsychotics, diuretics, and corticosteroids). Affected individuals may present with severe dehydration, seizures, progressive confusion, or altered consciousness. Swelling of the brain tissues (**cerebral oedema**) during fluid replacement is the most serious problem. The cerebral oedema can cause permanent neurological damage and death in severe cases. Dehydration and associated low blood pressure (**hypotension**) is more frequent and severe than that seen in DKA. Older patients and African and Afro-Caribbean populations are most commonly affected and HHS may be the first presentation of type 2 diabetes to a doctor.

CASE STUDY 13.3

A 21-year-old female was recently seen by her family doctor and prescribed Canesten cream for a vaginal thrush infection. She had noted a significant weight loss in the past month but was not dieting. For the previous week she had been feeling tired, thirsty, and had drunk a lot of water. She was admitted to the hospital in a coma with deep sighing respiration. She had a faint, rapid pulse and acetone was detected on her breath. Biochemical tests revealed the following results (reference ranges are given in brackets):

Blood glucose	35 mmol/L	(fasting 3.0–6.0)
Arterial blood pH	7.15	(7.35–7.45)
Urine ketones	positive	(negative)
Urine glucose	positive	(negative)
Serum potassium	5.9 mmol/L	(3.5–5.2)

(a) What is the diagnosis?

(b) Why does she have a low pH and why does her breath smell of acetone?

(c) Why is the serum potassium raised?

(d) Why did she have thrush?

(e) Why was she thirsty and drinking a lot?

(f) Why had she lost weight?

(g) Why did she have a faint, rapid pulse on admission?

(h) How should she be managed?

Mortality rates can be as high as 30% in HHS, much higher than those seen in DKA. We can see the main differences between DKA and HHS in Table 13.4.

Treatment aims are to rectify the precipitating cause(s) and to address the dehydration and hyperglycaemia. Initially, normal saline (0.9%) is the most appropriate fluid replacement, but it is essential not to replace the fluid too quickly as this can cause brain damage and death.

TABLE 13.4 Differences between DKA and HHS.

Diabetic ketoacidosis	Hyperosmolar hyperglycaemic syndrome
Short onset (few days)	Long onset (many days)
Osmolality rarely >320 mOsm/kg	Osmolality frequently >320 mOsm/kg
Significant hyperketonaemia and acidosis	None or low hyperketonaemia and acidosis
Hyperglycaemia relatively modest (or rarely absent)	Hyperglycaemia severe (can be up to 50 mmol/L)
Occurs mostly in type 1 diabetes	Occurs mostly in type 2 diabetes
Mortality <6%	Mortality up to 30%

Fluid replacement can be accompanied by insulin infusion (or i.v. administration) according to various sliding scales—the aim being to lower the blood glucose by no more than 3–4 mmol/L per hour.

Key Points

It is important to note that point of care testing (POCT) glucose meters are not accurate when used with samples from dehydrated, slightly acidotic and hyperglycaemic patients, so laboratory glucose measurements must be used.

Severe hyperglycaemia is a prothrombotic state (i.e. favouring the formation of a thrombus, or blood clot), increasing the risk of myocardial infarction and stroke. Anticoagulant therapy such as heparin and/or low dose aspirin may be indicated.

Acute hypoglycaemia in diabetes

In this section we will briefly discuss hypoglycaemia as an acute complication of diabetes. Non-diabetic hypoglycaemia and its consequences are covered in Section 13.13 of this chapter. Hypoglycaemia in diabetes tends to occur in patients who are either just commencing insulin or sulphonylurea therapy, in those who have introduced changes to their regime, or in those who have mismatched their treatment to food intake or exercise. Current guidelines recommend keeping the blood glucose above 4.0 mmol/L. Most diabetic patients learn to recognize the onset of hypoglycaemic symptoms and can usually self-treat at home. Deaths from hypoglycaemia account for only 3–4% of deaths in young patients with diabetes, with psychiatric problems being a significant contributing factor. Hypoglycaemia in the diabetic population is probably under-diagnosed.

13.9 Long-term complications of diabetes

Several factors contribute to the long-term complications of diabetes. Hyperglycaemia and excess glucose within target tissues can lead to excessive non-enzymatic **glycation** of several molecules. Glucose can attach covalently to proteins due to its abundance and proximity. A prime example is the glycation of haemoglobin to produce **haemoglobin A_{1c}**. Other soluble proteins, lipoproteins, and structural proteins can also be glycated and can be further transformed into proinflammatory molecules called **advanced glycation endproducts** (AGEs). Advanced glycation end products bind to receptors called **receptors for advanced glycation endproducts**, or RAGE, on the surfaces of target cells, especially in the vascular endothelium. Post-receptor binding induces a proinflammatory state in several tissues, triggering the release of adhesion molecules and inflammatory cytokines, promoting vascular inflammation and cardiovascular risk.

Disordered lipid metabolism seen in diabetes can contribute to the laying down of plaques on blood vessel walls. These are made up of lipid and cell fragments (**atheroma**) and lead to a narrowing and further inflammation of the vessel at these points (**atherosclerosis**).

Hypertension, or high blood pressure, another common complication of diabetes, can also cause vascular damage. In larger blood vessels this is called macrovascular damage and in the smaller, finer blood vessels of the body it is known as microvascular damage. This can affect several tissue/organ systems as follows.

Macrovascular disease

Atherosclerosis, causing vascular disease of the larger blood vessels in the arms and legs, affects peripheral vessels which become gradually blocked, causing pain and limping due to insufficient delivery of oxygen to the lower extremities (called claudication). More central defects in the larger vessels give rise to cardiovascular and cerebrovascular disease, reducing the quality of life and greatly increasing the risk of heart attack (**myocardial infarction**, MI), stroke (**cerebrovascular accident**, CVA), and death.

Microvascular disease

Damage to the smaller blood vessels can impair the function of tissues and organs such as the kidney, the eye, and the nervous system. Kidney damage (**nephropathy**) is more common in people with long-term diabetes and/or hypertension and tends to develop slowly.

Cross reference
Chapter 3 Kidney disease

As the nephropathy progresses blood pressure increases, causing swelling, especially in the feet, and a build up of waste products in the blood. Tight control of blood glucose and blood pressure can reduce the risk of nephropathy, an early sign of which is the presence of albumin in the urine (**microalbuminuria**). Diabetic patients are often prescribed blood pressure medication such as **angiotensin converting enzyme** (ACE) inhibitors or **angiotensin II receptor antagonists** (AIIRA) to keep the blood pressure in check. Renal function can be improved by restricting the intake of protein, potassium, phosphate, and sodium in the diet. In the eye, retinal damage (**retinopathy**) is caused by blocked or leaky blood vessels in the retina. This obstructs light reaching the retina which, when severe, can impair vision. People with diabetes need regular (usually annual) retinal screening. Existing early retinopathy can be successfully treated by laser-based therapy. Nerve damage (**neuropathy**) can occur to nerves that transmit impulses to and from the brain, spinal cord, muscles, skin, blood vessels, and other organs, including the penis. Neuropathy can cause erectile problems, which is a relatively common finding in diabetic men. Hyperglycaemia and neuropathy combine to delay wound-healing in diabetes. This promotes the formation of diabetic ulcers, especially in the lower limbs and the feet, which are a major cause of disability.

Musculoskeletal damage

Musculoskeletal problems occur more frequently in patients suffering long-term diabetes. This is thought to be due to damage to connective tissue proteins, notably collagen, caused by persistent hyperglycaemia. It results in reduced flexibility in muscle, skin, tendons, and ligaments. **Limited joint mobility** is a form of rheumatism most commonly seen in the hands but it can affect wrists, shoulders, and elbows. Glucose control can improve this condition. **Dupuytren's contracture** is more prevalent in the diabetic population, as is **carpal tunnel syndrome**, **tenosynovitis**, and **frozen shoulder**. **Charcot joint** is also known as **Charcot foot** as this is where sufferers are most affected, but it can also affect the ankle, knee and, more rarely, the wrist and hand. It is most common in older patients with long-term diabetes (over 15–20 years' duration).

Rarer complications associated with diabetes include skin rashes (**necrobiosis lipoidica**) and hardening of the breast tissue (**mastopathy**), mostly in diabetic women but sometimes also in diabetic men.

13.10 Evidence-based medicine and diabetes treatment

The main arm of the Diabetes Control and Complications Trial (DCCT) was a long-term (ten-year) prospective study of conventional versus intensive management of patients with type 1 diabetes. This showed a reduction of 76% in eye disease, a 50% reduction in kidney disease, and a 60% reduction in neuropathy in the intensively managed arm. This developed into the Epidemiology of Diabetes Intervention and Complications (EDIC) study. This showed that intensive diabetes management reduced the risk of non-fatal myocardial infarction, stroke, or death from cardiovascular disease by 57%.

The 20-year United Kingdom Prospective Diabetes Study (UKPDS) of type 2 diabetes showed that intensive versus conventional management reduced the incidence of major diabetic eye disease by a quarter and the risk of early kidney disease by a third. Also, the effect of intensive versus conventional antihypertensive treatment was shown to reduce the risk of death from any diabetes related end point by a third, reduce renal impairment by a third, and reduce visual deterioration by a third.

The Collaborative Atorvastatin in Diabetes Study (CARDS) investigated the management of dyslipidaemia in diabetes. In this large prospective randomized study, **atorvastatin**, a cholesterol reducing drug, was given to a large number of patients with type 2 diabetes who had no previous history of cardiovascular disease. This reduced heart attacks by 36% and strokes by 48%.

13.11 Treatments for diabetes

There is a wide range of treatments for diabetes appropriate to the severity or time course of the disease. At the mild end there is lifestyle modification. As the condition progresses *combinations* of various medications can be used. Finally, at the more pronounced and severe stages of the condition, for example in type 1 diabetes and long-standing and gradually deteriorating type 2 diabetes, the last line of conventional treatment is insulin therapy.

Lifestyle modification

Lifestyle modification is usually the first line of therapy for newly diagnosed, relatively mild type 2 diabetes. It normally consists of a two-pronged approach of diet modification, usually a reduction in total calories (to reduce obesity), carbohydrate, and lipid intake, linked to the introduction of physical exercise. This can be especially beneficial in the classic, overweight 'couch potato' type of patient most frequently seen in industrialized countries, provided that patients stick to the regime.

β Cell secretagogues

Sulphonylureas, originally derived from sulphonamide antibiotics, stimulate pancreatic β cells to secrete more insulin, so they are only useful in treating type 2 diabetes. First-generation sulphonylureas, such as **chlorpropamide** and **tolbutamide**, have been superseded by second and third-generation compounds, such as **gliclazide**, **glipizide**, and **glimepiride**.

Prandial glucose regulators also work by stimulating pancreatic β cells (via the K_{ATP} channel) to secrete more insulin. They are normally short-acting so are useful in preventing a rise in blood glucose after a meal. **Repaglinide** and **nateglinide** are prandial glucose regulators.

GLP-1 as a therapeutic agent in type 2 diabetes

Exogenous GLP-1, given by either subcutaneous or intravenous injection is rapidly degraded to inactive GLP-1. Within minutes of administration, 80% of the active GLP-1 has been converted to the inactive form. Although human GLP-1 (7-36) amide lowers blood glucose and improves glycaemic control in patients with type 2 diabetes, the short half-life of the native molecule makes it impractical as a therapeutic agent. Drug development for potential incretin treatments for diabetes has focused on two main strands of interest, outlined in Figure 13.11. The first is the development of DPP-IV resistant GLP-1 analogues, the so-called incretin mimetics, and the second is the development of DPP-IV inhibitors. Both therapies will prolong the half-life of biologically active GLP-1 in the circulation, the

(a) Endogenous GLP-1 metabolism

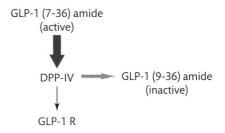

(b) Incretin mimetics

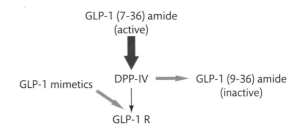

(c) DPP-IV inhibitors

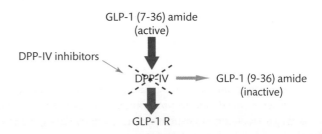

FIGURE 13.11
Alternative targets for incretin therapy in type 2 diabetes showing (a) Endogenous GLP-1 metabolism. (b) Incretin mimetics. (c) DPP-IV inhibitors.

former by being resistant to breakdown by DPP IV, and the latter by inhibiting the degrading activity of the enzyme.

Exenatide, isolated from the saliva of the hela monster, is a GLP-1 analogue which, like insulin, must be injected; **sitagliptin** and **vildagliptin** are DPP-IV inhibitors.

Insulin sensitizers

Biguanides inhibit gluconeogenesis in the liver and increase insulin sensitivity in muscle and adipose tissue. The most common biguanide is **metformin**. **Thiazolidinediones**, also referred to as **glitazones**, bind to the **peroxisome proliferator-activated receptor gamma** (PPARγ) found on target tissues and increase insulin sensitivity. Preparations are **rosiglitazone** and **pioglitazone**.

Other anti-diabetes medications

Acarbose and other α **glucosidase inhibitors** work by reducing the absorption of carbohydrate from the gut. They achieve this by inhibiting the activity of enzymes which digest starch and disaccarides, thus reducing post-prandial rises in glucose.

Amylin is deficient in established diabetes. Human amylin is chemically prone to forming amyloid, which can play a role in β cell destruction in type 2 diabetes. Stable analogues of amylin, especially if injected near mealtimes have beneficial effects on long-term glucose control. **Pramlintide** is an injectable, stable analogue of amylin, which inhibits glucagon release, delays gastric emptying, and signals satiety thereby reducing food intake.

Insulin

This hormone is given as the first line of therapy in type 1 diabetes as patients have no endogenous source. Most patients with type 2 diabetes also end up needing insulin therapy as their β cell mass gradually disappears. The most common mode of insulin administration is by daily injections but there are other forms. Some people choose to use insulin pumps which give a continuous supply of insulin. Insulin can be taken nasally (inhalation/absorption) and a method of getting insulin under the skin using a high pressure jet is under development for the small but significant number of patients who are terrified of needles. Human insulin is now manufactured in large quantities in vats using genetically modified bacteria. Porcine (pig) insulin is sometimes preferred by some older patients who started their treatment with animal insulin, although its use (and availability) is rapidly declining. Chemically modified forms of human insulin, the insulin analogues, are also in common use. The structures of these molecules have been altered to change their biological potency and half-life. Some of these preparations are short-acting, to be used around mealtimes to reduce post-prandial hyperglycaemia, some are medium to long-acting, and some are used as depot injections to give a background level of protection from hyperglycaemia. Slow-release depot insulins are especially useful overnight, during the peak time of hepatic glucose output.

Emerging treatments

In the field of transplantation joint pancreas/kidney transplants are now performed on diabetic patients with severe renal failure. The pancreas is included to give the transplanted kidney a less harmful environment to work in. Previous attempts to give diabetics just the endocrine part of the pancreas, that is, pancreatic islet grafts, usually met with failure due to the high rejection rates by recipients. However, recent advances in surgical, biological, and immunotherapy techniques have heralded the introduction of a more effective islet transplant procedure, known as the Edmonton protocol, pioneered by British surgeon James Shapiro in Toronto. The rate-limiting factor, as always in transplantation, is the scarcity of adequate organ donors. Experiments looking at the feasibility of harvesting and transforming stem-cells into insulin-secreting cells and placing them back into the same individual, thereby avoiding problems of rejection, are in progress but a long way from clinical use.

Due to the strong link between obesity and type 2 diabetes, weight loss surgery has been investigated as a treatment. The simplest procedure, known as **laparoscopic adjustable gastric banding** (LAGB), consists of placing a silicone band around the upper part of the stomach, reducing stomach volume and food intake. In clinical trials, obese type 2 diabetes patients having LAGB showed substantial, sustained weight losses and much improved blood glucose profiles.

In the early part of this millennium, the technique known as **gastric bypass surgery** is showing even more promise. In this procedure the size of the stomach is again greatly reduced and a part of the small intestine is bypassed (called a **Roux-en-Y bypass**). Of 22,000 patients who have had this procedure in the USA, 84% experienced a complete reversal of their type 2 diabetes. It was initially thought that this was due to simple weight loss but new research is pointing to modifications in intestinal hormone secretion following this procedure. Ghrelin, an intestinal hormone that signals hunger, is reduced post-bypass, whereas levels of GLP-1, which signals satiety, are increased. Currently, bariatric surgery is only approved for extremely obese subjects (BMI above 35), but this may change as research progresses.

Associated medications

Additional medications are commonly used by diabetic patients to treat co-existing conditions such as heart disease, stroke, and kidney disease. These may include anti-hypertensives for high blood pressure, statins and/or fibrates to improve the lipid profile, an ACE inhibitor to protect against kidney disease, and other drugs such as diuretics or low doses of aspirin.

SELF-CHECK 13.8

What effect does the drug tolbutamide have on insulin?

13.12 Monitoring and management of diabetes

Towards the latter part of the last century and the early part of this century there has been concerted commitment, throughout Europe and in the UK, to harmonize and improve standards of diabetes care and to strive for predefined goals in treatment. In Italy in 1989 the **Saint Vincent Declaration** was signed by government health services throughout all Europe. Early in this century the National Health Service introduced a **National Service Framework for Diabetes**, to stand alongside other national frameworks such as those for children and coronary heart disease. The standards of the framework were published in 2001 and the delivery strategy was published in 2003. This set out national targets against which services can be judged. Since then the National Institute for Health and Clinical Excellence (NICE) has published several excellent guideline reports for effective management of various forms of diabetes (see further reading).

In the UK, most people with diabetes will review and check their condition with their medical team at least once a year. This can be more frequent early in the diagnosis, or during changes in treatment or severity of disease. On an individual basis, some patients are happy to let others care for their condition whereas others are extremely independent and 'hands on' in the control of their diabetes. At a patient's review the following will be measured (at least once per year).

Assessing blood glucose control

A significant number of patients with diabetes will use point of care glucose monitors to regularly record their blood glucose concentration at any point in time. Figure 13.12 shows an example of a blood glucose meter often used by diabetic patients.

Whilst these meters are useful for the patient, they are prone to inaccuracies such as poor operator technique, variable storage conditions, and age of equipment, in addition to inbuilt inaccuracies in the machine if the patient's blood glucose concentration is very high or very low (Box 13.4).

Laboratory glucose measurements are performed regularly on patients but they only provide information regarding the state at the time of sampling. A measurement that assesses relatively long-term glucose control would be preferable. Glucose can attach to the amino groups of proteins non-enzymatically. This modification persists for the life of the protein and the amount of glycation is an accurate reflection of the concentration of glucose in the bloodstream during the lifespan of the protein. An estimation of the blood glucose control over the

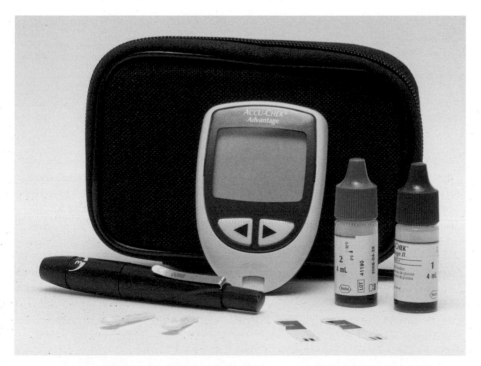

FIGURE 13.12
A portable blood glucose meter.

BOX 13.4 Point of care glucose meters

Ward-based glucose meters are not accurate at very low or especially, very high, glucose concentrations. For example the analytical ranges of three commercially available blood glucose meters were 0.6–33.3 mmol/L for the Bayer Ascensio Dex meter, 1.1–27.7 mmol/L for the Abbott precision PCX meter, and 0.6–33.3 for the Roche accu-chek Aviva meter. In addition, it is often very difficult to get a satisfactory finger-prick sample of blood from a hypoglycaemic or hyperglycaemic patient who may not be aware of their surroundings. However a meter will tell you if a sample is low in glucose, at which point it can be sent to the laboratory for accurate analysis.

last 6–8 weeks can be derived from measuring the **glycated haemoglobin** (HbA$_{1c}$), a glycated form of haemoglobin A, HbA. Using an approved laboratory method (see Box 13.5) HbA$_{1c}$ can be measured at clinic visits and a record of long-term glucose control accumulated for each patient. A percentage of haemoglobin that is glycated of 6.5 or less signifies good control. If the value is above this point then lifestyle and/or pharmaceutical intervention may be required. It should be noted that there are several genetic variants of haemoglobin which may give misleading results. In such circumstances other measurements such as **fructosamine**, a marker of total glycated protein, can be used to indicate glucose control, although it reflects a shorter time span (3–4 weeks).

Other tests

Cross references

Chapter 3 Kidney disease and Chapter 9 Abnormalities of lipid metabolism

Patients have the estimated glomerular filtration rate (eGFR) calculated along with blood and urine analysis for albumin, creatinine, and urea and action is taken accordingly.

Blood is checked for lipids, cholesterol, and triacylglycerols. Total cholesterol should be 4.0 mmol/L or less and fasting triacylglycerols should be 1.7 mmol/L or less.

BOX 13.5 HbA$_{1c}$ measurement and reporting

In 2007 an attempt was made to standardize laboratory HbA$_{1c}$ measurements around the world. A consensus of several groups, including Diabetes UK and the Association for Clinical Biochemistry, agreed that the International Federation of Clinical Chemistry (IFCC) reference measurement procedure should be the international method for calibrating all assays used for HbA$_{1c}$. This method reports values in mmol/mol (HbA$_{1c}$/Hb). Due to concerns that staff and patients would struggle with the new units and values as opposed to the old National Glycohaemoglobin Standardization Program (NGSP) units, which are given as a percentage of total haemoglobin, results are reported in both formats using the IFCC-NGSP master equation to convert between the two. This will continue until everyone is familiar with the new reference method and units. Plans to report an HbA$_{1c}$-derived average glucose value, along with the haemoglobin values, are also under way.

The medical team also record blood pressure, weight, and BMI, check legs, feet (chiropodist), and skin and circulation, and nerve supply. Retinal screening (fundoscopy) is regularly performed. There is also discussion as to depression, sexual function in men, and overall general health.

13.13 Hypoglycaemia

Hypoglycaemia is conventionally defined as a concentration of glucose in a venous capillary sample of blood collected into a fluoride oxalate tube (to prevent further glucose metabolism) of 2.5 mmol/L or less in a symptomatic patient. Diagnosis is made by demonstrating hypoglycaemia in a *laboratory* measured sample. Hypoglycaemic symptoms include tremor, sweating, anxiety, and numbness and/or tingling around the tongue and lips. Symptoms are due to the brain being starved of glucose; some patients experience mood swings, hallucinations, or erratic behaviour, which if left untreated can culminate in slurred speech, drowsiness, and coma. Brain cells do not need insulin stimulation to take up glucose, so the only thing that regulates the brain glucose level is the blood glucose concentration and the efficiency of the cerebral blood supply. These symptoms are collectively called the autonomic or neurogenic response. A drop in cerebral blood glucose concentration triggers the counter-regulatory response discussed in Section 13.4.

For a true diagnosis of hypoglycaemia, the clinical picture must resolve when glucose is given to the patient. This confirmed hypoglycaemia, with symptoms, which resolves after glucose is known as **Whipple's triad**.

Recovery from hypoglycaemia usually occurs within a few minutes following the intake of sugar. In serious cases, where the patient may be uncooperative or unconscious, intravenous glucose or intramuscular or subcutaneous glucagon may be given initially, which is usually sufficient to restore consciousness. Oral glucose can then be given. There are several subtypes of hypoglycaemia outlined below.

Fasting hypoglycaemia

The clinical investigation of hypoglycaemia in the fasting state entails sampling during either a series of three overnight fasts or during a prolonged 72-hour fast. Several blood samples are taken and sent to the laboratory for processing. When a blood glucose level of 2.5 mmol/L or less is obtained then the other samples obtained at the same time can be analysed for insulin (Box 13.6) and c-peptide.

Most hypoglycaemic attacks are associated with low circulating levels of insulin. The presence of an insulin-secreting tumour or insulinoma should be suspected if the insulin is inappropriately high when the glucose is very low. If the c-peptide is also increased during hypoglycaemia

BOX 13.6 Insulin assays

Some insulinomas secrete mainly proinsulin rather than insulin. Under these circumstances it can be useful to use an insulin assay that has some cross-reactivity to proinsulin (or one can measure proinsulin additionally). Most modern insulin assays utilize mouse monoclonal antibodies specific for human insulin and these will not detect proinsulin.

CASE STUDY 13.4

A 28-year-old man was taken to the casualty department of a local hospital because he was confused and disorientated. By the time he arrived he had lapsed into unconsciousness. On examination he was pale and clammy with a firm, rapid pulse.

A blood glucose estimation arranged via the emergency laboratory was 1.2 mmol/L. A further blood sample was taken and he was given 50 mL of 50% glucose solution intravenously. He 'woke up' almost immediately and had no recall whatsoever of recent events (reference ranges are given in brackets):

Urine drug screen: negative for hypoglycaemic drugs

Plasma cortisol	500 nmol/L	(150–700)
Fasting plasma insulin	52 mU/L	(2–10)
Fasting blood glucose	1.1 mmol/L	(3–6)

(a) Why is he pale and clammy, and why does he have a firm rapid pulse?

(b) What do the laboratory results suggest?

(c) What further investigations would you recommend?

then an insulinoma may be the cause. If, however, the c-peptide is low with a raised insulin level, then the insulin is likely to be exogenous.

Insulinomas can be localized using imaging and venous sampling techniques prior to surgical removal. **Multiple endocrine neoplasia** type 1 (MEN1) can be associated with multiple, mostly non-malignant, insulinomas. Other types of cancers can also cause hypoglycaemia. **Non-islet cell tumour hypoglycaemia** (NICTH) can occur in several types of non-pancreatic cancers, and is associated with normal to low insulin concentrations. The low blood glucose in these patients results from increased glucose uptake by the tumour, reduced hepatic glucose output and increased **insulin-like growth factor-2** (IGF-2) levels. Severe liver disease can cause hypoglycaemia due to compromised glycogen storage. The kidneys are the only other organ capable of gluconeogenesis and they also contribute to insulin degradation, which may explain hypoglycaemic episodes seen in end-stage renal failure. A lack of counter-regulatory hormones, such as cortisol, can also trigger hypoglycaemia. Alcohol excess is a common cause of hypoglycaemia via its suppressive action on hepatic glucose release. This is made worse in chronic alcoholics who have acquired ACTH deficiency, poor nutrition, and compromised liver function. Sepsis compromises renal function and also causes cytokine release, which can influence both insulin secretion and hepatic glucose output.

Reactive hypoglycaemia

Reactive hypoglycaemia can present in gastric surgery patients and usually occurs 90–150 minutes after a meal, especially if it is high in sugar or other carbohydrate. Because of the reduced stomach size there is a rapid transit of glucose into the small intestine where incretins are released leading to an excessive insulin response and hypoglycaemia. Symptomatic hypoglycaemia is sometimes recorded in normal individuals if an OGTT is sampled beyond two hours, but it must be stressed that an OGTT, consisting of rapidly ingesting a drink containing 75 g of glucose, is hardly physiological.

Factitious hypoglycaemia

Factitious (or non-physiological) hypoglycaemia occurs when insulin or sulphonylurea is taken or is administered deliberately. It is difficult to diagnose and is usually associated with psychological problems in the patient or carers involved. If a confirmed hyperinsulinaemic hypoglycaemic sample has a low c-peptide level then the insulin is likely to be exogenous. If the c-peptide level is high in such a sample, then the insulin is likely to be endogenous and a sulphonylurea screen should be performed. A positive screen indicates the raised insulin is due to sulphonylurea administration.

Neonatal and infant hypoglycaemia

Hypoglycaemia is relatively common in newborn babies. Premature babies and those that are small for their gestational age (called small for date) are at particular risk as they have feeding problems and relatively small livers with low glycogen stores.

Babies of mothers with diabetes can have increased β cell mass, with an associated risk of hypoglycaemia. This usually resolves rapidly. Inborn errors of metabolism, such as **glycogen storage disease type 1**, **galactosaemia**, **hereditary fructose intolerance**, and **fatty acid β oxidation defects** are all associated with hypoglycaemia. In some children, concomitant starvation and infection can precipitate hypoglycaemia. In **idiopathic ketotic hypoglycaemia**, insulin release falls and ketosis develops, usually in thin, small-for-date children. In **persistent hyperinsulinaemic hypoglycaemia of infancy**, PHHI (also called **congenital hyperinsulinism**, CHI), children present with persistent non-ketotic hyperinsulinaemic hypoglycaemia. This is caused by a genetic mutation resulting in the expression of a defective beta cell K_{ATP} channel, leading to unregulated insulin release. Short-term treatment can be achieved using diazoxide, but permanent treatment is by partial removal of the pancreas.

SELF-CHECK 13.9

When would you suspect factitious hypoglycaemia in a sample taken at the same time as a confirmed hyperinsulinaemic hypoglycaemic sample?

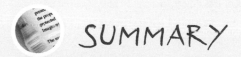

SUMMARY

- The incidence of obesity, metabolic syndrome, and diabetes is increasing at an alarming rate throughout the world. Estimates predict that by 2030, there will be 366 million people with diabetes worldwide.

- Diabetes is a consequence of failure of glucose and lipid handling, two of the main energy sources for the body. This is precipitated by either a complete lack (type 1), or a defective action (type 2), of insulin.

- The release of insulin by endocrine β cells of the pancreatic islet is controlled by the integrated interplay of glucose, lipid, and incretins.

- Wide excursions in the blood glucose concentration are normally prevented by insulin and its counter-regulatory hormones glucagon, cortisol, adrenaline, and growth hormone. This balance is disrupted in diabetes.

- The brain depends directly on the blood glucose concentration for its energy. Wide variations in blood glucose have significant effects on brain activity. Very low or very high blood glucose concentrations can cause significant physiological and behavioural changes and can be life threatening if left untreated.

- Long-term complications of diabetes, resulting from poor blood glucose control, are severely debilitating and include increased risk of heart attack and stroke, kidney disease, blindness, and limb amputations.

- There is now a wide range of therapies and treatments for diabetes, especially in richer, industrialized countries. New treatment strategies are emerging all the time.

- Several large, long-term studies have shown that the more intensive the control of blood glucose, lipid profile, and hypertension, the better the outcome in terms of increased lifespan, reduced cardiovascular risk and reduced risk of the long-term complications of diabetes. Even relatively minor improvements in diet and exercise can produce significant improvements.

- The treatment and monitoring of diabetes and its complications places a significant burden on healthcare resources and impact significantly on laboratory medicine workload.

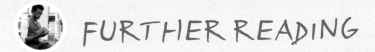

FURTHER READING

- **Ahmed N (2005) Advanced glycation endproducts: role in pathology of diabetic complications.** *Diabetes Research and Clinical Practice* **67**, 3-21.

- **Flatt PR (2008) Gastric inhibitory polypeptide (GIP) revisited: a new therapeutic target for obesity-diabetes? Dorothy Hodgkin lecture 2008.** *Diabetic Medicine* **25**, 759-64.

- **Frayling TM (2007) A new era in finding type 2 diabetes genes—the unusual suspects: RD Lawrence lecture 2006.** *Diabetic Medicine* **24**, 696-701.

- **Gromada J, Franklin I, and Wollheim CB (2007) Alpha cells of the endocrine pancreas: 35 years of research but the enigma remains.** *Endocrine Reviews* **28**, 84-116.

- **MacDonald P, Joseph JW, and Rorsman P (2005) Glucose sensing mechanisms in pancreatic β cells.** *Philosophical Transactions of the Royal Society: B.* **360**, 2211-25.

- **Meier JJ and Nauck MA (2005) Glucagon-like peptide 1 (GLP1) in biology and pathology.** *Diabetes Metabolism Research and Reviews* **21**, 91-117.

- **Scott A (2006) Hyperosmolar hyperglycaemic syndrome.** *Diabetic Medicine* **23**, 22-4.

- **Smith J and Nattrass M (2004)** *Diabetes and Laboratory Medicine*. **Marshall W and Horner J eds. London: ACB Venture Publications.**

- **Todd JF and Bloom SR (2007) Incretins and other peptides in the treatment of diabetes.** *Diabetic Medicine* **24**, 223-32.

- **Wallace TM and Matthews DR (2004) Recent advances in the monitoring and management of diabetic ketoacidosis.** *Quarterly Journal of Medicine* **97**, 773-80.

- **World Health Organization/International Diabetes Federation (2006)** *The Definition and Diagnosis of Diabetes Mellitus and Intermediate Hyperglycaemia*. **Report of a WHO/IDF Consultation. World Health Organization.**
  www.WHO.org

- **International Diabetes Federation. The IDF consensus worldwide definition of the metabolic syndrome, 2006.**
  www.idf.org

- **Diabetes UK website.**
  www.diabetes.org.uk

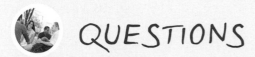

QUESTIONS

13.1 Which of the following are TRUE or FALSE?

(a) Serum triacylglycerols are often increased in type 2 diabetes

(b) Serum triacylglycerols are often low in type 2 diabetes

(c) Serum HDL-cholesterol is often high in type 2 diabetes

(d) Serum HDL-cholesterol is often low in type 2 diabetes

13.2 Type 1 diabetes is:

(a) More common in young people

(b) Associated with obesity

(c) Often accompanied by hypertension

(d) Not associated with the presence of antibodies to GAD

(e) None of the above

13.3 The following results were obtained from a two-hour oral glucose tolerance test:

(i) Fasting glucose 5.0 mmol/L and two-hour glucose 7.7 mmol/L

(ii) Fasting glucose 5.9 mmol/L and two-hour glucose 11.5 mmol/L

(iii) Fasting glucose 6.1 mmol/L and two-hour glucose 7.5 mmol/L

(iv) Fasting glucose 4.3 mmol/L and two-hour glucose 8.2 mmol/L

    (a) Which of these indicates normal glucose tolerance?

    (b) Which indicates impaired fasting glycaemia?

    (c) Which indicates diabetes?

    (d) Which indicates impaired glucose tolerance?

13.4 The anti-diabetes drug metformin:

(a) Directly stimulates insulin secretion

(b) Augments insulin action

(c) Is mainly used to treat type 1 diabetes

(d) Is not used to treat type 2 diabetes

(e) Is none of the above

13.5 Discuss the role incretins play in normal glucose homeostasis and their potential as therapeutic agents in the treatment of diabetes.

13.6 Outline the differences in DKA and HHS and give treatment strategies for both emergencies.

13.7 Define hypoglycaemia and give an outline of the different causes.

Answers to self-check questions, case study questions, and end-of-chapter questions are available in the Online Resource Centre accompanying this book.

 Go to www.oxfordtextbooks.co.uk/orc/ahmed/

14

Adrenal disease

John Honour

Learning objectives

After studying this chapter you should be able to:

- Describe the morphology, development, functions, and regulation of the adrenal glands
- Appreciate the value and limitations of measuring adrenal hormones in biological fluids
- Discuss the impact of changes in adrenal hormone concentrations under normal and abnormal circumstances, and the interpretation of endocrine tests
- Describe the screening and confirmatory tests for adrenal disorders
- Discuss the causes, outcomes, and treatments of adrenal disorders

Introduction

An adrenal gland is located above each kidney. Each gland is comprised of a **cortex** and **medulla** with distinct functions. The outer cortex produces steroid and the inner medulla produces catecholamine hormones. The most active steroids secreted by the normal adrenal cortex are **aldosterone** and **cortisol**. Their main functions are to help retain salt and regulate carbohydrate metabolism respectively. The adrenal cortex also produces relatively more **dehydroepiandrosterone sulphate** (DHEAS) that has weak male hormone-like testosterone activity. The adrenal medulla is part of the sympathetic nervous system producing the catecholamines **adrenaline** and **noradrenaline**. Renin production from the kidney leads to generation of **angiotensin II** in the circulation that stimulates synthesis of aldosterone. Cortisol is regulated by **adrenocorticotrophic hormone** (ACTH) release from the pituitary. The steroid hormones are produced in the adrenal cortex from cholesterol by a number of enzyme-catalysed steps. The diseased adrenal cortex may secrete abnormal amounts of intermediates, male sex steroids (androgens), and salt-retaining hormones. Disorders of the adrenal cortex are investigated by measurement of steroids in the blood, saliva, and urine and by measuring the regulatory hormones (ACTH and renin). Tumours of the adrenal medulla or **phaeochromocytomas** secrete high levels of catecholamines and measurements of the latter in blood and urine are needed to assist in their diagnosis.

14.1 **Adrenal glands**

The adrenal glands begin to form around the eighth week of gestation near the developing kidney. A foetal cortex is formed first and around this a definitive adrenal cortex begins. Cells from the neural crest migrate into the middle of the developing cortex to become the adrenal medulla and the whole gland becomes encapsulated. The medulla is part of the sympathetic nervous system. Neurotransmitters are released from the adrenal medulla into the blood so this organ, like the adrenal cortex, is a ductless gland and is part of the endocrine system.

The adrenal glands are triangular in shape and each is enclosed in a capsule surrounded by fat above the kidneys. At birth, the adrenal glands are larger than the kidneys but their relative sizes reverse during the first six months of life as the foetal adrenal zone disappears. During childhood the adrenal glands grow slowly. Beyond puberty until the sixth decade of life the two adrenal glands together are 8–12 g in weight.

Each adrenal gland is supplied with blood through three arteries to the ends and middle of the gland. From the outside of the gland the arteries divide and spread out across the surface then pass through the cortex to the medulla. Cortical arteries divide into arterioles and capillaries. Veins form near the cortico-medullary junction. Medullary arteries pass through the cortex without branching, then divide as capillaries through the medulla. Blood leaving the cortex and medulla drains into a single adrenal vein on each side. The right adrenal vein drains directly into the inferior vena cava but the left adrenal vein drains into the renal vein.

When the excised gland is dissected, the cortical zone looks somewhat yellow in colour whereas the medulla is reddish brown. Stained sections of the adrenal gland, when examined under the microscope, show three cortical zones (**glomerulosa**, **fasciculata**, and **reticularis**) between the capsule and the medulla. These are shown in Figure 14.1.

The outer zona glomerulosa under the capsule is recognized as clumps of cells that make up less than 10% of the cortex. The fasciculata cells are larger than cells of the glomerulosa and form long cords arranged radially with respect to the medulla forming 75–85% of the cortical volume. The zona reticularis is present from 5 to 8 years of age as a layer of cells at the boundary of the medulla and cortex. The reticularis cells have less smooth endoplasmic reticulum and lipid droplets than are seen in the fasciculata zone. By 13–15 years of age the histological pattern of the cortex ceases to change. In newborn infants about 80% of the cortex is the foetal zone that shrinks over the first six months after birth. The zona glomerulosa and fasciculata zones are small at that stage but grow as the foetal zone disappears. There is no reticularis in the adrenal cortex of newborn infants and younger children. If you were to look at sections of the adrenals under an electron microscope you would see that all the cells of the adrenal cortex have abundant smooth endoplasmic reticulum, many lipid droplets, and round or elongated mitochondria.

The adrenal medulla is composed mainly of hormone-producing chromaffin cells. Histologists noticed a long time ago that when an adrenal gland was fixed in a solution of chromium salts before sectioning this zone appeared brown when viewed through the microscope. This was due to oxidation of the catecholamines to melanin. The cells are modified neurons. The medulla is innervated from pre-ganglionic fibres. The various hormones secreted by the adrenal gland and their main functions are given in Table 14.1.

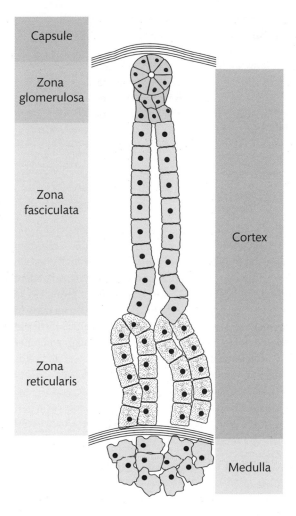

FIGURE 14.1
The adrenal gland showing the zona glomerulosa, fasciculata, and reticularis of the cortex and medulla within.

TABLE 14.1 The sources and functions of the adrenal hormones.

Region of adrenal gland	Zone	Major hormone	Functions of hormone
Cortex	Zona glomerulosa	Aldosterone	Regulation of sodium balance
	Zona fasciculata	Cortisol	Regulation of glucose, immunosuppressant
	Zona reticularis	Dehydroepiandrosterone sulphate	Weak androgen prohormone
Medulla		Adrenaline, noradrenaline	Increase heart rate, contract blood vessels, dilate air passages

Key Points

The adrenal glands have separate regions of outer cortex and inner medulla that produce steroids and catecholamines respectively. The sources of adrenal hormones and their actions are summarized in Table 14.1. There are differences in structure and activity in the foetus, young child, adult, and elderly. These issues need to be borne in mind when judging biochemical test results.

Functions of the adrenal gland

The most active hormones secreted by the adrenal cortex are cortisol and aldosterone and they respectively affect carbohydrate metabolism and electrolyte balance (glucocorticoid and mineralocorticoid effects). Steroids act at target sites in organs around the body through specific cytosolic receptors that become **transcription factors**, stimulating DNA-dependent synthesis of certain mRNAs in the nuclei of target cells. This in turn leads to formation of proteins that alter cell function. Cortisol has a number of actions that lead to an increase in the concentration of glucose in the plasma. In this respect cortisol is called a **glucocorticoid** and is an antagonist to insulin. An increase in glucose production occurs through a number of mechanisms, including an increase in protein catabolism, hepatic glycogenolysis, and keto-genesis. Cortisol is necessary for the actions of noradrenaline, adrenaline, and glucagon as well as the normal functions of blood vessels, water clearance by the kidneys, cardiac, and skeletal muscle, the nervous system, and gastrointestinal tract (permissive effects). Glucocorticoids also inhibit the inflammatory response to tissue injury. Aldosterone is a **mineralocorticoid** that increases sodium reabsorption from the urine, sweat, saliva, and gastric juice. Sodium ions are retained by exchange for potassium and hydrogen ions.

Steroid chemistry

The adrenal cortex synthesizes active steroid hormones from cholesterol. The 27 carbon atoms in cholesterol are numbered according to systematic nomenclature as shown in Figure 14.2.

If you study the structures of cortisol, aldosterone, and dehydroepiandrosterone (DHEA) there are a number of obvious differences. The numbers of carbon atoms in the steroids can be 21 (pregn-series) or 19 (androst-series). In cortisol the side chain of cholesterol is shortened at C-20 leaving a carbonyl group; the steroid nucleus is hydroxylated at C-11, C17, and C21. Cortisol and aldosterone have carbonyl groups at C-3 compared with a hydroxyl group in cholesterol, and the double bond at C-5 to C-6 in cholesterol is shifted so it is between C-4 and C-5 (isomerization). Aldosterone does not have a hydroxyl group at C-17 and has an aldehyde group at C-18. Dehydroepiandrosterone has only 19 carbon atoms (androstene) after loss of all of the side chain at C-17 in two stages (see later) to leave a carbonyl group at C17. Dehydroepiandrosterone has the same 3-hydroxyl-5-ene structure in the A and B-ring structures as cholesterol.

Steroid synthesis

Cortisol, aldosterone, and sex steroids are synthesized from cholesterol by a number of enzyme catalysed reactions that are seen in Figure 14.3.

Cholesterol

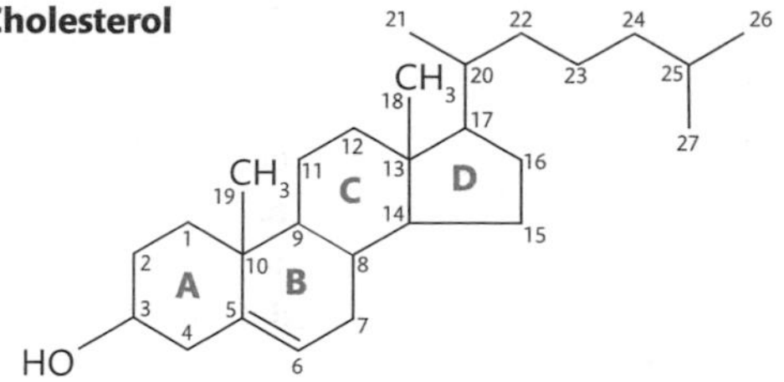

DHEA

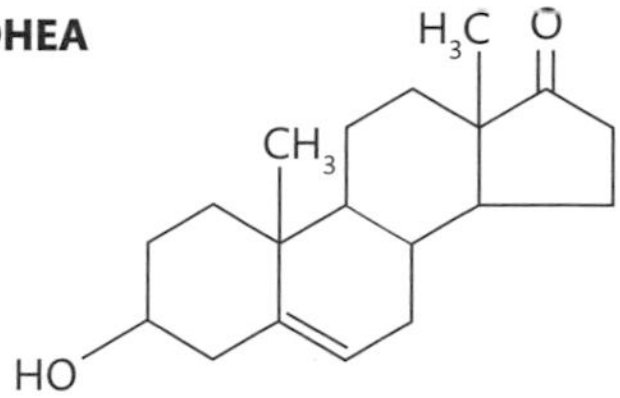

Cortisol

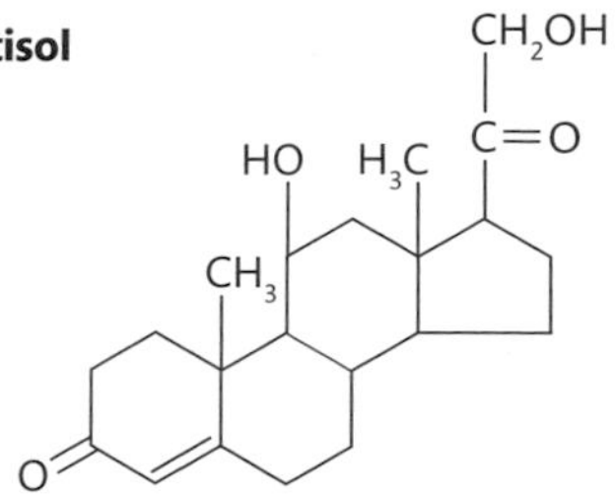

Aldosterone

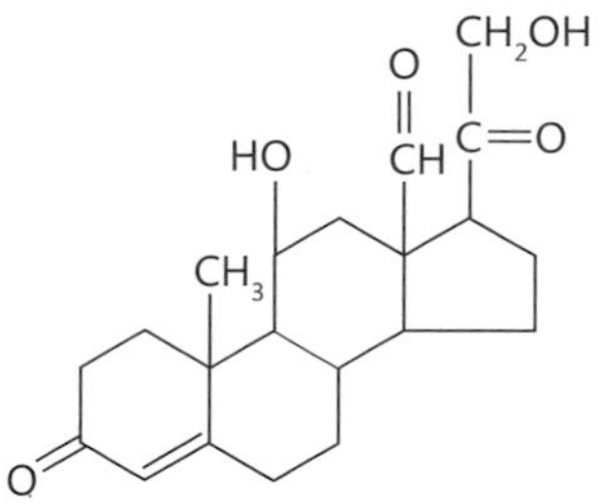

FIGURE 14.2

Cholesterol and the major adrenal steroids. Cholesterol has a total of 27 carbon atoms arranged in four conjoined rings plus two angular methyl groups at positions C10 and C13 with an eight carbon side-chain. The nucleus of cholesterol is a saturated phenanthrene (three fused benzene rings each of six carbon atoms: A, B, and C, with all double bonds saturated) coupled with a five-membered cyclopentane ring (ring D). Cholesterol is systematically named 3β-hydroxy-cholest-5-ene. The adrenal steroids have 19 carbons atoms (androstene) or 21 carbon atoms (pregnene). By comparing the structures you can spot the differences in structures of cholesterol and cortisol at carbons 3 to 4, 5 to 6, 11, 17, 20, and 21. Since cortisol has a double bond in the A-ring it is a pregnene steroid.

The scheme of adrenal steroidogenesis is drawn to show three distinct routes. In general terms that reflects the zonal differences in function and regulation and will help interpret the consequences of metabolic diseases affecting any of the enzymes. When you look later at diseases of the adrenal glands some of the substrates for a defective enzyme accumulate because they cannot be metabolized through the usual pathways to aldosterone and cortisol. When intermediates are present in large amounts they affect the clinical picture. Deoxycorticosterone, for example, is a mineralocorticoid and leads to salt retention and hypertension.

In the pathways there are hydroxylation steps, for example in the side chain cleavage process (SCC) and at C11, C17, and C21, that are catalysed by cytochrome P450 enzymes. All steroidogenic cytochrome P450s have a haem group at the active site of the enzyme that binds

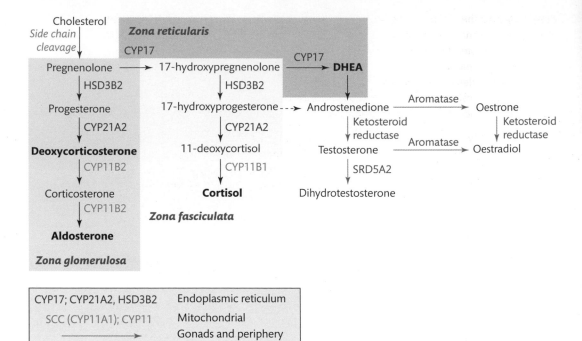

FIGURE 14.3
Synthesis of steroids from cholesterol. Enzymes are located in the endoplasmic reticulum (red) or mitochondria (green). Dehydroepiandrosterone is further metabolized to active androgens and oestrogens in gonads and periphery (orange). Clearly there is much trafficking of steroids in the cells of the adrenal cortex.

the substrate and molecular oxygen. Cytochrome P450 enzymes use a number of electron transporters. As seen earlier, the blood vessels in the adrenal cortex surround every adrenal cell with blood to maximize delivery of oxygen to the cells and speed delivery of the peptide hormones that stimulate adrenocortical action.

Cholesterol and cholesterol esters are taken up into the cells of the adrenal cortex. Cholesterol is released by hydrolysis of the esters. A transport protein is necessary for the initial mitochondrial uptake of cholesterol and is called **steroidogenic acute regulatory protein** (StAR). In the mitochondria cholesterol is converted to pregnenolone by SCC. Two hydroxylations at C-20 and C-22 precede cleavage of the C-20 to C-22 carbon bond leaving a carbonyl group at C20. The steroid hydroxylation steps involve an enzyme based on a cytochrome P450 for which the gene is abbreviated to *CYP11A1* because of similarity to other *CYP11* genes. In the endoplasmic reticulum, **pregnenolone** is oxidized to **progesterone** through the action of **3β-hydroxysteroid dehydrogenase** (gene *HSD3B2*). This enzyme can isomerize the C5-C6 double bond to the C4-C5 position. All active hormones have a 3-keto group with a C4-C5 double bond. The enzymes from HSD3B2 and CYP11A1 and the 21-hydroxylase (CYP21A2) are active in glomerulosa and fasciculata zones; 17-hydroxylase (CYP17) and 11β-hydroxylase (CYP11B1) are only expressed in the zona fasciculata.

Cortisol is produced from cholesterol in the zona fasciculata by the sequential action of five enzyme systems. The first steps to pregnenolone have been described. Three hydroxylation reactions involving positions 17, 21, and 11 then follow. In the adrenal cortex 3β-hydroxysteroid dehydrogenase acts to convert 17-hydroxypregnenolone to **17-hydroxyprogesterone** which is further converted to cortisol through **21-deoxycortisol** or **11-deoxycortisol**. In the zona glomerulosa and fasciculata progesterone can be hydroxylated at C-21 to deoxycorticosterone.

Aldosterone synthesis from cholesterol requires four enzymes as seen in Figure 14.3 in a path from progesterone, deoxycorticosterone, corticosterone, and 18-hydroxycorticosterone. CYP11B2 or aldosterone synthase has three activities that are only expressed in the zona glomerulosa; this enzyme catalyses hydroxylations at C-11 and then C-18 before oxidizing the C-18 hydroxyl to an aldehyde. The intermediates from deoxycorticosterone are corticosterone and 18-hydroxycorticosterone. CYP11B1 and CYP11B2 are located in the mitochondria of the fasciculata and glomerulosa cells respectively.

A 17,20 lyase enzyme acts on 17-hydroxypregnenolone to give androgens (that are male hormones). In the zona reticularis 17-hydroxylase is responsible for C-17 hydroxylation and cleavage of the C-17 to C-20 bond, thus leaving the steroid with 19 carbon atoms (androstene). Dehydroepiandrosterone is a substrate for sulphotransferase in the adrenal which therefore secretes high levels of DHEAS. From puberty the adrenal gland makes androgens (mainly DHEAS) in the zona reticularis. The action of 3β-hydroxysteroid dehydrogenase then leads to production of **androstenedione**.

Testosterone is not an important product of the adrenal cortex but is formed by reduction of androstenedione (ketosteroid reductase). The most active androgen dihydrotestosterone is formed by the action of 5α-reductase (SRD5A2). Oestrogen production requires aromatase (CYP19). The major androgen from the adrenal cortex from 5 to 8 years of age and throughout most of adult life is DHEAS. The adrenal cortex in the foetus has small islands of zona glomerulosa and fasciculata beneath the capsule. The large foetal adrenal zone, outside a poorly differentiated medulla, lacks the enzyme *HSD3B2*. Like the zona reticularis later in life, the foetal zone produces DHEAS. The foetal zone disappears over the first six months of life after birth. SCC can also act on cortisol to give 11-hydroxylated androgens (for example 11-hydroxyandrostenedione).

Key Points

The zones of the adrenal cortex produce steroids that have different functions according to their structure. Cytochrome P450 enzymes are required for the synthesis of steroid hormones in the cortex. A 3-keto-4-ene chemistry of the A-ring is a feature of all active adrenal hormones. 17-hydroxylation is a key reaction in the generation of cortisol and androgens through CYP17. Aldosterone has no 17-hydroxyl group and a unique aldehyde group at C18 but is otherwise structurally similar to cortisol.

SELF-CHECK 14.1

Which steroid is uniquely produced by the zona glomerulosa?

Regulation of adrenal steroid production

Cortisol production by the adrenal cortex is regulated by ACTH. This is a single polypeptide of 39 amino acid residues secreted from the anterior pituitary gland (adenohypophysis) at the base of the brain. Adrenocorticotrophic hormone is formed by specific proteolysis of a single precursor protein called pro-opiomelanocortin (POMC) with an M_r of 30,000 that also includes the sequences of melanocyte-stimulating hormones (MSH), lipotrophin (LPH), endorphin, and enkephalin, as shown in Figure 14.4.

Some of these peptides are neurotransmitters active in the CNS. They are not usually found at high concentrations in the peripheral circulation although certain tumours may secrete them along with ACTH, and high plasma concentrations can be found. An ACTH synthetic peptide

FIGURE 14.4

Adrenocorticotrophic hormone is a protein derived from pro-opiomelanocortin. Synacthen is a pharmaceutical product used for an adrenal stimulation test. Its activity is comparable with ACTH 1-39. MSH is melanocyte stimulating hormone; LPH is lipotrophin; CLIP is corticotrophin-like intermediate peptide; and Met-Enk is methionine-enkephalin. These peptides are secreted when pro-opiomelanocortin is processed excessively.

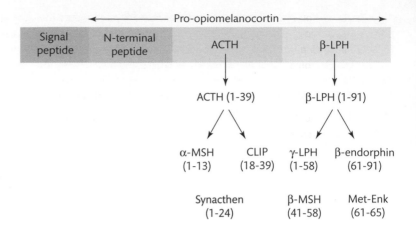

(synacthen) with the sequence of the first 24 amino acids of ACTH has biological activity similar to the natural ACTH hormone and a shorter half-life.

Adrenocorticotrophic hormone is secreted in regular pulses of variable amplitude over a 24-hour period. Most activity occurs at night-time between 2 am to 10 am. Cortisol is produced in response to the ACTH and the highest concentration is seen around 8 am, with lowest levels around midnight. This is the basis of the **circadian rhythm** of plasma cortisol concentrations. Trauma, emotional stress, and certain drugs initiate the secretion of corticotrophin-releasing hormone (CRH) by nerve endings in the hypothalamus and the median eminence of the brain. Corticotrophin releasing hormone is a peptide of 41 amino acids transmitted by the hypophyseal portal vessels to the adenohypophysis, evoking release of ACTH. Low circulating concentrations of cortisol lead to an increase in ACTH secretion. The locus of action of cortisol appears to be the hypothalamus, suppressing the release of CRH. High concentrations of cortisol inhibit ACTH secretion as shown in Figure 14.5.

The acute action of ACTH is to increase the transition of cholesterol through the steroidogenic pathway resulting in the rapid production of steroids. Adrenocorticotrophic hormone binds to

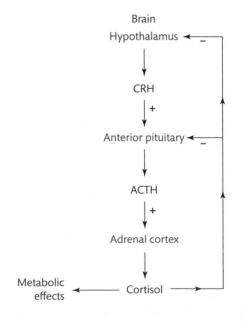

FIGURE 14.5

The hypothalamic-pituitary-adrenal (HPA) axis that regulates cortisol production. This is called a negative feedback; a biological mechanism like a central heating system. When the temperature in the room rises, the thermostat in the room turns off the boiler until the temperature drops and the boiler is turned on again. If the adrenal cortex cannot make cortisol (for example if there is a genetic defect of an enzyme in the pathway) then the ACTH levels will be very high. Continued stimulation of the adrenal gland will lead to cell growth and division and hyperplasia of the gland.

receptors on the plasma membranes of adrenal cells prior to activation of adenylate cyclase. The increase in cyclic AMP in the cytoplasm then activates phosphoprotein kinase and several other enzymes by phosphorylation. The newly activated enzymes stimulate several processes of which the hydrolysis of cholesterol esters is crucial.

SELF-CHECK 14.2

What is the impact on pituitary function of a low plasma cortisol concentration due to an adrenal defect in cortisol production?

Aldosterone production is principally regulated by the renin-angiotensin system. The process starts at the kidney which is responsive to changes in electrolyte balance and to plasma volume. Renin is released from the kidney in response to a reduced renal perfusion, by hyperkalaemia, or by reduced concentration of sodium. Renin is a glycoprotein of M_r of 42,000 produced in the kidney from an inactive proenzyme. Renin is a proteolytic enzyme that in the blood acts on renin substrate (angiotensinogen an α2-globulin synthesized in the liver) to release angiotensin I as outlined in Figure 14.6.

Angiotensin I has little biological activity until further hydrolysed to angiotensin II, an octapeptide, by the action of angiotensin converting enzyme (ACE). This converting enzyme is present in high concentrations in the lung, although widely distributed in the vasculature and other tissues. Angiotensin II has a short plasma half-life of 1–2 minutes, being rapidly destroyed by the action of angiotensinase.

SELF-CHECK 14.3

What is the response of the kidney to salt loss in the urine?

The synthesis of aldosterone is regulated by the actions of angiotensin II to increase the activity of enzymes at two steps in its biosynthetic pathway. The first regulation site is the conversion of cholesterol to pregnenolone. The second step is the conversion of corticosterone to aldosterone. Both sites are also stimulated by high concentrations of potassium ions. Adrenocorticotrophic hormone can induce a temporary rise in aldosterone synthesis but this is not sustained. Hyponatraemia can itself lead to an increase in aldosterone secretion; hyperkalaemia also stimulates aldosterone biosynthesis, whereas hypokalaemia inhibits aldosterone production.

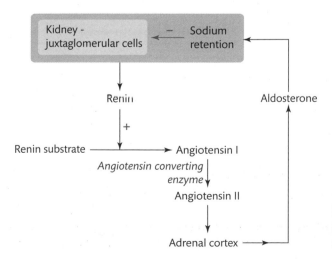

FIGURE 14.6
Regulation of aldosterone release by the renin-angiotensin-aldosterone (RAA) system.

Key Points

Adrenocorticotrophic hormone and the renin-angiotensin-aldosterone system control the rates of synthesis of cortisol and aldosterone respectively.

Steroid metabolism

Steroids are inactivated to metabolites prior to clearance from the body. Most steroids are excreted in the urine but some are lost through the intestinal tract. In general, steroid catabolism occurs by reduction mainly in the liver. Enzymes act principally on the A-ring and side chains. The double bond at C4-C5 can be reduced by the addition of two hydrogen atoms from 5β and 5α-reductases. The C-3 carbonyl groups are reduced by 3α-hydroxysteroid dehydrogenase found mainly in the liver. Since four hydrogen atoms have been added to the active steroid the metabolites are tetrahydro products of the systemic or trivially named hormones. Reduction of the 20-oxo group in the side chain of cortisol is catalysed by 20β and 20α-hydroxysteroid dehydrogenases producing isomeric hydroxyl groups at C-20 of which the 20α reduced steroids predominate in humans. Reduction at the C-20 position converts a tetrahydro reduced steroid to a hexahydro reduced steroid. These steroids have the trivial names cortol and cortolone from cortisol and cortisone respectively. Cortisol and cortisone are converted by side-chain cleavage to 11-oxygenated androgens. Aldosterone is catabolized in a similar fashion to cortisol with the formation of 3α,5β-tetrahydroaldosterone.

Cross reference

Chapter 15 Reproductive endocrinology

In the periphery and liver, DHEAS can be converted to DHEA (by a sulphatase enzyme) which is then acted on in the A-ring by the protein from HSD3B2 to produce androstenedione. This is acted upon by ketosteroid reductase to produce testosterone.

Androstenedione can be reduced with four hydrogens in the A-ring (in essence to tetrahydro-androstenedione) to the two important adrenal androgen metabolites, with trivial names androsterone and aetiocholanolone, as shown in Figure 14.7.

The majority of steroid metabolites are excreted in the urine as their sulphate and glucuronide conjugates, formed mainly in the liver through transferase enzymes (phosphoadenosine phosphosulphate and uridine diphosphoglucuronic acid). A highly polar 18-glucuronide conjugate of aldosterone is formed primarily in the kidneys and excreted in urine. Aldosterone can be liberated by hydrolysis at pH 1 which is a unique procedure incorporated into a urine test used to assess aldosterone production (see Section 14.2).

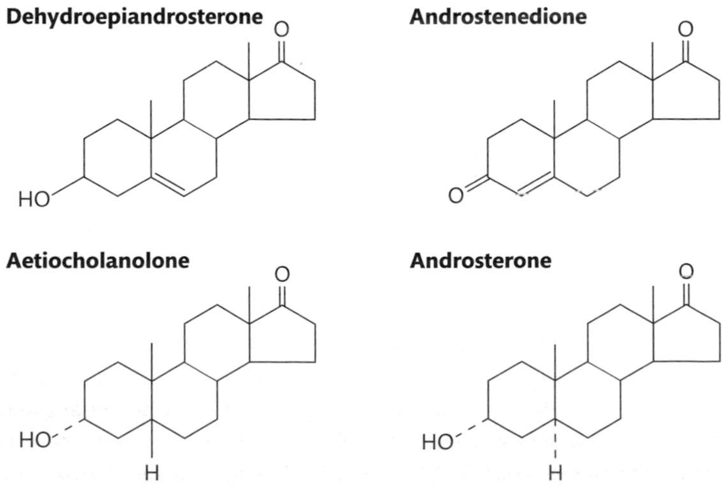

FIGURE 14.7

Androgen metabolites including dehydroepiandrosterone, androstenedione, aetiocholanolone, and androsterone.

In newborn children a number of steroid hydroxylases (1β-, 6α, and 6β, 15β, 16α, and 16β, 18) act on steroids possibly to increase the rates of steroid excretion. The metabolites are extremely polar and difficult to extract from urine. These enzymes become less active over the first year of life and apart from the induction of some of the enzymes by drugs (notably the 6-hydroxylase by anti-convulsants) do not influence the metabolism of steroids outside the newborn period.

Key Points

Steroids are inactivated before clearance of metabolites mainly in the urine. The enzymic steps are in general reductive before conjugation with glucuronic and sulphuric acid. In newborn infants further steroid hydroxylation is utilized to increase polarity and aid renal clearance.

Genetics of steroid production

The enzymes in steroid synthesis and catabolism are formed from genes expressed at many sites on several chromosomes. A little understanding of this will help make sense of some of the clinical disorders of the adrenal glands that are due to gene deletions, mutations, and conversions. The commonest defect of steroid synthesis is due to mutations in the 21-hydroxylase (*CYP21A2*) gene. The 21-hydroxylase protein has an M_r of 52,000 and is the product of a gene of 10 exons on the short arm of chromosome 6. Gene duplications are particularly common in the region of chromosome 6p21 coding for the human leucocyte antigen (HLA) compatibility locus, as can be seen in Figure 14.8.

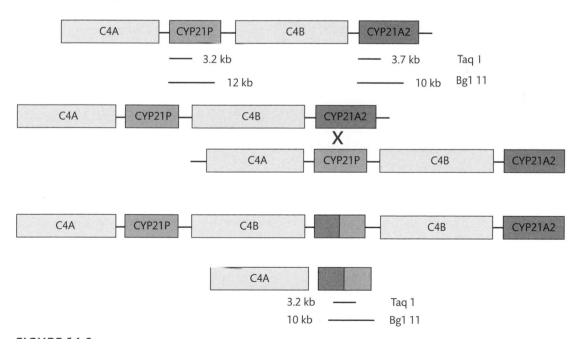

FIGURE 14.8

The arrangement of CYP21 genes on chromosome 6. Chromosome 6 has two related CYP 21 genes within a region coding for complement genes—C4A and C4B. CYP21A2 codes for 21-hydroxylase, CYP21P is a pseudogene with many sequence differences to CYP21A2. When the two copies of the gene are misaligned during meiosis, an unequal cross over leads to one chromosome with three copies of the C4/21-hydroxylase gene pairs and one chromosome with only one C4/21-hydroxylase gene pair that has no CYP21A2 gene and is thus an allele for 21-hydroxylase deficiency.

About 6 kb away from CYP21A2 there is a homologous gene (CYP21P) whose sequence has a number of differences from that of the active gene that prevents protein synthesis, which renders it non-functional. A gene can be deleted or duplicated by unequal crossing-over at meiosis. The changes in CYP21P can become incorporated into CYP21A2 by gene conversion, leading to a defective protein.

The gene for 11-hydroxylase in the cortisol pathway (called *CYP11B1*) is located in the chromosome 8q21-q22 region. About 45kb further along the chromosome is a homologous gene (*CYP11B2*) that codes for aldosterone synthase. The two genes are approximately 11 kilobases long and contain 9 exons. The genes are 95% identical in coding regions and 90% identical in introns. The enzymes are predicted to be 93% identical in amino acid sequence. CYP11B2 has 11-hydroxylase activity and will 18-hydroxylate corticosterone and further oxidize this to aldosterone. CYP11B1 only has 11-hydroxylase activity and is expressed in the zona fasciculata at significantly higher levels than CYP11B2 in the zona glomerulosa. A chimeric gene is the basis of a mutated protein in a condition called dexamethasone suppressible aldosteronism. The protein has high 18-hydroxylase activity for cortisol and ability to synthesize aldosterone in response to ACTH.

Adrenal medulla

The catecholamines, adrenaline (epinephrine) and noradrenaline (norepinephrine) are produced in the adrenal medulla. They increase the heart rate and blood pressure, constrict blood vessels, dilate bronchioles, and generally increase the rate of metabolism. Receptors for catecholamines are found on cells widely distributed throughout the body.

Adrenal medullary hormones

Adrenaline is synthesized from the amino acid tyrosine as shown in Figure 14.9.

Dihydroxyphenylalanine (DOPA) is formed by hydroxylation of tyrosine. On decarboxylation of DOPA, dopamine is produced. Hydroxylation of dopamine leads to production of noradrenaline which is N-methylated to adrenaline.

Regulation of catecholamines

Medullary cells release adrenaline into the blood in response to illness, exercise, trauma, and imminent danger. The effects of catecholamines are experienced when one is frightened. Heart rate increases, breathing is shallower, pupils dilate, and hair stands on end (goose pimples). Noradrenaline is predominantly derived from sympathetic nerve endings.

Catecholamine metabolism

Catecholamines have short half-lives of only a few minutes through the action of two enzymes. Catechol-O-methyl transferase is responsible for degradation of adrenaline and noradrenaline. Metanephrines are the products: 3-methoxymetanephrine from dopamine, normetanephrine from noradrenaline, and metadrenaline from adrenaline, as shown in Figure 14.10.

Monoamine oxidase converts adrenaline and noradrenaline to dihydroxymandelic acid, and metadrenaline and normetadrenaline to 4-hydroxy-3-methoxymandelic acid (vanillylmandelic acid VMA).

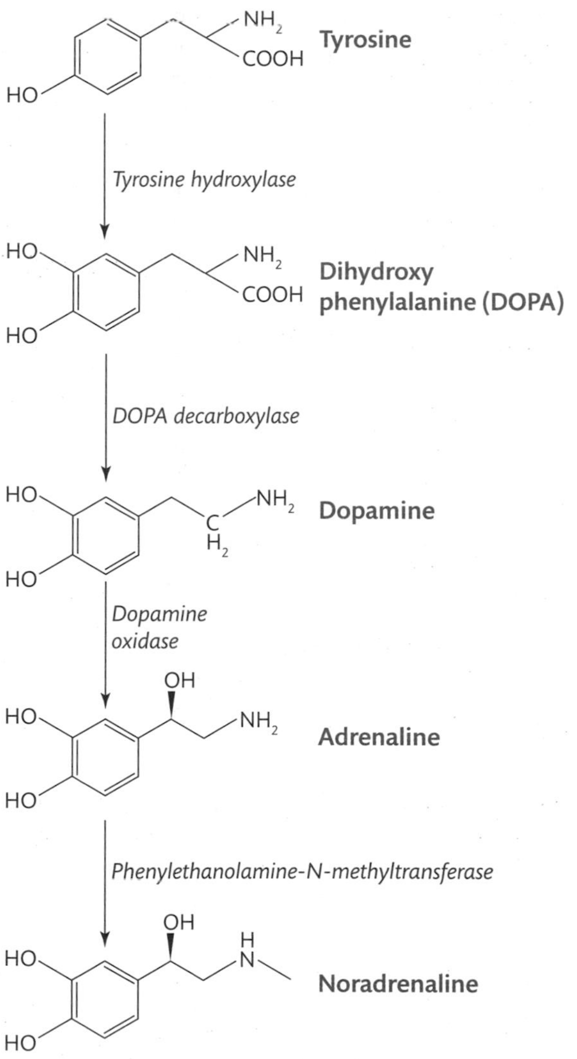

Tyrosine

Tyrosine hydroxylase

Dihydroxy phenylalanine (DOPA)

DOPA decarboxylase

Dopamine

Dopamine oxidase

Adrenaline

Phenylethanolamine-N-methyltransferase

Noradrenaline

FIGURE 14.9
Synthesis of the catecholamines adrenaline and noradrenaline from tyrosine.

Key Points

The adrenal medulla converts tyrosine into catecholamines. In response to exercise and imminent danger, cells of the medulla respond to neural signals and adrenaline (and to a lesser extent noradrenaline) is secreted into the blood. Heart rate and blood pressure increase. There is blood vessel constriction in the skin and dilatation of bronchial muscles. The catecholamine metabolites metadrenaline and normetadrenaline are excreted in the urine

SELF-CHECK 14.4

Where have you experienced the physical effects of catecholamines?

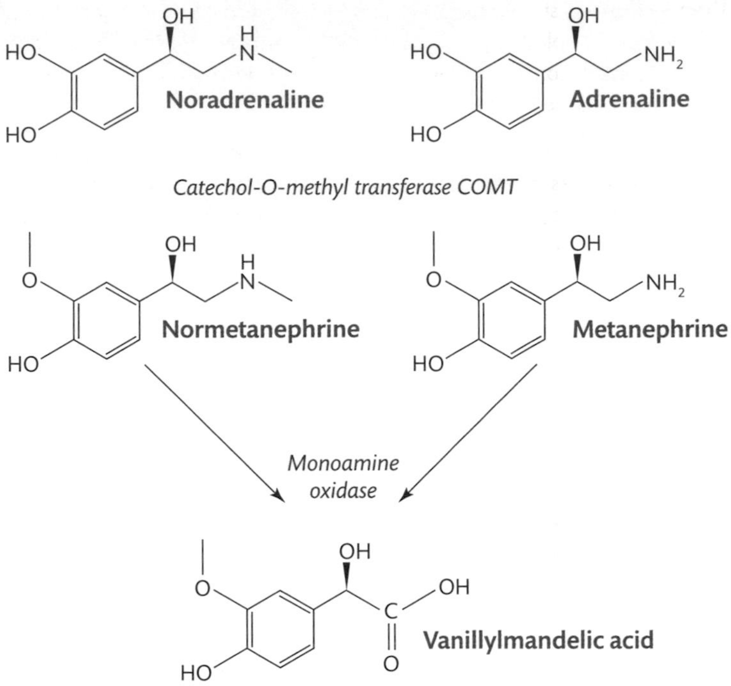

FIGURE 14.10
Metabolism of catecholamines where adrenaline and noradrenaline are converted to vanillylmandelic acid. Note COMT is expressed in adrenal medullary tumour cells so plasma metanephrines are markers for phaeochromocytoma.

14.2 Disorders of adrenal glands

A number of clinical problems of the adrenals are encountered and it is necessary to understand how laboratory tests are used to define the cause of each condition. In order to demonstrate any hormone abnormality in the pituitary, adrenal cortex, adrenal medulla, or elsewhere, many biochemical tests are required, sometimes supported by genetic tests. It is important to understand the means, limitations, and pitfalls of these determinations. It is also important to appreciate when a hormone concentration in a body fluid is outside normal limits and whether this is an artefact or whether it is clinically relevant.

Adrenal steroids in blood

Cross reference

Chapter 18 Immunological techniques in the *Biomedical Science Practice* textbook

Initially, steroids were measured by chemical methods and these were replaced by immunoassays around 1968. **Competitive immunoassays** for steroids in plasma or urine have been used since the early 1970s. The amount of label bound to the antibody is measured; this declines as the concentration of analyte increases. The plasma concentrations of adrenal regulatory peptides are determined with two-site immunoassays or **immunoradiometric assay** (IRMA).

The labels used in these assays were often radioactive but can now be fluorescent, chemiluminescent, or enzyme labels. In many of the immunological methods in use from the 1990s the antisera and the label were added directly to plasma samples. In some of these assays the antiserum is immobilized on the wall of a tube (coated tube) or particle that can be precipitated or sedimented. In direct assays, steroids like cortisol need to be displaced from binding proteins if the total serum concentration is to be determined. The concentrations of binding proteins are influenced by other factors like body weight, use of oestrogens (certain oral contraceptives, for example), and pregnancy.

The specificity of steroid measurement by immunoassay is a major challenge because within a biological sample there will be many structurally similar steroids which can cross-react. However, extraction and chromatography have been used to reduce such interferences. For more detail see the Method box below.

Care should be taken in the selection of assay methods for cortisol measurement in paediatric samples in the first six months of life. In the newborn infant, cortisol is readily converted to its inactive, C-11 oxidized product, cortisone, which occurs in plasma at concentrations up to 2,000 nmol/L, whereas cortisol can be less than 400 nmol/L. After six months of age, there are noticeable changes in the pattern of cortisol levels. First, the ratio of cortisol to cortisone is about one to four, as is found throughout the remainder of life. Second, the circadian rhythm of cortisol is also established around that time. A low cross-reaction of cortisone in the cortisol assay is necessary.

Chemical methods for steroid analysis based on **gas** and **liquid chromatography** (GC and LC), in some cases coupled to **mass spectrometry** (MS), have been used as **reference methods** for steroids since the 1970s. Gas chromatography coupled with mass spectrometry (GC-MS) is probably the most accurate analysis of steroids. Liquid chromatography coupled with tandem mass spectrometry is more amenable to routine hospital laboratory use and is being adopted by many laboratories. More details on tandem mass spectrometry are provided in the Method box below.

Cross reference

Chapters 11 and 13 Spectroscopy and Chromatography respectively in the *Biomedical Science Practice* textbook

METHOD *Extraction and chromatography of steroids*

Selective solvent extraction can partially take up certain steroids from the biological samples before immunoassay. This step reduces interference in steroid measurements in blood from newborn infants. The extraction step leaves in the plasma the steroid sulphates that have been secreted from the large foetal adrenal gland. The solvent containing the free steroids is evaporated and the dry extract dissolved in a buffer for assay. In many clinical circumstances, direct assays are functional but since the antisera used are rarely specific, chromatography may be necessary to purify steroids in an organic extract. This approach has the additional advantage in paediatric studies, where in a single plasma extract, several steroids can be extracted before measurement. It may be necessary to perform column or high performance liquid chromatography to collect fractions of the column elutes that contain the individual steroids prior to quantitative analysis of each by radioimmunoassay.

METHOD *Tandem mass spectrometry*

The mass spectrometer displays signals for the natural and stable isotope labelled steroids, and concentration of the steroid in the sample can be determined by comparison with the ratios determined with known mixtures or calibrants (isotope dilution analysis). Several steroids can be determined in one analysis by changing the mass spectrometer conditions during the duration of the liquid chromatography run, so as to detect individual steroids at their characteristic retention times.

Since 2005, there has been much criticism of the wide variations in reference ranges for steroid hormones from immunoassay data. Liquid chromatography coupled with tandem mass spectrometry will become available more widely for general use and there are likely to be standardized supplies of reagents that will reduce the imprecision seen between laboratories that perform these assays using reagents made in house.

Quantitative analysis is achieved by the addition of known amounts of stable-isotope labelled steroids to the biological sample.

In the blood, cortisol is largely bound to a specific binding protein called **cortisol binding globulin** (CBG) and to a lesser extent albumin. Less than 5% of the cortisol is free in the plasma. The concentrations of cortisol are usually determined as the total of bound and free steroid. This contrasts with investigations of thyroid function when plasma concentrations of the free thyroid hormones are determined. The times when blood samples for cortisol are taken are important. The adrenal gland secretes cortisol in what is described as a diurnal rhythm, with lowest plasma concentrations at midnight and highest levels at 8 am. In many textbooks this phenomenon is displayed as a sine curve but in fact there are in any individual a series of peaks between 3 am and 10.30 am at about 90-minute intervals of increasing amplitude. Since there are differences in the times of onset of this process the highest concentrations will be different in time and plasma concentration between individuals and this is the basis of the wide reference ranges for plasma cortisol in a population.

SELF-CHECK 14.5

(a) What are the ideal times for blood samples to be taken to assess cortisol production? (b) What are the limitations of your approach?

Aldosterone should be measured when electrolyte balance is disturbed or a patient has drug resistant hypertension. In most cases concomitant renin measurement is helpful. Adrenal and kidney diseases can be recognized using these results. Aldosterone should not be measured if aldosterone secretion is known to be low due to enzyme defects (see congenital adrenal hyperplasia) or adrenal failure (see Addison's disease). Aldosterone synthesis is affected by potassium concentrations and hypertensive drug treatments. Aldosterone concentrations in blood are most often measured by immunoassay. A few methods have been automated and tandem mass spectrometry is feasible.

Dehydroepiandrosterone sulphate is the major adrenal androgen and its secretion varies with age. It is secreted episodically with cortisol. Dehydroepiandrosterone sulphate has a long half-life of 10–20 hours and consequently the concentrations of DHEAS in plasma are relatively stable as opposed to the episodic secretion of androstenedione and DHEA. The dramatic increases in plasma concentrations of DHEAS before puberty are shown in Figure 14.11 and reflect remodelling of the adrenal cortex as the gland grows to adult size.

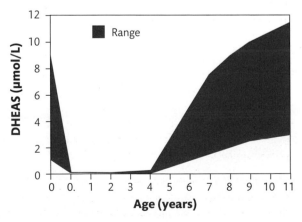

FIGURE 14.11
Plasma DHEAS concentrations with age during childhood.

The mechanism by which the additional zona reticularis develops has been controversial. The activity of 3β-hydroxysteroid dehydrogenase is markedly reduced. Growth factors may participate in intra-adrenal mechanisms that mediate steroidogenic changes that characterize the adrenarche. A pro-opiomelanocortin fragment (ACTH 79–96) as shown in Figure 14.4 was isolated as the putative androgen stimulating hormone but this action was not confirmed.

Intermediates in the path of cortisol synthesis (like 17-hydroxyprogesterone) become markers for the biosynthetic defects. The plasma concentration of the enzyme substrate is much higher than the normal ranges, whilst steroids that would be made further in the pathway are below the reference ranges when providing evidence for a defect in steroidogenesis. If immunoassays are replaced with tandem mass spectrometric methods it will be possible to determine several steroids in a single analysis.

Key Points

Most hospitals can offer a service for cortisol and DHEAS measurement. In many laboratories ACTH and 17-hydroxyprogesterone (and other relevant intermediates) will be referred elsewhere for analysis.

SELF-CHECK 14.6

Disorders of the adrenal cortex can be life threatening. (a) How should that influence the ways your laboratory handles samples? (b) If rapid turnaround times are not possible what procedures should be in place?

Adrenal steroids in urine

Free cortisol is present in the urine after filtration of free hormone in the blood at the kidney. Once plasma binding proteins are saturated, the free cortisol excretion rate will increase in proportion to the increased production of the hormone. Urine free cortisol excretion is therefore a useful measure of excess cortisol production as in Cushing's syndrome, but not for adrenal insufficiency. There are a number of drugs that interfere with its determination, for example synthetic steroids like prednisolone. Free cortisol in urine is usually measured by immunoassay preferably after solvent extraction which will not remove the many steroid conjugates from the urine sample. Antisera should be tested for their ability to bind metabolites of steroids since these are present in urine at relatively higher concentrations than cortisol itself.

Determining the rate of excretion of steroids in urine collected over 24 hours integrates the daily adrenal activity. The collection of a complete 24-hour urine sample cannot be relied upon, especially from children. Urine collection bags for babies are notoriously difficult to retain on an active newborn although some paediatric departments have special cots, in which the baby lies on a porous nylon sheet which allows urine to pass through, with flow directed to a collection vessel (preferably cooled). In older children, 24-hour collections are only possible with close supervision.

The **cortisol secretion rate** can be determined by stable isotope tracer techniques. The use of stable isotopes for this purpose has proved more accurate than the earlier radioactive tracers. The experiments can be conducted more physiologically and safely, even in children, but are only available in very specialized centres and are expensive. The secretion rate can be approximated from the combined excretion rates of all of the cortisol and cortisone metabolites.

Assessment of the total cortisol metabolite excretion rates has been used to determine if cortisol is produced in excess in obese children or suppressed in children with asthma taking inhaled steroids. A two-hour urine collection is essential for these studies because the excretion rates of cortisol metabolites vary, with highest concentrations between 10 am and 6 pm.

The excretion rate of the acid-labile conjugate of aldosterone (18-glucuronide) is a convenient index of aldosterone production. This conjugate is cleaved by acid at pH 1, whereas other glucuronides at C-3 and C-21 are not affected by this degree of acid exposure. After extraction of free steroids from urine and acid hydrolysis a second dichloromethane extract will contain aldosterone liberated by the acid hydrolysis and this can be measured by using an immunoassay. The excretion rate of aldosterone 18-glucuronide correlates with the sodium excretion in the urine. The excretion rate of 3α, 5β-tetrahydroaldosterone has proved useful in paediatric and hypertension studies. This assay usually requires the specialized facilities of a research laboratory or focused endocrine centre. Free aldosterone and 18-hydroxycorticosterone can be determined using HPLC and HPLC coupled with tandem mass spectrometry.

Key Points

Urine free cortisol is usefully measured to demonstrate high cortisol production but not adrenal suppression.

Profiles of urine steroids by gas chromatography and mass spectrometry

The term profile is often used when examining temporal changes of hormone concentrations in a fluid under basal conditions or in response to stimuli. A profile of urinary steroids refers to the chromatographic picture from the GC showing separation of steroids over time that gives information about the mix of all steroid metabolites in urine. For each steroid, several metabolites have to be identified and quantified.

The test is worth organizing when the results of local tests have proved normal or require confirmation using a different procedure. Some further information on the test is found in the Method box below.

 METHOD *Urinary steroid profile analysis*

Capillary column gas chromatography is used for the separation of individual steroids. The column outlet can be either a non-specific flame ionization detector (peaks characterized by absolute or relative retention) or a mass spectrometer where the steroid in each peak can be identified by its retention time and the mass spectrum which is a finger-print displaying molecular weight and a characteristic fragmentation pattern. In general, the urine steroid profile is an assessment of adrenal cortical function because the production of steroids by the gonads is insufficient to

influence the quantities of steroid metabolites in the urine. Analysis of steroids in urine, is in general, performed in specialist laboratories. The analysis is a general screen for abnormalities in amounts and ratios of steroids of wide diagnostic value. Despite the range of steroids in clinical use, few of them interfere with the analyses but can affect the steroid excretion rates. The drawback of urine profile analysis is that several metabolites arise from each hormone, complicating interpretation of the data. Steroid excretion rates in a 24-hour urine collection give indices of daily cortisol and androgen production. The ratios of various groups of metabolites can also be determined. For example the ratio of 5α to 5β reduced metabolites, and the ratio of cortisol to cortisone. Some drugs interfere with the assay, for example carbamazepine, cyproterone acetate, and megesterol give rise to peaks in the chromatogram. Some drug metabolites can be recognized by comparison of the mass spectra with those in a library that comes in most instrument software packages. Prednisone and prednisolone metabolites can be recognized from the mass spectra because the molecular weights are two mass units less than the equivalent endogenous cortisone and cortisol metabolites.

In the urine from a normal child the major steroid metabolites quantitatively are from cortisol and cortisone. The urine of a newborn infant up until about four to six months of age is an exception as only cortisone metabolites are detectable. For the sum of excretion rates of cortisol and cortisone metabolites there is a close agreement of absolute steroid excretion with body surface area. This information has been useful when judging adrenal excess in an obese child and in evaluating the effect of synthetic steroids in suppressing the hypothalamic-pituitary-adrenal axis. Blood sampling throughout the night to demonstrate low concentrations of cortisol through suppression of nocturnal cortisol pulses is not a practical alternative.

Androgen excretion rates are low for the first seven years then rise ten to twelve-fold over the next seven years to adult excretion rates. This is the adrenarche and the increased androgen metabolites that appear in the urine reflect the increased production of DHEAS. Premature adrenarche is a benign condition resulting from advanced adrenal growth with precocious differentiation of the zona reticularis. Androgen and cortisol metabolite excretion rates are marginally elevated for age and to a lesser extent body size.

Adrenocorticotrophic hormone and renin measurements

Adrenocorticotrophic hormone is usually measured by RIA, IRMA, or by use of an automated immunoassay analyser. Care must be taken to avoid stressing the patient when collecting blood samples for ACTH measurement. Adrenocorticotrophic hormone is the principal regulator of cortisol secretion and is usefully measured in plasma to see if secretion is high in adrenal failure or pituitary disease secreting ACTH autonomously, or is lowered, by the presence of high levels of glucocorticoids or pituitary failure.

Plasma renin activity (PRA) has been determined by immunoassay of angiotensin I; the enzyme product of renin action on endogenous angiotensinogen (renin substrate). The concentration of renin itself (PRC) can be measured by immunoassay. For PRA the blood sample after collection should not be kept cold but taken immediately to the laboratory for separation of the plasma that must be removed from the blood cells and frozen quickly. The plasma should then be kept frozen until thawed in the laboratory for the assay (see Method box below).

 METHOD *Plasma renin*

Inactive forms of renin in plasma are known to occur which are activated by acid, cold, or proteolytic treatment. Interpretations of renin results are complicated by the use of hypertensive drugs. These affect the production and actions of renin, aldosterone, angiotensinogen, and angiotensin converting enzyme. For example, treatment of a patient with a direct renin inhibitor will increase the plasma concentration of renin but block the renin action in the assay of activity. Renin activity is affected by posture, with an increase in activity on standing up from lying down due to a decrease in renal blood flow. Renin (through angiotensin II) is the principal regulator of aldosterone secretion and is usefully measured to distinguish primary and secondary hyper- and hypoaldosteronism.

> ## Key Points
>
> Renin and aldosterone assays are not particularly difficult to perform but are in general the tests best referred to specialist centres with experienced staff who can advise on interpretation of results in rare disorders. In the case of patients with hypertension, many drugs are available that impact on all stages in the production, regulation, and actions of these two hormones. The specialist centres should be provided with as much background information as possible to help them offer sound interpretation of the results.

Catecholamines

Catecholamines and their metabolites are measured in blood and urine samples and requested most often in patients with hypertension and adrenal or other masses. Requests to support diagnosis of phaeochromocytoma are increasing because of greater use of abdominal imaging. The secretion rate and plasma concentrations of catecholamines are affected by stress and renal failure so a 24-hour timed urine collection is required. Opaque polyethylene bottles are needed containing 10 mL of concentrated hydrochloric acid as preservative. This provides a pH range of 2 to 3 to ensure stability of the catecholamine metabolites. Measurements of VMA are becoming obsolete as drugs and certain foods (such as bananas and vanilla ice cream) can yield false positive results. Plasma metanephrines are now regarded as the best screening method because the tumours express catechol-O-methyltransferase activity but clinicians still request measurements of VMA, catecholamines, and metanephrines in urine. Enzymatic immunoassays based on microtitre plates are available but are often subject to cross-reaction and show poor agreement between manufacturers. Gas chromatography coupled with mass spectrometry is used as a reference method for metanephrines. Most routine methods now use HPLC with electrochemical detection but faster methods are available with tandem mass spectrometry. Sample preparation can be minimized using solid phase extraction and then the metanephrines are separated with a reverse-phase, cyano or hydrophilic interaction column. Normetanephrine in plasma typically above 1 nmol/L or metanephrine above 0.5 nmol/L (depending on the laboratory) are found in patients with phaeochromocytoma. Concentrations of metanephrines can however be raised due to stress and high levels of trauma so care must be taken with sample collection. False positive results can be caused by anti-hypertensive drugs such as methyl-DOPA, dopamine agonists, clozapine, calcium channel blockers, adrenergic blockers, tricyclic anti-depressants, and sympathomimetics.

Provocative testing

Measurements of hormones in blood and saliva samples provide assessment only of the situation at that point in time. However, the endocrine system is dynamic and abnormal results may need to be confirmed by further stimulating or inhibiting the endocrine gland.

Cortisol and ACTH stimulation tests

Four different stimulation tests are available all of which have limitations. The **synacthen** test is the most commonly used test of adrenal function usually as a high dose test. Before 250 µg of ACTH 1-24 (synacthen) are injected, a blood sample is taken for measurement of the plasma concentration of cortisol. Another blood sample for cortisol measurement is taken 30 minutes later and many protocols include a further sample at 60 minutes after the ACTH was injected.

The response to the stimulus is greater than from other stimulatory tests, but the results at maximum response are comparable with the other tests. Adrenocorticotrophic hormone stimulation tests should not be used in patients with a history of atopy because there is a danger of an anaphylactic response. A normal response was defined originally by fluorometric assays and those results remain within the literature. A peak plasma concentration of cortisol of 550 nmol/L and/or an increment of cortisol of 220 nmol/L are accepted to be the normal response, but these standards should be confirmed for the method in use. Lower peak plasma concentrations of cortisol or lesser increments are seen in adrenal insufficiency. A low dose test (1 µg is the most common dose for the stimulus, lower doses can be used as shown in Figure 14.12) achieves a cortisol response over 25 minutes similar to the high dose test and concentrations then fall. Most protocols, however, recommend bloods to be taken at zero and 30 minutes and may miss the highest concentrations achieved before that time, as shown in Figure 14.12.

A stimulation of ACTH release occurs in the **corticotrophin releasing hormone** (CRH) test. Corticotrophin releasing hormone (100 µg intravenously) stimulates ACTH secretion significantly over an hour only in patients with pituitary tumours. Plasma ACTH concentrations can then increase from a mean value of 35 ng/mL to more than three times the basal levels at 15 minutes, and cortisol concentrations 30–90 minutes after the CRF injection will double. In control subjects, plasma ACTH and cortisol concentrations may double during the test.

In the **metyrapone** test, cortisol synthesis is blocked at the 11-hydroxylase step, lowering plasma concentrations of cortisol and thus testing the ability of the pituitary gland to naturally stimulate ACTH release. The metyrapone test requires measurements of increased 11-deoxycortisol (the substrate for 11-hydroxylase, above 100 nmol/L) or ACTH (>75 ng/mL) on the morning after a night-time dose of the drug. Plasma cortisol should be <200 nmol/L. Failure to increase 11-deoxycortisol in the test indicates pituitary failure or adrenal failure. The responses observed with both of the above tests are more variable than seen with the synacthen test.

The **insulin induced hypoglycaemia test** also stimulates ACTH release and hence cortisol secretion. Plasma glucose concentration falls below 2.2 mmol/L. This procedure is potentially dangerous and should only be performed with medical supervision. A sugar rich drink needs to be available in order to raise blood glucose concentration to normal quickly. Blood samples are taken at 0, 10, 20, and 30 minutes into separate tubes for glucose and cortisol analysis.

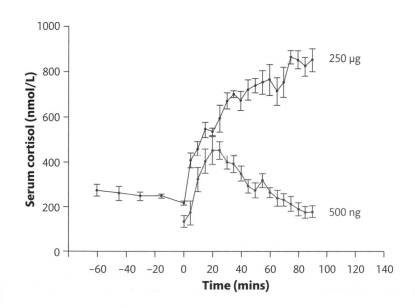

FIGURE 14.12
Serum cortisol concentrations after a low dose (250 µg) and a high dose (500 ng) synacthen test.

Failure to increase cortisol concentrations above 500 nmol/L in this test indicates pituitary failure or adrenal failure.

Dexamethasone suppression test

A high blood concentration or urine excretion rate of cortisol can occur due to stress, depression, or alcoholism. The synthetic corticosteroid, dexamethasone, acts like cortisol in the feedback loop to suppress ACTH from the pituitary. Plasma cortisol concentrations are then lowered. Failure to achieve suppression of cortisol to concentrations below 150 nmol/L with 0.5 mg of dexamethasone overnight or better four times a day for two days indicates that abnormal steroid production is autonomous. If the dose is increased to 2 mg four times a day for two days then plasma cortisol at 8 am will not be suppressed when there is an ectopic ACTH secreting tumour, whereas with a pituitary tumour plasma cortisol concentrations will be around half of the basal result.

Confirmation of phaeochromcytoma

The indications for an adrenal medullary tumour may need to be confirmed by repeat testing if plasma metanephrines are marginally raised. Analysis of urinary free metanephrines may also be helpful. The conjugates reflect metabolites formed in gastrointestinal tissue so may be less diagnostic. The condition is rare and experience of other tests is limited in most hospitals. A tumour may be visualized with an MRI scan. Additional functional scanning studies include injection of radioactive substances such as metaiodobenzylguanidine or octreotide. Some centres still use dynamic tests such as clonidine which acts like catecholamine in the brain, reducing activity of the sympathetic nerves to the adrenal medulla. If catecholamine concentrations are suppressed then a malignancy is unlikely. Blood pressure needs to be closely monitored during the test.

Key Points

It is not unusual in endocrine tests for results to be equivocal, bearing in mind the issues with age, gender, timing of sample, reference ranges, development, assay specificity, and pathology that can impact on the interpretation. Provocative tests are required not only to exclude pathology (raised cortisol that is normalized by a dexamethasone suppression test), but also to assist in the differentiation of causes of disease (Cushing's syndrome).

Reference ranges for adrenal cortical hormone

Aldosterone shows a circadian rhythm due to the transient influence of ACTH on aldosterone production in the morning. Concentrations of aldosterone in plasma increase on assuming an upright position after lying down quietly for 15 minutes. It is important to interpret aldosterone levels in relation to the plasma renin activity or concentration. Care must be taken in interpreting results in patients of different ages and against reference ranges in the literature (see Method box below).

At 8 am plasma cortisol can be in the range 200–800 nmol/L. This depends largely on the timescale of the pulsatile patterns in the individual. When concentrations below 100 nmol/L at 8 am can be demonstrated then adrenal failure is likely. Cortisol concentrations throughout most of the day are less than 200 nmol/L, with the lowest concentrations around midnight. Lack of this diurnal change, with high levels at midnight and in the morning, is diagnostic of autonomous cortisol secretion.

METHOD *Renin and aldosterone results*

High plasma concentrations of aldosterone and PRA are seen in the normal newborn infant. Both hormone concentrations decline over the first three years of life and adult values are reached at 3–5 years of age. Laboratories use many different units for reporting aldosterone concentrations (ng/dL, pmol/L) and plasma renin activity (ng/mL/h, pmol/L/min), or plasma renin concentration (ng/L, mU/L). Care must therefore be taken when interpreting a number for the aldosterone to renin ratio (ARR) for units, time, and volume. The cut-off point above which the ARR

indicates an adrenal tumour can vary numerically from 2.4 to 2,000. The ratios most frequently reported are:

>30 when aldosterone is in ng/dL and PRA is in ng/mL/h;

>750 when aldosterone is in pmol/L and PRA is in ng/mL/h.

Investigations are often undertaken when the patients are under some treatment for the underlying disease. Sometimes the aldosterone concentration can be in the normal range but above 350 pmol/L and the ARR is raised.

In neonates, high concentrations of DHEAS (up to 10 µmol/L) reflect the low activity of 3β-hydroxysteroid dehydrogenase in the foetal adrenal cortex. Plasma DHEAS concentrations are higher in preterm infants compared with those in full-term babies. The DHEAS concentrations decline over the first six months after delivery at term and later in a preterm baby. During infancy, from 6 months to 6–8 years DHEAS concentrations are below 1 µmol/L but by 8–10 years the levels are increasing as the zona reticularis develops. The increase in concentration of DHEAS (adrenarche) leads to pubic hair growth. In adults, plasma concentrations of DHEAS are normally between 2 and 10 µmol/L throughout the day and night until 50–60 years of age when the size of the zona reticularis decreases and the DHEAS concentrations decline.

The concentrations of 17α-hydroxyprogesterone (17-OHP) in the plasma are reported generally to be above 40 nmol/L during the first 48 hours of life. After that time the concentrations in normal children are less than 10 nmol/L. In very ill children, particularly preterm infants, persistently raised concentrations can be found. This may partly reflect the quality of the assays, particularly if assays are performed directly without extraction of steroid into an organic solvent. Steroids from the foetal adrenal cortex must be tested for cross-reactivity in any immunoassay used to measure 17-OHP in newborn infants. High concentrations of 17-hydroxypregnenolone sulphate have been found in newborn plasma and it binds to certain 17-OHP antisera. Plasma concentrations of 17-OHP by mass spectrometric methods are less than 3 nmol/L in normal infants.

Reichstein's compound S (11-deoxycortisol) is usually determined by RIA and the plasma concentrations are normally 5–20 nmol/L. High performance liquid chromatography has been used to reveal the high concentrations of 11-deoxycortisol in patients with congenital adrenal hyperplasis due to 11-hydroxylase defects and in patients receiving metyrapone medically to suppress cortisol production by blocking the 11-hydroxylase. Reference ranges for adrenal steroids are given in Table 14.2.

Reference ranges for catecholamines

Typical values are available for catecholamines as shown in Table 14.3 but the laboratory should establish values for the method in use.

Tests should be conducted without stress and after review of medications likely to affect results (for example tricyclic antidepressants or phenoxybenzamine).

TABLE 14.2 Reference ranges or upper limits for adrenal hormones in plasma and urine.

Hormone	Collection details	Concentration
Plasma aldosterone	Recumbent	100–500 pmol/L
Plasma aldosterone	Ambulant	600–1200 pmol/L
Plasma renin activity	Recumbent	0.5–2.2 nmol/L/h
Plasma renin activity	Ambulant	1.2–4.4 nmol/L/h
Plasma cortisol	8 am	200–800 nmol/L
Plasma cortisol	Midnight	50–200 nmol/L
Plasma DHEAS		2–10 μmol/L
Plasma ACTH		5–20 pmol/L
Urine free cortisol		<150 nmol/day
Urine adrenaline		<200 nmol/day
Urine noradrenaline		<680 nmol/day
Urine metanephrine		<2.1 μmol/day
Urine normetanephrine		<5.6 μmol/day
Urine vanillylmandelic acid		<40 μmol/day
Urine dopamine		<3 μmol/day

TABLE 14.3 Typical plasma concentrations of catecholamines under different circumstances.

	Adrenaline (pg/mL)	Noradrenaline (pg/mL)
Resting	40	200
Standing	40	500
Moderate exercise	130	1400
Heavy exercise	400	2200
Myocardial infarction	800	1400
Phaeochromocytoma	280	5500

SELF-CHECK 14.7

Reference ranges provide the clinician with a guide to interpretation of test results. Ideally reference ranges should be produced for the tests used in the laboratory but in some cases literature values are used. Do you think the laboratory and clinicians need to discuss any difficulties that arise from interpretation of abnormal results?

Key Points

The method used to measure the concentration of a hormone in a serum or plasma sample should have the required specificity and accuracy for clinical use. It is important to be aware of problems that can occur before the sample gets to the laboratory. The timing of sample collection, knowledge of drugs, and clinical information are all needed from the clinician to enable a dialogue when abnormal results are found.

14.3 Investigation of adrenal disorders

Cortisol synthesis may fail because of a defective enzyme in the pathway or destruction of the gland. Congenital adrenal hyperplasia (CAH) is one of the life-long diseases caused by inherited defects in enzymes of cortisol synthesis. In the absence of cortisol production from the adrenal gland, ACTH is released from the pituitary gland. Large amounts of ACTH in the blood lead to adrenal hyperplasia when cortisol synthesis is defective, as shown in Figure 14.13.

Congenital adrenal hyperplasia

The commonest form of CAH is due to a defect of 21-hydroxylase. This occurs in about 1 in 12,000 newborn infants each year in the UK so most hospitals will rarely see a case. Defects of other steroidogenic enzymes are at least ten times less common. In the absence of 21-hydroxylase, cortisol and sometimes aldosterone cannot be produced effectively. As a consequence of the defect 17-hydroxypregnenolone and 17-OHP accumulate and are available for the action of 17,20 lyase to produce excess androgens (male hormones). Females with 21-hydroxylase deficiency are born with ambiguous genitalia due to androgen exposure in foetal life. The genitalia can appear almost male with no testes. Some babies are wrongly assigned the male gender on the basis of a quick inspection of the genitalia. Males with 21-hydroxylase deficiency appear normal at birth, but 60% of boys and the affected girls will show signs of salt-loss from the fifth day of life and can die.

The affected boys may become sick at home after their early discharge from hospital. Affected girls with ambiguous genitalia are more likely to remain in hospital and can be monitored for electrolyte changes. The plasma potassium concentrations tend to rise before the plasma sodium falls.

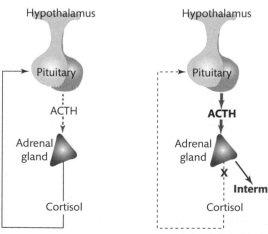

FIGURE 14.13
Schematic to show how adrenal hyperplasia occurs in congenital adrenal hyperplasia.

Boys and girls with mild forms of the defect (often called the non-salt losing or simple virilizing form) may present later with precocious puberty due to the excessive production of androgens. In adults, a non-classic form of CAH can occur. Basal plasma 17-hydroxyprogesterone concentrations are raised above the reference range and this can be further increased in a synacthen test. Synthesis of adrenaline is decreased in patients with defects in cortisol synthesis because normal medullary organogenesis is dependent on glucocorticoid secretion from the adrenal cortex. Patients with CAH can become hypoglycaemic during and after moderate exercise.

The investigation of potential cases of CAH is similar regardless of the enzyme defect. Finding low cortisol and high concentrations of the substrate of the defective enzyme characterizes the precise condition. Since 21-hydroxylase deficiency is the commonest cause, the concentration of the substrate (17-hydroxyprogesterone) needs to be determined in blood along with those of cortisol, ACTH, renin, and aldosterone. In the first three days, 17-OHP can be detected effectively in blood samples when measured specifically, such as with mass spectrometry or with an immunoassay performed on a solvent extract of the plasma to isolate free steroid from sulphated steroids that interfere in a direct immunoassay. Plasma concentrations of cortisol will be low and renin will be high because of the renal salt loss. A urine steroid profile should be requested because the urine analysis by gas chromatography and mass spectrometry is a definitive test that displays the nature of the excess steroids (gas chromatography retention time and mass spectrum of each of the abnormal steroids). Pregnanetriol is the main urinary metabolite of 17α-hydroxyprogesterone in children and adults and the excretion rate is high in CAH due to 21-hydroxylase deficiency. Using capillary column gas chromatography and mass spectrometry, the metabolites 17-hydroxyprogesterone and 21-deoxycortisol are markers for the disease. In the first three days of life, urine contains maternal and placental steroids so diagnosis of CAH on the basis of urine steroids should not be attempted in that early period.

If the blood concentration of 17-OHP is not elevated then other enzyme substrates need to be determined, such as 11-deoxycortisol for a 11-hydroxylase defect and DHEAS for a HSD3B2 defect. Tests may not be available for the full range of steroids in which case the urine steroid profile can pinpoint the defect in one cost effective analysis. Deoxycorticosterone is a potent mineralocorticoid and is produced excessively in this condition, causing sodium retention and hypertension. The sodium retention suppresses renin secretion from the kidney and aldosterone production is low in this condition, as shown in Figure 14.14.

A 17α-hydroxylase defect should be considered when investigations of a patient with primary infertility show persistently raised progesterone. The features of CAH according to the enzyme defects are summarized in Table 14.4.

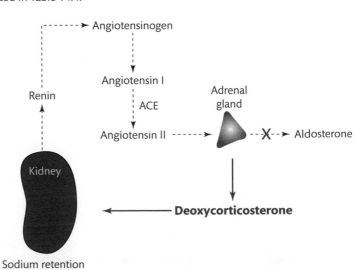

FIGURE 14.14

The renin angiotensin system in congenital adrenal hyperplasia, with low aldosterone production and high deoxycorticosterone production, due to 11-hydroxylase and 17-hydroxylase defects.

TABLE 14.4 Features of different types of congenital adrenal hyperplasia.

Defect	CYP21	CYP11B1	CYP17	HSD3B2	CAHo
Incidence	1 in 12,000	1 in 100,000	Rare	Rare	Rare
Ambiguous genitalia	Clitoromegaly (F)	Clitoromegaly (F)	Phenotypic female	Incomplete masculinization (M), mild clitoromegaly (F)	Incomplete masculinization (M), Phenotypic female
Adrenal failure	Hypoglycaemia, hyponatraemia (70%)	Not usually	Not usually	Hypoglycaemia, hyponatraemia	Hypoglycaemia, hyponatraemia
Hypertension		Yes	Yes	No	No
Puberty	Precocious	Precocious	No puberty	No puberty	No puberty
Potassium status	Hyperkalaemia	Hypokalaemia	Hypokalaemia	Hyperkalaemia	Hyperkalaemia
Cortisol	Low	Low unless 11-deoxycortisol cross reacts	Low	Low	Low
Aldosterone	Often low (defect in synthesis)	Often low (renin suppressed)	Often low (renin suppressed)	Often low (defect in synthesis)	Low
Androgens	Raised androstenedione	Raised androstenedione	Low	Raised DHEAS	Low
Diagnostic steroids	17-hydroxyprogesterone	11-deoxycortisol	Corticosterone, deoxycorticosterone and progesterone	DHEA, pregnenolone	None

To understand the implications of CAH consider the site of the enzyme block by referring back to Figure 14.3 and look at which of the three paths are affected and see where intermediates can be diverted to androgens and mineralocorticoids. The clinical picture reflects the lack of exposure of tissues to cortisol and or aldosterone, along with abnormally high or low production of male sex steroids. Hypertension will be found when deoxycorticosterone production is raised.

Key Points

The commonest of the adrenal disorders is congenital adrenal hyperplasia due to 21-hydroxylase deficiency. You may see samples for 17-hydroxyprogesterone assay from a newborn child with ambiguous genitalia (chances of this will depend on the number of deliveries in your local maternity unit). Consider the relevance of abnormal results and be prepared to draw these to the attention of reporting staff. If results for requested tests are normal, the laboratory needs to discuss other options with the clinician and look to refer work (urine steroid profile for example) to a specialist centre so as to support a precise diagnosis.

Cross reference

Chapter 16 Molecular biology techniques in the *Biomedical Science Practice* textbook

Gene deletions and large gene conversions can be detected by genomic Southern blot analysis. Restriction enzyme digestion, allele specific oligonucleotide hybridization, single-stranded conformation polymorphisms, and allele specific polymerase chain reaction (PCR) have been used for molecular diagnosis of point mutations. Care must be taken with genetic tests when more than one copy of a gene is present; this is quite common with *CYP21A2* genes.

In some cases the relevant gene will have to be sequenced. This testing is useful when families wish to have further children. Amniotic fluid can be taken at 11–12 weeks gestation for genetic analysis. If the foetus is an affected girl, then treatment with dexamethasone throughout

CASE STUDY 14.1

A three-week-old boy with failure to regain his birth weight was brought into hospital. He was dehydrated. Before he was given steroids and sodium supplements blood was taken for tests (reference ranges are given in brackets):

Sodium	110 mmol/L	(130–145)
Potassium	6.5 mmol/L	(3–5)
Chloride	70 mmol/L	(100–110)
Bicarbonate	19 mmol/L	(22–35)
Urea	6.5 mmol/L	(3–8)
Creatinine	50 μmol/L	(60–120)

Plasma 17-hydroxyprogesterone was 1600 nmol/L (<30 nmol/L) by immunoassay after extraction.

(a) What is the likely diagnosis?

(b) If the 17-OHP had been measured in a direct coated tube RIA, would your conclusion be the same, and if not state why?

(c) If the 17-OHP had not been elevated what further tests might the laboratory be asked to do and what might the results indicate?

pregnancy can reduce the virilization that might have been experienced with living girl siblings. The non-classic form of CAH is due to specific mutations amongst those seen in *CYP21A2*.

Key Points

Genetic testing in the context of adrenal disorders should be undertaken in specialist laboratories once there is reasonably clear biochemical evidence of the clinical problem. It is often necessary to send samples from the parents as well. Genetic testing is particularly useful when the parents wish to extend the family.

SELF-CHECK 14.8

In a newborn girl with ambiguous genitalia, which steroid should be measured first to support an adrenal defect of excess androgen production?

Addison's disease

In 1855 Thomas Addison described patients who typically presented with a range of symptoms including fatigue, muscle weakness, weight loss, vomiting, diarrhoea, and sweating. Tuberculous infection was often the basis of the adrenal failure and this cause may be on the increase. Adrenal enzyme antibodies are now known to block activity of 21-hydroxylase. Addison's disease is now known to be caused by cortisol deficiency. In the absence of cortisol production many patients have some signs of hyperpigmentation that can be in scar tissue or skin creases inside the cheek. This is due to increased production of melanocyte stimulating hormone as a consequence of increased pituitary processing of the ACTH precursor molecule pro-opiomelanocortin (POMC) following adrenal failure. The long-term use of corticosteroids and sudden withdrawal can also be responsible. In newborn infants there may be failure of adrenal development or a defect in

CASE STUDY 14.2

A 25-year-old woman had been vomiting over the past five days. She was brought unconscious to the hospital. Her blood pressure was 140/80 mm Hg lying and 110/60 on standing. She had not felt well for about four months. The following results were obtained for her blood tests (reference ranges are given in brackets):

Sodium	112 mmol/L	(130–145)
Potassium	6 mmol/L	(3–5)
Chloride	98 mmol/L	(100–110)
Bicarbonate	18 mmol/L	(22–35)
Urea	17 mmol/L	(3–8)
Creatinine	170 µmol/L	(60–120)

Urine was negative for glucose and ketones.

(a) What is the likely diagnosis?

(b) What further tests should be performed and how will the results be helpful?

cortisol synthesis. Rare forms of adrenal failure are due to lack of ACTH secretion or receptor action. The adrenal gland can also be destroyed by the high levels of very long chain fatty acids (VLCFA) that accumulate in the adrenal, brain, and myelin in adrenoleukodystrophy. Blood tests in adrenal failure will show hypoglycaemia, hyponatraemia, hyperkalaemia, and metabolic acidosis. Low plasma concentrations of cortisol will not be increased with synacthen.

A low cortisol (<100 nmol/L) at 8 am suggests adrenal insufficiency and a synacthen test with measurements of cortisol should be used to confirm this. If a biosynthetic defect is thought to be the cause then the basal and stimulated samples should be analysed to check the plasma concentrations of the intermediates before and after stimulation. Recent use of glucocorticoids may also lead to a state of adrenal hypoplasia lasting for some time. Herbal medicines can sometimes contain illegally added prescription medicines.

Cushing's disease and syndrome

In 1932, Harvey Cushing described a pituitary tumour in a patient with central obesity, a round, full-moon shaped face, reddened skin stretch marks (striae), and pigmentation. The clinical features of **Cushing's disease** are attributable to the production of cortisol by the adrenal cortex through the excess production of ACTH from the tumour as shown in Figure 14.15.

A similar syndrome is seen in the case of an ectopic (usually lung) ACTH secreting tumour or a tumour of the adrenal cortex secreting cortisol. **Cushing's syndrome** refers to the clinical picture from excess cortisol secretion. The administration of exogenous corticosteroids can also produce the picture of Cushing's syndrome (iatrogenic Cushing's). Patients with rheumatoid arthritis are often prescribed very large doses of glucocorticosteroids. When the cortisol excess is due to an autonomous or exogenous sources the concentrations of ACTH in blood will be reduced.

Cortisol excess will be reflected in an increase of free cortisol in the blood and since this is filtered at the kidney, a 24-hour urine free cortisol measurement is a good first test in a likely patient. Blood samples taken at 8 am and midnight will show high cortisol concentrations at midnight and loss of a diurnal rhythm of cortisol in affected patients. This is the next useful procedure and some advocate the use of saliva cortisol measurements for this purpose. Sensitive and specific cortisol assays will be needed for saliva samples. Very high plasma concentrations of ACTH can be found when tumours are ectopic. High cortisol concentrations may be found in situations of obesity, stress, depression, and alcoholism, as well as Cushing's syndrome. Provocative tests are essential. A low dose dexamethasone suppression test (0.5 mg four times a day for two days) will suppress cortisol in normal subjects and eliminate those where previous high cortisol concentrations were due to stress. High dose dexamethasone (2 mg four times a day) will not suppress cortisol in Cushing's syndrome. Further investigations are listed in Box 14.1.

SELF-CHECK 14.9

An obese male is found to have a raised 24-hour free cortisol excretion rate in his urine. On further investigation, cortisol excretion was still elevated after two days of dexamethasone treatment (0.5 mg four times a day for two days, 24-hour urine collected after fifth dose). The dexamethasone dose was increased to 8 mg per day. The urinary free cortisol decreased significantly.

(a) If you think there is a problem in cortisol production at what point in the hypothalamic-pituitary-adrenal (HPA) axis would you suggest this occurs?

(b) If plasma ACTH was low in the first investigations of this obese male what is the cause to eliminate?

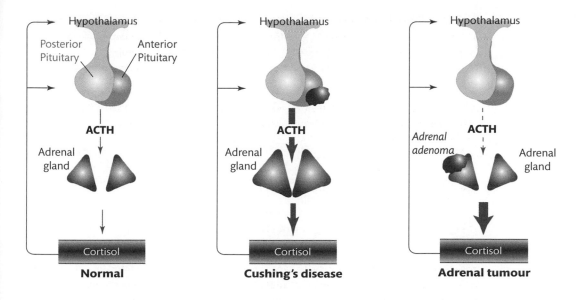

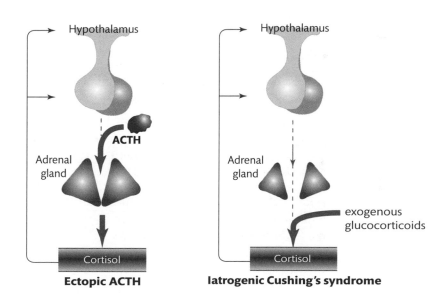

FIGURE 14.15
Causes of excess cortisol production.

Conn's syndrome

In 1955 Jerome Conn described an adrenal tumour secreting aldosterone (primary aldosteronism). A patient with this condition has high blood pressure and hypokalaemia and it can lead to muscle weakness, headaches, and alkalosis. At the time, it was hoped that this sort of tumour would be a common cause of hypertension treatable with adrenalectomy. The tumours vary in size from 0.5 cm and not all are visualized on imaging. In order to consider a patient for investigation, hypokalaemia was thought to be an important marker of the

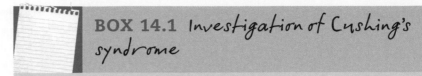

BOX 14.1 *Investigation of Cushing's syndrome*

Eye examination for visual field defects due to a pituitary tumour.

Adrenal and pituitary imaging for adenomas.

Blood sampling of the inferior petrosal sinuses (that drain from the pituitary) to lateralize the pituitary source of excess ACTH. The ACTH concentrations in blood samples from the right and left sinus are compared to predict tumour location. The plasma concentrations of cortisol in peripheral samples are taken at the same time. The secretion of ACTH is pulsatile so the injection of CRH can increase the diagnostic accuracy of the test. The highest concentrations of ACTH are seen at 5–10 minutes after the CRF injection on the side of the tumour.

Catheterization studies of veins to seek the source of ectopic ACTH may involve catheterization of major vessels around the abdomen since a tumour is often in the lung, but can be in the thyroid, pancreas, and elsewhere.

condition. On these criteria the incidence was low (less than 1% of hypertensive patients). Recent studies of patients with normokalaemia have revealed 10–20% of patients with familial and drug resistant hypertension have hyperaldosteronism.

The test of choice for primary aldosteronism is the aldosterone to renin ratio when blood pressure is controlled with drugs that do not affect the renin-angiotensin-aldosterone system.

CASE STUDY 14.3

A 60-year-old male was referred to outpatients with weight loss, muscle weakness, and pigmentation of a recent scar on the knee after a fall. Blood tests at midday showed the following results (reference ranges are given in brackets):

Sodium	144 mmol/L	(130–145)
Potassium	2 mmol/L	(3–5)
Chloride	90 mmol/L	(100–110)
Bicarbonate	>40 mmol/L	(22–35)
Urea	8.5 mmol/L	(3–8)
Creatinine	130 μmol/L	(60–120)
Cortisol	1500 nmol/L	(<200)

He was kept in hospital.

A 24-hour urine free cortisol measurement was 1600 nmol/24 h (<200 nmol/24 h). His plasma cortisol concentration was 1600 nmol/L at 8 am (200–800) and 1200 nmol/L at midnight (<150 nmol/L)

(a) What diagnosis is suspected?

(b) What further tests might be performed in the laboratory?

(c) How will the results help the diagnosis?

Since potassium ion concentrations in the adrenal cortex are important for the enzymes in the last step of aldosterone synthesis, patients with hypokalemia should be given potassium supplements prior to blood sampling for aldosterone and renin measurements.

Confirmation of primary aldosteronism

A high aldosterone to renin ratio in a screening test is suggestive of primary aldosteronism. As indicated above, you must be careful when offering an interpretation of the ratio from the literature because many units are used for both analytes. The aldosteronism can be due to a unilateral adenoma, bilateral adrenal hyperplasia, or dexamethasone suppressible types.

Catheterization studies will help localize a unilateral tumour but catheterization of adrenal veins requires a skilled radiologist. A catheter is inserted into a vein in the groin and moved up into the vena cava. For each blood sample taken as the catheter is repositioned, a peripheral blood sample should be taken for aldosterone concentrations. The right adrenal vein is difficult to access from the vena cava, whereas the left adrenal vein leads into the renal vein and is less problematic. The blood samples from the tumour are assayed for aldosterone and cortisol. The plasma aldosterone concentrations can be considerably higher than peripheral levels and will need to be assayed in dilution. Cortisol is measured to confirm that the catheter was in the vein.

CASE STUDY 14.4

A 50-year-old male with hypertension was referred to a specialist clinic because his blood pressure was not controlled despite treatment with an angiotensin II receptor blocker, a calcium channel blocker, and a diuretic. Conn's syndrome was a possibility, so following the laboratory guidelines his medications over a few weeks were withdrawn. Blood pressure was lowered with a beta blocker, prazosin, which does not affect the renin-angiotensin system. He then underwent a number of tests as follows (reference ranges are given in brackets):

Sodium	136 mmol/L	(130–145)
Potassium	2.9 mmol/L	(3–5)
Chloride	95 mmol/L	(100–110)
Bicarbonate	35 mmol/L	(22–35)
Urea	7.7 mmol/L	(3–8)
Creatinine	130 μmol/L	(60–120)
Cortisol 8 am	600 nmol/L	(200–800)
Cortisol 12 am	180 nmol/L	(<200)
Plasma aldosterone lying	1600 pmol/L	(100–500)
Plasma renin activity lying	0.2 pmol/mL/h	(0.5–2.2)
Aldosterone renin ratio lying	8,000 (>1200 suggestive of primary aldosteronism when aldosterone >350 pmol/L (note units for aldosterone and renin)	
Plasma aldosterone ambulant 12 am	690 pmol/L	(600–1200)
Plasma renin activity ambulant:	0.2 pmol/mL/h	(1.2–4.4)

(a) How would you interpret these results?

(b) What further investigations might be useful?

Clinical significance of other abnormal steroid results

Low aldosterone results can be found if liquorice and related confectionery is eaten excessively. Liquorice suppresses the inactivation of cortisol to cortisone so that cortisol remains available at high concentration in the kidney where it acts with the mineralocorticoid receptor. Sodium retention leads to hypertension and to suppression of renin. A similar clinical picture is associated with a defect of 11β-hydroxysteroid dehydrogenase type 2. Plasma aldosterone concentrations are low in CAH due to defects of 3β-hydroxysteroid dehydrogenase, 21-hydroxylase, and 11-hydroxylase, because of low production despite the high renin. However, in CAH with defects of 17-hydroxylase and aldosterone synthase, the plasma aldosterone is low due to suppression of renin activity through the action of deoxycorticosterone as a mineralocorticoid. Hyporeninaemic hypoaldosteronism is seen with kidney damage, particularly in patients with diabetes mellitus. Aldosterone results can be abnormal in renal disorders affecting electrolyte transport and can mimic defects of aldosterone production through high renin activity (Bartter's or Gitelman's syndromes) or low renin activity (Liddle's and Gordon's syndromes). High renin activity and aldosterone concentrations are seen in a number of situations (congestive heart failure, cirrhosis, and nephritic syndrome) and only rarely with defects of aldosterone action (defects of epithelial sodium transport channel or defects of the aldosterone receptor called pseudohypoaldosteronism).

Children with premature adrenarche have increased plasma concentrations of DHEA and DHEAS for age. Dehydroepiandrosterone sulphate concentrations are grossly elevated in children and adults with adrenal tumours and somewhat raised in CAH due to HSD3B2 deficiency. During childhood, testosterone and androstenedione production and their plasma concentrations are lower than in the first year and at puberty.

Key Points

An abnormal result needs to be considered in the broader context of hormone, production, regulation, action, and function as well as the other variables that have been referred to often in this chapter (age, timing, etc.).

Investigation of disorders of the adrenal medulla

A patient with excess catecholamine secretion can present with a range of signs and symptoms, including increased heart rate, high blood pressure with postural hypotension, palpitations, headaches, weight loss, and skin pallor. Plasma concentrations of metanephrines are the preferred screening test in samples from patients with drug resistant hypertension, especially when an adrenal mass has been seen. Tests should be conducted without stress and after review of medications likely to affect results, for example tricyclic antidepressants and phenoxybenzamine. A negative screen for metanephrines virtually rules out a catecholamine secreting tumour. In the methods for urinary metanephrines, the conjugated compounds are hydrolysed before analysis. Tandem mass spectrometry will probably replace methods based on HPLC with other detectors. Some genetic tests are being introduced. Plasma concentrations of chromogranin A may become an additional marker for phaeochromocytoma. Abdominal imaging with computed tomography or magnetic resonance are useful when localizing a tumour.

SELF-CHECK 14.10

An elderly female was found to have high blood pressure. She was anxious and sweating profusely.

(a) Should she be investigated for a disorder of the adrenal medulla or cortex?

(b) What tests would be most useful?

14.4 Management of adrenal disorders

In CAH and adrenal failure, cortisol treatment is essential for replacement and to suppress ACTH production. With current prescriptions it is difficult to replace cortisol to achieve the 24-hour profile of plasma concentrations seen normally. Usually 15–20 mg of cortisol (hydrocortisone, cortisone acetate) in two or three doses per day are prescribed, some physicians favour high doses at night, others favour high doses in the morning. Prolonged treatment with high doses of steroids can lead to adrenal atrophy. Steroids may be withdrawn gradually because recovery of function is slow. Cortisol secretion will slowly recover over 4–12 months. Plasma ACTH concentrations when low reflect over treatment but with time these values increase as the HPA axis returns to normal.

Fludrocortisone is a potent mineralocorticoid and typically 50 µg per day in an adult will control electrolytes and suppress renin production. Fludrocortisone does have glucocorticoid activity so doses must be carefully restricted. Renin is a good marker of replacement mineralocorticoid treatment, renin being normal or low when electrolyte balance is achieved.

The androgen production during childhood is largely from the adrenal cortex. Androgen excess can be counteracted by adrenal suppression (cortisol or dexamethasone) or anti-androgen treatment (cyproterone acetate, spironolactone).

Children with ambiguous genitalia from CAH will need surgery to improve appearance and function of the genitalia. For primary aldosteronism due to an adrenal adenoma and Cushing's syndrome, surgery will be required to remove neoplastic or abnormal tissue. Primary aldosteronism due to adrenal hyperplasia responds to treatment with mineralocorticoid receptor blockade (sprironolactone). The diagnosis of a phaechromocytoma is extremely rare but surgical treatment is very effective.

Catecholamine secreting tumours are largely in the adrenals but sometimes outside. Other tumours in tissues of the neural crest (paragangliomas) are also encountered that do not secrete such high levels of catecholamines. Familial phaeochromocytomas may occur in patients with multiple endocrine neoplasia syndromes and may be associated with medullary thyroid carcinoma, Von-Hippel Lindau disease, cutaneous neurofibromatosis, and mitochondrial disease due to mutations in succinate dehydrogenase subunits. Investigations will then extend beyond analysis of plasma metanephrines with broader imaging and genetic studies. Surgery is the treatment of choice for phaeochromocytomas but requires careful management before, during, and after tumour removal. Operations are performed laparoscopically and before and during the operation the patient is at risk of severe hypertensive crises and cardiac arrhythmias if catecholamine release from the tumour is provoked. Once the tumour vessels are clamped there is risk of hypotension. The patient should have plasma metanephrine concentrations monitored post-operatively.

SUMMARY

- The adrenal glands have an outer cortical region and an inner medulla with different functions. The adrenal cortex produces aldosterone, cortisol, and DHEAS, the medulla produces adrenaline and noradrenaline.

- The execution and interpretation of laboratory hormone tests need special consideration, particularly with regard to assay specificity, reference ranges for age, development, and in some cases body size. Timing of samples is also critical.

- Immunoassays have been the basis for measuring the concentrations of steroid hormones in biological fluids.

- Steroid hormone assays by tandem mass spectrometry are likely to replace most of the current steroid methods.

- The analysis of urine steroids by gas chromatography with mass spectrometry has an important place in assessing disorders of adrenal function in childhood and adults.

- A number of disorders of the adrenal glands are due to genetic defects in enzymes or neoplasms.

- Once recognized these disorders are treatable by surgery and/or hormone replacement therapy.

FURTHER READING

- **Antal Z and Zhou P (2009) Congenital adrenal hyperplasia: diagnosis, evaluation and management.** *Pediatric Reviews* **30**, 49–57.

  A review of congenital adrenal hyperplasia.

- **Dickstein G and Saiegh L (2008) Low dose and high dose adrenocorticotropin testing: indications and shortcomings.** *Current Opinion in Endocrinology, Diabetes and Obesity* **15**, 244–9.

  A meta-analysis of ACTH tests.

- **Findling JW and Rahh H (2006) Cushing's syndrome: important issues in diagnosis and management.** *Journal of Clinical Endocrinology and Metabolism* **91**, 3746–53.

  Presents a thorough account of this disorder.

- **Funder JW, Carey RM, Fardella C, et al. (2008) Case detection, diagnosis and treatment of patients with primary aldosteronism: an Endocrine Society clinical practice guideline.** *Journal of Clinical Endocrinology and Metabolism* **93**, 3266–81.

  A consensus document.

- **Fung MM, Viveros OH, and O'Connor DT (2008) Diseases of the adrenal medulla.** *Acta Physiologica* **192**, 325–35.

  This article covers the diseases affecting the adrenal medulla.

- Ghayee HK and Auchus RJ (2007) Basic concepts and recent development in human steroid biosynthesis. *Reviews in Endocrine and Metabolic Disorders* **8**, 289–300.

  Overview of steroid synthesis including some new concepts.

- Harding JL, Yeh MW, Robinson BG, Delbridge LW, and Sidhu SB (2005) Potential pitfalls in the diagnosis of phaeochromocytoma. *Medical Journal of Australia* **182**, 637–9.

  This article highlights important problems in diagnosis of phaeochromocytoma.

- Honour JW (2001) Urinary steroid profile analysis. *Clinica Chimica Acta* **313**, 45–50.

  Overview of this very useful procedure for defining most adrenal cortical disorders.

- Newell-Price J, Bertagna X, Grossman AB, and Niemann LK (2006) Cushing's syndrome. *The Lancet* **367**, 1605–17.

  A comprehensive account of Cushing's syndrome.

- Perry CG, Sawka AM, Singh R, Thabanes L, Bajnarek J, and Young WF (2007) The diagnostic efficacy of urinary fractionated metanephrines measured by tandem mass spectrometry. *Clinical Endocrinology* **66**, 703–8.

  Some laboratories will be moving to these techniques because of higher specificity and (it is hoped) accuracy.

- Riepe FG and Sippell WG (2007) Recent advances in diagnosis, treatment, and outcome of congenital adrenal hyperplasia due to 21-hydroxylase deficiency. *Reviews in Endocrine and Metabolic Disorders* **8**, 349–63.

  Presents a thorough account of this disorder.

- Ten S, New M, Maclaren N (2001) Clinical review 130: Addison's disease 2001. *Journal of Clinical Endocrinology and Metabolism* **86**, 2909–22.

  Presents a thorough account of this disorder.

- Whiting MJ (2009) Simultaneous measurement of urinary metanephrines and catecholamines by liquid chromatography with tandem mass spectrometric detection. *Annals of Clinical Biochemistry* **46**, 129–36.

  A critical examination of available tests.

- Young WM (2007) Adrenal causes of hypertension: phaeochromocytoma and primary aldosteronism. *Reviews in Endocrine and Metabolic Disorders* **8**, 309–20.

  Presents an overview of the adrenal causes of hypertension.

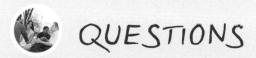

QUESTIONS

14.1 **(a)** How many enzymes are needed to convert cholesterol to cortisol?

(b) Discuss the changes to the cholesterol molecule that are required to synthesize cortisol and name the intermediates.

14.2 Is the activity of the adrenal foetal zone similar to the zona reticularis? If yes, what is the major steroid secreted and why?

14.3 In the following table link the hormone with its site of synthesis and regulatory mechanism

Hormone	Site of synthesis	Regulatory mechanism
Aldosterone	Adrenal medulla	ACTH
Adrenaline	Adrenal fasciculata zone	Angiotensin II
DHEAS	Adrenal glomerulosa zone	Neural
Cortisol	Adrenal reticularis	Unknown

14.4 If a patient has the commonest form of congenital adrenal hyperplasia would you expect the plasma aldosterone concentration to be increased in response to hyponatraemia?

14.5 **(a)** In a newborn girl with ambiguous genitalia and raised 17-hydroxyprogesterone, what other tests are available for this patient to be thoroughly investigated?

(b) What further tests can be performed in a patient with biochemical results that support CAH?

14.6 Some boys with congenital adrenal hyperplasia have no clinical problems in the first five years of life. Why can this be so?

14.7 State whether the following statements are TRUE or FALSE.

(a) Corticosterone is the marker for deficiency of the 17-hydroxylase enzyme

(b) A girl (46 XX chromosomes) with a deficiency of 3β-hydroxysteroid dehydrogenase will be born with some virilization, but a boy (46 XY) may appear to have genitalia more like a girl

(c) If cortisol is produced by an adrenal tumour, the ACTH concentrations would be high

(d) Tuberculosis is the commonest cause of adrenal failure

(e) Blood tests on a hypertensive patient taken on a hot summer afternoon would be falsely elevated if they had eaten ice-cream or chewed liquorice

Answers to self-check questions, case study questions, and end-of-chapter questions are available in the Online Resource Centre accompanying this book.

Go to www.oxfordtextbooks.co.uk/orc/ahmed/

15

Reproductive endocrinology

Ian Laing and Julie Thornton

Learning objectives

After studying this chapter you should be able to:

- Explain the role of the hormones involved in reproductive endocrinology in the female
- Provide an account of the hormonal changes in the menstrual cycle and the control mechanisms involved
- Describe how ovulation may be inhibited or induced for therapeutic purposes
- Describe the diagnosis of the menopause and the role of hormone replacement therapy
- Outline the causes and treatment of anovulatory infertility
- Discuss the polycystic ovary syndrome, its metabolic consequences, and its management
- Outline the hormonal interrelationships in male reproductive function
- Discuss the limitations of commonly used hormone assays

Introduction

Reproductive endocrinology in both sexes depends on a series of interrelated control systems involving the brain and higher centres, and hormones of the hypothalamus, pituitary gland, and the ovaries in the female, and the testes in the male. In the female, ovarian function is directed at producing one fertile oocyte each month, whilst in the male **spermatogenesis** is a continuous process. Whereas the reproductive lifespan of a woman is finite and ceases at the **menopause**, the male remains fertile throughout life. In both sexes reproductive function requires the individual to be in good health and both **ovulation** and spermatogenesis are compromised by acute and chronic illness. Disorders of reproductive function may occur at any level of the process involving the higher centres, the hypothalamus, pituitary, and the gonads, and the related abnormalities may be revealed by study of the hormones involved. Knowledge of the endocrinology of female reproductive function permits ovulation to be induced or suppressed for therapeutic purposes. Similarly, in the male testicular function and spermatogenesis can be suppressed, and in specific circumstances endocrine treatment can be made to stimulate spermatogenesis.

15.1 Reproductive endocrinology in the female

The foetal ovary is well established by ten weeks of gestation. By mid-gestation the pool of primordial follicles is at its highest and numbers about four million. The primordial follicle, from which an ovulatory follicle eventually develops, is an oocyte surrounded by a layer of flattened granulosa cells. The number of primordial follicles falls dramatically during the later part of gestation and this process continues throughout childhood into adult life. At puberty, about 400,000 follicles remain until at the menopause the numbers become exhausted. During the reproductive life of a woman, only about 400 primordial follicles will develop to a state where they can ovulate. During the process of folliculogenesis, primordial follicles develop into primary follicles and then into pre-antral and antral follicles. At the antral follicle stage a single follicle develops to become the dominant follicle and goes on to ovulate. The process of ovarian folliculogenesis is outlined in Figure 15.1.

A characteristic of all stages of follicular development is involution and atresia of the follicle, a process involving apoptosis, when follicles stop growing and shrink with subsequent follicle loss.

Hormones of the female reproductive system

The function of the gonads, the ovary in women and the testis in men, are controlled by the peptide hormones, follicle stimulating hormone (FSH) and luteinizing hormone (LH), which are secreted by cells in the pituitary gland called gonadotrophs. In both sexes, LH and FSH secretion are controlled by secretion of the hormone gonadotrophin releasing hormone (GnRH), sometimes called luteinizing hormone releasing hormone (LHRH), secreted by the hypothalamus. The principal endocrine secretions of the testis and ovary differ. In women they are the steroid hormones, oestradiol and progesterone, and in men the steroid hormone, testosterone.

Gonadotrophin releasing hormone

The hypothalamic derived GnRH is a peptide of ten amino acid residues. It is secreted by neurones in the hypothalamus into the pituitary portal circulation and acts on the gonadotrophin

FIGURE 15.1

Stages of follicular development in the ovary. Cohorts of primordial follicles develop under the influence of androgens and other cytokines. At the late antral stage, a dominant follicle is selected and goes on to ovulate. At all stages of development follicles are lost by a process involving apoptosis.

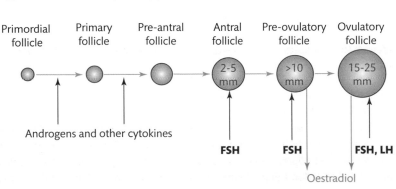

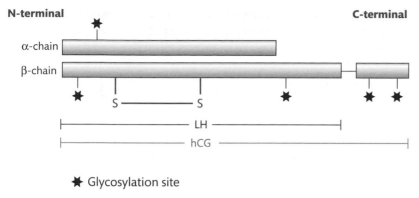

FIGURE 15.2
Schematic structure of human glycoprotein hormones. The α-polypeptide chain (M_r 10,000) is non-covalently associated with the β-chain (M_r 20,000). Both chains contain glycosylation sites (*). hCG shows the β-chain with additional glycosylation sites. Note the intra-chain disulphide bonds in the β-chain.

secreting cells of the pituitary to promote LH and FSH release. It acts on cell surface receptors. It is rapidly degraded by circulating peptidases in common with other molecules of this type and is not measurable in peripheral plasma by existing techniques. Gonadotrophin releasing hormone is secreted in pulses by the hypothalamus with a pulse frequency of approximately 90 minutes. The pulsatile release of GnRH is reflected by pulsatility in the secretion of LH and FSH.

Luteinizing hormone and follicle stimulating hormone

The gonadotroph cells of the pituitary secrete LH and FSH. These are glycoprotein hormones with molecular weights of approximately 30,000 Da. They act on specific cells within the ovary via G-protein coupled cell surface receptors. Both LH and FSH consist of two polypeptide chains (α, mw 10,000 Da, and β, mw 20,000 Da) held together by non-covalent linkage. In addition, the molecules are stabilized by intra-chain disulphide bonds within the β chain, providing the molecule with three-dimensional stability, as shown in Figure 15.2.

Both LH and FSH have high homology with thyroid stimulating hormone (TSH) and human chorionic gonadotrophin (hCG), with whom they share the common α subunit. The α subunit is identical in LH, FSH, TSH, and hCG. The specificity of action resides in the β subunit and there is some homology in the structure of the beta subunits of these hormones. In the case of LH and hCG, homology of the β subunit is virtually complete, but hCG has an extended carboxy-terminal β chain. All the pituitary glycoprotein hormones are glycosylated. There is a single glycosylation site on the beta chain and two glycosylation sites on the alpha chain. In the case of hCG, the C-terminal extension of the beta chain has additional glycosylation sites. One function of glycosylation is to inhibit degradation of the hormone and to increase its circulating half-life. Most importantly, glycosylation is required for biological activity. In the case of hCG, which is heavily glycosylated, this results in a longer half-life (8–12 h) than LH (20–30 minutes), with which it shares an identical profile of action. Follicle stimulating hormone also has a longer half-life than LH (approximately 8 hours). Under normal circumstances, small but measurable quantities of both free α and β subunits are secreted in measurable quantities by the pituitary. These are devoid of biological activity.

Prolactin

Prolactin is a peptide hormone consisting of a single polypeptide chain with a molecular weight of 23,000 Da, secreted by the pituitary. Prolactin secretion is controlled negatively by dopamine secreted by the hypothalamus, which anatomically sits just above the pituitary. Dopamine passes into the blood draining the hypothalamus into the pituitary via a portal circulation.

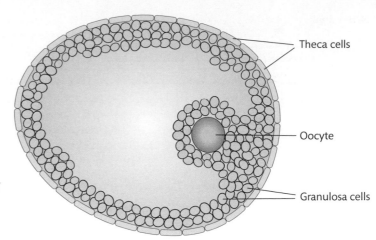

Theca cells

Oocyte

Granulosa cells

FIGURE 15.3
Schematic representation of mature pre-ovulatory ovarian follicle showing the oocyte surrounded by granulosa cells and the peripheral layer of granulosa and thecal cells.

High circulating levels of prolactin in plasma interfere with gonadotrophin action. Under normal circumstances, prolactin is secreted in a pulsatile manner at night, with lower values encountered during the day. Stimulation of the nipple, as in breastfeeding an infant, increases prolactin secretion significantly. Prolactin, together with adrenocorticotrophic hormone (ACTH), is a pituitary stress hormone and circulating prolactin levels increase with a number of stimuli, such as physical and psychological stress, trauma, and physical exercise.

Ovarian steroid hormones

The principal hormones secreted by the ovary are oestradiol and progesterone. Most circulating oestradiol is produced by the granulosa cells (Figure 15.3) of the developing dominant follicle in the early part of the menstrual cycle and from the luteinized granulosa cells in the corpus luteum after ovulation.

Extra-ovarian, or peripheral production of oestradiol from circulating precursor steroids in specific tissues is the main source of oestradiol in post-menopausal women and men. Oestradiol also has non-reproductive functions in both sexes, affecting the bones, the endothelium, and modulating lipoprotein metabolism. It also promotes wound-healing in men and women. The principal biologically active oestrogen is 17 β-oestradiol and is much more active than other related compounds such as oestrone and oestriol in binding to and activating the two intracellular oestrogen receptors ER alpha and ER beta. Within the ovary, FSH acts on the granulosa cells to induce the enzyme aromatase which converts locally produced testosterone to oestradiol as shown in Figure 15.4. Testosterone and its immediate precursor androstenedione are produced in the ovary principally by the theca cells stimulated by LH.

Both progesterone and oestradiol are produced by the corpus luteum under the influence of LH. In normal women, both testosterone and androstenedione are produced in, and secreted in measurable quantities by the ovary and the adrenals. The main biological functions of oestradiol are to promote development of secondary sexual characteristics in the female, and in terms of reproduction to induce proliferation of the endometrium in the early phase of the menstrual cycle. In addition, oestradiol acts on the cervix to induce secretion of clear, low-viscosity mucus which is easily penetrable by sperm. Oestradiol is also important in its action on the pituitary to initiate the mid-cycle surge of LH, responsible for producing ovulation of the developing follicle. Progesterone is responsible for the transformation of the proliferative (early) endometrium to the (late) secretory phase, which is required for implantation. It also changes the cervical mucus to a more viscous, less permeable form. When implantation of a

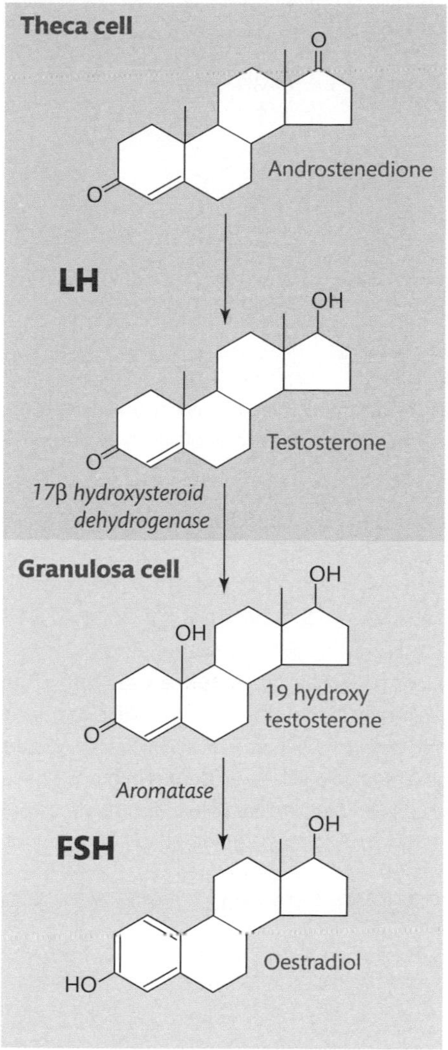

FIGURE 15.4
Biosynthesis of oestradiol in the mature ovarian follicle. Androstenedione produced in the theca cell under the influence of LH is metabolized to testosterone and enters the granulosa cell where it is converted to oestradiol by aromatase, which is induced by FSH.

fertilized egg occurs, progesterone maintains the endometrium in the secretory phase, allowing pregnancy to become established.

Anti-Mullerian hormone

The protein hormone anti-Mullerian hormone (AMH), a member of the transforming growth factor beta (TGF-β) family of hormones, is secreted by, and has important actions in the ovary. Ovarian AMH is also measurable in the peripheral circulation. This hormone was first recognized as a secretion of foetal Sertoli cells in the embryonic male testis. In early embryogenesis in both males and females two sets of ducts are present in the reproductive system, the Wolffian ducts (male) and the Mullerian ducts (female). Anti-Mullerian hormone secreted by the Sertoli cells of the foetal testis is responsible for regression of the Mullerian ducts during embryogenesis, resulting in differentiation of a male phenotype. In the female, AMH is produced in the ovary by granulosa cells of pre-antral and small antral follicles. Anti-Mullerian hormone has two principal actions. It inhibits the transition of primordial to primary follicles, and it inhibits the action of FSH on the growth of small antral and antral follicles. Its effect on

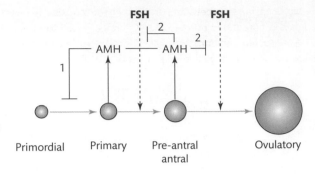

FIGURE 15.5
The role of AMH in ovarian folliculogenesis. Anti-Mullerian hormone inhibits early follicular development and inhibits late follicular development by opposing the actions of FSH (Carlsson IB, Scott JE, Visser JA, *et al.* (2006) *Human Reproduction* **19**, 2223–7).

antral follicles is to oppose the induction of FSH receptors, which are necessary for the expression of aromatase and consequently oestradiol synthesis. The functions of AMH in the ovary are summarized in Figure 15.5.

Thus, AMH has a key role in ovarian folliculogenesis. Both human and animal studies indicate that AMH is closely related to the ovarian follicle pool.

Inhibins

Two protein hormones, inhibin A and inhibin B, are produced by granulosa cells within the ovary. Inhibin B secretion from small antral follicles is maximal during the mid-follicular phase, and inhibin A is secreted in mid-cycle with a second peak in the luteal phase. Both are thought to have roles in the control of follicular development and follicle selection. While both have been shown to inhibit FSH secretion by pituitary cells *in vitro*, a clear role in modulating FSH secretion *in vivo* has not been established.

SELF-CHECK 15.1

Which gland secretes FSH and LH?

Hormonal changes from birth to puberty

From birth to puberty LH and FSH secretion in females is low (<1 IU/L) but measurable, and ovarian secretion of oestradiol <150 pmol/L) and progesterone (<3 nmol/L) is low. The onset of puberty is determined by the integration of metabolic and nutritional signals in the higher centres which result in an increase in the release of GnRH by the hypothalamus. This causes LH and FSH secretion to rise above the low levels encountered in childhood and promotes ovarian oestradiol production. Oestradiol is responsible for the physical signs of puberty. The first menstrual period is called the menarche. The importance of nutritional state in relation to the menarche is highlighted by the decrease in the mean age from 16 years to 13 years in the United Kingdom during the twentieth century, reflecting the improved nutrition of the population. Delayed puberty is considered to occur when pubertal signs are not apparent by 13 years. Precocious (early) puberty is considered when pubertal signs are present before eight years. Signs of puberty in girls are principally related to ovarian oestrogen and androgen secretion and are quantified clinically in terms of breast development and the growth of sexual hair. Following the menarche cycle lengths are variable and cycles without ovulation (anovulatory cycles) are common. In late puberty, a regular menstrual pattern develops, which is maintained until the menopausal transition. Once menstruation is established cycle lengths vary considerably between, and within, individuals and a cycle length of between 23 and 33 days is considered normal. **Amenorrhoea** is said to occur when menstrual function ceases. A practical definition of amenorrhoea is absence of menses for six months. Irregular menstruation is termed oligomenorrhoea.

The adrenarche

The **adrenarche** precedes puberty. It occurs when the adrenal glands begin to secrete steroids with androgenic activity. During childhood the adrenals secrete mainly cortisol. At the adrenarche there is an increase in the adrenal secretion of adrenal C-19 steroids (principally dehydroepiandrosterone sulphate, DHEAS) which are transformed to active androgens (testosterone) at the site of action, resulting in the growth of axillary and pubic hair, the development of apocrine sweat glands, and increased sebaceous gland secretion. In late puberty there is an increase in linear growth or a growth spurt, which is brought to an end when the epiphyses in the long bones fuse. Closure of the epiphyses is recognized to be an oestrogen dependent event in both sexes. The changes associated with puberty are well described and illustrated in detail in many paediatric texts and will not be further illustrated here.

15.2 **Hormonal changes in the menstrual cycle**

Normal menstrual function is a controlled process involving feedback control between the ovaries and the hypothalamus/pituitary axis. The hormonal interrelationships are outlined in Figure 15.6.

Both negative and positive feedback is involved in the process. When considering the menstrual cycle it is usual to regard day 1 of menstruation as the first day of the menstrual period. The cycle may be divided into three phases, called the follicular phase (days 1–12), the ovulatory phase (days 13 and 14), and the luteal phase (days 15–28). These hormonal changes are outlined in Figures 15.7 and 15.8.

Follicular phase of the menstrual cycle

In the early follicular phase, oestradiol levels are low and FSH levels high. The high FSH causes follicle development to take place and a dominant follicle starts to grow. As the follicle grows

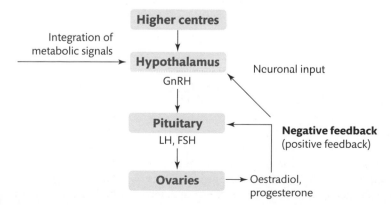

FIGURE 15.6
Hormonal interrelationships in the control of reproductive function in the female.

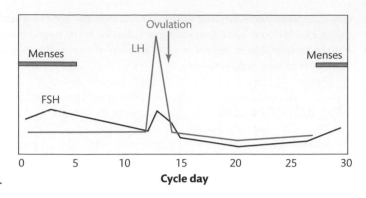

FIGURE 15.7
Hormonal changes in the menstrual cycle (LH and FSH).

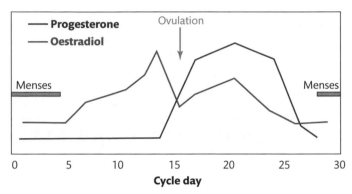

FIGURE 15.8
Hormonal changes in the menstrual cycle (oestradiol and progesterone).

the granulosa cells multiply and acquire more FSH receptors. The effect of FSH on the granulosa cells is to stimulate increased secretion of oestradiol. This has two effects. First, oestradiol feeds back negatively on the pituitary gonadotrophs to reduce FSH secretion. Second, as the oestradiol reaches a critical level its action changes to one of positive feedback on the gonadotrophs so that LH secretion rises dramatically. This is the mid-cycle LH surge. Follicle stimulating hormone also up-regulates the LH receptors in the granulosa cells. During the follicular phase, the effect of increasing oestradiol levels is to cause proliferation and thickening of the womb lining (endometrium).

Ovulatory phase of the menstrual cycle

The mid-cycle surge in LH production precedes and triggers release of the oocyte (ovulation), which occurs between 24 and 36 hours later. As the LH surge begins, progesterone secretion starts to rise and following ovulation the residual granulosa cells in the follicle accumulate lipid and luteinize to form a corpus luteum (yellow body). This secretes predominantly progesterone, although oestradiol release is maintained.

Luteal phase of the menstrual cycle

With the increase in oestradiol and progesterone secretion by the corpus luteum, LH and FSH secretion are suppressed by negative feedback and remain low during the remainder of the luteal phase. With increased progesterone, the endometrium changes from the proliferative to the secretory phase to facilitate implantation of a fertilized embryo. Progesterone secretion increases during the second half of the cycle (luteal phase), peaking at day 21 and declining thereafter. Late in the cycle, the progesterone and oestradiol fall to levels where the integrity of the endometrium is not maintained and it is shed, causing bleeding to occur. The combined effect of oestradiol and progesterone in the luteal phase is to lower LH and FSH secretion. The cycle is continuously repeated up until the menopause unless pregnancy intervenes. If the egg is fertilized, and **implantation** occurs, the early trophoblasts of the implanted embryo begin to secrete hCG, which has the same biological activity as LH. The action of hCG maintains progesterone and oestradiol secretion from the corpus luteum, thus maintaining the endometrium and the pregnancy.

Manipulation of the menstrual cycle

There are a number of circumstances when it is desirable to manipulate the menstrual cycle. The most common use is in the inhibition of ovulation for contraceptive purposes. The most widely used method is the combined oestrogen/progestagen oral contraceptive usually known as 'the pill'. Formulations contain the potent synthetic oestradiol derivative, ethynyl oestradiol, in combination with a synthetic progestogen. Progesterone although easily available cannot be used in oral preparations as it is quickly inactivated after absorption by first pass metabolism in the liver. The pill is usually administered daily for three out of four weeks. During the fourth week, oestrogen and progestogen levels fall, the endometrial lining is shed normally and a menstrual period occurs. Whilst taking the combined oral contraceptive pill, the oestrogen and progestogen levels mimic that of the luteal phase of the normal cycle and gonadotrophin secretion is reduced. This prevents the development of a dominant follicle and thus ovulation.

Endometriosis is a condition where there is abnormal proliferation of endometrial tissue outside the uterus. Endometriosis is an oestrogen dependent condition. It is an important cause of infertility and causes the patient great discomfort. It is useful to downregulate LH and FSH secretion in the treatment of endometriosis as this reduces oestrogen secretion. When oestrogen levels are suppressed, endometriosis regresses. This can be achieved by administration of long-acting GnRH analogues, such as buserelin, which gives steady circulating levels of a drug. Normally GnRH is secreted in pulses. Constant levels of GnRH or its agonists down-regulate GnRH receptors on the gonadotrophs in the pituitary. This causes FSH and LH secretion to fall, with reduction in circulating oestradiol. Long-acting LHRH analogues are also used to down-regulate pituitary function prior to induction of ovulation for *in vitro* fertilization (IVF) procedures.

SELF-CHECK 15.2

In the ovarian follicle, name three cells on which LH acts?

15.3 Menopause

The menopause is described as the last menstrual period. It occurs at an average age of 51 in women in the UK. The normal range for this is 42–57 years. Menopause below the age of 42 is considered to be premature. Approximately five years prior to the menopause the cycles become slightly shorter, less regular and frequently anovulatory. This period is known as the menopausal transition. The approach of the menopause is heralded between the age of 35–40 when a subtle increase in follicular phase FSH levels begins to occur, preceding irregularities of the cycle. This then extends to later phases of the cycles which become irregular. In addition

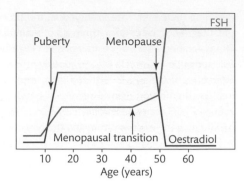

FIGURE 15.9
Age-related hormonal changes in the female.

to an increase in the follicular phase, the concentration of FSH becomes higher and more variable in the rest of the cycle. The result of this change in ovarian function is a considerable increase in the secretion of LH and particularly of FSH. Oestradiol concentrations fall, but post-menopausal oestrogen levels overlap with those found in the early follicular phase of the cycle in regularly menstruating women. You can see the age-related changes in pituitary and ovarian hormones in Figure 15.9.

The symptoms of the menopause include vaginal dryness and discomfort, and hot flushes due to oestrogen deficiency. This is also associated with an increase in body fat mass and a redistribution of fat to a more central (male-like) distribution. In terms of general health, the protective effect of the female hormones, oestradiol and progesterone, to the female gender in terms of cardiovascular disease (the major cause of death in UK women) is lost. Bone turnover increases at the menopause with a relative increase in bone resorption compared to deposition. Oestrogen prevents this and is critical in the maintenance of bone mass. Following the menopause, bone mass is lost at a steady rate, resulting in **osteoporosis**. This equates to an accumulating increase in the risk of osteoporotic fracture: a major cause of morbidity and mortality in elderly women. The general health consequences of the menopause are a major public health concern as increased life expectancy in women in the developed world means that the average woman will now spend 30 years of her life in the post-menopausal state.

Hormone replacement therapy

It is possible to effectively abolish the effects of the menopause by instituting hormone replacement therapy (HRT). Therapy may consist of administering oestrogen only or a combination of oestrogen and progesterone. Until comparatively recently, observational studies indicated that long-term HRT use was safe in terms of cardiovascular risk and in the development of breast and endometrial cancer. However, prospective randomized studies conducted in 2004 question this view and HRT use for more than ten years is not now considered advisable. Bisphosphonate drugs which are not oestrogenic are useful, safe, and effective in long-term osteoporosis management. Recently, selective oestrogen receptor modulators have been described for treating osteoporosis. These are oestrogen agonists in bone but not in the breast or endometrium.

SELF-CHECK 15.3

List three symptoms of menopause.

15.4 Causes of anovulatory infertility

When menstruation is absent this is called amenorrhoea. If there has never been a period it is called primary amenorrhoea and if menstruation ceases it is called secondary amenorrhoea. The principal causes of anovulatory infertility are ovarian failure, hypothalamic amenorrhoea, pituitary disease, **hyperprolactinaemia**, and **polycystic ovary syndrome** (PCOS). When **anovulation** is due to ovarian failure, this is called primary ovarian failure. In the other causes, ovarian function is intact and as the abnormality is elsewhere they are collectively termed secondary ovarian failure.

Primary ovarian failure

Ovarian failure occurs in all women at the menopause. However, when ovarian failure occurs in a woman aged less than 42 years it is regarded as premature. Chromosomal abnormalities such as Turner's syndrome, in which there is absence of an X chromosome, are associated with ovarian failure at an early age. Autoimmune processes can also cause premature ovarian destruction and both radiation and cytotoxic therapy as used in the treatment of common malignancies have this effect. You can see the changes in pituitary and ovarian hormone secretion associated with age in Figure 15.9.

Hypothalamic amenorrhoea

Several chronic processes can disrupt ovulation. Excessive weight loss as occurs in anorexia nervosa and bulimia are important examples. Most chronic debilitating diseases and prolonged trauma can induce amenorrhoea. These include chronic renal failure, poorly controlled diabetes, both hyperthyroidism and hypothyroidism, and chronic liver disease. Over-indulgence in physical exercise can also do this: 'jogger's amenorrhoea'. Collectively these phenomena are regarded as 'hypothalamic' amenorrhoea. In a few cases 'obvious' causes cannot be found. Some women with low weight without any obvious characteristics of eating disorders may also have amenorrhoea. In each individual there appears to be a critical weight below which menstruation ceases. Cessation of ovulation associated with disease probably represents an evolutionary defence mechanism, as in women the metabolic consequences of reproduction are high.

Hypogonadotrophic hypogonadism is a condition where there is absent secretion of LH and FSH by the pituitary. It may be diagnosed as a cause of delayed puberty in which LH and FSH secretion are found to be absent. The primary abnormality is hypothalamic and is due to GnRH deficiency due to abnormal developmental processes involving GnRH secreting neurones. When this occurs and the sense of smell is absent it is also known as Kallman's syndrome.

Pituitary disease

Anovulation, associated with amenorrhoea and reduced LH and FSH secretions is found in pituitary disease. Common causes are pituitary tumours including prolactinomas. Pituitary tumours secreting ACTH and human growth hormone, and non-secreting pituitary adenomas may cause amenorrhoea. Malignancies elsewhere in the body may also give rise to secondary tumour deposits in the pituitary, causing amenorrhoea. Sheehan's syndrome (post-partum pituitary necrosis) is caused by excessive blood loss in childbirth and is particularly common in developing countries, presenting with failure of **lactation**. Simmond's disease occurs when pituitary failure occurs following blood loss from other causes. The mechanism is a period of pituitary ischaemia induced by the poor blood circulation associated with blood loss. The pituitary is particularly susceptible to this, as the blood supply is portal. Autoimmune hypopituitarism (hypophysitis) occurs as an isolated entity and in association with other autoimmune processes such as hypothyroidism, and is more common in the post-partum period. Iron overload, frequently encountered in thalassaemia and less commonly haemochromatosis also cause pituitary destruction and amenorrhoea. Radiation therapy for pituitary tumours and intracranial neoplasms can also result in pituitary dysfunction, often after an interval of several years. Idiopathic hypopituitarism, when no obvious cause can be found, is also recognized.

Hyperprolactinaemia

An increased concentration of serum prolactin or hyperprolactinaemia is a frequent cause of anovulation and its associated infertility. Symptoms of hyperprolactinaemia are amenorrhoea and **galactorrhoea** (secretion of breast milk). However, not all patients with galactorrhoea show an increased prolactin. When hyperprolactinaemia is caused by a pituitary tumour, the patient may experience frontal headache and visual disturbances. Prolactin is a stress hormone so not surprisingly it is frequently increased in conditions of trauma, psychological, and surgical stress. It is also chronically increased in some common medical conditions such as chronic renal failure and cirrhosis of the liver. In normal pregnancies, prolactin is elevated and this occurs in tandem with a physiological increase in the number of normal prolactin secreting cells in the pituitary.

A number of drugs are associated with increased prolactin. Commonly used antipsychotic drugs of the phenothiazine class, such as chlorpromazine which are dopamine antagonists, together with tricyclic antidepressants such as imipramine, and also the selective serotonin re-uptake inhibitor drugs such as citalopram, are common offenders. Inhibitors of monoamine oxidase and the antihypertensive calcium channel blockers such as verapamil are also responsible for increasing prolactin levels. As pituitary prolactin secretion is negatively regulated by dopamine secreted into the portal circulation, interruption of this blood supply abolishes the dopamine effects. This can occur in pituitary trauma or in the presence of a pituitary tumour and is termed 'disconnection hyperprolactinaemia'. Hyperprolactinaemia occurs in hypothyroidism, particularly in severe disease. In hypothyroidism, hyperprolactinaemia occurs in conjunction with an increased number of prolactin secreting cells and demonstrable pituitary enlargement. Thyrotrophin releasing hormone (TRH) increases prolactin secretion. As circulating prolactin levels in hypothyroidism correlate to circulating TSH, the high prolactin in hypothyroidism is usually attributed to a direct TRH drive on the pituitary. Prolactin secretion falls when the hypothyroidism is treated. The degree of hyperprolactinaemia associated with different abnormalities is shown in Table 15.1.

Prolactinomas

Prolactin secreting pituitary tumours or prolactinomas are an important cause of hyperprolactinaemia. These are best visualized by magnetic resonance (MR) scanning which has a resolution of about 3 mm. At autopsy, pituitary tumours of greater than 3 mm are found in 11% of the population. When apparently healthy people are subjected to MR scans, 11% are found

TABLE 15.1 Prolactin values found in hyperprolactinaemia related to cause.

Cause	Concentration of prolactin (mU/L)
Pregnancy	Up to 8,000
Dopamine antagonists	Up to 5,000
Stress	Up to 2,000
Disconnection	Up to 8,000
Hyperthyroidism	Up to 2,500
Microprolactinoma	Up to 3,000–9,000
Macroprolactinoma	9,000–120,000
Renal failure	Up to 5,000
Macroprolactinaemia	Up to 6,000

to have small pituitary tumours. Prolactinomas of 3–8 mm are often referred to as microprolactinomas and those >8 mm as macroprolactinomas. This is a pragmatic definition, but useful in that macroprolactinomas are more likely to give local symptoms such as headache and visual impairment. As a rule, the larger the prolactinoma, the higher the circulating prolactin. Prolactinomas are almost invariably benign and slow growing.

Macroprolactinaemia

Prolactin is a monomeric protein which has a molecular weight of 23,000 Da and is the main circulating form, together with a much smaller amount of prolactin dimer also known as 'big prolactin'. A third form of prolactin is 'big big' prolactin or macroprolactin. Macroprolactin has a molecular weight of at least 150,000 Da and is a monomeric prolactin associated with an immunoglobulin molecule, usually an IgG. Macroprolactinaemia is a heterogeneous condition. There is evidence that macroprolactins occur as a conventional antibody-antigen complex, and also as a covalent complex of prolactin linked with immunoglobulin via disulphide bonds. You can see the chromatographic separation of macroprolactin in serum in Figure 15.10.

Because different prolactin immunoassays contain different prolactin antibodies, the ability of different assays to detect macroprolactin varies. All widely used commercial assays measure macroprolactin to some degree. Depending on the assay used, up to 15% of patients presenting with hyperprolactinaemia have macroprolactinaemia. Macroprolactinaemia is a persistent condition and all patients with hyperprolactinaemia should be screened for it at initial presentation. This is done by precipitating macroprolactin from serum using polyethylene glycol, which selectively precipitates macroprolactin. Whilst widely regarded as biologically inactive, macroprolactin has nonetheless been shown to have bioactivity in an *in vitro* assay system; long-term follow-up of many patients is needed to properly document the natural history of macroprolactinaemia.

The risks of having macroprolactinaemia for the patient are two-fold. First, as radiologically demonstrable pituitary tumours are fairly common, hyperprolactinaemia caused by macroprolactin can result in an unnecessary, potentially hazardous surgical procedure to remove

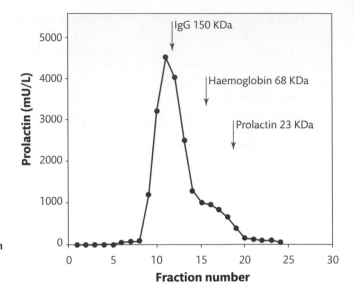

FIGURE 15.10
Gel chromatography of serum containing macroprolactin on Sephadex G-200. The majority of the prolactin elutes near IgG (molecular weight 150 KDa).

an insignificant tumour. Second, once the diagnosis of hyperprolactinaemia is established, treatment of the high prolactin is likely to dominate patient management so that, for example, in an infertile patient other more appropriate investigations and treatments such as IVF might not be properly considered.

CASE STUDY 15.1

Mary, aged 36, had recently married and wished to start a family. Since her periods started at age 13 they had been regular. After consulting her physician, he referred her to the reproductive medicine clinic after she had been trying, unsuccessfully to conceive for twelve months. Hormonal investigations were carried out on day 3 of her menstrual cycle (reference ranges are given in brackets):

LH	3.5 IU/L	(2–15)
FSH	5.3 IU/L	(1–15)
Prolactin	2070 mU/L	(50–700)

Progesterone on day 21 of her cycle was 58 nmol/L suggesting ovulation. A magnetic resonance scan of her pituitary showed no abnormality.

Mary was treated with the dopamine agonist bromocryptine and the raised prolactin normalized. She suffered postural hypotension and nausea with the bromocryptine and was unable to tolerate treatment. She was then treated with cabergoline, a second dopamine agonist, and she tolerated this much better. During the next four years she had several more courses of cabergoline and whilst on treatment her prolactin remained normal. Her periods remained regular throughout treatment.

(a) Was the high prolactin likely to have been the cause of the infertility in this case?

(b) Which additional test should have been carried out early in the investigation of this patient?

15.5 Polycystic ovary syndrome

In 1937, two Chicago gynaecologists, Stein and Leventhal described a syndrome of **hirsutism** and amenorrhoea in obese young women, associated with multiple cysts in enlarged ovaries. After removing wedges of ovarian tissue for biopsy they found that menstruation resumed and some patients, infertile due to lack of periods, went on to conceive. Thus the diagnostic procedure used, that is, wedge resection of the ovaries, was found to have therapeutic benefit. However, after a period of regular menstruation, irregular periods resumed in most patients. This syndrome is now more widely known as polycystic ovary syndrome (PCOS). It is the most common endocrine abnormality in women of reproductive age.

Depending on how it is defined, PCOS occurs in up to 20% of the population. The great majority of anovulatory infertility (as much as 80–90%) is associated with PCOS and it takes up much of the resources in the treatment of anovulatory infertility in the UK. It is about twice as common in women of south Asian origin in the UK, who tend to be affected at an earlier age. In the USA, American women of Hispanic origin are more frequently affected than Caucasian women or women of Afro-Caribbean descent. Although not uncommon in women of Chinese and Japanese descent, it is less frequent than in other ethnic groups in the USA.

Current definitions of polycystic ovary syndrome

The understanding of, and research into, PCOS has been hampered by the lack of a universally accepted definition of the syndrome. The 2003 consensus definition produced under the auspices of the American Society for Reproductive Medicine (ASRM) and the European Society for Human Reproduction and Embryology (EHSRE), although a topic for continuing debate, is gaining cautious acceptance.

According to the consensus definition, PCOS is indicated when two of three of the following criteria are present, and when other causes of the clinical abnormalities are excluded:

(1) the presence of **hyperandrogenism** and/or hyperandrogenaemia

(2) irregular or absent ovulation

(3) the presence of polycystic ovaries on ultrasound scanning

The abnormalities to be excluded in making a diagnosis of PCOS are androgen secreting adrenal and ovarian tumours, late-onset congenital adrenal hyperplasia (11-hydroxylase and 21-hydroxylase deficiency), and Cushing's syndrome.

Cross reference
Chapter 14 Adrenal disease

It is recognized that PCOS is associated with long-term health consequences in terms of diabetes and cardiovascular disease, and that pregnancy outcomes are also affected, with increased incidence of gestational diabetes and pregnancy associated hypertension.

Clinical presentation of polycystic ovary syndrome

Features of hyperandrogenism associated with PCOS are hirsutism, acne, and female androgenetic alopecia. Enlargement of the clitoris (clitoromegaly), male pattern balding, and development of a muscular body habitus are evidence of severe androgen excess and are described as **virilism**. Initially, PCOS presents to physicians in primary care. Most patients are referred at some stage to secondary care, usually to endocrinologists or gynaecologists. If the main complaint relates to hyperandrogenism, primary care physicians tend to refer the patients to endocrinologists and if it relates to **infertility** or irregular menstruation they are referred to gynaecologists. This results in different practitioners having totally different views of PCOS and the needs of individual patients. It is better for the patient to be evaluated and managed in secondary care in a combined clinic with care provided by both types of specialist.

Hyperandrogenaemia

No precise definition of hyperandrogenaemia is offered in the 2003 consensus guidelines. This leads to different local definitions of hyperandrogenaemia, which are further complicated by limitations in the currently used assays for testosterone. The suitability of reference ranges is recognized as a problem in the 2003 guidelines. In the UK, hyperandrogenaemia is usually defined in terms of circulating testosterone or as free androgen index (FAI). This is testosterone concentration divided by sex hormone binding globulin (SHBG) concentration, expressed as an index (FAI = T/SHBG × 100). The major circulating unconjugated C-19 steroids are testosterone, androstenedione, and dehydroepiandrosterone (DHEA), along with small quantities of 5α-dihydrotestosterone (DHT). Much larger quantities of the conjugated steroid dehydropepiandrosterone sulphate (DHEAS) also circulate. Of the circulating C-19 steroids only testosterone and DHT bind to, and activate, the androgen receptor. In general, circulating levels of androgens and precursor steroids relate poorly to the manifestations of hyperandrogenism such as hirsutism found in different patients. Other circulating C-19 precursor steroids such as androstenedione, DHEA, and DHEAS may be converted to biologically active androgens at the site of action. The circulating concentrations of the major C-19 steroids are given in Table 15.2.

An adrenal component is recognized in the hyperandrogenaemia of PCOS. The adrenal cortex secretes androstenedione as does the ovary, However, DHEA and DHEAS are secreted mainly by the adrenal cortex. When adrenal cortisol production rises, adrenal C1-19 steroid secretion also increases.

In androgen sensitive hair follicles, the inert steroid precursor DHEAS can be converted to both testosterone and dihydrotestosterone. Thus androgen action is contributed to both by circulating steroids and by precursor steroid conversion to active androgens at the site of action (Figure 15.11).

In the case of hirsutism the site of androgen action is the mesenchyme-derived dermal papilla at the base of the hair follicle (Figure 15.12).

TABLE 15.2 Circulating concentrations of reproductive steroid hormones and precursor steroids in men and women.

Steroid	Men	Women
Androstenedione (nmol/L)	2–12	2–8
Testosterone (nmol/L)	10–30	1–3
DHEA (nmol/L)	1–5	1–5
Oestradiol (pmol/L)	50–200	100–1500 (pre-menopausal) 50–150 (post-menopausal)
DHEAS (μmol/L)	2–10	3–8

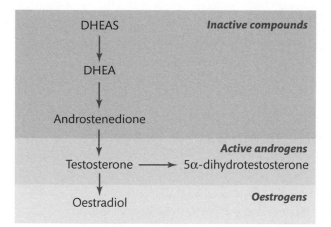

FIGURE 15.11

Transformation of inert steroids to active androgens at the site of action. The transformations above all take place in the dermal papilla of androgen dependent hair follicles.

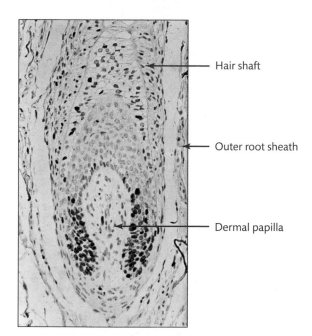

FIGURE 15.12

Cross-section through the base of an androgen dependent human hair follicle. The dermal papilla at the base of the hair follicle is the target for androgen action.

The origin of the hyperandrogenaemia in PCOS has long been debated. However, it is now recognized as having both ovarian and adrenal components. When ovarian function is suppressed either by using long-acting GnRH analogues or the oestrogen/progestogen contraceptive pill, circulating testosterone and androstendione concentrations in plasma fall significantly. A smaller decrease in DHEAS also occurs. Adrenal androgen secretion is reduced when suppressive doses of corticosteroids are administered.

Abnormalities of gonadotrophin secretion in polycystic ovary syndrome

When gonadotrophin immunoassays became available they confirmed earlier bioassay findings that LH concentration was increased in PCOS. The ratio of LH to FSH is also increased. There also appear to be qualitative differences in circulating LH, attributed to altered glycosylation.

In PCOS there appear to be more circulating basic LH glycoforms. This leads to the LH immunoreactivity being expressed differently in different immunoassays. About half of PCOS patients will have a raised LH or ratio of LH to FSH. This abnormality in gonadotrophin secretion is accompanied by increased pulse frequency of GnRH secreted by the hypothalamus. The precise reasons for these abnormalities are not defined, but androgen actions are thought to dysregulate the normal negative feedback mechanisms of both oestradiol and progesterone.

Ovarian ultrasound examination in polycystic ovary syndrome

The characteristic small cysts in the ovaries can be visualized by ultrasound examination. Transabdominal ultrasound has generally been used but this is complicated by the requirement for a full bladder (the ovaries sit just behind). In obese patients, abdominal fat spoils the image with the transabdominal approach and the ovaries may not be observed properly. Better visualization is given by the transvaginal approach. Patients with PCOS tend to have larger ovaries than normal (ovarian volume >10 mL). The 2003 consensus criteria require 12 follicles of between 2–9 mm in each ovary for an ultrasound classification of PCOS using modern equipment and trained operators. The distribution of the cysts tends to be peripheral and the central portion of the ovary may have a characteristic bright ultrasound appearance. Improvements in ultrasound methodology have led to the realization that many women have measurable small cysts in their ovaries, but not in abnormal numbers. Thus quantitative rather than qualitative ultrasound evaluation of ovarian morphology is appropriate, as shown in Figure 15.13.

Obesity and insulin resistance in polycystic ovary syndrome

Obesity is a feature of PCOS and about half of the patients presenting in secondary care have a body mass index of greater than 30. A further 25% are 'overweight', with a body mass index of greater than 25. Ideal body mass index is considered to be between 20 and 25. Body mass index (BMI) is calculated as the weight in kg divided by the height in metres squared (kg/m^2). All features of PCOS are made worse by increased weight and are ameliorated by weight loss. Insulin resistance is considered to be present when the body requires more insulin than normal to regulate glucose homeostasis. In the general population, a relationship between obesity and insulin resistance has been recognized for many years. The more obese a patient is the more insulin resistant they are likely to be. Insulin resistance is well recognized as a prominent

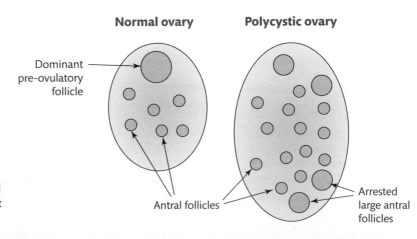

FIGURE 15.13
Schematic representation of normal polycystic ovaries. The polycystic ovary is larger, with an increased number of antral and large antral follicles the 'cysts' without a dominant pre-ovulatory follicle.

association of PCOS and affects 40–50% of patients. Whatever method is used to measure the insulin resistance, not all women with PCOS can be shown to be insulin resistant. However, insulin resistance, when present in PCOS is greater than can be accounted for by obesity alone. Insulin resistance precedes, and is a strong predictor for, type 2 diabetes, which is more common in young women with PCOS. In PCOS the incidence of type 2 diabetes and hypertension also increases with age. Insulin resistance is also associated with specific atherogenic abnormalities of lipoprotein metabolism which are raised triacylglycerols and low HDL cholesterol. Altered endothelial function is also recognized. These abnormalities are important features of the metabolic syndrome, which predicts higher risk of later diabetes and cardiovascular disease in middle age. In spite of many studies showing increased cardiovascular risk markers in PCOS, strong evidence for increased cardiovascular mortality, a hard end point, is lacking.

Pregnancy outcomes in polycystic ovary syndrome

Patients with PCOS have an increased risk of gestational diabetes. There is also an increased risk of pregnancy associated hypertension and **pre-eclampsia**. Babies tend to be delivered early and are more likely to be admitted to a neonatal intensive care unit. Perinatal mortality is higher. Whether these abnormalities are specifically associated with PCOS remains to be established as similar problems are found in obese women without PCOS.

Insulin and androgen action

Insulin, in addition to its essential role in regulating glucose homeostasis, is a hormone of many actions. Some of these are particularly relevant to the abnormalities found in PCOS. Insulin, together with LH, acts on the ovary to increase the ovarian secretion of androstenedione and testosterone. Insulin directly increases androstenedione secretion by ovarian theca cells acting via its own receptor. Measures which decrease insulin secretion by reducing insulin resistance, such as dietary manipulation, and treatment with insulin sensitizing agents such as metformin and thiazolidenedione drugs, which reduce circulating insulin, also reduce circulating androgens. Sex hormone binding globulin, which is lower in patients with PCOS, is secreted by the liver, a major target of insulin action. Insulin down-regulates the hepatic production of SHBG, independent of its actions on glucose regulation. Therefore SHBG is a marker of insulin action on the liver and low SHBG is a marker for insulin resistance. In the circulation, SHBG has a direct role in controlling the concentrations of circulating non-protein bound or free testosterone and DHT. These are regarded as the biologically active fraction. As testosterone and DHT are not involved in a feedback control with the hypothalamus and pituitary in women, rising SHBG levels reduce their biological availability and vice versa (see Figure 15.14).

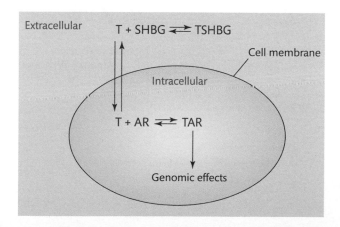

FIGURE 15.14
Modulation of androgen action by SHBG. Unbound testosterone (T) passes freely across the cell membrane. By binding to testosterone, SHBG limits the amount which can bind to the androgen receptor (AR) and elicit metabolic effects.

Thus SHBG is an important controlling factor in the expression of androgen action, especially in women. Therefore insulin has two main effects on androgen action. First, it increases ovarian androgen secretion and second, it enhances androgen bioavailability.

Disordered follicular maturation in PCOS

The principal cause of infertility due to PCOS is abnormal development of ovarian follicles with the accumulation of small follicles 'the cysts' (see Figure 15.15).

There is failure of selection of a dominant follicle from the cohort of smaller follicles. The ovary is not regarded as a 'classical' androgen target. However, androgen levels in the ovary are high and act within the ovary in a paracrine manner. Theca cells in the PCOS ovary secrete significantly higher amounts of androgens compared to normal ovaries. Androgen receptors occur in small primary ovarian follicles. The effect of androgens in early follicle development is to accelerate the development of small primary follicles from primordial follicles.

Circulating levels of AMH are higher in PCOS. Small pre-antral and antral follicles produce AMH. Granulosa cells from patients with PCOS appear to secrete more AMH than normal. Circulating AMH is closely associated with the antral follicle count in both normal women and in women with PCOS. The two major effects of AMH are to inhibit early stages of follicular recruitment and to inhibit the actions of FSH on antral follicles. The latter is brought about by inhibition of the induction of FSH receptors. In the absence of FSH action, a dominant follicle cannot be selected, the enzyme aromatase and LH receptors cannot be induced. Thus the mid-cycle LH surge, which is dependent on increasing follicular oestradiol secretion that depends on the induction of aromatase, cannot be initiated. Excessive intra-ovarian androgen action, together with increased production and actions of AMH contribute to the dysregulated folliculogenesis in PCOS.

SELF-CHECK 15.4

What lipid abnormalities are seen in patients with insulin resistance?

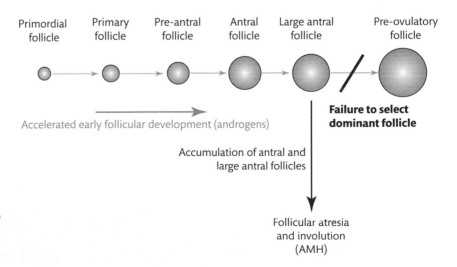

FIGURE 15.15
Dysregulated folliculogenesis in PCOS. There is an accelerated early follicular development with an accumulation of antral and large antral follicles. Increased secretion of AMH by the increased number of antral follicles prevents selection of a dominant follicle and ovulation.

15.6 Investigations in female reproductive endocrinology

The determination of hormones involved in reproductive function gives important information in women with absent or irregular menstruation. The hormones in question are progesterone, LH and FSH, oestradiol, prolactin, testosterone (with SHBG), and AMH.

Progesterone

Normal menstrual cycles vary from 23–33 days. Most of the cycle variation is in the follicular phase. In regularly menstruating women ovulation is assumed to occur. This can be confirmed by measuring luteal phase progesterone. When a sample is taken seven days before the next period a value of more than 25 nmol/L may be considered to confirm ovulation. In practice a window of 5–9 days before the next period is considered acceptable. The progesterone result cannot be interpreted without knowledge of the date of the subsequent period. In patients with irregular menstruation, samples can be taken 2–3 times weekly until menstruation occurs and interpreted retrospectively.

Luteinizing hormone and follicle stimulating hormone

In women with absent menstruation LH and FSH are measured at any time. In regularly menstruating women, serum gonadotrophins should be measured on days 3–5 of the cycle. In the case of ovarian failure or the menopause FSH is the analyte of importance. In regularly menstruating women under 30, day 3–5 FSH values are <9 IU/L. Day 3–5 FSH begins to rise over the age of 35. Values of FSH >15 IU/L in the follicular phase of the menstrual cycle strongly suggest a perimenopausal state. Elevated FSH (>20 IU/L) in the context of absent menstruation strongly predicts ovarian failure or an established menopause. Low gonadotrophins (LH and FSH <2 IU/L) in the absence of menstruation suggest a hypothalamic or pituitary cause for the amenorrhoea. However, values in this range are often found in patients on 'the pill' and in the luteal phase of the normal menstrual cycle. In a woman with amenorrhoea, an LH/FSH ratio of more than 2 with or without an elevated LH (>20 IU/L) is suggestive of PCOS. A raised LH, however, can be associated with an ovulatory peak and needs to be interpreted in the light of the subsequent menstrual history. In a normal menstrual cycle LH can be measured as a predictor of ovulation. When daily serum LH values are measured a doubling in LH indicates the start of the mid-cycle LH surge. This is important for optimum timing of insemination.

Oestradiol

Measurements of serum oestradiol are surprisingly unhelpful in the investigation of anovulatory infertility. Day 3–5 oestradiol is unchanged during the menopausal transition and bears little relation to ovarian reserve. Post-menopausal oestradiol values overlap with those encountered in the follicular phase of the normal cycle, whilst after the menopause FSH values are reliably elevated. Oestradiol has an important use in the monitoring of post-menopausal hormone replacement therapy when this is being given via oestradiol implants. Oestradiol implants have a variable period of effectiveness from a few months to over a year. If repeat implants are given on the grounds of recurrence of symptoms, oestradiol concentrations can rise to abnormally high levels. This is called tachyphylaxis. Although effective for symptom relief, this greatly increases the risk of thrombotic episodes. In post-menopausal patients treated with oestradiol implants, oestradiol should be measured before a repeat implant is considered. If the oestradiol is above 400 pmol/L reimplantation should be deferred. Oestradiol should be measured during induction of ovulation with exogenous FSH and used in conjunction with ultrasound measurements for the timing of hCG injection to produce ovulation and to avoid the occurrence of ovarian hyperstimulation.

Prolactin

It is appropriate to measure prolactin in any woman with absent or irregular menstruation, or who has galactorrhoea, or in whom there is evidence of subfertility. It should be remembered that prolactin is a stress hormone and that any elevated result should be confirmed on a repeat sample. An elevated result should be considered in terms of the causes of raised prolactin. A macroprolactin screen should be performed once when hyperprolactinaemia has been established. Prolactin should be measured at subsequent clinic visits to assess response to treatment.

Testosterone and sex hormone binding globulin

In any woman who is amenorrhoeic, has irregular periods, is subfertile, or shows signs of androgen excess, measurement of testosterone and SHBG is indicated. As circulating SHBG varies widely and it binds the majority of circulating testosterone, testosterone is best interpreted in conjunction with SHBG. The results may be expressed as a free androgen index (T/SHBGx100) usually abbreviated FAI. Except where SHBG is low, the FAI bears a satisfactory relationship to the free testosterone. In the diagnosis of female hyperandrogenaemia and in the context of PCOS, FAI is more sensitive and specific than testosterone alone. If an elevated serum testosterone is found using a direct (non-extraction) immunoassay, the sample should be re-analysed after extracting the serum with diethyl ether and the extracted result interpreted.

Additional steroid investigations

Further investigations carried out when a raised FAI is found depend largely on local guidelines. The aim is to exclude other pathologies related to hyperandrogenism and hyperandrogenaemia, which are androgen secreting ovarian and adrenal tumours and late onset congenital adrenal hyperplasia (CAH). Often when FAI is elevated then androstenedione is also measured and if this is raised, 17-hydroxyprogesterone and DHEAS are also measured. These three tests are also done if the extracted testosterone is greater than 5 nmol/L. Measurement of urinary cortisol, or performance of an overnight dexamethasone suppression test for the diagnosis of Cushing's syndrome is optional and depends on the presentation and clinical evaluation.

Anti-Mullerian hormone

Serum AMH is a useful investigation in the context of anovulatory infertility. Circulating AMH reflects the primordial follicle pool, closely reflects ovarian ageing, and is a useful predictor

CASE STUDY 15.2

Nasreen, aged 27 years, had been married for 18 months and wished to start a family.

She had been trying to conceive without success for 12 months. Her periods had been irregular for the last two years and she had noted increased growth of facial hair on her chin and upper lip and also on her arms and legs. Her grandmother had type 2 diabetes and her mother had high blood pressure. She was obese, with a BMI of 31 kg/m².

On ultrasound examination both ovaries were enlarged with many peripheral cysts. Investigations were as follows (reference ranges are given in brackets):

Testosterone	3.1 nmol/L	(0.5–2.5)
Sex hormone binding globulin	15 nmol/L	(23–120)
Free androgen index	20.6	(1–8)

(a) Which common condition is associated with Nasreen's symptoms and laboratory findings?

(b) Which additional investigation is required in this case?

of ovarian reserve. Serum AMH is also closely related to antral follicle count (ovarian follicles which are 2–5 mm in diameter), an independent measure of ovarian reserve. Unlike LH, FSH, oestradiol, and progesterone, AMH secretion is independent of the stage of the menstrual cycle. Unlike these hormones its concentrations are not affected by the contraceptive pill. It also shows reduced day-to-day variability. Anti-Mullerian hormone is useful in predicting response to ovarian stimulation with gonadotrophins in induction of ovulation. High AMH is associated with increasing risk of hyperstimulation and low AMH with inadequate gonadotrophin response. It performs better than FSH in identifying poor responders. Unlike FSH, however, AMH is also useful in identifying those patients at risk of hyperstimulation. At the menopause, circulating AMH is undetectable. Anti-Mullerian hormone is also elevated in patients with PCOS. It has significant potential as a supplementary investigation in making the diagnosis. Granulosa cell tumours are uncommon ovarian tumours. Anti-Mullerian hormone is secreted by some granulosa cell tumours and may have use as a tumour marker in these cases.

15.7 Management of anovulatory infertility

In a patient presenting with amenorrhoea, a full history and examination should be carried out to rule out underlying causes of absent menstruation. It is appropriate to begin investigations when periods have been absent for six months. It is common practice to start investigations in couples who have been trying unsuccessfully to conceive for one year. Young healthy couples trying to conceive after ceasing contraception do so at a rate of 22% per cycle.

Weight-related amenorrhoea

Amenorrhoeic women with low BMI (typically <19 kg/m²) should be encouraged to gain weight. In a majority of cases satisfactory weight gain causes menstruation and fertility will return. Although ovulation can be induced by GnRH and exogenous gonadotrophins (see below) pregnancies in low-weight women are considered to be high risk. Overweight women (BMI> 30 kg/m^2) should be encouraged to lose weight. Pregnancies are also high risk in overweight women.

Induction of ovulation

Various therapeutic approaches are possible to induce ovulation when there are abnormalities at the hypothalamic and pituitary level. The approaches used depend on the site of the abnormality. In hypothalamic amenorrhoea there is a resetting of the pituitary-ovarian estradiol negative feedback threshold which results in reduced FSH secretion. When this happens insufficient FSH is produced to initiate growth of a dominant follicle. Under these circumstances the selective oestrogen receptor modulator clomiphene citrate can be used. It stimulates FSH secretion by antagonizing the negative feedback of oestradiol. The use of clomiphene citrate is associated with a modest increase in the risk of multiple pregnancies and it should be used under the supervision of a reproductive medicine clinic. In hypothalamic amenorrhoea when clomiphene citrate is ineffective, GnRH can be employed. Ovulation can be induced by pulsatile administration of GnRH via a subcutaneous line attached to a pump. The pump may be worn for several months at a time. Following treatment, pituitary LH and FSH secretion normalize, the ovary is stimulated in a physiological manner and a regular menstrual pattern is established. Treatment with GnRH is not associated with an increased risk of multiple pregnancies.

When the cause of anovulation is related to the pituitary or the hypothalamus, exogenous treatment with FSH can be used to induce ovulation. When ovulation is induced by exogenous gonadotrophins it is imperative that this is done under careful monitoring with ultrasound and oestradiol measurements to avoid the high risk of ovarian hyperstimulation and multiple pregnancy. Multiple pregnancies represent a major hazard to both the mother and the foetuses. A particular danger of exogenous gonadotrophin treatment is the ovarian hyperstimulation syndrome. Gross ovarian enlargement occurs, liver function is compromised, ascites may develop, and in rare cases death may ensue from multi-organ failure.

Induction of ovulation in PCOS

Weight management is crucial in achieving a successful outcome in induction of ovulation in PCOS. Initially, the patient should be encouraged to achieve a weight loss of 10% of body weight. Earlier enthusiasm for ovulation induction with the insulin sensitizing drug metformin has not been borne out by controlled trials. Clomiphene citrate (see above) is also useful in PCOS. Wedge resection is no longer widely carried out. It has been superseded in clomiphene resistant PCOS by ovarian 'drilling' where a controlled thermal injury to the ovary is carried out using diathermy. In diathermy a controlled electric current is used to inflict heat injury to the ovary.

In patients who fail to conceive after clomiphene citrate and after ovarian drilling, induction of ovulation with exogenous FSH may be used. When these treatments are unsuccessful, patients with PCOS may be treated by IVF using a controlled ovarian stimulation (COS) procedure provided they can achieve a BMI of less than 30 kg/m^2. As they are at increased risk of ovarian hyperstimulation, they are carefully managed with lower doses of exogenous gonadotrophins.

Ovarian failure

Established ovarian failure associated with the menopause, premature menopause, or resulting from chromosomal disorders is not amenable to the above treatments which are absolutely dependent on a viable pool of oocytes in the ovary. In ovarian failure the only possible course to pregnancy is egg donation, where fertilized eggs from another patient are implanted in the woman's uterus.

Ovulation induction for *in vitro* fertilization

In vitro fertilization (IVF) is a major advance in the treatment of subfertility. It can be used in women with normal cyclical menstrual function who suffer pathology of the Fallopian tubes, and in those where the abnormalities arise at the hypothalamic and pituitary level. Women with the PCOS, who do not respond to other fertility treatments can also be treated with IVF. In addition it is useful when there is a male factor associated with the subfertility.

As originally developed, IVF was used on women during natural menstrual cycles. It is possible to retrieve eggs for fertilization after the mid-cycle surge of LH has commenced. As single eggs were retrieved and the conception rates encountered, 8–10% per cycle, many treatment cycles were needed to achieve satisfactory conception rates. This compared with conception rates of about 20–25% per cycle in young healthy couples attempting to conceive with regular unprotected intercourse. This led IVF practitioners to develop methods to induce ovulations in which many oocytes are produced and fertilized. The resulting surplus embryos can then be frozen for use in subsequent cycles if the first implantation is not successful. Multiple ovulation induction in IVF is referred to as COS. The aim is to produce several (10–12) oocytes for fertilization while avoiding the consequence of severe ovarian hyperstimulation. The process begins in the mid-luteal phase of the menstrual cycle. Daily injections of a long-acting GnRH analogue are given, which down-regulates LH and FSH production, and are continued during the procedure. This enables control of the cycle to be established which is independent of the natural feedback controls.

Following the resulting menstrual period after GnRH down-regulation, daily injections of exogenous FSH are given. After four days, serum oestradiol concentration is measured in the serum and ovarian ultrasound performed. Depending on these results, the dose of FSH used may need to be increased or decreased. Treatment with FSH is continued with daily oestradiol measurements and ultrasound monitoring of the resulting follicles carried out on day 8 and day 10 of FSH treatment. Several outcomes are possible in a COS cycle. First of all there may be absence of stimulation as judged from ultrasound and oestradiol monitoring and the cycle abandoned due to lack of response. Second, response may be considered satisfactory and an injection of hCG given to cause ovulation and eggs retrieved 24–36 hours later. Third, the cycle may be abandoned due to hyperstimulation. An intermediate approach is used when modest hyperstimulation occurs. In these circumstances when embryos are replaced in the stimulation cycle there is still an increased risk of hyperstimulation syndrome developing. Eggs are retrieved and fertilized with the embryos frozen for replacement in subsequent cycles.

Hyperprolactinaemia

Most patients with hyperprolactinaemia and those with prolactinomas are managed without pituitary surgery with dopamine agonist drugs, such as the short-acting bromocryptine and the

CASE STUDY 15.3

Sharon, aged 26 years, had been married for two years. Menarche was at age 14 with regular periods. When she was 24 her periods became irregular and she was started on the combined oral contraceptive pill. Wishing to become pregnant, she had stopped taking the pill one year earlier but her periods did not return. Her partner's sperm count was normal. The following investigations were carried out (reference ranges are given in brackets):

LH	21 IU/L	(2–15)
FSH	42 IU/L	(1–15)
Prolactin	350 mU/L	(50–700)

(a) What do the high gonadotrophins suggest in this case?

(b) What is Sharon's principal option to achieve a pregnancy?

long-acting cabergoline. When a patient with a prolactinoma is treated with a dopamine agonist the tumour shrinks, prolactin concentrations fall over a period of weeks to months and can be reduced into the normal range in 80–90% of patients. Dopamine agonists have disturbing and unacceptable side effects in many patients, including depression and postural hypotension. Once prolactin has been normalized, it is worthwhile considering stopping the drugs with follow-up estimations of prolactin. With successful treatment of hyperprolactinaemia, periods return in amenorrhoeic women and irregular cycles become more regular and ovulatory. In Table 15.1 degrees of hyperprolactinaemia associated with various pathologies are summarized.

When large prolactinomas are encountered, expansion of the tumour upwards from the pituitary causes pressure on the optic chiasma with resulting visual loss. Dopamine agonist treatment is usually promptly effective in relieving symptoms in these cases. Pituitary surgery is only carried out when large prolactinomas cause visual field loss which is not improved by dopamine agonists.

15.8 Reproductive endocrinology in the male

The testis is active before birth and secretes testosterone in response to pituitary LH. Both LH and testosterone are secreted for the first 6–12 months of life, as shown in Figure 15.16.

During the remaining part of childhood, gonadotrophin and testosterone concentrations remain low. The secretion of LH and FSH begins to rise just before puberty and the testes begin to grow. At the onset of puberty, nocturnal pulses of LH and FSH are secreted and these gradually extend into the daytime as puberty progresses. With advancing puberty testosterone concentrations rise and spermatogenesis commences. Puberty is also accompanied by a growth spurt which comes to an end in terms of growth when the epiphyses of the long bones close.

Testosterone is secreted by the Leydig or interstitial cells of the testis. It stimulates the development of the secondary sex characteristics in men which include growth of the external genitalia, development of a muscular body shape, deepening of the voice, facial and body hair

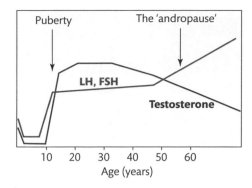

FIGURE 15.16
Age-related changes in reproductive hormones in men during life.

growth, and, in susceptible males, loss of hair in middle age. Testosterone is also responsible for development and maintenance of the sex drive. In adult men there is a diurnal rhythm of testosterone secretion, with higher circulating levels in the morning than in the afternoon or evening and this needs to be considered in the interpretation of testosterone results. The diurnal rhythm is lost or reduced in elderly men.

Gonadal function is maximal and circulating testosterone highest from the early twenties to the late thirties. Testosterone then shows a gradual decline. As testosterone concentration falls, both LH and FSH concentrations rise gradually with age. This is not a rapid transition as occurs in the female menopause but a much more gradual process. It is usually referred to as the 'andropause'. The fall in testosterone concentration is masked to some degree by a gradual age-related rise in SHBG concentration. The fall in free testosterone with increasing age is more noticeable than the fall in total testosterone.

Hormonal control mechanisms in the male

As in the female, GnRH stimulates the pituitary to release LH and FSH in the male. Follicle stimulating hormone and LH are also regulated by negative feedback inhibition in the male. Testosterone is produced in the Leydig cells of the testis and negatively regulates LH secretion. Follicle stimulating hormone acts on the Sertoli cells to cause secretion of inhibin B which has a negative regulatory effect on the production of FSH by the pituitary. The process is summarized in Figure 15.17.

Testosterone has local actions in the testis where its main role is to maintain spermatogenesis. The concentrations of testosterone are higher in the testis than elsewhere in the body and the

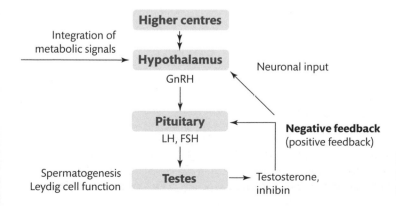

FIGURE 15.17
Hormonal control and interrelationships in the male.

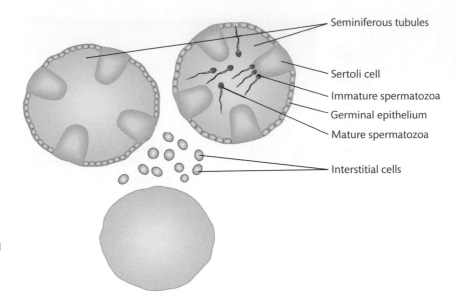

Seminiferous tubules

Sertoli cell

Immature spermatozoa

Germinal epithelium

Mature spermatozoa

Interstitial cells

FIGURE 15.18
Schematic representation of testicular tissue showing germinal epithelium, seminiferous tubules, Sertoli cells, interstitial cells, and spermatozoa.

testis is a distinct area of local high testosterone concentration. Testosterone also up-regulates the production of androgen binding protein (ABP) in the testis, which has a role in modulating local testosterone levels. The action of FSH on the germinal epithelium is to assist spermatogenesis. Spermatogenesis can continue in the absence of FSH provided testosterone concentrations are adequate.

Spermatogenesis

Spermatozoa originate in the germinal epithelium and gradually mature and develop as they migrate to the centre of the tubules. Sertoli cells assist in the maturation process. The process of spermatogenesis takes about two and a half months from start to finish. If hormonal input, that is, LH and FSH secretion ceases, spermatogenesis winds down slowly over this period. The process of spermatogenesis is outlined in Figure 15.18.

Locally produced testosterone from the interstitial cells is essential to spermatogenesis.

Testicular failure

Testicular failure involving the germinal epithelium results in impaired or absent spermatogenesis. Under these circumstances FSH is usually raised. When testicular damage includes the Leydig cells, LH concentrations also rise. Raised FSH and LH concentrations are indicators of primary testicular failure. Leydig cell damage is also accompanied by reduced testosterone. Testicular failure may occur after mumps infection when this occurs following puberty. The seminiferous tubules tend to be affected and the Leydig cells remain intact. Treatment with cytotoxic drugs for a range of malignancies also causes testicular failure, which is often irreversible. This is also encountered after radiation treatment. Testicular failure may be encountered in Klinefelter's syndrome, a chromosomal abnormality with an XXY rather than a normal male XY complement. This usually presents about the time of puberty. In the testis, the interstitial cells are reduced in number and testosterone secretion although impaired is not usually absent. Secondary sex characteristics are less well developed and **gynecomastia** may be present.

TABLE 15.3 Normal values for sperm evaluation.

Evaluation	Value
Ejaculate	2–5 mL volume
Sperm concentration	>20 million/mL
Motility	>50% progressive
Morphology	>30% normal morphology
White blood cells	<1 million/mL

The seminiferous tubules are abnormal and sperm production is absent. Low sperm counts are also found in men with partial deletions of the Y chromosome.

Evaluation of reproductive function in the male

The normal values for sperm evaluation are given in Table 15.3. Sperm are evaluated in terms of count morphology and motility. A low sperm count is termed **oligospermia** and absent sperm **azoospermia**. Possession of a normal sperm count implies satisfactory endocrine status regarding spermatogenesis. It also rules out ductal obstruction. Endocrine investigations are warranted when sperm counts are low. Sperm production is variable and can be markedly affected by intercurrent illness. Serum testosterone is likewise affected by systemic illness. A diagnosis of oligospermia or azoospermia should not be made on the basis of a single count but on a least two counts separated by at least three months. Smoking, alcohol abuse, and recreational drug use compromise reproductive function in both sexes. Male smokers have on average 15% lower sperm counts than non-smokers.

Oligospermia

Oligospermia is not an absolute bar to fertility. Successful natural fertilization may occur with sperm counts of less than one million per mL. When sperm counts are low, intra-uterine insemination of the partner may be effective and it is usual to time the fertilization to coincide with ovulation by monitoring LH in the partner. Some cases of oligospermia, however, require intracytoplasmic sperm injection (ICSI) into the oocyte to achieve fertilization.

Azoospermia

Azoospermia is said to occur when no sperm are present in the ejaculate. In many men with azoospermia (absent sperm production) Leydig cell function remains intact and testosterone is normal. Azoospermia may occur due to a blockage in the ducts which is frequently amenable to surgical treatment (obstructive azoospermia). In these cases the concentration of serum FSH is usually normal. However, the cause is often severely compromised spermatogenesis; when this is the case FSH levels are usually raised. In the past this was a total bar to male fertility. However, in many cases a few sperm can be obtained by careful centrifugation of the ejaculate. In other cases sperm can be surgically aspirated from the epididymis, adjacent to the testis or extracted from a surgical biopsy of testicular tissue. The few sperm obtained by these techniques can be used in conjunction with ICSI to achieve fertilization.

Intracytoplasmic sperm injection

Intracytoplasmic sperm injection is a major advance in the treatment of male infertility. It is used in conjunction with IVF. In this procedure a single sperm is directly injected into an oocyte to achieve fertilization. It is effective when only a very few sperm are obtainable and is also effective with immotile sperm. Its use in conjunction with IVF is increasing to include men with only moderately decreased spermatogenesis. Currently in IVF units 40–70% of all cycles now use ICSI. There are no universally agreed criteria as to when ICSI should be employed for IVF.

Donor insemination

When no sperm can be obtained, donor insemination can be employed to achieve a pregnancy. Young healthy anonymous donors are used. They are fully screened for communicable disease. Sperm for donation is quarantined frozen for at least three months before use to allow possible incubating infections in the donor to be revealed. Frozen sperm is less effective than fresh sperm in terms of insemination effectiveness. If women are inseminated on the basis of menstrual history for timing of ovulation, the results are poor, approximately 6% conception per cycle inseminated. The results are much improved when the insemination is done on the day of the mid-cycle LH surge and conception rates of 22% per inseminated cycle can be obtained. With this technique, pregnancy rates approach those obtained in young healthy couples conceiving spontaneously. With improvements in the treatment of male infertility related to ICSI the need for donor insemination continues to decrease.

Endocrine causes of male infertility

Endocrine causes of male infertility are quite uncommon. Infertility is associated with severe manifestations of hypo- and hyperthyroidism, congenital adrenal hyperplasia, and hyperprolactinaemia. Treatment of the underlying condition is the appropriate course of action. When associated with gonadotrophin insufficiency as in hypogonadotrophic hypogonadism or pituitary disease, spermatogenesis can be induced by exogenous gonadotrophin treatment or pulsatile administration of GnRH. When endogenous gonadotrophin production is absent, spermatogenesis may be induced by hCG alone or if this is unsuccessful by adding in FSH.

Manipulation of testicular function

As reproductive function in women can be manipulated by using synthetic steroid hormones these can be used in men in a similar way. Prostate cancer in men is androgen responsive. In metastatic disease useful remission is obtained by reducing circulating testosterone following pituitary down-regulation of FSH and LH secretion with GnRH agonists, which causes circulating testosterone to fall. Antiandrogens, such as flutamide are compounds which bind to but do not activate the androgen receptor. They inhibit androgen action at the receptor level and are also useful for this situation. They may be used alone or combined with GnRH agonists. Hormonal 'escape' of the tumour after 1–2 years is usual. Spermatogenesis in men is inhibited by synthetic progestogens. This happens because they inhibit LH and FSH production. Testosterone production also falls, accompanied by a negative effect on libido. 'The male pill' is a combination of a synthetic progestogen to inhibit LH, FSH, and testosterone to maintain libido. However, once starting treatment at least two months must elapse before spermatogenesis is reduced to levels low enough to ensure effective contraception. Although shown to be effective and comparatively safe, the male pill seems less generally acceptable to couples than the female alternative.

SELF-CHECK 15.5

List three endocrine causes of infertility in males?

15.9 Limitations in assays for steroid and polypeptide hormones

We have seen how useful information regarding pathological and normal changes in the pituitary gonadal axis can be obtained by measurement of the hormones involved. It is important to be aware of some of the limitations of the assays used and the influence of these limitations on the interpretation of the results.

Luteinizing hormone and follicle stimulating hormone

Luteinizing hormone and FSH are now almost universally measured by two-site labelled antibody assays on commercial immunoassay instruments. It is important for purposes of comparison between assays that they are standardized in terms of the agreed international standards. The biological specificity for hormone action resides in the beta chain as do the immunological determinants of specificity. Standard preparations of gonadotrophins, which are heavily glycosylated proteins, are heterogeneous in terms of different glycoforms. Pituitary derived standards have different glycoform distribution to circulating forms. Different reagent manufacturers use different antibody combinations and this affects the results obtained. Although the carbohydrate portions of the molecule are not considered as participating directly in the antibody binding, different glycated forms of gonadotrophins show subtle differences of immunoreactivity with various antibodies. As various clinical circumstances may be associated with different circulating glycoforms of the parent hormones, this is a potential source of assay variability.

Prolactin

As with LH and FSH, prolactin is usually assayed with two-site labelled antibodies on immunoassay instruments. Although most prolactin assays have the required sensitivity to measure prolactin in a wide range of clinical circumstances, different prolactin assays detect macroprolactin to different degrees due to the use of different antibodies by different manufacturers.

Oestradiol

Oestradiol circulates mainly bound to SHBG and albumin, with a small amount in the free form. Universally total estradiol is measured. The values encountered are usually less than 1,000 pmol/L so the assay of oestradiol is a considerable analytical challenge. Whereas in the late follicular phase of the menstrual cycle values of between 200–500 pmol/L are encountered, in the early follicular phase values are less than 200 pmol/L. In a clinical context, most samples for oestradiol are assayed in unextracted serum on automated immunoassay analysers, almost invariably using competition type immunoassays. It is questionable whether most direct non-extraction oestradiol assays can perform adequately at values less than 100–200 pmol/L. Improvements in sensitivity and specificity can be obtained on extracted serum using diethyl ether for this purpose. Low oestradiol values are encountered in males, prepubertal boys and girls, in the early follicular phase of the menstrual cycle, and after the menopause. Various oestrogen containing medications contain compounds which cross-react variably with antisera used in oestradiol assays.

Testosterone

In a similar manner to oestradiol, testosterone circulates mainly bound to SHBG and albumin, with a small amount in the free form. The free form is regarded as being the biologically active fraction. Because of the wide range of values for SHBG which binds testosterone strongly, it is advisable to measure SHBG with testosterone when assessing androgen status in both sexes. The majority of testosterone assays are carried out in unextracted serum on commercial immunoassay analysers. These assays are usually suitable for measuring testosterone in men, although they may not be entirely reliable at lower levels.

Key Points

In women and children, testosterone circulates in much lower concentrations and direct (non-extraction) assays do not have sufficient specificity to measure testosterone adequately.

The specificity of the assays can be improved by prior solvent extraction but even with this step, problems of specificity remain. A solution to this problem, now generally applicable in clinical laboratories, is to use techniques involving liquid chromatography-mass spectrometry (LC-MS). Using LC-MS for the assay of testosterone gives satisfactory sensitivity and specificity in all clinical applications.

Progesterone/anti-Mullerian hormone assays

Currently progesterone immunoassays appear to have adequate performance for their main purpose, which is as a test for ovulation. Anti-Mullerian hormone assays are presently available in kit form in 96-well plate ELISA format. The available kits have appropriate characteristics for the measurement of AMH in the context of ovarian reserve and in the investigation of PCOS.

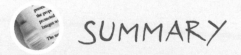

SUMMARY

- Reproductive function in both sexes is a controlled process involving hormones of the hypothalamus, pituitary, and the gonads.

- The processes in the male and the female have common endocrine features. These are the hypothalamic GnRH, which acts on the pituitary gland, and the pituitary hormones LH and FSH, which act on the ovary and the testis respectively.

- The main functions of the ovary are to produce a single egg (oocyte) for ovulation each month and to secrete oestradiol and progesterone which regulate the uterine cervix and the endometrium to permit the passage of motile sperm and to provide a suitable environment for implantation of a fertilized oocyte.

- The main targets of LH in the ovary are the theca cell of the ovarian follicle that secretes testosterone and androstenedione, which are converted into oestradiol by follicular granulosa cells under the influence of FSH.

- Following ovulation, the corpus luteum secretes progesterone and oestradiol under the influence of LH. Ovarian function ceases abruptly at the menopause with the exhaustion of the ovarian follicle pool when most women still have a considerable life expectancy.

- The main function of the testis is to provide a continuous supply of very large numbers of healthy motile sperm by continuous meiosis for fertilization of a single oocyte.

- The reproductive function of the testis remains intact during the life of the individual although there is a gradual age-related decline in gonadal function.

- The LH target in the testis is the interstitial or Leydig cell and the FSH target is the Sertoli cell.

- Spermatogenesis is a continuous process in the male and the cyclical changes in gonadotrophin production are not evident. In both the male and female, pathological processes involving the hypothalamus are responsible for cessation of gonadal function.

- Intercurrent illness also impairs both ovulation and spermatogenesis. In both sexes negative feedback control of gonadotrophin secretion is exerted by steroid hormones secreted by the gonads and cessation of gonadal function is accompanied by increases in circulating LH and FSH.

- By making use of exogenous steroid hormones it is possible to regulate the function of both the ovaries and the testis.

FURTHER READING

- Broekmans FJ, Knauff EAH, te Velde ER, Macklon NS, and Fauser BC (2007) Female reproductive ageing: current knowledge and future trends. *Trends in Endocrinology and Metabolism* **18**, 58–65.

- Beltran L, Fahie-Wilson MN, McKenna TJ, Kavanagh L, and Smith TP (2008) Serum total prolactin and monomeric prolactin reference intervals determined by precipitation with polyethylene glycol: evaluation and validation on common immunoassay platforms. *Clinical Chemistry* **54**, 1673–81.

- European Society for Human Reproduction (ESHRE) and the American Society of Reproductive Medicine (ASRM) (2004) Revised 2003 consensus on diagnostic criteria and long-term health risks related to polycystic ovary syndrome. Rotterdam ESHRE/ASRM-sponsored PCOS Consensus Workshop Group. *Fertility and Sterility* **81**, 19–25.

- European Society for Human Reproduction (ESHRE) and the American Society of Reproductive Medicine (ASRM) (2008) Consensus on infertility treatment related to polycystic ovary syndrome. ESHRE/ASRM-sponsored PCOS Consensus Workshop Group, Thessaloniki. *Human Reproduction* **23**, 462–77 (Erratum in: *Human Reproduction* (2008) 23, 1474).

- Hadley ME and Levine JE (2006) **Hormones and male reproductive physiology in** *Endocrinology.* **New Jersey: Pearson Prentice Hall. pp. 387–407.**

Useful websites

- http://www.netdoctor.co.uk/menshealth/facts/malefertility.htm
- http://www.nhs.uk/Conditions/Infertility
- http://www.nice.org.uk/Guidance/CG11

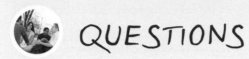 QUESTIONS

15.1   FSH acts on the ovary to promote:

   (a)   Progesterone secretion

   (b)   Oestradiol secretion

   (c)   Testosterone secretion

   (d)   Insulin secretion

   (e)   None of the above

15.2   Which of the following are TRUE or FALSE in the early follicular phase of the menstrual cycle?

   (a)   FSH is low

   (b)   FSH is high

   (c)   Oestradiol is high

   (d)   Oestradiol is low

15.3   At mid-cycle the LH surge is triggered by:

   (a)   Falling FSH

   (b)   Falling oestradiol

   (c)   Rising progesterone

   (d)   Rising oestradiol

   (e)   None of the above

15.4   Which of the following are TRUE or FALSE during the luteal phase of the menstrual cycle?

   (a)   FSH is increased

   (b)   FSH is suppressed

   (c)   LH is increased

   (d)   LH is suppressed

15.5   Which ONE of the following is correct for the luteal phase of the menstrual cycle?

   (a)   FSH secretion is stimulated by oestradiol

   (b)   LH and FSH are negatively regulated by oestradiol and progesterone

   (c)   LH secretion is stimulated by progesterone

   (d)   FSH secretion is stimulated by progesterone

   (e)   None of the above

15.6 What happens to the hormones LH, oestradiol, and FSH during menopause?

15.7 What effect does the combined oestrogen-progestogen contraceptive pill have on serum FSH and LH concentrations in women?

15.8 Which ONE of the following is correct for the polycystic ovary syndrome?

   (a) Obesity is always present

   (b) IVF is the first-line treatment for induction of ovulation

   (c) Weight management is a key element in treatment

   (d) Diabetes is less common

   (e) None of the above

Answers to self-check questions, case study questions, and end-of-chapter questions are available in the Online Resource Centre accompanying this book.

 Go to www.oxfordtextbooks.co.uk/orc/ahmed/

16

Biochemical nutrition

Pat Twomey and William Simpson

Learning objectives

After studying this chapter you should be able to:

- Discuss nutritional considerations in the management of health and disease, both in hospital and in the community at large
- Describe the assessment of nutritional status
- Outline the contribution of the laboratory to nutritional assessment
- Explain the effect of illness on the interpretation of tests used to assess nutrition
- Appreciate the value of nutritional intervention strategies

Introduction

Nutrition (from the Latin for nurse, *nutrix*) is the science that deals with nutrients and nourishment. **Nutrients** are the chemical components of food which are required for tissue function (both cellular metabolism and physical activity of the intact organism) and structure (growth and repair). **Nourishment** is the process by which organisms obtain, assimilate, and utilize nutrients. This chapter will examine different types of nutrients and the consequences of their deficiencies and excesses. The chapter will also consider assessment of nutritional status, in particular the role of laboratory tests. Features of disordered eating patterns and the value of nutritional interventions will also be described. Box 16.1 gives some historical highlights of clinical nutrition.

16.1 Types of nutrients

Nutrients can be divided into two broad categories which can be further subdivided into categories as given below:

- Macronutrients:
  - organic macronutrients:

BOX 16.1 Notable history of clinical nutrition

AD 30 (approx) Aulus Cornelius Celsus writes in *De Medicina* about the role of food in maintaining health.

1611 John Woodall describes the value of lemon juice in protecting against scurvy.

1747 James Lind undertakes the first 'controlled clinical trial' to investigate the ability of citrus fruits to prevent scurvy.

1838 Jöns Jakob Berzelius first describes the term 'protein'.

1896 Christiaan Eijkman identifies dietary factors as the cause of Beri Beri in chickens.

1906 Sir Frederick Gowland Hopkins describes the need for 'accessory food factors' in the diet.

1912 Kazimierz Funk coins the term vitamine ('vital amine') to describe these factors.

1932 Sir David Cuthbertson describes the metabolic response to trauma.

1940 McCance and Widdowsons describe the chemical composition of foods.

1968 Stanley Dudrick describes long-term total parenteral nutrition.

1963 UK Department of Health constitutes the Committee on Medical Aspects of Food and Nutrition Policy.

- o carbohydrates (sugars and starches)
- o lipids (fats and oils)
- o proteins
- — inorganic macronutrients:
  - o electrolytes
  - o minerals
- Micronutrients:
  - — organic micronutrients:
    - o fat-soluble vitamins
      - vitamin A (retinoids)
      - vitamin D (calciferols)
      - vitamin E (tocopherols)
      - vitamin K (phylloquinone and related compounds)
    - o water-soluble vitamins
      - vitamin B group
      - vitamin C (ascorbic acid)
  - — inorganic micronutrients:
    - o trace elements

Macronutrients are those nutrients required in amounts of a gram or more per day. The organic (carbon containing) macronutrients are primarily sources of energy. An excess of these macronutrients can lead to chronic disease states such as obesity, while a general deficiency can lead to **starvation**. The inorganic (non-carbon containing) macronutrients include major electrolytes and minerals, for example sodium, chloride, potassium, calcium, magnesium, phosphorus, and sulphur. Whilst not being sources of energy, excesses or deficiencies of these can lead to alterations of metabolism, including effects on the metabolism of other nutrients.

Micronutrients are those nutrients required in amounts of less than a gram per day. They are necessary, for example, as cofactors for specific enzymes in metabolic pathways. Excess amounts can lead to toxicity, whereas deficiencies of individual micronutrients can lead to classical deficiency diseases such as scurvy.

The organic micronutrients are usually referred to as **vitamins**—these compounds were initially believed to be 'vital amines', and although it was later shown that they were not actually amines, the derived term 'vitamin' remained. Individual vitamins were named as they were discovered using letters of the alphabet; A, B, and so on. Subsequent research revealed that vitamin B was actually a group of several different molecules, often co-existing in the same foods, but with different roles. The group term was retained, but the individual vitamins were subclassified as B_1, B_2, B_3, etc. In addition, it transpired that many other vitamins existed in isomers and other chemical forms, for example vitamin A refers to a group of compounds known as retinoids.

The inorganic micronutrients are inorganic elements required by the human body in minute quantities. Approximately a century ago, scientists began to detect trace amounts of several elements in human tissue. These substances were called **trace elements**, although this term has various definitions. In biology, trace elements are those present at levels of less than 100 ppm in tissue. In human nutrition, however, the term essential trace element is generally defined as an element required in mg quantities in the diet.

Currently, there is evidence, such as a defined biochemical function, for up to 15 elements required in trace amounts by humans or animals, although it is likely that not all the trace elements essential for humans have been identified. In addition, it is probable that many biochemical functions of the known trace elements remain to be identified. Some trace elements have other metabolic effects when taken in quantities higher than usual dietary amounts, and so specific trace elements may be used as medicines, for example lithium is sometimes used to treat severe depression. All trace elements are, however, toxic when large amounts are ingested.

16.2 **Specific nutrients: macronutrients**

Carbohydrates are energy rich molecules that contain solely carbon, hydrogen, and oxygen, usually in the ratio of 1:2:1. They can be classified into those that can be digested (sugars and starches, which are 16 kJoules or 4 kcal energy per gram) or not (fibre) by the human small intestine. The basic unit is a monosaccharide (sugar) such as glucose, fructose, or galactose. Two monosaccharides form disaccharides, for example glucose plus fructose produce sucrose, glucose and galactose produce lactose respectively. Oligosaccharides are composed of 3–10 monosaccharides while polysaccharides contain more than ten monosaccharides: they include starch and cellulose from plants and glycogen from animals. The enzyme amylase acts on polysaccharides containing glucose, mainly starch, converting them to disaccharides. These are then broken down by specific disaccharidases, for example lactase hydrolyses lactose to galactose and glucose.

In clinical nutrition, glucose is measured routinely to assess glucose intolerance or diabetic control, but is also sometimes useful to indicate contamination of blood samples with intravenous nutrition fluid, which contains concentrated glucose.

Proteins are polymers of (>50) amino acids, joined by a peptide bond (-C(=O)NH-). In adults, there are eight essential amino acids (necessary in the diet) and 12 non-essential amino acids (not required in the diet). However, the latter are *so* essential to life that the healthy human metabolism has preserved the ability to synthesize them. In times of ill-health this ability can be impaired, and so certain amino acids become essential in the diet. For this reason glutamine and arginine, for example, are sometimes referred to as 'conditionally essential'. For infants, histidine is essential. Further classification depends on the presence of functional groups: hydrophobic, hydrophilic, dicarboxylic, and basic amino acids. Protein provides 16 kJoules (4 kcal) energy per gram.

Specific amino acids are, however, rarely assessed in routine clinical nutrition. Albumin and total protein are routinely measured, but concentrations of these in blood reflect underlying disease rather than providing information on protein nutrition status. Blood albumin concentration does not fall even in starving patients, unless their condition is complicated by, for example, sepsis. It had previously been suggested that the reason albumin concentration does not fall in starvation is because of its relatively long half-life in blood of around 30 days, and so various other proteins of shorter half-life in blood have been assessed. The blood concentrations of these other proteins were, however, also affected by the **acute phase response**, and so they are not being routinely assessed.

Key Points

Acute phase response is characterized by release of proteins in the plasma in response to inflammation. Some of these proteins have a protective function, that is, to destroy microbes such as C-reactive protein and complement, whereas others have a negative feedback effect on the inflammatory response, for example serpins.

CASE STUDY 16.1

A 75-year-old man was admitted to the respiratory ward with a severe chest infection and poor oral intake. Clinical and X-ray examinations revealed a lobar pneumonia. Results of routine biochemistry tests were as follows (reference ranges are given in brackets):

Sodium	134 mmol/L	(137–144)
Potassium	4.8 mmol/L	(3.5–4.9)
Urea	12.4 mmol/L	(2.5–7.0)
Creatinine	146 µmol/L	(60–110)
Alkaline phosphatase	90 IU/L	(45–105)
Albumin	28 g/L	(37–49)

(a) Comment on these results.

(b) What test might aid interpretation?

(c) What intervention (if any) does this patient require?

Lipids are molecules which dissolve in alcohol but not water. They contain carbon, hydrogen and oxygen, but have proportionally less oxygen than carbohydrates. The major energy store in animals is triacylglycerols (commonly referred to as triglycerides). Animal sources of dietary lipids predominantly contain saturated fatty acids whilst those from plants and fish often contain some unsaturated fatty acids; monosaturated fatty acids contain one C=C double bond whilst polyunsaturated fatty acids contain several. These double bonds are named by counting their position from the terminal methyl group. This position is referred to as omega (written ω or n), thus ω3 fatty acids contain a double bond at position 3. Furthermore, these double bonds have two potential isomers, called *cis* and *trans*; *cis* is the naturally occurring form, with *trans* only occurring with hydrogenation of unsaturated fatty acids. These man-made *trans* fatty acids confer a longer shelf-life and greater palatability to foods, so became popular commercially, but they are now known to be more atherogenic than other forms of fat and so are currently being phased out. Lipids provide 37 kJoules (9 kcal) energy per gram. Assessment in routine clinical nutritional practice is, however, limited to assessing tolerance of artificial nutritional supplements, for example it may be necessary to reduce the fat content of a feed if the patient develops hypertriglyceridaemia.

Ethanol or ethyl alcohol (C_2H_5OH) is an energy rich alcohol that is commonly found in nature due to the fermentation of sugars. Whilst minimally toxic in small amounts, ethanol does have a concentration dependent toxicity. Ethanol is the main psychoactive ingredient in alcoholic drinks. The majority of alcohol is catabolized by hepatocytes under the action of alcohol dehydrogenase; the remaining alcohol is eliminated unchanged through the lungs and in urine. The rate of hepatic catabolism varies from each individual; if an individual consumes more alcohol per unit time than the liver can break down, then the alcohol concentration in the blood increases. One consequence of excess alcohol is hypoglycaemia which can then result in further nutrition being sought. Ethanol provides 29 kJoules (7 kcal) energy per gram. Ethanol concentration is not routinely assessed in clinical nutrition, although a history of excess ethanol intake increases the suspicion that there may be other nutritional deficiencies.

Cross reference

Chapter 10 Disorders of calcium, phosphate, and magnesium homeostasis and Chapter 5 Fluid and electrolyte disorders

Inorganic macronutrients, sodium, chloride, potassium, calcium, magnesium, and phosphorus in the form of inorganic phosphate are dealt with in detail in other chapters.

Their importance in the context of nutrition, however, cannot be overstressed, and they are assessed frequently in routine practice.

SELF-CHECK 16.1

What is the difference between dietary lipids from animal and plant sources?

CASE STUDY 16.2

A 55-year-old male with a known history of alcohol abuse attended hospital complaining of pins and needles. On examination, a positive Chvostek's sign was noted (hyperexcitability of the facial nerve; when the skin over the nerve is tapped, the facial muscle goes into spasm). Routine biochemistry tests showed the following results (reference ranges are given in brackets):

Sodium	134 mmol/L	(137–144)
Potassium	3.1 mmol/L	(3.5–4.9)
Urea	10.4 mmol/L	(2.5–7.0)

Creatinine	135 µmol/L	(60–110)
Calcium (corrected)	1.36 mmol/L	(2.20–2.60)
Phosphate	0.65 mmol/L	(0.8–1.4)
Alkaline phosphatase	90 IU/L	(45–105)
Albumin	38 g/L	(37–49)

Despite intravenous calcium, 12 hours later the serum calcium remained low at 1.34 mmol/L.

(a) What nutrient needs to be measured?

(b) How does alcohol affect this nutrient?

(c) How does this nutrient affect his calcium results?

16.3 Specific nutrients: organic micronutrients

The organic micronutrients include vitamins, which are organic substances required in small amounts in order to regulate cell function and maintain life.

In general, vitamins cannot be synthesized endogenously (although there are a few exceptions) and must be provided in the diet. We as humans require 13 different vitamins. Of these vitamins, four are fat-soluble and nine are water-soluble.

Fat-soluble vitamins

The solubility of a vitamin can affect how it is absorbed, transported, and stored in the body. The absorption of fat-soluble vitamins is via a complex absorptive process in the gastrointestinal tract (GIT). After absorption, these vitamins are transported in the blood bound to carrier proteins and stored either in the liver or in fat tissue. When required, these vitamins can be mobilized from their storage sites. The fat-soluble vitamins include vitamins A, D, E, and K.

Vitamin A

Vitamin A is commonly known as retinol. Figure 16.1 demonstrates its structure.

Retinol and related compounds have three main functions in humans. These are:

(1) Vision: retinal is present in the rods and cones of the retina; it changes state when hit by a photon of light, causing membrane potential changes which are then relayed to the brain.

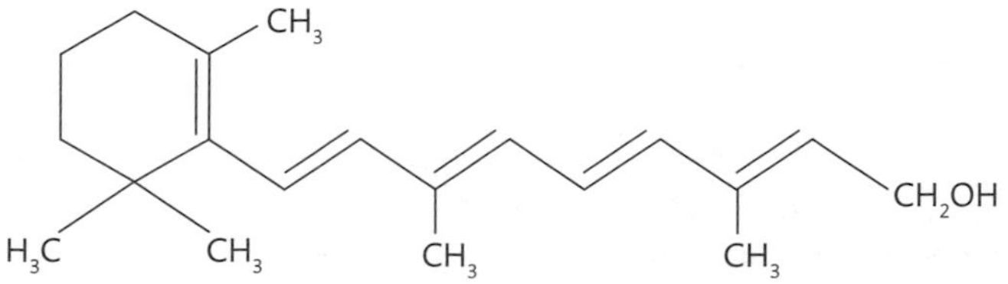

FIGURE 16.1
Retinol.

(2) Control of differentiation and proliferation: retinoids are required by certain types of cells, notably epithelial and bone cells.

(3) Glycoprotein synthesis: retinyl phosphate is a cofactor in the synthesis of certain glycoproteins.

Retinol is derived mostly from retinol esters in food, with about a quarter being made in the body from beta carotene. A specific transport protein, retinol binding protein exists; it circulates in a complex with prealbumin, and as both of these proteins are affected by the acute phase response, serum retinol concentrations are of limited value in illness. Dietary deficiency of retinol leads to **night blindness**, which is reversible, but its deficiency can also cause irreversible corneal degeneration ('xeropthalmia'), which is a major cause of blindness in the world.

Vitamin D

Vitamin D_2 is also known as ergocalciferol and is the semi-synthetic form used to fortify certain foodstuffs, whereas vitamin D_3 commonly known as cholecalciferol is the naturally occurring form found in dairy produce and eggs. The main source of cholecalciferol is, however, from endogenous production in skin exposed to ultraviolet radiation. You can see the chemical structure of cholecalciferol in Figure 16.2.

Both forms of vitamin D are rapidly converted in the liver to 25-hydroxy forms, which undergo further conversion in the kidneys to 1,25-dihydroxy forms. For example in the skin, UV light acts on 7-dehydrocholesterol and converts this to cholecalciferol which in turn is converted to 25-hydroxycholecalciferol in the liver and then to 1,25-dihydroxycholecalciferol in the kidneys (1,25-DHCC) or calcitriol; the latter has a key role in regulation of calcium homeostasis.

Cross reference

Chapter 10 Disorders of calcium, phosphate, and magnesium homeostasis

The serum concentrations of the 25-hydroxy forms can be differentiated by modern analytical techniques such as HPLC and mass spectrometry. The concentrations of the 1, 25 forms are, however, much lower, making analysis more difficult. Deficiency of vitamin D in children results in the classical bone deformity disease known as **rickets**, whereas adults with deficiency develop **osteomalacia**, where the defective bone mineralization can lead to pathological fractures rather than deformity. For this reason, there is widespread advocation for the supply of vitamin D supplementation (combined with calcium) for the elderly, to reduce risk of fractures. Hyperphosphataemia suppresses the renal conversion to the 1,25 forms.

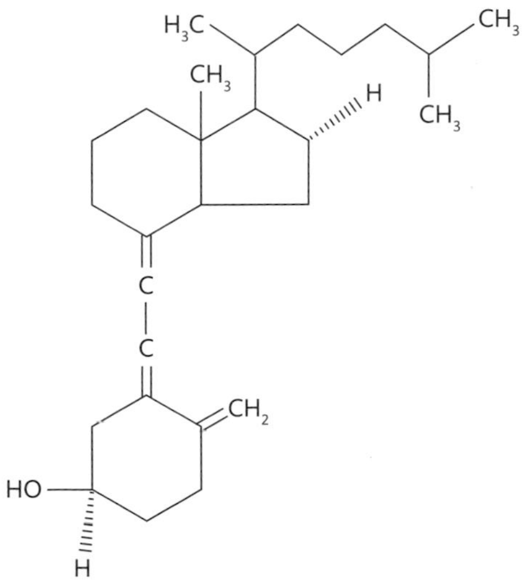

FIGURE 16.2
Cholecalciferol.

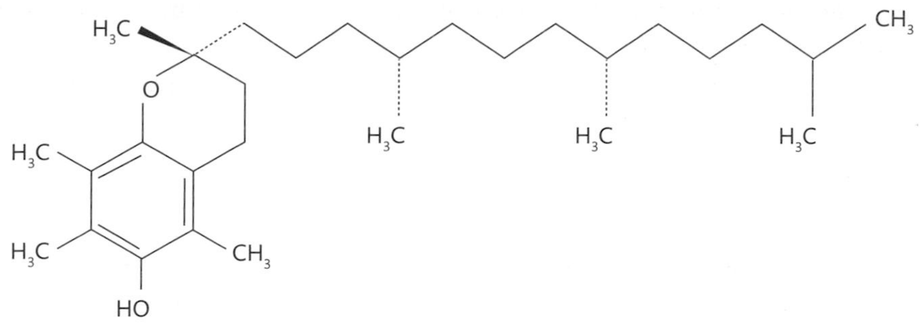

FIGURE 16.3
α-tocopherol.

This may occur in chronic renal failure, leading to 'renal rickets'. Inadequate 1,25-DHCC causes hypocalcaemia, resulting in a secondary hyperparathyroidism. Excessive intake can lead to **hypervitaminosis D**. This could, theoretically, cause hypercalcaemia, but there is little evidence to suggest that this occurs in clinical practice.

Vitamin E

Like vitamin C, vitamin E or tocopherol does not have a role as a specific cofactor. The structure of α-tocopherol is shown in Figure 16.3.

It acts as an **antioxidant** and **free radical** scavenger, principally protecting lipids, particularly in cell membranes, from oxidation. Tocopherols are ubiquitous in food and so their deficiency is rare, partly because they can be regenerated *in vivo* using vitamin C as a reducing agent. A relatively low intake had been associated with increased risk of cardiovascular disease, but clinical trials of supplements have not shown any benefit. As vitamin E is fat-soluble, its concentration is usually expressed relative to the total cholesterol concentration.

Vitamin K

Vitamin K_1 (phylloquinone and phytomenadione) and K_2 (menaquinone) occur naturally; K_3 (menadione) is a synthetic compound used therapeutically. The chemical structure of vitamin K_1 (phylloquinone) is shown in Figure 16.4.

The principal role of vitamin K is in blood clotting; it is essential for the function of clotting factors II, VII, IX, and X, along with the inhibitory factors protein C and protein S. Deficiency of vitamin K is rare unless fat malabsorption is present. Clinically, deficiency of vitamin K presents as a bleeding tendency in affected individuals. Warfarin inhibits the regeneration of vitamin K, and the resulting inhibition of clotting is used therapeutically.

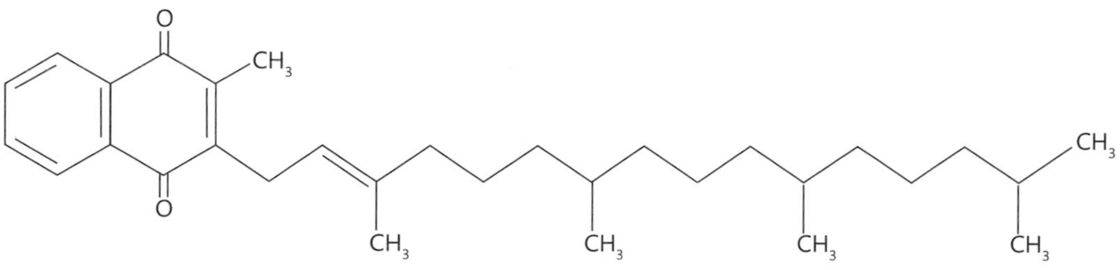

FIGURE 16.4
Phylloquinone.

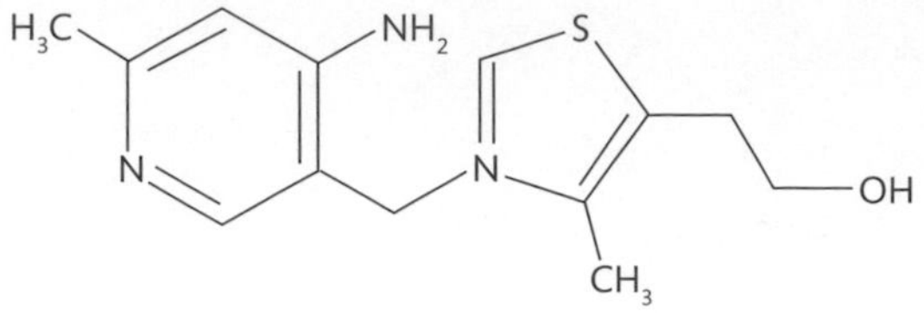

FIGURE 16.5
Thiamine.

Water-soluble vitamins

The water-soluble vitamins are absorbed from the GIT and enter the blood, in which they are transported to the sites where they are required. Any excess water-soluble vitamins are detected by the renal system and removed from the body by excretion in the urine. The water-soluble vitamins include vitamin B_1, B_2, B_3, B_5, B_6, B_7, B_9, B_{12}, and C.

Vitamin B_1

Vitamin B_1 commonly known as thiamine is a cofactor for a number of metabolic processes, most notably the conversion of pyruvate to acetyl CoA. You can see its chemical structure in Figure 16.5.

Thiamine deficiency leads to impaired carbohydrate metabolism, and potentially to life-threatening **lactic acidosis** as you can see in Figure 16.6.

Care must always be taken, therefore, to ensure that a starved individual has adequate thiamine replacement before giving macronutrients. Apart from the systemic lactic acidosis, irreversible neurological damage (**Wernicke-Korsakoff syndrome**) can be precipitated. The classical deficiency syndrome is **Beri-Beri**, which may be either 'dry' or 'wet'. The term Beri-Beri translates literally as 'man who walks like a sheep', the effect produced by the peripheral neuropathy of dry Beri-Beri. In wet Beri-Beri, the main finding is heart failure, leading to the development of **oedema**. Despite sounding exotic, Beri-Beri does present in the UK, and should be considered in any unexplained metabolic acidosis. Thiamine is also a cofactor for transketolase, and the activity of transketolase in blood has been used to assess body status.

Vitamin B_2

Vitamin B_2 commonly referred to as riboflavin is also required for numerous metabolic processes involving redox reactions. You can see the chemical structure of riboflavin in Figure 16.7.

Although there is no specific syndrome described in riboflavin deficiency, a number of clinical signs may be present, including angular stomatitis and glossitis (inflammation of the tongue). It is noteworthy that a number of these clinical signs have poor specificity, as they can be found in association with a number of the B vitamins. Deficiencies of B vitamins are often multiple, as many are found in similar types of food, and so therapy is usually given as a complex of B vitamins.

Vitamin B_3

Vitamin B_3 also known as niacin, is part of nicotinamide adenine dinucleotide (NAD^+) and NADP, is involved in many redox reactions in human metabolism, as well as being a common component of clinical chemistry reagents. The chemical structure of niacin is shown in Figure 16.8.

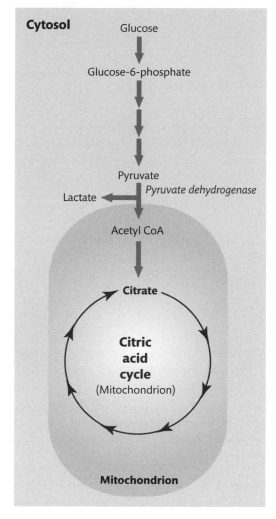

FIGURE 16.6

Role of thiamine in metabolism of carbohydrates. Pyruvate enters the mitochondrion, where it is converted to acetyl CoA by the enzyme pyruvate dehydrogenase. The acetyl CoA can then enter the citric acid cycle and be used to release more energy. The enzyme pyruvate dehydrogenase is dependent on thiamine, and during thiamine deficiency a build up of pyruvate occurs. Excess pyruvate is converted to lactic acid and the patient will suffer from lactic acidosis.

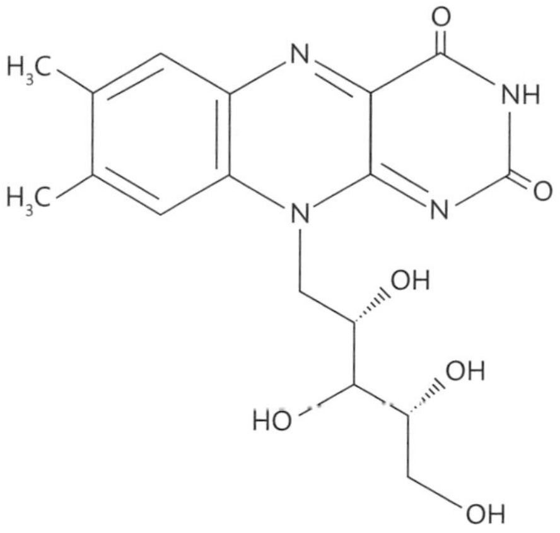

FIGURE 16.7
Riboflavin.

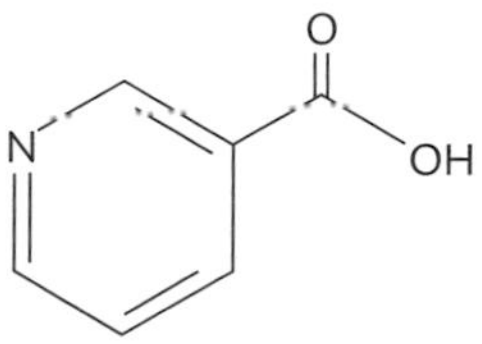

FIGURE 16.8
Niacin.

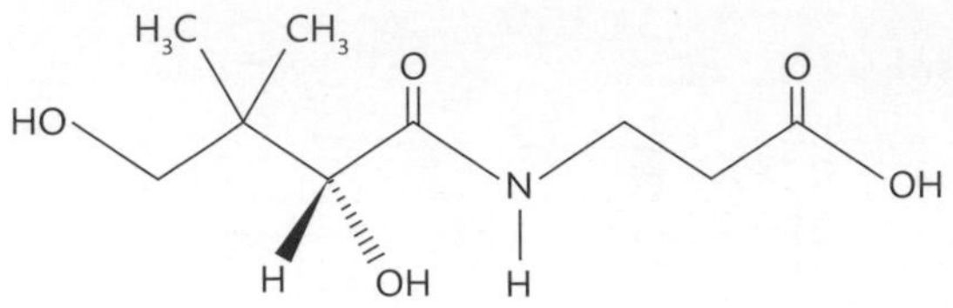

FIGURE 16.9
Pantothenic acid.

Severe deficiency of niacin results in the deficiency syndrome **pellagra**, where patients present with dermatitis, diarrhoea, and dementia, but milder deficiencies may present with non-specific signs such as glossitis. In pharmacological doses (of the order of one hundred times the typical daily intake of niacin), niacin is used as a treatment for dyslipidaemia.

Vitamin B_5

Vitamin B_5 also known as pantothenic acid is part of coenzyme A and plays a fundamental role in carbohydrate metabolism, but despite this, it is so ubiquitous in food that no classical deficiency syndrome has been described. The structure of pantothenic acid is shown in Figure 16.9.

Vitamin B_6

Vitamin B_6 or pyridoxine is so ubiquitous in foodstuffs that deficiency is rare, but clinical signs such as glossitis and neuropathy can occur, and a sideroblastic anaemia has been described. You can see the chemical structure of pyridoxine in Figure 16.10.

Pyridoxine is used, along with B_{12} and folate, in the treatment of homocystinuria. Pharmacological doses of pyridoxine (of the order of one hundred times the usual intake of pyridoxine) have also been used to treat premenstrual tension.

Vitamin B_7

Vitamin B_7 or biotin is sometimes referred to as 'vitamin H'. Biotin is a cofactor in some carboxylase reactions involved with fatty acid metabolism and gluconeogenesis. The chemical structure of biotin is shown in Figure 16.11.

It binds strongly to avidin (a phenomenon utilized in some analytical reagent systems) which is found in raw egg white; consumption of significant quantities of raw egg has therefore been associated with biotin deficiency, the main clinical sign being a form of dermatitis, but otherwise deficiency is rare.

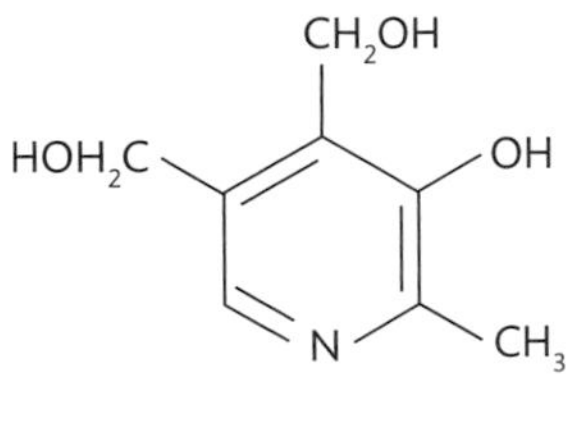

FIGURE 16.10
Pyridoxine.

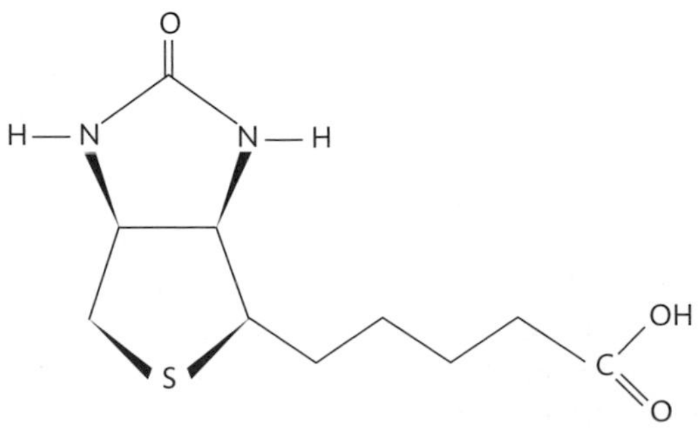

FIGURE 16.11
Biotin.

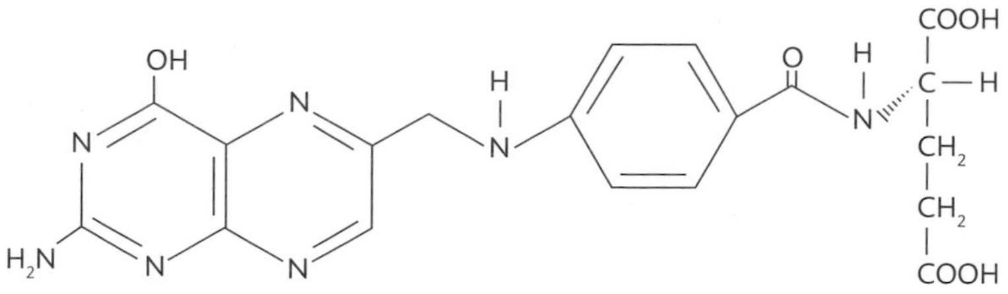

FIGURE 16.12 Folic acid.

Vitamin B_9

Vitamin B_9 or folic acid, in the form of tetrahydrofolic acid, is involved in single carbon transfer reactions, including a key role in DNA synthesis. Figure 16.12 shows the structure of folic acid.

Deficiency of folic acid results in macrocytosis, and **megaloblastic anaemia** in severe cases. Megaloblastic anaemia refers to the abnormal red cells found in this type of anaemia— megaloblasts are large, functionally deficient forms. They are normally found in bone marrow, as they are one of the primitive maturation stages of erythrocyte formation, but they are not usually found in peripheral blood.

A *relative* deficiency in early pregnancy and prior to conception has been associated with neural tube defects in the foetus, and so supplementation is often recommended. Folate (along with B_6 and B_{12}) is also used to help reduce the homocysteine level in patients with homocystinuria, and this helps reduce the high rate of vascular disease otherwise experienced in these patients. Trials using supplements of these vitamins to improve cardiovascular disease outcomes in the wider population have shown no benefit (despite significant reductions in homocysteine levels being achieved).

Vitamin B_{12}

Vitamin B_{12} or cobalamin (shown in Figure 16.13) is a cofactor for methionine synthase and methylmalonyl coenzyme A.

In the nineteenth century, a megaloblastic anaemia was identified that was invariably fatal and thus called **pernicious anaemia**. The first effective treatment for this disease was to eat a pound (approximately 0.5 kg) of raw liver daily. In 1948, cobalamin, the factor in liver which had this anti-pernicious anaemia effect, was isolated. It was later found that '**intrinsic factor**', a protein released into the lumen of the stomach by the parietal cells, binds with B_{12} to facilitate its absorption in the terminal ileum. Addisonian pernicious anaemia refers to a specific autoimmune disease, where antibodies to intrinsic factor prevent the binding of B_{12}. Production of intrinsic factor can, however, be impaired by a number of mechanisms, including atrophic gastritis, *Helicobacter pylori* infection, and bacterial overgrowth. Furthermore, B_{12} is only found in foods of animal origin, and absolute vegetarianism will lead to deficiency of B_{12} after 5–10 years.

In addition to pernicious anaemia, the pathological consequences of severe vitamin B_{12} deficiency include peripheral neuropathy, memory loss, dementia, an irreversible neurological disease called subacute degeneration of the spinal cord, and in the most severe cases, death. Vitamin B_{12} deficiency also raises the concentration of homocysteine in blood. Because B_{12} deficiency is usually related to impaired absorption, replacement is usually given by parenteral injection. Whilst up to one year's requirements can be retained in the liver, B_{12} is water-soluble and excessive intakes are efficiently excreted in the urine. As such, B_{12} toxicity has not been described and doses up to 10,000 times the minimal daily adult human requirement do not appear to have adverse effects.

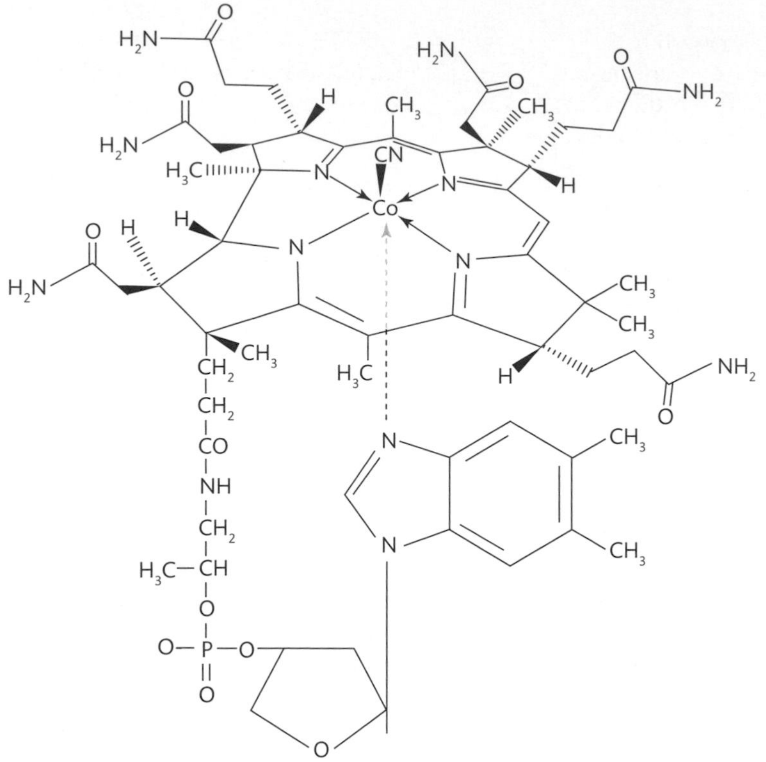

FIGURE 16.13
Cyanocobalamin.

Vitamin C

Vitamin C, also known as ascorbic acid, as shown in Figure 16.14, does not have a role as a specific cofactor, but acts instead as an antioxidant and free radical scavenger in aqueous fluids.

This includes maintaining iron in the Fe II state, which is the state required for absorption of dietary iron. Certain enzymes are also dependent on iron being in the Fe II rather than the Fe III state of oxidation, including those involved in collagen crosslinking. The classical deficiency disease is **scurvy**, which results from defective collagen synthesis. Many individuals self-medicate with pharmaceutical doses of vitamin C to 'prevent colds' or for cardiovascular disease prevention, but clinical trials have not shown any objective benefit in these situations. Large doses of vitamin C have, however, been associated with the formation of kidney stones (due to high levels of oxalate in the urine).

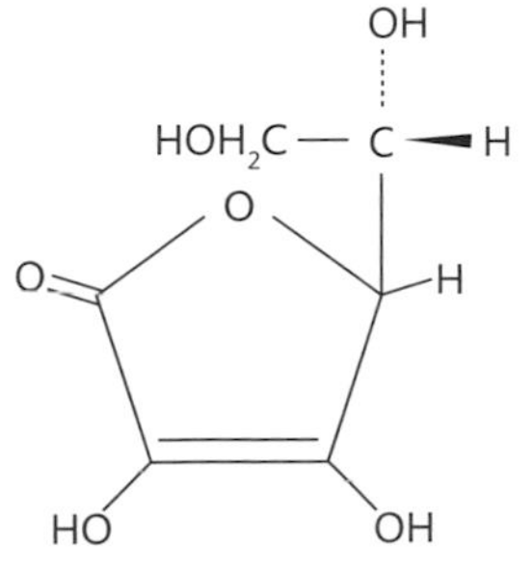

SELF-CHECK 16.2

What does deficiency of vitamin D cause in (a) children and (b) adults?

FIGURE 16.14
Ascorbic acid.

CASE STUDY 16.3

A 42-year-old woman with anorexia nervosa went to see her doctor complaining of blood-blisters at the back of her mouth which had ruptured. The doctor examined the oral mucosa to find two ruptured blisters, considered vitamin C deficiency, and contacted the clinical biochemistry department for advice on vitamin C analysis. She was informed that the best approach was simply to give vitamin C for several weeks and assess the response clinically.

(a) Why did the laboratory decline analysis for vitamin C?

(b) What condition does vitamin C deficiency cause?

(c) What are the manifestations of vitamin C deficiency?

(d) What foods need to be consumed to avoid vitamin C deficiency?

16.4 Specific nutrients: inorganic micronutrients

Copper is a cofactor for a number of oxidase enzymes which are involved in the stabilization of matrixes of connective tissue, oxidation of ferrous iron, synthesis of neurotransmitters, assurance of immune system competence, generation of oxidative energy, and protection from **reactive oxygen species**. Although copper is a well-established essential trace element, its practical nutritional importance is a subject of debate. Well-established pathological consequences of copper deprivation in humans have been described primarily for premature and malnourished infants and include a hypochromic, normocytic, or macrocytic anaemia; bone abnormalities resembling scurvy by showing osteoporosis, fractures of the long bones and ribs, epiphyseal separation, and fraying and cupping of the metaphyses with spur formation; increased incidence of infections; and poor growth. The consequences of the genetic disorder **Menkes disease** (copper deficiency caused by a cellular defect in copper transport) in children include 'kinky-type' steely hair, progressive neurological disorder, and death. Nutritional copper toxicity is not a major health issue. The ingestion of fluids and foods contaminated with high amounts of copper can cause nausea. Because their biliary excretion pathway is immature, accumulation of toxic amounts of copper in the liver could be a risk for infants if intake is chronically high; this apparently caused cases of childhood liver cirrhosis in India. However, genetic copper toxicity (**Wilson's disease**) occurs in about 1:30,000 individuals. Its identification is important as it can be treated; one such treatment is very high doses of zinc, which competitively reduces gastrointestinal copper absorption.

Intestinal absorption is a primary homeostatic mechanism for copper. Copper enters epithelial cells of the small intestine by a facilitated process that involves specific copper transporters, or non-specific divalent metal ion transporters located on the brush-border surface. Then the copper is transported to the portal circulation where it is taken up by the liver and resecreted in plasma bound to caeruloplasmin. Transport of copper from the liver into the bile is the primary route for excretion of endogenous copper. Copper of biliary origin and non-absorbed dietary copper are eliminated from the body via the faeces. Only an extremely small amount of copper is excreted in the urine. The absorption and retention of copper varies with dietary intake and status. For example, the percentages of ingested copper absorbed were 56%, 36%, and 12%, with dietary intakes of 0.8, 1.7, and 7.5 mg/day, respectively. Moreover, tissue retention of copper is markedly increased when copper intake is low. The best nutritional sources of copper are legumes, whole grains, nuts, offal, seafood, peanut butter, chocolate, mushrooms, and ready-to-eat cereals.

Iodine has one known function in higher animals and humans—it is a constituent of thyroid hormones. Thyroxine (T4) is converted to tri-iodothyronine (T3), which functions as a regulator of growth and development by reacting with cell receptors, resulting in energy production and the activation or inhibition of synthesis of specific proteins. Iodide, an anion, is rapidly and almost completely absorbed from the stomach and upper GIT. Most other forms of iodine are changed in the GIT to iodide and completely absorbed. Absorbed iodide circulates in the free form; it does not bind to proteins in blood. Iodide is rapidly removed from circulation by the thyroid and kidney. Urinary excretion is a major homeostatic mechanism. If iodine intake has been adequate, only about 10% of absorbed iodide appears in the thyroid, the rest appears in the urine. However, if iodine status is inadequate, a much higher percentage, up to 80%, can appear in the thyroid. The thyroid gland is essentially the only storage site for iodine, where it appears mostly as mono-, di-iodotyrosine, and T4, with a small amount of T3.

Cross reference

Chapter 12 Thyroid disease

The necessity for iodine in the diet was recognized in the 1920s when it was found that iodine prevented goitre and decreased the incidence of the sequelae of iodine deficiency. In the western world, iodized salt has been the major method for assuring adequate iodine intake since that time. Other sources of iodine are seafood and foods from plants grown on high-iodine soils. The consequences of iodine deficiency do, however, remain a major public health problem in other parts of the world. Iodine deficiency is the most prevalent global cause of preventable mental retardation. The spectrum of iodine deficiency disorders is large and includes foetal congenital anomalies and perinatal mortality. Although homeostatic mechanisms allow for a substantial tolerance to high intakes of iodine, iodine-induced hyperthyroidism has been recognized for nearly two centuries. People who have had a marked iodine deficiency and are then given high amounts of iodine as part of a preventative program are at risk of getting hyperthyroidism, with clinical signs including weight loss, tachycardia, muscle weakness, and skin warmth.

Iron is a component of molecules that transport oxygen, and numerous enzymatic oxidation and reduction (redox) reactions use iron as the agent through which oxygen is added, hydrogen is removed, or electrons are transferred. Among the trace elements, iron has the longest and best described history. By the seventeenth century, drinking wine containing iron filings was a recognized treatment for iron deficiency anaemia. Despite the extensive knowledge about its treatment and prevention, and the institution of a variety of effective interventions, iron deficiency is the primary mineral deficiency in the USA and the world today. The physiological signs of iron deficiency include anaemia, glossitis (smooth atrophy of tongue surface), angular stomatitis (fissures at the angles of the mouth), koilonychia (spoon nails), blue sclera, lethargy, apathy, listlessness, and fatigue. Pathological consequences of iron deficiency include impaired thermoregulation, immune function, mental function, and physical performance. Complications in pregnancy include increased risk of premature delivery, low birth weight, and infant mortality, and possibly increased risk of osteoporosis. There is no effective mechanism for excretion of excess iron, and concerns have been expressed about high intakes of iron being a health issue. This has come about through epidemiologic observations associating high dietary iron or high body iron stores with cancer and coronary heart disease. Further experimental studies, however, are required to confirm whether the high intakes of iron increase the risk for these diseases. The toxic potential of iron arises from its biological importance as a redox element that accepts and donates electrons to oxygen; this can result in the formation of reactive oxygen species or free radicals that can cause damage to cellular components.

An iron overload disease known as **hereditary haemochromatosis** is caused by a defective regulation of iron transport with excessive iron absorption and high transferrin (transport form) iron in plasma. Clinical signs appear when body iron accumulates to about ten times normal and include cirrhosis, diabetes, heart failure, arthritis, and sexual dysfunction. Haemochromatosis also increases the risk for hepatic carcinoma. The treatment for hereditary haemochromatosis is repeated phlebotomy.

Control of absorption from the GIT is the primary homeostatic mechanism for iron. Dietary iron exists generally in two forms, haem and non-haem, that are absorbed by different mechanisms. Haem iron is a protoporphyrin molecule containing an atom of iron; it comes primarily from haemoglobin and myoglobin in meat, poultry, and fish. Non-haem iron is primarily inorganic iron salts provided mainly by plant-based foods, dairy products, and iron-fortified foods. Haem iron is much better absorbed and less affected by enhancers and inhibitors of absorption than non-haem iron. Iron absorption is regulated by mucosal cells of the small intestine, but the exact mechanism of regulation has yet to be established. Both iron stores and blood haemoglobin status have a major influence on the amount of dietary iron that is absorbed. Under normal conditions, men absorb about 6% and menstruating women absorb about 13% of dietary iron. However, with severe iron deficiency anaemia (functionally deficient blood low in haemoglobin), absorption of non-haem iron can be as high as 50%. Iron loss from the body is very low, about 0.6 mg/day. This loss is primarily by excretion in the bile and, along with iron in desquamated mucosal cells, eliminated via the faeces. Menstruation is a significant means through which iron is lost for women. It should be noted, however, that non-physiological loss of iron resulting from conditions such as parasitism, diarrhoea, and enteritis account for half of iron deficiency anaemia globally. Excess iron in the body is stored as ferritin and hemosiderin in the liver, reticuloendothelial cells, and bone marrow.

The recommended intakes of iron for various age and sex groups are shown in Table 16.1, which shows that the recommended dietary allowance (RDA) for adult males and post-menopausal women is 8 mg/day; and that for menstruating adult women is 18 mg/day. Meat is the best source of iron, but iron-fortified foods (cereals and wheat-flour products) also are significant sources.

Manganese is an essential trace element, although deficiency is rare. It is mentioned here, however, as it has recently been shown to accumulate in the basal ganglia of the brain, potentially causing Parkinson-like symptoms in patients receiving long-term intravenous feeding.

Selenium is a component of enzymes that catalyse redox reactions including glutathione peroxidase, iodothyronine 5-deiodinase, and thioredoxin reductase. Selenium was suggested as being essential in 1957 but a biochemical role was not identified for selenium until 1972. The first report of human selenium deficiency appeared in 1979; the subject resided in a low-selenium area and was receiving total parenteral nutrition (TPN) after surgery. The subject and other selenium-deficient subjects on TPN exhibited bilateral muscular discomfort, muscle pain, wasting, and cardiomyopathy. Subsequently, it was discovered that **Keshan disease**, prevalent in certain parts of China, was prevented by selenium supplementation. Keshan disease is a multiple focal myocardial necrosis resulting in acute or chronic heart function insufficiency, heart enlargement, arrhythmia, pulmonary oedema, and death. Other consequences of inadequate selenium include impaired immune function and increased susceptibility to viral infections. Selenium deficiency can also make some non-virulent viruses become virulent.

Recently, however, not only selenium deficiency, but effects of supranutritional intakes of selenium have become of great health interest. Several supplementation trials have indicated that selenium has anti-carcinogenic properties. For example, one trial with 1,312 patients supplemented with either >200 µg selenium/day or with a placebo found the selenium treatment was statistically associated with reductions in several types of cancer, including colorectal and prostate cancers. Selenium is a relatively toxic element; intakes averaging 1.2 mg/day can induce changes in nail structure. Chronic selenium intakes over 3.2 mg/day can result

TABLE 16.1 Recommended dietary allowances for selected trace elements established by the Food and Nutrition Board, Institute of Medicine, National Academy of Sciences

	Copper (μg/day)	Iodine (μg/day)	Iron (mg/day)	Selenium (μg/day)	Vitamin B_{12} (μg/day)	Zinc (mg/day)
Age						
0–6 months	–	–	–	–	–	–
7–12 months	–	–	11	–	–	3
1–3 years	340	90	7	20	0.9	3
4–8 years	440	90	10	30	1.2	5
9–13 years	700	120	8	40	1.8	8
14–18 years	890	150	11 M/15 F	55	2.4	11 M/9 F
19–50 years	900	150	8 M/18 F	55	2.4	11 M/9 F
51 years and greater	900	150	8	55	2.4	11 M/8 F
Pregnancy						
18 years or less	1,000	220	27	60	2.6	13
19–50 years	1,000	220	27	60	2.6	11
Lactation						
18 years and less	1,300	290	10	70	2.8	14
19–50 years	1,300	290	9	70	2.8	12

Abbreviations: F, female; M, male.

in the loss of hair and nails, mottling of the teeth, lesions in the skin and nervous system, nausea, weakness, and diarrhoea. Selenium, which is biologically important as an anion, is homeostatically regulated by excretion, primarily in the urine, but some is also excreted in the breath. Selenate, selenite, and selenomethionine are all highly absorbed by the GIT; absorption percentages for these forms of selenium are commonly found to be in the 80–90% range. The recommended intakes for selenium are shown in Table 16.1, which shows that the RDA for adults is >55 μg/day. Food sources rich in selenium are fish, eggs, and meat from animals fed abundant amounts of selenium and grains grown in high-selenium soil.

Zinc is the only trace element that is found as an essential component in enzymes from all six enzyme classes. Zinc also functions as a component of transcription factors known as zinc fingers that bind to DNA and activate the transcription of a message, and impart stability to cell membranes. Signs of human zinc deficiency were first described in the 1960s. Although it is generally thought that zinc deficiency is a significant public health concern, the extent of the problem is unclear because there is no well-established method to assess the zinc status of an individual accurately. The physiological signs of zinc deficiency include: depressed growth; anorexia (loss of appetite); parakeratotic skin lesions; diarrhoea; and impaired testicular development, immune function, and cognitive function. Pathological consequences of zinc deficiency include dwarfism, skin rash, delayed puberty, failure to thrive (**acrodermatitis enteropathica** infants), impaired wound-healing, and increased susceptibility to infectious disease. It has also been suggested that low zinc status increases the susceptibility to osteoporosis and to pathological changes caused by the presence of excessive reactive oxygen species or free radicals.

Zinc is a relatively non-toxic element. Excessive intakes of zinc occur only with the inappropriate intake of supplements. The major undesirable effect is an interference with copper metabolism that could lead to copper deficiency. Long-term high zinc supplementation can reduce immune function and high density lipoprotein (HDL) cholesterol (the 'good' cholesterol). These effects are seen only with zinc intakes of 100 mg/day or more. A primary homeostatic mechanism for zinc is absorption from the small intestine. Absorption involves a carrier-mediated component and a non-mediated diffusion component. With normal dietary intakes, zinc is absorbed mainly by the carrier-mediated mechanism. Although absorption can be modified by a number of factors, about 30% of dietary zinc is absorbed. The efficiency of zinc absorption is increased with low zinc intakes. The small intestine has an additional role in zinc homeostasis by regulating excretion through pancreatic and intestinal secretions. After a meal, more than 50% of the zinc in the intestinal lumen is from endogenous zinc secretion. Thus, zinc homeostasis depends upon the reabsorption of a significant portion of this endogenous zinc. Intestinal conservation of endogenous zinc is apparently a major mechanism for maintaining zinc status when dietary zinc is inadequate. The urinary loss of zinc is low and generally not markedly affected by zinc intake.

The RDA for adult males is 11 mg/day and for adult females is 8 mg/day. The best food sources for zinc are red meats, organ meats (for example liver), shellfish, nuts, whole grains, and legumes. Many breakfast cereals are fortified with zinc.

SELF-CHECK 16.3

Which one of the following molecules acts as a potent antioxidant protecting the cell membranes from the harmful effects of free radicals: iodine, vitamin D, vitamin E, or vitamin K?

16.5 Assessment of nutritional status

Nutritional status is an all-embracing term used to describe existing body stores, current intake (and the ability of the body to use this intake), and ongoing requirements. As such, nutritional status has been described by the Inter-Collegiate Group on Nutrition as 'what you are, what you eat, and what you (can) do'. These three aspects are obviously inter-related: the extent of your stores (your size and composition) will dictate what you need to eat; what you eat dictates what you can do with the nutrients; what you do both physically and metabolically will dictate your body's ability to store nutrients, and so on. Prolonged disturbances of the normal relationship of these will lead to malnutrition.

Malnutrition is a state of impaired function, growth, or development resulting from the failure to utilize one or more nutrients, either through altered supply ('what you eat'), altered demand ('what you are'), or altered metabolism of nutrients ('what you can do'). This impairment may be due to a lack of food, insufficient amounts of a nutrient in the diet, problems with digestion or absorption, an inability to utilize the nutrient once absorbed, or increased requirements. Complexities arise in illness, for example in active inflammatory bowel disease, problems with absorption are obvious, but the acute illness leads to a diminished appetite which decreases food availability, whilst increasing metabolic demands and altering the metabolic processing of absorbed nutrients. Malnutrition is a result, therefore, of an imbalance, for whatever reason, between nutrient availability and requirements. It may represent an excess or a deficiency, either of food in general or of one or more specific nutrients.

Key Points

Malnutrition occurs not only due to nutritional deficiencies but also excesses.

Detection of malnutrition is vitally important because such patients have disturbances in function at the organ and cellular level which can manifest as:

- altered partitioning and impairment of normal homeostatic mechanisms
- muscle wasting and impairment of skeletal muscle function
- impaired respiratory muscle function
- impaired cardiac muscle function
- atrophy of smooth muscle in the GIT
- impaired immune function
- impaired healing of wounds and **anastomosis**

Even recently, despite an increasing understanding of the importance of nutrition in health, studies have shown that up to 40% of hospitalized patients in the UK can be classified as being malnourished; unfortunately, in many of these patients, the problem is not recognized. The result of these malnutrition-induced changes is an increased risk of morbidity and mortality, particularly in patients undergoing surgery. To identify patients at increased risk, it is therefore necessary to determine whether their needs are being met, and to do this the patient's nutritional status must be assessed first.

A variety of methods have been advocated as useful in the assessment of a patient's nutrition. These include:

- Taking a history and examination of the patient
- Clinical assessment:
  - anthropometric measurements:
    - height and weight
    - body mass index
    - skinfold thickness
    - bioelectrical impedance
    - others
  - biochemical measurements
    - serum proteins
      - albumin
      - other proteins
    - nitrogen balance
    - specific nutrients
  - functional measurements
    - immune competence
    - muscle function

Assessment of the nutritional status of a patient begins with a review of the patient's history and a clinical examination. Alteration of supply of dietary nutrients would seem relatively straightforward to assess, by means of diet history, intake charts, etc., but in some circumstances it is less easy: dietary recall is subjective and notoriously unreliable, and even objective intake charts are prone to considerable error. The patient's perspective should be noted—feelings of hunger or of fullness will have an effect on intake, but will also affect the accuracy of reported intake, particularly with regard to portion size. Alteration of demand is even more subtle. Demand varies with activity, but also depends on body size, more body mass requiring more energy even at rest, referred to as the 'resting energy expenditure' (REE). Demand also depends on general metabolic state (for example fever will markedly increase the REE). Possibly the most complicated area is altered metabolism of micronutrients, as this will affect the body's ability to handle macronutrients. Such effects on the metabolism differ in different conditions of altered nutrient supply, such as fasting or starvation, but also vary in the face of trauma and sepsis, and these effects all have to be taken into account when assessing nutritional status and planning nutritional interventions.

Clinical assessment forms the mainstay of nutritional status evaluation. The specific methods that have been used in clinical practice include **anthropometric**, functional, and biochemical measurements.

Anthropometric measures

Height and weight are two of the most commonly used indices of nutritional status. A history of weight loss in a patient is particularly important information, although it is often difficult to determine accurately. Weight loss is usually determined by subtracting the current weight from the weight the patient recalls having when they were 'well'. Although a patient's recall of their previous weight can often be inaccurate, loss of more than 10% of body weight, or more than 4.5 kg, is associated with a significantly poorer outcome following surgery. In addition, more rapid weight loss, is associated with a higher risk of complications. It is often desirable to know the lean body mass (or conversely, the total body fat), and although total body weight can be compared with values standardized for age and sex, this approach takes no account of frame size. The **body mass index** (BMI) takes some account of frame size and so has been suggested as the most practical anthropometric indicator of total body fat from deficient to excess. The BMI is defined as weight in kilograms divided by the square of the height in metres.

Total body fat has been estimated more specifically by a number of other anthropometric measures, including skinfold thickness and bioelectrical impedance, or directly assessed using underwater weighing or by imaging techniques. Approximately 50% of total body fat is in the subcutaneous layer, and so the thickness of a fold of skin measured using calipers can be compared with published correlations to give an estimate of total body fat and calculating the body fat using published correlations (approximately 50% of total body fat is in the subcutaneous layer). Skinfold thickness of the triceps is most commonly measured, but assessment of skinfolds at multiple sites is better. However, measurements of skinfold thickness are susceptible to intra- and inter-observer variability, which limits their use clinically. **Bioelectrical impedance** has also been used to characterize body composition. Electrical current passes through the body by means of the movement of ions, and so depends principally on total body water and electrolyte content, which correlates with lean body mass. Movement of ions is limited by cell membranes, and so cells become polarized, but cannot sustain the flow of current in a single direction. If the current is alternated, however, the current can be sustained because the cells do not become so polarized. In this way, cells act as miniature capacitors, and the restriction of current is referred to as impedance. By measuring the flow of current at various frequencies, the proportion of intracellular to extracellular fluid can be estimated. Unfortunately, although bioelectrical impedance can give an accurate estimate of body composition in stable subjects, it becomes less reliable in patients with oedema and electrolyte shifts, and so its value in critically ill patients is limited. Specialized equipment is required for underwater weighing and for imaging, and so these techniques are usually limited to research use.

Biochemical measurements in the assessment of nutritional status

In comparison with the anthropometric techniques, it is relatively simple to send a sample to a laboratory. Assessing a patient's nutrition in this way might seem ideal, and clinical laboratories often receive requests related to nutritional assessment. The biggest problem, however, is that the constituents of the sample do not accurately reflect the whole body composition.

Plasma proteins

Albumin is the most abundant protein in plasma, and many studies show that a low concentration of albumin in plasma is associated with an increased risk of complications and mortality. As a result, for many years the albumin concentration was erroneously thought to reflect nutritional status. Other studies have shown that, despite significant loss of body weight and long periods of semi-starvation, albumin concentration in plasma may even be increased. Plasma albumin acts as a negative marker of the acute phase response, and so its concentration is lowered in malignancy, trauma, and sepsis, even in the presence of an adequate nutritional intake. As such, low levels point to the increased nutritional risk associated with underlying disease, but it should *not* be used to assess nutritional status. It had also been suggested that the half-life of albumin in plasma (approximately 30 days in healthy individuals) is too long to reflect recent changes in nutritional status. Other plasma proteins have, therefore, been suggested as alternatives to albumin in assessing nutritional status. These include transferrin (half-life of seven days), retinol-binding protein (half-life of 1–2 hours), and pre-albumin (half-life of two days). The plasma levels of these proteins are, however, also altered in stress, sepsis, and cancer, and so, as for albumin, they are not useful for assessing nutritional status in routine clinical practice. C-reactive protein (CRP) is usually measured, not as a nutrient *per se*, but as an assessment of any ongoing acute phase response (APR), aiding the interpretation of measurements of other protein concentrations.

Nitrogen balance

Of the nitrogen lost from the body, most is excreted in urine. Urea accounts for approximately 80% of the total urinary nitrogen, and so urea alone may be measured as an approximate indicator of losses. Total urinary nitrogen may also be measured, although this latter technique is not widely available. In addition, nitrogen is lost from the skin and in stools (approximately 2–4 g per day). One method for assessing nitrogen balance uses the following equation:

Nitrogen balance (g) = (dietary protein (g) × 0.16) − (urine urea nitrogen (g) + 2 g stool + 2 g skin)

(where urine urea nitrogen (g) = urine urea (mmol) × 28).

Care should be taken using such equations if the patient's plasma urea is changing, however, as this reflects changes in the proportion of urea being cleared in urine. As an approximation, in an adult human, one mmol/L increase in plasma urea equates to one extra gram of nitrogen lost (or one gram less if the plasma urea falls by one mmol/L). Although nitrogen balance has not been shown to be a prognostic indicator, the method is widely used to assess a patient's protein requirements, and to assess the response to the provision of nutritional support.

Specific nutrients

Serum electrolyte and mineral measurements are routinely performed as part of nutritional assessment and monitoring of nutritional therapy. These are discussed in the relevant chapters, but it should be stressed that serum measurements rarely reflect whole body status, for example the serum sodium concentration is more dependent on water status than sodium. Urinary electrolyte output may give a more useful indicator of requirements.

Cross reference
Chapter 5 Fluid and electrolyte disorders

Assessment of micronutrients

In clinical nutrition, assessment of vitamin or mineral status is only required if specific deficiency is suspected, or to confirm adequacy of supply in long-term artificial nutrition. Vitamins A, E, B_1, B_2, and B_6 are measured using HPLC techniques, although historically these B vitamins were measured using red cell enzyme activation analyses. Vitamin C may be measured by HPLC, but the concentration in blood reflects recent intake rather than body status. Intracellular vitamin C (for example in leukocytes) may better reflect body stores, but this is technically difficult and rarely performed. Folate and B_{12} are measured by immunoassay, and vitamin D by immunoassay, HPLC, and LC-MS/MS. Vitamin K status is assessed by undertaking a clotting screen rather than measuring the vitamin concentration. The remaining B vitamins are not measured routinely.

Trace elements are bound to plasma proteins which are affected by intercurrent illness, and so to aid interpretation of measured concentrations, CRP is usually measured in order to assess the APR. In addition, the specific binding protein is measured. Iron is measured routinely colorimetrically, and its binding proteins transferrin and ferritin are measured by immunoassay. Copper and zinc are also measured routinely by atomic absorption spectroscopy (AAS), inductively coupled plasma atomic emission spectroscopy (ICP-AES), or colorimetrically. Caeruloplasmin (specific binding protein for copper) is measured by immunoassay. Selenium is also measured, although glutathione peroxidase (GSH-PX) activity provides a better assessment of body selenium status. Manganese is assessed in patients receiving home parenteral nutrition, but other inorganic micronutrients are not measured routinely.

The haematinics (folate, B_{12}, iron, transferrin, ferritin) are widely available; other micronutrients are specialized tests usually only performed in centralized laboratories.

A suggested scheme for the testing of specific nutrients is given in Box 16.2.

BOX 16.2 Suggested scheme of laboratory measurements

Before commencing nutritional support:

- Electrolytes and renal function, liver function tests, albumin, total protein, CRP, minerals, and glucose.

- Also consider lipids, thyroid function, and haematinics (folate, B_{12}, iron, transferrin, ferritin).

In addition, in the presence of clinical evidence of specific or general nutritional deficiency, it is worthwhile taking samples to check other micronutrients (vitamins A, D, E, B_1, B_2, B_6; copper and caeruloplasmin, zinc, selenium, and GSH-PX). Different micronutrient deficiencies often occur together, and so there may be value in knowing their pre-treatment levels.

After commencing nutritional support:

- Daily assessments (until stable).
- Electrolytes, renal and liver function tests and glucose.
- CRP if baseline was raised.
- Plus twice weekly assessments (until stable) of minerals.

Longer term artificial nutritional support:

Electrolytes and renal function, liver function tests, albumin, total protein, CRP, minerals, glucose, micronutrients (vitamins A, D, E, B_1, B_2, B_6, folate, B_{12}, iron, transferrin, ferritin, copper, caeruloplasmin, zinc, selenium, and GSH-PX).

These can be assessed monthly in hospital in-patients, or three-monthly in outpatients. Patients receiving home parenteral nutrition should also have blood manganese concentrations assessed every three months.

Functional tests in the assessment of nutritional status

Functional changes during impairment of nutrition have been used by examining changes in immune competence and changes in muscular function such as the ability to do exercise.

Immune competence

In malnutrition there is a reduction in the total number of circulating lymphocytes and impairment in a wide variety of immune functions. However, these alterations in immune function are non-specific and can be affected by trauma, surgery, anaesthetic and sedative drugs, pain, and psychological stress and as a result are not generally applicable to clinical practice.

Muscle function

Malnutrition can adversely affect both the structure and function of skeletal muscle. Measurements of handgrip strength are cheap and easy to perform in patients who are able to cooperate, and can be used to predict those patients who develop post-operative complications. However, grip strength may be influenced by other factors such as the patient's motivation and cooperation. Furthermore, such tests may be difficult to apply to patients who are critically ill who may be unable to cooperate. Alternatively, stimulation of the ulnar nerve at the wrist, with a variable electrical stimulus, results in contraction of the adductor pollicis muscle, the force of which reflects nutritional intake. The function of the respiratory muscles is also impaired in malnutrition. This deterioration can be detected by various indices of standard respiratory function tests, in particular vital capacity. As for skeletal muscle, however, whilst the influence of nutritional status should be taken into account when performing these tests, the non-specific nature of changes limits their clinical usefulness as tests of nutritional status.

Nutrition risk index

A nutrition risk index of nutritional status is based on a combination of variables. Several indices exist, but one often used, in surgical practice particularly, is calculated as follows:

$$\text{Nutrition risk index} = 1.519 \times \text{serum albumin (g/L)} + 0.417 \times (\text{patient's current weight/patient's usual weight}) \times 100$$

The score obtained is used to categorize the patient's nutritional risk:

- <83.5 'severe'
- 83.5–97.5 'mild'
- 97.5–100 'borderline'

It is important to note that this index does not assess the patient's nutritional status: it is purely a *prognostic* index. It can, however, help identify patients who may require nutritional intervention.

Assessment in everyday practice: MUST

The malnutrition universal screening tool (MUST) tool is a simple screening tool developed by the British Association for Parenteral and Enteral Nutrition (BAPEN). The procedure is outlined in Box 16.3 and details are available from the website (www.BAPEN.org.uk). The tool was developed because of the difficulties encountered with detailed assessments, whilst recognizing the prevalence of malnutrition in hospital patients. It has been recommended by the National Institute for Health and Clinical Excellence (NICE) and by Quality Improvement Scotland (QIS) for routine screening of all hospital patients. It should be noted that the MUST tool is simply a low-specificity, high-sensitivity, screening tool. Individuals identified as being at risk should then be referred to the specialist team for more detailed assessment and formation of a nutritional action plan.

SELF-CHECK 16.4

What is the value of measuring serum CRP when assessing the nutritional status of a patient?

BOX 16.3 The malnutrition universal screening tool (MUST)

Essentially, MUST consists of a series of five steps:

- Step 1: measure height and weight to obtain body mass index (BMI in kg/m²) and this is then given a numerical score (>20 = 0; 18.5–20 = 1; <18.5 = 2).
- Step 2: note percentage unplanned weight loss in the previous 3–6 months and give this a numerical score (<5% = 0; 5–10% = 1; >10% = 2).
- Step 3: establish the 'acute disease effect' and also give this a numerical score (if patient is acutely ill and there is or will be no nutritional intake for more than five days = 2).

- Step 4: add scores from steps 1, 2, and 3 together to obtain the 'overall risk of malnutrition'.
- Step 5: a decision is taken as to what to do depending on the resultant score.

Significance of the resultant score and clinical management:

- Score 0 (low risk): repeat the screening process subsequently.
- Score 1 (medium risk): observe by noting the patient's dietary intake for the next three days. If this improves then there is little concern. However, if there is no improvement then this is of clinical concern and one should follow the local policies for what to do next, for example referral to nutrition support team/dietician.
- Score 2 or more (high risk): these patients should be referred to the nutrition support team/dietician to try to increase their nutritional intake and there should be policies in place for the nutritional support given to these patients.

16.6 Disordered eating patterns

Individual food intake, and thus energy consumption, depends on hunger, **appetite**, and **satiety**. Physiologically, we all have a basic need to eat (hunger) while psychologically we may desire to eat (appetite). Thus, appetite is a learned behaviour that is affected by the sensory characteristics of food (seeing, smelling, and thinking about food) as well as social factors. With abundant food supplies, appetite triggers eating rather than hunger. A point is arrived at for individuals when sufficient food is consumed such that there is a sense of feeling full (satiety). This sense of fullness depends on both physiological and psychological factors, such as the amount of food and the characteristics of the food such as water, fibre, and macronutrient content. Blood levels of glucose, amino acids, fatty acids, and hormones all play a part in the regulation of food intake by the hypothalamus.

Some illnesses including inflammation, infection, malignancy, and pain can cause anorexia (loss of appetite), while others such as some gastrointestinal problems and malignancies can produce a feeling of fullness. Other patients have obsessive and compulsive behaviour and cognitive processes that drive them to eat in a way that disturbs their psychological, mental, and physical health. Such conditions are psychiatric and are called eating disorders. The disturbance may often be extreme and life threatening, ranging from restrictive and limited eating as in **anorexia nervosa**, to excessive eating, as in compulsive over-eating and binge eating disorder; and can include intermittent episodes of purging (**bulimia nervosa**) punctuated with episodes of normal eating. Unsurprisingly, patients with diagnosed eating disorders are at risk of nutrient deficiencies. As well as **protein-energy malnutrition**, patients who have anorexia nervosa often have iron, zinc, and vitamin D deficiencies, while those who vomit often have depleted potassium stores as well as being dehydrated.

An imbalance over time between appetite and satiety can result in **obesity**, a condition where there is excess body weight due to an abnormal accumulation of fat (obēsus, which means 'stout', 'fat', or 'plump' in Latin). The objective definition is a BMI of 30 kg/m^2 or more, that is, the weight in kg divided by the square of the height in metres. Thus, a 1.7 metre man with a weight of 100 kg would have a BMI of 100 / (1.7 × 1.7) or 34.6 kg/m^2. In addition, there are other measures of the degree of fatness, including the waist circumference. While the BMI cut-offs are the same for both genders, waist circumference cut-offs differ between men and women; however, both measures have different cut-offs based on ethnicity. The prevalence of obesity varies between countries and as a general rule, the less developed the country the lower the prevalence of obesity. However, even within developing countries, the prevalence of obesity is increasing, in particular, in those countries where there is a rural to urban population shift. Over the last few decades, the prevalence of obesity in the United Kingdom has been increasing. The 2003 Health Survey for England specifically examined the prevalence of obesity among adults and found that 22% and 23% of adult men and women were obese with 43% and 33% being overweight respectively. It was estimated that there would be an increase of 55% in the number of male obese patients by 2010.

The consequences ('what you (can) do') of being overweight and having obesity include a greater risk of atherosclerosis due to dyslipidaemia, hypertension, and type 2 diabetes mellitus; degeneration of weight-bearing joints due to osteoarthritis; gallbladder disease; oesophagitis; urinary incontinence; sleep apnoea and respiratory problems; and some malignancies (breast, colon, and endometrial). The ideal treatment ('what you are, what you eat') is long-term lifestyle change. Obesity does not occur overnight and thus weight loss takes time. In simplistic terms, there is a need for a reduction of input relative to output such that more calories are burnt up than supplied to the body. This is easier said than done, especially over the longer term; once a patient loses weight, then the next step is maintaining the weight loss. Clearly, the ideal treatment is not to become overweight in the first place, that is, the education of children is essential to reduce the future prevalence. The ideal weight is not a BMI of 25 kg/m^2 but a weight loss of 5–10% in the first instance. This is because such a weight loss can undo many of the metabolic complications of obesity. Where lifestyle measures have failed, those patients who have obesity related co-morbidities may be offered medications that help them lose weight. However, many patients do not lose weight when prescribed a weight-lowering medication and of those that do, many regain weight upon cessation of the medication. Another option is **bariatric surgery** which is generally reserved for those who are quite obese. Such techniques are more likely to maintain long-term weight loss and as a result, more obese patients are undergoing such surgery.

SELF-CHECK 16.5

Calculate the BMI of a man weighing 86 kg who has a height of 1.75 m?

CASE STUDY 16.4

A 40-year-old man underwent work-related medical screening. His weight was 108.2 kg and his height was 1.72 m. Clinical biochemistry tests gave the following results (reference ranges are given in brackets):

Cholesterol	6.6 mmol/L	(<3.0)
Triacylglycerols	4.8 mmol/L	(<1.7)
HDL-C	0.8 mmol/L	(1.0–2.0)
Cholesterol: HDL-C ratio	8.3	
Fasting plasma glucose	8.9 mmol/L	(<6.0)

(a) Calculate the BMI and comment on it.

(b) What symptoms may be associated with a raised glucose?

(c) Describe and comment on his lipid profile

16.7 **Nutritional intervention**

The form of nutritional intervention depends on the nutritional and clinical status of the patient. It can take several forms, including dietary changes, oral supplementation, enteral nutrition, including nasogastric tube, percutaneous endoscopic gastrostomy tube, etc., and parenteral (intravenous) nutrition using either a peripheral or central venous line. While it is beyond the scope of this chapter to discuss in detail, an otherwise well patient who has a minor single micronutrient deficiency may require either dietary change or an oral supplement, while a patient who cannot absorb macro- and micro-nutrients for whatever reason may require **parenteral feeding** via a central venous line. The more complex the intervention, the greater the clinical risk of complications and thus the greater need for a more specialist input. It is recommended that parenteral nutrition involves multidisciplinary nutrition teams.

The **refeeding syndrome** is one important complication of nutrition support. It may occur when nutrients are supplied after a period of starvation. The types of patients who are at risk of developing the refeeding syndrome are given in Box 16.4. The metabolism adapts to starvation by suppressing glycolysis, and there may also be specific vitamin deficiencies, notably thiamine. As a result the handling of nutrients becomes impaired. Clinically, the syndrome is a heterogeneous group of signs and symptoms which may manifest as abnormal serum electrolytes (potassium, inorganic phosphate, calcium, and magnesium), vitamin deficiencies, or fluid retention. Identification of patients at risk and subsequent reductions in initial calorific intake with a gradual increase to full calorific requirements reduce the likelihood of its occurrence. Deficiencies of micronutrients, especially thiamine, should also be addressed before restoring macronutrient intake. For this reason, it is common to give water-soluble vitamins at an early stage, particularly in patients with habitual excess alcohol intake.

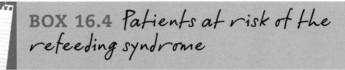

BOX 16.4 *Patients at risk of the refeeding syndrome*

- prolonged fasting, chronic malnutrition/underfeeding
- prolonged intravenous hydration (especially >7 days)
- anorexia nervosa
- some post-operative patients
- some patients on chemotherapy
- morbid obesity with massive weight loss
- classic kwashiorkor
- classic marasmus
- chronic alcoholism
- hyperemesis gravidarum

CASE STUDY 16.5

A 75-year-old woman had a malignant tumour of her sigmoid colon surgically removed. In the three months prior to the operation, she had unintentionally lost more than 10% of her body weight, and latterly she had not been eating anything substantial at all. The day following her surgery, her bowel motions returned, and she was started on sips of water. Over the next three days she was allowed to drink more water; her appetite returned so she was allowed to eat. However, over the next few days she developed a significant weakness in both upper and lower limbs; routine biochemistry tests showed the following (reference ranges are given in brackets):

Sodium	134 mmol/L	(137–144)
Potassium	3.1 mmol/L	(3.5–4.9)
Urea	10.4 mmol/L	(2.5–7.0)
Creatinine	135 µmol/L	(60–110)
Calcium (corrected)	2.16 mmol/L	(2.20–2.60)
Phosphate	0.25 mmol/L	(0.8–1.4)
Alkaline phosphatase	150 IU/L	(45–105)
Albumin	33 g/L	(37–49)

(a) What nutritional risk factors does this patient have?

(b) Describe and comment on the clinical biochemistry results.

(c) How should this patient be treated?

In view of the low phosphate result and her history of significant and unintentional weight loss, refeeding syndrome was presumed.

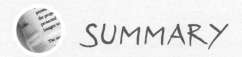

SUMMARY

■ Clinically, effective nutritional care requires assessment and monitoring of nutritional status and an understanding of the effects of illness, in order to ensure appropriate provision of nutrients to patients.

■ The laboratory has an important analytical role in assessment and monitoring of specific nutrients, but results of these analyses must be interpreted appropriately, with special regard to the acute phase response.

■ Highly specialized methodologies may be required for the assessment of some nutrients because of their very low concentrations in patient samples. As a result, assays for many specific nutrients will not be available immediately in the majority of hospitals.

■ It must also be realized that the samples available for analysis in the laboratory rarely provide information on whole body nutrient status.

■ Measurements are most often performed in serum or urine because of the relative analytical ease, but these measurements can only offer limited information.

■ Results of all nutritional analyses need to be put into a clinical context, and knowledge of the clinical nutritional status is a prerequisite for full interpretation of the laboratory results.

■ Prolonged disturbances of nutritional status will lead to malnutrition, and knowledge of specific history and examination is required to determine which analyses should be performed, and how the results are interpreted.

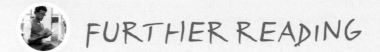

FURTHER READING

● Ayling R and Marshall W (2007) *Nutrition and Laboratory Medicine*. Cambridge, UK: ACB Venture Publications.

A comprehensive guide to laboratory aspects of clinical nutrition.

● Beastall G and Rainbow S (2008) **Vitamin D reinvented: implications for clinical chemistry.** *Clinical Chemistry* **54**, 630–2.

● Department of Health (1991) *Department of Health Report on Health and Social Subjects 41. Dietary Reference Values for Food Energy and Nutrients for the United Kingdom. The definitive reference list of dietary nutrients for the United Kingdom.* London: The Stationery Office.

● Jeejeebhoy KN (2000) **Nutritional assessment.** *Nutrition* **16**, 585–90.

● Shenkin A (2006) **Micronutrients in health and disease.** *Postgraduate Medical Journal* **82**, 559–67.

A useful overview of micronutrients and their use clinically.

● Shenkin A (2006) **Biochemical monitoring of nutrition support.** *Annals of Clinical Biochemistry* **43**, 269–72.

A review of laboratory assessment of micronutrients.

● World Health Organization (2004) *Vitamin and Mineral Requirements in Human Nutrition*, 2nd edition. Geneva: World Health Organization.

● World Health Organization (2006) *Guidelines on Food Fortification with Micronutrients*. Geneva: World Health Organization.

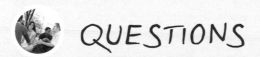

QUESTIONS

16.1 Which of the following are regarded as components of nutritional status? State whether TRUE or FALSE for each statement.

(a) The food you are able to ingest

(b) The air you breathe

(c) Your income

(d) Your body composition

(e) What you are able to do

16.2 Which ONE of the following is a micronutrient?

(a) Protein

(b) Zinc

(c) Glucose

(d) C-reactive protein

(e) Phospholipid

16.3 Which ONE of the following analyses is routinely performed in small general hospital laboratories?

(a) Vitamin A

(b) Folate

(c) Vitamin D

(d) Rhubarb

(e) Manganese

16.4 Which of the following analyses are performed as part of the nutritional assessment of patients on parenteral nutrition at home? State whether TRUE or FALSE for each analysis.

(a) Renal function

(b) Vitamin A

(c) Vitamin K

(d) Manganese

(e) Insulin

16.5 Which of the following changes are associated with a raised CRP? State whether TRUE or FALSE for each analysis.

(a) A fall in serum iron

(b) A fall in serum zinc

(c) A rise in serum ferritin

(d) A fall in serum copper

(e) A rise in serum magnesium

Answers to self-check questions, case study questions, and end-of-chapter questions are available in the Online Resource Centre accompanying this book.

Go to www.oxfordtextbooks.co.uk/orc/ahmed/

17

Gastrointestinal disorders and malabsorption

Gordon Brydon

Learning objectives

After studying this chapter you should be able to:

- Explain the different functions of acid in the stomach and bile acids in the small intestine
- Identify the secretion and absorption processes within the small intestine
- Explain the functions of secretin and cholecystokinin in nutrient digestion
- Describe the symptoms and laboratory diagnosis of gut hormone secreting tumours
- Discuss the advantages and limitations of the faecal elastase test as a measure of exocrine pancreatic disease
- Outline the use of simple tests for initial investigation of patients with chronic watery diarrhoea
- Discuss the biochemical tests used to investigate malabsorption
- Outline the factors which have contributed to some gastrointestinal tests becoming obsolete

Introduction

The use of biochemical tests for the diagnosis of gastrointestinal disease has increased over recent years, as a result of several analytical developments.

Many older laboratory tests have become obsolete because of their poor predictive value and have been replaced by improved ones, which if used appropriately can facilitate or exclude a diagnosis. The advent of procedures such as **endoscopy**, where a flexible tube is passed through the mouth or the anus (**colonoscopy**) to visualize and sample tissues within the gastrointestinal tract, have transformed patient investigation. However, these are invasive, expensive procedures and some are not without risk.

Laboratory diagnosis of *Helicobacter pylori*, the microorganism which is the major cause of duodenal and gastric **ulcers**, has significantly reduced the upper endoscopy workload. Serological testing for **coeliac disease**, caused by sensitivity to **gluten** in the diet, has resulted in more frequent investigation for this disease and has revealed a much higher incidence than previously thought.

The importance of tests in faeces, the ultimate receptacle of gastrointestinal secretions, is now recognized. Some proteins released from fluids, sites of inflammation, or from bacteria, are stable in faeces and have significant diagnostic potential, providing direct information about gastrointestinal pathology. The introduction of specific commercial immunoassays to identify and quantify these proteins has greatly facilitated and standardized this process. In this chapter, you will be introduced to the basic biochemical and, where relevant, physiological processes in the gut, their optimization for digestion and absorption, and how diagnostic tests have evolved to investigate the breakdown of gut homeostasis.

17.1 **Gastrointestinal biochemistry and physiology**

We digest all food mostly as an extracellular process within the gastrointestinal tract (GIT) otherwise known as the alimentary canal. The GIT is a hollow continuum of several organs extending from the mouth to the anus. Unlike other internal body organs, the GIT, at either extremity, is in contact with the skin via **peri-oral** and **peri-anal** regions. In normal human adults, the GIT is about 7–8 metres long and consists of upper and lower gastrointestinal tracts. The upper tract consists of the mouth, pharynx, oesophagus, and stomach. The mouth contains the **buccal mucosa** where the openings to our salivary glands are found and also the tongue and teeth. Behind our mouth is the pharynx which leads to a hollow muscular tube, the oesophagus (or gullet). Food is passed down the oesophagus into the stomach by muscular contractions referred to as **peristalsis**. The lower intestinal tract consists of the small and large intestine. The small intestine is about five metres long and has three parts: the duodenum, jejunum, and ileum. The large intestine has three parts: the caecum, colon, and rectum, where the faeces are finally passed through the anus. An outline of the GIT is shown in Figure 17.1.

The main functions of the GIT are the ingestion, digestion, and absorption of nutrients, conservation of secretions, and elimination of unabsorbed waste products. This requires the coordinated function of the GIT and other associated organs. Digestion involves physical and biochemical processes where food is mixed with a series of secretions containing enzymes at optimal pH, and moved along the gut in a controlled way to allow maximal absorption of digestion products. In addition to the organs of the GIT, both the liver and **exocrine** pancreas produce secretions which are essential for digestion.

Complex unconscious neural and **humoral** processes are involved in the secretory and motility processes and the controlled movement of the milieu from one alimentary compartment to the next, with entry to each being controlled by circular sphincter muscle groups.

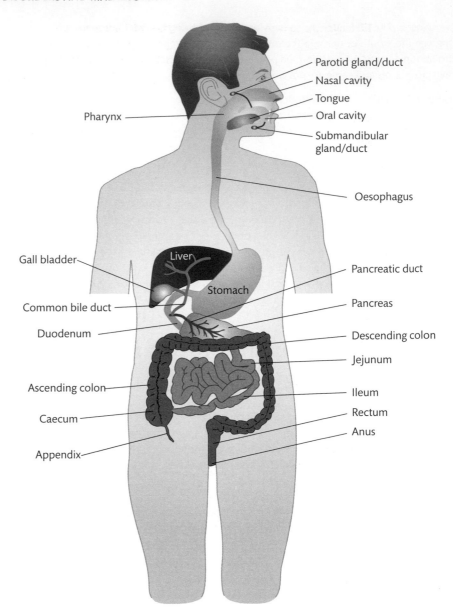

FIGURE 17.1
The gastrointestinal tract showing the major organs involved.

The swallowing reflex is the last motility process which is under conscious control in the GIT until **defaecation**. There are two major absorption processes:

- Absorption of dietary water, minerals, and digestion products, required for the maintenance of body structure and function; this occurs mainly in the jejunum.
- Conservation of water, minerals, and cofactors secreted during the digestive processes to maintain body fluid or physiological homeostasis.

Up to eight litres of fluid can be secreted in 24 hours, of which more than 98% is reabsorbed. Further information on fluid absorption and secretion is shown in Table 17.1, and humoral and **neurocrine** processes in Table 17.2.

The final process in the distal colon involves storage and elimination of undigested food, bacteria, bile pigments, and epithelial cells shed into the gut.

TABLE 17.1 Normal daily secretions and absorption of fluids in the GIT.

Source of fluid	Secreted (or ingested) fluids per day (L)	Absorbed fluids per day (L)
Diet	2	—
Saliva (mouth)	1.5	0
Gastric juice	2.5	0
Pancreatic juice and bile	2	0
Small intestine	1.5	7.9
Colon	0	1.5
Faeces	0.1	—

TABLE 17.2 Major humoral and neurocrine regulators of digestion and absorption.

Organ	Source	Factor	Function
Oral cavity	Vagal reflex	Acetylcholine	Saliva production
Stomach	Vagal reflex	Acetylcholine	HCl production
	Vagus/G cells	Gastrin	HCl production
	Enterochromaffin like cells	Histamine	HCl production
Duodenum	Mucosal osmo- and chemo-receptors	Gastro-inhibitory peptide	Closes pyloric sphincter; regulates volume and tonicity of duodenal contents when these increase after gastric emptying
	Mucosal chemoreceptors	Secretin	Stimulates pancreatic fluid and bicarbonate production in presence of acid and peptides (gastric acid production inhibited)
		Pancreozymin/cholecystokinin	Stimulates pancreatic enzyme production when amino acids/peptides present (gastric emptying inhibited); gall bladder contraction releases bile acids and bicarbonate into duodenum

Key Points

The main functions of the GIT are the ingestion, digestion, and absorption of nutrients, conservation of secretions, and elimination of unabsorbed waste products.

SELF-CHECK 17.1

Which organs outside the gastrointestinal tract are required for digestion?

Oral and gastric digestion

Digestion begins in the mouth, where food is ground or chewed into smaller fragments and mixed with saliva, which is released from the parotid and submandibular salivary glands. This fluid contains **mucins** which help lubricate food to aid swallowing and prevent aspiration into the lungs, amylase that initiates starch digestion, and bicarbonate ions for both optimal enzyme action and for neutralizing the acid which has been refluxed from the stomach. In addition, the presence of lingual lipase initiates triacylglycerol hydrolysis in the mouth, and secretory IgA provides mucosal immunity.

The sight, smell, and taste of food, as well as chewing, initiate release of acid via the **vagus nerve** reflex. This innervates the intestinal organs from the brain stem (cephalic phase of digestion) and is followed by passage of food through the oesophagus to the stomach. The stomach initially acts as a food reservoir and as it fills with food, both stretch and **chemoreceptors** are activated which in turn stimulate release of the hormone **gastrin** into the blood circulation. Gastrin acts on the **parietal cells** of the stomach and stimulates release of hydrochloric acid (gastric phase of digestion). The hydrochloric acid causes protein denaturation by disruption of hydrogen bonds, and unfolding of the protein tertiary structure.

The high hydrogen ion concentration also causes the release of pepsinogen from **G cells** and its activation to pepsin which initiates proteolysis. The hydrogen ion concentration in the stomach is about a million times greater than that of the plasma. Such a concentration gradient is created by the primary active transport of hydrogen ions by an H^+/potassium ion **ATPase**. This 'proton pump', is located in the luminal membrane of parietal cells, and transfers H^+ ions released from carbonic acid (H_2CO_3) by the action of the enzyme carbonic anhydrase into the stomach. There are a number of drugs which are **proton pump inhibitors** specifically designed to reduce gastric acid synthesis, for example omeprazole. The secretion of hydrogen ions into the lumen is balanced by the secretion of an equal amount of bicarbonate into the blood, creating an increase in the pH of venous blood leaving the stomach, known as the 'alkaline tide'. In the fasted state, gastric secretion is predominantly isotonic saline, whilst in the fed state the active transport of hydrogen ions predominates, causing the pH of gastric contents to fall to between 1.5 and 2.

Epithelial cells secrete mucin and bicarbonate. This solution resists digestion and protects the gastric mucosa from acid, pepsin, and mechanical trauma. The mixing of food with secretions is aided by the contraction and relaxation of the stomach smooth muscles.

The contents of the stomach are delivered in a controlled manner through the relaxed **pyloric sphincter** muscles and into the duodenum to initiate the intestinal phase of digestion.

Key Points

Sphincter muscles are groups of circular muscles which regulate flow along ducts by contracting to close and relaxing to open them.

Key Points

The three phases of digestion in the upper gut are: (1) cephalic, response to sight, smell of food, causes release of acid via vagal stimulation; (2) gastric, release of acid from parietal cells in response to vagal mediated release of gastrin and initial denaturation/ unfolding of protein tertiary structure and early digestion by pepsin; and (3) intestinal, stomach contents delivered through pyloric sphincter into the duodenum with release of humoral agents from duodenal epithelium initiating intestinal digestion.

SELF-CHECK 17.2

Which phase of digestion is associated with the sight and smell of food and which nerve initiates this?

Intestinal digestion and absorption

The transfer of food into the duodenum triggers the epithelium to release the hormones secretin and cholecystokinin into the circulation. These stimulate the release of pancreatic and biliary secretions which pass via the common bile duct into the duodenum through the relaxed **sphincter of Oddi**. The contents of these secretions are responsible for digestion in the small intestine.

The inactive precursor trypsinogen is converted to trypsin by enteropeptidase secreted from the duodenal mucosa. Pancreatic juice is alkaline (due to bicarbonate) and contains inactive pro-enzymes which are activated by trypsin within the duodenum. The exocrine pancreas has the highest rates of protein synthesis and secretion of any organ in the body. Phospholipases, cholesterol esterases, and nucleases are also present in exocrine pancreatic fluid and hydrolyse their relevant substrates to release fatty acids, cholesterol, and oligonucleotides.

Absorption from the jejunum is optimized by the tissue structure of the **enterocytes**, where the surface area is hugely increased by finger-like **villi** folds which project into the lumen (20–40 per mm² of mucosa). The **apical** membrane of each enterocyte is folded into **microvilli** which form the brush border where nutrients are absorbed. These microvilli are shown in Figure 17.2.

Enterocytes have a short lifespan and are formed in the crypts between the villi, migrate up to the tips, and are sloughed off (released) into the lumen. **Goblet cells** are scattered among the enterocytes and secrete mucins which have a predominantly protective function against shear stress, chemical agents, and bacteria.

Digestion and absorption of carbohydrates

Amylase converts carbohydrates (polysaccharides) to smaller oligosaccharides and disaccharides in the gut lumen; these are further converted to monosaccharides by enzymes within the mucosal brush border. Monosaccharides like glucose and galactose are absorbed with sodium by a specific transport protein (sodium-glucose transporter). Disaccharides are digested as follows: maltose and isomaltose are converted to glucose, sucrose is converted to fructose and glucose, whereas lactose is converted to glucose and galactose. Fructose is absorbed independently of sodium by facilitated diffusion via a separate transporter, and pentoses by simple diffusion. Up to 20% starch may escape absorption from the small bowel as it can be associated with indigestible complex carbohydrates (**dietary fibre**).

Digestion and absorption of proteins

In the newborn, immunoglobulins in the **colostrum** (first food from mother prior to breast milk) can be absorbed without prior digestion. However, intact proteins cannot be absorbed by adults. Proteins are metabolized to small peptides and dipeptides within the jejunum by

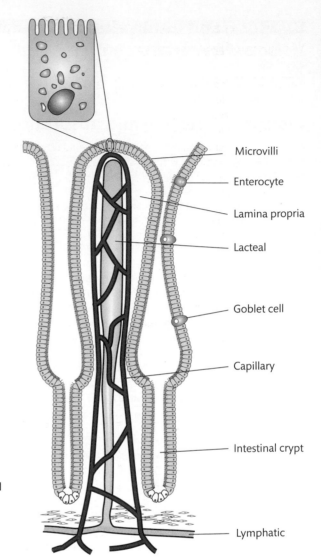

FIGURE 17.2
Vertical section through an intestinal villus showing the microvilli. Adapted with permission from Ayling R and Marshall W (2007) *Nutrition and Laboratory Medicine.* **London, UK: ACB Venture Publications.**

Labels: Microvilli, Enterocyte, Lamina propria, Lacteal, Goblet cell, Capillary, Intestinal crypt, Lymphatic

several proteolytic enzymes specific for different sites on the protein molecule, for example **exopeptidases** (carboxypeptidases) and **endopeptidases** (trypsin, chymotrypsin, and elastase). Several transport proteins within the jejunal brush border act on both free amino acids and small peptides. For example, aminopeptidases will remove N-terminal amino acids. Some di- and tripeptides have been shown to be absorbed into enterocytes more rapidly than amino acids.

Food only accounts for some of the amino acids absorbed, with about half in total being derived in roughly equal amounts from digestive juices and desquamated cells; less than 5% escape absorption and pass to the colon.

SELF-CHECK 17.3

Can intact proteins be absorbed from the healthy gut?

Digestion and absorption of fats and fat-soluble vitamins

Fats are emulsified into droplets by the mechanical action of peristalsis and the detergent action of both bile acid conjugates (bile salts) and phospholipids. Bile salts aggregate to form disc-shaped **micelles** where hydrophobic domains are projected inwards and hydrophilic domains outwards. Lipids enter these micelles, with cholesterol and fat-soluble vitamins being drawn into the central hydrophobic region. In addition, phospholipids and **monoacylglycerols** align with both the interior hydrophobic and exterior surface hydrophilic regions. The formation of these mixed micelles is essential for both digestion and absorption of lipids. This is seen in Figure 17.3.

Pancreatic lipase preferentially attacks the 1 and 3 bonds in **triacylglycerols** to produce 2 monoacyl glycerols. Co-lipase is essential for this process, acting as a co-enzyme, displacing surface bile salts from fat droplets, binding lipase, and facilitating lipolysis.

At the brush border of the enterocytes, products of lipolysis diffuse out of the mixed micelles down a concentration gradient into the cells. In the enterocytes, free fatty acids and monoacylglycerols with more than 10–12 carbons (long chain) are esterified to form triacylglycerols. These are associated with **apolipoproteins** to form **chylomicrons** (very large lipoproteins) before being secreted into the **lacteals** (intestinal lymphatic vessels) and then the thoracic duct

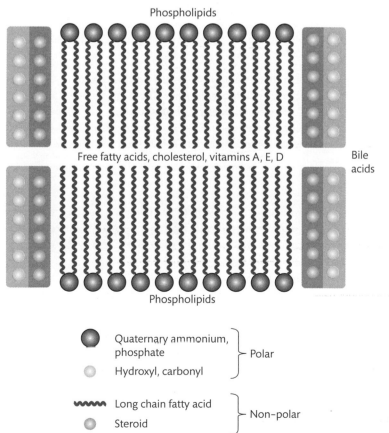

FIGURE 17.3
A schematic showing the mixed micelles required for digestion and absorption of lipids.

before transfer into the bloodstream. Chylomicrons also contain cholesterol, cholesterol esters, and the fat-soluble vitamins A, D, E, and K. Fat digestion and absorption are efficient, with >95% absorption. Most of the fat in faeces is derived from bacteria and desquamated cells.

Key Points

Since fat is not water-soluble, it has to be emulsified into droplets which are stabilized within micelles in the presence of bile salts and phospholipids; these structures provide a platform for effective fat digestion and also facilitate absorption at the brush border epithelium.

Enterohepatic circulation of bile acids

The primary bile acids, cholic and chenodeoxycholic acid, are synthesized from cholesterol in the liver, where they react with the amino acids glycine or taurine to form conjugated bile acids (bile salts). This is followed by their secretion, concentration, and storage as constituents of bile in the gall bladder. Cholecystokinin induced gall bladder contraction results in the transfer of bile, which contains bile salts, cholesterol, and phospholipids, into the duodenum, facilitating lipid digestion and absorption.

More than 95% of bile acid conjugates are actively reabsorbed at the terminal ileum by the apical sodium dependent bile acid transporter (ASBT). Two specific proteins, intestinal bile acid binding protein and organic solute transporter, complete the transfer across the cell cytoplasm and the **basolateral** membrane into the portal circulation, which returns bile acids to the liver.

Only a small percentage of the total body pool of bile acids is lost per day in faeces as the secondary free bile acids deoxycholic and lithocholic acids. These are formed by bacterial metabolism in the distal ileum and colon by bacterial **deconjugation** and removal of the hydroxyl group in the 7 position of the steroid nucleus. Some deoxycholic acid is absorbed passively and re-secreted into bile following conjugation. This is shown in Figure 17.4.

Key Points

Bile salts are synthesized in the liver, and stored in the gall bladder, from where they are released into the duodenum to facilitate lipid digestion. They are then 'rescued' by active absorption from the terminal ileum and returned to the liver by the hepatic portal vein, which completes their enterohepatic circulation.

SELF-CHECK 17.4

What is the difference between a bile acid and a bile salt?

Absorption of minerals and water

Sodium is transported out through the basolateral membrane by active transport, facilitated by the Na/K ATPase pump. This ensures that the enterocyte cytoplasm has a low sodium concentration and a negative charge.

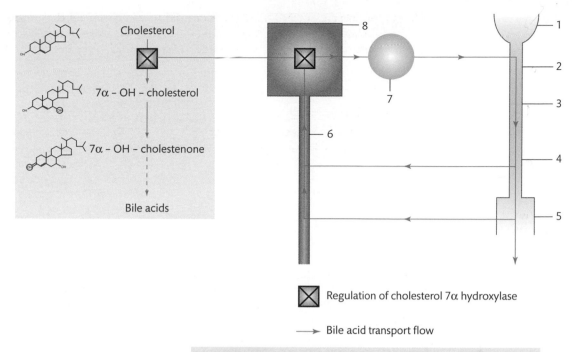

⊠ Regulation of cholesterol 7α hydroxylase

⟶ Bile acid transport flow

1 **Stomach**
2 **Duodenum** – emulsification and digestion of fats in micelles
3 **Jejunum** – fat absorption from micelles
4 **Terminal ileum** – active absorption of bile salts
5 **Colon** – passive absorption of secondary bile acids
6 **Hepatic portal vein** – returns absorbed bile salts/acids to liver
7 **Gall bladder** – stores bile salts between meals
8 **Liver** – bile salts synthesized by acinar cells

FIGURE 17.4

Enterohepatic circulation of bile acids and their regulation of synthesis.

The resultant electrical and osmotic gradient created by this pump drives most intestinal transport processes.

In the fed state, sodium absorption in the jejunum is coupled to the transport of monosaccharides and amino acids at the villous brush border membrane. Different proteins are responsible for the transport of a range of amino acids with sodium. In the jejunum, water can pass freely along osmotic gradients created by the Na/K ATPase pump through the intercellular junctions. This is illustrated in Figure 17.5.

In the colon, the intercellular junctions between enterocytes are much less permeable, thus minimizing osmotic gradient driven permeability. This allows sodium to be absorbed even when luminal concentrations are low, resulting in the formation of a low volume, formed stool and conservation of electrolytes.

In the terminal ileum and colon, sodium absorption is also linked by exchange with hydrogen ions and potassium, and chloride is linked with bicarbonate. Sodium absorption and potassium secretion are enhanced by **aldosterone**. Where the rate of delivery of fluid increases, potassium secretion is increased in the proximal colon. This can result in potassium deficit (hypokalaemia) in severe **diarrhoea**, which in combination with bicarbonate loss can result in a metabolic acidosis.

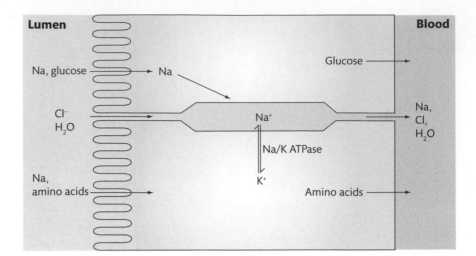

FIGURE 17.5

A schematic showing the movement of water from the gut lumen into the blood. See text for details.

SELF-CHECK 17.5

What are the forces driving absorption across the jejunal mucosa?

Secretion of minerals and water

Jejunal secretion is driven primarily by the active secretion of chloride at the villous crypts. Chloride is transported into the enterocytes coupled with sodium by the Na-K2Cl transporter on the basolateral membrane. Increases in cyclic adenosine monophosphate (cAMP) and calcium result in the opening of the chloride channels on the luminal enterocyte membrane, with resultant movement of chloride down the electrochemical gradient into the lumen, followed by sodium via the ATPase pump on the basolateral membrane, and water. Sodium secretion is required for jejunal glucose absorption (see Figure 17.6). This chloride channel is defective in cystic fibrosis and is now referred to as the **cystic fibrosis transmembrane conductance regulator** (CFTR). Normal jejunal secretion is stimulated by distension of the duodenum, and modulated by neural and humoral mechanisms.

Cholera toxin causes adenyl cyclase activation, resulting in chloride channel opening with jejunal secretion of chloride, followed by sodium and water through the intercellular spaces.

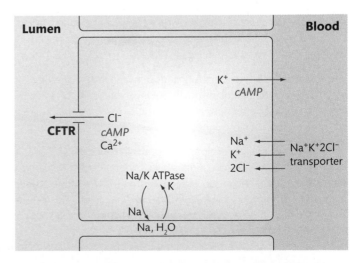

FIGURE 17.6

A schematic showing secretion of chloride in the gut. See text for details.

Uncontrolled diarrhoea ensues, with luminal and faecal electrolytes resembling those of plasma. The symptoms are treated using oral hydration therapy, which is a simple mixture containing glucose and salt solutions with sodium, chloride, potassium, and bicarbonate that facilitates glucose stimulated sodium absorption followed passively by water and chloride. This simple solution counteracts the severe and often fatal dehydration caused by uncontrolled secretion and saves millions of lives, especially children, in the developing world.

Absorption of calcium and iron

Calcium absorption is dependent on the concentration of **calcitriol** (activated vitamin D). This stimulates the synthesis of calbindin, a calcium binding protein present in enterocytes. Calcitriol synthesis and calcium absorption are stimulated by calcium deficiency.

Calcium can form insoluble salts with dietary phosphate, oxalate, and **phytates**, thus limiting its absorption. This also happens when calcium forms complexes with free fatty acids in fat **malabsorption**.

Iron (inorganic) is absorbed mainly in the duodenum. The reduced ferrous (Fe^{2+}) form is more readily absorbed than the oxidized ferric (Fe^{3+}) form. Organic iron as **haem** is more readily absorbed than the inorganic forms.

Gastric acid facilitates iron absorption, preventing insoluble complex formation and assists ascorbate (vitamin C) induced reduction to the ferrous form. Iron binds to a brush border receptor and is transported into the cell cytoplasm where it then binds to **mobilferrin**; this transports iron across the cell preventing toxicity. **Transferrin** receptors on the basal membrane then facilitate passage of iron to transferrin, the main carrier protein in the circulation.

Apoferritin can also bind iron irreversibly within enterocytes to form **ferritin**. This cannot be transported to the circulation and is excreted when the mature epithelial cells are released into the lumen. Iron deficiency results in an increase in mobilferrin and a decrease in apoferritin. Iron overload causes the reverse, thus regulating iron absorption.

Absorption of water-soluble vitamins

Most water-soluble vitamins are absorbed in the duodenum and jejunum by diffusion, carrier mediated diffusion, or secondary to active transport of sodium. Vitamin B_{12} is bound to protein in food and released by gastric acid during digestion. It then combines with **intrinsic factor** (IF) produced by parietal cells to form a complex which resists digestion. This complex is absorbed intact by specific receptors in the ileal mucosa and released into the bloodstream. About 80% is bound in an inactive form to the transport proteins **transcobalamin** I and 3, and 20% in a biologically active form to transcobalamin II (TC 2), also known as holotranscobalamin (HoloTC).

Colon function

The major function of the large intestine is the absorption of fluid and salts, and the storage of faeces until voiding. Normally the colon receives less than 1 L of fluid per day, but this can increase greatly where malabsorption or secretion from the small bowel occurs, and where it exceeds 5 L per day then diarrhoea will occur. The large intestine has a huge population of bacteria. In fact, the number of bacterial cells in the colon exceeds the total number of other cells in the body. Some vitamins, notably vitamin K, are derived from colonic bacterial synthesis. Bacteria also contribute to digestion and nutrition through **anaerobic** fermentation, producing short chain fatty acids from starch and complex carbohydrates, predominantly acetate, and also butyrate, which is used as a source of energy by the colonic mucosa. Fermentation also results in gas (flatus) containing hydrogen, methane, carbon dioxide, and hydrogen sulphide. A diet rich in complex carbohydrates (dietary fibre) can lead to a physiological increase in flatus production.

SELF-CHECK 17.6

What are the two major components of faeces, excluding water?

Other gastrointestinal functions

The GIT has an important barrier function, which prevents entry of foreign proteins and microorganisms. Any breakdown in the mucosa can result in bacterial invasion and systemic infection. The GIT has immunological functions and contains **gut associated lymphoid tissue** (GALT). Diverse areas of endocrine tissue secrete hormones directly into the circulation to target tissues. These hormones regulate fluid and enzyme secretion, gut motility, sphincters, and also appetite. Osmoreceptors respond to nutrient concentrations and regulate fluid secretion.

17.2 Biochemical investigation of gastrointestinal tract diseases

The major GIT tract diseases investigated using biochemical tests are listed in Table 17.3. Disorders of the GIT can affect the stomach, pancreas, gall bladder, and small and large intestine. Some diseases can affect more than one area of the GIT. Most of the conditions affecting the small intestine will cause malabsorption (see Table 17.4).

TABLE 17.3 Biochemical tests used to investigate gastrointestinal disorders.

Gastrointestinal disorder	Biochemical test
Duodenal, peptic ulcer disease	C^{13} urea breath test, serum antibodies to *Helicobacter pylori*, faecal *Helicobacter pylori* antigen
Gastrinoma (Zollinger Ellison syndrome)	Plasma gastrin
Acute pancreatitis	Serum amylase, lipase
Exocrine pancreatic insufficiency	Faecal elastase
Coeliac disease	Tissue transglutaminase antibodies
Bacterial colonization of the small intestine	Glucose breath hydrogen test
Bile acid malabsorption	C^{14} glycocholate breath test, serum 7α-OH-cholestenone
Gastrointestinal inflammation (Crohn's, ulcerative colitis)	Faecal calprotectin, lactoferrin
Factitious diarrhoea	Laxative screen

TABLE 17.4 Causes of malabsorption.

Types of malabsorption	Disease
Mucosal	<ul><li>Coeliac disease</li><li>Dissacharidase deficiencies</li><li>Cows' milk intolerance</li><li>Crohn's disease</li><li>Lymphangiectasia</li><li>Radiation enteritis</li><li>Post-infectious malabsorption, including tropical sprue, Whipple's disease, giardiasis, nematode parasites</li><li>Immunodeficiency (congenital or acquired)</li></ul>
Intraluminal	<ul><li>Pancreatic insufficiency as in chronic pancreatitis, cancer of pancreas and cystic fibrosis, causing defective nutrient digestion</li><li>Cholestatic liver disease: causing reduced bile secretion, fewer bile salts in the duodenum and defective fat digestion</li><li>Gastrinoma causing increased acid production and steatorrhoea</li><li>Intestinal hurry: hyperthyroidism, drugs</li></ul>
Structural	<ul><li>Intestinal hurry: dumping syndrome after gastric surgery</li><li>Bacterial colonization caused by stasis (blind loop syndromes), after gastric surgery, or secondary to scleroderma or diabetic neuropathy</li><li>Short bowel syndrome (resection of small bowel segments, for example in severe Crohn's disease) resulting in reduced mucosal surface</li><li>Fistulae, diverticuli, strictures resulting in colonization and/or reduced mucosal surface</li></ul>

Diseases of the stomach

Diseases or conditions of the stomach that are investigated by clinical biochemistry laboratories include:

- peptic ulcers (mainly gastric or duodenal)
- gastrinomas
- vitamin B_{12} deficiency

Peptic ulcers

The stomach produces hydrochloric acid which is required for activation of digestive enzymes and kills bacteria. However, acid is corrosive in action and so the cells lining the stomach (and duodenum) produce mucus which has a protective effect. There is a balance between the amount of acid produced and the mucus defence barrier. An ulcer develops if this balance is disrupted, for example by excessive production of hydrochloric acid and pepsin in gastric juice or a reduction in the protective mechanisms of the mucosa. An ulcer is a break (hole) in the lining of the GIT that fails to heal and is often accompanied by inflammation. Ulcers can affect the oesophagus, stomach, and duodenum. Clinical features of peptic ulcers include a chronic upper abdominal pain that is relieved by foods or antacids. About eight out of ten cases of peptic ulcers are due to a *Helicobacter pylori* infection. This infection weakens the stomach (or duodenal) lining by causing inflammation which results in disruption of the mucus layer, allowing the hydrochloric acid and pepsin to cause damage producing ulceration.

The metabolism of urea to ammonia and carbon dioxide by microbial urease, protects these mucosal bacteria from gastric acidity. This reaction forms the basis of the urea breath test (UBT) which detects gastric *Helicobacter pylori*. Where ulceration has been shown not to be associated with *Helicobacter pylori* infection, plasma gastrin concentrations may assist in diagnosis (see below). There are currently three types of *Helicobacter pylori* tests:

- urea breath test
- serological *Helicobacter pylori* IgG antibody
- faecal *Helicobacter pylori* antigen

In the UBT, patients are given C^{13} labelled urea with a fruit drink and a breath sample is taken after 30 minutes. This is assayed by mass spectroscopy for C^{13} labelled carbon dioxide which diffuses across the gastric mucosa and is transported to and released from the lungs. High concentrations of labelled breath $^{13}CO_2$ indicate the presence of *Helicobacter pylori* in gastric mucosa.

This test has the advantage in that it can be used both to establish the presence of *Helicobacter pylori* where suspected and also monitor eradication of the infection. Serological antibody tests can be used to establish infection, but cannot be used to monitor treatment because of residual antibody present after treatment. Faecal *Helicobacter pylori* antigen tests are a direct measurement of *Helicobacter pylori* and improved assays have now been described where predictive values are equivalent to that of the urea breath test.

SELF-CHECK 17.7

What is the disadvantage of serological testing for *Helicobacter pylori*?

Gastrinoma

Gastrin is synthesized and released from the G cells in the **gastric antrum** (distal stomach) and duodenum, and stimulates secretion of gastric acid from parietal cells in the stomach. Where gastric pH is low, gastrin release is inhibited. Conversely, release is increased in conditions or therapies resulting in low or negligible gastric acid concentration (hypo- or **achlorhydria**). Thus, plasma gastrin may increase in response to hypo- or achlorhydria caused by chronic or **atrophic** gastritis, pernicious anaemia, previous vagotomy, or treatment with H_2-**receptor antagonists** such as cimetidine (by competitive inhibition of histamine at histamine H_2 receptor sites) or proton pump inhibitors such as omeprazole (by inhibiting acid secretion from parietal cells). The most frequent reason for measurement of plasma gastrin concentration is in the investigation of **gastrinoma** (as in the **Zollinger Ellison syndrome**). This is a rare disorder characterized by severe, multiple, and recurrent peptic ulcers that arise due to excessive gastric acid secretion. This can be caused by neoplasia (gastrinoma) arising within the pancreas (most frequent) or within the duodenum. Further information on gastrinoma is given in Table 17.5.

About 60% of these tumours are malignant and 30% arise as part of the **multiple endocrine neoplasia** (MEN) syndrome type 1. Chronic overproduction of gastric acid can lead to fat malabsorption and **steatorrhoea** caused by acid inhibition of the pancreatic lipase.

The diagnosis of gastrinoma is based on the detection of a high plasma gastrin concentration in the presence of symptoms of acid hyper-secretion. Patients should not be receiving acid inhibiting drugs at the time of testing. The basal plasma gastrin concentration may be normal or only slightly increased in some patients with gastrinoma.

TABLE 17.5 Neuroendocrine tumours: incidence, symptoms, and biochemical tests for diagnosis (adapted from Ramage J K, et al. (2005) Guidelines for the management of gastroenteropancreatic neuroendocrine (including carcinoid) tumours, *Gut* 54:iv1–iv16, with permission).

Tumour	Annual incidence	Symptoms	Biochemical test	Result
Carcinoid	1 in 50,000	Flushing, diarrhoea		
Foregut			24-hr 5HIAA	Sometimes raised
Midgut			24-hr 5HIAA	Usually raised
Hindgut			24-hr 5HIAA	Not raised
Gastrinoma	1 in 500,000	Severe peptic ulceration, malabsorption and diarrhoea	Fasting gastrin Gastric acid secretion	Raised basal gastrin; raised basal acid output
Insulinoma	1 in 500,000	Confusion, weakness, sweating, unconsciousness; relief with eating	Fasting insulin, glucose and C-peptide	Raised fasting insulin/ glucose ratio, proinsulin or C-peptide
Glucagonoma	<1 in 10 million	Erythema, weight loss, diarrhoea, diabetes mellitus	Fasting gut hormones	Raised serum glucagon and enteroglucagon
VIPoma	<1 in 10 million	Profuse watery diarrhoea and hypokalaemia	Fasting gut hormones	Raised fasting vasoactive intestinal peptide
Ppoma (Pancreatic polypeptide)	<1 in 10 million		Fasting gut hormones	Raised fasting pancreatic polypeptide
Somatostatinoma	<1 in 10 million	Cholelithiasis, weight loss, diarrhoea and steatorrhoea	Fasting gut hormones	Raised fasting somatostatin
All NETs			Plasma chromogranin A	Raised in most cases

Vitamin B_{12} deficiency

Vitamin B_{12} deficiency should be suspected in all patients with unexplained anaemia and/or neurological symptoms, as well as patients at risk of developing vitamin B_{12} deficiency such as the elderly, and patients with intestinal diseases. Patients at risk include the following groups:

- Elderly patients with **hypochlorhydria**, unable to release vitamin B_{12} from dietary proteins.
- Pernicious anaemia: an **autoimmune** disease preventing absorption of vitamin B_{12} is associated with atrophic gastritis, achlorhydria, and an increased risk of gastric cancer. Most patients have parietal cell antibodies, and about 50% patients have antibodies to IF. These are of two types: one prevents binding of vitamin B_{12} to IF and the other prevents binding of IF/vitamin B_{12} complex to the ileal receptor. The diagnosis is based on the demonstration of a **macrocytic anaemia** with low serum concentrations of vitamin B_{12} in the absence of dietary deficiency. Treatment is with intramuscular hydroxycobalamin.

Furthermore, it is worth nothing that:

- Patients with total gastrectomies: unable to produce IF.
- Patients with terminal ileal disease or resection: unable to absorb the IF/vitamin B_{12} complex.
- Patients with **blind loop syndrome** or parasitic infestations will also result in malabsorption.

Disease of the pancreas

The main conditions investigated by clinical biochemistry laboratories are **acute** and **chronic pancreatitis**, **cystic fibrosis**, and carcinoma at the head of the pancreas.

Acute pancreatitis

The main causes of acute pancreatitis are **gallstones**, other gallbladder (biliary) disease, and alcohol use. Viral infection (mumps, coxsackie B, mycoplasma), traumatic injury, congenital, pancreatic or common bile duct surgical procedures, certain medications (for example oestrogens, corticosteroids), or **idiopathic** are other causes. Acute pancreatitis is initiated by auto-digestion of the pancreatic tissue by activated pancreatic enzymes and can result in haemorrhage. The patient can often experience severe abdominal pain. Confirmation of the clinical diagnosis usually depends on the plasma amylase activity. In severe cases, plasma calcium concentration may fall following the formation of insoluble calcium salts of fatty acids in areas of fat necrosis.

Increased amylase concentration in the plasma normally arises from the pancreas (P-isoamylase) and the salivary glands (S-isoamylase).

Plasma P-isoamylase is a more sensitive and specific test for acute pancreatitis than total amylase, but total amylase is usually greatly increased in acute pancreatitis, with values >10 times the upper limit of normal considered almost diagnostic for acute pancreatitis. Values of up to five times above normal may indicate acute pancreatitis, but can also be present in conditions such as mesenteric infarction (occlusion of blood supply to tissue), acute biliary tract disease, and acute **parotitis** (salivary gland inflammation).

Smaller increases in amylase concentration may occur in almost any acute abdominal condition (for example perforated peptic ulcer or injection of drugs which cause spasms in the sphincter of Oddi). Moderate increases in plasma amylase can also occur in diabetic ketoacidosis.

In the rare condition of **macroamylasemia**, part of the plasma amylase exists as a high molecular weight form which cannot be cleared by the kidneys. This can be demonstrated by a persistently high amylase concentration in the plasma in the presence of a normal amylase concentration in the urine. Serum lipase is a more specific measure of acute pancreatitis than total amylase and has a diagnostic power similar to p-isoamylase, but is seldom performed.

Key Points

Acute pancreatitis can be a life threatening disease; serum amylase of more than ten times above normal is considered almost diagnostic of this condition.

SELF-CHECK 17.8

Name two non-pancreatic diseases where serum amylase is raised.

The clinical conditions, chronic pancreatitis, cystic fibrosis (CF), or cancer of the head of the pancreas can result in reduced digestion and consequent malabsorption, of protein, fat, and carbohydrates.

Chronic pancreatitis

This is a progressive condition leading to inflammation and calcification of exocrine pancreatic ducts and the subsequent reduction in the volume and enzymes of pancreatic secretions. This may not be evident until the disease is advanced and can then result in malabsorption and steatorrhoea. The main causes are alcohol related, or more rarely genetic and autoimmune.

Cystic fibrosis

This is an inherited disease (autosomal recessive) and is caused by a mutation in the CFTR gene. The product of this gene is a chloride ion channel protein important in creating sweat, digestive juices, and mucus. This condition is characterized by thick mucus production and early exocrine pancreatic insufficiency.

Cancer at the head of the pancreas

This can lead to occlusion of the pancreatic duct with much reduced output of pancreatic secretions.

Key Points

Conditions leading to inflammation or occlusion of the pancreatic ducts will result in reduction in digestive enzymes in the small bowel; maldigestion is usually a late symptom of pancreatic disease.

Tests for assessing pancreatic insufficiency can be either invasive, requiring duodenal intubation, or non-invasive, avoiding intubation.

Invasive tests of pancreatic insufficiency: these are seldom used now because of expense and discomfort to the patient, and a high degree of technical expertise in siting the tube is required. However, they have a high sensitivity for detection of chronic pancreatitis of >90%, and can detect some cases of mild, as well as moderate and severe cases. The secretin/cholecystokinin-pancreozymin (CK-PZ) test requires measurement of bicarbonate and amylase or trypsin in duodenal fluid following stimulation of the pancreas/gall bladder with intravenous secretin and CK-PZ.

The Lundh test assesses the pancreatic secretory response to a standard test meal of glucose, corn oil, and protein but is also dependent on extra-pancreatic factors, such as gastric and vagal function, and secretin and cholecystokinin secretion. Abnormal results are obtained in most cases of chronic pancreatitis, with bicarbonate and enzymatic activity tending to fall before pancreatic fluid volume.

Non-invasive tests for pancreatic insufficiency: these involve measuring a pancreatic specific enzyme called **faecal elastase**. This enzyme, unlike faecal chymotrypsin, is resistant to degradation during gut transit, thus directly reflecting exocrine pancreatic function. Indeed in normal subjects, it reaches concentrations in faeces which are 5–6 times higher than those of duodenal fluid. Low levels are associated with pancreatic insufficiency (<200 µg/g). Another benefit of this test is that levels in faeces are unaffected by enzyme substitution therapies using pancreatin, which lack human elastase, allowing this test to be used for monitoring patients on treatment. One drawback is that watery diarrhoea samples can cause falsely low, pathological results. Elastase 1 is measured by ELISA using a monoclonal antibody. Only a single random stool sample is required in a universal container, which is stable for mail delivery and then at 4°C for five days, or six months stored frozen at −20°C.

Two other tests, the fluoroscein dilaurate (pancreolauryl) test and the N benzoyl-L-tyrosyl para aminobenzoic acid (BT-PABA) test, which require the patient to ingest synthetic substrates for pancreatic digestion and where the products of digestion are excreted in urine and quantified, have now mostly been withdrawn from use. All non-invasive tests are only reliable in diagnosing moderate to severe disease.

Key Points

Elastase passes through the gut without being degraded and faecal elastase reflects pancreatic exocrine secretion. Levels are unaffected by replacement therapies but watery diarrhoea can result in false positive results.

SELF-CHECK 17.9

Are low or high levels of faecal elastase associated with exocrine pancreatic insufficiency?

Disease of the small intestine

Disease of the small intestine can affect the duodenum, jejunum, and ileum. Since absorption of most nutrients occurs in the small intestine, disorders affecting this region of the GIT often cause malabsorption. However, malabsorption of nutrients may also occur due to disease affecting other parts of the GIT. Diseases of the small intestine may arise from:

- abnormal luminal function
- bacterial overgrowth
- abnormal jejunal mucosa (mostly coeliac disease)
- disaccharidase deficiency particularly lactase deficiency
- Crohn's disease (mostly ileum resulting in bile acid malabsorption)
- neuroendocrine tumours, in particular carcinoid tumours

Abnormal luminal function

Gastric surgery can cause impaired mixing and intestinal hurry (reduced intestinal transit time). In the resultant **dumping syndrome**, food and secretions from the stomach move to the small intestine in an unregulated, abnormally fast manner. This may develop when the opening (pylorus) between the stomach and duodenum has been damaged or removed during an operation. Thyrotoxicosis can also cause intestinal hurry.

Bacterial overgrowth

Bacterial colonization of the small bowel can lead to fat maldigestion and steatorrhoea as a result of bile salt de-conjugation, and the resultant reduction in micelle formation.

The most common causes arise from gut surgery, resulting in anatomical changes which predispose to bacterial growth, for example partial gastrectomy with construction of a blind duodenal loop.

Other causes of colonization include pathological processes such as jejunal **diverticulae** (outpouch from the main gut lumen) or internal **fistulae** (an abnormal connection or passageway between two epithelium-lined organs or vessels that normally do not connect). This can occur in **Crohn's disease**. Neuropathies (nerve damage) which lead to stasis in the gut such as diabetic **neuropathy** and **systemic sclerosis** also predispose to colonization.

The gold standard method for detection is aspiration and culture of fluid from the small intestine, but this requires intubation of the patient and a high level of technical expertise. False negatives due to sampling errors and false positives due to non-pathological bacteria normally present in the small bowel also make this a less attractive procedure to use. The most commonly used tests now are breath tests of which three have been described:

- C^{14} glycocholate breath test: isotopically labelled glycocholate is given orally. Bacterial metabolism of glycocholate leads to release of glycine and further metabolism to $C^{14}O_2$ in the liver. This is then measured in expired air between two and six hours following intake. This test is sensitive but is also positive in patients with ileal disease.

- C^{14} xylose breath test: this results in small bowel bacterial metabolism of xylose and release of $C^{14}O_2$ in breath.

- Glucose breath hydrogen test: is the most frequently used test. Patients should be on a low fibre diet for two days prior to the test to reduce baseline values of hydrogen production from colonic bacteria. They are then given 50 g glucose solution and breath samples (end alveolar air) measured at 15-minute intervals for two hours. An increase above baseline of >20 parts per million within the first hour indicates bacterial colonization. Later rises should be interpreted with caution as these may indicate malabsorption. This can be done as a point of care test.

Key Points

Conditions resulting in gut stasis predispose to bacterial overgrowth and subsequent fat maldigestion/malabsorption; the breath hydrogen test is most frequently used for diagnosis.

SELF-CHECK 17.10

Why should neuropathy in the GIT result in the bacterial overgrowth syndrome?

Abnormal intestinal mucosa

Generalized conditions such as radiation enteritis and **lymphoma**, resulting in infiltration and destruction of the small intestine, and short gut syndrome where more than two-thirds of the small bowel have been surgically removed, will reduce the viable surface area for nutrient and vitamin absorption.

Mucosal disorders such as coeliac disease, post-infectious malabsorption (including **tropical sprue**), Whipple's disease (bacterial disease affecting intestinal mucosa), *Giardia lamblia* (protozoan parasite affecting intestinal mucosa), nematodes (worms), immunodeficiency (congenital and acquired), and Crohn's disease, can all result in a decreased area for absorption. Crohn's disease of the terminal ileum may lead to loss of bile salts and vitamin B_{12}, while more extensive disease can also affect the jejunum.

Lymphatic obstruction can be caused by a rare condition called intestinal **lymphangiectasia**, resulting in fat malabsorption. On a global scale, acute and chronic infections such as tropical sprue are important causes of malabsorption, but in the UK by far the commonest cause is coeliac disease.

No specific clinical disorder of protein malabsorption has been described. Some disorders causing mucosal disease, such as Crohn's disease and non-steroid anti-inflammatory drug use, result in loss of protein across the damaged mucosa, without malabsorption occurring.

Two inherited disorders are described, which affect intestinal absorption of amino acids (as well as their reabsorption from the proximal tubules in the kidney). These are **cystinuria** (affecting dibasic amino acids) and **Hartnup disease** (affecting tryptophan and other neutral amino acids). There is no deficiency syndrome associated with cystinuria, since affected amino acids can be absorbed as constituents of di- and tripeptides.

Key Points

Conditions resulting in infiltration, infection, destruction, or surgical removal of the small intestine will reduce the viable surface area for nutrient and vitamin absorption. Protein loss can occur across the damaged mucosa from plasma. Some rare inherited amino acid malabsorption disorders are described.

Coeliac disease is an autoimmune disease triggered by a sensitivity to **gliadin**, a fraction of gluten which is present in the cereals, wheat, barley, and rye. The incidence in the Caucasian population is now thought to be about 1 per 100. Genetic factors contribute to the pathogenesis. The condition is found in up to 15% of first-degree relatives of affected individuals and there is a strong association with the **HLA-DQ2** antigen of the **human leukocyte antigen** (HLA) **complex**. It has a strong association with other autoimmune disorders, notably type 1 diabetes and autoimmune thyroid disease. Coeliac disease was first described as far back as the second century AD. Cereals were only recognized as the source of the problem during World War II, when a Dutch paediatrician observed that during a cereal famine some children's health actually improved, only to deteriorate once more when cereal availability was restored. Subsequently, the gluten fraction of wheat flour was shown to be the causative agent.

Endomysial and tissue transglutaminase antibodies (TTGA) are demonstrable in almost all affected patients. Tissue transglutaminase (TTG) is associated with the **endomysium** and is the target of a **T cell mediated** autoimmune response.

Tissue transglutaminase reacts with gliadin so as to enhance the T cell response in susceptible individuals. Enterocytes exposed to gliadin undergo destruction and subtotal villous atrophy ensues, with an associated chronic inflammatory infiltrate in the **lamina propria.** The age of presentation varies from infancy to late adulthood and the symptoms vary in presentation from relatively minor tiredness and malaise to small bowel malabsorption and steatorrhoea. Many patients remain without symptoms for years.

Serum anti-endomysial IgA and IgA TTGA are very sensitive (>90%) and specific (>95%) for the condition, and should now replace anti-gliadin and anti-reticulin antibodies which have a lower predictive value.

Total IgA should be determined in cases where coeliac disease is suspected, and where IgA TTGA gives a normal but very low result. Where IgA is deficient (1 in 500 of normal population, and 1 in 50 of patients with coeliac disease), the IgG TTGA should be measured. However, the predictive value of IgG antibodies is inferior to those of IgA antibodies, since increased levels are less specific. Although the gold standard diagnosis is from small bowel histology, treatment will often be commenced with a typical presentation and high positive antibody result, with response to a gluten free diet confirming the diagnosis. The concentration of IgA TTGA returns to normal in serum with successful response to diet.

Patients with coeliac disease have an increased risk of osteoporosis even with strict dietary adherence. Incidence of lymphoma and **adenocarcinoma** of the small intestine also increase but are reduced by adherence to a gluten free diet. **Dermatitis herpetiformis**, a blistering condition of the skin, is associated with coeliac disease and responds to a combination of oral dapsone (an anti-infective drug also used in treatment of leprosy) and a gluten free diet.

Clinically, coeliac disease is thought of as an iceberg, with the majority of non-symptomatic cases (latent and silent) remaining undiagnosed, with only the 'tip' of symptomatic subjects being tested.

Key Points

Coeliac disease is an autoimmune disease triggered by a sensitivity to gliadin; the incidence is much higher than previously thought; it has a strong association with other autoimmune disorders, for example diabetes mellitus. Serum IgA TTGA is a sensitive marker of this disease and returns to normal with successful response to diet.

SELF-CHECK 17.11

If coeliac disease is suspected and IgA TTGA gives a very low result, what else should you measure?

Disaccharidase deficiency

Disaccharidase deficiency, particularly **lactase deficiency**, is a common cause of malabsorption. Approximately 5% of Caucasian adults are deficient as are the majority of adults in Africa and Asia. However, this only results in lactose intolerance where other criteria exist, such as colonic sensitivity to lactose, presence of specific colonic bacteria, and possibly inability to metabolize intraluminal gases. The severity of symptoms varies considerably; nausea, abdominal distension and cramps, diarrhoea, and increased flatus following milk ingestion are usual. Diagnosis is usually made on the basis of symptoms, the link to milk ingestion, and resolution by abstention from milk.

Lactose tolerance tests are available where the diagnosis requires to be confirmed and these measure lactose malabsorption rather than lactose intolerance. A 50 g lactose meal is given in solution and the blood glucose is measured: where this fails to rise by more than 1.1 mmol/L above baseline over a three-hour period, then lactose malabsorption may be present. The breath hydrogen is measured at 30-minute intervals over a three-hour period and where this increases by >20 parts per million (ppm) above baseline then lactose malabsorption may be present. This is the more sensitive test but has poor specificity. It is dependent on anaerobic bacterial metabolism of lactose in the colon, and can be easily carried out as a point of care test. In addition to primary lactase deficiency, this can occur secondary to coeliac disease, Crohn's disease, and following gastroenteritis.

Congenital deficiencies of **sucrase-isomaltase glucosidase** and **maltase** have been described but are rare and can be investigated using a breath hydrogen test with the appropriate substrate given orally. Substrates should be given on a per kilogram body weight basis, especially

CASE STUDY 17.1

A 43-year-old female with a history of type 1 diabetes mellitus presented with symptoms of lack of energy and mild recurrent abdominal pain. On investigation, standard biochemistry/haematological tests revealed the following results (reference ranges are given in brackets):

Haemoglobin	11.4 g/dL	(11.5–16.5)
Glucose (fasting)	8.0 mmol/L	(4.0–5.9)

Other tests such as urea, electrolytes, liver function tests, calcium, and serum albumin were all within their reference ranges.

(a) What do the clinical history and results indicate and what further tests might be performed?

Further biochemical investigations were carried out with results as follows:

Serum ferritin	10 mg/L	(20–150)
Serum TTGA	26 U/L	(0.1–7.9)

(b) What is the likely diagnosis?

in young children, to minimize dehydration where there is the possibility of an acute diarrhoeal response to the dietary load.

Cows' milk allergy is much more common than lactose intolerance in babies and is caused, not by lactose, but by an allergic reaction of the baby's immune system to proteins in the milk.

Bile acid malabsorption

This is classified into three types:

- Type 1: caused by mucosal disease or resection of the terminal ileum, the final segment of the small intestine.

- Type 2: cause unknown, possible defect in bile acid transporter mechanism, or disordered ileal signalling affecting hepatic synthesis and bile acid pool size.

- Type 3: caused by non-ileal disease with secondary bile acid malabsorption, such as bacterial colonization where deconjugation of bile acids reduces absorption, cholecystectomy, diabetes mellitus where the ileal absorption may become compromised by rapid transit, or overload of the **enterohepatic** circulation.

Mucosal disease of the terminal ileum caused by Crohn's disease, radiation damage, or ileal resection following obstruction, results in reduction in active absorptive capacity and loss of bile acids into the colon.

At higher concentrations in the colon (>1.5 mmol/L) of the dihydroxy bile acids, chenodeoxycholic and deoxycholic acid, can inhibit sodium absorption and cause water secretion and diarrhoea. The water secretion mechanism is complex, involving both intracellular calcium and cAMP. This effect only occurs at luminal pH >6.5 where bile acids exist as anions.

The following are used as diagnostic procedures:

- C^{14} glycocholate breath test, as described previously for bacterial overgrowth syndrome.

- Whole body retention of selenium labelled homotaurocholic acid test (WBR SeHCAT). Here the patient is given a capsule of a gamma emitting conjugated bile acid analogue, SeHCAT, with 40 kBq activity. The 100% value for whole body retention measurement is determined after 30 minutes with a whole body scanner or **gamma camera** and repeated after seven days. Results of <10% WBR SeHCAT at seven days are considered pathological, and such patients have a good response to bile acid binding agents which remove bile acids from the soluble phase of the lumen. The SeHCAT is mainly resistant to deconjugation and **dehydroxylation** and is absorbed solely at the terminal ileum and as such is a more specific marker of ileal disease than other tests. These tests are carried out in medical physics departments, are expensive, time consuming, and expose the patients to radiation.

- Serum 7 α-hydroxycholestenone (7-HCO) has recently been shown to correlate well with the WBR SeHCAT test. This is a test of bile acid turnover (rate of synthesis) and directly relates to the activity of the enzyme liver cholesterol 7α hydroxylase, the rate limiting step in bile acid synthesis. The 7 α-hydroxycholestenone is an intermediate in the synthesis of bile acids from cholesterol and is formed by the oxidation of 7α-OH-cholesterol. As such, it is also a marker of bile acid malabsorption, increasing as bile acids are lost from the entero-hepatic circulation. Further details are given in Figure 17.4. This is a non-invasive test done by analysis of serum using high performance liquid chromatography, and is a simpler and cheaper alternative to the other procedures.

SELF-CHECK 17.12

Irritable bowel syndrome and diabetes mellitus are conditions which may cause which type of bile acid malabsorption?

Neuroendocrine tumours

The incidence of neuroendocrine tumours (NETs) is extremely rare and all other possible diagnoses should be excluded before their investigation. They may be associated with multiple endocrine neoplasia 1 (MEN1).

The most common NETs are **carcinoid tumours**, often found in the ileo/caecal region of the GIT or less commonly in tissues derived from embryological foregut, for example the thyroid and bronchus. These tumours secrete vasoactive peptides, mainly serotonin, which pass directly to the liver by venous drainage and are inactivated there. Symptoms are only likely to occur when the tumour has spread to the liver, when serotonin is released into the systemic circulation causing episodic flushing and diarrhoea—the carcinoid syndrome.

Most carcinoid tumours secrete excessive amounts of 5-hydroxytryptamine (5-HT, serotonin). This neurotransmitter is synthesized from tryptophan and subsequently metabolized and excreted in urine as 5-hydroxyindole acetic acid (5-HIAA), which can be measured for diagnosis. Values of <40 μmol/24 hours are considered normal and values >120 μmol/24 hours are considered diagnostic of carcinoid tumour.

Foregut tumours can lack the enzyme amino acid decarboxylase and consequently secrete 5 hydroxytryptophan rather than serotonin and this can lead to negative results for 5-HIAA. Urine collections should be carried out when patients are symptomatic. False positive results can occur in patients consuming foods rich in serotonin such as bananas and avocados. The urinary excretion of 5-HIAA correlates well with tumour size and can be used to monitor therapy. Plasma chromogranin A is another tumour marker for this disease where metastasis has occurred and is raised in >90% of all metastatic neuroendocrine tumours.

Disease of the large intestine

The large intestine consists of the caecum, appendix, colon, and rectum. The major disorders affecting the large intestine are Crohn's disease, **ulcerative colitis**, **irritable bowel syndrome** (IBS), **diverticular disease**, polyps, and carcinoma.

Ulcerative colitis is a form of inflammatory bowel disease (IBD) which affects the colon, often forming characteristic ulcers or open sores. The main symptoms of ulcerative colitis are diarrhoea mixed with blood and lower abdominal pain. Ulcerative colitis is frequently abbreviated as IBD and thus often confused with the less serious IBS condition. Ulcerative colitis has similarities to Crohn's disease, another form of IBD. Patients with ulcerative colitis (and Crohn's disease) have periods of exacerbated symptoms and then periods in remission where they are symptom free.

Crohn's disease is an autoimmune inflammatory disease of the intestines (a type of IBD) that may affect different parts of the GIT. The regions affected become inflamed, thickened, and ulcerated. It usually affects the terminal ileum and colon but can also affect other regions of the GIT including the jejunum and may cause obstruction of intestines leading to pain, diarrhoea (which may contain blood), and malabsorption. Evidence has been presented suggesting a genetic link for Crohn's disease in that siblings of affected individuals are at higher risk. Furthermore, smokers are more likely to develop Crohn's disease.

Diverticulosis is characterized by the presence of small pouches in the colon (called diverticula) that bulge outwards. The condition becomes more common with age. Often these pouches become infected or undergo inflammation and the condition is then called diverticulitis. Together diverticulosis and diverticulitis describe the different types of diverticular disease. Symptoms of diverticular disease include abdominal pain sometimes accompanied by fever, nausea, vomiting, and constipation. Diverticular disease can lead to bleeding in the stools, infections, perforations, or blockages of the GIT.

Many of the conditions affecting the large intestine are inflammatory conditions often accompanied by blood loss in the stool. Blood loss in the faeces and its detection are mentioned in the Method box below. Inflammatory reactions occurring in the GIT can be detected by measurement of markers such as faecal **calprotectin**.

Faecal calprotectin

Calprotectin is a heterocomplex protein expressed by myeloid cells and released from the cytoplasm of activated neutrophils during inflammation. Faecal concentrations of calprotectin are elevated when pathology resulting in an inflammatory process occurs in the GIT.

Highest levels are found in active inflammatory bowel disease and bacterial infections, but can also be increased in cancer of the colon and stomach, and in **enteropathy** caused by nonsteroidal anti-inflammatory drugs.

Since it is resistant to degradation in the gut, faecal calprotectin is a direct measure of mucosal inflammatory activity and may be detected at a level insufficient to cause an increase in the serum erythrocyte sedimentation rate (ESR) and C-reactive protein (CRP). Furthermore, faecal levels seem to be unaffected by a variety of non-intestinal conditions which may result in a systemic elevation of these inflammatory markers.

More recently the test has been investigated as a tool to distinguish organic from non-organic disease, and it has the potential to be used to separate younger patients with irritable bowel disease (no significant pathology) from those with more serious illness, such as inflammatory bowel disease, who require invasive investigations such as colonoscopy for diagnosis. Patients with IBS characteristically have no organic lesion within the gut and would be expected to have a normal faecal calprotectin.

A tool capable of screening patients in the outpatient setting or in the community and avoiding unnecessary invasive and expensive investigations has important implications for both waiting lists and hospital expenditure, and warrants further evaluation.

This test can also be used to monitor activity of disease in patients with inflammatory bowel disease. Faecal lactoferrin is another stable protein released from activated leucocytes which has also been used as a surrogate marker of gut inflammation.

Diarrhoea

Diarrhoea is defined as passage of stool weight of >200 g/day. Stool frequency or liquidity are also used to describe diarrhoea, but both can increase without stool mass, for example in the functional disorder, irritable bowel syndrome (IBS), which accounts for up to half of all visits to gastroenterology outpatient departments. Increases in stool mass of 500–1,000 g/day are highly significant of disease and require medical attention; however, diarrhoea has to be more severe to result in dehydration.

Significant malabsorption of nutrients does not automatically cause diarrhoea, since the normal colon can absorb up to 5 L fluid per day. Although malabsorption of carbohydrate will result in greater bacterial metabolism to short chain fatty acids and increased absorption from the colon, osmotic diarrhoea will only occur when the absorptive capacity is overwhelmed. Patients may have profuse diarrhoea in the absence of nutrient malabsorption or mucosal disease, for example when a bacterial toxin triggers secretory diarrhoea. Many patients with diarrhoea have

METHOD Faecal blood loss

The most accurate method of measuring blood loss in faeces is by injecting chromium labelled white cells into the bloodstream and measuring their appearance in faeces. This is expensive and involves administration of radiochemicals and is little used in practice.

Various slide tests based on the ability of haem to split peroxide and cause impregnated guaiac to turn blue have been used extensively, but are now used only to screen the well population, where despite their low sensitivity and specificity, they have been shown to be positive in 10–15% of adults aged between 50 and 70 who have undiagnosed colon cancer.

Animal and vegetable peroxidases can cause false positive and reducing substances such as vitamin C can cause false negative results. These tests are no longer used in the investigation of symptomatic patients.

Other tests used include those based on immunological detection of the globin fraction of human haemoglobin, which are unaffected by dietary components such as meat or animal blood and have improved sensitivity.

Both slide tests and immunological tests are only useful in detecting colonic blood loss since degradation of blood released is higher in the gut by bacteria and/or digestive enzymes renders this undetectable.

The 'Haemoquant' test measures protoporphyrin in faeces and as such measures haemoglobin and metabolites, and so can measure blood loss from any region in the gut, but is affected by animal haem products, is labour intensive, and little used.

a mixture of mechanisms contributing to their diarrhoea. An osmotic component, inhibition of active transport, motility changes, and secretory processes may all contribute to a different degree in most forms of diarrhoea. Further information is found in Table 17.6.

TABLE 17.6 Types and causes of chronic diarrhoea.

Types	Mechanism
Osmotic diarrhoea	Non-absorbable sugars (lactulose, sorbitol) or antacids (magnesium sulphate) retain water in the small bowel and colon.
	Malabsorbed sugars (lactase deficiency) or carbohydrate (coeliac disease, pancreatic insufficiency) result in a large increase in short chain fatty acid production in the colon.
Secretory diarrhoea	Cholera toxin.
	Hormone secreting neoplasms and neural mediators.
	Inflammatory bowel disease: in both ulcerative colitis and in Crohn's disease inflammatory mediators such as prostaglandins stimulate colonic secretion. Cytokines arising from inflamed mucosa may also down-regulate fluid absorptive mechanisms.
	Bile acid malabsorption.
Mixed secretory and osmotic diarrhoea	Tropical sprue and coeliac disease.
	Unabsorbed solutes cause increased osmotic component as do increased levels of IFN-γ and TNF-α in the lamina propria which down-regulate nutrient absorption.
	Crypt **hyperplasia** and villi destruction cause secretory processes to dominate.
	Unabsorbed bile acids and fatty acids stimulate fluid secretion in the colon, as do the inflammatory mediators.
	Diabetes of long-standing causes degeneration of adrenergic nerves which are proabsorptive in function and this can result in an imbalance of fluid homeostasis and diarrhoea.
	Bacterial colonization: deconjugated bile acids result in fat malabsorption with associated steatorrhoea and colonic secretory diarrhoea from bacterial production of hydroxy fatty acids.
Active transport defects	Many different active transport processes involving Na-ATPase enyzmes can be affected by mucosal disease processes.
	Congenital chloride losing diarrhoea causes severe watery diarrhoea from birth and a high concentration of chloride ions in stool water. The deletion in normal chloride/bicarbonate exchange in the mucosa results in a metabolic alkalosis.
	In glucose-galactose malabsorption, the jejunal sodium glucose/galactose transport protein is defective, leading to chronic osmotic diarrhoea after the infant's first feed.
	Both these conditions are life threatening and though very rare need to be diagnosed promptly.
Motility disorders	Increased motility in thyrotoxicosis and opiate withdrawal.
	Fluctuating motility in functional bowel disease.
	Decreased motility due to large diverticula, smooth muscle damage (scleroderma, amyloidosis, radiation injury) or autonomic neuropathy (diabetes).

A systematic approach to investigation of chronic diarrhoea is outlined in the Method box below.

METHOD *A systematic approach to biochemical investigations of chronic diarrhoea*

Collect stool for 48 hours
↓
Assess weight
if >200g on standard ward diet: diarrhoea;
if > 500g on standard ward diet: watery diarrhoea;
if <200g: NOT diarrhoea.
↓
Unusual appearance/colour of stool should assist in direction of diagnostic process
↓
Any blood; possible infection by amoeba, schistosomiasis, *Colostridium difficile*
Pale colour; steatorrhoea indicates fat malabsorption
↓
Check medical history for predisposing factors
↓
Foreign travel; antibiotic use; drug use; alcohol abuse; immune compromised;
hyperthyroid.
↓
Basic investigations
↓
Full blood count, urea and electrolytes, liver function tests, calcium, vitamin B_{12}, folate,
iron status, thyroid function, coeliac screen i.e. TTG antibodies, CRP. For weight loss, low
ferritin, low calcium, or vitamin B_{12}, or elevated inflammatory parameters; investigate
further for malabsorption and/or inflammatory bowel disease.
↓
Where TTG antibodies are normal, consider bacterial colonization (glucose hydrogen
breath test) or exocrine pancreatic disease (faecal elastase). Do faecal calprotectin,
a sensitive test for active intestinal inflammation.
↓
Where stool weight is greater than 500g and diagnosis is still unclear, response to
fasting should be assessed. Where response is positive, secretory diarrhoea is unlikely
and a faecal osmotic gap undertaken to confirm this with follow up of magnesium
measurements or lactose malabsorption tests. Where there is no response to fasting
or faecal osmotic gap suggests secretory diarrhoea, then investigate further for
factitious diarrhoea (laxative screen)/gut hormone screen/urinary 5-HIAA, and bile
acid malabsorption.
↓
Where intestinal pathology is suspected, definitive diagnoses are made from histology of
biopsies taken at endoscopy (e.g. coeliac disease) or colonoscopy (e.g. Crohn's disease,
ulcerative colitis, and colonic cancer).

Factitious (self-induced) diarrhoea: laxative abuse

A significant number of patients referred to gastrointestinal outpatient departments with chronic diarrhoea have been found to be abusing **laxatives**. A random urine sample taken when the patient has diarrhoeal symptoms should be screened for stimulant laxatives which can be bought over the counter without prescription, including bisacodyl and senna. Magnesium sulphate can also be taken inappropriately and will cause an osmotic diarrhoea. This can be measured in faecal supernatants used for **faecal osmotic gap** (FOG) calculations.

Faecal osmotic gap

This may be useful in distinguishing between an osmotic and a secretory diarrhoea and is derived from measurement of faecal sodium and potassium concentrations. This can be calculated as follows:

$$\text{Faecal osmotic gap (FOG)} = 290 - 2x \text{ (faecal sodium + faecal potassium)}$$

where 290 is the assumed serum osmolality.

Faecal osmolality *per se* is not used since this can rise quickly in faecal samples due to microbial short chain fatty acid production at room temperature.

Faecal creatinine should be determined to exclude urine contamination of the faecal sample. This test should be undertaken prior to testing for bile acid malabsorption, laxatives in urine, gut hormone screen, and faecal magnesium, and is most useful when faecal output is >500 g per day. Below this output, interpretation is less clear. Further information is supplied in Table 17.7.

Many tests that were used to investigate gastrointestinal disorders have become obsolete and been replaced by more convenient tests as shown in Table 17.8.

17.3 Management of gastrointestinal disease

The principles underlying the treatment of gastrointestinal disorders depend on the cause of the initial problem and have been outlined in Table 17.9.

TABLE 17.7 Faecal osmotic gap in secretory and osmotic diarrhoea.

Faecal osmotic gap	Interpretation	Possible further investigations
<75 mOsm/kg	Excludes osmotic diarrhoea (for example lactose intolerance and magnesium induced diarrhoea)	Laxative screen Gut hormone screen Bile acid malabsorption
>75 mOsm/kg	Excludes secretory diarrhoea (for example caused by laxative abuse, hormonal diarrhoea)	Response to low lactose diet; lactose breath test; faecal magnesium analysis

TABLE 17.8 Biochemical tests for gastrointestinal disorders that are now obsolete or done only in specialist centres.

Tests now obsolete or done in specialist centres	Function measured	Other comments
D-xylose absorption test	Carbohydrate absorption	Not sensitive
Sugar permeability tests	Intestinal absorption	Very sensitive, requires high performance liquid chromatography
Jejunal disaccharidases	Sugar absorption, for example lactase	Requires small bowel biopsy
Secretin/pancreozymin test Lundh test meal: tube tests	Exocrine pancreatic function	Very sensitive; invasive and require high degree of nursing/medical skill
Basal and maximal acid output	Resting and stimulated gastric acid output	Invasive and mostly unnecessary since introduction of *Helicobacter pylori* testing
Triolein breath test	Fat malabsorption	Now largely unused; uses radioisotopes
Acid steatocrit (like haematocrit using acidified faeces)	Fat malabsorption	Has been used in paediatrics
Stool microscopy after lipid staining	Fat malabsorption	Sometimes used in monitoring patients with cystic fibrosis
3–5 day faecal fat test	Fat malabsorption	Not sensitive, impractical

TABLE 17.9 Management of patients with various gastrointestinal disorders.

Pathology	Treatment
Duodenal/gastric ulcer	*Helicobacter pylori* eradication with combination of antibiotics and proton pump inhibitor
Neuroendocrine tumours	Surgical removal where no metastases have occurred; otherwise treatment with octreotide, a somatostatin analogue which suppresses hormone secretion
Chronic pancreatitis/exocrine pancreatic disease	Replacement of exocrine pancreatic enzymes with pancreatin to restore digestion; pain control with analgesics
Gallstones	Surgery, usually keyhole, which involves very small incisions (laparoscopic cholecystectomy)
Coeliac disease	Diet control; patient put on a gluten free diet and compliance monitored with regular coeliac serology
Small bowel bacterial overgrowth	Antibiotics and treatment of predisposing cause where possible
Crohn's disease	Biological therapies, analgesics, anti-inflammatory agents, elemental diets
Ulcerative colitis	Biological therapies, anti-inflammatory agents, analgesics
Lactose malabsorption	Reduce or exclude foods containing lactose from diet
Bile acid malabsorption	Bile acid sequestering agents such as cholestyramine bind bile acids out of solution
Irritable bowel syndrome	Stress reduction, behavioural therapy, dietary changes, antispasmodic drugs, anti-diarrhoeal drugs
Factitious diarrhoea	Psychological assessment

The range of treatments used include antibiotics (for example *Helicobacter pylori*, small bowel colonization); nutritional (for example exclusion of gluten in coeliac disease, and **elemental diets** in Crohn's disease); **biological therapies** and anti-inflammatory agents (Crohn's disease, ulcerative colitis), **antispasmodic** drugs for IBS; surgery for removal of gallbladder (cholecystectomy), and ileum/colon (for Crohn's disease, ulcerative colitis or cancer), as appropriate.

CASE STUDY 17.2

A 50-year-old woman was admitted to a medical ward with a history of chronic profuse watery diarrhoea and muscle weakness. Standard biochemical investigations were normal except for the serum potassium measurement (reference range given in brackets):

Potassium   2.6 mmol/L   (3.6–5)

The diarrhoea did not improve with fasting. The stool volume was in excess of 1 L per day

(a) What type of diarrhoea does this patient probably have?

A diarrhoeal sample was centrifuged and sodium, potassium, and creatinine measured with results as follows:

Sodium   125 mmol/L
Potassium   31 mmol/L

Faecal osmotic gap of 290–312 = −22

Creatinine was negligible relative to urine concentration, indicating no urine contamination. This confirmed a secretory diarrhoea (FOG <75 mOsm/L). A urine laxative screen was negative on two separate occasions.

(b) What further tests should be undertaken to investigate this patient?

CASE STUDY 17.3

A 60-year-old man with a previous partial gastrectomy was referred by his doctor to the hospital. He had a recent history of chronic diarrhoea and some weight loss. The following tests gave results within their reference ranges: urea and electrolytes, liver function tests, calcium and albumin, free T4 and TSH, serum CRP, serum IgA TTGA, and faecal elastase.

(a) What conditions might this patient develop secondary to the partial gastrectomy?

Results for further tests are given below (reference ranges are given in brackets):

Haemoglobin   11.3 g/dL   (13.0–18.0)
Serum vitamin B12   180 ng/L   (>250)

Blood film showed a macrocytic anaemia. A glucose breath hydrogen test was undertaken and results were as follows;

Zero	2 ppm
+15 min	3 ppm
+30 min	8 ppm
+45 min	12 ppm
+60 min	+25 ppm
+75 min	+24 ppm
+90 min	+15 ppm
+105 min	+12 ppm
+120 min	+6 ppm

(b) Comment on these results.

CASE STUDY 17.4

A 38-year-old woman had a long history of abdominal pain, bloating, and bouts of diarrhoea followed by constipation but no weight loss. The following tests were all within their reference ranges: urea and electrolytes, liver function tests, free T4 and TSH, calcium and albumin, and CRP.

What further tests of gastrointestinal function should now be undertaken?

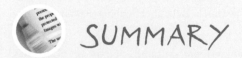

SUMMARY

- Normal gut function allows optimal digestion and absorption of nutrients and conservation of fluid and minerals.

- Disease within the gut can result in breakdown of normal homeostasis leading to inefficient digestion and absorption of nutrient with inappropriate loss of fluid and minerals.

- Laboratory tests of gastrointestinal function have been designed to measure gastric, exocrine pancreatic, jejunal, and ileal function, which are the main sites of digestion, absorption, and conservation.

- Several gastric tests now detect infection with the bacterium *Helicobacter pylori*, by far the major cause of duodenal, gastric ulcer, and ultimately atrophic gastritis and gastric cancer. Serum gastrin assays are of occasional value, where patients are *Helicobacter pylori* negative or where Zollinger Ellison syndrome is suspected.

- Pancreatic exocrine function tests investigate the major digestive capacity of the gut. These have evolved from the original 'tube' tests where tubes were swallowed, sited by radiography and used for aspiration of secretions, to non-invasive tests involving urine analysis, mostly now unavailable, to ELISA measurement of faecal elastase 1 which is stable throughout the gut and in faeces.

■ Bacterial colonization of the small bowel is usually secondary to stasis and can result in lipid maldigestion through bile acid deconjugation. The glucose hydrogen breath test is the most widely used test for diagnosis.

■ Coeliac disease with an incidence approaching 1 per 100 in the UK population is a major cause of jejunal dysfunction and can be virtually diagnosed with a high serum TTGA result. Lactose intolerance, also common, can in most cases be diagnosed by response to dietary restriction.

■ Investigation of chronic profuse watery diarrhoea should only be performed after establishing whether diarrhoea is secretory or osmotic in origin, by testing response to fasting or by calculating faecal osmotic gap. Tests for bile acid malabsorption, laxative abuse, or gut hormones can then be performed as appropriate.

■ Faecal calprotectin is evolving into a test which can be used to distinguish between organic and non-organic disease and as such may reduce the need for expensive colonoscopic tests on younger patients with lower gut symptoms.

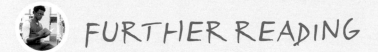

FURTHER READING

● **Ayling R and Marshall W (2007)** *Nutrition and Laboratory Medicine*. **London: ACB Venture Publications. pp. 79–108.**

A clear presentation of this topic.

● **Duncan A and Hill PG (2007) A review of the quality of gastrointestinal investigations performed in UK laboratories.** *Annals of Clinical Biochemistry*, **44**, 145–58.

This is a comprehensive article outlining the quality of the current gastroenterology biochemistry services available in the UK. It is written in an unequivocal manner and is highly recommended.

● **Field M (2003) Intestinal ion transport and the pathophysiology of diarrhoea.** *Journal of Clinical Investigation* **111**, 931–43.

Excellent comprehensive review of a complex subject.

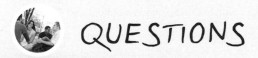

QUESTIONS

17.1 Which ONE of the following statements concerning gastric acid is correct?

  (a) It kills bacteria in the stomach

  (b) It reduces conversion of pepsinogen to pepsin

  (c) It digests carbohydrates

  (d) It synthesizes mucus in the stomach

  (e) None of the above

17.2 Which intestinal cells secrete sodium chloride solution and what does this facilitate?

17.3 What role do bile acids play in lipid digestion?

17.4 Briefly outline why you think faecal calprotectin is a useful diagnostic test.

17.5 What are the limitations of the faecal elastase test?

17.6 Where a patient presents with severe watery diarrhoea, which preliminary investigations should be carried out?

17.7 Where a patient suspected of having coeliac disease has a very low serum IgA TTGA result, what further tests should be carried out?

17.8 Why has FOB testing been discontinued in patients who have gastrointestinal symptoms?

17.9 What conditions predispose to bacterial colonization of the small bowel?

Answers to self-check questions, case study questions, and end-of-chapter questions are available in the Online Resource Centre accompanying this book.

Go to www.oxfordtextbooks.co.uk/orc/ahmed/

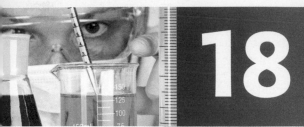

18

Specific protein markers

Gerald Maguire

Learning objectives

After studying this chapter you should be able to:

- List the principal proteins found in blood plasma
- Describe the functions of the principal proteins found in plasma
- Discuss the changes in the concentrations of major plasma proteins during disease
- Describe the principal methods used in the analysis of plasma proteins

Introduction

The plasma of blood is rich in numerous proteins that perform a variety of different functions. Figure 18.1 shows the numerous proteins that can be separated in a sample of plasma using the high resolution technique of two-dimensional (2D) electrophoresis.

The proteins present in plasma include a number of enzymes, transport proteins, protein hormones, cytokines, clotting factors, and complement proteins. This chapter will describe only a few of the individual proteins for which the measurement of their concentration has proved useful in the diagnosis and management of diseases. These proteins are sometimes called specific proteins, a rather illogical term given that all proteins have individual structures and specific function(s). This chapter will examine the various proteins in plasma and the value of measuring their concentrations in relation to disease investigation.

18.1 Plasma proteins

If you examine Figure 18.1, it is apparent that plasma has a complex composition. However, if plasma or serum (Box 18.1) is subjected to cellulose acetate electrophoresis, a technique with less resolving ability than 2D electrophoresis, fewer proteins are resolved and the results are

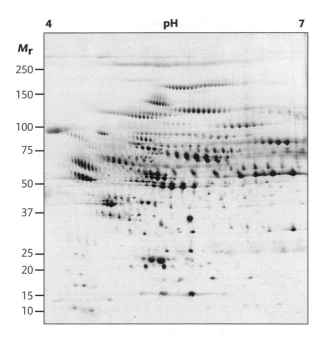

FIGURE 18.1

Two-dimensional electrophoresis of plasma. Each of the stained spots represents a plasma protein or one of a group of related plasma proteins. Courtesy of Dr PM Palagi, Swiss Institute of Bioinformatics, Geneva, Switzerland.

BOX 18.1 What's in a word: serum or plasma?

Plasma is the physiological fluid that cells of the blood float in. Serum is the fluid which is obtained after a blood sample has been allowed to clot. In practice most measurements of albumin are performed on serum rather than plasma.

much easier to interpret. Indeed many of the specific proteins mentioned in the Introduction are easy to distinguish (Figure 18.2).

Specific proteins comprise approximately 80% by weight of the 70 g/L of total protein in plasma. They are listed in Table 18.1.

With the exception of immunoglobulins, the liver is the principal site of the synthesis of all plasma proteins. Their concentrations in the plasma are maintained by a balance between the rates of synthesis and release from the liver and the speed at which they are degraded or lost. Changes in the concentrations of proteins in plasma can thus result from alterations in their rate of synthesis or their rate of loss.

The major cause of a change in the rate of synthesis is the **acute phase response**, which is a rapid and coordinated change in the concentrations of many plasma proteins (Table 18.1) that is instigated in situations where damage to host tissues can occur. Examples of these situations include infection, trauma, and a number of inflammatory diseases. The effect is to divert synthesis towards those plasma proteins that are required to cope with the new situation. The concentrations of acute phase proteins do not all increase and decrease in concert. Some, for example C-reactive protein (CRP), are part of the innate immune system and its concentration increases rapidly to a thousand-fold the concentration found in the absence of disease. It can also decrease just as rapidly. C-reactive protein has a direct antibacterial activity.

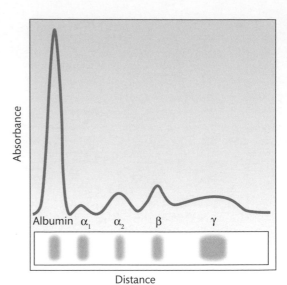

FIGURE 18.2

Cellulose acetate electrophoresis of a sample of serum. It is readily apparent that albumin is the major protein constituent. The α_1 fraction is mainly α_1-antitrypsin, while the α_2 fraction includes haptoglobin. The β globulin fraction includes a number of lipoproteins, transferrin, plasminogen, and complement proteins. The γ globulins are antibodies. These proteins are listed in Table 18.1.

Other acute phase proteins, such as α_1-antitrypsin (AAT) increase in concentration more slowly and are thought to function in preventing hydrolytic damage by proteases released at the site of inflammation. Albumin is a *negative* acute phase protein, that is, its concentration falls in an acute phase reaction. Figure 18.3 shows the changes found in one patient during an exacerbation of Crohn's disease an inflammatory condition of the intestine.

An increased loss of protein can occur in kidney and intestinal diseases. In these cases, the proportional loss of low M_r proteins is greatest. A less common cause of a lower concentration

TABLE 18.1 **The major proteins in the plasma.**

Protein	Normal concentration in serum (g/L)			Response to an acute phase reaction
	Mean	Lower limit	Upper limit	
Total protein	70	63	83	No effect
Albumin	40	32	48	Decrease
α_1-antitrypsin	1.5	0.9	1.8	Increase
α_1-acid glycoprotein (orosomucoid)	1.0	0.4	1	Increase
Haptoglobin	1.5	0.5	2.6	Increase
Caeruloplasmin	0.3	0.2	0.6	Increase
Transferrin	2.5	1.9	3.5	Decrease
Complement C3	1.0	0.8	2.14	Increase
Complement C4	0.3	0.13	0.6	Increase
IgG	10	6	13	No effect
IgA	2.0	0.8	3.7	No effect
IgM	1.5	0.4	2.2	No effect

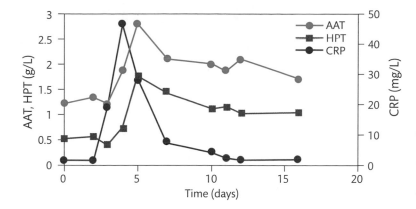

FIGURE 18.3

An example of an acute phase response in a patient suffering from Crohn's disease (abbreviations AAT, α_1-antitrypsin; HPT, Haptoglobin; CRP, C-reactive protein). Note the different scale for CRP.

of a plasma protein is a reduction in the rate of its synthesis by the liver. For example, in cirrhosis of the liver the capacity of the organ to synthesize proteins is reduced. However, for many specific proteins, genetic polymorphisms may be a major cause of reduced concentrations in plasma. The following sections will consider each of the major plasma proteins in turn.

Cross reference
Chapter 10 Disorders of calcium, phosphate, and magnesium homeostasis

18.2 **Albumin**

Albumin is the most abundant protein in blood plasma (Table 18.1). It consists of a single polypeptide of 584 residues with a M_r of 69,000. Albumin has two major functions: it is a transport protein *par excellence* and it also serves to maintain the **oncotic pressure** of the blood. Many sparingly soluble substances can bind to albumin (Table 18.2) and be transported in the plasma.

The poor solubility in water of some of these, for example bilirubin and fatty acids, is due to their hydrophobic nature. Others, such as bilirubin are toxic in free solution; binding to albumin renders them less harmful. Some are only physiologically active when free in solution. This includes calcium, whose active concentration is reduced on binding to albumin, as described in Chapter 10.

Albumin is largely responsible for the oncotic pressure of the blood, that is, the osmotic pressure of the plasma, which arises from the presence of proteins. The oncotic pressure prevents fluid leaking out of capillaries into the tissue fluid (Figure 18.4).

Albumin is synthesized and released directly into the blood at a rate that is sufficient to balance its loss by degradation and leakage into urine and other fluids. It is lost in small amounts into the urine but in health this loss is less than 30 mg a day.

A number of pathological conditions result in low concentrations of albumin in the blood. These conditions include:

- an acute phase reaction
- liver disease
- malnutrition
- malabsorption
- nephrotic syndrome
- protein losing enteropathy

TABLE 18.2 Compounds that bind albumin.

Fatty acids
Bilirubin
Drugs
Salicylate
Warfarin
Clofibrate
Phenytoin
Divalent cations
Calcium
Copper
Selenium
Zinc
Hormones
Thyroxine
Cortisol
Aldosterone

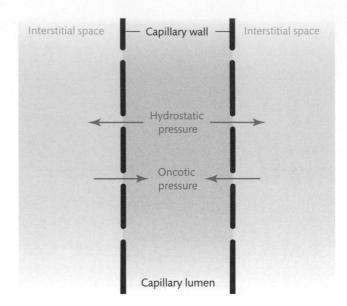

FIGURE 18.4

Leakage of capillary fluid. Contraction of the heart results in hydrostatic pressure, which pumps blood around the body. This pressure could also force the fluid across the capillary membrane. This is prevented by the oncotic pressure of the blood, which is mainly due to albumin.

Any disease that leads to an acute phase reaction will eventually result in a low plasma albumin concentration. However, this may take some weeks to occur and so will be observed only if the duration of the disease is prolonged. Albumin is synthesized in the liver, thus low plasma albumin concentrations are associated with liver disease. In prolonged starvation, the lack of protein in the diet and the consequent reduced supply of amino acids restricts the synthesis of albumin. The production of albumin also falls in malabsorption, where the problem is not a lack of protein in the diet but the absorption of amino acids by the gastrointestinal tract (GIT) following its digestion. In the nephrotic syndrome and protein losing enteropathy, proteins are lost in the urine or from the intestine respectively. The rate of loss is greater than the synthetic capacity of the liver and a low concentration of albumin in the plasma is the result.

A plasma albumin concentration lower than the lower limit of the reference range, that is, **hypoalbuminaemia** does not usually produce any symptoms unless it becomes very low indeed: less than 15 g/L. At this point, the oncotic pressure is insufficient to prevent fluid leaking from blood capillaries into the interstitial fluid and the result is oedema.

Clinical use of serum albumin measurements

A serum albumin concentration lower than the lower limit of the reference range is indicative of disease but is not diagnostic for any specific ailment. Nevertheless, the finding of a low concentration of albumin in a serum sample would prompt further investigations of its underlying cause (disease). In clinical laboratory practice, most determinations of serum albumin concentrations are made not for diagnostic reasons but to aid in the estimation of serum calcium. Similarly, its measurement can be useful in interpreting the concentrations of some drugs, for example, phenytoin a drug used in treating epilepsy, which can also bind to albumin. A value for the concentration of phenytoin in serum that has been adjusted for albumin concentration is a more accurate predictor of whether the epileptic fits will be controlled than one unadjusted.

Clinical use of urine albumin measurements

Albumin is the principal protein present in urine and is the one detected by dipstick tests for urine protein. Concentrations lower than about 30 mg/L can be found in the urine of healthy individuals but higher values are a feature of renal disease. The limit of sensitivity for urine dip stick testing for protein is 250 mg/L. However, values lower than 250 mg/L may be associated with disease, a condition termed **microalbuminuria**, which describes urine albumin concentrations 30–250 mg/L. Measuring the concentration of urine albumin is particularly useful in monitoring diabetes. Diabetics are susceptible to a number of complications, one of which is kidney disease. This is referred to as diabetic nephropathy. The finding of microalbuminuria in a diabetic is an indication of an increased risk of diabetic nephropathy. Diabetics therefore have regular urine albumin measurements made and when microalbuminuria is found, appropriate therapeutic measures must be taken to delay its progression.

Urine output varies to a great extent depending on the amount of fluid a particular individual drinks. The concentration of albumin in urine will reflect this, being lower if urine output is high (the urine is dilute) or high when urine output is low (the urine is concentrated). To compensate for this the albumin: creatinine ratio is used. The urine creatinine concentration reflects urine flow as the creatinine output per day of an individual is quite constant. An alternative is to measure the albumin output in a 24-hour urine collection by multiplying the albumin concentration by the urine volume. However, collecting urine for 24 hours is a lot less convenient for the patient.

SELF-CHECK 18.1

A patient with cirrhosis of the liver has an albumin concentration of 25 g/L. Is this likely to cause oedema?

18.3 Alpha 1-antitrypsin

The serine protease inhibitor AAT has an M_r of approximately 52,000. Inhibitors of serine proteases are often called serpins. Serine proteases get their name because they contain a particularly reactive serine residue in their active site, which is essential for enzymic activity. All serine proteases hydrolyse other proteins to form smaller fragments. Over 200 serine proteases are known and they participate in numerous physiological activities, including digestion, clotting, dissolution of clots, control of blood flow, lung elasticity, cell lysis in immune responses, and spermatozoa penetration of the ovum. Thus, for example, the serpins antithrombin III and C1 esterase inhibitor are necessary for the control of coagulation and complement activation respectively. The inhibitory activity of AAT was first observed when blood was found to inhibit the action of trypsin. However, the natural substrate of AAT is elastase, an enzyme released from neutrophils. Release occurs when neutrophils are activated at the site of an infection. Thus AAT serves to restrict the activity of elastase to sites near the centre of infection.

The mechanism of action of serpins had been likened to that of a mousetrap. The 'bait' is a reactive centre in the inhibitor which has a unique amino acid sequence that confers specificity for the complementary serine protease. The sequence in AAT is Met-Ser-Ile-Pro-Pro-Glu. The reactive centre docks with the active site of the enzyme and is hydrolytically cleaved by it. However, in the case of serpins, the products are not released from the active site and the serpin remains covalently attached to the enzyme. This cleavage results in a conformational change of the serpin in which a loop containing the active centre (with the protease still attached) swings out and distorts and denatures the protease against the opposite end of the serpin.

There are many genetic variants of AAT which are inherited in a Mendelian fashion (see Box 18.2). Most variants result in normal concentrations of the serpin in plasma and are not associated with disease. Of the common variants, only S and Z are associated with low plasma levels of AAT and may predispose to liver and/or lung disease. The general features associated with each are shown in Table 18.3.

In the Z variant, a lysine residue is substituted for a glutamate at position 342. This alters the structure of the serpin opening its main β sheet. The reactive loop of another Z molecule can become inserted into the β sheet. This process can repeat, allowing the Z molecules to stack up on one another forming an aggregate of protein molecules. Aggregation occurs in the endoplasmic reticulum of liver cells and prevents the newly synthesized protein from reaching the circulation and thus low concentrations (15% of normal) result. Indeed, the levels are so low that it may not be detectable by serum protein electrophoresis (Figure 18.5) and the Z phenotype is often referred to as AAT deficiency. The aggregation and precipitation of serpins in the cell causes liver disease. Whether this mechanism also causes lung disease or whether

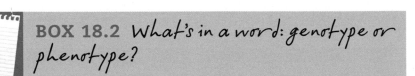

BOX 18.2 What's in a word: genotype or phenotype?

We can use the word genotype when we know what variant each of the AAT genes contain. Often we do not know this. If we observe only M protein, this could be the result of two genotypes M or M null. Thus we refer to the M phenotype. In the heterozygous state we can infer the genotype from the phenotype, thus the MZ phenotype can only result from the MZ genotype.

TABLE 18.3 Common variants of α_1-antitrypsin.

AAT variant	Percentage of population	Mean serum concentration (g/L)	Risk of liver disease	Risk of lung disease
M	90	1.5	No increased risk	No increased risk
MZ	3	0.8	Small increase	No increased risk
Z	<1	<0.3	Large increase	Large increase
S	<1	0.8	No increased risk	No increased risk
MS	8	1.0	No increased risk	No increased risk
SZ	<1	0.6	Small increase	Small increase

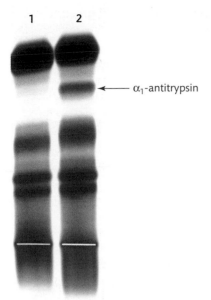

α_1-antitrypsin

FIGURE 18.5
Serum protein electrophoresis, where lane 1 is from a patient with α_1-antitrypsin deficiency and lane 2 is from a healthy patient.

it is due to the low level of serpins circulating that allows unrestricted action of neutrophil elastase is unclear.

Liver disease associated with the ZZ phenotype may be one of two types. About 15% of individuals present with neonatal hepatitis. This resolves in about half of those affected, but 30% progress to cirrhosis and death by the age of 20 years. Lung disease usually presents in early middle age as **emphysema**. This is a chronic condition in which alveoli are destroyed leading to shortness of breath. Not every individual with the Z variant progresses to overt lung disease. Factors other than AAT deficiency contribute. The most significant being tobacco smoking, which doubles the rate of decrease in lung function. The phenotype of AAT deficiency cannot always be deduced from measuring its concentration in blood alone. Normal levels may be found in MZ individuals during an acute phase response. To identify the phenotype, isoelectric focusing (IEF) is required (Figure 18.6).

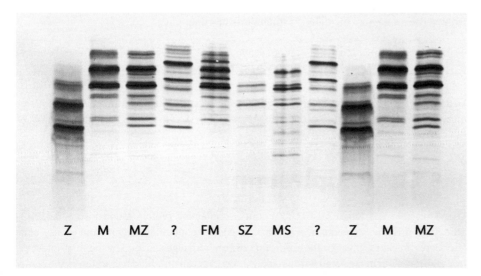

Z   M   MZ   ?   FM   SZ   MS   ?   Z   M   MZ

FIGURE 18.6
Photograph showing an isoelectric focusing gel with different α_1-antitrypsin variants.

Clinical use of α_1-antitrypsin measurements

Measurement of AAT is conducted in babies presenting with **neonatal jaundice** to investigate any possible deficiency, and in patients with liver disease or emphysema. No specific treatment is available for AAT deficiency other than advice on stopping smoking and treating the resulting organ damage.

SELF-CHECK 18.2

A baby presents with neonatal jaundice and the serum AAT concentration is 0.8 g/L. (a) Is the jaundice likely to be due to AAT deficiency? (b) How would you confirm your suggestion?

18.4 Haptoglobin

Haptoglobin is a haemoglobin-binding protein found in plasma. Haptoglobin is a tetramer, that is, it consists of four subunits: 2α and 2β polypeptides. The situation is slightly complicated in that there are two types of α subunts, α_1 of M_r 9,000 and α_2 with an M_r of 17,300 but only a single type of β with an M_r of 40,000. These can combine to form three subtypes of haptoglobin designated 1.1 ($2\alpha_1 + 2\beta$), 2.1 ($2\alpha_2 + 2\beta$) and 2.2 ($\alpha_1\alpha_2 + 2\beta$). The proportions of each show remarkable racial differences.

Haemoglobin is found at a high concentration in blood of 130 g/L. However, in normal circumstances, all of the haemoglobin is packaged within erythrocytes, with little occurring *free* in plasma. Free haemoglobin in plasma is dangerous as it can precipitate in kidney tubules and damage them. A number of diseases occur in which erythrocytes can lyse within the blood vessels: a process called **intravascular haemolysis**. For example, the inherited condition glucose 6-phosphate dehydrogenase deficiency causes erythrocyte fragility, which leads to intravascular haemolysis and the associated clinical problems. However, the free haemoglobin complexes with haptoglobin to form a haemoglobin-haptoglobin complex that is subsequently removed from the circulation when it binds to specific receptors on the surfaces of macrophages.

Inherited deficiency states of haptoglobin leading to disease are unknown. Studies have shown both positive and negative associations of haptoglobin subtypes with some diseases. However, the data from the different studies is conflicting.

Clinical use of haptoglobin measurements

Determining the concentration of haptoglobin can be useful when investigating anaemia or unexplained jaundice. In both cases, a low haptoglobin (less than 0.3 g/L) suggests that haemolysis may be the cause.

18.5 Caeruloplasmin

Caeruloplasmin is a large (M_r 132,000) copper-containing protein found in plasma. In its pure form, it is blue in colour as a result of its copper content: *caerulo* is Latin for sky blue. Caeruloplasmin is essential for the absorption of iron from digested food by the small intestine. Iron is absorbed in the reduced ferrous ($Fe(II)^{2+}$) state but can only bind to transferrin, the *carrier* of iron in plasma (see below) in the oxidized ferric (Fe(III)$^{3+}$) state. Caeruloplasmin is a ferro-oxidase which is responsible for this conversion. Caeruloplasmin contains over 90% of the copper content of the plasma; one of its functions may be to transport copper around the body.

The pathology of caeruloplasmin is intimately related to the pathology of copper. Low concentrations of which are associated with two rare inherited diseases of copper metabolism: Wilson's and Menkes diseases. There are no primary inherited deficiency states of caeruloplasmin but there are some inherited diseases which can lead to a secondary deficiency of caeruloplasmin. **Wilson's disease** results from the deficiency of a copper transport protein that affects both the transport of copper out of the liver into bile and the incorporation of copper into caeruloplasmin. The result is an accumulation of copper within the liver and low concentrations of caeruloplasmin in the plasma. The accumulated copper in the liver leaks out in a free form, that is, *not* within caeruloplasmin, and it is deposited at various sites in the body, the principal ones being the liver and brain. Thus Wilson's disease can present with both liver and neurological disease. Treatment of Wilson's disease is with copper chelating agents such as penicillamine and liver transplantation.

Menkes disease is an X-linked inherited condition in which copper absorption from the small intestine is impaired, leading to low concentrations of copper in the plasma. This deficiency impairs the activities of copper containing enzymes and results in neurodegeneration and death within the first decade of life. There is no effective treatment for Menkes disease.

Clinical use of caeruloplasmin measurements

The only use for caeruloplasmin measurement is in the investigation of possible cases of Wilson's or Menkes disease. When its concentration is determined, it is best accompanied by the measurements of copper concentration. In Menkes disease, concentrations of copper and caeruloplasmin in serum samples are both pathologically low. In Wilson's disease, the serum copper and caeruloplasmin concentrations are usually (but not always) low. In contrast, concentrations of copper in solid body tissues are always high.

18.6 **Transferrin**

Transferrin (M_r approximately 77,000) is an iron transport protein found in plasma. Each molecule of transferrin is able to bind two molecules of iron, which must be in their oxidized ferric state. There are more than 30 genetic variants of transferrin but little evidence that these are functionally different.

The concentration of transferrin in plasma is about 2 g/L, so it can bind about 50 µmol/L of iron. This is about three times the normal iron concentration in blood (about 20 µmol/L). The affinity of transferrin for iron is so high that free iron is not found in the blood. The amount of transferrin (or its concentration) can be estimated by determining the iron-binding capacity of a clinical sample of serum: there is an exact correlation between the concentration of transferrin in serum and the iron binding capacity.

Only eight patients worldwide have been reported to suffer a transferrin deficiency. The major feature in these individuals was iron deficiency anaemia. The pathology of transferrin is bound up with the pathology of iron. In iron deficiency, the liver synthesizes increased amounts of transferrin, therefore its concentration in serum increases. In iron overload states the transferrin concentration does not change but its percentage saturation, that is, the amount of iron bound is increased.

Transferrin is a negative acute phase protein: a transferrin less than the lower reference limit (less than 2 g/L) may be a non-specific feature of disease.

Clinical use of transferrin measurements

Measurements of serum transferrin are only of use in calculating the transferrin saturation value, which is a sensitive test for iron overload.

$$\text{Transferrin saturation (\%)} = 4 \times \text{serum [iron] } (\mu mol/L) / \text{serum [transferrin] } (g/L)$$

Values over 45% indicate an iron overload. When investigating the iron status of a patient, however, the most appropriate single test is determining the concentration of ferritin in serum. The concentration of plasma ferritin increases in iron overload but is reduced in iron deficiency.

SELF-CHECK 18.3

A patient with slightly abnormal liver function tests was referred to hospital for investigation. His serum transferrin concentration was 2.0 g/L and his serum iron concentration was 30 µmol/L. Is iron overload a likely cause of his liver disease?

18.7 C-reactive protein

C-reactive protein is an acute phase protein whose concentration increases dramatically (up to one thousand-fold) following an inflammatory event and falls rapidly, returning to normal values within a few days of the end of the inflammation. C-reactive protein was discovered as the active agent that could precipitate a cell wall component of pneumococci called fraction C. This binding is dependent on binding of CRP to phosphocholine, which is found in many bacterial and fungal polysaccharides.

C-reactive protein consists of five subunits with a combined M_r of 105,000. Binding to susceptible organisms is restricted to one site of the protein. Once bound to bacteria, CRP activates the immune system to attack and, it is hoped, destroy the bacteria.

There are no deficiency states described for CRP but the gene for the protein shows considerable polymorphism. Over 40 different single nucleotide polymorphisms have been described and, in some instances, these correlate with concentrations of CRP in serum samples. The clinical relevance of these polymorphisms is yet to be established.

Clinical use of C-reactive protein measurements

Given its functional properties, CRP is an ideal marker of inflammation. Indeed, the most significant reason for determining the concentration of CRP in clinical samples is to assess the inflammatory response. After albumin, it is arguably the protein most often investigated in clinical laboratories. It is used clinically to assess infection, to detect and monitor acute and chronic inflammatory conditions and, more recently, it is used along with other markers to assess the risk of cardiovascular disease. Figure 18.7 shows the changes in concentration of CRP in an acute (myocardial infarction) and a chronic (Crohn's disease) condition.

The assessment of coronary risk require assays for CRP that are highly sensitive and are able to accurately measure concentrations as low as 0.5 mg/L. The relationship between concentrations of CRP and the risk of a coronary event is demonstrated in Table 18.4. An individual in

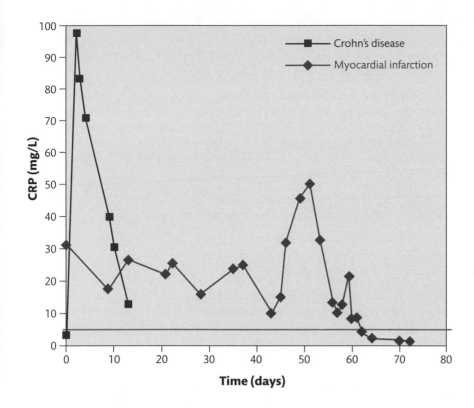

FIGURE 18.7
Changes in C-reactive protein in an acute myocardial infarction and a chronic (Crohn's disease) condition. The orange line indicates the upper limit of normality.

TABLE 18.4 Risk of a coronary event in relation to the concentration of CRP.

Median CRP in each quartile (mg/L)	Relative risk	Confidence interval of relative risk
0.6	1.0	
1.9	2.1	1.0–4.5
3.8	2.1	1.0–4.4
8.5	4.4	2.2–8.9

the highest quartile for CRP concentration has about four times the risk of a coronary event compared to a normal healthy individual.

Unfortunately, although it is possible to estimate the increased risk, unlike the situation with cholesterol, there are no medical interventions to reduce the concentration of plasma CRP. Thus, the use of serum CRP measurements to assess coronary risk is not routine.

18.8 **Immunoglobulins**

The immune system is a complex and integrated system of organs, tissues, cells, and cell products, including antibodies and complement proteins. Specialized organs and tissues of the system include lymph nodes, spleen, and bone marrow. In blood, the immune system consists of white blood cells (neutrophils, lymphocytes, and monocytes), antibodies, and another group of plasma proteins called complement proteins. Antibodies or immunoglobulins (Box 18.3.) and complement proteins are circulating soluble proteins. The immune system recognizes and fights foreign substances and cells, and so protects the body against diseases by killing pathogens such as viruses, bacteria and fungi, and tumour cells.

An understanding of the mechanism of **antibody** action and production is needed to understand the diseases with which they are associated. The molecule to which the antibody binds is called an **antigen**. The binding site on the antibody is only large enough to bind to a small part of the surface of its cognate antigen and is roughly equivalent to the area occupied by eight amino acid residues. The site on the antigen to which the antibody binds is called an **epitope**.

Immunoglobulins are divided into five classes: IgG, IgA, IgM, IgD, and IgE. Of these, only IgG, A, and M are present in plasma at appreciable concentrations (Table 18.5).

All immunoglobulin molecules consist of the same basic unit of two heavy and two light chains (Figure 18.8). The heavy chain consists of four domains, three of which are constant and one that is variable. The light chain consists of two domains, one is constant and one variable.

The heavy and light chains associate to form three functional sections: two 'arms' each containing a binding site that is unique for a specific epitope and a 'tail' that is common to all immunoglobulins within a class.

BOX 18.3 What's in a word: antibodies or immunoglobulins?

These words antibodies or immunoglobulins are often used interchangeably. They refer to the same molecules. 'Immunoglobulin' is often used when referring to their structure and 'antibody' when referring to their function. The word 'antiserum' refers to serum that contains antibodies. For example, if a rabbit is injected with human IgG, it will develop antibodies against human IgG and a serum sample from it is referred to as an antiserum against human IgG.

TABLE 18.5 Classes of immunoglobulins.

Class	Approximate concentration in serum (mg/L)	No. of Ig subunits	No. of heavy chains	No. of light chains	Approximate M_r	Adult reference range (g/L)
IgG	10,000	1	2	2	150,000	6.0–13.0
IgA	2,000	1 or 2	2 or 4	2 or 4	150,000 or 300,000	0.8–3.7
IgM	1,000	5	5	5	900,000	0.4–2.2
IgD	50	1	2	2	170,000	
IgE	0.1	1	2	2	190,000	

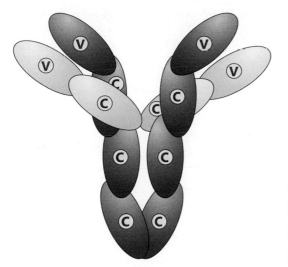

FIGURE 18.8
Structure of an IgG molecule. There are two heavy chains (shown in red) each consisting of four domains and two light chains (shown in green) each consisting of two domains. C and V refer to constant and variable respectively.

Antibodies have three ways of protecting an individual from a potential invading organism such as a virus or a bacterium (Figure 18.9).

Antibodies are made by specialized cells called plasma cells. Each individual plasma cell makes antibodies that are specific to a single epitope. A group of identical plasma cells will make and secrete an identical immunoglobulin molecule, and is called a **clone**. The body contains many different clones of plasma cells. Normally, the immunoglobulins in plasma are therefore referred to as **polyclonal**, given that they arise from many clones (Figure 18.10). Collectively, immunoglobulins have the capacity to bind to a vast number of different epitopes on protein and carbohydrate molecules.

Plasma cells are derived from B lymphocytes which circulate around the body. Each B cell has a unique receptor on its cell surface, which is a surface bound form of IgM. The receptor has the

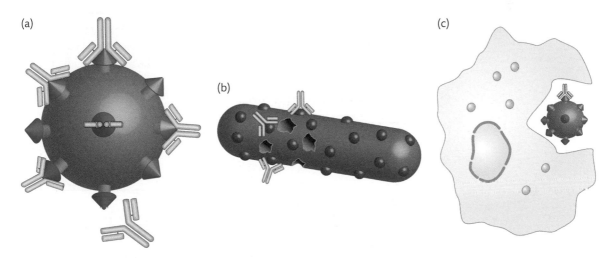

FIGURE 18.9
The ways in which an antibody can protect against infection. (a) Masking a surface molecule of a virus can prevent its entry. (b) Antibody binding to a bacterium can activate the complement system which punches holes in the bacterial cell membrane. (c) Antibody binding can 'tag' the invader, marking it out for destruction by neutrophils.

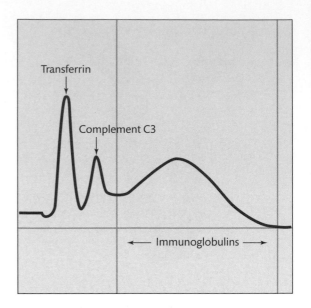

FIGURE 18.10

Detail from an electrophoretogram of a normal serum sample showing the polyclonal nature of immunoglobulins. The immunoglobulin peak is very broad as it consists of many different proteins each with a different amino acid sequence and therefore with a different mobility. This should be contrasted with the sharp peaks of transferrin and complement C3 which each consist of a single protein.

same binding site as the antibody that the plasma cell derived from the B lymphocyte is destined to produce. Following an encounter with a foreign antigen in the spleen or lymph node, B lymphocytes differentiate and multiply further, first producing plasma cells that make IgM and then 'class switching' to produce plasma cells making IgG or IgA (or IgE). The plasma cells making IgM remain in the lymph node in which they were first produced, whereas the plasma cells that make IgA and IgG migrate elsewhere: IgG plasma cells mainly to the bone marrow and IgA plasma cells mainly to intestinal lymphoid tissue. In a normal individual, plasma cells make up less than 5% of the total cells within the bone marrow. The bone marrow is also responsible for the production of red blood cells, platelets, and leukocytes, and the majority of cells within the bone marrow are blood cells in different stages of differentiation.

A foetus *in utero* is unable to make immunoglobulins and, indeed, the immune system is immature at birth. Any immunoglobulins a baby possesses at birth were transferred from the mother during development in the womb. Gamma immunoglobulins are the only type that can cross the placenta and at birth are the only class present in babies. The circulatory half-life of IgG is about six weeks so its concentration decreases until the baby's own immune system develops and begins synthesis. Immunoglobulins do not reach adult levels until the age of about one year for IgM, two years for IgG, and eight years for IgA (Figure 18.11).

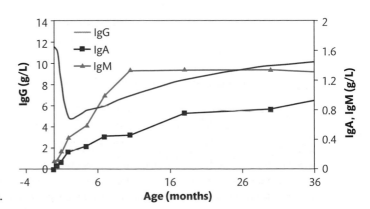

FIGURE 18.11

Development of immunoglobulins with age.

Immunoglobulin deficiency

Immunoglobulin deficiency leads to susceptibility to infections; particularly those caused by bacteria. Most commonly, immunoglobulin deficiency is secondary to another disease, either ones that interfere with antibody production, such as leukaemia or lymphoma, or those causing an excessive loss of immunoglobulins, for example nephrotic syndrome when an excessive loss of proteins occurs in the urine. Immunodeficiency can also affect the cellular arm of the immune system without affecting the production of antibodies. This occurs in acquired immunodeficiency syndrome (AIDS); indeed, the concentrations of immunoglobulin are generally high in patients with AIDS.

Primary immunodeficiencies (PID) are much less common and collectively affect about 1 in 10,000 of the population. This is on a par with phenylketonuria. Over 100 different PIDs with different characteristics have been reported. They can affect one or more arms of the immune system, including immunoglobulins, white blood cells, or complement. **Common variable immunodeficiency** (CVID) is a common immunodeficiency. It affects only immunoglobulin production and is characterized by recurrent bacterial infections. It usually presents at ages from the late teens to early 30s. It can be treated by regular infusions of immunoglobulins. **Bruton's disease**, named after its first describer, is a rarer form of immunoglobulin deficiency, which presents with recurrent infections in boys from about the age of six months when the maternally transferred IgGs have been lost. Bruton's disease is caused by the lack of an X-linked gene necessary for normal B lymphocyte development.

SELF-CHECK 18.4

A five-month-old boy is being investigated following an episode of bacterial meningitis. His serum immunoglobulin concentrations are as follows: IgG 3.0 g/L, IgA 0.2 g/L, and IgM 0.2 g/L. Do these immunoglobulin concentrations suggest that the boy may have an immunodeficiency?

Immunoglobulin excess

A normal response to infection is to produce immunoglobulins to fight the infection so concentrations of immunoglobulins in the blood are often increased in bacterial infections. Indeed, in areas of endemic infection, the 'normal' concentrations of immunoglobulins in samples of sera are often higher than areas with low infections. As immunoglobulins are polyclonal in nature, this increase is polyclonal. Increases in the polyclonal concentrations of serum immunoglobulins are also seen in a number of inflammatory conditions, such as **rheumatoid arthritis** and **systemic lupus erythematosus** (SLE), and particularly liver disease. A polyclonal increase in IgM is characteristic of the liver disease, primary biliary cirrhosis, and IgA is often raised in alcoholic liver disease. However, these changes are not specific to any particular disease and the values do not change predictably during the course of disease.

As discussed above, a normal increase in serum immunoglobulins is polyclonal. However, if one particular clone of plasma cells grows large, then it will synthesize and secrete a correspondingly large amount of a single immunoglobulin. This particular immunoglobulin will therefore be described as **monoclonal** (Figure 18.12).

The presence of monoclonal antibodies (and other proteins) in serum samples can be detected as 'bands' using electrophoretic techniques. However, the presence of a monoclonal band in

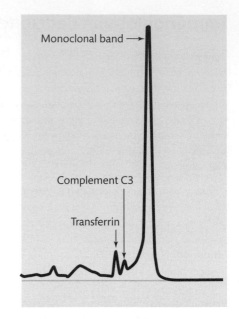

Monoclonal band →

Complement C3

Transferrin

FIGURE 18.12
Detail from an electrophoretogram of
a serum sample showing a monoclonal
immunoglobulin band.

a sample of serum need not necessarily be of any clinical significance, although a number are associated with clinical conditions, as you can see in Table 18.6.

When a serum monoclonal band is found in the absence of overt disease, it is called a monoclonal band of undetermined significance (MGUS); these are common and are found in 1% of the over 50s and 3% of those over 70 years. However, approximately 1% per year of patients showing an MGUS progress to malignant disease (either **multiple myeloma** or **Waldenstrom's macroglobulinaemia** (WM)), so annual follow-ups are required.

Multiple myeloma affects about 50 per million of the population. In about 80% of cases, the excessive antibodies are IgG; in about 10% of cases they are IgA, and in about 10% only light chains are produced (light chain myeloma). As light chains have an M_r of about 25,000, they are freely filtered at the glomeruli and are lost into the urine. Urinary free light chains are called **Bence-Jones protein** (BJP) after their nineteenth-century discoverer. In about 1% of cases, the myeloma may not produce heavy or light immunological chains (non-secreting myeloma). Free light chains are present at a low concentration in normal serum but can be present at high concentration in myeloma.

TABLE 18.6 Monoclonal gammopathies.

Condition	Ig affected	Features
MGUS	G, A, M	Symptomless
Multiple myeloma	G, A (rarely D, extremely rarely M, E)	Bone pain/lesions Anaemia Hypercalcaemia Renal insufficiency
Waldenstrom's macroglobulinaemia	M	Enlarged liver and spleen Anaemia, hyperviscosity
Solitary plasmacytoma	G, A	Bone pain/lesions

Waldenstrom's macroglobulinaemia affects about 10 per million of the population. It is associated with the presence of a monoclonal band of the IgM type in the serum. Clinically, WM is a very different disease to multiple myeloma because the clones of IgM producing cells reside in the lymph nodes and not in the bone marrow. Thus bone lesions are not found in WM patients.

A **plasmacytoma**, unlike multiple myeloma which affects multiple bones, affects only a single site. As there is only a small mass of cells, antibody production is low and the monoclonal bands are usually small and in about half of affected individuals may be too small to detect.

The diagnostic criteria for the different **monoclonal gammopathies** have been defined by an international myeloma group (Table 18.7). The presence of BJP in the urine although characteristic of myeloma can be found in MGUS.

Myeloma related disease is a consequence of the proliferation of the abnormal clone. Occupation of the marrow space by the myeloma suppresses the production of the other marrow cells including erythrocytes, leading to anaemia and also normal plasma cells leading to suppression of the production of normal immunoglobulin with consequent susceptibility to bacterial infection. The abnormal amount of free light chains is damaging to the kidney, leading to renal impairment, and extension of the marrow space leads to lytic lesions in bone, susceptibility to fractures, and hypercalcaemia.

So far we have discussed immunoglobulins in serum but they are also present in other bodily fluids including the cerebrospinal fluid (CSF). Most of the proteins in CSF, including immunoglobulins, are derived from plasma. Some proteins are made in the central nervous system (CNS). Immunoglobulins can be made in the CNS in response to CNS infections and also in an autoimmune condition called **multiple sclerosis**. Thus the presence in CSF of locally derived immunoglobulin is a marker for CNS infections and multiple sclerosis. The symptoms of CNS infection are quite different from those of multiple sclerosis, so usually there is no diagnostic confusion. Locally produced IgG in the CNS is oligoclonal, that is, consisting of many individual bands (Figure 18.13).

Clinical use of immunoglobulin measurements

Measuring the concentrations of immunoglobulins in samples of serum is useful in investigating suspected immunodeficiency, when, of course, low values are found. Susceptibility to bacterial infections is increased at low concentrations of IgG, particularly concentrations less

TABLE 18.7 Diagnostic criteria for diseases associated with monoclonal gammopathies.

MGUS	Asymptomatic myeloma	Symptomatic myeloma	Waldenstrom's macroglobulinaemia
Monoclonal band <30 g/L	Monoclonal band <30 g/L	Monoclonal band in serum or urine	IgM monoclonal band
<10% clonal plasma cells in bone marrow	>10% clonal plasma cells in bone marrow	Clonal plasma cells in bone marrow or plasmacytoma	Bone marrow infiltration of plasmacytoid/plasma cells
No myeloma related disease	No myeloma related disease*	Myeloma related disease	Characteristic surface staining pattern of lymphocytes

* Myeloma related disease includes anaemia, hypercalcaemia, renal impairment, bone fractures, lytic lesions, and bacterial infections.

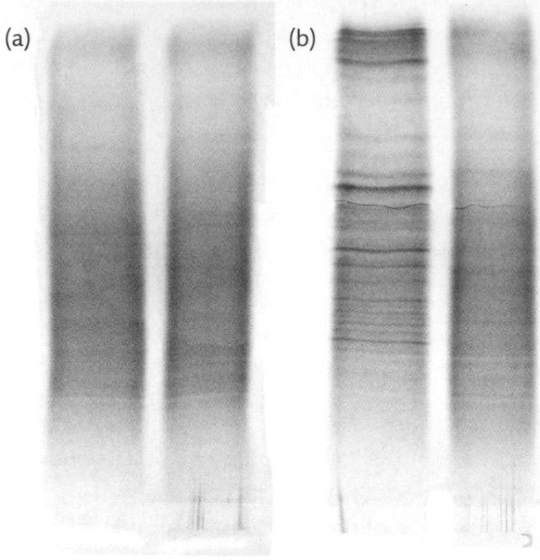

(a)   (b)

CSF   Serum   CSF   Serum

FIGURE 18.13

IgG oligoclonal banding in cerebrospinal fluid. Isoelectric focusing and staining for IgG in paired CSF and serum samples from: (a) A healthy patient. (b) A patient with multiple sclerosis. Note the bands in the CSF from the patient with multiple sclerosis.

than 3.0 g/L. IgA deficiency is relatively common (about 1 in 700) but in most patients is not associated with an increased susceptibility to infection. In immunodeficient patients being treated with immunoglobulin therapy, measurements of immunoglobulin concentrations are required to monitor its efficacy. If the concentration of IgG falls below 6.0 g/L, the treatment regime may need to be altered by increasing the amount or frequency of IgG infusions.

Immunoglobulin measurements are of some use when investigating liver disease and rheumatoid disease when high concentrations may occur. Repeated measurement in these conditions is not warranted as changes in concentration are not correlated with the clinical disease.

Immunoglobulin concentrations are determined in the investigation and management of monoclonal gammopathies. Despite their clinical differences, from the biochemistry laboratory perspective the investigation and monitoring of WM and myeloma are very much the same: tumour burden is reflected in the concentration of the monoclonal band. Usually this means monitoring the band in serum, but with light chain myeloma the concentration of BJP in urine is usually monitored. Alternatively, the concentration of the free light immunoglobulin chain in the serum may be measured.

18.9 Measuring concentrations of specific proteins

The concentrations of specific proteins in serum and urine can be measured using immunological and non-immunological techniques. The major non-immunological techniques are:

- dye-binding for albumin
- biuret for serum total protein

In addition, there are techniques to qualitatively assess proteins in serum and urine. These are electrophoretic methods using a variety of different support media (gels):

- agarose gel electrophoresis
- capillary zone electrophoresis

Cross reference

Chapter 14 Electrophoresis in the *Biomedical Science Practice* textbook

Immunological methods

The concentrations of most individual proteins are measured in a clinical laboratory using antibodies specific to them. There are a variety of immunological methods, but they all employ antibodies and rely on the formation of an insoluble antibody-antigen complex. Insoluble complexes only form at appropriate concentrations of antibody and protein (Figure 18.14). At relatively low concentrations of the protein compared to its antibody, the excess of antibody means there is little chance of an antibody molecule reacting with more than one protein molecule so a complex cannot form.

At high concentration of protein relative to that of antibody, the excess of protein molecules means there is little chance of a protein molecule reacting with more than one antibody molecule so, again, a complex formation cannot occur.

The major immunological techniques encountered in clinical laboratories are:

- radial immunodiffusion
- electroimmunoassay
- immunonephelometry
- immunoturbidimetry

We shall briefly discuss each in turn. In radial immunodiffusion (RID), antisera specific to the protein of interest is mixed with a solution of 1% agarose at 56°C. At this temperature, the agarose is liquid and can be poured onto a glass plate and as it cools allowed to set as a gel. Small wells are punched out of the gel and samples of serum are added to the wells (Figure 18.15).

The gels are incubated for one or more days. During the incubation, the protein of interest diffuses out of the well in all directions. Close to the well the protein concentration is high and is in excess of the antibody, so insoluble complexes are not formed. However, the protein in the sample becomes more dilute as it diffuses out from the central well. At some distance from the well, a concentration of protein occurs at which insoluble complexes can form. At this point, the protein stops diffusing. The higher the initial concentration of protein in the sample, the further it will need to diffuse before it reaches the point at which complex formation can occur. Thus, the diameter of the ring of precipitation will be greater. The gel is then washed

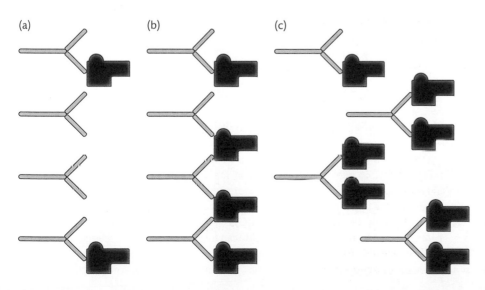

(a) (b) (c)

FIGURE 18.14
Effect of protein concentration on the formation of antibody-antigen complexes where: (a) Small amount of antigen = no insoluble complexes. (b) Large amount of antigen = insoluble complexes. (c) Very large amount of protein = no insoluble complexes.

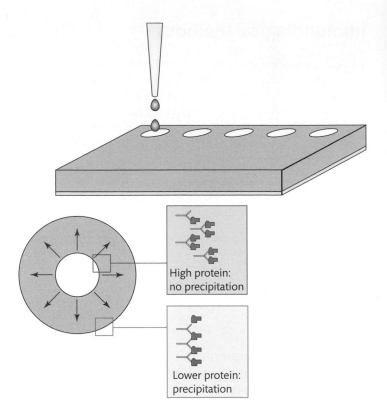

FIGURE 18.15
Schematic illustrating the principle of radial immunodiffusion. See text for details.

Labels within figure:
High protein: no precipitation
Lower protein: precipitation

to remove soluble protein and stained with a protein specific dye, usually Coomassie blue. The concentration of the protein in the samples can be estimated by constructing a standard curve. A series of standards of known concentrations are added to separate wells and the diameters (d) of the resulting rings measured. You can see an example of this in Figure 18.16. A graph of the square of the diameter (d^2) against the concentration of protein is then plotted. The concentration of protein in a sample can then be estimated from the graph. Radial immunodiffusion assays are still used for measuring the concentrations of some proteins, especially those which are infrequently determined.

SELF-CHECK 18.5

Use Figure 18.16 to calculate the concentrations of the protein in samples 1–15. The protein concentrations of the standards are as follows, St1 = 100, St2 = 50, St3 = 25, St4 = 12.5, St5 = 6.25 g/L.

In rocket electroimmunoassays, antisera specific to the protein of interest is, again, mixed with a solution of 1% agarose at 56°C. The agarose is then poured onto a glass plate and allowed to set. Wells are punched into the agar, and samples added to them. However, rather than allowing the proteins to simply diffuse into the agar, the plate is subjected to electrophoresis, usually for 30 to 60 minutes but occasionally for longer periods of time. The sample and standard concentrations of protein in the wells are negatively charged at physiological pH and migrate towards the positive electrode or anode. In each case, as the protein migrates from its well, it also diffuses in all directions. As with RID, as it diffuses sideways it will eventually reach a concentration at which an insoluble complex can form. Thus, there is a gradual loss of protein from the migrating mass and eventually this reduces its concentration to the level where it also forms an insoluble complex, so at that position electrophoretic migration will cease.

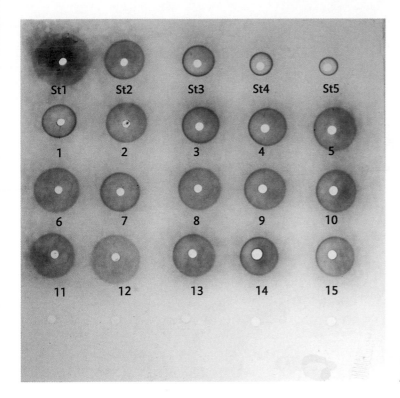

FIGURE 18.16
An example of a radioimmunodiffusion plate.

The more protein initially present in the well, the further the distance it will migrate before it is all precipitated. The protein from each well gives rise to a roughly rocket-shaped pattern of precipitation; hence the name of the technique. The height of each rocket is determined by the concentration of protein (Figure 18.17).

In immunoturbidimetric and immunonephelometric assays, an antibody specific to the protein of interest is diluted in buffer containing polyethylene glycol (PEG) and mixed with diluted patient sample (serum, plasma, urine, or CSF). If the concentration of antibody has been chosen correctly, complexes will form between it and the antibody. Polyethylene glycol is a protein precipitant and at a concentration of 60 g/L enhances the formation of complexes. The size and/or number of complexes formed will depend on the concentration of protein in the sample. Complex formation causes turbidity, which then absorbs and scatters light. The greater the concentration of protein, the greater the amount of light absorbed and scattered (Figure 18.18).

In immunoturbidimetric assays, the amount of light *absorbed* is measured. In immunonephelometric assays, it is the amount of light *scattered* which is determined.

One drawback of these methods is that at high protein concentrations, the protein will be present in such an excess that complexes cannot form. The concentration at which this occurs will differ with different assays, depending on the conditions used. The larger the concentration of antibody used, the higher the antigen concentration before it is present in excess ('antigen excess'), as shown in Figure 18.19.

However, the more antibody used, the more expensive the assay will be. If antigen excess has been reached even a high protein concentration will give a low turbidity and lead to an underestimation of the true concentration. For most proteins this is not a problem as the normally encountered range of protein concentrations, even in extreme pathological situations, is not great.

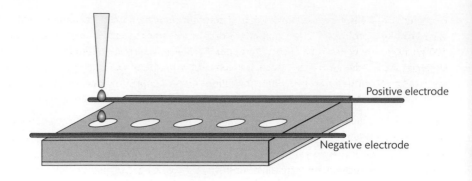

Positive electrode

Negative electrode

FIGURE 18.17
Schematic illustrating the principle underlying electro-immunoasay. See text for details.

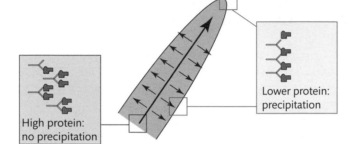

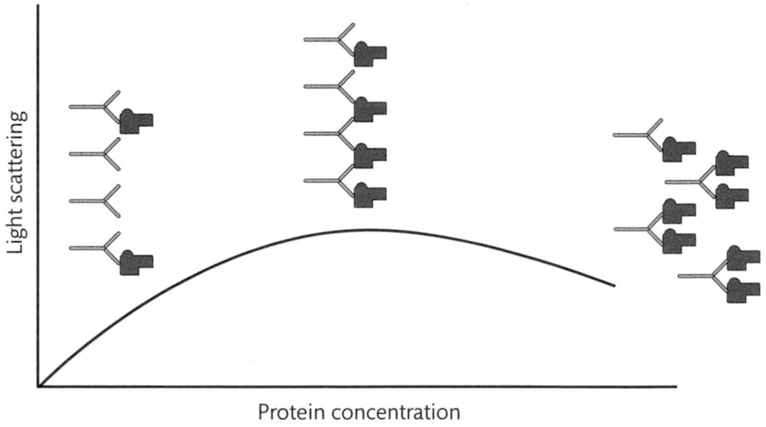

High protein: no precipitation

Lower protein: precipitation

FIGURE 18.18
Schematic illustrating the principles underlying immunoturbidimetric and immunonephelometric assays.

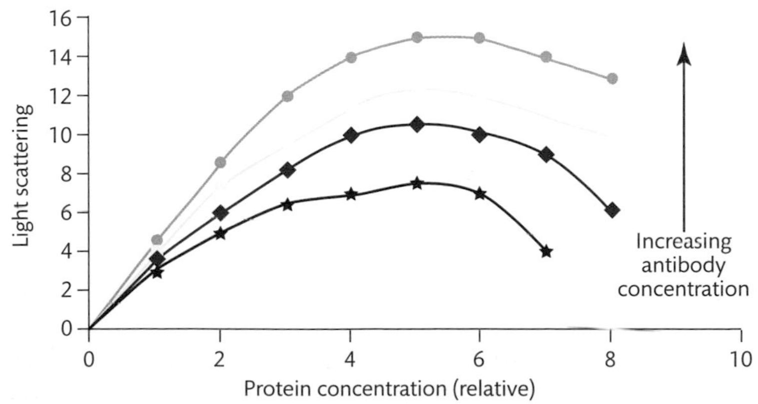

Light scattering

Protein concentration

FIGURE 18.19
Effect of increasing protein concentration on light scattering at different antibody concentrations.

Light scattering

Increasing antibody concentration

Protein concentration (relative)

However, if there is a large concentration of serum monoclonal immunoglobulins, antigen excess can be a problem. The concentrations of these proteins can sometimes be as great as 100 g/L in serum, which is fifty times the upper limit of normality for serum IgM. Obviously, strategies are needed to overcome this 'antigen excess' problem. One strategy is to measure the protein concentration at two different dilutions of the sample. Another is to add the sample in two stages. Most automated instruments are programmed to detect antigen excess.

The limit of sensitivity using turbidimetric and nephelometric techniques is about 10 mg/L. To increase sensitivity, the antibody used in detection can be coated onto a latex bead. This technique increases the size of the complexes formed and therefore increases their ability to scatter light. The subsequent increase in sensitivity means this technique can detect concentrations as low as 0.5 mg/L.

Non-immunological methods

Before the widespread availability of antibodies, the measurement of specific proteins was difficult as all proteins had essentially the same chemical behaviour. They had, therefore, to be measured on the basis of a property specific to the particular protein. Albumin could be measured on the basis of its property as a transport protein where dyes would bind to the transport sites and change colour upon so doing. Transferrin could be measured on the basis of its iron-binding properties, and caeruloplasmin on its oxidase activity. However, these methods (with the exception of dye-binding for albumin) are rarely used and will not be described.

Albumin in serum or plasma

The commonest methods used to determine the concentration of albumin in samples of serum rely on the change in absorption shown by certain indicator dyes when they bind to albumin (Figure 18.20). The two dyes principally used are bromocresol purple (BCP) and bromocresol green (BCG).

Both have the advantage that the binding to proteins occurs quickly (unlike antibody based methods) and the reagents are inexpensive. Both dyes are adequate for measurements in serum; however, they are not specific for albumin. Bromocresol green can bind to other blood proteins. This is not usually a problem as the interference is small. However, when the concentration of albumin is low, the interference can give a significant overestimation of the concentration of albumin in the sample and be clinically misleading. Bromocresol purple does not suffer from this problem but is more prone to negative interference, that is, it can

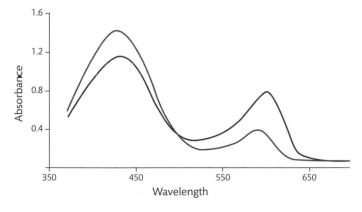

FIGURE 18.20

Spectrophotometric scan of bromocresol purple in the presence (red line) and absence (blue line) of albumin. Note the large increase in absorbance at 600 nm.

underestimate the true concentration of albumin in patients in renal failure. The most accurate means of determining the concentration of albumin are immunoturbidimetric or immunonephelometric methods but these are more expensive and, in practice, the BCP method agrees well with the immunologically based methods in all but a few patients.

Total protein in serum or plasma

Proteins differ in their amino acid compositions (indeed, each has a unique amino acid sequence). Thus, any method used to measure the concentration of total protein in a sample should be independent of amino acid composition. In effect, this means the method should detect only the peptide bonds in a protein. In addition, if it is to be used routinely, the method should be easily automated and inexpensive. In practice, only one method meets these requirements and is used universally for the determination of total protein in serum or plasma: the biuret method. The method gains its name because biuret (a product formed when urea is heated) has a similar structure to the peptide bond and reacts with alkaline copper solution in a similar manner to peptides and proteins to produce a violet coloured complex. The biuret method is not specific to proteins. Smaller peptides will react although they are not present in great concentrations in serum so this is not a big problem.

Total protein in urine

There are several methods in common use to determine the total protein concentration in urine. Unfortunately, none of them is ideal. The biuret method is not applicable to use with urine samples because it contains peptides at up to fifty times the concentration of its proteins. The biuret assay can be used if the proteins are precipitated using a reagent, for example 10 g/L of trichloroacetic acid (TCA), but which leaves the peptides in solution. The test can then be performed once the proteins have been resuspended in a buffered solution. However, this adds to the complexity of the assay and makes it difficult to automate.

Three methods for determining urine protein concentrations in common use rely on:

- pyrogallol red-molybdate (PRM)
- Coomassie brilliant blue (CBB)
- benzethonium chloride (BEC)

The first two are dye-binding methods, but in both cases, the dyes react to differing extents with different proteins. Thus, urine which contains 500 mg/L of albumin reacts differently to one that contains 500 mg/L of globulins. Hence, different results would be recorded for the total protein concentration in the urine sample.

The BEC method is based on turbidimetry. The proteins are precipitated in strongly alkaline solution in which the peptides are soluble. Hence, the turbidity formed is due only to the proteins; further albumin and globulins produce about the same amount of turbidity on a weight for weight basis. Thus, different results for the total protein concentration in urine samples can be obtained depending which method is used for analysis. It is therefore essential to use only *one* method to monitor a patient's urine protein concentration.

Protein electrophoresis

Electrophoresis is the movement of ions in an electric field. Protein molecules carry a net charge, with most proteins being negatively charged. Thus, when a sample of protein is placed in an electric field, it is attracted to the positive electrode and moves towards it. The rate of

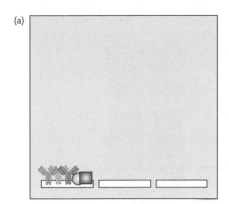

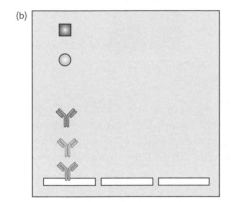

FIGURE 18.21
Schematic illustrating the principle underlying electrophoresis. (a) A mixture of proteins is added to a slit cut in an agarose gel. (b) After electrophoresis, the proteins are separated.

migration of the protein depends on its overall charge (the more negatively charged it is, the faster it will move) and its mass (the lighter it is, the faster it will move). Thus, the determining factor in how fast a protein moves is its mass-to-charge ratio. A mixture of proteins in an electric field will thus separate if they possess different mass-to-charge ratios (Figure 18.21).

Electrophoresis is normally performed in a support medium that minimizes the diffusion of the protein, which would lead to poor separation. The original media were paper, but this is no longer used, and cellulose acetate, which is still used in some circumstances (see Figure 18.2). Currently, the most common support media are gels of agarose or polyacrylamide. If electrophoresis is conducted in a tube with a tiny diameter, diffusion is not a problem so a support medium is not needed. This is the basis of a high resolution form of electrophoresis called capillary zone electrophoresis (CZE). Instruments are now available that fully automate the process of CZE.

In agarose gel electrophoresis, a small volume of the serum sample is added to a well cut into the gel and an electrical potential is applied across the gel for a fixed time. This process separates the proteins in the sample. The proteins are then fixed in position by precipitating them with strong acid and their positions on the gel located by staining them with a protein adherent dye, usually Coomassie blue, as shown in Figure 18.22.

This procedure separates the proteins in the serum into six major bands of proteins. However, when paper is used as the support medium, only four bands are observable. You may also wish to re-examine Figure 18.2, which shows the separation of serum proteins using cellulose acetate as a support medium. Since paper was the support medium originally used when separating serum proteins, the protein peaks were named on the basis of this system and

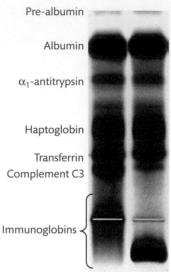

Pre-albumin

Albumin

α_1-antitrypsin

Haptoglobin

Transferrin
Complement C3

Immunoglobins

Monoclonal band

FIGURE 18.22
An example of agarose gel electrophoresis.

those names have been retained. The most rapidly migrating protein is albumin. The three other groups of proteins were referred to as α, β, and γ bands. The γ region contained the immunoglobulins, hence they are sometimes referred to as γ globulins. The major proteins of the α region are AAT and haptoglobin. The major proteins in the β region are transferrin and the complement protein, C3.

During CZE (Figure 18.23), the proteins migrate along a capillary tube and are detected by their absorbance at 214 nm as they pass an optical window at the end of the capillary.

Absorption at this wavelength is due to the presence of peptide bonds. Thus, the absorbance at 214 nm is proportional to the number of peptide bonds in the protein which, in turn, is proportional to the M_r of the protein (bigger proteins will have more peptide bonds). A graph of the absorbance at 214 nm versus time is automatically displayed and printed as seen in Figure 18.24.

The main clinical use of protein electrophoresis is to detect monoclonal immunoglobulin proteins (Figure 18.25). Once found, it is necessary to determine the concentration of the protein and to identify its immunoglobulin class.

Quantitation of monoclonal bands

If a monoclonal band has been identified on agarose gel electrophoresis, its concentration can be determined by densitometry. The gel is placed in a gel scanner which moves it through a narrow beam of light and the amount of light transmitted through the gel measured.

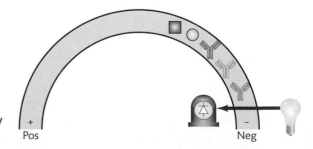

FIGURE 18.23
Schematic diagram of capillary zone electrophoresis.

Pos

Neg

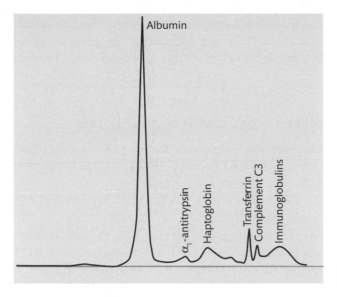

FIGURE 18.24
Example of capillary zone electrophoresis of normal serum.

More light is absorbed and therefore less transmitted when the beam passes through a band on the gel; the denser the band, the more light is absorbed. The output of a densitometric scanner is similar to that of a CZE trace. If the amount of dye bound to the protein is proportional to the quantity of protein present in the band, then the amount of light absorbed (equivalent to the area under the peak of the densitometric trace) will be proportional to the concentration of the protein. Thus, a complete scan of the gel will give the relative amounts of each protein in the different bands but not their absolute amount. To determine the absolute amount present in any one band, it is necessary to measure the total protein of the sample using a separate method, usually the biuret method. If the relative proportion of the band is multiplied by the known total protein of the sample, its absolute value can be calculated.

In CZE, quantitation of a monoclonal band does not require any additional steps as the proportion of each band can be calculated directly from the signal output. A separate measurement of total protein in the serum sample is, however, required.

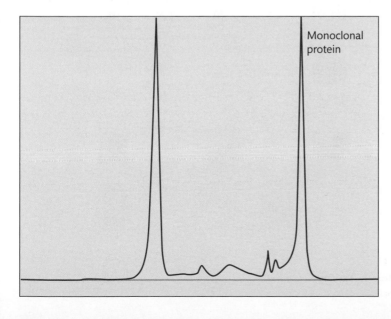

FIGURE 18.25
Example of capillary zone electrophoresis of serum with a monoclonal band.

SELF-CHECK 18.6

In Figure 18.25, the area under the monoclonal band is 40% of the total area under all the bands. The total protein concentration of the sample is 80 g/L. What is the concentration of the monoclonal band?

Identification of monoclonal heavy and light chain bands

When a monoclonal band has been found, it is essential to identify its heavy and light chain class. If a gel technique has been used to detect the band, it is usually characterized using immunofixation (Figure 18.26). In this technique, five separate aliquots of the serum are applied to adjacent lanes in a gel and, following electrophoresis, a different specific antiserum to IgG, IgA, IgM, kappa (κ) or lambda (λ) is layered over each lane.

The specific antisera will bind to and fix the protein to which it is specific. The unfixed protein is washed away and the fixed protein is stained (usually with Coomassie blue). However, if CZE is used, the monoclonal band is characterized using immunosubtraction (Figure 18.27).

In this case, five aliquots of the serum are mixed with a different specific antiserum to human G, A, or M heavy chains or to human κ or λ light chains. The mixtures are incubated for sufficient time to allow the antiserum to bind and remove its complementary immunoglobulin from the aliquot of serum. Each aliquot, together with one not treated with antigen, is subjected to electrophoresis. Following electrophoresis, the electrophoretogram of the untreated serum is compared with that treated with antiserum. The monoclonal band can be identified because it will 'disappear' in the aliquot in which it was treated with its complementary antiserum.

Detection of urinary Bence-Jones protein

Urinary BJP is a monoclonal free light chain in urine. However, the presence of free light chain in urine is not always pathological; it can be present in healthy individuals. Urinary free light chain in health is polyclonal in nature and occurs at low concentrations (less than 10 mg/L). In some clinical conditions in which there is an increase in the total concentration of polyclonal

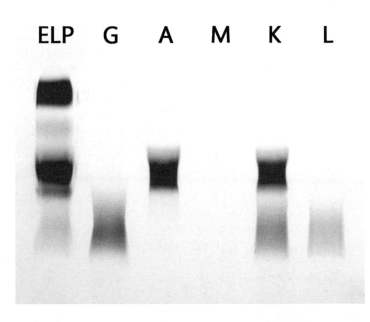

FIGURE 18.26

Serum immunofixation. The monoclonal band shown in the first lane (stained for all serum proteins) is fixed only by antiserum to A and to kappa. Hence it is an A kappa monoclonal band.

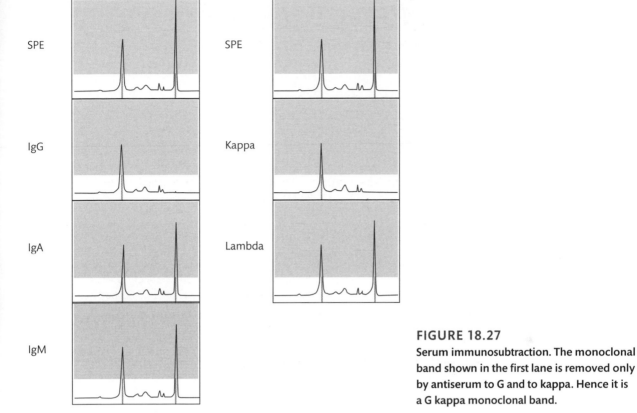

FIGURE 18.27
Serum immunosubtraction. The monoclonal band shown in the first lane is removed only by antiserum to G and to kappa. Hence it is a G kappa monoclonal band.

IgGs in the blood, it is sometimes possible for the concentration of free polyclonal light chains in urine to be increased. This would not constitute BJP, which is an appreciable concentration (greater than 30 mg/L) of a monoclonal free light chain in urine.

Bence-Jones protein is detected by electrophoresis of samples of urine followed by the fixing and staining of proteins as for serum protein electrophoresis. Small amounts of albumin and transferrin in urine are not unusual; large amounts in excess of 1 g/L may be found in some kidney diseases where glomeruli 'leak' proteins into the urine. In some urine samples, protein constituents of the kidney tubules may also be detected. Thus, an electrophoretic analysis of a urine sample may indicate the presence of several monoclonal bands, but it is not possible by a simple visual inspection to determine if they are free immunoglobulin light chains. To do this requires immunofixation. Only a urine sample that contains a monoclonal band which is fixed by antisera to κ or λ light chains and not by antibodies to any immunoglobulin heavy chain can be regarded as positive for BJP (Figure 18.28).

Isoelectric focusing

Isoelectric focusing is a type of electrophoretic separation of high resolving power. It can separate proteins that exhibit only minor differences in their charges. *In vitro* it is possible to alter the charge on a protein by altering the pH of the solution in which the protein is dissolved. At lower values of pH, carboxy groups of the protein are more likely to be protonated and therefore lose their negative charge:

$$-COO^- + H^+ \rightarrow -COOH$$

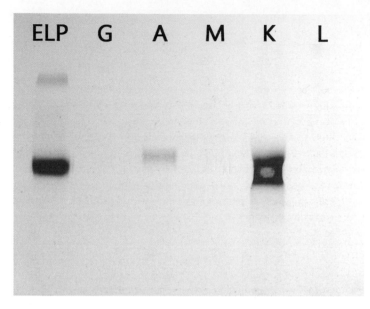

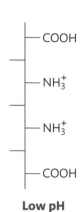

Low pH
Overall charge +2

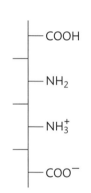

pH = pI
Overall charge zero

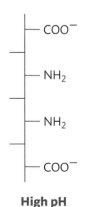

High pH
Overall charge –2

FIGURE 18.29

Effect of pH on the overall charge of a protein.

FIGURE 18.28

Urine immunofixation. The monoclonal band shown in the first lane is fixed only by antiserum to kappa. Hence it is a free kappa monoclonal band (note the slight reaction to A; this is the result of a small amount of A kappa monoclonal band being present in the urine).

Similarly, at lower pH value, amino groups are also more likely to gain an H$^+$ but they would become positively charged:

$$-NH_2 + H^+ \rightarrow -NH_3^+$$

If the pH were gradually increased, at some point both groups would lose their H$^+$. Thus, if the pH of the solution were gradually varied, it would eventually reach a pH at which the net charge on the protein was zero (Figure 18.29).

The pH at this point is called the **isoelectric point** (pI) of the protein. If a protein was placed in a solution that had a pH equal to the pI of the protein, the protein would be uncharged and therefore would not move in an electric field. Different proteins, however, have different values for their respective isoelectric points. This is the basis of IEF. A stable pH *gradient* is formed in a gel by applying an electric potential to a gel containing a mixture of **ampholytes**. Ampholytes are ions of low M_r that contain different numbers of positive and negative charges. Thus, they migrate through the gel at speeds determined by their charge and therefore become aligned in the gel in the order of the values of their isoelectric points. For instance, the ampholyte with the lowest pI will be the most negatively charged. A range of ampholytes must be chosen that have pI values which span those of the proteins of interest. When a mixture of proteins is added to a gel with an appropriate set of ampholytes and subjected to an electric field, the different proteins will migrate at rates determined by their charges. At the high pH end of the gel, most proteins are negatively charged and will therefore migrate towards the anode. The anodic side of the gel is of low pH, hence as the proteins move towards the anode they enter a region of the gel of lower pH and become less negatively charged. When a particular protein reaches the part of the gel which has a pH equal to its pI it will stop moving. Thus, each protein will migrate to the position on the gel at which it is uncharged, that is the pH is equal to its pI. Even if a protein begins to *diffuse* from this position, the change in pH will cause it to become charged, hence it will migrate back to its pI position. Thus, IEF separates proteins into very narrow bands indeed.

CASE STUDY 18.1

A 10-year-old boy was referred to hospital by his doctor following recurrent infections. The family had just moved into the area and the child had previously been investigated for 18 months at another hospital where a low concentration of albumin had been noted. Albumin measurements at the new hospital gave a value of 25 g/L (reference range 35–45) but this dropped to 18 g/L a few weeks later. The child was then referred to a teaching hospital where severe hypoalbuminaemia was found (albumin 5 g/L). This prompted urgent investigation and a high urine protein excretion of 4.7 g/L (reference range <0.25 g/L) was found. The serum IgG concentration was 3.5 g/L (reference range 6–13 g/L). A renal biopsy showed 'minimal change glomerulonephritis' and treatment with prednisolone was commenced. Within three months, the albumin and immunoglobulins were back to normal concentrations. The local hospital had used a BCG-based method to measure the concentration of serum albumin. The teaching hospital used one based on BCP.

(a) What is the relevance of the low Ig G concentration to the patient's symptoms?

(b) What can be deduced from the use of different methods to measure the concentration of albumin in two hospitals?

Isoelectric focusing has high resolving powers and separates proteins that have only very small differences in their charges. For instance, the carbohydrate side chains found attached to a specific glycoprotein may differ in number and structure between different protein molecules with identical amino acid sequences. The protein is said to show **microheterogeneity**. The different molecular species of a protein exhibiting microheterogeneity may not be separated by simple electrophoretic techniques but can be resolved by IEF. Thus, a protein which appears as a single band on normal electrophoresis (for example α_1-antitrypsin in Figure 18.5) may be separated into several distinct types by IEF (Figure 18.6).

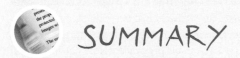

SUMMARY

■ Plasma contains many different proteins each with a different function. Their concentrations in health are fairly constant but change in disease because of excessive rates of loss or altered rates of synthesis. The acute phase reaction affects the concentrations of several proteins: some increase and some decrease.

■ Measurements of a number of serum proteins such as albumin, α_1-antitrypsin, haptoglobin, caeruloplasmin, transferrin, and C-reactive protein are of value in investigation of certain disease states.

■ Low concentrations of serum antibodies can result from secondary or primary causes and lead to susceptibility to bacterial infections. High serum concentrations of antibody can be polyclonal or monoclonal in nature. The presence of monoclonal serum immunoglobulin is not uncommon in healthy elderly people, but if accompanied by disease associated symptoms, suggests a diagnosis of myeloma or Waldenstrom's macroglobulinaemia. Oligoclonal IgG in the cerebrospinal fluid is found in multiple sclerosis.

- Concentrations of proteins are usually measured in samples of serum. Except for albumin, where a dye-binding method is usually used, the concentrations of individual serum proteins are normally measured using antisera specific to the individual protein. The techniques used include radial immunodiffusion, electroimmunoassay ('rocket') techniques, immunoturbidimetric, and immunonephelometric assays.

- The total concentration of proteins in serum is usually measured by the biuret method and in urine using dye-binding, with pyrogallol red molybdate or by turbidimetry with benzethonium chloride.

- Agarose or capillary zone electrophoresis is used to qualitatively assess serum and urine proteins. These techniques can detect Bence-Jones protein in urine and a serum monoclonal band in myeloma or Waldenstrom's macroglobulinaemia. Monoclonal bands can be typed by immunofixation or immunosubtraction using antisera specific to each of the heavy and light chain classes. Isoelectric focusing, which is a high resolution technique can also be used to separate proteins.

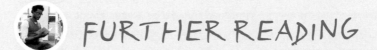

FURTHER READING

- **Beetham R (2000) Detection of Bence-Jones protein in practice.** *Annals of Clinical Biochemistry* **37**, 563–70.

  A comprehensive review of determination of Bence-Jones protein.

- **Camper WD and Osicka TM (2005) Detection of urinary albumin.** *Advances in Chronic Kidney Disease* **12**, 170–6.

- **Carrell RW and Lomas DA (2002) Alpha-1-antitrypsin deficiency—a model for conformational diseases.** *New England Journal of Medicine* **346**, 45–53.

  A good review of α_1-antitrypsin deficiency.

- **Hill PG (1985) The measurement of albumin in serum and plasma.** *Annals of Clinical Biochemistry* **22**, 565–78.

- **Whicher J and Spence C (1987) When is serum albumin worth measuring?** *Annals of Clinical Biochemistry* **24**, 572–80.

- **Whicher JT, Calvin J, Riches P, and Warren C (1987) The laboratory investigation of paraproteinaemia.** *Annals of Clinical Biochemistry* **24**, 119–32.

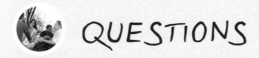

QUESTIONS

18.1 State whether the following are TRUE or FALSE for albumin.

(a) It is made in the liver

(b) It is an acute phase protein

(c) When present at high concentrations in the blood it can lead to oedema

(d) Can be overestimated by the BCG method of measurement

(e) Binds bilirubin with high affinity

18.2 State whether the following are TRUE or FALSE for α_1-antitrypsin.

(a) It is a serpin

(b) It inhibits neutrophil elastase

(c) When found at low concentrations in the blood it can be associated with lung disease

(d) It is a negative acute phase protein

(e) It is commonly measured by a dye-binding method

18.3 State whether the following are TRUE or FALSE for haptoglobin.

(a) It is a serpin

(b) It binds iron in the ferrous state

(c) It is found at high concentrations in the blood in conditions of intravascular haemolysis

(d) It is a negative acute phase protein

(e) Runs in the alpha 2 position on serum protein electrophoresis

18.4 State whether the following are TRUE or FALSE for caeruloplasmin.

(a) It is green in colour

(b) It contains two molecules of iron

(c) It is found at low concentrations in the blood in Wilson's disease

(d) It oxidizes iron from the ferrous to the ferric state

(e) It is a negative acute phase protein

18.5 State whether the following are TRUE or FALSE for transferrin.

(a) It transports iron in the ferric state

(b) It binds two molecules of iron

(c) It is a negative acute phase protein

(d) It is fully saturated with iron in healthy individuals

(e) It is found at high concentrations in the blood in iron deficiency

18.6 State whether the following are TRUE or FALSE for CRP.

(a) It can be found in blood at thousand-fold its normal concentration in response to an infection

(b) It stands for carbohydrate receptor protein

(c) Its concentration in blood can be used to monitor the disease activity of inflammatory conditions

(d) When bound to bacteria it activates the immune system

(e) It is present in the blood of healthy individuals at concentrations similar to that of albumin.

18.7 State whether the following are TRUE or FALSE for IgG.

(a) It is made in the liver

(b) It is composed of two heavy and two light chains

(c) It is present in blood at birth

(d) It is usually monoclonal in healthy individuals

(e) When found at low concentrations in blood it leads to an increased risk of bacterial infections

18.8   Why does albumin migrate further than transferrin in gel electrophoresis?

18.9   What are plasma cells?

18.10 Which immunoglobulin classes can be elevated in multiple myeloma?

18.11 C-reactive protein is present in blood at concentrations of up to 10 mg/L. What method would you use to measure its concentration?

18.12 What determines the rate of migration of a protein in isoelectric focusing?

Answers to self-check questions, case study questions, and end-of-chapter questions are available in the Online Resource Centre accompanying this book.

 Go to www.oxfordtextbooks.co.uk/orc/ahmed/

19

Cancer biochemistry and tumour markers

Joanne Adaway and Gilbert Wieringa

Learning objectives

After studying this chapter you should be able to:

- Describe some of the causes of cancer
- List the commonest types of cancer
- Discuss the effects of cancer on some organs
- Understand the different methods of cancer treatment
- Describe the attributes of an ideal tumour marker
- List the different types of molecules that can be used as tumour markers
- Determine the most appropriate tumour marker for different types of cancer
- Discuss possible future advances in cancer biochemistry and molecular diagnostics

Introduction

Cancer is a group of diseases characterized by uncontrolled growth and division of cells in the body which may invade surrounding tissues. The abnormal mass of tissue that results from uncontrolled cell growth is called a **tumour**. Cancer can have many adverse effects on the body, for example destruction of healthy tissues surrounding the tumour or obstruction of ducts or hollow organs such as the urethra, which can lead to renal failure. Tumours sometimes produce hormones which upsets the body's normal hormonal balance, causing a variety of symptoms.

Each year in the UK, approximately 277,600 people are diagnosed with cancer and at any one time 1,207,000 people have the disease. Cancer is the second most common cause of death in the UK after cardiovascular disease. In 2005 more than 138,000 people died from cancer. This makes cancer a significant healthcare issue that consumes a lot of resources for diagnosis, monitoring, and treatment.

Cancer may cause many biochemical abnormalities, therefore a diagnostic clinical biochemistry service has wide-ranging roles in monitoring and managing patients, and guiding their treatment. In addition to the usual repertoire of investigations, a specialist cancer service needs to provide a range of tests for tumour markers, hormone investigations and molecular diagnostic testing techniques are increasingly being used.

This chapter will discuss the role of the clinical biochemistry service in the diagnosis and monitoring of patients with cancer, the contribution of cancer-specific investigations, and the unique elements of service provision compared to that seen in non-specialist units.

19.1 Causes of cancer

Cancer develops as a consequence of multiple genetic mutations which usually occur over a long period of time. These mutations are often in genes that disrupt the cell cycle, differentiation, or apoptosis (programmed cell death) pathways, and the genetic alteration is inherited by successive generations of daughter cells. These pathways are usually under strict control, ensuring a balance between division, differentiation, and death of cells in tissues, to maintain tissue structure. If control of one or more of these pathways is lost, there may be uncontrolled division, lack of cellular differentiation, or accumulation of abnormally long-lived cells. Figure 19.1 demonstrates this imbalance between cellular proliferation and cell death in cancer.

Tumour: abnormal mass of tissue that results from uncontrolled cell growth.

These cells tend to have a growth or survival advantage over normal cells as they are not subject to the usual strict control steps, and their numbers accumulate, forming a tumour.

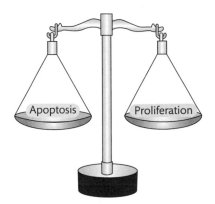

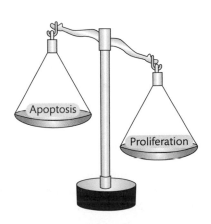

FIGURE 19.1

The delicate balance between apoptosis and cell proliferation in a tissue. In health, cell proliferation and apoptosis are in balance, maintaining tissue structure. This is maintained by a host of regulatory factors. If regulation is lost, the tissue gains a growth advantage and proliferation is favoured over apoptosis, leading to the development of a tumour.

The tumour may be **benign** and remain localized to the area in which it was formed, or it may be cancerous. Cancerous tumours are **malignant** and have the potential to become **metastatic** and spread to distant sites.

There are two broad classes of genes that are mutated in cancer. These are oncogenes and tumour suppressor genes. Oncogenes are mutated versions of normal genes (proto-oncogenes). These genes code for proteins that are involved in control of the cell cycle or cellular differentiation, such as growth factors or transcription factors. When these genes develop 'gain-of-function' mutations, control is lost, the protein is overexpressed, and the affected cell has a growth advantage. Tumour suppressor genes are inactivated in cancers. These genes are often involved in cell growth regulation pathways, for example p53, which suppresses growth of cells which have damaged DNA. Loss of function of these genes also confers a growth advantage.

As we said above, cancer is caused by multiple gene mutations. If a gene mutation is inherited and therefore present in every cell of the body, fewer additional mutations are needed to cause cancerous transformation and cancer may develop at an earlier age. This is a rare occurrence, however, and most cancers are caused by acquired mutations. These mutations are more likely to occur in rapidly dividing cells or if cells are repeatedly exposed to carcinogens, such as toxins in tobacco smoke.

There are many factors that increase the risk of developing cancer. Excess body weight, especially around the abdomen, has been found to increase the chance of developing colon or thyroid cancer in men, endometrial and gallbladder cancer in women, and renal and oesophageal cancer in both sexes. The reason for this increased risk is not yet fully understood. Other lifestyle factors may increase cancer risk, for example alcohol consumption may increase the likelihood of breast and oesophageal cancer, and smoking is associated with a higher risk of stomach, renal, bladder, larynx, and throat cancer, as well as lung cancer. Some infections are associated with a greater possibility of developing certain cancers, for instance infection with human papilloma virus is associated with cervical cancer, and girls aged 12–13 years are now vaccinated against this virus to help prevent cervical cancer.

Benign: a tumour that does not invade and destroy the tissue in which it originates or spread to distant sites in the body.

Malignant: a tumour that can invade and destroy the tissue in which it originates and can spread to distant sites in the body.

Metastatic: tumour that has spread to a site in the body distant from its origin.

SELF-CHECK 19.1

(a) What is an oncogene? (b) How does an inactivating mutation of an oncogene confer a growth advantage on cells?

19.2 **Types of cancers**

Cancers may be classified into different types depending on the tissue from which they originate:

- carcinomas
- sarcomas
- lymphomas
- leukaemias
- blastomas
- germ cell tumours
- other tumours

The most common type of a cancer is a carcinoma. Carcinomas arise from epithelial cells that line the skin and internal organs of the body. Sarcomas are cancers of connective tissue, such as muscle, fat, bone, and cartilage. Lymphomas are malignant tumours of lymph nodes of the immune system. Leukaemia is a group of malignant diseases in which the bone marrow produces increased numbers of white blood cells, many of which are abnormal or immature. The production of normal white cells, red cells, and platelets is suppressed, leaving patients susceptible to infection, anaemia, and bleeding. Blastomas are rare cancers of immature undifferentiated cells. Blastoma is often used as part of the name of the tumour, for instance a retinoblastoma is a rare cancer of the retina that occurs in childhood, and osteoblastoma is a blastoma of the bones. Germ cell tumours are tumours of cells that can form sperm or oocytes. These tumours often develop in the gonads but can be found elsewhere in the body. Some rare tumours are not named according to their tissue of origin, for instance an insulinoma is an insulin-secreting tumour of the pancreatic islet cells and a phaeochromocytoma is a tumour of the chromaffin cells of the adrenal medulla or extra-adrenal sympathetic nervous tissue.

Although the same number of men and women develop cancer each year, the incidence of the various types of cancer in the two sexes is different. Figure 19.2 illustrates this.

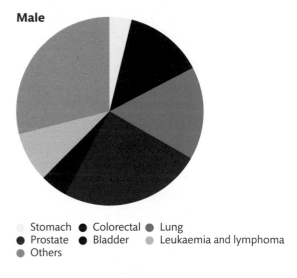

Male

● Stomach ● Colorectal ● Lung
● Prostate ● Bladder ● Leukaemia and lymphoma
● Others

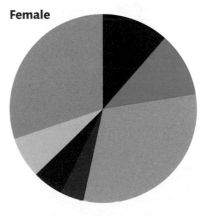

Female

● Colorectal ● Lung ● Breast
● Uterus ● Ovary ● Leukaemia and lymphoma
● Others

FIGURE 19.2

The incidences of different types of cancer in male and female patients. See text for details.

The most common type of cancer in men is prostate cancer, with around 32,800 new cases each year in the UK. The incidence of prostate cancer is increasing by 10–15% every five years in developed countries. About 68% of patients survive for at least five years after diagnosis, and this number is increasing as treatments improve. Breast cancer is the most common type of cancer in women, with around 43,100 new cases each year. The incidence of breast cancer is increasing in developed countries and its incidence increases with age. However, 80% of patients survive for at least five years after diagnosis and 72% survive for at least ten years.

Lung cancer is the next commonest cancer in men, followed by colorectal cancer. This trend is reversed in women, with slightly more developing colorectal rather than lung cancer. Most (90%) cases of lung cancer are associated with smoking and the lag time between exposure to cigarette smoke to the development of cancer (the latency period) is at least 20 years. The incidence of lung cancer in males below 74 years of age has been falling since the 1970s. However, the peak incidence in women aged 55–64 occurred in the late 1980s and the incidence in women 65–84 years old is still increasing, reflecting the different trends in cigarette smoking in the sexes. The mortality rate of lung cancer is high, with only 5% of patients surviving for at least five years after diagnosis.

The incidence of colorectal cancer increases with age and the overall incidence has been slowly rising since the 1970s. About 42% of men and 43% of women survive for at least 5 years after diagnosis, and the survival rates are increasing by around 6% every 5 years.

SELF-CHECK 19.2

(a) Which is the commonest type of cancer in male patients? (b) What is the most common contributory factor in lung cancer?

19.3 **The effects of cancer**

Patients with cancer can show many different abnormalities caused by direct effects of the tumour, such as obstruction, tissue destruction or tumour growth, or by substances produced by the tumour, such as hormones or cytokines. These substances can cause effects that are not directly related to the presence of the tumour and this is called the paraneoplastic syndrome.

Obstruction

Tumour growth can lead to blockage and obstruction of hollow organs such as the oesophagus, gastrointestinal tract (GIT), or bile duct. Depending on their location, some tumours cause biochemical changes, which give an early indication of the likely site of obstruction. This allows the opportunity to start appropriate treatment. For example, raised bilirubin, alkaline phosphatase (ALP), and γ-glutamyl transferase (GGT) levels may indicate bile duct obstruction (for example caused by cancer of the head of the pancreas) that leads to **obstructive jaundice**. Another example may be an increase in the protein content of cerebrospinal fluid (CSF), which can occur when the central canal of the spinal cord is blocked by a **primary** or **secondary tumour** that interrupts the flow of CSF. Plasma proteins may then pass through the walls of meningeal capillaries into the stagnant CSF increasing its protein content below the tumour site. This high protein concentration can be identified by analysing CSF collected following a lumbar puncture. Symptoms of cancer are often non-specific, and the detection of biochemical changes may provide an early diagnostic indicator that warrants further investigation.

Obstructive jaundice: increased serum bilirubin concentration caused by the obstruction of the flow of bile into the duodenum.

Primary tumour: tumour that is at the site from which it first developed.

Secondary tumour: tumour that has developed as a result of metastasis from the original tumour site.

Once a patient is suspected of having cancer, disciplines such as endoscopy, radiology, and histopathology typically become involved in making a definitive diagnosis and identifying the site of the tumour. This information is then used to inform on likely prognosis and the appropriate treatment for the patient. Endoscopy involves using a small camera on a flexible tube to look inside the GIT to see if there are any abnormal growths causing a blockage. A biopsy of any suspicious areas can be taken and this is examined by a histologist to see if the tissue is cancerous.

SELF-CHECK 19.3

What biochemical changes might you see in a patient with obstructive jaundice due to carcinoma of the head of the pancreas?

Tissue destruction

Tissue invasion by tumours may cause significant functional damage which may lead to release of enzymes such as lactate dehydrogenase (LD), aspartate aminotransferase (AST), and alanine aminotransferase (ALT) from the cells. However, destruction often follows a slow, inexorable process in which moderate but persistently increased activity of these enzymes occurs, for example the liver is the organ that is most commonly affected by tumour invasion but many patients show only minimal biochemical abnormalities. Cholestasis, which is characterized by raised ALP and GGT activities in serum with normal or slightly raised bilirubin levels, may indicate deposits in the liver that are causing local disruption to the flow of bile. With more generalized tissue destruction, the same rise in ALP and GGT activities may be measured in serum samples but the bilirubin may also be markedly raised.

Cross reference

Chapter 14 Adrenal disease

Secondary tumour deposits, for example lung or breast tumours metastasizing to the adrenal glands may occasionally lead to adrenal hypofunction.

Cross reference

Chapter 5 Fluid and electrolyte disorders

Decreased adrenal function may cause non-specific symptoms such as weakness and weight loss if it has a slow onset. However, in situations of increased stress such as infection, the patient may develop severe hypovolaemia, shock, and hypotension and require emergency medical treatment. The biochemical features of hyponatraemia with hyperkalaemia may develop and provide a clue to the diagnosis.

Tumour invasion of the pituitary gland can cause destruction of the anterior or posterior gland or the pituitary stalk. Figure 19.3 shows the basic structure of the pituitary gland and the hormones that are produced.

Cross reference

Chapter 11 Abnormal pituitary function

The tumour may be primary, often a benign adenoma, or a secondary deposit, most commonly from breast or lung carcinomas. Symptoms of hypopituitarism, such as weakness and weight loss, may not appear until there is considerable tissue destruction and may not be recognized due to the overlap with symptoms of advanced malignancy. The commonest symptom of metastatic pituitary destruction is thirst and the production of large amounts of dilute urine due to failure of antidiuretic hormone (ADH), also known as arginine vasopressin (AVP) secretion by the posterior pituitary gland. This is known as cranial diabetes insipidus and may lead to life-threatening dehydration if water losses are not replenished. Synthetic ADH may be used to treat the condition.

The anterior pituitary gland can also be affected by tumour invasion. Decreased secretion of growth hormone is an early feature of such invasion, followed by reduced release of luteinizing hormone (LH) and follicle stimulating hormone (FSH). Subsequent loss of

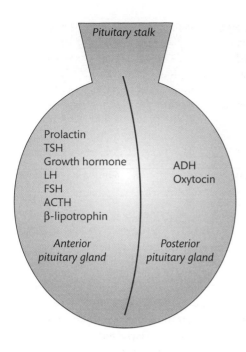

FIGURE 19.3

The pituitary gland. The anterior pituitary gland produces hormones such as TSH, prolactin, and growth hormone; the posterior pituitary produces ADH and oxytocin. The pituitary stalk leads to the hypothalamus, which releases hormones that control production of pituitary hormones.

adrenocorticotrophic hormone (ACTH) and thyroid stimulating hormone (TSH) may also occur. If the pituitary stalk is affected, hyperprolactinaemia may develop. Prolactin secretion is inhibited by dopamine, which is released from the hypothalamus and enters the pituitary gland through the pituitary stalk. If the stalk is destroyed or obstructed by a tumour, prolactin secretion is not inhibited and becomes autonomous, leading to hyperprolactinaemia.

Tumour growth and cell turnover

> ## Key Points
>
> Rapid tumour growth and cell turnover can cause biochemical changes due to breakdown or leakage of cell contents from immature tumour cells.

Lactate dehydrogenase (LD) concentrations in the serum have been shown to correlate with tumour mass in some solid tumours such as those of lung, ovaries, and breast. However, raised LD concentrations can also be measured in serum from patients with liver disease and after myocardial infarction, so its lack of specificity limits its clinical usefulness. Five **isoenzymes** of LD, called LD1, LD2, LD3, LD4, and LD5, are widely distributed across different tissues yet all are detected by conventional activity based measurement methods and cannot be distinguished. Figure 19.4 shows the different isoenzymes.

Isoenzymes are physically distinct forms of a given enzyme.

Isoenzyme analysis may improve specificity (for example LD1 and LD2 levels are raised in ovarian/testicular cancer) but more specific tumour markers are available that are simpler to measure, therefore LD isoenzyme analysis is rarely used in the management of these malignancies.

Uric acid is the major breakdown product of the purines adenosine and guanosine that form part of DNA and RNA. Its concentration may be increased in plasma when there is increased cell turnover, for example during rapid tumour growth. High serum uric acid concentrations

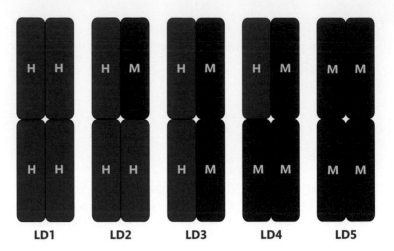

LD1   LD2   LD3   LD4   LD5

FIGURE 19.4

Lactate dehydrogenase isoenzymes. LD1 consists of four H subunits (H4) and is the major isoenzyme in cardiac muscle and red blood cells. LD2 is H3M and is also found in cardiac enzyme and muscle cells. LD3 is H2M2 (lung tissue), LD4 is HM3, found in white blood cells and lymph nodes, and LD5 has four M subunits and is found in skeletal muscle and liver.

may be associated with an increased mortality risk in cancer, especially in malignancies of digestive organs and the respiratory system.

Paraneoplastic syndrome

Paraneoplastic syndrome includes all the systemic signs and symptoms of cancer that are not directly related to the physical presence of a tumour. In most cases, the syndrome is caused by something that is secreted by the tumour. This could be a hormone, cytokine, protein, or neurotransmitter. In most cases, removal of the tumour leads to improvement or disappearance of the syndrome.

Tumour related hypoglycaemia

Hypoglycaemia can be caused by excess insulin production from a pancreatic islet cell tumour or insulinoma. This can cause severe hypoglycaemia that is difficult to control. Some tumours may produce antibodies which can bind to and activate the insulin receptor. This has been seen in patients with Hodgkin's lymphoma. Hypoglycaemia can also be caused by non-islet cell tumours such as lung or stomach tumours. This condition is known as non-islet cell tumour hypoglycaemia (NICTH) and is caused by an increase in insulin-like growth factor-2 (IGF-2). Insulin-like growth factors are structurally related to insulin and can cross-react with the insulin receptor. The concentration of the IGFs in the circulation is much higher than the concentration of insulin so IGFs could cause hypoglycaemia by binding to the insulin receptor. This does not occur as IGF-1 and 2 are bound to IGF-binding protein 3 (IGFBP3) and an acid labile subunit (ALS), forming a complex that is too large to cross the capillary endothelial barrier; the complex remains in the intravascular space and therefore cannot gain access to the insulin receptor. However, some tumour cells seem to over-express the IGF-2 gene, producing large amounts of an IGF-2 precursor protein known as 'big IGF-2', which overflows into the circulation. This suppresses growth hormone and IGFBP3 production. Big IGF-2 cannot form

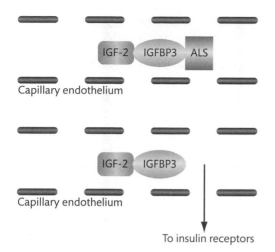

Capillary endothelium

To insulin receptors

FIGURE 19.5

The mechanism of non-islet cell tumour hypoglycaemia. See text for details.

the normal large complex with IGFBP3 and ALS and instead circulates as free IGF-2 or as a smaller complex with IGFBP3. In this form, it is able to cross the capillary endothelial barrier and bind to insulin receptors of liver and muscle. This leads to decreased release of glucose into the blood from the liver and an increased rate of glucose utilization by tissues such as muscle, causing hypoglycaemia. Figure 19.5 describes this process.

Patients with NICTH have low insulin, C-peptide, IGF-1, and IGFBP3 concentrations, and a diagnosis can be made if patients have an IGF-2: IGF-1 ratio of 10 or above. Tumour-related hypoglycaemia can be difficult to treat, with patients requiring repeated intravenous glucose infusions. The most effective treatment is removal of the tumour, although if this is not possible, glucocorticoids seem to suppress the production of big IGF-2 and can be used to control the **hypoglycaemia**.

Cross reference
Chapter 13 Diabetes mellitus and hypoglycaemia

Hypoglycaemia: low blood glucose concentration.

SELF-CHECK 19.4

List the three different causes of tumour-related hypoglycaemia?

Syndrome of inappropriate antidiuretic hormone

The syndrome of inappropriate antidiuretic hormone (SIADH) is a condition where the concentration of ADH in the circulation is inappropriately high, causing water retention, low plasma sodium concentrations and concentrated urine production. This occurs when the hormone ADH is secreted from tumour cells, most commonly in small cell lung cancer. In a healthy individual, ADH is released from the posterior pituitary gland in response to an increase in the osmolality of the extracellular fluid (ECF) and renders the renal collecting ducts more permeable to water, allowing water resorption and a decrease in the ECF osmolality.

Cross reference
Chapter 5 Fluid and electrolyte disorders

When ADH is produced in excess, as in SIADH, water is reabsorbed in the kidneys and the extracellular fluid becomes diluted, leading to hyponatraemia. The diagnosis can be made if the following criteria are met:

- serum sodium must be less than 135 mmol/L
- serum osmolality must be less than 270 mOsm/kg
- urine osmolality must be greater than 100 mOsm/kg
- urine sodium excretion must be greater than 30 mmol/L
- the patient must have no clinical evidence of hypo- or hypervolaemia

- the patient must have normal renal, adrenal and thyroid function
- the patient must not be on diuretics

The SIADH can occur in patients without malignancy; it can also be caused by lung or central nervous system diseases or by certain drugs. The immediate treatment is restricting the amount of fluid the patient is allowed to drink so as to prevent severe hyponatraemia. Removal of the cause is required for cure.

Hypercalcaemia of malignancy

Up to 20% of patients with cancer develop hypercalcaemia at some stage in their disease. It is often seen in patients with lung, breast, ovarian, or haematological malignancies and is a poor prognostic indicator. Hypercalcaemia can develop in two ways in cancer patients and the patients may have either osteolytic hypercalcaemia or humoral hypercalcaemia. Osteolytic hypercalcaemia is caused by factors produced by the tumour which lead to increased reabsorption of bone and therefore release of calcium. The factors that cause this are not yet fully understood, but cytokines such as interleukin 1 and 2 and transforming growth factor-α may be involved. Patients have raised serum calcium, normal phosphate, and a decreased parathyroid hormone (PTH) concentration. This type is often seen in patients with haematological malignancies such as multiple myeloma. Humoral hypercalcaemia of malignancy is caused by a peptide hormone called parathyroid hormone related peptide (PTHrp). As the name suggests, it has a similar structure to parathyroid hormone and has 60% homology to PTH in the first 13 amino acid residues at the N terminal end of the hormone. Parathyroid hormone related peptide (PTHrp) can bind to the PTH receptor in kidney and bone to mimic the action of PTH, causing calcium release from bone and reabsorption of calcium at the renal tubule.

Cross reference

Chapter 10 Disorders of calcium, phosphate, and magnesium homeostasis

CASE STUDY 19.1

A 79-year-old lady with a non-small cell lung cancer was admitted to hospital presenting with acute confusion.

Biochemical investigations on serum and urine gave the following results (reference ranges are given in brackets):

Serum

Sodium	114 mmol/L	(135–145)
Potassium	4.4 mmol/L	(3.5–5)
Urea	2.3 mmol/L	(2.7–7.5)
Creatinine	59 μmol/L	(50–120)
Osmolality	256 mOsm/kg	(275–295)

Urine

Osmolality	479 mOsm/kg

The patient was not on any diuretic treatment and had no clinical signs of hypo- or hypervolaemia.

The clinicians suspected that this patient had SIADH.

(a) What other biochemical tests should be carried out to support this diagnosis?

(b) What results would you expect if this patient did have SIADH?

SELF-CHECK 19.5

Name the two different types of hypercalcaemia of malignancy.

19.4 Cancer treatments

There are three main modes of treatment for cancer. These are:

- surgery
- chemotherapy
- radiotherapy

Surgery is widely used to treat cancers and is usually the first-line treatment. During surgery a histopathologist will often be present to confirm that cancerous tissue has been fully excised. Unfortunately, surgery is unable to treat cancer that has widely metastasized through the body. In this case, chemotherapy and/or radiotherapy will be needed. Sometimes, tumours are too large or inaccessible to be completely removed. In these cases, part of the tumour may be removed to reduce pressure on organs and improve the quality of life for the patient. This is called tumour debulking, and may be followed by other modes of treatment.

Chemotherapy is the use of chemicals to destroy cancerous tissue. Cancer cells divide more rapidly than non-cancerous cells so they are more vulnerable to these drugs. However, these drugs are usually delivered systemically via the bloodstream and they can also destroy normal cells, thus causing side effects such as nausea and vomiting, hair loss, or bone marrow suppression. Different types of chemotherapy drugs are used, depending on the type and stage of the cancer. All the classes of drugs interfere with DNA replication to prevent cell division. Alkylating drugs such as cyclophosphamide, which is used in chronic lymphocytic leukaemia and lymphoma, form covalent bonds with molecules in DNA to prevent DNA replication. Cytotoxic drugs interfere with enzymes involved with DNA replication and include drugs such as doxorubicin and idarubicin, which are used in haematological malignancies such as leukaemias. Antimetabolites such as 5-fluorouracil become incorporated into new DNA and RNA or interfere irreversibly with cellular enzymes to prevent cell replication and are used to treat tumours of the GIT and breast. Some chemotherapeutic agents have been derived from plant materials. Vinca alkaloids interfere with microtubular function by acting as a spindle poison. Such drugs include vincristine, which is used to treat a variety of cancers, including leukaemias, breast, and lung cancer.

Newer, more targeted chemotherapeutic agents are now becoming available. These include trastuzumab, which is a humanized monoclonal antibody directed against the Her2 protein in breast cancer cells. Her2 is the receptor for human epidermal growth factor. Trastuzumab binds Her2, preventing binding of epidermal growth factor and therefore preventing cell division. This therapy is only effective in patients who overexpress Her2, so analysis of Her2 expression in tumour cells must be carried out before any treatment is given. Imatinib is a tyrosine kinase inhibitor which was developed as a treatment for chronic myeloid leukaemia, which is characterized by abnormal tyrosine kinase signalling. The drug has since been found to be useful in some gastrointestinal stromal tumours and acute lymphoblastic leukaemia. Immunotherapeutic reagents are also being developed to stimulate the patient's own immune system to recognize and attack cancer cells. Interleukin-2 is used in metastatic renal carcinoma and has been shown to shrink tumours in some patients but has not been proven to increase survival. Rituximab is a monoclonal antibody which causes lysis of B lymphocytes and can be used in the treatment of B cell lymphomas.

Current chemotherapeutic agents are toxic to healthy cells as well as cancerous cells, so it is important to monitor patients for evidence of organ damage. Renal function must be monitored during treatment as most chemotherapy drugs are renally cleared. A decline in renal function can cause a toxic build-up of the drug in the body, so the glomerular filtration rate must be regularly measured and the dose of the drug decreased if the renal function deteriorates. Many chemotherapeutic agents are especially toxic to the liver, as the liver is involved in drug metabolism and is thus exposed to high concentrations of the drugs. Liver function tests such as AST, ALT, GGT, and bilirubin should be carried out to detect if damage is occurring.

Radiotherapy is the use of X-rays to destroy cancerous cells. The radiation causes structural damage to DNA. The affected cell attempts to repair the damage, and, if this is not successful, programmed cell death is initiated. Figure 19.6 shows this process.

Healthy cells are more successful at repairing DNA damage than tumour cells, so healthy cells surrounding the tumour are minimally affected. In addition, the treatment is targeted at the

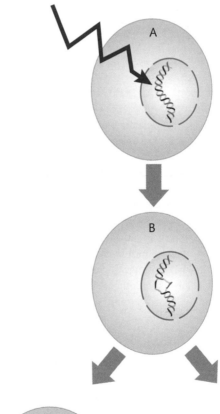

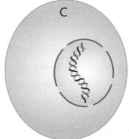

FIGURE 19.6

Mechanism of action of radiotherapy. Irradiation of a cell (A) causes structural damage to the DNA (B). The affected cell attempts to repair the DNA damage (C). If this is not successful, programmed cell death is initiated (D). Healthy tissue is more effective at DNA repair than cancerous cells, so radiotherapy kills more cancer cells than healthy ones.

affected area and careful planning is needed to ensure that the radiotherapy beam is directed at the correct area and a minimum of healthy tissue is affected. Radiotherapy is usually given externally using a linear accelerator, a machine used to deliver uniform doses of high energy X-rays to the region of the patient's tumour. The course of radiotherapy is usually given over a number of days, with each treatment known as a fraction. This minimizes the damage to healthy tissue. Internal radiotherapy may be given for some tumour types. Brachytherapy is the use of internal radiotherapy implants inserted near the site of the tumour. This type of radiotherapy is used for prostate, uterine, and head and neck cancer as it enables more direct treatment of tumours that are not close to the surface of the body and therefore are inaccessible to external irradiation. Some patients may receive radioisotope treatment, which is the use of radioactive liquids, given orally or by injection. Radioactive iodine is given orally to treat thyroid cancers and benign thyroid conditions such as thyrotoxicosis, a disease where excess amounts of thyroid hormones are released into the bloodstream. Iodine is taken up by the thyroid gland, so small doses of radioactive iodine given orally are taken up and concentrated in this tissue, destroying cells in the gland but sparing other cells in the body. This is effectively a treatment that is targeted only at the thyroid gland. Other forms of radioisotope treatment are also available, for instance radioactive strontium is used in patients with secondary bone cancers.

SELF-CHECK 19.6

(a) Name the three main methods of cancer treatment. (b) Why is it important to monitor renal function in patients on chemotherapy?

19.5 Tumour markers

By definition, a marker is a molecule, substance, or process whose deviation from normal can be detected by an assay. This deviation may be a change in the serum concentration of the marker or an alteration in the structure or site of production of the molecule. The ideal tumour marker should:

- be undetectable in health
- be produced only by malignant tissue
- be specific to the site of malignancy
- circulate in concentrations proportional to the tumour mass

The ideal tumour marker could be used in screening, diagnosis, prognostic assessment, staging, managing, and monitoring of cancer. No tumour marker currently in use fulfils all the criteria of an ideal tumour marker. Different classes of substances have been developed as tumour markers and we will now explore some of these further. The determination of sensitivity and specificity for a test are given in the Method box below.

Oncofoetal proteins

Oncofoetal proteins are proteins that are normally produced during embryonic development and are not usually found in adult tissues. High concentrations are only found in adults when genes that control cellular growth are reactivated in malignancy. Carcinoembryonic antigen (CEA) is an oncofoetal protein and is mainly used as a marker for colorectal cancer. Another member of this group is α-fetoprotein (AFP). Alpha-fetoprotein is a glycoprotein that is produced in the yolk sac, epithelial cells of the GIT, and the liver during embryonic development.

METHOD *Determination of test sensitivity and specificity*

The analytical sensitivity of a test is the lowest concentration of the analyte that can be accurately measured by the assay. The analytical specificity of the test refers to its ability to measure the analyte of interest, rather than any other in the sample. We need to know this information before we introduce a new test into the laboratory, but we also need to know the clinical sensitivity and specificity of the test, to show us how useful the test will be.

The clinical or diagnostic sensitivity describes the percentage of people with a particular disease who test positive for the disease, that is, the number of true positive results that are obtained. This is calculated using the following equation:

Sensitivity = true positive/(true positive + false negative)

The higher the sensitivity of the test, the more cases of the disease will be identified.

The clinical specificity of a test describes the percentage of people without a disease who have a negative result, that is, the number of true negative results that are obtained. This can be calculated using this equation:

Specificity = true negative/(true negative + false positive)

The higher the specificity of the test, the more likely it is that a patient without the disease will obtain a negative result. Ideally, the sensitivity and specificity of a test should be 100%, but unfortunately very few tests are ideal.

In healthy non-pregnant adults, low concentrations of AFP are found in the blood. However, AFP is produced by hepatocellular carcinoma and non-seminomatous germ cell tumours and is used as a tumour marker in these cancers.

Hormones

The synthesis and secretion of certain hormones can be altered in some types of cancer. Large quantities of some hormones can be produced if a tumour develops in the tissue of endocrine glands and these hormones can be used as tumour markers for these particular types of cancer. Insulin levels can be raised in insulinomas of the pancreas, serum prolactin is increased in patients with prolactinomas, and catecholamine levels are high in patients with phaeochromocytomas. In some cancers, cells of some organs such as the lung or central nervous system (CNS) may start producing hormones that they do not normally secrete. This is called **ectopic hormone** secretion and measurement of the hormones produced can again give information about the type of tumour that is present; for example ACTH can be produced by some small cell lung cancers, causing high levels of cortisol to be secreted by the adrenal glands.

Ectopic hormone: a hormone formed by a tissue outside its normal endocrine site of production.

Tumour-associated antigens

Tumour-associated antigens are antigenic markers on cell membrane surfaces. Although not necessarily unique to tumour type, the serum concentration of a marker is sometimes associated with tumour size or burden so that changes in levels can help guide and monitor treatment and provide prognostic information. Diagnosis remains the domain of histology. Specific monoclonal antibodies have been produced against antigenic structures of different types of tumour cells to produce more specific tumour marker assays. Markers of this group are all named with the CA (carbohydrate antigen) prefix.

Tumour markers justify their use if a more favourable clinical outcome is achieved. In practice, however, their introduction has been poorly controlled and the evidence base of their applicability (as determined by outcomes of randomized controlled trials) is often lacking or still emerging. We shall see evidence of this in the next section as we discuss specific tumour markers.

Carbohydrate antigen 15.3

The CA 15.3 assay measures the mucin MUC1, a glycoprotein of high M_r with multiple glycosylation sites expressed by epithelial cells in bladder, stomach, pancreas, ovary, respiratory tract, and breast. In normal breast tissue, MUC1 is shed into breast milk. In breast tumours, the loss of normal tissue organization causes MUC1 to be shed into the circulation, where it can be detected by immunoassay techniques.

In malignancy, changes in the activities of the enzymes that catalyse glycosylation produces altered polysaccharide chains and hence a multiplicity of epitope recognition sites. In turn, a plethora of assays such as CA 15.3, CA 549, CA 27.29, and EMCA have been developed to detect MUC1.

> ## Key Points
>
> **CA 15.3 is the most widely used serum assay for MUC1 in breast cancer.**

The CA 15.3 assay is typically a sandwich immunoassay using two monoclonal antibodies, one directed against human milk fat globular proteins and one against a membrane enriched fraction of metastatic breast carcinoma. This assay is widely available on automated immunoassay platforms, making it a simple and convenient assay.

Lack of sensitivity and specificity makes CA 15.3 unsuitable for screening for breast cancer. Up to 5% of healthy individuals may have slightly raised CA 15.3 levels; raised levels may also be seen in liver disease and in patients with other types of advanced adenocarcinomas. Mammography and histopathology are currently the established technologies for, respectively, screening and diagnosis of breast cancer.

In breast cancer, CA 15.3 has been used as a prognostic marker. At least ten different studies involving more than 4,000 patients have shown that high concentrations of serum CA 15.3 at diagnosis predict poor patient outcome. Unfortunately, these different studies used different cut-offs of CA 15.3 concentrations in their analysis, so more work is now needed to examine all these data and relate CA 15.3 concentrations to patient outcomes, such as disease-free survival times and overall survival. Some research groups have also studied CA 15.3 concentrations in patients who have been treated and have disease recurrence. Their research has shown that levels of CA 15.3 at the time of disease recurrence can give prognostic information, and that the time interval between diagnosis and first abnormal CA 15.3 concentration can indicate likely prognosis. Measuring serum concentrations of CA 15.3 can therefore give useful information about prognosis, both at the time of disease diagnosis and during patient follow-up.

It is unclear whether CA 15.3 is useful in monitoring response to treatment. Several studies have shown that serial changes in CA 15.3 concentration correlate with response to treatment. However, in a significant number of patients, serial measurement of CA 15.3 concentration does not provide information about treatment response and, because of this, expert groups such as the American Society for Clinical Oncology do not recommend the use of CA 15.3 in the monitoring of response to treatment except in cases such as patients with bony metastases when it is difficult to monitor the patient clinically.

SELF-CHECK 19.7

What role does CA 15.3 have in the management of breast cancer?

Carbohydrate antigen 125

The CA 125 assay uses the OC-125 antibody to detect a mucin-like glycoprotein present in tissues derived from foetal coelomic epithelium. In adults, this mucin is found mainly in the peritoneum, pleura, GIT, and female reproductive tract.

> ## Key Points
>
> **CA 125 is often used in the management of ovarian cancer.**

In 90% of patients with advanced ovarian cancer, CA 125 is raised mainly due to tissue destruction, vascular invasion, and inflammation. Its best recognized and most substantiated role is in monitoring the response to treatment (typically chemotherapy). Serial decreases of 75% over three samples or a 50% decrease sustained over a minimum of four samples have been widely substantiated as representing response to treatment. Increasing serum CA 125 concentrations may predate clinical relapse in around 70% of patients, in turn allowing earlier instigation of salvage chemotherapy on the assumption that enhanced clinical outcomes can then be achieved.

Raised serum concentrations of CA 125 are also found during menstruation and pregnancy, endometriosis, peritonitis, or cirrhosis. Serum concentrations can be particularly high in patients with ascites and they are also raised in 40% of women with intra-abdominal malignancies and 28% of patients with non-ovarian malignancies. The lack of specificity limits its role in aiding differential diagnosis.

Evidence for CA125's role in screening is still awaited, in part based on arguments as to the cost effectiveness of screening programmes for ovarian cancer, the seventh most common cancer in women in the UK. This relatively low prevalence means that screening the population as a whole would give a large number of false positive results. A more useful approach could be to screen a selected (high-risk) population. Annual ultrasound screening is reported to offer a sensitivity close to 100% compared to 80% for CA 125 based screening, but the former's false positive rate of 1.2–2.5% compared to 0.1–0.6% for the tumour marker may lead to a higher recall and hence distress to otherwise healthy women. Modelling studies suggest that annual CA 125-based screening may provide lower overall benefits but be more cost-effective at detecting early stage cancers than annual ultrasound screening.

The best recognized role for CA 125 is in monitoring response to treatment. Levels may rise within two weeks of treatment due to release of CA 125 from damaged tissue, but outside this time period, changes in CA 125 concentrations can give useful information about the effectiveness of treatment. Carbohydrate antigen 125 can also be used in the follow-up of patients following treatment for ovarian cancer. In around 70% of patients, a rising CA 125 level may be the first indication of relapse, predating clinical signs of relapse by a median of four months. In order for this to be of clinical use, it must be proven that early treatment of relapse improves patient outcome. The Medical Research Council has carried out a randomized controlled trial to investigate whether giving chemotherapy based on rising CA 125 concentrations improves survival rates compared to the current practice of administering chemotherapy once there is

CASE STUDY 19.2

A 45-year-old woman was diagnosed with ovarian cancer and was referred to a surgeon for oophrectomy (removal of the ovaries). Prior to surgery, her CA 125 concentration was 25,392 mU/L. Three days after successful surgery, her CA 125 concentration was measured again and found to be 28,795 mU/L.

(a) Why had the CA 125 concentration increased despite the successful removal of this woman's tumour?

(b) Should the clinician still ask for CA 125 measurements in this patient?

clinical evidence of recurrence. The results suggest that withholding treatment in isolated cases of increasing CA125 levels does not impact negatively on overall survival for these patients.

Carcinoembryonic antigen

Carcinoembryonic antigen is a highly glycosylated cell surface glycoprotein detectable in foetal colon and adenocarcinomas but not in healthy colon. Its wide deployment as a tumour marker in colorectal cancer also allowed evidence to emerge of its distribution elsewhere—raised concentrations are also found in breast, lung, gastric, and ovarian cancer, as well as alcoholic liver cirrhosis, gastrointestinal inflammatory diseases, and in smokers. Carcinoembryonic antigen serves as a prime example of a tumour marker whose evidence of efficacy played little part in its introduction and subsequent popular demand, as it was introduced as a specific marker for colon cancer but was later found to be raised in many other diseases. Once a test is introduced for a defined purpose, it is very difficult to subsequently re-educate users about its drawbacks, so CEA remains in use for patients with colorectal cancer.

Key Points

Carcinoembryonic antigen's best established value is in the pre-clinical detection of recurrence after surgical treatment for colorectal cancer, particularly the early detection of liver metastases.

Intensive follow-up that includes CEA monitoring for at least two years is now recognized as being associated with a significant reduction in all causes of mortality and a modest improvement in outcome.

Its most popular use is in the monitoring of colorectal cancer treatment, given that levels are increased in 80% of patients with distant metastases. Several studies have shown that decreases in levels while on chemotherapy are associated with better overall survival whilst increases generally predict progressive disease. Evidence of the impact of CEA testing on quality of life or cost of care is still awaited.

Carcinoembryonic antigen's low sensitivity (36%) and specificity (87%) in detecting the earliest stages of colorectal cancer (Dukes stage I and II) limits its use as a screening test and as a diagnostic aid. Its low specificity is related to the elevated levels (usually <10 μg/L) seen in benign conditions such as GIT diseases, fibrocystic breast disease, and renal failure. Smoking is associated with a doubling of underlying CEA concentrations and men tend to have higher levels than women. As a prognostic indicator, numerous studies have shown that patients with a pre-operative CEA concentration above 5 µg/L level have a worse outcome, but evidence that adjuvant chemotherapy leads to improved outcome is awaited. After successful surgical treatment of cancer, CEA concentrations should return to normal levels within 4–6 weeks. Failure to fall to normal levels is associated with early recurrence of disease.

Carbohydrate antigen 19-9

Carbohydrate antigen 19-9 (CA 19-9) is a carbohydrate antigen that was originally derived from the culture medium of a colorectal cancer cell line but later found to be a sialylated form of the Lewis A blood group antigen.

> ### Key Points
> Although initially described as a tumour marker for colorectal cancer, CA 19-9 has since found a role as a marker of pancreatic cancer.

Early symptoms of pancreatic cancer are non-specific, including weight loss and back pain. As a result, diagnosis is often not made until the disease is advanced and treatment options are limited. Carbohydrate antigen 19-9's contribution to earlier detection is limited by low median sensitivity (79%) and specificity (82%). Raised concentrations are seen in extrahepatic biliary obstruction and in chronic liver disease. In addition, 5–10% of the population are Lewis genotype negative and therefore do not express CA 19-9. The low prevalence of pancreatic cancer also ensures a low positive predictive value for the test, meaning it has little use as a screening tool.

Carbohydrate antigen 19-9 could be used to give prognostic information. One group adjusted CA 19-9 concentrations by dividing them by the bilirubin concentration to try and improve the specificity of the test. They found that patients with a pre-operative adjusted CA 19-9 concentration above 50 mU/L had twice the risk of cancer recurrence than patients with concentrations below 50 mU/L. Another group discovered that patients whose post-operative non-adjusted CA 19-9 fell to within the normal range survived longer than patients whose CA 19-9 did not. This information could be used to determine which patients will need adjuvant treatments, although larger studies are needed to confirm the results.

Some studies have shown that CA 19-9 may be used to monitor recurrence of pancreatic cancer post-surgery, with rising concentrations giving a lead time of several months. However, there is currently no effective treatment for advanced pancreatic cancer so the clinical use of CA 19-9 in this setting is limited.

SELF-CHECK 19.8
Why does CA 19-9 only play a limited role in the diagnosis of pancreatic cancer?

Prostate specific antigen

Prostate specific antigen (PSA) is a serine protease with a M_r of 34,000–35,000. Most is bound to α_1-antichymotrypsin, and only a small amount is found in serum as free PSA. It was originally thought to be synthesized solely by epithelial cells in the prostate, but is now also recognized to be present in sweat glands, endometrium, salivary glands, and breast tissue. Serum PSA is raised in patients with prostate diseases such as prostate cancer.

Screening for prostate cancer using PSA has the potential to detect disease at least five years before it is clinically evident. However, there is some doubt as to whether screening for prostate cancer detects clinically significant disease and whether it improves patient outcome; the role of PSA in screening for prostate cancer is therefore controversial. In the USA, supporters of screening point to the rise and then fall in incidence and mortality as a result of the large uptake in PSA testing during the 1980s. In the UK, a risk management programme based on informed choice was introduced in 2003, reflecting concerns over needlessly detecting men who die with, rather than from, prostate cancer, the risks of prostate biopsy to confirm a diagnosis, the low number of individuals thought to benefit from treatment, and questions over the cost effectiveness of screening programmes.

Benign prostatic hypertrophy (BPH) accounts for 80% of cases with increased PSA concentrations detected during screening. Acute urinary retention, acute prostatitis, and prostatic ischaemia are also associated with increased levels. To improve discriminatory power, a variety of (as yet) poorly substantiated strategies have been proposed. Age-specific reference ranges have been proposed to account for physiological increases with age. Measurement of free and bound PSA in serum has also been proposed. In healthy individuals, including those with BPH, around 82% of PSA in serum is bound to α_1-antichymotrypsin. In malignancy, the bound proportion increases to around 90%.

A third option, PSA doubling time, seeks to differentiate slower linear increases in PSA seen in BPH from more rapid rises seen in prostate cancer. Finally, the ratio of serum PSA to the volume of the prostate gland can be used to calculate PSA density (PSAD), which is raised in prostate cancer. However, this requires trans-urethral ultrasound to be carried out as an additional, invasive, relatively expensive, and poorly tolerated diagnostic test.

A key role for PSA in a hospital setting is the detection of disease recurrence after radical prostatectomy. Patients with a PSA concentration below 0.4 µg/L 3-6 months after radical prostatectomy are less likely to develop recurrent disease. A subsequent rise in PSA may be detected 3-5 years before clinical relapse. Prostate specific antigen is also widely used to follow the success/failure of treatment of advanced prostatic cancer. Prognostically, pre-treatment PSA concentrations have been shown to be inversely proportional to survival, with progression-free survival of greater than 36 months for patients with PSA levels 0-99 µg/L, falling to 7.4 months with PSA concentrations greater than 500 µg/L. Overlap in concentrations prevents use of PSA in staging of prostate cancer on PSA alone.

Paraproteins

A paraprotein is a monoclonal immunoglobulin or immunoglobulin light chain in the blood or urine resulting from a clonal proliferation of plasma cells or B-lymphocytes.

Cross reference

Chapter 18 Specific protein markers

Key Points

Measurement of paraproteins in patients with suspected myeloma or Waldenström's macroglobuinaemia provides an example of a central role for a biochemical marker in diagnosis.

The diagnosis of the haematological malignancy multiple myeloma relies on two out of three diagnostic criteria being met:

- **osteolytic lesions** on skeletal X-rays
- an increase in plasma cells in the bone marrow to >10%
- detection of a paraprotein-intact immunoglobulin (two heavy and two light chains) or a Bence-Jones protein (a monoclonal light chain in urine)

Osteolytic lesions: areas of bone loss that can be seen as holes on X-rays caused by infiltration of cancer cells into the bone.

Approximately 80% of myelomas secrete an intact immunoglobulin (light chain plus heavy chain) and 20% produce light chains only. These light chains are filtered by the kidney and detected in the urine. Free light chains in the urine are called Bence-Jones protein. Less than 1% of patients are non-secretors. The symptoms of multiple myeloma and Waldenström's macroglobulinaemia are very vague, with bone pain, fatigue, and recurrent infections, so the detection of a paraprotein or Bence-Jones protein is often the first indication of B cell malignancy and the sensitivity of the test is around 98%. Urine and serum electrophoresis and immunofixation should be carried out on all patients in whom a diagnosis of myeloma is suspected.

The specificity of paraprotein detection is around 61% as paraproteins are also found in patients with monoclonal gammopathy of undetermined significance (MGUS), amyloidosis, and chronic lymphocytic leukaemia. Monoclonal gammopathy of undetermined significance is a pre-malignant disorder characterized by the presence of a paraprotein in the serum but no end-organ damage. The prevalence of MGUS is 2% in people over 50 years old and 3% of people over 70 years old, and in order to distinguish between patients with MGUS and those with myeloma or Waldenström's macroglobulinaemia, further investigations must be carried out. These include a bone marrow aspirate to look at the percentage of plasma cells present in the bone marrow and skeletal survey to look for lytic lesions. Paraprotein and Bence-Jones protein measurement could therefore be classed as screening tests for patients in whom there is a high suspicion of the disease.

Paraproteins may be typed as either immunoglobulin G (IgG), IgA, or IgM (rarely IgD or IgE) heavy chains and either kappa or lambda light chains by immunofixation of serum and/or urine. Figure 19.7 shows the structure of an immunoglobulin.

Typing may provide prognostic information, with IgA myelomas associated with longer survival than IgG myeloma and IgM paraproteins being more commonly associated with Waldenström's macroglobulinaemia (a disorder characterized by an IgM paraprotein that is classed as a low grade non-Hodgkin's lymphoma) that may require different treatment.

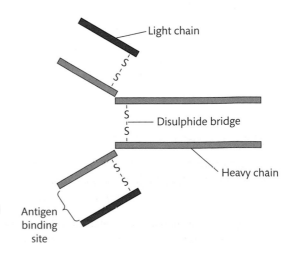

FIGURE 19.7

The structure of immunoglobulins. Immunoglobulins consist of two heavy and two light polypeptide chains joined by disulphide bridges. The antigen binding site is formed by the amino-termini of the heavy and light chains. The amino-termini have very variable amino acid sequences to enable many different antigens to be recognized.

There is no proven benefit in treating patients with MGUS or asymptomatic myeloma. The only proven prognostic factor for progression to symptomatic myeloma is serum paraprotein level, and the percentage risk of progression within ten years equates to the paraprotein concentration, for example a paraprotein concentration of 20 g/L equates to a 20% risk. Guidelines from the British Committee for Standards in Haematology state that all patients with MGUS or asymptomatic myeloma should therefore be monitored indefinitely using clinical assessment and measurement of serum paraprotein and urine Bence-Jones protein.

Quantitation of paraprotein levels by densitometric scanning is widely used to monitor the success of treatment. A 50% fall in concentration is typically associated with disease regression and improved clinical outcome. Typically, successful treatment with chemotherapy such as melphalan/prednisolone or vincristine/adriamycin/dexamethasone may improve overall survival. The last 20 years has also seen wide uptake of peripheral stem cell transplantation following bone marrow chemotherapy as radical treatment for the disease.

SELF-CHECK 19.9

Name the different types of paraprotein that can be identified using serum immunofixation.

Alpha-fetoprotein and human chorionic gonadotrophin

Alpha-fetoprotein is a glycoprotein of M_r 70,000 that is synthesized by foetal yolk sac, liver, and intestine. In the foetus it acts as an albumin-like carrier protein and its concentrations peak at 12–14 weeks of gestation. Alpha-fetoprotein levels are undetectable in healthy individuals after the first year of life.

> ### Key Points
> Alpha-fetoprotein has been found to be of use as a tumour marker in hepatocellular carcinoma and germ cell tumours.

Alpha-fetoprotein is also measured in pregnancy as high maternal levels may indicate foetal neural tube defects.

Human chorionic gonadotrophin (hCG) is a glycoprotein of M_r 45,000 that is secreted by placental trophoblasts. Human chorionic gonadotrophin consists of an α and a β subunit. The α subunit is identical to the α subunit of the pituitary hormones LH, FSH, and TSH. The β subunit is unique to hCG. Human chorionic gonadotrophin is raised in ovarian and testicular germ cell tumours, extragonadal germ cell tumours, and choriocarcinoma, as well as in pregnancy.

> ### Key Points
> Human chorionic gonadotrophin has been found to be a very useful tumour marker for choriocarcinoma, and, along with AFP, for germ cell tumours.

Hepatocellular carcinoma is associated with chronic liver diseases such as hepatitis B and C. The tumour tends to grow rapidly and treatment is more likely to be successful if the disease is found at an early stage when the lesion is small and amenable to surgical removal. Alpha-fetoprotein has a limited role in the diagnosis of early hepatocellular carcinoma. The normal range for AFP is <20 µg/L and values of greater than 400 µg/L have been found to be virtually diagnostic for hepatocellular carcinoma. However, two-thirds of patients with tumours less than 4 cm have serum AFP concentrations less than 200 µg/L, and up to 20% of patients never have a raised AFP, even when their tumours are very large. In addition, raised AFP levels are found in benign diseases such as hepatitis B and C, and cirrhosis, which decreases the specificity of the test. Such patients are at higher risk of developing hepatocellular carcinoma and need to be regularly screened for hepatocellular carcinoma. Ultrasound can detect hepatocellular carcinoma with very high sensitivity and specificity and it has been found that combining ultrasound with AFP measurement increases detection rates. Alpha-fetoprotein can be used to give prognostic information in hepatocellular carcinoma and to detect recurrence following treatment.

Germ cell tumours arise from primitive cells. They occur predominantly in men and are the most common cancers of young adult men. Over 90% arise in the testes and are classified as seminomas or non-seminomas after histopathological examination. Ovarian germ cell tumours are very rare. Alpha-fetoprotein or hCG are elevated in 80% of metastatic and 57% of stage 1 non-seminomatous germ cell tumours. Both markers must be measured as either or both may be secreted by the tumour. In contrast, hCG is raised in less than 20% of seminomas and AFP is raised only if there are liver metastases.

Alpha-fetoprotein and hCG cannot be used as a population screening tool for germ cell tumours as their sensitivity is too low. They are both raised in a variety of other conditions, thus decreasing their specificity for germ cell tumours. Alpha-fetoprotein and hCG may aid in diagnosis but again, their sensitivity and specificity are not good enough for use as sole diagnostic tools. If biopsy is not possible in a patient with a suspected germ cell tumour, elevated AFP and hCG may be used to give a provisional diagnosis and allow treatment to start, as elevation of both AFP and hCG is rarely seen except in germ cell tumours.

Alpha-fetoprotein and hCG can be used as prognostic markers in germ cell tumours. The degree of elevation of AFP immediately before chemotherapy is closely related to prognosis. After surgery for stage I disease, serial measurement of the markers can be used to calculate their half-lives. A half-life for AFP of less than seven days and for hCG of less than three days suggests elimination of the tumour; 89% of patients with satisfactory marker half-lives have complete remission, compared to only 9% of patients with longer marker half-lives. Measurement of hCG and AFP has been incorporated into the International Germ Cell Cancer Collaborative Group's prognostic classification system for non-seminomatous germ cell tumours. Studies have also been carried out on the use of AFP and hCG in monitoring treatment of germ cell tumours. Raised serum AFP and/or hCG were the first indicators of relapse in 55% of patients in one study, but it was also noted that some metastases may behave differently to the primary tumour and may not secrete AFP and hCG even if the primary tumour did. In spite of this, many expert panels now recommend the use of hCG and AFP in the follow-up of patients with treated germ cell tumours.

Choriocarcinoma is a rare form of cancer that originates in the outermost of the membranes surrounding a foetus (the chorion). It is a highly malignant tumour which can rapidly spread to the lungs, but treatment with chemotherapy is highly successful in such tumours. The risk of developing choriocarcinoma after a normal pregnancy or following abortion is around 1 in 50,000 but the risk is increased following a molar pregnancy. This is where the chorion surrounding the embryo degenerates in early pregnancy and fluid-filled sacs develop. Molar pregnancies occur in around 1 in 1,500 pregnancies and 20% of such patients will go on to develop choriocarcinoma.

Serial measurement of hCG is used to monitor patients for choriocarcinoma following a molar pregnancy. Three hospital centres in the UK, namely Charing Cross in London, Sheffield, and Dundee have been set up to carry out this screening, and samples from all over the UK are sent to these centres to ensure the results produced are consistent and reliable. Most cases of choriocarcinoma occur within six months of evacuation of molar pregnancy; therefore hCG levels are monitored weekly until undetectable for three weeks, then monthly until undetectable for six months. Rising hCG levels are diagnostic of choriocarcinoma once a new pregnancy has been ruled out. In addition, abnormal bleeding for more than six weeks following pregnancy should be treated as suspicious and patients should have hCG measurement carried out to exclude a new pregnancy or choriocarcinoma. Patients who have had previous molar pregnancies have a ten-fold increased risk (1-2% incidence) of future molar pregnancies, therefore such patients should be monitored with early ultrasounds in any subsequent pregnancies.

Human chorionic gonadotrophin measurement can be used as part of a clinical classification system to assign patients with choriocarcinoma to good prognosis and poor prognosis groups. Patients with metastatic choriocarcinoma, an hCG level below 40,000 mU/L, and no other risk factors, such as prior chemotherapy or brain or liver metastases, have a good prognosis and are likely to respond to single agent chemotherapy. Patients with risk factors, including a pre-chemotherapy hCG concentration greater than 40,000 mU/L are classed as poor prognosis, with an increased risk of failure of single-agent chemotherapy and increased risk of death.

Human chorionic gonadotrophin is also used to guide treatment of choriocarcinoma and monitor patients for recurrence. Chemotherapy is continued until hCG values have normalized. Two or three courses of maintenance chemotherapy are then given in the hope of eradicating all viable tumours. Once treatment is complete and patients are in remission, serial hCG measurements are carried out every two weeks for the first three months, then monthly until the patient has been in remission for one year. The risk of disease recurrence after one year of remission is less than 1%; therefore frequent hCG measurement is not required after this time.

Alpha-fetoprotein and hCG are versatile tumour markers. Alpha-fetoprotein can be used in the diagnosis of hepatocellular carcinoma and in screening of high-risk patients, although the lack of specificity of the test is a problem. Alpha-fetoprotein can also be used to give prognostic information and monitor patients for recurrence of hepatocellular carcinoma. Alpha fetoprotein is used along with hCG in the diagnosis of germ cell tumours although the sensitivity of the tests is too poor to allow their use in screening. Alpha-fetoprotein and hCG can give prognostic information in germ cell cancers and detect recurrence of the disease. Human chorionic gonadotrophin is the only tumour marker that is used in choriocarcinoma. It is used to screen patients following molar pregnancies, gives prognostic information, and is also used to guide treatment duration and detect disease recurrence. Human chorionic gonadotrophin in the context of choriocarcinoma is the closest marker we have to an ideal tumour marker.

SELF-CHECK 19.10

What are the different types of cancer in which AFP has found use as a tumour marker?

19.6 The future of clinical biochemistry in cancer diagnostics

As most of the tumours markers currently in use are far from ideal, work is being carried out to develop new types of tumour marker. New technology is also being developed to try to tailor cancer treatment to individual patients. Emerging tumour markers and technologies include:

- serum free light chains
- Her2/neu
- molecular diagnostics
- pharmacogenomics

Serum free light chains

Assays for serum free light chains have recently been developed as an alternative to the measurement of urinary Bence-Jones protein in multiple myeloma. The excretion of Bence-Jones protein is dependent on the resorptive capacity of the renal tubules as well as the amount of free light chains that are produced by the B cells. This means that Bence-Jones protein is not detected in the urine until the concentration exceeds the resorptive capacity of the renal tubules and so limits the sensitivity of urine investigation. Measuring the serum free light chain concentration in serum does not rely on renal function, therefore the sensitivity for detecting myeloma is increased. In addition, the serum test could be used in the differential diagnosis of the 1% of multiple myeloma patients previously diagnosed as having non-secretory myeloma. The increased sensitivity of the serum free light chain measurement has allowed 70% of such patients to be identified as paraprotein secretors. However, as it has been shown that treating patients with non-symptomatic myeloma offers no benefit, it remains to be seen whether the increased sensitivity of this test offers a clinical advantage. Serum free light chains can also be used to detect disease recurrence once patients have been treated and it has been shown that patients with an abnormal ratio of free kappa: free lambda light chains are more at risk of progressing from MGUS to multiple myeloma. The tests may also be of value in earlier prediction of response to treatment given free light chain half-life is 6 days compared to 20 and 15 days for IgG and IgA respectively. We shall see in years to come whether this assay replaces urine electrophoresis in the clinical biochemistry laboratory.

Her2/neu

Her2/neu is an oncogene that is over-expressed in late stages of an aggressive form of breast cancer. The extracellular domain (ECD) of Her2/neu is shed from cancer cells and is detectable in the circulation of breast cancer patients who over-express Her2/neu. Her2/neu expression can be assessed using immunohistochemistry and/or fluorescent *in situ* hybridization (FISH) of tumour cells. Over-expression of Her-2/neu identifies women who may benefit from treatment with trastuzumab (Herceptin®). This is a monoclonal antibody that binds with high affinity to the extracellular domain of Her2/neu and inhibits proliferation of the tumour cells that over-express the oncoprotein.

The extracellular domain of Her2/neu may be shed from cancer cells and is detectable in the circulation by immunoassay. The assay provides a means of monitoring treatment and early clinical research indicates that rising serum Her2/neu concentrations may predict non-responders to Herceptin®. This in turn could allow an earlier decision to stop ineffective treatment with a very expensive drug.

SELF-CHECK 19.11

What uses has Her2/neu found in the management of breast cancer?

Molecular diagnostics

Molecular diagnostics is the use of diagnostic techniques to understand the molecular mechanisms of an individual patient's disease. A wide array of techniques such as Southern blotting, fluorescence *in situ* hybridization (FISH), **cytogenetics**, polymerase chain reaction (PCR), and DNA microarrays are used to provide information about diagnosis, prognosis, and therefore aid treatment decisions. The Method box below describes the procedure of FISH. Molecular diagnostics is widely used in the diagnosis and classification of haemopoietic malignancies. Cytogenetic techniques provide information about gene translocations to, for instance, diagnose acute promyelocytic leukaemia, where the PML gene on chromosome 15 fuses with the retinoic acid receptor-α gene on chromosome 17. Most patients with this chromosome translocation respond to treatment with all-trans retinoic acid, therefore it is vital to know whether this particular translocation is present. Molecular diagnostic techniques are also being developed for use in solid tumours such as breast cancer. Gene expression profiling of 78 women with breast cancer identified 70 genes that could predict whether these women would develop metastatic disease within the next five years. Gene expression microarray studies have also been carried out to try and predict sensitivity to chemotherapeutic agents. This information may in future allow us to tailor chemotherapy regimes to individual patients, decreasing the number of patients who undergo unsuccessful and unnecessary therapies.

> **Cytogenetics**: the study of the structure of chromosomes.

Pharmacogenetics and pharmacogenomics

Pharmacogenetics is the study of inherited differences in drug metabolism and response of a single gene or phenotype. Pharmacogenomics is the study of genes that determine drug behaviour. These are new fields that seek to discover the genetic basis for variation in drug response between patients and to use this information to improve the safety and efficacy of drug prescribing and dosing in individual patients. Inherited variations in drug targets, drug metabolizing enzymes, or drug transporters can have a major impact on a patient's response to a drug and their chance of experiencing toxic side effects or adverse reactions. The genetic variations include nucleotide deletions, insertions, and single nucleotide polymorphisms, all

METHOD *Fluorescence in situ hybridization*

Fluorescence *in situ* hybridization (FISH) is a technique used in cytogenetics to look at a specific gene or chromosome abnormality that is beyond the resolution of conventional cytogenetic techniques. A fluorescently labelled probe is used that consists of the DNA sequence complementary to the DNA sequence of interest. The sample DNA, consisting of metaphase chromosomes or interphase nuclei is denatured using heat to separate the complementary strands of the DNA. The probe is added to the mixture and allowed to hybridize to the DNA. The sample is then viewed under the fluorescent microscope and presence or absence of the fluorescent probe is looked for.

One use of FISH is to look for gene rearrangements that are specific for certain types of leukaemias. This information can be used to give an accurate diagnosis and place a patient in a particular prognostic group to enable the clinician to decide on the appropriate treatment for each patient. The probe is designed to hybridize to the specific gene rearrangement so that the fluorescent signal is only seen in samples which are positive for the gene rearrangement.

of which can alter the amino acid sequence of the encoded proteins and therefore their function. Once these variations have been identified, patients can be screened before the drug in question is prescribed, and it also allows the dose to be individually tailored for maximal clinical effect and minimal side effects.

An important example of a genetic polymorphism which alters the metabolism of a drug is thiopurine methyltransferase (TPMT). This is an enzyme involved in the metabolism of azathioprine, a purine antimetabolite that is used in the treatment of leukaemia. Patients with TMPT deficiency caused by mutations in the TPMT gene fail to metabolize azathioprine at the normal rate; therefore high levels build up in the blood. This can lead to suppression of bone marrow activity, leaving the patient anaemic and susceptible to infection. Analysis of TPMT activity should be carried out before azathioprine treatment is started, to identify patients with low TPMT activity and allow lower drug doses to be given, to try to avoid toxic build-up of the drug in the patient.

SELF-CHECK 19.12

How may we be able to use pharmacogenomics in cancer patients in the future?

Cancer can develop in many different tissue types and different organs. The type and site of the cancer may affect the biochemical changes that are seen, for instance, some types of cancer may cause obstruction of hollow organs such as the liver, or destroy organs such as the pituitary gland. Cancer treatment can also damage organs such as the liver or kidneys, or cause specific biochemical abnormalities such as hypomagnesaemia. Tumour markers have been developed for use in certain types of cancer such as CA 19-9 in pancreatic cancer and PSA in prostate cancer. Ideally, we would like to use these for screening, diagnosis, monitoring of disease, and to give prognostic information, but none of the tumour markers currently available is good enough to use in all these situations. New tumour markers such as serum free light chains for myeloma and Her2/neu for breast cancer are being developed which may have more widespread use in the future. Fields such as cytogenetics and pharmacogenomics are coming to the fore to try and tailor treatments to individual patients, improving their efficacy and decreasing the toxic effects of the drugs. In future, these techniques may play an even larger role in the diagnosis and treatment of patients with cancer.

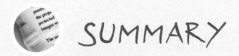

SUMMARY

- Cancer is a group of diseases and the incidence of cancer is the same in both sexes but the incidence of the different types of cancer differs.

- Survival rates for many types of cancer are improving as the disease is detected earlier and treatments become more effective.

- Malignancy can have wide-ranging effects on the body, depending on the location and size of the tumour and the substances it is secreting.

- Routine biochemical tests can give clues as to the presence of cancer, but further investigations are required to confirm the diagnosis.

- The ideal tumour marker could be used in all aspects of cancer management, from screening and diagnosis to monitoring recurrence, but unfortunately the ideal tumour marker does not exist.

- Different tumour markers have been developed which are useful in certain types of cancer.

- Care must be taken in the interpretation of tumour marker results as the majority suffer from a lack of specificity which can limit their usefulness.

- New, more specific tumour markers such as Her2/neu and serum free light chains are being developed and could soon be in widespread use.

- Fields such as molecular diagnostics and pharmacogenomics are increasingly being used to improve cancer diagnosis and tailor treatment to the individual patient.

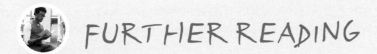

FURTHER READING

- **Ceschi M, Gutzwiller F, Moch H, Eichholzer M, and Probst-Hensch NM (2007)** Epidemiology and pathophysiology of obesity as cause of cancer. *Swiss Medical Weekly* **137**, 50–6.

  A review of research into the link between obesity and cancer.

- **Dear R, Wilcken N, and Shannon J (2008)** Beyond chemotherapy—demystifying the new 'targeted' cancer treatments. *Australian Family Physician* **37**, 45–9.

  Clear overview of new chemotherapy treatments.

- **Duffy MJ (2004)** Evidence for the clinical use of tumour markers. *Annals of Clinical Biochemistry* **41**, 370–7.

  A good overview of the clinical utility of a range of tumour markers currently in widespread use.

- **Lee W, Lockhart C, Kim RB, and Rothenberg ML (2005)** Cancer pharmacogenomics: powerful tools in cancer chemotherapy and drug development. *Oncologist* **10**, 104–11.

  A detailed review of the potential uses of pharmacogenomics in cancer treatment.

- **Pannall P (1992)** The clinical biochemistry of malignancy. *Clinical Biochemistry Reviews* **13**, 142–51.

  Presents a very clear account of the biochemical effects of cancer.

- **Smith A, Wisloff F, and Samson D (2006)** Guidelines on the diagnosis and management of myeloma. *British Journal of Haematology* **132**, 410–51.

  A detailed overview on the current guidelines for diagnosis and treatment of multiple myeloma.

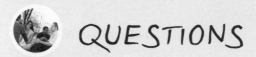

QUESTIONS

19.1 Lung cancer is the most common cancer in male patients: TRUE or FALSE?

19.2 Are routine biochemistry tests (excluding tumour markers) useful in the diagnosis of cancer? Which tests could be helpful in a patient with cancer?

19.3 Which of the following may be a factor in the development of cancer? More than one answer may be correct.

    (a) Inactivation of a tumour suppressor gene

    (b) Overexpression of a tumour suppressor gene

    (c) Inactivation of a proto-oncogene

    (d) Gain of function mutation of a proto-oncogene

19.4 No current test can be classed as an ideal tumour marker. What are the characteristics that are required of the ideal tumour marker? Should we be using the currently available tests in view of the fact that they all have drawbacks?

19.5 Tumour related hypoglycaemia can only be caused by excess production of insulin from tumour cells: TRUE or FALSE?

19.6 Which three of the following are used as diagnostic signs for myeloma?

    (a) Osteolytic lesions

    (b) Kidney failure

    (c) Presence of a paraprotein

    (d) Fractured bone

    (e) Anaemia

    (f) Recurrent infections

    (g) An increase in bone marrow plasma cells to greater than 10% of the total

Answers to self-check questions, case study questions, and end-of-chapter questions are available in the Online Resource Centre accompanying this book.

Go to www.oxfordtextbooks.co.uk/orc/ahmed/

Inherited metabolic disorders and newborn screening

Mary Anne Preece

Learning objectives

After studying this chapter you should be able to:

- Explain modes of inheritance
- Explain the effects of a metabolic block
- Give examples of treatment for inherited metabolic disorders
- Describe, with examples, the groups of inherited metabolic disorders
- Know which investigations can be used to diagnose inherited metabolic disorders
- Explain the strategies for antenatal diagnosis of inherited metabolic disorders
- Describe the disorders detected by newborn screening

Introduction

Inherited metabolic disorders (IMDs) are a heterogeneous group of genetic conditions that present mostly in childhood. Although individually rare, collectively they are numerous and cause substantial morbidity and mortality. It is estimated that the prevalence of IMDs in the UK is more than 1 in 1,000 live births. Some IMDs present acutely and lead to death in the neonatal period, whereas others are benign. Some may be diagnosed in infancy, and need complex, often dietary treatment requiring lifelong monitoring. Many cause significant health problems for affected individuals. A specific diagnosis enables **genetic counselling** to be offered to the family and subsequent **antenatal diagnosis** in future pregnancies. As girls with IMDs reach adulthood, care during pregnancy becomes an additional issue, to achieve optimum outcome for the baby without compromising the mother's health. The field of IMDs is vast and continually expanding. This chapter is an introductory overview of IMDs with illustrative examples.

20.1 Modes of inheritance

Humans have 23 pairs of chromosomes. For each pair of chromosomes, one is inherited from the mother and one from the father. One pair is the sex chromosomes; females have two X chromosomes, males have one X and one Y chromosome. The non-sex chromosomes are called **autosomes**. Inherited metabolic disorders are due to mutations in the genes (DNA) on the chromosomes. Autosomal recessive, autosomal dominant, and X-linked recessive are forms of Mendelian inheritance, as shown in Figure 20.1.

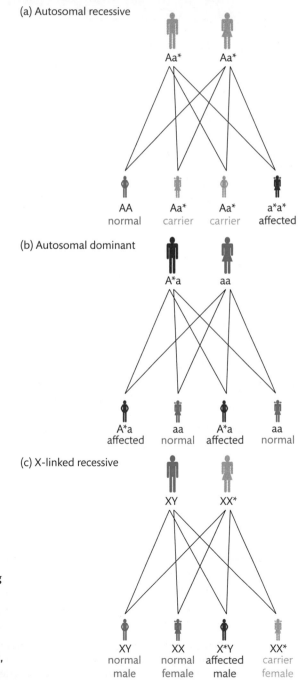

FIGURE 20.1

Modes of inheritance. The chromosome carrying a mutation is indicated with an asterisk. A green individual is normal, an orange individual is a carrier, and a red individual is affected with the disorder. The modes of inheritance shown are: (a) autosomal recessive, (b) autosomal dominant, (c) X-linked recessive.

Autosomal recessive inheritance

This situation is shown in Figure 20.1(a). Both parents are **heterozygotes**, that is, they carry an abnormal recessive gene. Carriers are usually asymptomatic and most families are unaware that they carry a disorder until they have an affected child. For each pregnancy there is a 1 in 4 risk of the child inheriting one abnormal gene from each parent and thus being affected with the disorder. On average, 25% of offspring will be affected (abnormal **homozygotes**), 25% will be normal, and 50% will be carriers (heterozygotes).

Key Points

Most IMDs have an autosomal recessive mode of inheritance.

Autosomal dominant inheritance

This situation is shown in Figure 20.1(b). An affected parent has one copy of an abnormal dominant gene (and is affected with the disorder). On average, half of that parent's offspring will inherit the abnormal gene and will be affected with the same disorder as their parent. Half of the parent's offspring will not inherit the gene and so they will not be affected with the disorder, nor will they be a carrier. Some of the **porphyrias** are inherited in an autosomal dominant fashion.

X-linked inheritance

X-linked disorders are due to mutations on the X chromosome. Males inherit their one X chromosome from their mother. If it carries a mutation they will be affected with the corresponding disorder, that is, they will be **hemizygote**. This situation is shown in Figure 20.1(c).

In females, a process called **lyonization** randomly inactivates one of the X chromosomes in each cell of the body. Thus in some female heterozygotes there is a chance that more abnormal than normal X chromosomes will be active. This can result in a female having symptoms of the disorder. If affected, females generally have a milder form of the disorder, but nevertheless they may have significant clinical problems. Examples of X-linked disorders are **adrenoleukodystrophy** and the urea cycle disorder ornithine transcarbamylase deficiency.

Maternal inheritance

Mitochondrial DNA (mtDNA) defects are transmitted from the mother. Mitochondria have their own DNA, each mitochondrion has several copies of DNA, and each cell has many mitochondria. There are approximately 700 copies of mtDNA per cell in fibroblasts and over 200,000 copies per cell in oocytes. When a cell divides mitochondria are shared randomly between the two new cells, and then multiply to form the appropriate number of mitochondria for the cell as shown in Figure 20.2. This phenomenon in which cells have different proportions of normal and abnormal mitochondria is known as **heteroplasmy**.

At fertilization all the mitochondria (and mtDNA) come from the ovum, the sperm does not contribute any mitochondria to the offspring.

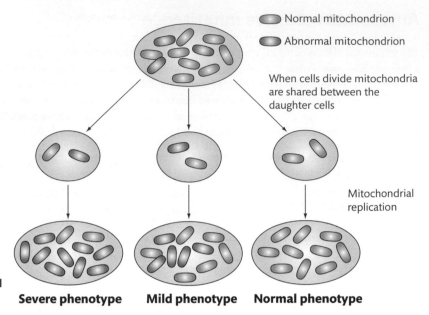

FIGURE 20.2
A schematic illustrating mitochondrial
replication during cell division.

Severe phenotype **Mild phenotype** **Normal phenotype**

> ## Key Points
>
> **If there is a mtDNA mutation, symptoms will depend upon the percentage of abnormal mtDNA in each cell.**

SELF-CHECK 20.1

(a) A couple have had a child affected with an autosomal recessive disorder. What is the chance of their next child being affected? (b) Which disorders are transmitted by maternal inheritance?

20.2 Effects of a metabolic block

The effects of a metabolic block can be considered in simplistic terms by looking at a hypothetical pathway. Figure 20.3 shows an example where DNA codes for an enzyme, which may require a cofactor to function.

If there is a block in the pathway between B and C due to an enzyme deficiency shown by the red square, then B, and to a lesser extent A, will accumulate. These may be toxic, or may be converted to other toxic metabolites, F and G. C and D and any metabolites of C and D, such as H, may become deficient, leading to other clinical consequences. Approaches to treatment are

FIGURE 20.3
Effects of an enzyme defect in a
hypothetical metabolic pathway.
See text for details.

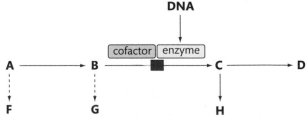

TABLE 20.1 Strategies for treatment of IMDs, with specific examples.

Treatment goal	Principle used	Examples
Substrate reduction	Dietary restriction	Low phenylalanine diet in phenylketonuria
		Galactose free diet in galactosaemia
		Use of medium chain fatty acid as a fat source in disorders of long chain fatty acid oxidation
	Decreasing flux through the pathway leading to the block	NTBC in tyrosinaemia type 1 blocks the pathway at a stage prior to the enzyme defect
		Avoiding lipolysis in fatty acid oxidation disorders
		Haem arginate in porphyrias exerts negative feedback on the first step of the haem biosynthetic pathway
	Removal of substrate	Exchange transfusion, dialysis, haemofiltration used acutely in many disorders
		Carnitine therapy in organic acidurias (will conjugate with the organic acid and can be excreted)
		Betaine in homocystinuria
Correcting product deficiency	Supply of the product	Continuous enteral glucose or uncooked corn starch in glycogen storage disease
		Cystine supplementation in homocystinuria
	Increasing substrate supply	Carnitine in carnitine transporter defect
Decreasing toxicity of metabolites	Removal of toxic metabolites	Sodium benzoate and sodium phenylbutyrate for treatment of hyperammonaemia
Stimulating residual enzyme	Vitamin treatment (pharmacological doses)	Pyridoxine in homocystinuria
		Vitamin B_{12} in methylmalonic aciduria
	Drug treatment	Phenobarbitone in Crigler Najjar syndrome type 2
Enzyme replacement	Organ transplantation	Liver transplantation for some disorders
		Bone marrow transplantation for some disorders
	Enzyme therapy	Enzyme available for several lysosomal storage disorders (for example Gaucher's disease) with more under development
	Gene therapy	Not yet in routine clinical use

considered in more detail below, but fall into three categories: removal of the substance and/or metabolites that accumulate before the block (i.e. A, B, F, G), improving the function of or replacing the defective enzyme, or replacement of the deficient products of the blockage (i.e. C, D, H).

20.3 Approaches to treatment

As the understanding of IMDs has advanced, so has the range of treatments available. This, together with improvements in diagnostic testing and in general medical care, has meant that IMD patients are surviving longer. Examples of treatment are shown in Table 20.1, and some are further discussed below. Remember that symptomatic treatment is also necessary, for example anticonvulsants for seizures.

20.4 Laboratory testing for inherited metabolic disorders

Routine laboratory tests may be useful as first-line investigations in patients suspected of an IMD. These are shown in Table 20.2.

TABLE 20.2 Tests that may be useful when investigating an IMD.

Test (blood unless indicated otherwise)	Comments
Glucose	Hypoglycaemia is a presenting feature of many disorders, notably organic acidurias, fatty acid oxidation disorders, and carbohydrate disorders.
Ammonia	Hyperammonaemia (ammonia >200 µmol/L) is a medical emergency and needs prompt treatment and investigation. The main disorders with hyperammonaemia are organic acidurias and urea cycle defects. Beware of artefactual increases of ammonia.
Acid-base status	Metabolic acidosis is found in many disorders. Respiratory alkalosis is an early finding in the urea cycle disorders.
Lactate	Lactate may be increased due to hypoxia, shock, or seizures. If secondary causes of lactic acidaemia are excluded, primary causes should be sought.
Liver function tests including clotting	Some disorders have major abnormalities of liver function, for example galactosaemia.
Creatine kinase	Increased in long chain fatty acid oxidation defects, muscle glycogen storage diseases.
Lipids	Increased in some glycogen storage diseases.
Urate	Increased in glycogen storage disease type 1, decreased in molybdenum cofactor deficiency.
CSF glucose	Inappropriately low in glucose transporter deficiency (GLUT 1).
CSF lactate	May be increased in electron transport chain disorders. Useful as a pointer to proceed to further invasive testing.

The basic repertoire of tests for IMDs is shown in Table 20.3.

More specialist tests may also be needed; a useful directory is available on the Metabolic Biochemistry Network website (www.metbio.net). With the rapid growth of molecular biology techniques, many defective genes for IMDs are known. However, analysis of DNA does have pitfalls. If a base change is found in a gene it may not necessarily be disease causing. Some patients do not have a detectable mutation. If a **mutation** is not detected a specific IMD cannot be excluded, it just becomes less likely.

Who to investigate

Inherited metabolic disorders are complex and may seem bewildering to those who are not working in the field. Many of the specialist investigations are complex and expensive so it is helpful to use a structured approach to investigation. Useful information can be obtained from:

- family history (has there been a previously affected sibling, is there an extended family history suggesting X-linked inheritance?)
- clinical features (is there evidence of hepatomegaly, dysmorphic features, hypotonia?)
- evidence of regression (loss of developmental milestones that have previously been reached)
- results of the initial tests—see Table 20.2
- age when the patient first developed symptoms
- relationship between feeding and symptoms

TABLE 20.3 Specialist tests for investigation of IMDs.

Test	Principle of method	Comments
Amino acids (plasma, urine, CSF)	Based on ninhydrin reaction with amino and imino groups. Analysed by qualitative, for example thin layer chromatography, or quantitative techniques, for example HPLC, ion-exchange chromatography. Also analysed by tandem mass spectrometry.	Some disorders diagnosed better in plasma and others in urine. Qualitative techniques will not detect low concentrations of amino acids. Some drugs react with ninhydrin, for example antibiotics.
Homocystine Total homocysteine	Ion-exchange chromatography, HPLC or tandem mass spectrometry.	Special sample handling required.
Organic acids (urine)	Solvent extraction from urine, derivitization, for example trimethylsilyl then separation and identification by gas chromatography-mass spectrometry.	Will detect most organic acids and other compounds, for example some amino acids, purines and pyrimidines, drugs and drug metabolites.
Acyl carnitines (blood, bile)	Analysed by tandem mass spectrometry.	Useful for diagnosis of organic acidurias, particularly fatty acid oxidation defects. Carnitine depletion leads to false values for some metabolites.
Very long chain fatty acids (plasma)	Fatty acids are transmethylated and measured by GC-MS.	Useful first-line test for peroxisomal disorders.
Glycosaminoglycans (urine)	Isolation from urine and separation by 1- and 2-dimensional cellulose acetate electrophoresis.	Spot tests are unreliable.
Oligosaccharides (urine)	Separation by thin layer chromatography.	Abnormal in glycogen storage disorders and some lysosomal storage disorders. May find diet or drug artefacts.
Sugar chromatography (urine, faeces)	Separation by thin layer chromatography.	Faeces—may detect disaccharidase deficiencies. Urine—may detect carbohydrate disorders. Results depend upon dietary intake.
Free fatty acids and 3-hydroxybutyrate (plasma)	Useful test to assess the physiological response to hypoglycaemia.	Results may suggest hypoketotic hypoglycaemia or hyperinsulinism.
Chitotriosidase (plasma)	Fluorimetric enzyme assay.	Increased in some lysosomal storage disorders.
Leukocyte enzymes (blood)	Isolation of leukocytes from whole blood. Various enzyme assays used for diagnosis of lysosomal disorders.	Relatively large amount of blood required. Some assays available on dried blood spots.
Fibroblast assays	Fibroblasts cultured from skin biopsy. Enzyme assays or radiochemical 'whole pathway' assays can be carried out.	Fibroblasts can be stored permanently in liquid nitrogen.
DNA testing	Mutation analysis or sequencing is available for many disorders but for some the primary enzyme defect is not known.	Not all mutations are causal, some are polymorphisms. Gene expression required to establish significance of some mutations.

For most patients suspected of an IMD a reasonable set of first-line investigations is urine for amino acids and organic acids, together with blood for amino acids and acyl carnitines. If there are features that suggest a storage disease then consider urine oligosaccharide and mucopolysaccharide analysis. If there are dysmorphic features with hypotonia, then consider plasma very long chain fatty acid measurement. Specialist laboratories will be happy to advise on tests for a particular situation. The Metabolic Biochemistry Network website (www.metbio.net) has guidelines for investigation of certain presentations which are intended for those who work in non-specialist laboratories.

There are two situations when it is important to obtain specimens. These are in an acutely presenting patient who is likely to die and in sudden unexplained death in infancy (SUDI). The following specimens should be considered:

- blood collected in lithium heparin tubes for amino acids and acyl carnitines
- blood collected in EDTA tubes for DNA analysis
- urine for organic acids and amino acids
- skin biopsy for fibroblast culture

20.5 Groups of metabolic disorders

There are several ways of classifying IMDs, which perhaps tells us that none are really satisfactory. Pathways have a range of potential disorders due to defects at each step (for example Figure 20.4).

The groups of metabolic disorders will be considered in this chapter in terms of their primary abnormality in body fluids, that is, the test by which they are most likely to be detected.

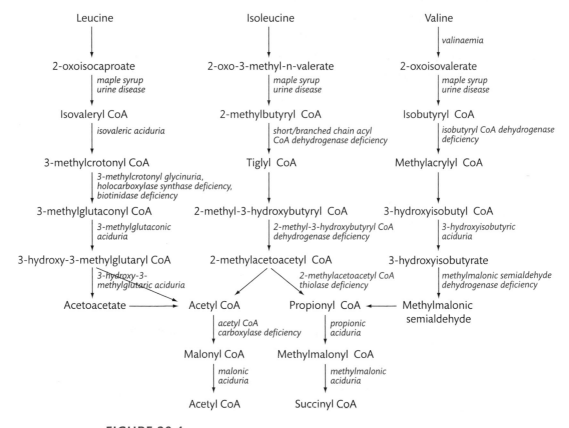

FIGURE 20.4
Disorders affecting the metabolism of branched chain amino acids.

Amino acid disorders

Amino acids are the building blocks of proteins, and disorders of amino acids can result from defects in the breakdown of amino acids or their metabolism. The build-up of certain amino acids in the blood can cause serious medical complications in the newborn. Selected amino acid disorders will be discussed in this section and include phenylketonuria, tyrosinaemia, homocystinuria, cystinuria, and cystinosis.

Phenylketonuria

Phenylalanine hydroxylase converts phenylalanine to tyrosine as shown in Figure 20.5.

Phenylketonuria (PKU) is caused by a deficiency of the enzyme phenylalanine hydroxylase. This leads to increased phenylalanine concentrations in the body. If untreated, severe developmental delay occurs. In affected neonates, phenylalanine concentrations increase rapidly after the introduction of milk feeds. In the UK, all babies are screened for PKU usually 5–8 days after birth by measurement of phenylalanine concentration in dried blood spots, so that treatment can be commenced as soon as possible, leading to optimal outcome. The children are treated with a low protein diet, supplemented with an amino acid mixture from which phenylalanine has been removed. The diet is difficult and requires expert dietetic supervision. Regular monitoring of blood phenylalanine concentrations is required to guide treatment, usually at least weekly throughout childhood. During adolescence and adulthood, the diet can be relaxed but most centres now recommend diet for life. Females must return to a strict diet during pregnancy and ideally this should be started pre-conceptually. If females are not well controlled during pregnancy, there is a high risk of foetal abnormalities, for example **microcephaly**, congenital heart disease and learning difficulties.

Some patients (1–3%) with increased blood phenylalanine concentrations detected by newborn screening do not respond to low phenylalanine diets. They have defects in

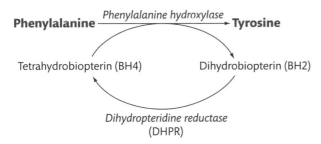

FIGURE 20.5
Enzyme block in phenylketonuria showing the involvement of biopterin. See text for details.

CASE STUDY 20.1

A male baby had a dried blood spot phenylalanine concentration determined in the newborn screening specimen collected at five days of age (reference range given in brackets):

Phenylalanine   2050 μmol/L   (<100)

(a) What is the most likely diagnosis?

(b) What is the treatment?

dihydropteridine reductase or in biopterin synthesis, which cause defective phenylalanine hydroxylase function. In biopterin defects, tyrosine hydroxylase and tryptophan hydroxylase are also affected leading to a deficiency of neurotransmitters. These patients require supplementation with neurotransmitters as well as a low phenylalanine diet.

Tyrosinaemia type 1

Fumarylacetoacetate lyase converts fumarylacetoacetate to fumarate and acetoacetate. **Tyrosinaemia** type 1 is caused by a deficiency of fumaryl acetoacetate lyase (FAL). The disorder can present in several ways:

- acute liver failure and renal tubular dysfunction
- hypophosphataemic rickets and chronic liver disease
- abdominal and neurological crises thought to be caused by accumulation of 5-aminolevulinate

Patients usually have markedly increased plasma α-fetoprotein in addition to abnormal liver function tests. Plasma amino acids show increased tyrosine and sometimes methionine and phenylalanine. These findings are non-specific and may be secondary to liver dysfunction. Urine 5-aminolevulinate is increased due to inhibition of porphobilinogen synthase by succinyl acetone.

Key Points

The diagnosis of tyrosinaemia type 1 is made by demonstration of succinyl acetone in urine.

The disorder is treated using 2-(2-nitro-4-trifluoromethylbenzoyl)-1,3-cyclohexanedione (NTBC) and this has been in use since 1992. This compound blocks the metabolic pathway several steps before FAL. It prevents the production of succinyl acetone and the toxic effects thereof. Figure 20.6 shows that NTBC creates an additional block in the pathway, so patients need a low tyrosine diet (similar to that used in PKU).

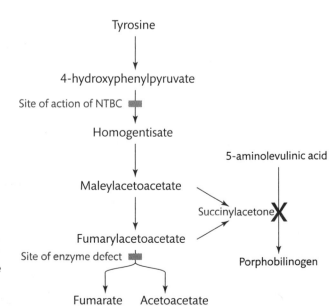

FIGURE 20.6

Metabolic pathway of tyrosine showing the enzyme block in tyrosinaemia type I and the site of action of NTBC. Also note the inhibitory effect of succinyl acetone on porphobilinogen synthase.

There is a long-term risk of development of **hepatoma** so patients are regularly monitored to detect this.

Homocystinuria

Homocystinuria is due to a defect of methionine metabolism. The commonest form is known as classical homocystinuria and is due to a deficiency of the enzyme cystathionine-β-synthase (CBS) which converts methionine to homocystine. Classical homocystinuria is a multisystem disorder. Clinical features include dislocated lenses, developmental delay, thromboembolic episodes, bony abnormalities, osteoporosis, and psychiatric disorders. The enzyme blockage causes both homocystine and methionine to accumulate in body fluids. Cystathionine-β-synthase requires pyridoxine as a cofactor. Approximately half of the patients with classical homocystinuria respond to pharmacological doses of pyridoxine (250–500 mg/day). Unresponsive patients are treated with a low methionine diet (similar to that used in PKU) or with the methyl donor betaine which acts by converting homocystine to methionine as shown in Figure 20.7. The Method box below describes the measurement of homocysteine.

Some patients may need diet and betaine to achieve good biochemical control.

There are two further forms of homocystinuria, known as remethylation defects. In contrast to classical homocystinuria they have low (rather than high) methionine concentrations. They present at any age, usually with neurological symptoms. Figure 20.7 shows the steps catalysed by 5,10-methylene tetrahydrofolate reductase and methionine synthase. Methionine synthase deficiency causes methylmalonic aciduria as well as homocystinuria, (see below) and treatment is with pharmacological doses of vitamin B_{12}. Deficiency of 5,10-methylene tetrahydrofolate reductase is treated with betaine.

Cystinuria

Cystinuria is an amino acid transport disorder in which there is defective transport of cystine and basic amino acids across the renal tubule and the small intestine. Urine amino acids show grossly increased concentrations of cystine and basic amino acids (ornithine, lysine, and arginine), whereas plasma amino acids are normal. The only clinical symptom is the formation of

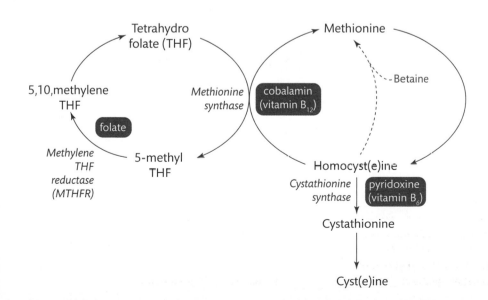

FIGURE 20.7
The metabolic pathway of homocystine, together with the defects which cause homocystinuria.

METHOD Measurement of Homocysteine

Homocystine is a sulphur containing amino acid. Two forms exist in equilibrium; with a sulphydryl group it is known as homocysteine (hcys), and with a disulphide bridge it is known as homocystine (hcys₂).

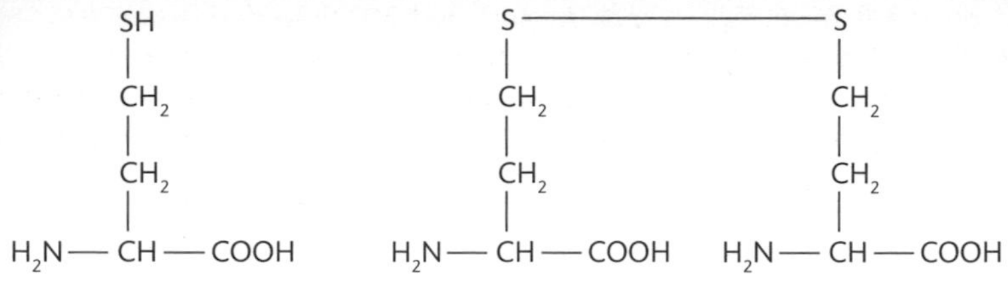

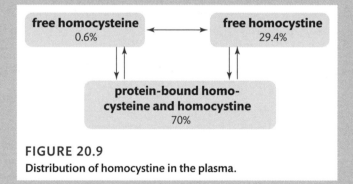

FIGURE 20.8
Structures of homocysteine and homocystine.

About 70% of homocystine is protein bound.

free homocysteine 0.6%	⟷	**free homocystine** 29.4%

protein-bound homo-cysteine and homocystine 70%

FIGURE 20.9
Distribution of homocystine in the plasma.

Plasma for free hcys₂ measurement (by an amino acid analyser) should be separated and deproteinized promptly. Total homocysteine measures all the fractions of hcys₂ and hcys—free and protein bound, but blood should still be separated within 30 minutes of venepuncture. Homocystine can be difficult to detect in urine by qualitative techniques and measurement of total homocysteine in blood is recommended if homocystinuria is suspected.

cystine stones in the kidney, due to its low solubility. Treatment is aimed at preventing stone formation by maintaining a high fluid intake, alkalinization of the urine to increase cystine solubility, or the use of drugs that form soluble disulphide compounds with cystine.

Key Points

Diagnosis of cystinuria is by measurement of amino acids in the urine.

Cystinosis

Cystinosis is a lysosomal storage disorder. Cystinosin is a lysosomal membrane protein that transports cystine out of the lysosome. Cystinosis is due to a defect in cystinosin and results in the formation of cystine crystals in the lysosome. In infantile nephropathic cystinosis, the commonest form of the disorder, children present with failure to thrive, growth retardation and hypophosphataemic rickets at a few months of age. Slit lamp examination of their eyes shows pathognomonic crystals of cystine in the cornea, which cause photophobia from about two years of age. They have a Fanconi syndrome, that is, suffer glycosuria, generalized amino-aciduria, acidosis, hypouricaemia, hypokalaemia, and phosphaturia. If untreated, end-stage renal failure occurs by 5–12 years of age. Treatment is with the drug cysteamine, a weak base that is able to enter the lysosome. It forms disulphides with cystine that can pass out of the lysosome via the lysine carrier. Treatment can slow the progression towards renal failure.

Key Points

Diagnosis of cystinosis is by detection of increased cystine concentration in leukocytes.

SELF-CHECK 20.2

(a) What is the basis for treatment of PKU? (b) What are the clinical features of classical homo-cystinuria? (c) What is the difference between cystinuria and cystinosis?

Urea cycle disorders

The urea cycle is a series of enzyme-catalysed reactions responsible for formation of urea to enable elimination of waste nitrogen from the body, as shown in Figure 20.10.

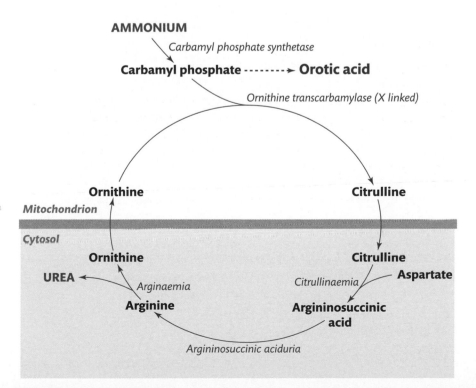

FIGURE 20.10
The urea cycle showing the sites of inherited metabolic disorders.

Deficiency of any of the enzymes involved can lead to **hyperammonaemia**, which is a medical emergency. Some causes of hyperammonaemia are listed in Box 20.1. Ammonia is toxic to the brain and hyperammonaemia can cause coma. Neonatal onset urea cycle disorders present with severe hyperammonaemia after an initial 24–48 hour period of having been well. The clinical features are non-specific and may be confused with sepsis, but babies deteriorate rapidly and without treatment may go into a coma and die within days. Urea cycle defects can, however, present at any age. In particular, female heterozygotes for X-linked ornithine transcarbamylase deficiency can have a wide range of phenotypes.

BOX 20.1 Causes of hyperammonaemia

Artefactual

- delay in transit/analysis
- haemolysis
- capillary blood
- difficult venepuncture

In neonates/infants

- sick preterm neonates
- severe perinatal asphyxia
- infection
- parenteral nutrition
- transient hyperammonaemia of the newborn

In infants/children/adults

- valproate therapy
- Reye's syndrome
- liver disease
- urinary tract infection
- leukaemia

IMDs

- urea cycle disorders
- organic acid disorders, especially methylmalonic and propionic acidurias

Miscellaneous other IMDs

- hyperornithinaemia, hyperammonaemia, and homocitrullinaemia syndrome
- lysinuric protein intolerance
- hyperinsulinaemia due to glutamate dehydrogenase defect
- neonatal ornithine aminotransferase deficiency

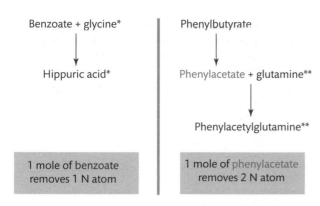

Benzoate + glycine*

↓

Hippuric acid*

Phenylbutyrate

↓

Phenylacetate + glutamine**

↓

Phenylacetylglutamine**

| 1 mole of benzoate removes 1 N atom | 1 mole of phenylacetate removes 2 N atom |

FIGURE 20.11
Mode of action of drugs used to achieve alternative nitrogen excretion.

Key Points

Plasma ammonia should be measured in any patient with unexplained encephalopathy, and if genuinely increased an underlying IMD should be sought.

Emergency treatment of hyperammonaemia includes use of drugs that utilize alternative pathways of nitrogen excretion, as shown in Figure 20.11.

If ammonia concentration is not controlled by drug treatment then haemofiltration techniques will be necessary. If a patient has hyperammonaemia, appropriate follow-up tests for urea cycle defects include quantitative plasma amino acids and urine orotate, which will give the following results:

- carbamyl phosphate synthetase (CPS) deficiency—low citrulline, normal orotate
- ornithine transcarbamylase (OTC) deficiency—low citrulline, high orotate
- citrullinaemia—very high citrulline, high orotate
- argininosuccinic aciduria—slightly increased citrulline and orotate, plus the presence of argininosuccinate
- argininaemia—increased orotate, normal citrulline, increased arginine

Some organic acidurias can also present with severe hyperammonaemia so urine organic acids and blood acyl carnitines should also be measured.

SELF-CHECK 20.3

What is the toxic compound that accumulates in urea cycle disorders?

Organic acid disorders

Organic acidurias are due to defects which cause the accumulation of low M_r carboxylic acids in the body. The defects primarily affect intermediary metabolism of small molecules, particularly the catabolism of the branched chain amino acids valine, leucine, and isoleucine, and the steps of propionate metabolism (Figure 20.4). Classical organic acidurias present in the

CASE STUDY 20.2

A female neonate presented at 48 hours of age with jitteriness and abdominal distension. Initial investigations including glucose, lactate, urea and electrolytes, septic screen, chest X-ray, and cranial ultrasound were all normal. She deteriorated and developed seizures. Biochemical investigation showed the following (reference range is given in brackets):

Ammonia   1250 µmol/L   (<100)

(a) What group of disorders could present in this way?

(b) What investigations should be carried out urgently?

neonatal period with severe metabolic acidosis accompanied by hypoglycaemia, hyperammonaemia, hypocalcaemia, and ketosis. However, in common with other IMDs the clinical spectrum is wide and patients may present in infancy or childhood with a less severe or episodic clinical picture. In some organic acidurias metabolic acidosis is not a major feature. The first organic aciduria described was isovaleric aciduria, which is due to a defect of an enzyme in the leucine catabolic pathway. The disorder was initially called the sweaty sock syndrome because the volatile isovalerate accumulates in body fluids and imparts a distinctive odour.

Urine organic acid analysis by gas chromatography-mass spectrometry and blood acyl carnitine analysis by tandem mass spectrometry are the most important diagnostic investigations.

Disorders of propionate metabolism—propionic aciduria and methylmalonic aciduria

Propionyl CoA is formed by the catabolism of the amino acids valine, isoleucine, threonine, and methionine, and also that of thymine, uracil, cholesterol, and odd chain fatty acids. Some propionate originates from bacterial action in the gastrointestinal tract. Methylmalonic aciduria (MMA) and propionic aciduria (PA) are relatively common organic acidurias due to defects in the propionate pathway. Patients present early in life with metabolic acidosis, hyperammonaemia, and hypoglycaemia. They may also have hypocalcaemia and **neutropenia**. Clinically, the two disorders are indistinguishable, but they can be differentiated by urine organic acid analysis. Adenosyl cobalamin is an essential cofactor for methylmalonyl CoA mutase and some later presenting cases of MMA respond biochemically and clinically to pharmacological doses of vitamin B_{12} as shown in Figure 20.12.

There are also defects in the intracellular metabolism of vitamin B_{12}. Methylmalonic acidurias can also arise due to defects that impair production of adenosylcobalamin (required for the metabolism of methylmalonate) and/or methylcobalamin (required for the metabolism of homocystine). Depending upon where in the metabolic pathway of vitamin B_{12}, a defect is located, a patient will have MMA and/or homocystinuria. Indeed, patients with dietary vitamin B_{12} deficiency will have increased homocystine and methylmalonate. Measurement of these compounds is used as a marker of vitamin B_{12} deficiency.

Severe PA and MMA are treated with protein restriction and carnitine supplementation. Carnitine acts by combining with organic acids to form esters that are excreted in the urine. All patients need an emergency regime for use during times of metabolic decompensation.

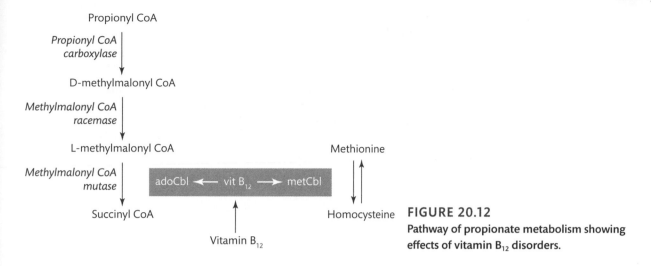

FIGURE 20.12
Pathway of propionate metabolism showing effects of vitamin B_{12} disorders.

Some patients have chronic hyperammonaemia that requires treatment. Patients with vitamin responsive methylmalonic aciduria may not require protein restriction.

Fatty acid oxidation disorders

Fatty acid oxidation is a major source of energy in the body. Long chain fats (C16-C20) are stored as triacylglycerols in adipose tissue. Under fasting conditions fatty acids are mobilized and oxidized in the mitochondria. Long chain fats are transported across the mitochondrial wall by a shuttle process involving carnitine. Once inside the mitochondria the fatty acid acyl CoA undergoes ß-oxidation. The acyl CoA molecule is progressively shortened by two carbon atoms that are released as acetyl CoA molecules. Acetyl CoA is further metabolized to form ketones (acetoacetate and 3-hydroxybutyrate). Many enzymes are involved in fatty acid oxidation. They have varying activity towards chain lengths of acyl CoAs, for example there are three acyl CoA dehydrogenases—very long chain (VLCAD), medium chain (MCAD), and short chain (SCAD). Any blockage in fatty acid oxidation will result in impaired production of ketones under fasting conditions.

The commonest fatty acid oxidation disorder (FAOD) is medium chain acyl CoA dehydrogenase deficiency (MCADD). This classically presents with hypoketotic hypoglycaemia, reflecting the inability to form ketones under fasting conditions. **Encephalopathy** occurs due to increased free fatty acids in the body. The peak age of presentation is during infancy, usually following a prolonged period of fasting, for example as diarrhoea and vomiting. Approximately 25% of patients die during their presenting episode. Some patients do not present until adulthood, and others do not present at all but are detected by family screening. Once diagnosed the disorder is easily treatable by avoidance of fasting and use of glucose polymer drinks or drips during episodes of illness. Many newborn screening programs now include MCADD.

The clinical severity of the other FAODs is variable. Some disorders, particularly long chain FAODs may have muscle disease and adults may present with **rhabdomyolysis** (breakdown of muscle tissue). Fatty acid oxidation disorders are a cause of sudden explained death. It is possible to provide energy in the form of medium chain triacylglycerols to patients with long chain FAODs as this bypasses the metabolic block.

Common mutations can be tested for, but their absence does not exclude a disorder. Fibroblast or muscle fatty acid oxidation studies may be necessary in some patients.

Key Points

Blood acyl carnitine analysis is the single most useful investigation for diagnosis of FAOD but may show no abnormalities in adult onset cases.

Carbohydrate disorders

Carbohydrates are required for energy metabolism in our bodies. Indeed, metabolism of carbohydrates such as glucose, fructose, and galactose are linked through interactions between different enzymatic pathways. Disorders that affect these pathways produce clinical features that can be mild to severe or even life threatening. Most carbohydrate disorders can be treated, or controlled with dietary intervention. The carbohydrate disorders discussed in this section include those of galactose and fructose metabolism.

Disorders of galactose metabolism

There are three disorders of galactose metabolism:

- galactose-1-phosphate uridyl transferase (GALT) deficiency (classical **galactosaemia**)
- galactokinase deficiency
- uridine diphosphate galactose-4-epimerase deficiency

The steps these enzymes catalyse are outlined in Figure 20.13.

Galactose and glucose are the monosaccharide residues in lactose. Lactose is the main carbohydrate in milk, providing 40% of energy requirements in neonates. Symptoms in classical galactosaemia occur early in life, when lactose is introduced into the diet. Babies usually present with jaundice and liver failure during the first week of life. Clinical symptoms are thought to be due to the accumulation of galactose-1-phosphate. Patients also develop cataracts, due to accumulation of galactitol, which exerts an osmotic effect in the lens of the eye. The diagnosis may be suspected by the presence of a non-glucose reducing sugar in urine (positive with reducing substances and negative Clinistix™) and can be confirmed by the measurement of GALT activity. There are pitfalls in these tests, which are shown in Table 20.4.

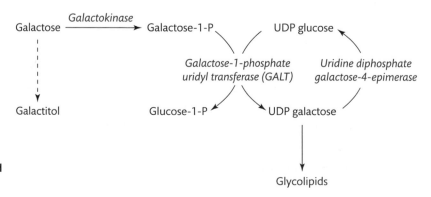

FIGURE 20.13
Metabolism of galactose and the enzymes associated with inherited metabolic disorders.

TABLE 20.4 Pitfalls in diagnosis of classical galactosaemia.

Test	Expected result in classical galactosaemia	Pitfalls
Urine sugars	Galactose present in high concentration but *no* glucose	Baby may be so ill that it is not taking milk feeds, i.e. lack of ingestion of galactose leading to lack of excretion.
		Galactose present in the urine due to hepatic dysfunction (the liver is the only organ in the body that metabolizes galactose).
		Positive with reducing substances and positive Clinistix™ may be found due to the presence of galactose and glucose in the urine (glucose may be present due to renal tubular dysfunction).
GALT measurement	GALT undetectable	Baby has had a blood transfusion for jaundice. The red cells in a blood specimen will be from the donor and will give a normal enzyme result.
		Glucose-6-phosphate dehydrogenase deficiency can cause a false abnormal test when using some screening tests to assess GALT activity.

Restriction of galactose intake (i.e. all milk and milk products) leads to prompt resolution of hepatic dysfunction. Despite treatment, some patients have learning difficulties and a high proportion of females have ovarian dysfunction.

Disorders of fructose metabolism

Fructose is found in fruits and the disaccharide sucrose which is composed of glucose and fructose. Disorders of fructose metabolism are rarer than those of galactose and do not present

CASE STUDY 20.3

A female baby aged three days had a normal birth and went home at 48 hours of age. The midwife noted increasing jaundice so the baby was readmitted at four days of age. On examination the baby had cataracts. Biochemical investigations showed the following (reference range is given in brackets):

Total bilirubin   450 µmol/L   (< 200)

Liver function tests were deranged, prolonged clotting times and the urine showed positive reducing substances but negative glucose.

(a) What can cause the positive results in the urine specimen?

(b) Which IMD could present in this way?

(c) Which investigation should be carried out urgently?

until fructose is introduced into the diet, generally at weaning. Patients with **hereditary fructose intolerance** (HFI) avoid sweet foods and because of this, in the past, had a conspicuous absence of dental caries. They will have fructose in the urine if they are ingesting it. Diagnosis is made by investigation of enzyme activity in liver biopsy or by DNA analysis.

SELF-CHECK 20.4

(a) Which of the following are reducing sugars: glucose, galactose, fructose, ribose, sucrose?
(b) Which disorders should be suspected when a non-glucose reducing sugar is present in urine?

Glycogen storage disorders

Glycogen storage disorders affect the synthesis and breakdown of glycogen. Glycogen is a highly branched polymer of glucose containing up to 55,000 glucose residues. The glucose residues are linked with α-1,4 linkages except at the branch point, where there are α-1,6 links. Branching of the molecule increases its compactness. At the heart of the molecule is a protein called glycogenin, which initiates the process of forming a new molecule. Glycogen is an effective way of storing glucose because:

- it prevents glucose from diffusing back out of the cell and being lost
- it reduces the osmotic burden on the cell (if the glucose molecules were not converted to glycogen, its concentration would be approximately 400 mmol/L)
- it provides a readily accessible store of glucose for use when required

Glycogen is stored as particles in the cytoplasm; the amount present at any one time will depend upon the fasting status of the individual. In the liver, glycogen constitutes about 10% of wet weight and is used to maintain blood glucose concentration. This glycogen is depleted by 24–36 hours of fasting. In muscle, glycogen constitutes about 2% of wet weight and is used to provide energy. **Glycogenesis** and **glycogenolysis** are outlined in Figure 20.14 and occur by the following processes:

- Glycogenesis:
  - glucose-1-phosphate is converted to UDP-glucose, which acts as a donor of glucose molecules
  - eight glucose molecules are added to glycogenin using UDP-glucose as the donor
  - glycogen synthetase adds further glucose molecules, forming α-1,4 linkages using UDP-glucose as the donor
  - branch points are introduced by branching enzyme as α-1,6 linkages
- Glycogenolysis:
  - phosphorylase b kinase activates phosphorylase b to phosphorylase a
  - phosphorylase cleaves one glucose residue at a time to form glucose-1-phosphate until there are four glucose residues left before a branch point (limit dextrin)
  - debranching enzyme removes the outer glucose residues and then the glucose residue at the branch point
  - phosphorylase is able to act on the debranched glycogen
  - phosphoglucomutase converts glucose-1-phosphate to glucose-6-phosphate
  - glucose-6-phosphatase converts glucose-6-phosphate to glucose which passes into the bloodstream
  - in lysosomes, 1–3% of cellular glycogen is broken down by acid α-glucosidase

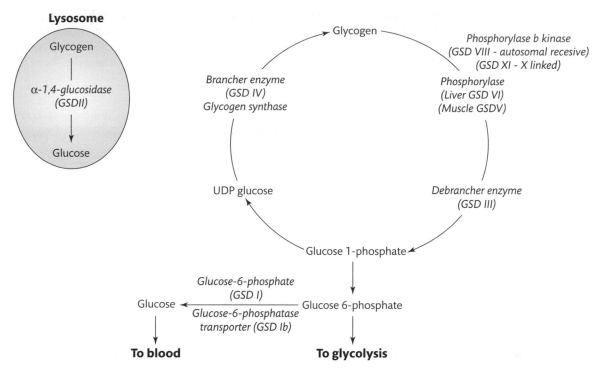

FIGURE 20.14

Pathways of glycogenesis and glycogenolysis showing the defects in glycogen storage disease.

Glycogen storage diseases that affect the liver present with hepatomegaly and may also have hypoglycaemia and lactic acidosis. Glycogen storage disease I (GSD I) is the most severe form and presents with hypoglycaemia and lactic acidosis after a fasting interval of as little as two hours. Triacylglycerols and urate may show striking increases. Hypoglycaemia in GSD III is less severe than GSD I, but some patients also have muscle involvement, with increased plasma creatine kinase. Glycogen storage diseases VI and IX are milder and hepatomegaly may be the only finding. Enzyme analysis for GSD III, VI, and XI are available in blood cells. The enzyme in GSD I can only be measured in the liver. Analysis of DNA is a practical alternative for patients suspected of GSD I.

Treatment of the hepatic GSDs includes the following:

- assessment of fasting tolerance (how long after a feed until the patient becomes hypoglycaemic)
- regular feeds approximately 2–4 times every hour
- continuous overnight nasogastric feeds of glucose polymer (approximately five residues of glucose per molecule)
- uncooked cornstarch as a form of 'slow release' glucose
- allopurinol to prevent hyperuricaemia

The GSDs which affect muscle are GSD V, VII, and some forms of GSD III. Glycogen storage disease V is known as McArdle's disease and presents with muscle cramps and rapid fatigue during exercise. These patients are at risk of rhabdomyolysis. Creatine kinase is chronically increased. A definitive diagnosis can be made by measurement of phosphorylase activity in a muscle biopsy. Furthermore, DNA analysis may reveal a disease causing mutations that would avoid a the need for muscle biopsy.

Glycogen storage disease II is a lysosomal storage disease. Patients do not have hypoglycaemia since the majority of the glycogen is broken down normally. The infantile form presents with severe cardiomyopathy and hypotonia, whereas the adult form is milder and presents with muscle weakness. Enzyme replacement therapy is available for this disorder (see below).

Glycogen storage disease IV is a rare form of GSD in which glycogen structure is abnormal. It presents with liver cirrhosis and treatment is by liver transplantation.

SELF-CHECK 20.5

What biochemical abnormalities are present in glycogen storage disease type 1?

Gluconeogenetic disorders

Gluconeogenetic disorders affect the pathways involved in the synthesis of glucose from pyruvate. The metabolic pathway involved is shown in Figure 20.15.

The defects are:

- glucose-6-phosphatase deficiency (glycogen storage disease type 1)
- fructose-1, 6-bisphosphatase deficiency
- phosphoenolpyruvate carboxykinase (PEPCK) deficiency
- pyruvate carboxylase (PC) deficiency

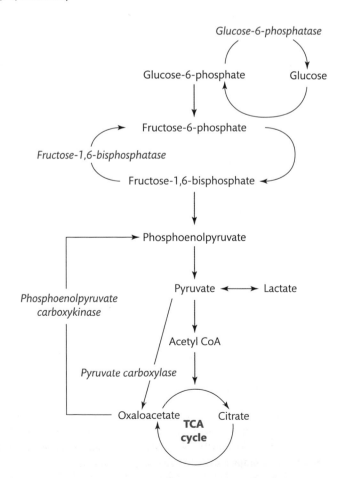

FIGURE 20.15

Pathways of glycolysis and gluconeogenesis indicating the enzymes which are deficient in gluconeogenetic disorders.

Key Points

Gluconeogenetic defects cause hypoglycaemia and lactic acidosis.

Patients have recurrent lactic acidosis and hypoglycaemia due to their inability to convert lactate and pyruvate to glucose. Patients with PEPCK and PC deficiencies also have progressive neurodegeneration.

Lysosomal storage disorders

Lysosomes are organelles within the cell where breakdown of a variety of complex compounds occurs. Lysosomes contain hydrolytic enzymes that work best at an acid pH. Each enzyme is responsible for the breakdown of a specific substrate. If a lysosomal enzyme is defective, its substrate accumulates within the lysosome and eventually causes cellular damage. More than 40 individual lysosomal storage disorders (LSDs) are known. They do not generally have acute problems. Most patients appear normal at birth and develop symptoms progressively over the first few years of life, depending on the organs and storage material involved. Some do not present until adulthood. Most LSDs are autosomal recessive, with the exception of Fabry disease and mucopolysaccharidosis type 2, which are X-linked.

The affected areas of the body may include:

- connective tissues, leading to coarse facial features, bone abnormalities (dysostosis multiplex)
- nervous tissue, causing developmental regression, myoclonus, spasticity, peripheral neuropathy
- major organs, for example hepatosplenomegaly, cardiomyopathy
- other suspicious features are corneal clouding, cherry red spot on the retina, and angiokeratoma

Investigation of these disorders requires a structured approach. Urine mucopolysaccharide analysis will detect the mucopolysaccharidoses (but spot tests are unreliable). Biochemical investigation is by specific enzyme analysis in plasma, leucocytes, or cultured fibroblasts. Consideration of the clinical picture is essential so that the appropriate test or group of tests is requested. These disorders are listed in Table 20.5.

Treatment for lysosomal storage disorders

Enzyme replacement therapy is available for several of the LSDs. Treatment is usually started in hospital with weekly enzyme infusions but eventually can be performed at home. This is an expensive treatment and there are strict criteria that must be met before it is funded. Storage material in the brain is not cleared because infused enzymes cannot cross the blood-brain barrier. Substrate deprivation is an alternative approach used in some milder cases. An artificial compound inhibits the synthesis of the natural substrate of the defective enzyme. In mild cases there may be sufficient residual enzyme to break down the smaller amount of substrate.

Newborn screening for LSDs is being considered now that enzyme replacement therapy is available. Multiplex assays have been developed to measure the activity of several LSD enzymes in a single blood spot using tandem mass spectrometry. If diagnosed early, enzyme replacement therapy may result in a better outcome.

TABLE 20.5 Some of the groups of LSDs with the tests required for diagnosis.

Disorder group	Disorder	Investigations
Mucopolysaccharidoses	MPS type 1 Hurler/Scheie MPS type 2 Hunter MPS type 3 San Filipo MPS type 4 Morquio MPS type 6 Maroteaux-Lamy MPS type 7 Sly	Urine mucopolysaccharides, then specific enzyme assay or DNA
Oligosaccharidoses	Aspartylglucosaminuria Fucosidosis α-mannosidosis β-mannosidosis Sialidosis Schindler's disease	Urine oligosaccharides may be abnormal, otherwise proceed to specific enzyme assay
Sphingolipidoses	Niemann Pick's disease A and B Fabry's disease Farber's disease Gaucher's disease GM_1 gangliosidosis GM_2 gangliosidosis (Tay-Sachs, Sandhoff) Galactosialidosis Krabbe's disease Metachromatic leucodystrophy Multiple sulphatase deficiency	Chitotriosidase increased in some disorders, otherwise proceed to specific enzyme assay
Glycogen	Glycogen storage disease II (Pompe)	Specific enzyme assay
Amino acids	Cystinosis	Leukocyte cystine

Peroxisomal disorders

Peroxisomes are organelles responsible for a range of biochemical reactions, including oxidation of very long chain fatty acids (VLCFA), synthesis of plasmalogens, and synthesis of bile acids. The commonest peroxisomal disorders are Zellweger syndrome (ZS) and adrenoleukodystrophy (ALD). Plasma VLCFA concentrations are abnormal in both of these disorders. Zellweger syndrome is a peroxisomal biogenesis disorder in which the assembly of the peroxisome is defective, leading to deficiencies in several enzyme functions. Patients present in the neonatal period or in infancy with severe congenital abnormalities, including dysmorphic facies, severe psychomotor retardation, profound hypotonia, seizures, and liver dysfunction.

Adrenoleukodystrophy is a peroxisomal disorder where the structure of the peroxisome is intact, but an ABC transporter membrane protein is defective. Patients have neurological symptoms and/or adrenal failure. A third of patients present between the ages of five to ten years and progressively deteriorate to a vegetative state within a few years. Some patients have

TABLE 20.6 Schematic showing the complexes of the respiratory chain and where the subunits are coded in the DNA.

Complex number	Complex name	Number of subunits coded for by	
		Nuclear DNA	Mitochondrial DNA
I	NADH dehydrogenase	Approx 34	7
II	Succinate dehydrogenase	4	0
III	Cytochrome bc$_1$ complex	10	1
IV	Cytochrome c oxidase	10	3
V	ATP synthase	11 or 12	2

a milder form that presents in adulthood and is known as adrenomyeloneuropathy. Some patients develop adrenal failure before the onset of neurological signs.

Key Points

Consider adrenoleukodystrophy in any male diagnosed with adrenal failure.

Respiratory chain disorders

Respiratory chain disorders are also known as mitochondrial myopathies, OXPHOS disorders, and electron transport chain disorders. The respiratory chain is located in the inner mitochondrial wall. It consists of five complexes each made up of a number of subunits. Some subunits are coded in nuclear DNA; the remainder in mtDNA. Table 20.6 shows the complexes of the respiratory chain and where the subunits are coded in the DNA.

The function of the respiratory chain is to make ATP. Electrons feed into the pathway at the level of complex I or complex II and are transferred to oxygen which is reduced to form water. Simultaneously a proton gradient is created across the membrane by complexes I, III, and IV. This gradient is used to make ATP by activating ATP synthase (complex V).

Respiratory chain disorders affect tissues that have a high ATP requirement, for example brain, skeletal muscle, heart, liver, kidney, gut, and endocrine glands. They often have increased lactate concentrations in blood and cerebrospinal fluid. In children defects are likely to be due to nuclear encoded defects and mtDNA analysis is usually unrewarding. Often a diagnosis can only be made by obtaining a muscle biopsy and assaying biochemically the complexes of the respiratory chain. Usually patients have neurological features but some will present with acute liver failure. Respiratory chain defects should be excluded in patients with acute liver failure of unknown cause since this diagnosis is a contraindication to transplantation. Respiratory chain disorders can cause lactic acidaemia as can a number of other disorders. These are outlined in Box 20.2.

Key Points

Respiratory chain disorders can give rise to any symptom in any organ at any age, and with any mode of inheritance.

BOX 20.2 *Causes of lactic acidaemia*

Acquired

- hypoxia/hypoperfusion
- drugs
- sepsis
- seizures

Secondary

- organic acidurias
- fatty acid oxidation defects

Primary

- glycogen storage disorders
- gluconeogenetic defects
- pyruvate disorders
- respiratory chain disorders

Mitochondrial DNA abnormalities

Mitochondria are the only organelles in the cell to contain their own DNA, which is called mitochondrial DNA (mtDNA). It contains 37 genes, 13 of which code for subunits of enzymes in the respiratory chain (see Table 20.6). The remainder code for transfer RNA and ribosomal RNA molecules used within the mitochondria. There are a number of clinical conditions associated with specific mtDNA point mutations or deletions that include:

- MELAS—Mitochondrial Encephalomyopathy, Lactic Acidosis, and Stroke-like episodes
- MERRF—Myoclonic Epilepsy with Ragged Red Fibres
- NARP—Neurogenic muscle weakness, Ataxia and Retinitis Pigmentosa
- FBSN—Familial Bilateral Striatal Necrosis and Leigh syndrome
- DEAF—Maternally inherited antibiotic induced deafness
- MMC—Maternally inherited Myopathy and Cardiomyopathy
- LHON—Leber's Hereditary Optic Neuropathy

Cholesterol biosynthesis and bile acid synthesis

The normal functions of cholesterol are:

- component of biological membranes, especially neurones
- precursor for bile acid synthesis
- precursor for steroid hormone synthesis
- activator of hedgehog proteins

Defects of cholesterol synthesis are multisystem disorders, usually with dysmorphism, and sometimes with skeletal dysplasias. The Smith-Lemli-Opitz (SLO) syndrome is the most common defect of cholesterol biosynthesis. Prior to the discovery of the underlying defect in 1994 it was thought to be a syndrome due to a chromosomal abnormality. The defect is in the final step of cholesterol biosynthesis and results in low cholesterol with grossly increased 7-dehydrocholesterol concentrations. The clinical features of SLO are dysmorphic features;

mental retardation; **syndactyly**; ambiguous genitalia in males; and brain, cardiac, and renal malformations. However, it has been shown that treatment with cholesterol can improve behaviour in some children.

Bile acids (BAs) are synthesized from cholesterol in the liver to form the two primary BAs, cholate and chenodeoxycholate. Peroxisomal enzymes catalyse some of the steps. Bile acids are conjugated with glycine or taurine to form conjugated BAs (for cholate these are glyco-cholate and taurocholate respectively). The conjugated BAs pass via the gall bladder to the intestine where they emulsify fats. Disorders of BA synthesis present with neonatal cholestasis, progressive liver disease, diarrhoea, and fat-soluble vitamin deficiency due to malabsorption. Notably, these patients tend to have a low plasma γ-glutamyl transferase activity. Some BA synthesis disorders are treatable with chenodeoxycholic or cholic acid. These disorders are diagnosed by analysis of bile acids in blood or urine.

SELF-CHECK 20.6

Which disorder causes decreased plasma cholesterol and increased 7-dehydrocholesterol concentrations?

Porphyrias

The porphyrias arise from defects in the haem biosynthetic pathway. This is illustrated in Figure 20.16.

Haem is synthesized in all tissues, but mainly in the erythroid cells (where it is further converted to haemoglobin) and in the liver (where it is further converted to cytochrome P450

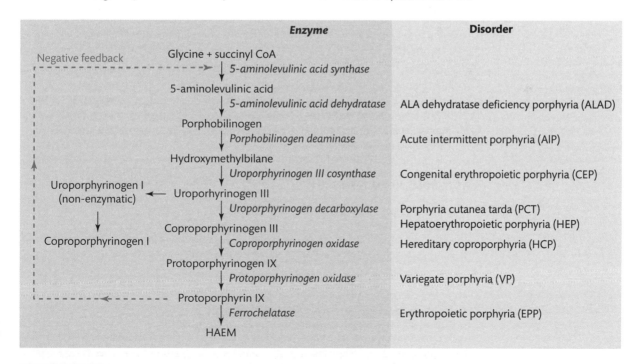

FIGURE 20.16

Pathway of haem biosynthesis showing the enzymes involved and the porphyrias (together with the abbreviation used) caused by the various enzyme defects.

TABLE 20.7 Classification of the porphyrias.

Disease	Hepatic	Erythropoietic	Acute	Cutaneous	Acute and cutaneous
ALA dehydratase deficiency porphyria (ALAD)			Yes		
Acute intermittent porphyria (AIP)	Yes		Yes		
Congenital erythropoietic porphyria (CEP)		Yes		Yes	
Porphyria cutanea tarda (PCT)	Yes			Yes	
Hepatoerythropoietic porphyria (HEP)	Yes	Yes		Yes	
Hereditary coproporphyria (HCP)	Yes		Yes (72%)	Yes (7%)	Yes (21%)
Variegate porphyria (VP)	Yes		Yes (20%)	Yes (59%)	Yes (21%)
Erythropoietic porphyria (EPP)		Yes		Yes	

containing enzymes). There are eight steps in the pathway. If there is a blockage, porphyrinogens accumulate and are oxidized to porphyrins which are detectable in urine, blood, or faeces. If haem synthesis is impaired, there is a lack of negative feedback on the first enzyme of the pathway and so the flux through the pathway continues with over-production of porphyrin intermediates. Porphyrias may be classified by the location of the defect (hepatic or erythropoietic) or by clinical presentation (acute or cutaneous) and the classification is outlined in Table 20.7. Some of the porphyrias are autosomal dominant.

Acute porphyric attacks

These attacks are potentially life threatening. The main symptoms are severe colicky abdominal pain usually with vomiting, neuropathy (which may be severe enough to cause paralysis) and mental and psychiatric disturbances. Drugs or alcohol may precipitate attacks. They are more common in females, although rarely occur before puberty. Once diagnosed an affected individual will be given information about which drugs are unsafe. Apart from removing precipitating factors, treatment of acute episodes is symptomatic, together with intravenous haematin which exerts negative feedback on 5-aminolevulinic acid (ALA) synthase and prevents further production of abnormal metabolites.

Cutaneous porphyrias

Two different forms of skin symptoms occur in the cutaneous porphyrias. Bullous porphyrias have fragile skin, vesicles and bullae (blisters), and hypertrichosis (excess body hair). In some porphyrias, the predominant symptoms are of acute photosensitivity. Excess porphyrins in the skin are excited by sunlight and this is thought to lead to the skin damage. Most patients with cutaneous porphyrias are advised to avoid sunlight.

Laboratory investigation for porphyrias

Laboratory testing is guided by clinical presentation. Random urine porphobilinogen (PBG) measurement is available as an emergency test in many hospitals with an Emergency Department. If PBG is increased, further investigation should include urine and faecal porphyrins to delineate the underlying defect. If urine PBG is normal at the time of acute symptoms, then an acute porphyria can be excluded (with the exception of the vanishingly rare ALA dehydratase deficiency porphyria). For patients with cutaneous symptoms porphyrin analysis should be carried out in urine, blood, and faeces. Abnormal results should be followed with enzyme or DNA analysis where available. At risk family members should be counselled.

There are several secondary causes of increased porphyrins, including liver disease (increased urine coproporphyrin), lead poisoning (increased blood zinc protoporphyrin), and tyrosinaemia type 1 (increased urine 5-aminolevulinate).

SELF-CHECK 20.7

Which test is abnormal in acutely presenting porphyrias?

Creatine metabolism

Creatine phosphate is found mainly in the brain and muscle. It is formed by the action of creatine kinase on creatine. Three disorders of creatine synthesis have been described; all have similar symptoms of psychomotor retardation, speech impairment, and epilepsy. Low creatine concentration in the brain can be demonstrated by proton magnetic resonance spectroscopy (MRS). Specialist biochemical testing is required to differentiate between the possible defects. Some forms of creatine deficiency syndromes are treatable with dietary creatine.

Congenital disorders of glycosylation

Many proteins in the body are glycosylated. This process adds side chains of oligosaccharides called 'glycans' that are essential to the function of the protein. Glycans are linked to proteins either by the amide group of asparagine (N-glycans) or the hydroxyl group of serine or threonine residues (O-glycans). More than twenty different defects of protein glycosylation have been described. There is a broad spectrum of clinical presentations. The only routinely available diagnostic test is isoelectric focusing of transferrins in plasma (see Box 20.3 on transferrin). This will detect disorders of N-glycosylation associated with deficiency of sialic acid (SA) residues in the glycan side chain, such as congenital disorder of glycosylation (CDG) type 1a. There may be other apparent protein abnormalities in these patients.

Key Points

The test used for carbohydrate-deficient transferrin as a marker of excessive alcohol consumption will NOT detect CDG.

BOX 20.3 Transferrin

Transferrin is a polypeptide of 679 amino acid residues with two branched N-linked oligosaccharides that terminate in SA residues. Normally up to 85% of transferrin carries four SA residues (termed tetrasialotransferrin), 5% carries five (pentasialotransferrin), and 5% carries three (trisialotransferrin). Patients with CDG have lower levels of the tetrasialotransferrin in their plasma, with increases in di- and a-sialotransferrin.

Neurotransmitter disorders

Patients with neurotransmitter disorders present with neurological problems such as:

- severe early onset epileptic encephalopathies
- progressive extrapyramidal movement disorders

Most possible sufferers require specialist analysis of cerebrospinal fluid (frozen in liquid nitrogen at the bedside).

Purine and pyrimidine disorders

Purine and pyrimidine metabolism is complex. There are pathways for synthesis, catabolism, and salvage. Disorders have neurological, immunodeficiency, or renal stone symptoms depending upon which part of the pathway is affected:

- Adenosine deaminase and purine nucleoside phosphorylase (PNP) deficiencies lead to immunodeficiency. Plasma and urine urate concentrations are decreased in PNP deficiency.
- Xanthine oxidase (XO) deficiency causes low urate concentrations in plasma and urine. Renal stones are present in ~30% of cases. The molybdenum cofactor defect causes deficiency of XO and sulphite oxidase.
- Lesch-Nyhan syndrome is an X-linked disorder with deficiency of hypoxanthine: guanine phosphoribosyl transferase (HPRT) activity. Plasma and urine urate concentrations are high.

Some disorders can be detected by urine organic acid analysis, but most purine and pyrimidine disorders will need purine and pyrimidine analysis in blood or urine, followed up by specific enzyme or DNA analyses.

Cross reference

Chapter 4 Hyperuricaemia and gout

20.6 Antenatal diagnosis

Many IMDs are severe conditions and families may wish for antenatal testing in future pregnancies. This is usually carried out by chorionic villus sampling as early as ten weeks gestation. Tests are performed directly on the chorionic villus cells, or the cells may be cultured for analysis. Amniotic fluid sampling (amniocentesis) is an alternative method usually carried out at 16–18 weeks gestation. Analyses are carried out on cultured amniotic fluid cells or on the amniotic fluid itself. The disadvantage of amniocentesis is that if the foetus is affected a later termination may be required, which is more distressing to the family. Occasionally foetal blood sampling or foetal liver biopsy is required. Antenatal testing is usually by measurement of enzyme activity or DNA analysis. It is essential that the diagnosis has been proven in the index case. For antenatal diagnosis by DNA, parents should be tested to confirm that they carry the mutation. Maternal contamination must be excluded. Antenatal diagnosis is a complex area and close liaison must take place between the laboratory, the clinical geneticist, and the obstetrician.

20.7 Newborn screening

Newborn screening aims to test all babies for certain treatable disorders. The aim is to diagnose disorders before the onset of symptoms so that treatment can be initiated and outcome

TABLE 20.8 Criteria for a screening program.

The condition	Important health problem
	Natural history and epidemiology should be understood
	Latent presymptomatic phase
The test	Simple, safe, precise, validated, and acceptable to the general population
	Agreed policy for diagnostic follow-up of screen positive cases
The treatment	Effective
	Evidence that early treatment leads to improved outcome
The program	Benefits outweigh harm
	Value for money
	Auditable quality standards
	Quality assurance

improved. In 1968 Wilson and Jungner set out principles for screening programs; these have been updated by the UK National Screening Committee in March 2003. The key criteria are shown in Table 20.8.

A capillary blood specimen is collected from a heel prick at 5–8 days of age, usually by the community midwife. Blood is spotted onto cards made of absorbent filter paper to form dried blood spots. The cards are sent to the screening laboratories by post. Most newborn screening laboratories in England test >50,000 babies per year. The disorders screened for are shown in Table 20.9.

Phenylketonuria

In the UK there has been newborn screening for PKU since 1969. The disorder is tested for by demonstration of increased phenylalanine in dried blood spots. Most laboratories use tandem mass spectrometry for PKU screening.

TABLE 20.9 The following disorders are tested for in England by the newborn screening program.

Disorder	Date of commencement of universal newborn screening	Approximate incidence
Phenylketonuria	1969	1 in 14,000
Congenital hypothyroidism	1981	1 in 3,500
Sickle cell disorders	2006	1 in 2,500
Cystic fibrosis	2008	1 in 2,500
Medium chain acyl CoA dehydrogenase deficiency	2009	1 in 10,000

Congenital hypothyroidism

Cross reference

Chapter 12 Thyroid disease

Congenital hypothyroidism occurs when the thyroid gland is unable to produce thyroxine and therefore the serum thyroid stimulating hormone (TSH) concentration increases as more is released in an attempt to stimulate the thyroid gland to produce thyroxine.

Newborn screening is carried out by demonstration of increased TSH in dried blood spots. In normal babies there is a surge of TSH production in the first 24 hours and a return to normal by five days of age. Most babies with congenital hypothyroidism have serum concentrations in excess of 200 mU/L (normally <10 mU/L). Children affected by congenital hypothyroidism are treated by thyroxine replacement therapy.

Sickle cell disorders

Adult haemoglobin (HbA) is made up of 2α and 2β polypeptide chains. Sickle cell disease is due to an abnormality in the haemoglobin molecule caused by an amino acid change in the β chain. This haemoglobin is called HbS. Individuals who are homozygous for HbS are affected with sickle cell disease. They may develop painful life-threatening 'sickling' crises at times of infection. The different haemoglobins are examined following separation by HPLC or isoelectric focusing.

Cystic fibrosis

Newborn screening is carried out for this disorder by measurement of immunoreactive trypsin (IRT) in dried blood spots. A second tier test is included in the screening program whereby DNA testing for a panel of mutations is carried out on the same blood spot card.

Medium chain acyl CoA dehydrogenase deficiency

Medium chain acyl CoA dehydrogenase deficiency (MCADD) is a fatty acid oxidation defect and is tested for by measurement of octanoyl carnitine in blood spots by tandem mass spectrometry (simultaneously with phenylalanine).

Other conditions and future developments

Most developed countries screen for PKU and congenital hypothyroidism. Coverage of other disorders around the world is variable, partly reflecting the local population. Many countries screen at an earlier age than the UK so different cut-offs may be required. Tandem mass spectrometry has revolutionized newborn screening because the tests for PKU and MCADD can also detect a whole range of amino acids and acyl carnitines and therefore diagnose a wide range of IMDs. Some countries now operate extended newborn screening programs, although not all of the disorders tested for fulfil the newborn screening criteria.

CASE STUDY 20.4

A male infant aged four months presented to the emergency department acutely unwell. On examination he had massive hepatomegaly. Initial investigations of his blood showed (reference ranges given in brackets):

Glucose   <0.6 mmol/L   (2.0–5.5)
Lactate   8.0 mmol/L   (0.5–2.0)

(a) Which abnormality requires urgent treatment and why?

(b) Why might the infant have hepatomegaly (enlarged liver)?

(c) As well as inherited metabolic disorders, which group of disorders should be considered in patients with hypoglycaemia?

(d) Why is medium chain acyl CoA dehydrogenase deficiency (MCADD) unlikely?

SELF-CHECK 20.8

(a) Which disorders does the newborn screening program in England test for? (b) Which laboratory technique is used to test for PKU and MCADD?

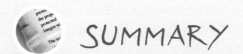

SUMMARY

- Inherited metabolic disorders can be inherited by autosomal recessive, autosomal dominant, or X-linked modes of inheritance.

- Inherited metabolic disorders arise due to a defective enzyme causing deficiency of a product, accumulation of a substrate, or product from a minor pathway.

- Inherited metabolic disorders include those affecting amino acids, urea cycle, organic acids, fatty acids, carbohydrates, glycogen storage, lysosomal storage, and peroxisomes.

- Defects of enzymes in the haem biosynthetic pathway cause a group of disorders called the porphyrias.

- Antenatal diagnosis is used to detect IMDs prior to birth by measuring enzyme activity or DNA analysis.

- A nationwide screening program is available for detection of certain IMDs, for example phenylketonuria and congenital hypothyroidism.

FURTHER READING

- Fernhoff PM (2009) Newborn screening for genetic disorders. *Pediatric Clinics of North America* **56**, 505–13.

- Green A, Morgan I, and Gray J (2003) *Neonatology and Laboratory Medicine.* Cambridge: ACB Venture Publications.

   Practical chapters on testing for IMDs that present in the neonatal period and on newborn screening.

- Hoffman GF, Nyhan WL, Zschocke J, Kahler SG, and Mayatapek E (2002) *Inherited Metabolic Diseases.* Philadelphia: Lippincott Williams & Wilkins.

   This little book is good value and provides a comprehensive overview of inherited metabolic disorders.

- Williams RA, Mamotte CD, and Burnett JR (2008) Phenylketonuria: an inborn error of phenylalanine metabolism. *Clinical Biochemistry Reviews* **29**, 31–41.

- www.metbio.net. Website of the National Metabolic Biochemistry network. Includes an assay directory for specialist metabolites and enzymes and best practice guidelines written for non-specialists.

- www.newbornscreening-bloodspot.org.uk. Website of the UK Newborn Screening Programme Centre, with a range of information.

QUESTIONS

20.1 For each statement, state whether TRUE or FALSE.

   (a) Phenylketonuria has an autosomal dominant mode of inheritance

   (b) Phenylketonuria often arises due to a deficiency of the enzyme phenylalanine hydroxylase

   (c) Mitochondrial DNA disorders often have maternal inheritance

   (d) The urea cycle disorder ornithine transcarbamylase deficiency is an example of an X-linked disorder

20.2 Describe, using examples, the effects of a metabolic block.

20.3 Give three examples of how IMDs may be treated using named disorders.

20.4 Highlighting the pitfalls, explain how to test for homocystinuria.

20.5 Highlighting the pitfalls, explain how to test for classical galactosaemia.

20.6 Describe the approaches which can be used for antenatal diagnosis.

20.7 Describe the disorders tested for by the newborn screening program.

Answers to self-check questions, case study questions, and end-of-chapter questions are available in the Online Resource Centre accompanying this book.

 Go to www.oxfordtextbooks.co.uk/orc/ahmed/

Therapeutic drug monitoring

Robin Whelpton, Nigel Brown, and Robert Flanagan

Learning objectives

After studying this chapter you should be able to:

- Explain, with the aid of suitable examples, what a drug is
- Describe how drugs get into the body, how they are metabolized, and how the drug or their metabolites are removed from the body
- Explain elimination half-life and how it relates to clearance, volume of distribution, and attainment of 'steady-state' conditions
- List and discuss the analytical methods used in therapeutic drug monitoring
- Explain the reasons for performing therapeutic drug monitoring and the drugs for which it is useful

Introduction

Therapeutic drug monitoring (TDM) is the term used to describe the measurement of the plasma concentrations of a **drug** and/or any pharmacologically active metabolites attained during treatment. The aim of TDM is to help optimize treatment by, for example, selecting a dose that maximizes the desired effects a drug has whilst minimizing the risk of **toxicity**. Therapeutic drug monitoring may also be performed to assess if a patient is taking a drug as prescribed. For most drugs, there is a wide margin (so-called margin of safety) between the clinically-effective dose and the dose associated with marked adverse effects. However, for some drugs the difference between a clinically-effective dose and a

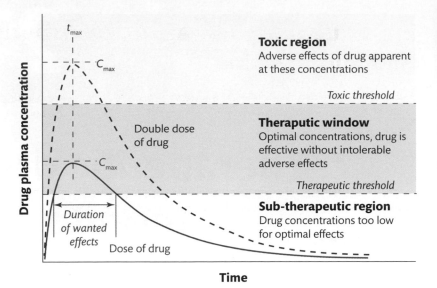

Toxic region
Adverse effects of drug apparent at these concentrations

Toxic threshold

Theraputic window
Optimal concentrations, drug is effective without intolerable adverse effects

Therapeutic threshold

Sub-therapeutic region
Drug concentrations too low for optimal effects

FIGURE 21.1
Concept of a 'therapeutic window' during therapeutic drug monitoring. See text for details.

potentially toxic dose is small. These drugs are said to have a small **'therapeutic range'** or narrow **'therapeutic window'**, and the dose of the drug must be adjusted carefully for each patient. Figure 21.1 demonstrates how a drug with a narrow therapeutic window is more difficult to use effectively and safely compared to one with a wide therapeutic window.

At low plasma concentrations dosage is **sub-optimal**, this is the **sub-therapeutic** region. Above the therapeutic threshold concentration the drug is clinically effective, but if the concentration rises above the toxic threshold, adverse effects may be apparent. Concentrations between the therapeutic and toxic thresholds are in the 'therapeutic window'. In the example shown in Figure 21.1, when a single dose of drug is taken, the concentration rises as the drug is absorbed into the plasma, until the rate at which the drug is absorbed equals the rate at which it is being eliminated, this occurs at t_{max}, the plasma concentration at this time being C_{max}. The duration of action will depend on where the drug concentration curve crosses the therapeutic threshold. In this example, doubling the size of the dose, results in the higher concentrations entering the toxic region. Clearly the larger the therapeutic window the easier it will be to prescribe a drug safely. To maintain the drug in the therapeutic window, smaller divided doses are given at regular intervals.

In order to appreciate the reasons for TDM and the decisions that may have to be made in dose adjustment, it is necessary to understand what is meant by a 'drug', how drugs are given, what the body does to a drug, and how the effect of dose adjustment on the plasma concentration of a drug may be predicted.

Key Points

The therapeutic range for a drug is the plasma concentration range over which it shows clinical benefit with minimal toxicity for the vast majority of patients.

21.1 Drug administration, distribution, and elimination

Before considering the value of TDM, it is necessary to appreciate the nature of drugs, their routes of administration, and distribution within, and **elimination** from, the body.

Nature of drugs

A drug is a substance that has a physiological effect on the person taking it. Thus, drugs may be either medicines, or 'recreational drugs'. This latter category includes compounds such as caffeine, ethanol, and nicotine, and also drugs that are so dangerous if used incautiously that access to them is controlled, for example cocaine and lysergic acid diethylamide (LSD). Table 21.1 shows that drugs come from many sources, and include chemicals that are normally present in the body, such as adrenaline (epinephrine) and insulin. Note that some inorganic salts are used as drugs. Lithium carbonate, for example, is used in treating bipolar disorder, a condition in which the patent's mood swings from a state of over-activity (mania) to severe depression. Also note that drugs may be used for several, sometimes seemingly unrelated, purposes. Ethanol, for example, is commonly used as a recreational drug, but it may also be used as an antidote to methanol and to ethylene glycol poisoning, and to sterilize skin prior to giving an injection.

TABLE 21.1 Some examples of compounds used as drugs.

Example	Origin	Obtained by	Examples of usage
Adrenaline	Animal/synthetic	Prescription	Treating shock (very low blood pressure). Given with local anaesthetics to constrict blood vessels and maintain anaesthetic in area of injection.
Aspirin	Synthetic	Prescription/OTC*	Treating pain, fever, inflammation.
Cocaine	Plant	Prescription/illicit	Local anaesthetic. Recreational.
Ecstasy	Synthetic	Illicit	Recreational.
Ethanol	Microorganisms, for example yeast	Prescription/recreational	Treating methanol and ethylene glycol poisoning. Cleansing skin prior to venepuncture/injection.
Insulin	Animal/genetically engineered	Prescription	Treating diabetes.
Lithium carbonate	Inorganic salt	Prescription	Treating manic depressive bipolar illness.
Morphine	Plant	Prescription/illicit	Treatment for pain, diarrhoea, and coughing. Recreational.
Penicillin	*Penicillium* moulds	Prescription	Treating bacterial infections.
Sodium bicarbonate	Inorganic salt	Prescription/OTC*/grocers	Used as an antacid and treating overdose, e.g. aspirin.

* Over-the-counter, usually purchased from a pharmacy without prescription.

Administration of drugs

Drugs can be administered in a number of different ways. The commonest route is by mouth (orally, Latin *per os*, abbreviation: p.o.). Drugs can also be injected: directly into a vein (**intravenous**, i.v.), into a muscle (**intramuscular**, i.m.), or under the skin (**subcutaneous**, s.c.). Drugs prescribed to treat asthma are generally inhaled as they go straight to their site of action when taken in this way. **Sublingual**, meaning 'under the tongue', describes when a drug is held in the mouth and is absorbed across the buccal membranes. Drugs may also be administered as **suppositories**, that is, inserted into the rectum, or as **pessaries** inserted into the vagina. Such drugs may be given for local effect, whilst others are given for distribution to other parts of the body. Increasingly, drugs are applied to skin, from where they are absorbed into the bloodstream, for example nicotine patches used in smoking withdrawal.

Oral administration of drugs has several advantages, including:

- it is convenient and usually acceptable to the patient
- sterile conditions are not required
- no special medical skills in giving the drug are required
- tablets, capsules, or liquids may be used, so that patients who have difficulty swallowing tablets, for example, may take their medicine as a solution or a suspension in an appropriate fluid

Assuming that an orally-administered drug has been given for effect in the body, that is, systemic effect, rather than a local effect in the gastrointestinal tract (GIT), it has to be absorbed into the blood and then transported to its site(s) of action. If a drug is not absorbed and transported in this way then the oral route cannot be used. Drugs such as penicillin, which is unstable in acid, and insulin, which is digested by proteolytic enzymes, are destroyed by gastric contents. Generally, ionized drugs such as quaternary ammonium compounds are poorly absorbed because ionized molecules diffuse through cell membranes poorly. Basic drugs that are protonated in the acidic environment of the stomach cannot be absorbed until they have passed into the intestine where the pH is higher and protonation is much reduced. However, aspirin (acetylsalicylic acid) is less than 1% ionized in the stomach and can be absorbed across the stomach wall. Some drugs are formulated with a special **enteric coating** that prevents them releasing their contents until they reach the intestine.

Generally, once in the small intestine a drug will be in a form suitable for absorption into the hepatic portal circulation. However, the body has developed a number of systems for keeping ingested poisons or potential poisons from reaching the general (systemic) circulation. Orally-administered drugs may be metabolized by: (1) the intestinal flora (bacteria living in the gut lumen); (2) the intestinal wall; and (3) the liver (Figure 21.2). This process is termed presystemic or **first-pass metabolism**.

Also, efflux pumps (proteins that act to 'pump' organic molecules from one side of a biological membrane to the other) in the intestinal mucosal cells may return absorbed drugs to the intestinal lumen, or the drug may be excreted from the liver via the bile. Apart from drugs that are absorbed from the mouth or the lower part of the rectum, drugs absorbed across the GIT are carried via the hepatic portal vein to the liver where they may undergo extensive first-pass metabolism. Acidic drugs (AH) are protonated in gastric acid and may be absorbed from the stomach. On the other hand, protonated basic drugs (BH^+) are ionized in gastric acid and cannot be absorbed from the stomach. Extensive inactivation or **excretion** of the drug may result in little, if any, of the dose reaching the systemic circulation and in such cases administration by the oral route is impractical.

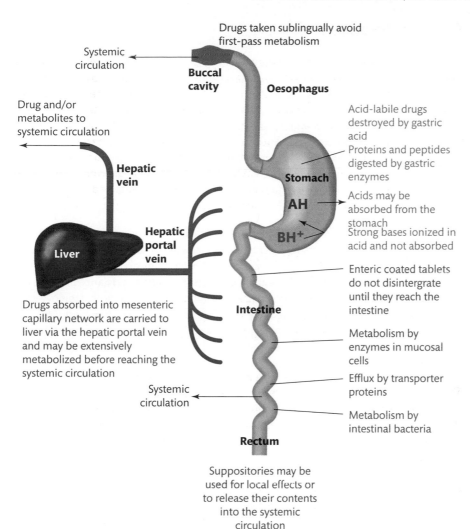

FIGURE 21.2
A schematic showing absorption of drugs from the gastrointestinal tract. See text for details.

The pharmacological activity of a drug may not be reduced by **metabolism** if the metabolite(s) are pharmacologically active; indeed **prodrugs** are compounds designed to be easily absorbed and then metabolized into an active drug once in the body.

First-pass metabolism can be circumvented by giving the drug by either injection, or sublingually because absorption from the mouth is directly into the systemic circulation. Intravenous injection will give immediate delivery of the drug to the circulation and hence an almost immediate effect. Intramuscular or subcutaneous injections require the drug to be absorbed from the injection site. Some drugs are formulated to give a steady release of drug into the blood from an injection site.

Intravenous injections are given by trained professionals, requiring the patient to be in hospital, to visit their general practitioner, or to have a health professional visit them at home. Less skill is required for i.m. and s.c. injections and patients, for example diabetics or their carers, can be taught to give s.c. injections. As injections can be painful and have to be sterile they are not as convenient as taking a drug orally.

First-pass metabolism for a drug refers to its removal from the plasma after absorption and before it reaches the systemic circulation

Drug distribution

After a drug has been absorbed it is distributed to its site of action. If the drug is heparin, an **anticoagulant**, then the blood is its site of action. However, for most drugs their site of action is elsewhere in the body, for example in the case of a 'sleeping-pill' it is the brain. The process of drug movement from blood to tissues is termed distribution. Obviously, drugs do not 'know' where to go; rather they are carried in the bloodstream and may cross membranes and enter tissues provided the conditions are appropriate. Generally, unionized, lipid-soluble drugs are more extensively distributed into tissues than ionized, water-soluble drugs because uncharged molecules can cross cell membranes more easily. The driving force for movement is usually the concentration gradient, that is, drugs pass from an area of high concentration to areas of low concentration.

Drugs may also be transported across membranes by carrier proteins. These are specialized transmembrane proteins that transport endogenous substances such as sugars, amino acids, and so on. However, if the molecular structure of the drug is similar to an endogenous substance it, too, may be transported across the membrane. **Multidrug resistance** proteins, also known as P-glycoproteins (PGP), function as transmembrane *efflux* pumps that remove drugs from cells. They influence the absorption, distribution, and elimination of some drugs. In the gut they may act to remove poisons that have been absorbed from the lumen, as noted above. They prevent some drugs from entering the brain or crossing the placenta. They, along with other transport proteins, actively secrete some drugs into renal tubular fluid and bile. As the term multidrug resistance protein implies, these compounds are implicated in the development of resistance to **cytotoxic** drugs because the ability to pump these latter agents out of tumour cells is often increased when compared with normal cells. This is a major factor limiting the efficacy of **chemotherapy**.

Plasma protein binding can affect the distribution, and thus the pharmacological activity, of a drug. Drugs in the systemic circulation may bind to one or more types of plasma protein. Acidic drugs are often bound to albumin. Basic drugs may bind to albumin and also to α_1-acid glycoprotein. Binding to plasma proteins is a mechanism by which many drugs are transported in blood and can markedly reduce the amount filtered out of the plasma by the kidney.

Lipophilic drugs tend to bind extensively and the concentrations in plasma, as protein bound and non-bound or 'free' drug, can exceed the amount that is possible to dissolve in water. Drugs that are highly bound to plasma protein have a low 'free fraction' that is available to diffuse to their site(s) of action in tissue(s). Disease conditions where the concentrations of plasma proteins are low can lead to a marked increase in the free fraction. Another factor that can influence the free fraction is competitive displacement from plasma binding sites by other drugs or by endogenous molecules such as bilirubin. A high free fraction may lead to toxicity, but increased distribution of non-bound drug in tissues often reduces the *total* concentration of drug in the blood that is measured. Thus, on some occasions it may be better to measure the concentration of 'free drug', provided of course that this is technically feasible and that one can correlate the 'free' drug concentration with clinical effect.

Cerebrospinal fluid (CSF) contains little protein and for some drugs their concentrations in CSF are often similar to those of non-protein bound drugs in plasma, provided that the drug is able to diffuse across the **blood-brain barrier**. This is the term used to describe an arrangement of cells that excludes many blood components from the central nervous system (CNS). It consists of a specialized capillary network in which the endothelial cells are tightly-packed together. Some drugs may also be excluded from the CNS by PGP, as explained above. Thus, the concentrations of some drugs in the brain bear little relationship to their total concentrations in plasma.

The extent to which a drug is transferred from the plasma to tissues is indicated by the **apparent volume of distribution**, V. To calculate V it is necessary to measure the *plasma* concentration of the drug and to know the amount of drug in the body at the time the blood sample was taken. Immediately after a rapid i.v. injection the amount of drug in the body will be the dose injected and if the plasma concentration at time zero (C_0) is measured:

$$V = \frac{Dose}{C_0}$$

This equation shows that the lower the plasma concentration, the greater the apparent volume of distribution. Table 21.2 shows that some drugs are distributed into anatomical volumes that can be measured using suitable markers such as the dye Evans blue, which binds so avidly to plasma albumin it is restricted to plasma. Inulin, a very water-soluble molecule that cannot penetrate cells, is distributed in extracellular fluid. Isotopically-labelled water is used to measure the volume of total body water.

Drugs that are almost entirely distributed in plasma have low apparent volumes of distribution, but at the other end of the scale some lipid-soluble drugs have relatively high apparent volumes of distribution, even greater than the total amount of water in the body. This is because these drugs become extensively concentrated in tissues, leading to very little being in the plasma, and hence the high *apparent* volumes of distribution. The apparent volume of distribution is important because it can have a large influence on the amount of drug needed to achieve the intended clinical effect. Furthermore, it has a major effect on the rate of elimination of a drug.

TABLE 21.2 Some examples of apparent volumes of distribution (V).

Substance	V (L/kg of body weight)	V in 70 kg subject (L)
Heparin	0.06	4.2
Evans blue*	0.05	3.5
Amoxycillin	0.2	14
(+)-Tubocurarine	0.2	14
Inulin**	0.21	14.5
Ethanol	0.65	45
Phenazone	0.6	42
Deuterium oxide (^2H_2O)***	0.55–0.7	38–50
Digoxin	5	350
Chlorpromazine	20	1400
Quinacrine	500	35,000

Anatomical volumes: * plasma, ** extracellular fluid, *** total body water.

Key Points

The apparent volume of distribution for a drug is used to quantify the distribution of a drug between the plasma and the rest of the body. It is often defined as the volume in which the drug would have to be uniformly distributed to give the observed concentration in the plasma.

Drug elimination

Removal of drugs from the body is termed elimination and drugs are eliminated primarily by metabolism and excretion. Generally drugs are metabolized by the liver to make them more suitable for excretion either in urine or in bile. Some drugs may be excreted without prior metabolism. Drug metabolism is usually considered as two phases: 1 and 2. Phase 1 reactions are oxidation, reduction, or hydrolytic reactions that insert or reveal a functional group. They are sometimes referred to as *functionalization* reactions. Phase 2 reactions are conjugations, usually with highly water-soluble compounds such as glucuronate or sulphate. A phase 1 reaction may be necessary before phase 2 can take place, but if suitable functional groups are present, the drug may be conjugated without the need for prior modification. The term phase 3 metabolism is used sometimes to describe further metabolism of phase 2 metabolites.

Most phase 1 reactions are oxidations and because the enzymes metabolize a wide variety of substrates, they are known as **mixed function oxidases** (MFOs). Many, but not all, MFOs are from the **cytochrome P450** (CYP450) family. These enzymes contain a haem group, and need NADPH and molecular oxygen as cofactors. The CYP450 isoforms involved in drug metabolism are typically found in the endoplasmic reticulum, and may also be involved in the metabolism of hormones such as steroids and prostaglandins. The amounts or activities of some of these enzymes can be increased or decreased in response to changes in drug treatment (including use of some herbal medications), cigarette smoking, or dietary components. Typically these enzymes will hydroxylate or N-demethylate a compound to facilitate a phase 2 reaction.

The main phase 2 metabolizing enzymes are the uridine diphosphate glucuronosyltransferase (UDPGT) family. Like the CYP450 enzymes there are a large number of these proteins, but all will attach glucuronic acid, a glucose derivative, to a primary metabolite, or drug if it contains a suitably reactive group, to increase the water solubility of the final metabolite. Sulphotransferases catalyse sulphation of suitable molecules, again producing water-soluble metabolites.

Several aspects of drug metabolism are shown in Figure 21.3.

Although phenacetin is no longer used as a drug in western countries, its active metabolite, paracetamol is widely used as an analgesic. Looking at Figure 21.3 you can see that phase 1 O-deethylation of phenacetin produces the active metabolite, paracetamol (acetaminophen) which undergoes phase 2 conjugation with glucuronate and sulphate. These water-soluble metabolites are excreted in the urine. Paracetamol can undergo phase 2 conjugation without a prior phase 1 reaction. A proportion of a dose of paracetamol is oxidized to a very reactive and toxic metabolite, N-acetyl-*para*-benzoquinoneimine (NAPQI) which is normally conjugated with reduced glutathione (GSH). This illustrates that drug metabolism should not be considered solely as detoxification. Glutathionases catalyse further reactions of the paracetamol-glutathione conjugate to produce paracetamol mercapturate, which is excreted in the urine.

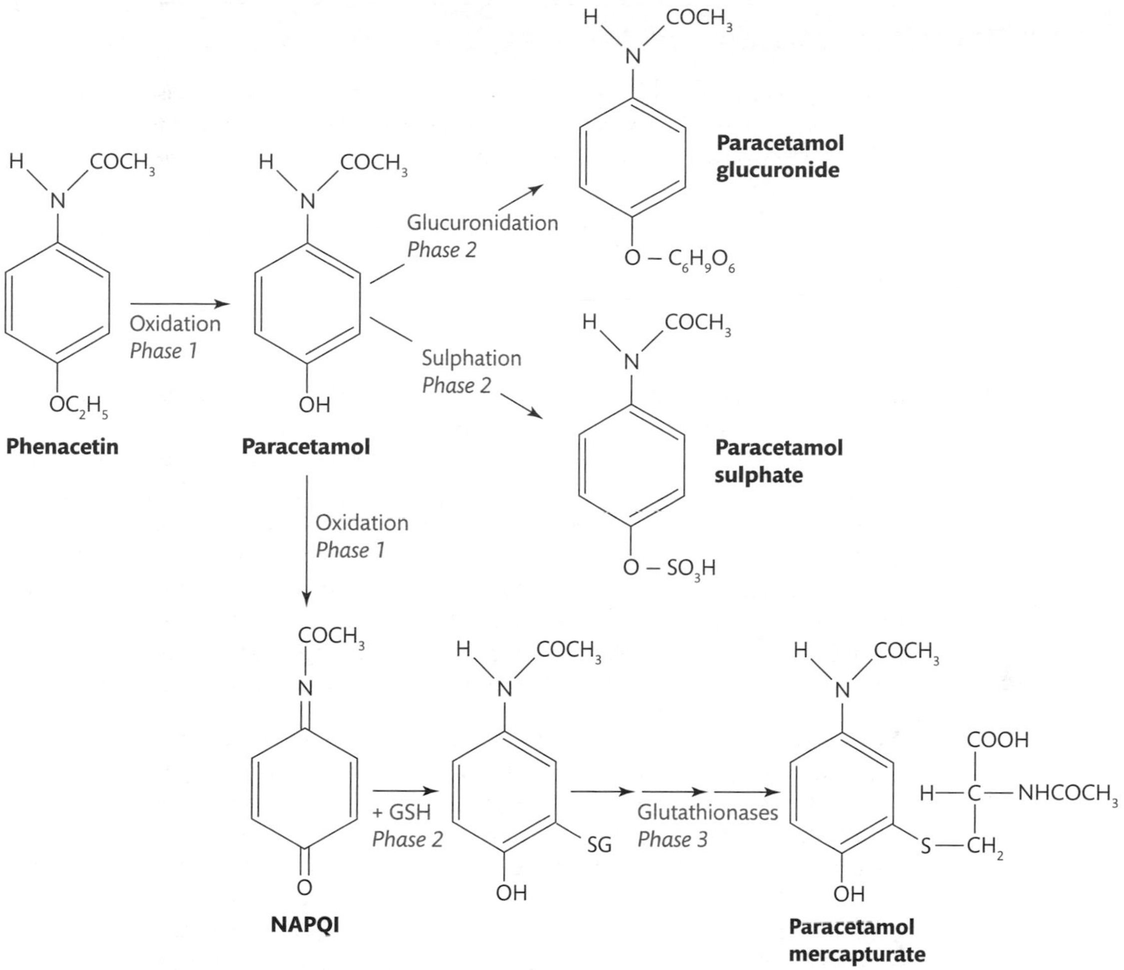

FIGURE 21.3
Example of phases of drug metabolism. See text for details.

Explain why some drugs can undergo phase 2 conjugation without first undergoing a phase 1 metabolic reaction.

21.2 Pharmacokinetics

Pharmacokinetics is the study of drug concentration-time relationships in the body. Often plasma concentrations are used (see Method box below), but drug concentrations in other fluids such as urine or saliva may be studied.

The aim of pharmacokinetics is to describe the shape of concentration-time curves as equations and to calculate values for parameters, such as **half-life**, apparent volume of distribution, and **bioavailability**. It may then be possible to predict the effect of changing the size of the dose given, the route of drug administration, and so on.

METHOD *Collection of blood and blood fractions for drug analysis*

Plasma

(1) Obtain blood by venepuncture of the median cubital vein of an arm, after ensuring that the sampling site is not near the site of any drug injection or infusion.

(2) A hypodermic needle and syringe, or a commercially available vacuum system may be used.

(3) A tourniquet may be used to distend the vein prior to venepuncture, but should be released just prior to sampling.

(4) If using a needle and syringe, remove the needle from the syringe and, so as to avoid haemolysis, slowly transfer the blood to a sample tube containing an appropriate anticoagulant such as heparin or EDTA.

(5) If the analyte is sensitive to light, wrap the tubes with aluminium foil.

(6) Cap and gently rotate the tube to ensure the anticoagulant is mixed with the blood.

(7) Centrifuge to separate the plasma—typically 15 minutes at 2,000 g, at 4°C if necessary.

(8) Transfer plasma, the upper straw-coloured aqueous layer, which is approximately 50% by volume of the total, to a second sample tube.

(9) Store the plasma sample as appropriate, usually 4, −18 or −70°C, depending on the nature of the analyte and how long the sample is to be kept before analysis.

(10) Prior to analysis thaw if necessary and shake the tubes to ensure the contents are homogenous before removing a portion for analysis.

Serum

(1) Collect blood into a plain tube.

(2) Allow the blood to clot; 15–30 minutes standing at room temperature is usually sufficient.

(3) Centrifuge to help separate the serum from the clot.

(4) Transfer the serum to a second tube and store as for plasma.

Whole venous blood

(1) Collect blood into anticoagulant tubes and mix carefully as above.

(2) Store as appropriate, remembering that freezing will haemolyse the blood.

Note

The anticoagulant, the type of tube (glass or plastic), and the storage conditions will depend on a number of factors, including the nature of drug, the assay method, and when the assay is to be performed.

Key Points

The bioavailability of a drug is the fraction of the given dose that reaches the systemic circulation.

Order of reaction and half-life

Most drugs, given at therapeutic doses, are removed from the body at a rate that is proportional to the plasma concentration or the amount of drug in the body. That is, the greater the amount of drug present, the faster the rate at which it is removed. This relationship is known as **first-order elimination**. As you can see in Figure 21.4, a feature of first-order kinetics is that the plasma half-life ($t_{1/2}$), the time taken for the initial concentration to decrease by half, is constant. This means that the proportion of drug removed or remaining is easily calculated

(a) First-order elimination

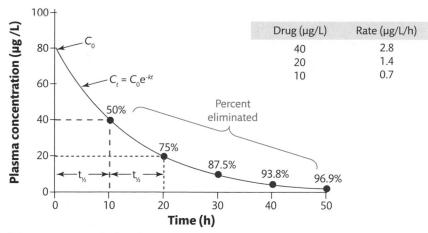

Drug (µg/L)	Rate (µg/L/h)
40	2.8
20	1.4
10	0.7

(b) Zero-order elimination

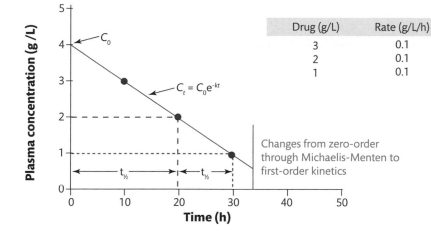

Drug (g/L)	Rate (g/L/h)
3	0.1
2	0.1
1	0.1

FIGURE 21.4

Comparison of: (a) First-order elimination of a drug after a rapid intravenous injection. (b) Zero-order elimination of a drug after a rapid intravenous injection. See text for details.

from knowledge of $t_{1/2}$. A graph of the natural logarithm of concentration ($\ln C_t$) against time is a straight line of slope $-k$ (the first-order rate constant). Such plots were used to obtain values for C_0 and k, but today desktop computers and inexpensive software can obtain these values from the C_t versus t data.

If you look at Figure 21.4, you can see that because the entire drug dose is introduced into the bloodstream almost immediately, only the decline of the drug in the plasma is observed. Extrapolation of the curves to $t = 0$ gives C_0, the 'initial' concentration, that is the value that would be obtained if the drug is instantaneously mixed with the plasma. In Figure 21.4 (a) a constant *proportion* of drug is removed per hour. The elimination half-life ($t_{1/2}$) is the time it takes for C_0 to fall by half, which in this example is ten hours. The concentration falls again by 50% in the next half-life and so on. Irrespective of the size of the dose, after five half-lives less than 4% of the dose remains. The rate of elimination is proportional to the concentration and declines as the concentration falls (see the inset box in Figure 21.4 (a)). In Figure 21.4 (b) the reaction is zero-order, the rate of elimination is independent of the plasma concentration, and a constant *amount* of drug is removed per hour, in this example 0.1 g/L per hour (see inset), and therefore, the larger the C_0, the longer it will take to reach $C_0/2$. Thus $t_{1/2}$ is not constant and is proportional to the concentration of the drug.

The plasma half-life of a drug is directly proportional to the apparent volume of distribution and indirectly proportional to **systemic clearance**. Clearance is an indicator of how easily organs such as the liver and kidney can metabolize and excrete a drug. For example, the renal clearance of creatinine, which is filtered at the glomerulus, is 125 mL/min, whereas benzyl-penicillin, which is actively secreted into the urine, has a renal clearance of approximately 400 mL/min. Plasma clearance, also known as whole body clearance is the total clearances of all the eliminating organs. The higher the plasma clearance, the faster the drug is eliminated and so the shorter will be its half-life. However, the drug has to be carried in the blood to the eliminating organs, and if a drug has a large volume of distribution its plasma concentration will be low. Thus, the larger the volume of distribution, the longer the half-life.

Cross reference

Chapter 3 Kidney disease

The induced synthesis of some drug metabolizing enzymes, for example in cigarette smokers or by some drugs such as phenobarbital (phenobarbitone), a well-known enzyme inducer, may increase hepatic clearance. Changing the pH of urine may increase the excretion of weakly acidic or basic drugs. For example, the aspirin metabolite, salicylic acid is more ionized in alkaline renal tubular fluid and, because the ionized form is not readily reabsorbed back into the body, more salicyclate remains to be excreted in the urine. Inhibition of drug metabolizing enzymes may reduce clearance. Liver or kidney diseases can have similar effects, but there may be accompanying changes in the apparent volume of distribution, so it can be very difficult to predict the effect of changes in organ clearance on the plasma half-life of the drug. It is not normally possible to measure plasma clearance directly, but it can be obtained from the concentration-time data following an i.v. injection of the drug. This is explained in the Method box below.

SELF-CHECK 21.2

How would an increase in apparent volume of distribution affect the elimination half-life?

METHOD *Measuring the pharmacokinetic parameters of a drug*

(1) Inject the dose of drug as a rapid i.v. injection.

(2) Collect timed blood samples into tubes containing an anticoagulant.

(3) Centrifuge the blood to separate the plasma.

(4) Measure the plasma concentrations of drug using a suitable analytical method, such as high performance liquid chromatography.

(5) Draw a graph of the natural logarithm of measured concentrations ($\ln C_t$) versus time (t).

(6) Extrapolate the line to the concentration axis to obtain the intercept, I, ($= \ln C_0$).

(7) Calculate C_0 ($= e^I$).

(8) Calculate the slope of the line; this is numerically equal to the rate constant, k.

(9) Calculate the elimination half-life, $t_{\frac{1}{2}} = 0.693/k$.

(10) Calculate the apparent volume of distribution, $V = Dose/C_0$.

(11) Calculate the plasma clearance, $CL = V \times k$.

Note

V has units of volume and k of time^{-1}, so CL has units of flow-rate, for example mL/min.

An i.v. injection places all of the dose of the drug into the bloodstream. However, with other routes of administration there is no guarantee that the entire dose will reach the general circulation. This is particularly true of oral administration when poor absorption, first-pass metabolism, and efflux by PGP can have a marked effect on the fraction of the dose (F) that enters the general circulation. Because the area under the drug plasma concentration-time curve (AUC) is proportional to the *amount* of drug that enters and leaves the plasma, we can estimate F by comparing the AUC values after i.v. AUC_{iv} and oral AUC_{po} doses:

$$F = \frac{AUC_{iv}}{AUC_{po}}$$

To measure AUC a series of plasma samples is collected and a graph of plasma concentration against time is constructed. Usually F, a quantitative estimate of bioavailability, is measured in the same individual. The same dose of drug is given i.v. and orally on two separate occasions, ensuring that the entire first dose has been eliminated before the next one is given. Clearly, the size of F can have a major influence on the clinical activity of drugs administered orally and this must be considered when interpreting TDM results.

When drugs are given as repeated doses at regular intervals, before all of the previous dose has been eliminated, the plasma concentration will rise due to accumulation of the drug in the body. However, the concentration will not rise indefinitely. If drug elimination is first-order then the rate of elimination increases with increasing concentration until the amount being removed in each dosage interval equals the dose given. This is known as **steady-state**. Figure 21.5 shows how:

(1) The plasma concentrations fluctuate between *peak* concentrations after each dose and *trough* concentrations just before the next dose.

(2) The peak-to-trough fluctuations are greater for the short half-life drug. This is because a greater amount of drug is eliminated during the dosage interval (Figure 21.5 (a)).

(3) Both the peak and trough concentrations are levelling off with time to steady-state values.

(4) The plasma concentrations of the short half-life drug approach steady-state more quickly than the one with the longer half-life (Figure 21.5 (b)). The plasma concentrations reach 96.9% of their steady-state values (C_{ss}) within five plasma half-lives.

If you look at the example in Figure 21.5, the rate of elimination of the drugs increases as the concentration increases and the average concentration (- - - -) approaches a steady-state concentration, C_{ss} (average). If the drugs were given as a constant rate i.v. infusion the plasma concentrations would be superimposed on the average concentration curves. If you look at Figure 21.5 (a), it is more difficult to maintain a drug with a short half-life in the therapeutic window (shaded area). Peak-trough fluctuations can be reduced by using a sustained-release preparation of drug, as with lithium carbonate, for example, and make it easier to maintain the concentrations in the therapeutic window. In Figure 21.5 (b) drugs with long half-lives take longer to reach steady-state but the peak-trough fluctuations are less.

To avoid delay before therapeutic concentrations are reached, a higher initial dose, known as a *loading dose*, can be given to achieve almost instantaneous steady-state conditions. The subsequent smaller doses are referred to as *maintenance doses*. Digitoxin, a cardioactive drug, has an elimination half-life of approximately seven days, so it would take about five weeks to reach steady-state if a loading dose was not used.

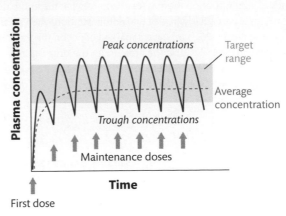

(a) Short half-life

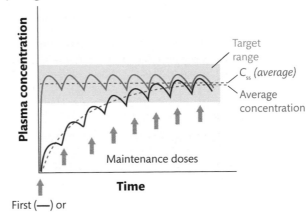

(b) Long half-life

FIGURE 21.5
Multiple-dosing of drugs with: (a) Short half-life. (b) Long half-life. See text for details.

Most drugs show first-order elimination, making dosing and interpretation of TDM relatively straightforward. For such drugs:

- the half-life is constant
- the time to peak concentration is constant
- clearance is constant
- AUC is directly proportional to the dose
- C_{max} is directly proportional to the dose
- C_{ss} is directly proportional to the dose

However, a few drugs when prescribed at therapeutic doses do not show first-order elimination. Such drugs tend to be present in plasma at relatively high concentrations and are subject to extensive metabolism. Consequently, the enzymes responsible for metabolizing the drug become progressively saturated and the elimination kinetics are no longer first-order. In such cases the elimination kinetics can be explained by the Michaelis-Menten equation. At very high concentrations the kinetics approach **zero-order**, which is when the rate of elimination remains constant and independent of the amount of drug present. This is explained in more

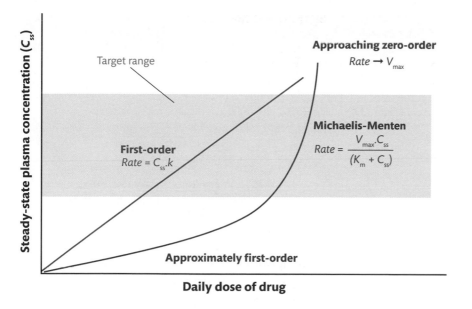

FIGURE 21.6
Comparison of drugs eliminated according to first-order and Michaelis-Menten kinetics. See text for details.

detail in the Method box below. Figure 21.6 shows how drugs that exhibit first-order kinetics differ from those that exhibit Michaelis-Menten kinetics.

The steady-state plasma concentrations of drugs eliminated according to first-order kinetics are directly proportional to the daily dose of drug. However, for drugs eliminated according to Michaelis-Menten kinetics, the steady-state concentration increases disproportionately with increasing daily dose. Small increments in dose may lead to large increases in plasma concentrations, making it difficult to find the correct daily dose to maintain the concentration in the therapeutic range (Figure 21.6, shaded area). The classic example of this type of behaviour is that shown by the **anticonvulsant** drug phenytoin. It is very difficult to assess the correct phenytoin dose, especially as there are large inter-individual variations in the shapes of the concentration-dose curve, and TDM of phenytoin is used to help ascertain the appropriate daily dose for a given patient. Phenytoin exhibits Michaelis-Menten kinetics when given at therapeutic doses. Furthermore, individuals show large variations in apparent K_m and V_{max} for this drug.

21.3 Analytical methods

A wide variety of analytical methods are used in TDM. The choice of which method to use is made on the basis of workload, available equipment, and the suitability of the method to the measurement of drug or metabolite in question in a particular patient group. A clinical decision is often based on the results of an assay, therefore it is usually necessary to select assays with the potential for a short turnaround time.

Immunoassays

Immunoassays are based on the interaction of a drug with an antibody that binds selectively to that drug or to molecules of similar structure to the drug. The principle of the assay is that the drug being measured competes with another molecule, known as the 'label', for antibody binding. The label may be a radioactive drug, or drug attached to an enzyme or a fluorescent molecule making the label easy to measure. Radioactive labels are less common than they

METHOD Michaelis-Menten equation applied to drug elimination

The Michaelis-Menten equation relates the velocity (v) of an enzyme reaction to the substrate concentration (S):

$$v = \frac{V_{max}\, S}{K_m + S}$$

where V_{max} is the maximum velocity, and K_m is the Michaelis constant, and equals the substrate concentrations at $V_{max}/2$.

For most drugs given at therapeutic doses, the concentration is very small compared to K_m and so S contributes little to the denominator, $K_m + S$. Thus the denominator approximates to K_m and:

$$v \approx \frac{V_{max}\, S}{K_m} \approx kS$$

where $k = V_{max}/K_m$. In the equation above the velocity, or rate, is directly proportional to substrate concentration and therefore, is a *first-order* equation; k is known as the (first-order) elimination rate constant.

When the drug concentration is very large compared to K_m, then $K_m + S$ approximates to S and so:

$$v \approx \frac{V_{max}\, S}{S} \approx V_{max}$$

Under these conditions the enzyme is saturated with substrate, and the reaction cannot proceed any faster. The rate of reaction is independent of the substrate concentration and therefore it is a *zero-order* reaction.

When a patient takes a drug, we cannot measure the drug concentration in equilibrium with the drug-metabolizing enzymes, but we can measure the plasma concentration. Furthermore, when a drug is given repeatedly at regular intervals, the plasma concentration does not rise indefinitely, but 'levels off' to a constant value known as the steady-state concentration, C_{ss}. When steady-state is reached, the rate at which the drug is removed equals the rate at which it is being taken. If we use the daily-dose as the rate at which the drug is being given and eliminated, the Michaelis-Menten equation is:

$$Daily\ dose = \frac{V_{max}\, C_{ss}}{K_m + C_{ss}}$$

Ethanol is removed from the body largely by metabolism. At high concentrations its elimination can be described as zero-order, at low concentrations as first-order, but the kinetics at intermediate concentrations can be described by Michaelis-Menten kinetics.

used to be because of the problems of shelf-life and waste disposal inherent in the use of radioactive materials.

The advantages of immunoassay methods include:

- commercially available kits
- ease of automation
- high throughput
- fast turnaround

With immunoassay kits, factors such **selectivity**, sensitivity, and precision (reproducibility) will have been investigated by the manufacturer, at least to an extent. However, the kits may be expensive and may not be readily applicable to specimens other than those for which they were developed, usually plasma or serum in the case of TDM assays.

Some immunoassays require the antibody-bound drug to be separated from the unbound drug before one of them is measured. These are heterogeneous assays. Homogeneous assays, where no separation is required, are more easily automated. Three commonly used homogeneous immunoassays for TDM are:

- enzyme-multiplied immunoassay technique (EMIT)
- fluorescence polarization immunoassay (FPIA)
- cloned enzyme donor immunoassay (CEDIA)

The EMIT assay has two main components. The first is an enzyme, glucose-6-phosphate dehydrogenase, from a bacterial source with the drug chemically attached to it, and the second is an antibody to the drug. The drug in the sample being analysed competes with the drug bound to the enzyme for the antibody binding site. Binding of the antibody to the enzyme prevents the enzyme from working, so the concentration of drug in the sample is inversely related to the enzyme activity. The enzyme activity is monitored by the change in absorbance at 340 nm produced by the conversion of the cofactor NAD^+ to NADH. The nature of the method means it can be used easily by high-throughput clinical chemistry analysers with minimal operator involvement. There is extensive experience of EMIT for TDM. It is simple and it has adequate sensitivity for compounds with an M_r of less than 200 if present in biological fluids at moderate concentrations.

With FPIA, the label is a fluorescent derivative of the drug. These molecules rotate freely in solution, and when they are irradiated with polarized light of the appropriate wavelength they emit light in different planes. However, when bound by antibody the molecules rotate more slowly and emit more light in a similar plane to that of the incident light, which is measured using a polarizing filter as shown in Figure 21.7 (a). The plane polarized light emitted by the label is in the same plane as that produced by polarizing filter 1 and passes through polarizing filter 2 to be measured by the detector. However, any analyte (drug) in the sample competes with the fluorescent-labelled drug for the antibody and so the numbers of slowly rotating fluorescent molecules decrease as the drug concentration increases (Figure 21.7 (b)). The fluorescent light from the freely-rotating label is emitted in all planes and so there is less in the same plane as the incident light. Thus, less plane polarized light passes through polarizing filter 2 and the detector signal is lower than it was in the absence of analyte. Therefore, the amount of polarized light emitted from the reaction cell is inversely related to the drug concentration. This technique is not as widely applicable to clinical chemistry analysers and hence requires a dedicated instrument, but has the advantage of improved precision and reagent stability over EMIT.

Cloned enzyme donor immunoassay exploits antigen-antibody binding to influence enzyme activity. Beta-galactosidase from *Escherichia coli* is supplied as inactive fragments. The large fragment (some 95% of the enzyme) is termed the enzyme acceptor (EA), and the smaller fragment is termed the enzyme donor (ED). By conjugating analyte to the ED fragment, antibodies to the analyte can prevent the formation of intact, active enzyme. Any analyte present in the sample competes for binding sites on the antibody, hence an increase in analyte concentration will decrease binding of antibody to the ED fragment and increase enzyme activity, which can be monitored by production of chlorophenol red (CPR) from CPR-β-galactoside. Like EMIT, this assay can be used by the high-throughput clinical chemistry analysers with minimal operator involvement, but the enzyme fragments are not inherently stable and the complex can be disrupted by sample components.

Interferences and assay failures with immunoassays

A potential problem with immunoassays is poor selectivity. Other, not necessarily related, drugs or constituents of the sample may be measured as well as any drug that might be present. Metabolites and other structurally-related compounds often cross-react in immunoassays. This

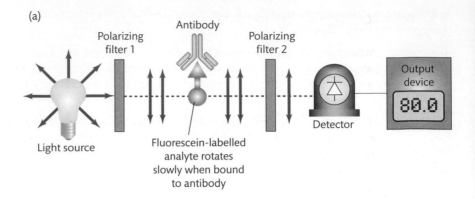

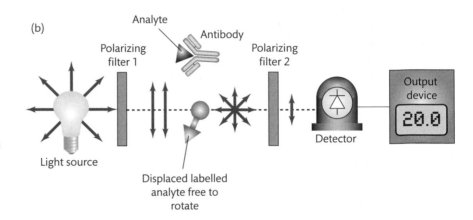

FIGURE 21.7

Schematic illustrating the principle of the fluorescence polarization immunoassay in: (a) The absence of analyte. (b) The presence of the analyte. See text for details. Adapted from Flanagan RJ, Taylor A, Watson ID, and Whelpton R (2007) *Fundamentals of Analytical Toxicology*. Chichester, UK: Wiley

can be helpful in qualitative work such as drug abuse screening, but is generally undesirable in quantitative work. Up-to-date information for interferences with any immunoassay should be available on the kit insert or they can generally be obtained from the manufacturer's website.

Digoxin, a cardioactive drug extracted from the yellow foxglove (*Digitalis lanata*), provides an example of problems that have occurred in immunoassays. Other digitalis glycosides (digoxin-like drugs), digitoxin, for example, cross-react in many assays and this cross-reactivity has been exploited to measure these other glycosides. Endogenous substances, called digoxin-like immunoreactive substances (DLIS), can also react with anti-digoxin antibodies leading to an overestimate of the digoxin concentration. Digoxin-like immunoreactive substances occur in patients with diabetes, uraemia, essential hypertension, liver disease, and pre-eclampsia. They may be a natural digoxin-like hormone, or an abnormal metabolite produced in the above diseases. Unlike digoxin these substances are highly protein bound and it is possible to separate them from digoxin by **ultrafiltration** before measuring the digoxin in such samples. A further complication is when digoxin poisoning is treated with anti-digoxin antibody fragments (F_{ab} fragments) that bind digoxin in plasma *in vivo*. This means that plasma 'free' digoxin concentrations cannot be measured in plasma once the fragments have been given without removing the bound fraction by dialysis.

Ultrafiltration: a method of separating macromolecules such as proteins from smaller molecules by filtration through a membrane, often under pressure.

Cross reference

Chapter 15 Immunological techniques in the *Biomedical Science Practice* textbook

SELF-CHECK 21.3

Explain the role(s) of antibodies in immunoassay.

Chromatographic methods

Gas liquid chromatography (GLC) has been used to measure anticonvulsant, **antidepressant**, and some **antipsychotic** drugs, as well as cardioactive compounds. High-performance liquid chromatography (HPLC) with ultraviolet (UV) or fluorescence detection may be used for TDM (Figure 21.8).

In HPLC, samples are introduced via the injection valve and carried to the column where the components are separated and measured as peaks as they pass through the detector. Typically, ultraviolet or fluorescent detectors are used, or the column outlet can be connected to a mass spectrometer, which may give more selective and potentially more sensitive detection.

Liquid chromatography-mass spectrometry (LC-MS or LC-MS/MS also known as LC-tandem MS) techniques are becoming widely used TDM applications, particularly if immunoassays are either not available or unsuitable for the analyte(s) in question. The analytes are separated on an LC column and passed to the MS where they are ionized and fragmented. The ions are separated according to their mass to charge ratios (m/z) and either quantified, or in the case of LC-MS/MS a selected ion is further fragmented, and then a selected fragment ion is quantified. Because the fragments are selected as characteristic of the particular drug, this

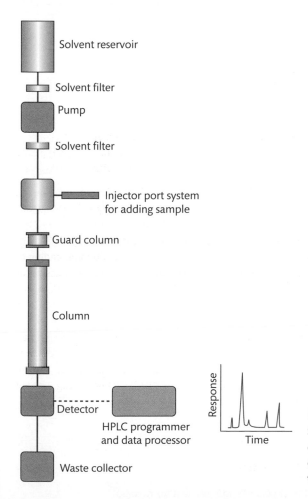

FIGURE 21.8
Schematic showing components of a high-performance liquid chromatography system. See text for details.

technique can be highly selective and thus sensitive. This allows short LC columns to be used, which in turn means that large numbers of samples can be assayed in a given time. However LC-MS and LC-MS/MS have high capital cost, experienced operators are required, and the technique is not devoid of interferences and other problems, most notably ion suppression enhancement (the presence of co-eluting components that either suppress, or enhance the ionization of an analyte).

Liquid chromatography-mass spectrometry can be used to measure a wide range of drugs, often in the same sample. For example, it has been reported that the antipsychotics amisulpride, bromperidol, clozapine, droperidol, flupentixol, fluphenazine, haloperidol, melperone, olanzapine, perazine, pimozide, risperidone, sulpiride, zotepine, and zuclopenthixol, as well as norclozapine, clozapine N-oxide, and 9-hydroxyrisperidone, can all be measured in plasma after appropriate sample pre-treatment steps using the same method. One would not expect all the drugs listed above to be present in a single sample. However, being able to detect the drug and its metabolites and other related drugs is a potential advantage LC-MS has over immunoassays, when it is not always clear what is being measured. Figure 21.9 shows how the combination of LC and MS imparts selectivity to an assay.

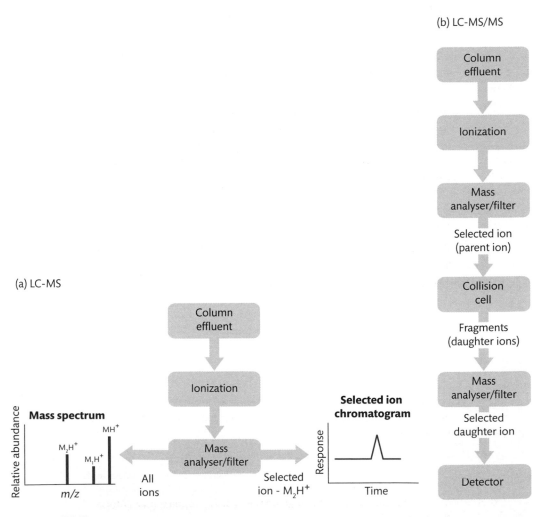

FIGURE 21.9
Schematic to illustrate the principles of: (a) LC–MS. (b) LC-MS/MS. See text for details.

Figure 21.9 (a) shows how the eluent from the LC column is introduced into the MS interface and the analytes ionized. Various modes of ionization are available and generally the molecules are ionized by the addition of a proton to give a pseudomolecular ion, MH^+. A number of fragments or adducts of the analyte may be produced. These are separated according to their mass to charge ratio, m/z, to give a mass spectrum. A particular ion can be chosen and monitored to give a selected ion chromatogram. Figure 21.9 (b) shows that in LC-MS/MS, an analyte ion sometimes known as a parent ion, from the first mass analyser is selected and passed to a 'collision cell' to induce fragmentation of this ion. The fragments, sometimes known as product or daughter ions, are again separated according to their m/z ratios. This approach imparts a high degree of selectivity because even structurally similar molecules are unlikely to give identical parent and daughter ions.

Cross reference
Chapter 13 Chromatography and Chapter 11 Spectroscopy, both in the *Biomedical Science Practice* textbook

Key Points

Selectivity is used to describe the likelihood that a method measures the analyte in question and not interferences.

Quality control and quality assurance in therapeutic drug monitoring

As with any analysis performed in a clinical laboratory, TDM assays must be fit for purpose and must be used with appropriate internal quality control (IQC) and external quality assurance (EQA) in place.

Many chromatographic assays are often developed in-house, and therefore guidelines for method implementation and validation adopted by the pharmaceutical industry, notably those provided by the US Federal Drug Administration (FDA), should be adopted. Unlike automated immunoassays, where the calibration curve may be stable for many weeks, each chromatographic assay requires its own calibration curve (consisting of 6–8 standards across the assay range) to be analysed with each batch of samples, together with IQC samples at low, medium, and high concentration. However, more robust assays can be calibrated with fewer standards if this has been shown to provide satisfactory results.

Unlike most routine clinical chemical assays, chromatographic methods may have a number of manual sample preparation stages, thus increasing the likelihood of human error in the assay and emphasizing the need for robust IQC procedures.

Cross reference
Chapter 1 Biochemical investigations and quality control

21.4 Therapeutic drug monitoring: practicalities

For some drugs it may be possible to measure a clinical response to indicate when a dose of drug needs adjusting. For example, blood glucose concentrations can indicate changes in dosage of an anti-diabetes drug.

Key Points

Therapeutic drug monitoring can be used to guide the clinician to maximize efficacy and minimize the risk of toxicity.

TDM will be of benefit when there:

- is no direct method of measuring the pharmacological effect of the drug
- is a poor or non-existent correlation between dose and effect
- is a narrow margin between 'therapeutic' and 'toxic' plasma concentrations
- are large inter- and intra-individual differences in drug absorption, distribution, and elimination
- is a correlation between the plasma concentration and pharmacological effect of the drug
- is reliable methodology for measuring plasma drug and/or active metabolite concentrations in a clinically relevant timeframe

Adherence: also known as compliance, is when patients take their medication as advised, the correct doses at the correct times.

There may be reasons for TDM even if the above criteria are not fulfilled. One is to check for **adherence**. For example, patients with serious mental illness may say they are taking their medication when in fact they are not. Measuring the plasma concentration of the drug in question may reveal this and knowing that drugs can be measured in this way may encourage patients to take medicines as advised.

Detailed knowledge not only of the limitations of the analytical method(s) used, but also of the clinical pharmacology, toxicology, and pharmacokinetics of the drug(s) monitored is often required when interpreting the results of TDM measurements.

Therapeutic and reference ranges

A key requirement of TDM is that the clinical effects of a drug are in some way related to its plasma concentration even if the relationship between the plasma concentration and the concentration of drug at its site of action is complex. Extreme examples are the **immunosuppressants** used to prevent organ rejection after transplantation. Too little drug may result in loss of the new organ, but too much drug may result in infection, which is often life threatening. In the long term, use of these drugs may result in development of malignancy.

Patients vary considerably in how they respond to drugs, therefore the term therapeutic range is usually applied to an individual, that is, it is the range of concentrations applicable to *that* patient. Furthermore, the patient's treatment must be based on sound clinical judgement and not solely on the basis of laboratory TDM results. To reflect this, the term **reference range** is recommended when helping interpret TDM data, although the expressions 'therapeutic range' or 'target range' are also encountered.

Time of sample collection

As discussed above, plasma drug concentrations fluctuate between doses unless the patient is undergoing an i.v. infusion, when the infusion rate should balance the rate of elimination, thus maintaining the plasma drug concentration effectively constant. Generally, sufficient time from the start of drug treatment, or dose adjustment, should be allowed to ensure steady-state conditions before blood sampling.

The AUC is a way of expressing the total amount of drug in the body during the period between each dose. However, it is impractical to take a large number of samples routinely, so in some cases one (or occasionally two) samples are taken and used to estimate AUC. Typically, a pre-dose sample provides a better estimate of the AUC than samples collected at other times. Also, peak concentrations are likely to be affected by differences in absorption, perhaps due to the presence of other drugs, or of food in the GIT. Therefore, pre-dose samples are considered more consistent, although some protocols may require samples to be collected at some other specified time after the dose.

Type of sample collected

Typically, drug concentrations are measured in plasma or serum. The use of blood tubes containing barrier gels to facilitate plasma or serum collection is not recommended as some of these gels have been shown to absorb some drugs, for example antidepressants and benzodiazepines. For drugs that are bound extensively to red blood cells, ciclosporin for example, use of haemolysed whole blood (EDTA anticoagulated) is mandatory. When transport and storage of liquid samples is impractical, in the tropics for example, it is possible to measure some drugs in dried blood collected onto filter paper, provided the original volume of blood absorbed is known.

Saliva is a plasma ultrafiltrate and the salivary concentrations of some drugs reflect the free fraction. However, a great degree of cooperation is required from the patient to collect this sample reliably, and contamination with food or the remains of the drug tablet are possible. Moreover, some diseases may inhibit production of saliva, and drug treatment may also have this same effect. Saliva may be a useful alternative to plasma for children, many of whom are happy to spit into a container.

Keratinaceous samples (hair and nail) can be used to give a record of drug exposure, but the analytical procedures are complex and thus expensive, and are normally only used in a forensic context in order to establish prior drug use.

Type of measurement

Generally, the parent drug is measured, or in the case of a prodrug, the active metabolite. Aspirin is rapidly hydrolysed to salicylate *in vivo*, and it is usually salicylate that is measured. For some drugs, an active metabolite may make a significant contribution to the overall clinical and/or toxicological effects. Some TDM protocols may advocate using a total value, parent drug plus active metabolite(s), others may consider the concentration of metabolite separately. An immunoassay may measure parent drug and metabolite, and the result will depend on the relative concentrations of drug and metabolite, and the cross-reactivity of the antibody for the metabolite. A different immunoassay may give a different result and this is one of the reasons why laboratories, based upon their experience, may define their own reference ranges.

Measuring the concentration of an active metabolite as well as the concentration of parent drug will probably require a chromatographic method. A further advantage of knowing the relative concentrations of drug and metabolite is that they may give an indication of when the sample was taken relative to time of the last dose of drug. A higher than expected drug: metabolite ratio may indicate that the time between the dose having been taken and the sampling is shorter than the prescribed interval.

The importance of time of sample collection, conditions, and interference by metabolites is stressed in the case history in Box 21.1

Units of measurement in therapeutic drug monitoring

Generally the results of plasma drug measurements are reported in mass concentration, mass units, for example mg/L, rather than molar concentration, for example mmol/L, except

BOX 21.1 A case that highlights the importance of sample collection

A young patient who received a liver transplant for juvenile haemochromatosis was being maintained on a regimen of low dose ciclosporin and mycophenolic acid (MPA). The patient responded well and was successfully discharged from hospital. After six months the family moved house and a further set of monitoring samples were taken and analysed.

Date	MPA (mg/L)	Time post dose (hrs)
03 Sep 2007	0.7	12.6
05 Sep 2007	0.7	13.5
05 Sep 2007	Patient discharged from hospital	
10 Dec 2007	0.5	17.9
12 Mar 2008	0.4	16.5
20 Apr 2008	Family moved house	
06 Jun 2008	Sample collected	
20 Jun 2008	Sample received by transplant laboratory	
20 Jun 2008	1.7	16.5

The patient was still stable and the paediatrician queried the increase in MPA. The suggested therapeutic range is 1–3.5 mg/L. The concentrations of MPA were sub-therapeutic, but this is not a cause for concern as the patient was also receiving ciclosporin and showed no signs of rejection. Immunosuppressants often show serious side effects, so clinicians minimize doses (and hence concentrations) whenever possible.

The age of the sample on the 20 June 2008 was important. MPA has a major metabolite in the plasma (a glucuronide) that is present in concentrations of 10–100 times that of MPA. This metabolite is unstable in plasma and will slowly break down to give MPA, particularly in whole blood at room temperature. The parents were contacted to arrange a repeat sample and the referring laboratory was asked to ensure the sample was separated on the day of receipt and stored at 4°C before postage. The repeat sample gave an MPA concentration of 0.6 mg/L.

This case highlights the importance of disseminating data for the handling of samples to referring laboratories. In this case this was the first sample for MPA that had been seen by the laboratory. Also of note in this case was the good education that had been provided to the parents regarding dose and sample times, so that they were able to insist that the phlebotomist recorded all the relevant information.

for lithium, methotrexate, and thyroxine. You may encounter other units such as parts per million or µg/mL.

$$[\text{parts per million}] = \mu g/g = \mu g/cm^3 = \mu g/mL = mg/L = g/m^3$$

The conversion of mass concentration (ρ) to amount concentration (c), 'molar units', and *vice versa* is simple if the molar mass, M_r, of the compound of interest is known. Thus, using the example of a compound with a molar mass of 151.2 g/mol:

$$c = \rho \div M_r \qquad \text{For example: } (1 \text{ mol/L}) = (151.2 \text{ g/L}) \div (151.2 \text{ g/mol})$$

$$\rho = c \times M_r \qquad \text{For example: } (151.2 \text{ g/L}) = (1 \text{ mol/L}) \times (151.2 \text{ g/mol})$$

Cross reference

Chapter 6 Preparing and measuring reagents in the *Biomedical Science Practice* textbook

However, such conversions always carry a risk of error. Especial care is needed in choosing the correct M_r because the drug could be supplied as a salt, or in a hydrated form. This can cause large discrepancies, especially if the mass of the accompanying anion or cation, or water of crystallization, is relatively high. Most analytical measurements are reported in terms of the free acid or base, and not the salt.

21.5 Therapeutic monitoring of specific drugs

This section summarizes information on the main groups of drugs where TDM may play a part in patient management. Unless stated otherwise, reference ranges refer to pre-dose (trough) concentrations. It must be emphasized that reference ranges can be different with different analytical methods, or with differing patient populations. As with any analyte, it is good practice to ascertain the reference range in the population served by the laboratory with the method used by that laboratory.

Analgesics, including non-steroidal anti-inflammatory drugs

There is very little routine monitoring of non-steroidal anti-inflammatory drugs (NSAIDs). Salicylates used to treat rheumatoid arthritis have been monitored, in part because of the availability of simple methodology. Salicylates react with ferric ions, Fe(III), in aqueous solution to produce a purple colour that can be used to quantify the amount of salicylate present. The reference range is 250–300 mg/L. Monitoring of other **analgesics** and NSAIDs that have largely replaced salicylates in rheumatoid arthritis is rare and is mainly used to diagnose, and if possible minimize, toxicity.

Antiasthmatic drugs

There is usually no reason to monitor **bronchodilators**, such as salbutamol and ipratropium that are commonly taken by inhalation, as the clinical effect is easily monitored and the drugs are relatively non-toxic even if taken in overdose. The exception is theophylline, which is given by mouth and is monitored to minimize the risk of adverse effects at higher plasma concentrations. The adult reference range is 8–20 mg/L. When theophylline and caffeine are used to treat **neonatal apnoea** the reference ranges are 6–12 and 10–30 mg/L respectively.

Anticonvulsant drugs

Therapeutic drug monitoring is vital for anticonvulsant drugs. The aim is to find a dose of drug that prevents convulsions, but because the severity and frequency of seizures varies so much between patients, treatment is usually guided by the plasma anticonvulsant concentration (pre-dose sample). Further complications are that some of the drugs show non-first order absorption or elimination, or saturation of protein binding at higher concentrations. In addition, many drugs are used in combination as some patients with epilepsy do not respond to one drug alone. Carbamazepine induces not only the metabolism of some other drugs, but also its own metabolism, and thus the half-life of this drug decreases with continued usage.

Some of the metabolites of anticonvulsant drugs, for example carbamazepine-10, 11-epoxide, have marked anticonvulsant activity and plasma concentrations of these metabolites should be monitored as well as those of the parent compound. In addition, some of these drugs are used for other indications and the target ranges for these conditions may be markedly different from that for the anticonvulsant activity.

Phenytoin has a relatively narrow therapeutic window, the reference range being generally taken as 7–20 mg/L, and exhibits Michaelis-Menten elimination kinetics (Figure 21.6). The situation is further complicated by inter-individual variations in apparent V_{max} and K_m. However, if C_{ss} is measured at two dosage rates, it is possible to calculate these constants for that patient. Substitution of the appropriate values into the Michaelis-Menten equation allows a suitable dose to be calculated for that patient.

Antimalarial drugs

Quinine was the first **antimalarial** drug. The plasma total (bound and free) quinine concentrations associated with effective antimalarial therapy are in the range 10–15 mg/L. However, in non-infected subjects plasma concentrations above 10 mg/L are commonly associated with clinical features of toxicity such as visual disturbance, leading in some cases to permanent visual deficit or blindness. Normally, quinine is 70–90% bound to plasma proteins, notably α_1-acid glycoprotein, the concentration of which is increased in severely infected patients. Because quinine protein binding is also increased, to 93% or so, higher plasma total quinine concentrations can be tolerated with no apparent toxicity.

Chloroquine and hydroxychloroquine are antimalarial drugs that again bind to α_1-acid glycoprotein. The reference concentration when they are taken to prevent malaria is less than 0.05 mg/L. These drugs are also used at higher doses as second-line agents to treat rheumatoid arthritis, the reference ranges being 0.2–0.3 and 0.4–0.5 mg/L, for chloroquine and hydroxychloroquine respectively. Chloroquine also has anti-HIV-1 activity. Chloroquine is used as a malaria **prophylactic** at a single adult oral dose of 300 mg free-base once weekly. Taking the drug daily by accident often leads to toxicity in non-infected subjects.

Antimicrobial drugs

Therapeutic drug monitoring of aminoglycoside **antibiotics** such as gentamicin and tobramycin has been well-established for many years. Interpretation of the measured concentration is best provided in conjunction with a microbiology laboratory. Peak (two-hour post-dose) concentrations of isoniazid, rifampin, pyrazinamide, and ethambutol give information as regards effective oral dosage, but an additional sample at six hours may help differentiate between delayed absorption and generally poor absorption, which results in ineffective treatment as well as giving further information such as an indication of plasma half-life. Prompt, effective treatment minimizes the risk of bacteria developing resistance to the drugs.

The observed wide inter-individual variation in the pharmacokinetics of antiretroviral drugs, chiefly those used to treat HIV, has led to increasing interest in the use of TDM to help optimize dosing with these drugs.

More recently TDM of drugs such as fluconazole, itraconazole, and its active metabolite hydroxyitraconazole, posaconazole, and voriconazole that are used in the prophylaxis and treatment of fungal infections of the lung such as Aspergillosis in immunosuppressed patients has been advocated.

Antineoplastic drugs

Therapeutic drug monitoring has the potential to improve the clinical use of **antineoplastic** (anti-cancer) agents, most of which have very narrow therapeutic windows and highly variable pharmacokinetics in different patients. Plasma concentration-effect relationships have been established for 5-fluorouracil, 6-mercaptopurine, which is the active metabolite of the drug azathioprine, and methotrexate. Interpretation is complicated by the use of different agents simultaneously. The *AUC* may give more useful information than measurements at a single point in time. Relationships between plasma concentration and dose-limiting toxicities for epipodophyllotoxins, platinum-containing compounds, camptothecin, anthracyclines, and antimetabolites have also been described, but these are seldom monitored outside specialist laboratories.

The advent of a range of tyrosine kinase inhibitors following the success of imatinib in treating chronic myeloid leukaemia (CML) is opening up a whole new field of application for TDM. It has been suggested that trough plasma imatinib concentrations of 1 mg/L or more are associated with an increased likelihood of maintaining a good response in CML.

Cardioactive drugs

Therapeutic drug monitoring of digoxin is well established, although its clinical relevance is often obscure. The clinical features of overdose can be confused with the features of under-dosing. Sometimes TDM can be useful in the case of amiodarone to monitor adherence and assess toxicity, and to monitor adherence to sotalol and to other β-adrenoceptor blockers such as atenolol and propranolol. Use of the calcium channel blockers verapamil and diltiazem is normally assessed by monitoring haemodynamic effects. Diltiazem and *N*-desmethyldiltiazem, desacetyldiltiazem and *N*-desmethyldesacetyldiltiazem are unstable in plasma; all may be pharmacologically active and there is no rationale for attempting to measure plasma concentrations of diltiazem. Therapeutic drug monitoring is also useful in assessing dose requirements for flecainide usage in children.

Immunosuppressants

Therapeutic drug monitoring of ciclosporin, mycophenolic acid (the active metabolite of mycophenolate mofetil), sirolimus, and tacrolimus is well established, with TDM a requirement of drug licensing within the European Union for sirolimus. Haemolysed whole blood is the preferred specimen for all but mycophenolic acid. All these drugs have narrow therapeutic windows and show considerable pharmacokinetic variability, and TDM is essential to avoid adverse effects such as nephrotoxicity while maximizing efficacy. All is not straightforward, however, as some patients experience acute rejection episodes or post-operative complications despite blood concentrations within the reference range. As in the case of antineoplastic

CASE STUDY 21.1

A 65-year-old man was on digoxin therapy of 0.13 mg per day. He went to see his family doctor at 9 am and complained of feeling tired and suffering from nausea. A blood specimen was collected and sent to the laboratory for analysis (target range given in brackets):

Serum digoxin   3.2 µg/L (0.8–2.0)

(a) Can you provide a possible explanation for this result?

(b) What further tests would be useful?

agents, *AUC* calculations may give more useful information than measurements at a single time point, particularly for mycophenolic acid, but collection of such samples in the outpatient setting is often impractical. Peak, two-hour post-dose, sampling may be a better indicator of optimal ciclosporin dosage than pre-dose or four hour post-dose sampling, but this is disputed and in any event peak sampling is difficult to achieve in practice.

Interpretation of either 'trough' or 'peak' results is complicated because:

(1) There may be considerable differences between the results obtained with immunoassays as compared to chromatographic methods.

(2) Immunosuppressants are often used in combination to reduce the risk of toxicity from individual compounds hence the concentrations attained during optimal treatment are lower than when the drugs are used alone.

(3) The amount of immunosuppression required for maintenance treatment varies widely depending on the engrafted organ.

Psychoactive drugs

Lithium carbonate is used to treat bipolar disorder (also known as manic depression). The optimal plasma concentration range for treatment (Li^+ ion, 12-hour post-dose sample) is 0.6–1.0 mmol/L, with mild to moderate toxicity often occurring between 1.5–2.0 mmol/L. However, patients with mania seem to have an increased tolerance to the drug, and in some patients unwanted effects are apparent at concentrations as low as 1 mmol/L. Lithium is excreted in urine, and renal excretion of the drug decreases with age and when dietary sodium intake is lowered. The excretion rate also varies widely between individuals. Unlike many of the other drugs discussed in this chapter, lithium can be measured easily using an ion selective electrode or by colorimetric assays on high-throughput clinical chemistry analysers. Serum lithium should be monitored during initiation of therapy and every three to four months thereafter to ensure continued optimal dosage. The risk of interference from lithium heparin anticoagulant is avoided if serum is used rather than plasma.

Generally there is little need for TDM of tricyclic antidepressants, that is, drugs such as imipramine and nortriptyline, as patients become tolerant to some of the adverse effects of these

CASE STUDY 21.2

A sample for tacrolimus analysis was received from an external hospital which gave a result of 20.5 µg/L but there was no dosage data on the request. The referring laboratory promised to chase the sample and three days later provided a repeat sample marked 'urgent', but again with no dosage data. The sample was analysed and the tacrolimus concentration was reported as 22.6 µg/L. The patient's doctor was contacted and asked to check that the specimen provided was a genuine trough sample. Furthermore he was to ensure a further sample was sent with full dosage data (sample and last dose time, last dose amount). A further sample was received a week later, again with no dosage data. The tacrolimus concentration in this sample was 1.3 µg/L, and even lower (<1.0 µg/L) in a further sample received a week later. When the hospital was contacted again, it was learnt that the patient had developed severe (life-threatening) rejection.

What message can be learnt from this case study?

drugs on chronic dosage. However, TDM reference ranges have been suggested and may be useful in dose adjustment in individual patients. Therapeutic drug monitoring of antidepressants in general is mainly concerned with assessing whether treatment failure is due to poor adherence, ultra-rapid metabolism, or drug interactions leading to induction or inhibition of drug metabolizing enzymes.

The antipsychotic drugs fall into two categories: 'typical' or first generation, which include chlorpromazine and haloperidol, and 'atypical' or second generation. The former have been in use for many years and there is little benefit to be gained from TDM. However, monitoring of one of the atypical drugs, clozapine, is required as there is a 50-fold inter-patient variation in the rate of clozapine metabolism. Alteration in smoking habit can have a dramatic effect (on average ± 50%) on clozapine dose requirement, and the plasma concentrations associated with serious toxicity, most notably convulsions, are little different from those associated with effective therapy. In addition the clinical features of clozapine overdosage can mimic those of the underlying disease. There is also some requirement to monitor another atypical antipsychotic drug, olanzapine, most notably in assessing adherence.

SELF-CHECK 21.4

Having read the above, what do you think is a necessary criterion for TDM to be successful?

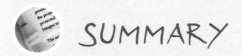

SUMMARY

- This chapter has introduced a complex field and has made no attempt to discuss the diseases for which the drugs are prescribed.

- Knowledge of how drugs are absorbed, distributed, and eliminated is required to understand the basis for sample collection in TDM.

- It is important to ensure that the analytical method used gives accurate and reliable results in a clinically relevant timescale since patient treatment may be dependent on the data generated.

- Once an appropriate measurement has been performed it is crucial to know how the measured concentration relates to pharmacological effect in order to give appropriate clinical interpretation of the results.

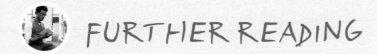

FURTHER READING

- **Birkett DJ (2002)** *Pharmacokinetics Made Easy*, **2nd edition. Sydney: McGraw-Hill.**

  Originally published as a series of articles in the *Australian Prescriber*. Gives further explanation of clearance and elimination, and is a useful source of equations.

- **Flanagan RJ, Brown NW, and Whelpton R (2008) Therapeutic drug monitoring (TDM) CPD Bulletin.** *Clinical Biochemistry* **9**, 3–21.

  More advanced account of TDM and useful source of reference ranges for drugs.

- Flanagan RJ, Taylor A, Watson ID, and Whelpton R (2007) *Fundamentals of Analytical Toxicology*. Chichester: Wiley.

  Details of analytical methods, pharmacokinetics, and drug metabolism.

- Gibson GG and Skett P (2001) *Introduction to Drug Metabolism*, 3rd edition. Cheltenham: Nelson Thornes.

  A classic textbook on drug metabolism.

- Gross AS (2001) Best practice in therapeutic drug monitoring. *British Journal of Clinical Pharmacology* 52, 5S–10S.

- Hallworth M and Watson I (2008) *Therapeutic Drug Monitoring and Laboratory Medicine*. London: ACB Venture Publications.

  Details of TDM in laboratory medicine.

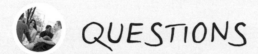

QUESTIONS

21.1 State whether the following statements are TRUE or FALSE:

(a) Enteric coating of drugs allows them to be released immediately in the mouth

(b) Prodrugs are easily absorbed and metabolized to active components in the body

(c) Movement of drugs from the blood to the tissues is referred to as distribution

(d) N-acetyl-*para*-benzoquinoneimine is a toxic product derived from the metabolism of aspirin

(e) The plasma half-life of a drug is directly proportional to its clearance

21.2 State whether the following statements are TRUE or FALSE:

(a) Heparin is an anticoagulant

(b) Pharmacokinetics is the study of drug concentration-time relationships

(c) Radioactive labels are now more commonly used in immunoassays

(d) The abbreviation FPIA stands for fluorescence polarization immunoassay

(e) Salicylate is an anti-asthmatic drug

21.3 Which one of the following statements concerning drugs is incorrect?

(a) Salbutamol is a bronchodilator

(b) The therapeutic window for use of phenytoin is relatively small

(c) Quinine is an anticonvulsant drug

(d) Lithium is used to treat patients suffering from manic depression

(e) Hair can be used to assess drug exposure

21.4 Describe briefly the steps required to set up an assay for measuring the concentration of a drug using HPLC.

Answers to self-check questions, case study questions, and end-of-chapter questions are available in the Online Resource Centre accompanying this book.

 Go to www.oxfordtextbooks.co.uk/orc/ahmed/

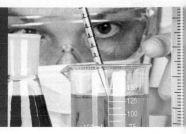

22

Chemical toxicology

Gwendolen Ayers

Learning objectives

After studying this chapter you should be able to:

- List the most common poisons, describe the circumstances in which poisoning occurs and the factors that influence the toxicity of a substance
- Describe how poisons enter and are removed from the body, and the mechanisms of poisoning as a basis for rational treatment of poisoning
- Describe with examples the analytical methods used to identify and measure poisons
- Select appropriate laboratory investigations based on clinical features and history

Introduction

Toxicology is the scientific study of **poisons** and poisoning, and has applications in many areas, including the industrial, agricultural, veterinary, environmental, forensic, and medical fields.

In clinical biochemistry we use the term toxicology to refer to the identification and measurement of the concentrations of poisons or biomarkers of their effects, in body fluids, tissues, and sometimes other materials. To appreciate why toxicology investigations are performed and the way results are used we need to know how poisoning occurs, how a diagnosis of poisoning is made and how poisoning is treated. In this chapter, we will consider the different types of poisons, their clinical and biochemical features, their laboratory investigation and management. Selected examples of common poisons will be considered in detail.

> A **poison** is a substance capable of causing harm to a living organism.

22.1 Types of poisons and poisoning

We can think of poisoning as a medical condition caused by a substance that is not usually present in the body, for example methanol, or is present at a much higher concentration than usual, for example iron. In the latter case, for the condition to be a poisoning, the high concentration must be due to intake of an excessive amount from an external source. Poisons can be referred to as **xenobiotics**.

> A **xenobiotic** is a compound that is foreign to life.

The word is derived from the Greek *xenos* meaning *foreign* and *bios* meaning *life*. We have seen that the term poison can be defined as any substance that causes harm to a living organism.

Key Points

Any substance can be a poison if the body is exposed to an amount that exceeds the body's capacity to detoxify it.

For convenience poisons can be divided into seven main groups:

- anions
- corrosives
- gases and volatiles
- metals and metalloids
- toxins
- pesticides
- drugs

These vary in their chemical structures and properties, and the ways in which they affect the human body. This relates mainly to their effects and to some extent how they can be identified and measured, but the classification is not entirely satisfactory.

Toxic anions include cyanide, fluoride, nitrite, and oxalate. They often bind to metal ions in metalloproteins and thus interfere with their function. For example, cyanide causes tissue **hypoxia** by binding to iron in cytochrome c reductase and inhibiting electron transfer. Corrosives cause destruction of body tissues on contact. They can be strong acids or bases such as sulphuric acid and sodium hydroxide, organic acids for example acetic acid, heavy metal salts or strong detergents. Gases and volatiles include products of combustion such as carbon monoxide, solvents, and gases used in manufacturing or the home, and alcohols including ethanol. Metals include iron, lead, and arsenic. They can cause acute toxicity, for example ingestion of an overdose of iron tablets can cause liver damage, but more often toxicity is due to accumulation of the metal as a result of exposure to relatively low amounts over a prolonged period of time. **Toxins** are biological compounds produced by plants, animals, bacteria, and fungi, for example digitalin from the foxglove, snake venom, botulinum toxin, and the phallotoxins and amatoxins of the death cap mushroom. **Pesticides** are chemicals used to kill pests and include organophosphate insecticides and the herbicide paraquat. **Drugs** are substances used to produce a desired effect but can also have unwanted toxic effects. This is especially likely in non-accidental self-poisoning, when medicines are taken in excessive amounts, and in drug misuse.

A toxic exposure describes the amount of a substance that will cause features of poisoning if it enters the body.

Because of the dependence on amount, poisoning is also referred to as toxic exposure. Toxic exposure may be intentional (non-accidental) or unintentional (accidental), acute, or chronic. Misuse of substances including drugs comes somewhere between these in that exposure to the poison is intentional but the toxic effects are an unwanted consequence. Intentional poisoning, where the exposure is set up deliberately with the intent of harming oneself or others is the most likely type of poisoning to cause serious harm. If the poison is self-administered this is referred to as suicide if the result is death or parasuicide (previously known as attempted suicide) if the person survives. These definitions remove the need to make a judgement as to whether the toxic exposure was a deliberate attempt to end life or made to seek attention. While the distinction may be important for the future care of survivors, laboratory investigation cannot usually help with this. The circumstances and the medical history are more useful guides, for example in attempts to end life steps are often taken to avoid discovery. **Munchausen's syndrome**, where an individual inflicts injury to themselves in order to attract medical attention, is a rare reason for intentional self-poisoning. Munchausen's by proxy where the sufferer poisons another, for example a parent giving an unnecessary medicine to a child, is another form of intentional poisoning. Other examples are homicide (murder), judicial (execution by lethal injection), euthanasia, chemical warfare, and terrorist attacks.

Unintentional or accidental poisoning is more common than intentional poisoning and is particularly common in children, where the child's natural curiosity leads them to eat or drink cleaning products or medicines. Exposure to toxic chemicals in the workplace (**occupational exposure**) is a potentially serious cause of poisoning. Developed countries usually have legislation to limit the amount of substance that can be handled or is present in the workplace environment and to check the effectiveness of this by personal and/or environmental monitoring. Occupational exposure is often chronic, due to exposure to relatively low levels of the toxic substance over a prolonged period, as is environmental exposure, for example due to contaminated air or water supplies. This is an example where laws have been put in place to prevent poisoning by limiting pollution of the environment.

Iatrogenic (medically induced) toxicity is also a significant cause of unintentional poisoning. Medication errors occur, such as prescribing an incorrect dose or combination of drugs, dispensing the wrong tablet strength or administering medication daily instead of weekly, or by the wrong route, for example giving an oral dose by intravenous injection. Healthcare organizations and the individuals providing the care have a duty to put systems in place to identify problems and their causes and to learn from experience in order to avoid these occurrences.

As we have seen, poisoning can take place in a variety of circumstances with a huge range of different substances. In practice, poisoning accounts for a small proportion of medical presentations and deaths. The majority of clinically significant poisonings are caused by only 20–30 substances. Factors that affect the statistics on which these statements are based include the inherent toxicity of the substance, the availability of the substance, which will differ for different age groups, occupations, and geographical areas, and whether exposure is intentional or not. The quality of the scientific evidence for poisoning depends on the extent, sensitivity, and specificity of laboratory investigation. The major poisons causing death in England and Wales in 2005 are listed in Table 22.1.

CASE STUDY 22.1

Children living in Birmingham close to a motorway were found to have high blood concentrations of lead. Analysis of surface dust samples in the locality also showed high lead concentrations.

Where do you think the lead in the dust was coming from?

TABLE 22.1 Poisons causing death in England and Wales. Based on Registrar General's Annual Report for 2005, Office of National Statistics.

Poisons causing death ranked with most frequent first	Number of deaths
Opioids, including heroin and methadone	376
Carbon monoxide	257
Alcohol	164
Tricyclic antidepressants	145
Paracetamol	134
Other antidepressants and antipsychotic drugs	109
Cocaine and other psychostimulants	61
Benzodiazepines	32
Drugs, medicaments, and biological substances, including those specified above	1783
Substances chiefly of non-medicinal source including alcohol	610

22.2 Clinical features of poisoning

Poisoning may present with neurological, respiratory, cardiovascular, and gastrointestinal features. The eyes, skin, and muscle may also show signs which aid diagnosis. Neurological features can be broadly grouped into those associated with central nervous system (CNS) depression and those associated with CNS stimulation. Features of CNS depression range from relatively minor, for example confusion and impaired muscle coordination causing unsteadiness, abnormal gait (**ataxia**), and involuntary eye movements, to life threatening. Impaired consciousness, reduced muscle tone and reflexes, decreased heart rate (bradycardia) and respiratory drive, hypotension, and **hypothermia** can lead to cardiovascular collapse, hypoxic damage to the brain and other tissues, seizures, coma, and death.

The Glasgow Coma Score (GCS) is widely used for assessing CNS function. It involves assessing eye opening (score 1–4), verbal response (score 1–5) and motor response (score 1–6). A score of 1 indicates no response and the highest score in each case indicates normal function. The GCS is the sum of the scores and ranges from 3 to 15, with a score of 8 or less indicating coma.

SELF-CHECK 22.1

An overdose patient is assessed with a GCS of 6 on admission. Are they (a) fully conscious (b) mildly confused or (c) comatose?

The features of CNS stimulation are essentially the opposite of those of CNS depression. There is restlessness and agitation, increased muscle tone and reflexes, increased body temperature ranging from mild fever to **hyperpyrexia**, increased respiratory and heart rate (tachycardia), and hypertension. Severe cases may present with delirium, mania, seizures, and coma.

Any agent that impairs consciousness may cause hypoventilation (reduced depth of respiration) by depression of the respiratory centre. Some compounds, for example salicylate and sympathomimetics have a direct stimulatory effect on the respiratory centre causing hyperventilation. Hyperventilation may also occur as a compensatory effect in metabolic acidosis. Respiratory rate may be increased or decreased. Irritants cause cough, breathlessness, and increased sputum production. There may be aspiration of gastric contents, pulmonary oedema, or other direct evidence of lung injury.

Poisons can affect the cardiovascular system by altering the rhythm at which the heat beats causing tachycardia, bradycardia, or other **dysrhythmia** (irregular heart beat). Hypotension may be a result of hypovolaemia caused by vomiting and diarrhoea, depressed muscle contractility, induced either by a direct action of the toxin or acidosis, or dysrhythmias. Hypertension can also be a feature.

Ocular features of poisoning include **miosis** (pin point pupils) or **mydriasis** (dilated pupils), decreased visual acuity and loss of control of eye movements (**nystagmus**). The skin may be flushed or pale, blistering may be present and needle marks may suggest intravenous drug use. **Rhabdomyolysis** is associated with CNS stimulant use and in the unconscious patient the presence of blisters on areas of maximum pressure can suggest this diagnosis.

Damage to the lips, mouth, or tongue, increased or decreased salivation, unusual breath odour, and **dysphagia** may suggest poisoning. Vomiting and diarrhoea are less specific gastrointestinal features of poisoning.

22.3 **Biochemical features of poisoning**

Disturbances in acid-base balance are common in poisoning, the most common being metabolic acidosis. This can come about due to ingestion of acids, for example aspirin (acetylsalicylic acid) or compounds that are metabolised to acids, for example ethylene glycol.

Nephrotoxicity may also cause metabolic acidosis. Lactic acidosis due to fits or **ischaemia**, and ketoacidosis due to starvation or diabetic ketoacidosis are other causes of metabolic acidosis associated with, but not directly due to, poisoning. Metabolic alkalosis, respiratory acidosis, and respiratory alkalosis may also occur. A mixed picture may be seen as in salicylate poisoning, where stimulation of the respiratory centre causes respiratory alkalosis alongside the metabolic acidosis mentioned above.

Hyponatraemia is rarely seen in the acute situation but Ecstasy users who drink water rather than isotonic fluids may suffer from water overload. Hypernatraemia may be due to water loss or excess sodium intake as in bleach ingestion. Hypokalaemia can occur as potassium redistributes into erythrocytes in alkalosis or in response to insulin. Release of potassium from damaged cells in metabolic acidosis or in **digoxin** toxicity can cause hyperkalaemia.

Acute hypocalcaemia may result from complex formation, in poisoning with anions, for example fluoride, or with compounds that give rise to anions, for example ethylene glycol, which is metabolized to oxalate and glycollate. Loss of magnesium in diuretic or laxative abuse may lead to chronic hypocalcaemia.

An increased osmolar gap may indicate poisoning by alcohols or ethylene glycol. The osmolar gap is calculated by subtracting the estimated plasma osmolality from the measured plasma osmolality. The estimated plasma osmolality is obtained by multiplying the plasma sodium by two then adding the plasma urea and glucose concentrations. An osmolar gap greater than 10 mmol/L suggests the presence of unmeasured osmotically active species.

Cross reference
Chapter 6 Acid-base disorders

Cross reference
Chapter 5 Fluid and electrolyte disorders

Key Points

The osmolar gap is calculated by subtracting the estimated plasma osmolality from the measured plasma osmolality and can be used to indicate the presence of a foreign substance in the plasma if the difference between the two values is greater than 10 mmol/L.

Calculation of the anion gap may also be helpful. This is calculated by subtracting the sum of the plasma chloride and bicarbonate concentrations from the plasma sodium concentration and is normally between 8–18 mmol/L. A low anion gap is unusual and indicates the presence of an unmeasured cation. In an overdose situation the possibility of lithium should be considered. Unmeasured anions give rise to an increased anion gap and if there is no obvious explanation lactate and ketones should be measured.

High plasma concentrations of urea and creatinine indicate renal impairment. A ratio of urea to creatinine of >100 when both are measured in mmol/L is suggestive of pre-renal uraemia due to hypovolaemia. Hyperglycaemia may be present due to the anti-insulin actions of catecholamines. Hypoglycaemia is seen in overdose with hypogycaemic agents and occasionally salicylates, ethanol, and β-blockers. Changes in plasma enzyme activities are usually a reflection of the underlying pathological changes caused by the poisoning. Creatine kinase (CK) will be raised in rhabdomyolysis. Liver damage, for example in paracetamol toxicity, will cause raised transaminases (alanine aminotransferase (ALT) and aspartate aminotransferase (AST). An exception is the **hyperamylasaemia** seen in morphine overdose which is due to a direct action preventing release of pancreatic secretions into the gastrointestinal tract (GIT).

22.4 Management of the poisoned patient

In order to understand the rationale underlying the medical management of poisoning we need to know how patients are exposed to poisons, how poisons exert their effects, and how they are removed from the body. Poisons can enter the body by contact with skin or mucous membranes, through the lungs by inhalation, via the GIT by ingestion, and from tissue, for example muscle, or directly into the bloodstream by injection, animal bites, and stings. The toxicity of the substance often varies according to the route of exposure, for example the amount of substance reaching the site of action when taken orally is generally much less than if the same amount is injected due to 'first-pass' metabolism. Poisons can be removed from the body by excretion in urine and faeces and volatile substances can be lost in the breath.

Cross reference

Chapter 21 Therapeutic drug monitoring

The relationship between time from exposure and the concentration of substance in the blood depends on the pharmacokinetic processes of absorption, distribution, metabolism, and excretion.

Provided the blood concentration of the poison relates to that at its site of action, pharmacokinetic calculations based on the size of the exposure or measurement of timed blood concentrations may be helpful in predicting the time course of appearance of toxic effects and recovery from them.

Some poisons, namely corrosive liquids, irritant gases, and particulates produce direct (local) toxicity by damage to membranes on contact. Clinical features will be present at the sites of exposure, for example the skin or eyes and the severity relates both to the concentration of the substance and the degree of ionization. More commonly toxic effects are systemic and result from absorption into the circulation and distribution to tissues. Systemic effects may be structurally specific, due to interaction with tissue receptor proteins and related to the chemical structure of the poison, or non-specific. In some cases it is a metabolite rather than the substance itself which is responsible for toxicity. The process by which this occurs is known as metabolic activation and the metabolism of paracetamol to a free radical product, N-acetyl-*para*-benzoquinoneimine (NAPQI) that causes liver damage is an example of this.

In the UK, guidance for the emergency treatment of poisoning is published in the British National Formulary (BNF). If the patient has features of poisoning or is well but has a history of exposure to a poison of which the action is delayed, admission to hospital is advised. The National Poisons Information Service (NPIS) provides a database of toxicology information for healthcare professionals (Toxbase), accessible via the Internet. This includes monographs on a large number of substances. These give the clinical and biochemical features of poisoning for each substance, together with treatment advice. Regional Poisons Information Centres provide a telephone service to give advice specific to an individual case.

Most poisoned patients will survive with supportive care. The ABC of resuscitation (airway, breathing, circulation) is followed. Many poisons suppress breathing and ventilatory support may be needed. Hypotension is common and the cause should be determined. Hypoxia and acidosis should be corrected and treatment of arrhythmias may be required. Body temperature can be corrected by warming or cooling. Intravenous diazepam is given to control seizures if these are prolonged.

Minimizing absorption, enhancing elimination and the use of **antidotes** are more specific approaches to the management of poisoning. There is a lack of evidence to support many traditional treatments but if a patient presents within one hour of ingestion and a large amount has been taken, gastric lavage may be effective in removing the poison from the GIT. The dangers must be weighed against the benefits and it is not suitable if the patient is drowsy or comatose, or if petroleum products or corrosives have been ingested. Activated charcoal can be given by mouth to prevent absorption but should not be given if petroleum products or corrosives have been ingested and it is not effective for removal of alcohols, dichlorodiphenyltrichloroethane (DDT), malathion insecticides, or metal salts, for example iron or lithium.

CASE STUDY 22.2

A 3-year-old girl was brought to hospital because she had been found in her grandmother's bedroom playing with tablets from a pack of aspirin. The child's mother had brought the pack with her and tablets from the floor were found. The pack contents were aspirin 300 mg, 16 tablets, of which two were still in place. A further 8 tablets were recovered from the floor. The girl weighed 20 kg. On examination, she appeared well.

(a) Calculate the maximum amount of aspirin ingested by the child and the amount per kg of body weight.

(b) Toxbase states that if no other substances have been taken and if less than 125 mg/kg of aspirin has been ingested and there are no symptoms then hospital admission is not required. Based on this information, does she need her blood salicylate concentration measured?

If large amounts of sustained release drugs have been taken whole bowel irrigation may be considered.

Key Points

Gastric lavage involves passing a tube via the mouth or nose down into the stomach, which is then followed by sequential administration and removal of small volumes of either warm water or saline. This procedure is continued until the returning liquid shows no gastric contents and the ingested substance has effectively been washed out.

Alkalinization of the urine is effective in enhancing the elimination of acidic drugs, such as salicylates, and the chlorphenoxy herbicides. Multiple dose activated charcoal is useful for removing some drugs, for example carbamazepine and theophylline. Dialysis will remove low molecular weight (<350 D), hydrophilic substances with a low **volume of distribution**, for example lithium, methanol, ethylene glycol and salicylates. **Haemodialysis** is more efficient than **peritoneal dialysis**. For lipophilic substances, for example theophylline, chloral hydrate, and meprobamate haemoperfusion is effective.

Antidotes are medicines that prevent the toxic effects of a specific poison or group of poisons. The most commonly used is N-acetylcysteine for the treatment of paracetamol poisoning. This provides a source of sulphydryl groups to react with the NAPQI metabolite preventing the binding to cellular components, which results in damage to cells and is an example of substrate supplementation. **Chelating agents** are used to bind metal ions and thus prevent toxicity, for example desferrioxamine for acute iron overdose and calcium disodium edetate for lead poisoning. Competitive inhibition of metabolism to prevent the formation of toxic metabolites is another mechanism of antidote action and the basis for the use of ethanol or fomepizole (4-methylpyrazole) in the treatment of methanol and ethylene glycol poisoning. Naloxone acts as an antidote for poisoning with opiates or opioids by competing for binding to opiate receptors. Digoxin toxicity can be prevented by giving digoxin-binding antibody (Digibind).

22.5 Laboratory investigation of poisoning

For diagnosis, identification of the poison(s) involved, that is, qualitative analysis, may be sufficient, but for monitoring treatment measurement of the concentration in the blood will usually be required.

SELF-CHECK 22.2

Why is qualitative analysis useful in toxicology?

Toxicology testing is carried out for five reasons:

- confirmation of suspected poisoning
- differential diagnosis of coma
- diagnosis of brain death
- to influence active therapy
- assessment and treatment of drug misuse

In the last area, testing is used to identify which drugs are being used and to monitor compliance with treatment programs. During treatment, negative results are expected for substances that should not be taken and positive results are used to confirm that prescribed substitutes, for example methadone, are being taken and not diverted.

Figure 22.1 shows the steps involved in the laboratory investigation of poisoning. Notice how both analytical expertise and knowledge of the clinical aspects of poisoning are required in order to provide an effective toxicology service. The importance of correct identification of the patient and samples, and observing the necessary criteria for sample collection cannot be overemphasized. Good communication with requesting clinicians is essential. Analyses may have legal implications so a properly functioning quality management system with particular emphasis on accurate and comprehensive record keeping is essential.

Routine biochemistry tests such as plasma electrolytes, arterial blood gases, and osmolar and anion gaps are helpful in managing the supportive treatment of poisoning and may give an indication of severity. It might be expected that identifying the agent and the extent of exposure

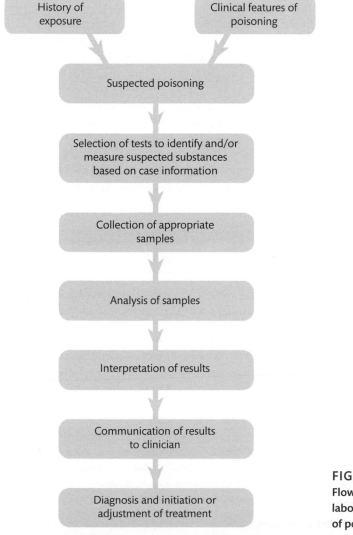

FIGURE 22.1
Flow diagram for the laboratory investigation of poisoning.

TABLE 22.2 Emergency toxicology investigations that should be available at all times within two hours.

Investigation	Reason
Quantitative in blood or serum:	
Paracetamol	Antidote available
Salicylate	Specific treatment
Ethanol	Specific treatment and monitoring treatment when used as antidote
Carboxyhaemoglobin	Antidote available
Methaemoglobin	Antidote available
Iron	Antidote available
Digoxin	Antidote available
Lithium	Specific treatment
Theophylline	Specific treatment
Qualitative:	
Urine paraquat	Prognosis

would be a prerequisite for treatment. In practice this is only true when specific treatment such as antidote administration is available. Recommendations (NPIS and ACB, 2002) have been published that indicate the ten or so substances for which quantitative analysis within two hours of request is required on a 24-hour, 7 days a week basis because they are common, treatable poisonings where the circulating concentration is related to toxicity. These are shown in Table 22.2. Turnaround times for identifying or measuring other important agents that occur frequently are also suggested.

SELF-CHECK 22.3

Why is rapid measurement of paracetamol required at all times?

Cross reference

Chapter 1 Biochemical investigations and quality control

The three main factors affecting the choice of method are analytical sensitivity, analytical specificity, and whether it is necessary to know the amount present in the sample.

The analytical sensitivity required depends on the amount of specimen available and the concentration to be measured. Toxic concentrations are by definition higher than normal and generally higher than therapeutic in the case of drugs. This means that the analytical sensitivity needed for a toxicology assay is usually less than for a clinical biochemistry or therapeutic drug application. It is, however, still necessary to determine the limit of detection as part of the validation of toxicology methods. The presence and nature of interfering substances, for example components of the sample matrix or metabolites in the case of drugs, determines the requirements for analytical specificity. If it is sufficient to identify a substance, qualitative or semi-quantitative, possibly non-instrumental, methods may be adequate. If quantitative analysis is required some form of instrumentation is usually needed in order to meet the analytical sensitivity and specificity requirements using the volumes of sample available in the clinical situation.

Specificity can be improved by including some form of sample preparation prior to instrumental analysis. This also provides the opportunity to improve sensitivity by concentrating the analytes of interest. Sample preparation can also extend the life of chromatography columns. Diffusion, as in head space analysis, can be used to extract gases and volatiles. Protein precipitation by acid/base or organic solvent is widely applicable. Acid or enzymic digestion can be used to release components from tissues and also to hydrolyse conjugated metabolites.

Liquid-liquid solvent extraction is used for non-volatile organic compounds as shown in the Method box below. A cleaner extract can be produced by back extraction of the analyte(s) from the organic phase into an aqueous phase of suitable pH, then re-extracting instead of immediately proceeding to the evaporation stage. The aqueous phase can be discarded or the pH can be adjusted and it can be re-extracted to select other compounds. This type of extraction is labour intensive and there are health and safety hazards associated with use of glassware and handling significant amounts of acids, alkalis, and organic solvents (Box 22.1).

A wide variety of devices are now available for solid phase extraction with a range of different selectivities and the process can be partly or fully automated. Solid phase extraction has safety benefits as smaller volumes of organic solvents are used and the use of glassware is decreased. The devices have a significant consumable cost and there may also be a capital cost, for example for a vacuum unit to speed up elution from columns.

Any form of sample preparation may have less than 100% recovery and recovery may also vary between samples. To compensate for this, it is usual in quantitative chromatographic analyses to add a fixed amount of an **internal standard** to all test, calibrator, and control samples at the start of sample preparation and to express sample response as a ratio to the internal standard response.

METHOD Liquid-liquid solvent extraction

The pH of the sample is adjusted (acidic buffer for acids, alkaline buffer for bases) to suppress the ionization of the compound(s) of interest.

The resulting solution is then mixed with between 1 and 10 times its volume of organic solvent.

After mixing, the organic phase containing the analyte(s) is usually separated off and the solvent extract is evaporated to dryness.

The concentrated extract is then reconstituted with a suitable solvent before further analysis.

BOX 22.1 Health and safety

Hazards associated with liquid-liquid extraction:

- use of glassware
- acids and alkalis
- organic solvents

Where the requirement is to identify an unknown poison the traditional approach has been to perform a number of qualitative, often comparatively non-specific, spot tests in order to direct more complex analyses of increased specificity. In clinical work this approach is mainly of historic interest but an example of a qualitative colour test still in general use is the detection of paraquat in urine.

Spot colour tests are limited in their sensitivity by the capacity of the human eye to detect the colour change. Their specificity is limited because the colour reaction depends on the presence of a functional group which may be present in a number of compounds. The spot test for paraquat is given in the Method box below.

These limitations can be overcome to an extent by using instrumental techniques. Visible and ultraviolet (UV) spectrophotometry are of limited use except to detect or quantify (bio) chemical reactions. The essential prerequisite is a characteristic absorption spectrum and this is fulfilled by the haemoglobin derivatives. **Carboxyhaemoglobin** for carbon monoxide poisoning and methaemoglobin for exposure to oxidizing agents are measured by difference spectrometry. Modern blood gas analysers usually incorporate a co-oximeter, which by measuring absorbance at multiple wavelengths on a haemolysate of the blood sample, will measure total haemoglobin and calculate the percentages of oxy-, carboxy-, met-, and reduced haemoglobin.

In dilute solutions the intensity of fluorescence is directly proportional to the concentration of flurophore. Fluorescence spectrometry has better sensitivity than spectrophotometry because measurement is made against a dark background. Specificity is also better because both excitation and emission wavelengths are compound specific. Its application is limited to compounds with natural fluorescence, for example quinine, or for which a fluorescent derivative can be prepared. Specialist instrumentation (fluorimeter) is required and there are problems with both background fluorescence and quenching associated with the analysis of biological samples.

Cross reference
Chapter 11 Spectroscopy in the *Biomedical Science Practice* textbook

Atomic absorption spectrometry is used for the measurement of metals and metalloids. In clinical laboratories ion selective electrode technology is now widely used for lithium, and induction-coupled plasma mass spectrometry (ICPMS) is increasingly used for metal analysis. Technological advances both in engineering and information handling have led to increasing numbers of applications for mass spectrometry, both in toxicology and in other areas of clinical biochemistry.

Chromatography separates compounds based on partitioning between a stationary and a mobile phase due to differences in physico-chemical properties. Detection can be visual, particularly for planar chromatography, as in viewing a thin-layer chromatogram under UV light or after staining. For column chromatography some form of instrumental detection is used. Identification is by retention factor (R_f), the distance run relative to that of the mobile phase, in planar chromatography or by retention time for column chromatography. In toxicology,

 METHOD Spot test for paraquat

An aliquot of urine is mixed with an equal volume of alkaline sodium dithionite solution (0.1% dithionite in 1 mol/L sodium hydroxide).

The development of a blue colour indicates the presence of the herbicide paraquat. Diquat, a chemically related compound, also reacts giving a green colour.

some form of sample pre-treatment is generally used to prevent interference from the biological matrix and to maintain the integrity of the stationary phase. Thin layer chromatography (TLC), high performance liquid chromatography (HPLC), and gas liquid chromatography (GLC) are widely used to identify compounds. For quantification the column techniques (HPLC and GLC) are generally used, as the effluent from the column is transferred directly to the detector which is an integral part of the instrumentation. Chromatography, by its separative nature, particularly if preceded by a selective extraction procedure, has inherent specificity. Similar compounds can often be resolved so that a number of compounds can be identified and/or quantified in a single run. A single chromatographic spot or peak, however, does not necessarily mean a single compound and quite unrelated compounds may have the same extraction and chromatographic characteristics in any particular system. This may be overcome by running the sample through a number of chromatographic systems and/or using a detection system that has a degree of specificity.

Planar chromatography in the form of TLC is an important technique for identification. Compounds can be visualized non-destructively and can be recovered from the TLC plate for further analysis. The sequence of colour changes with different stains can be used to improve specificity. There is a commercial system (Toxilab®) that uses this type of identification approach following chromatography on a glass fibre matrix which is impregnated with drug standards for comparison. An example of drugs analysed using the Toxilab® procedure is shown in Figure 22.2.

The system also includes solvent extraction tubes for acidic and basic drugs. The extracts are concentrated onto discs which are inserted into spaces on the plate in order to minimize variability between operators due to application technique.

Gas liquid chromatography is the method of choice for separating volatile compounds. It is unsuitable for compounds that are not stable at the temperature required for their conversion to the gaseous phase. An example of a chromatogram for determination of ethanol and methanol using gas chromatography is shown in Figure 22.3.

The flame ionization detector (FID) is non-selective and response generally relates well to the mass injected into the chromatograph. Detectors with improved selectivity for nitrogen containing (alkali-FID), or halogen containing (electron capture), are available. Gas liquid chromatography can also be combined with mass-spectrometry (GC-MS) or tandem mass spectrometry (GC-MS/MS). The sample pre-treatment requirements are often less rigorous and the range of compounds that can be analysed is wider for HPLC than for GLC. Detectors available include visible-UV, diode array, fluorescence, electrochemical, MS, and tandem MS.

Cross reference
Chapter 13 Chromatography in the *Biomedical Science Practice* textbook

An approach to improving specificity and sensitivity that does not rely on extensive sample preparation is to use biochemical reactions which have inherent selectivity, such as enzyme catalysed or protein-binding reactions. Enzymatic assays for ethanol and paracetamol are widely used. Immunoassay is an application of protein-binding assay where one or more antibodies are prepared which are able to bind to specific epitopes.

Cross reference
Chapter 21 Therapeutic drug monitoring

In toxicology the ability of antibodies to cross-react with different molecules of similar chemical structure can be used to advantage, as in urine testing for drugs of abuse. Using antibodies with broad specificity can enhance detection times by picking up metabolites as well as the parent compound and can also avoid the need to apply more complex tests, as the negative immunoassay test will exclude the presence of a number of chemically related compounds of interest.

The need for rapid testing in the emergency situation might seem to justify point of care testing (POCT). As mentioned above, carboxyhaemoglobin and methaemoglobin are readily available on co-oximeters associated with the blood gas analysers commonly available in critical care

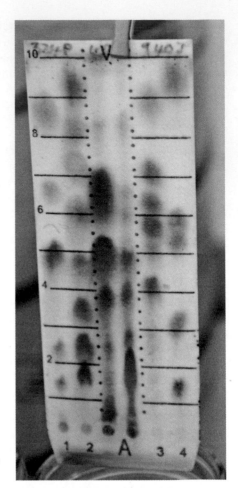

FIGURE 22.2
Photograph to demonstrate visualization of basic drugs following planar chromatography using the Toxilab A procedure.

situations. Apart from this, the need for specific, sensitive, quantitative analysis when testing is required in the emergency situation means that testing is generally better carried out in the laboratory. Breath alcohol meters are used by the police and are sometimes used in critical care situations. Point of care testing may be helpful in non-acute situations. Breath carbon monoxide analysers are often used in health promotion work to demonstrate the immediate biochemical effect of smoking. A variety of analytical systems and devices are used for urine drugs of abuse testing in mental health care and addiction treatment, where timely results are important so that compliant behaviour can be rewarded and clients who have deviated from agreed care plans can be confronted.

Cross reference

Chapter 18 Point of care testing in the *Biomedical Science Practice* textbook

As with clinical chemistry procedures, it is important to have appropriate internal quality control and external quality assessment procedures in place for all analyses, qualitative as well as quantitative. Salicylate, paracetamol, and lithium are included in the UK National External Quality Assessment Service (UKNEQAS) for clinical chemistry and there are separate schemes for drugs of abuse in urine, toxicology cases, and various therapeutic drug monitoring (TDM) schemes, including antidepressant and antipsychotic drugs.

Communication between laboratory staff and clinicians over toxicology requests is even more important than for other laboratory tests. Effective laboratory investigation of poisoning needs a selective approach to analysis for which a detailed clinical history is required.

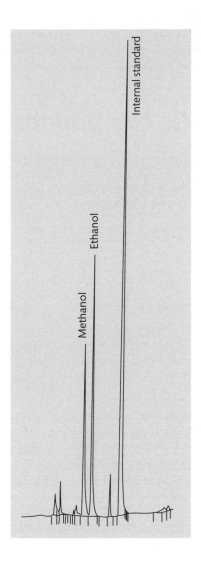

FIGURE 22.3
Chromatogram for the gas chromatographic determination of ethanol and methanol.

Further, because the effects of poisoning are difficult to distinguish from symptoms of other organic disease states, a high index of suspicion is needed to make the diagnosis, particularly for chronic toxicity. The laboratory should, therefore, be prepared to be active in raising awareness of clinicians to situations where poisoning may be a possible diagnosis.

An appropriate sample is always a prerequisite for effective testing. Blood is required for quantitative testing. Blood, urine, or less often, gastric contents may be used for qualitative testing. Usually serum or plasma can be used but some analytes, for example heavy metals like lead, are present at higher concentrations in erythrocytes so whole blood collected into an appropriate anticoagulant is required. Preservative, for example fluoride/oxalate for ethanol, or a specific anticoagulant may be required, or a method may be susceptible to interference from compounds present in gel tubes or to plasticizing agents in collection tubes and closures. Information on sample type, collection container, sample volume, and appropriate sampling times must be readily available to requesting clinicians.

The interpretation of toxicology results tends to be less straightforward than for other clinical chemistry tests and usually depends on the time of sampling relative to exposure, which may

Cross reference

Chapter 21 Therapeutic drug monitoring

not be known. Valuable information on the effectiveness of active treatment can be gained by collecting two or more samples at suitable time intervals and using toxicokinetic calculations. The same principles and equations apply as in pharmacokinetics, but toxin concentration is used instead of drug concentration.

22.6 Toxicology of specific compounds

This section will consider the toxicology of selected substances commonly encountered in clinical biochemistry laboratories. The pathophysiology underlying toxicology of the substance, its clinical features, and aspects of investigation will be covered. The substances we will look at include carbon monoxide, ethanol, paracetamol, salicylate, and drugs of abuse.

Carbon monoxide

Carbon monoxide gas is the non-medicinal poison most frequently associated with death due to poisoning in the UK. It is formed by incomplete combustion of hydrocarbons and other organic compounds. Sources therefore include fumes from vehicles or malfunctioning gas appliances, smoke from fires, and cigarette smoke. Metabolism of organic solvents, for example dichloromethane is another source of particular importance for those who are exposed to solvents in the course of their work. Poisoning can be acute, sub-acute, or chronic. The latter is often occult and may not be diagnosed, as symptoms (headache, dizziness, nausea, malaise, and a flu-like feeling) are vague and non-specific.

Following inhalation, carbon monoxide is rapidly absorbed into the blood, where it binds to haemoglobin to form carboxyhaemoglobin, thus reducing oxygen delivery to tissues. It is excreted via the lungs. The half-life of carboxyhaemoglobin (COHb) depends on the oxygen tension of inspired air. After removal from the source of exposure, the half-life of COHb is 5–6 hours in room air and one hour with 100% oxygen. In **hyperbaric oxygen** (2–3 atmospheres) the half-life is further reduced to 25 minutes. Hyperbaric oxygen is sometimes used in treatment of carbon monoxide poisoning but the evidence of additional benefit compared to 100% oxygen at atmospheric pressure is poor. Carboxyhaemoglobin is usually measured by spectrophotometry although it can be measured by gas chromatography.

Blood carboxyhaemoglobin relates best to carbon monoxide exposure and toxicity when blood is collected soon after exposure and before oxygen treatment. The blood COHb depends on both the amount of carbon monoxide in the atmosphere and the duration of exposure. Various causes of carbon monoxide poisoning, COHb levels, and resultant toxic effects are given in Table 22.3.

Features of acute carbon monoxide poisoning include metabolic acidosis, normal PO_2 with reduced oxygen saturation, features of hypoxic damage to organs and tissues, for example kidney, skeletal muscle, and a high blood COHb.

Carbon monoxide is just one of the noxious gases that, together with particulates, constitute smoke from fires. Of deaths occurring as a result of fires, more are due to smoke inhalation than to burns.

Ethanol

Ethanol (CH_3CH_2OH) is an aliphatic alcohol produced by fermentation of sugars. It is widely available in the form of alcoholic drinks and is also used as a solvent in many industrial and

TABLE 22.3 Relationship between carbon monoxide exposure, blood carboxyhaemoglobin, and toxic effects. Note: heart or respiratory disease increase the susceptibility to carbon monoxide poisoning so that toxic effects are seen at lower COHb levels.

Reason for high carbon monoxide concentration in air	Exposure time	Blood carboxyhaemoglobin %	Toxic effects
City air	Continuous	<5	
Tobacco smoke	Intermittent	<10	
		10–20	Headache, nausea
Faulty boiler	Intermittent	20–50	Headache, nausea, weakness, impaired vision, fainting , vomiting, diarrhoea
Faulty boiler Car exhaust into sealed car	Hours Minutes	>50	Bradycardia, hypotension, respiratory depression, coma convulsions and death

household products. In developed countries, excessive alcohol drinking is second only to tobacco smoking as the cause of overall damage to health. Ethanol use is also linked to violence and other crime.

Following ingestion, some ethanol is absorbed from the stomach but the small intestine is the main site of absorption. Peak blood concentrations of ethanol occur 0.5–3 hours after ingestion. The rate of absorption increases with the concentration of ethanol ingested, up to a concentration of 20% alcohol by volume (ABV). Higher ethanol concentrations cause gastric irritation and slower absorption. Absorption also depends on recent food intake, the rate being slower if ethanol is taken with food. Ethanol distributes throughout the body water. Up to 90% of the amount of ethanol ingested is metabolized in the liver by alcohol dehydrogenase to acetaldehyde, then to acetate then carbon dioxide via the tricarboxylic acid (TCA) or Kreb's cycle. Ethanol is expired in the breath and is also excreted unchanged in the urine. The rate of ethanol metabolism normally ranges from 120–500 mg/L/hour and does not depend on the blood ethanol concentration. This means that on average it takes about one hour to eliminate one unit of alcohol.

Acute ethanol intoxication is common and is an increasing problem in the early teen years. In patients who have taken a deliberate overdose of drugs 60% of males and 40% of females have also consumed ethanol. Ethanol acts directly on the brain, initially causing disinhibition of brain activity and behaviour, and subsequently CNS depression, with decreases in respiration, blood pressure, and body temperature. It potentiates the effects of other CNS depressants, for example carbon monoxide and drugs. The acute toxic effects increase with blood concentration (see Table 22.4).

The clinical features of ethanol toxicity include a sense of enhanced well-being, dizziness, nystagmus, ataxia, dysarthria, nausea, vomiting, drowsiness, hypotension, hypothermia, respiratory depression, convulsions, and coma. Inhibition of reflexes can lead to aspiration (inhalation) of vomit. Laboratory features include raised plasma osmolality, mild metabolic acidosis,

TABLE 22.4 Relationship between blood ethanol concentration and effects. Note: In the UK the legal blood ethanol limit for driving is 800 mg/L (80 mg/100mL).

Blood ethanol mg/L	Effects
0–500	Mild euphoria, decreased fine motor function
500–1500	Euphoria, slowed reactions, impaired judgement, reduced motor function
1,500–3,000	Obvious intoxication with impaired balance, speech, vision, comprehension, reaction time, and emotional instability
3,000–5,000	Marked intoxication with impaired consciousness, hypotension, hypothermia, and respiratory depression, leading to seizures, coma and death

ketosis, and hypoglycaemia in children and some adults. Supportive treatment is usually sufficient but ethanol can be readily removed by haemodialysis.

Prolonged excessive ethanol use causes physical disease including liver damage, cardiovascular disease, diabetes, upper gastrointestinal cancer, and brain damage. It also leads to tolerance (larger amounts are required to achieve desired and unwanted toxic effects) and addiction. 'Safe' drinking limits of 21 units/week for men (14 for women) with a minimum of two alcohol free days each week, where one unit of alcohol is approximately 10 g of ethanol, have been suggested. To calculate the number of 'units' in a drink work out the alcohol concentration in the drink in g/L by multiplying the percentage of alcohol by volume (ABV) by the density of ethanol in g/mL multiplied by ten. The density of ethanol is approximately 1 g/mL. The ethanol concentration in g/L is then multiplied by the volume of the drink in L to give the amount of ethanol in grammes, then divided by ten to give the number of units. This is equivalent to multiplying the ABV by the volume of the drink in litres (see Method box below).

SELF-CHECK 22.4

How many alcohol units are there in a can (440 mL) of beer (3.5% ABV)?

Approaches to the detection of problem drinking by laboratory tests that have been proposed include the combination of raised γ-glutamyl transferase, raised mean cell volume and the presence of ethanol in a morning blood sample, serum carbohydrate deficient transferrin, and the measurement of ethyl glucuronide in serum or urine.

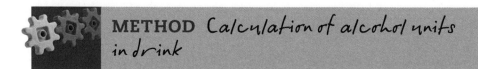

METHOD Calculation of alcohol units in drink

Alcohol units = % alcohol by volume (ABV%) × volume of drink (L)

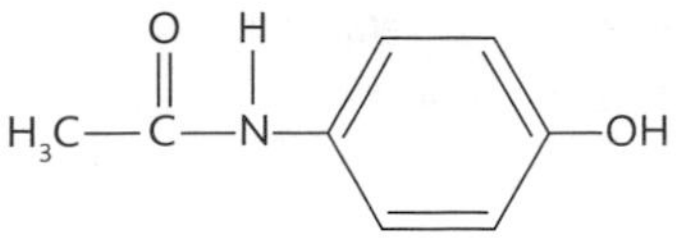

FIGURE 22.4
Chemical structure of paracetamol.

Paracetamol

Paracetamol is the second most common drug involved in poisoning death in the United Kingdom. Figure 22.4 shows its chemical structure.

It is readily available as an over-the-counter analgesic and as a component in many compound cold and cough remedies. It is readily absorbed, with peak blood concentrations occurring 1–2 hours after therapeutic ingestion. In overdose the time to peak concentration is increased and may be as long as four hours. Paracetamol is metabolized in the liver to sulphate and glucuronide conjugates and the toxic metabolite NAPQI. The latter binds to thiol (–SH) groups. Glutathione provides a store of –SH groups and is protective but in overdose, stores are depleted and NAPQI binds to proteins. This disrupts their normal function, leading to damage to tissues and organs, most notably the liver and kidneys. The toxicity of paracetamol is outlined in Figure 22.5.

The normal plasma half-life for paracetamol is about two hours but this is extended to as much as 12 hours in an overdose. Paracetamol is usually measured spectrophotometrically following chemical or enzymatic reaction. It can also be measured by immunoassay, GLC or HPLC.

The toxic effects of paracetamol overdose are delayed, with hepatic necrosis leading to liver failure and death after 2–3 days if untreated. About 15% of cases are complicated by acute renal failure. N-acetylcysteine, given by intravenous infusion, prevents liver damage by supplying SH groups. There is a risk of serious **anaphylactic** reaction to acetylcysteine, so confirmation of paracetamol ingestion is required before it is administered. Oral methionine can be used as an alternative provided the patient is not vomiting. The risk of **hepatotoxicity** is related to the plasma concentration of paracetamol measured in samples collected at least four hours after ingestion. The risk is higher for patients previously exposed to liver enzyme inducing agents such as ethanol or the anticonvulsant drugs carbamazepine and phenytoin. A nomogram relating the risk of hepatic damage to plasma paracetamol concentration at a given time post-ingestion is used, with concentrations above the line indicating the need for antidote treatment. Figure 22.6 shows that the nomogram has two lines, with the lower line applying to patients at 'high risk', as mentioned above.

Early/mild clinical features of paracetamol toxicity include vomiting, abdominal pain, and jaundice. Laboratory features include raised serum alanine and aspartate aminotransferases, with peak activities occurring at 72–96 hours after the overdose, raised bilirubin, prolonged prothrombin time, haematuria, and raised serum creatinine.

Salicylate

Aspirin (acetylsalicylic acid) is another analgesic that is available over the counter and like paracetamol is also included in compound formulations. Poisoning is still common, usually by oral overdose but also via dermal absorption in the treatment of psoriasis and through buccal absorption from teething gels in infants. In overdose, absorption is delayed, with peak concentrations occurring at about four hours after oral ingestion compared to 1–2 hours for therapeutic doses. Sustained release preparations can form concretions (bezoars) which delay

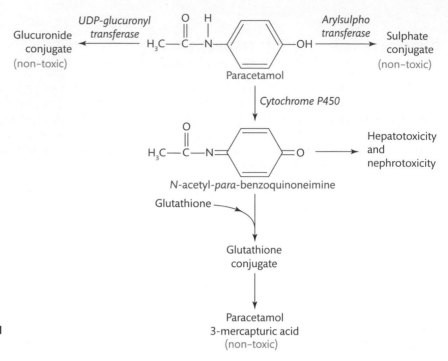

FIGURE 22.5
The pathophysiology of paracetamol poisoning. See text for details.

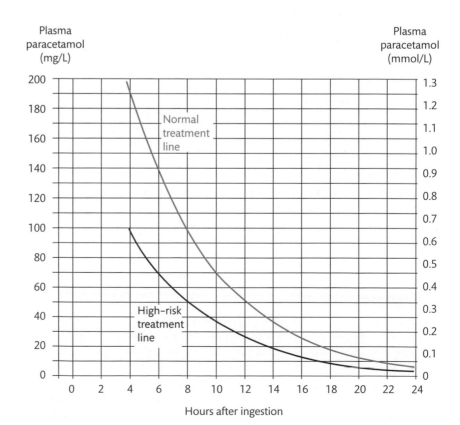

FIGURE 22.6
Nomogram for interpretation of serum paracetamol concentrations in cases of suspected toxicity.

CASE STUDY 22.3

A 21-year-old female was brought to hospital by her boyfriend at 9 pm. He had gone out after they had an argument earlier that day and when he returned she was feeling sick and vomiting; she told him that she had taken two packets (16 tablets per pack) of paracetamol at about 3.30 pm. Blood was collected immediately for paracetamol, urea and electrolytes, creatinine, liver function tests, and prothrombin time. Her serum paracetamol concentration was 250 mg/L.

Should she be given N-acetylcysteine to prevent liver damage? Explain your answer and refer to Figure 22.6 to decide.

absorption even more. Aspirin is metabolized in the liver to salicylic acid commonly referred to as salicylate as shown in Figure 22.7 and other metabolites.

The plasma half-life is 2.5–19 hours and may be longer in an overdose due to saturation of elimination mechanisms. Salicylate can be measured colorimetrically using the Trinder reaction (salicyalte reacts with ferric ions to give a purple colour). Enzymatic assay, immunoassay, and chromatography can also be used to measure plasma salicylate.

Salicylate has direct effects on the inner ear and GIT. It stimulates the respiratory centre and also causes uncoupling of oxidative phosphorylation. The pathophysiology of salicylate poisoning is outlined in Figure 22.8.

Clinical features include nausea and vomiting, vasodilatation, sweating, tinnitus, impaired hearing, pulmonary oedema, renal failure, convulsions, and arrhythmias. Impaired consciousness occurs rarely. Laboratory features include a raised anion gap, acid-base disturbance (mixed respiratory alkalosis with metabolic acidosis), hypo- or hyperkalemia, hypoglycaemia, and increased prothrombin time. Specific treatment, that is, alkalinization of the urine with intravenous sodium bicarbonate, is indicated if the plasma salicylate exceeds 350 mg/L in a child, or 450 mg/L in an adult. Haemodialysis is indicated if the plasma salicylate is above 700 mg/L and in severe poisoning intravenous glucose is given to avoid **neuroglycopenia** which may occur even in the absence of hypoglycaemia. You can see the procedure for haemodialysis outlined in Figure 22.9.

The haemodialysis procedure relies on the principle of dialysis. Blood taken from the patient's artery is circulated through a dialyser on one side of a selective or semi-permeable membrane, whereas a solution of normal electrolytic composition circulates on the other side of the membrane. Waste products, poisons, and small molecules cross the membrane and the purified dialysed blood is returned to the body via a vein.

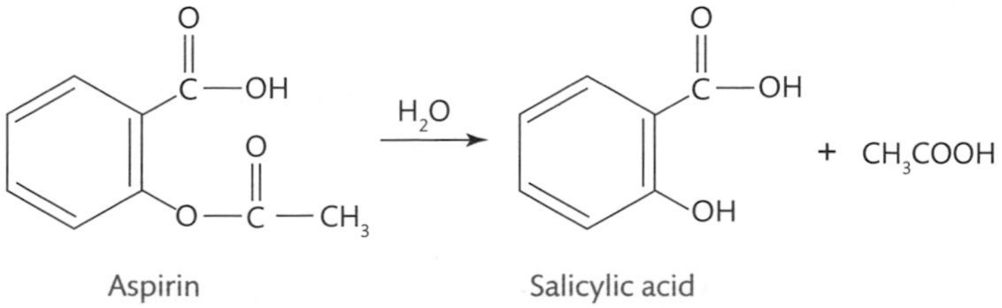

FIGURE 22.7
Conversion of aspirin (acetylsalicylic acid) to salicylic acid.

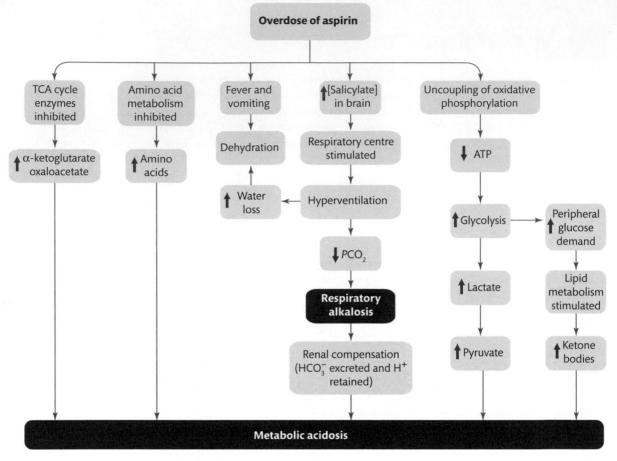

FIGURE 22.8
The pathophysiology of salicylate poisoning. See text for details.

Key Points

Salicylate poisoning can cause a mixed acid-base disorder. A respiratory alkalosis due to the effect of salicylate on the respiratory centre is often accompanied by a severe metabolic acidosis due to increased production of acids.

Drugs of abuse

'Drugs of abuse' is a collective term used for several classes of drugs that are used for non-medical purposes. Some are used medically, for example opiate analgesics, but others have no medical use, for example methylenedioxymethamphetamine (MDMA), often referred to as Ecstasy. The chemical structure of MDMA is shown in Figure 22.10.

These drugs are usually manufactured illicitly and often contain contaminants that may be more toxic than the drugs themselves. These may be the present in the materials used in manufacture, by-products of the manufacturing process, or added to the drugs as 'cutting agents'

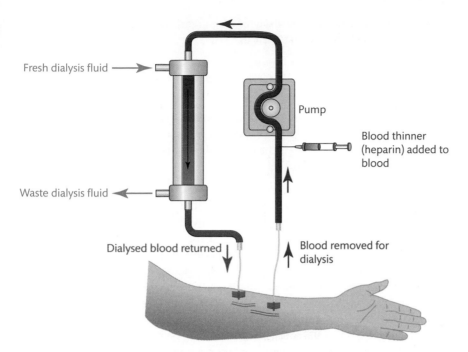

Fresh dialysis fluid

Pump

Blood thinner (heparin) added to blood

Waste dialysis fluid

Dialysed blood returned

Blood removed for dialysis

FIGURE 22.9
An outline of the procedure for haemodialysis. See text for details.

prior to sale. Accidental overdose can occur when the substance is used in what seems to be the habitual amount, but is of greater purity than usual. Another common circumstance of accidental overdose is when use is recommenced at the previous habitual dose after a period of abstinence during which tolerance has been lost.

SELF-CHECK 22.5

Look at Case study 22.4. Why does this man have a high risk of drug toxicity?

For clinical purposes drugs of abuse testing is mainly done on urine. This is advantageous as drugs and their metabolites are often present in urine at higher concentrations than in blood and urine is essentially protein-free so simpler sample preparation methods can be used. Intravenous drug users are at high risk of blood-borne infections, for example human immunodeficiency virus (HIV) and hepatitis due to needle sharing. Using urine as the analytical sample rather than blood is therefore safer for laboratory staff. Urine concentrations do not relate closely to blood concentrations so quantitative analysis does not confer any benefit over qualitative analysis. Immunoassays run on clinical chemistry analysers are commonly used as first-line tests to detect the presence of a drug class, for example opiates. Positive results are then followed up with further analysis to identify the individual compound(s). In the emergency situation testing for drugs of abuse is used to confirm a diagnosis of overdose with a particular drug or drugs. Other uses are to confirm claims of drug use at entry to treatment programs, to confirm compliance with abstinence agreements, and/or the use of prescribed

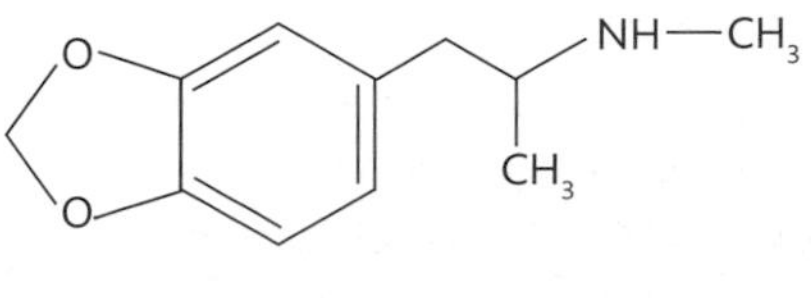

FIGURE 22.10
Methylenedioxymethamphetamine.

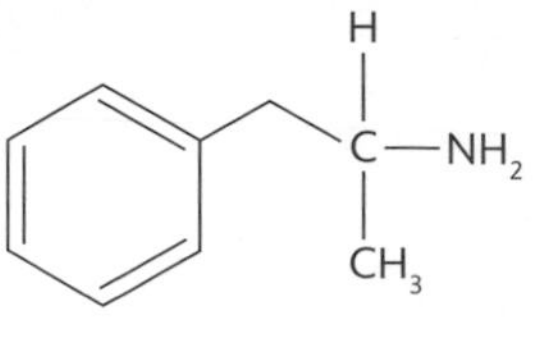

FIGURE 22.11
Amphetamine.

substitutes and to investigate the possibility of drug misuse in the differential diagnosis of mental health problems or behavioural changes.

Amphetamines are basic compounds (aromatic amines) and are most often seen clinically in the UK as amphetamine and Ecstasy (3,4-methylenedioxymethamphetamine, MDMA). The chemical structure of amphetamine is shown in Figure 22.11.

They are usually used orally as tablets but in powder form they can be snorted or smoked, and solutions are injected. Intravenous use can cause delusions and paranoia, and cardiomyopathy is seen with long-term use. Outside the UK, smoking a form of crystalline methamphetamine known as 'ice' is a common form of misuse.

Amphetamines are CNS stimulants and common clinical features in overdose include euphoria, increase in energy with agitation and excessive activity, self-confidence and extrovert behaviour, loss of appetite, dilated pupils, tachycardia, hyperthermia, and hypertension. Less commonly seen are aggressive, violent behaviour, hyperpyrexia, convulsions, coma, and rhabdomyolysis. Biochemical features of poisoning include metabolic acidosis, raised creatine kinase activity, and increased alanine and aspartate aminotransferase activities. Hyponatraemia due to inappropriate antidiuretic hormone (ADH) secretion and self-induced water intoxication may complicate Ecstasy use. It is important to be aware of the selectivity of the method(s) used as some urine screens for amphetamines do not detect methamphetamines and vice versa. Treatment is supportive and cooling may be needed. Diazepam may be used for sedation and to control convulsions. Acidification of the urine will increase renal elimination of methamphetamines but is seldom necessary. Ecstasy should not be confused with 'liquid ecstasy', a term used for gamma-hydroxybutyrate (GHB) which is a sedative.

Cocaine (cocaine hydrochloride) is typically used nasally ('snorted') but is also used intravenously, sometimes in combination with heroin. 'Crack' cocaine (cocaine base) is smoked. The chemical structure of cocaine is shown in Figure 22.12.

Cocaine is commonly smuggled by 'body packing'. Signs of cocaine toxicity in a body packer indicate the presence of leaking packages and the need for their urgent surgical removal.

Cocaine is initially a CNS stimulant but high doses cause CNS depression. It also causes intense vasoconstriction. Clinical features in acute overdose are similar to those described above but more intense. Additional common features include sweating, rapid breathing, ataxia, and hallucinations. Less common features are myocardial infarction, cerebral infarction, subarachnoid haemorrhage, and hypokalaemic paralysis. Laboratory features include metabolic acidosis and hypokalaemia, or hyperkalaemia with a raised creatine kinase if rhabdomyolysis occurs. Diagnosis is readily confirmed by a positive urine screen for the cocaine metabolite benzoylecgonine. Essentially treatment is as above but a higher level of support may be required, for example mechanical ventilation, intravenous sodium bicarbonate to correct metabolic acidosis, and pharmacological intervention to reduce body temperature and control hypertension.

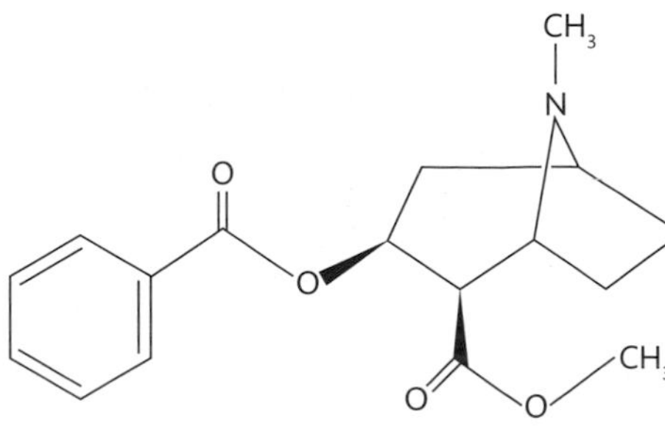

FIGURE 22.12
Cocaine.

Chronic use of cocaine leads to cardiomyopathy and psychiatric disorders, including paranoia and intense depression when use is ceased. Use of cocaine with ethanol results in the formation of the cardiotoxic metabolite, ethylcocaine.

Opiates are a class of alkaloid drugs that get their name from opium, the dried latex from the seed capsule of the opium poppy, *Papaver somniferum*. They can be extracted from opium (e.g. morphine) or synthesized. Opioids are synthetic, basic compounds which act at opiate receptors and have similar effects. Opiates are most commonly misused in the form of heroin (diacetylmorphine) the structure of which is shown in Figure 22.13.

Diacetylmorphine is highly soluble in water making it ideal for injection. The pharmaceutical form, diamorphine, has a purity of 99.5%, whereas the purity of street heroin, a white or brown powder, often 'cut' with other drugs, is typically about 40% but can range from 1–98%. This variation can easily lead to accidental overdose. Heroin is commonly smoked or snorted as well as injected.

Opiates and opioids are used medically as analgesics for the control of acute and chronic pain and are sometimes used as compound preparations with paracetamol. The opioids methadone and buprenorphine are used as substitutes in the treatment of heroin addiction. The mechanism of opiate/opioid toxicity is CNS and respiratory depression, which is enhanced if ethanol is also taken. Clinical features include nausea, vomiting, pin point pupils, a very low respiratory rate, drowsiness progressing to coma, hypothermia, hypotension, cardiovascular collapse, and pulmonary oedema. Laboratory features are a low PO_2 with a high PCO_2 in severe cases and raised CK if skeletal muscle has been damaged. A positive urine opiates immunoassay screen confirms the diagnosis but a negative result does not exclude opioid overdose as opioids, for example methadone, do not cross-react. Treatment consists of oxygen with assisted ventilation if required and administration of the opioid antagonist naloxone. Naloxone does not suppress convulsions and cardiac arrhythmias, which need to be treated conventionally after correction of hypoxia and acidosis.

Barbiturates are derivatives of barbituric acid (2,4,6-trioxohexahydropyrimidine) which was first synthesized in 1864. They have been widely used in medicine as sedatives, hypnotics, anaesthetics, and antiepileptics, but have largely been replaced by the benzodiazepines. They are CNS depressants and long-term use is associated with dependency and addiction. Misuse, including deliberate self-poisoning, was common in the mid-twentieth century but is now rare due to decreased accessibility.

Benzodiazepines are a group of synthetic, lipophilic compounds that act as CNS depressants. They are used as anxiolytics, hypnotics, muscle relaxants, and anti-epileptics depending on their

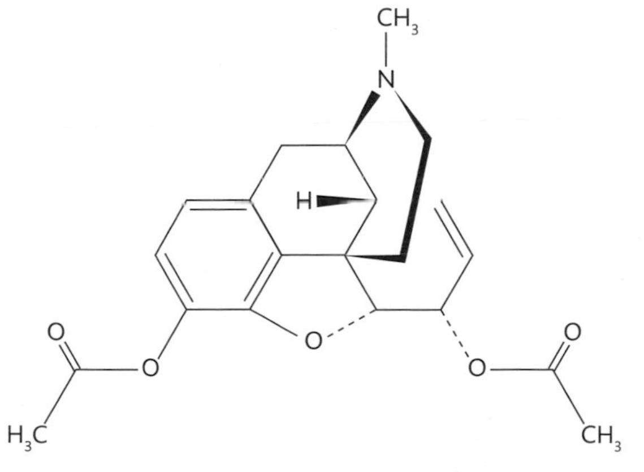

FIGURE 22.13
Heroin.

CASE STUDY 22.4

A 30-year-old man recently released from prison collapsed on arrival at his hostel, apparently drunk after going to a pub with friends.

Examination in hospital showed a GCS of 7, bradycardia, hypotension, shallow breathing, and pinpoint pupils. A recent needle puncture mark was noted in his groin. He rapidly regained consciousness when given naloxone and subsequently recovered with supportive care.

What toxicology tests would be appropriate?

individual chemical structures, usually in the form of tablets or capsules for ingestion. They can alleviate withdrawal symptoms and are prescribed as part of drug dependency treatment regimes. They are also illicitly manufactured or diverted and are widely misused often in combination with other drugs, for example heroin. They have a wider therapeutic margin than opiates and barbiturates. If taken alone in overdose they cause drowsiness, ataxia, slurred speech, and occasionally short-lived depression of consciousness. They potentiate the effect of other CNS depressants. Flumazenil is a benzodiazepine antagonist. Its use as an antidote in benzodiazepine overdose is rarely justified and can be hazardous in mixed overdoses involving tricyclic antidepressants and in benzodiazepine dependent patients. Benzodiazepine immunoassays tend to have wide selectivity but cross-reactivity to conjugated metabolites may be low. The urine concentrations of newer drugs, for example lorazepam, tend to be significantly lower than for the earlier ones and it may be advisable to include a hydrolysis step at the beginning of the analytical procedure.

Cannabis is usually smoked. It is derived from the plant *Cannabis sativa* and contains lipophilic tetrahydrocannabinols in amounts that depend on the form. Tetrahydrocannabinol content is lowest in the herbal form (marijuana), intermediate in resin (hashish), and highest in cannabis oil. The chemical structure of tetrahydrocannabinol is illustrated in Figure 22.14.

Cannabis is a mild hallucinogen which gives feelings of relaxation, sleepiness and lack of concentration. Problems associated with long-term use and/or high doses include apathy and low energy, hallucinations, panic attacks, and psychosis. Cannabinoids are often included as part of a drugs of abuse screening panel but are seldom of importance in acute overdose.

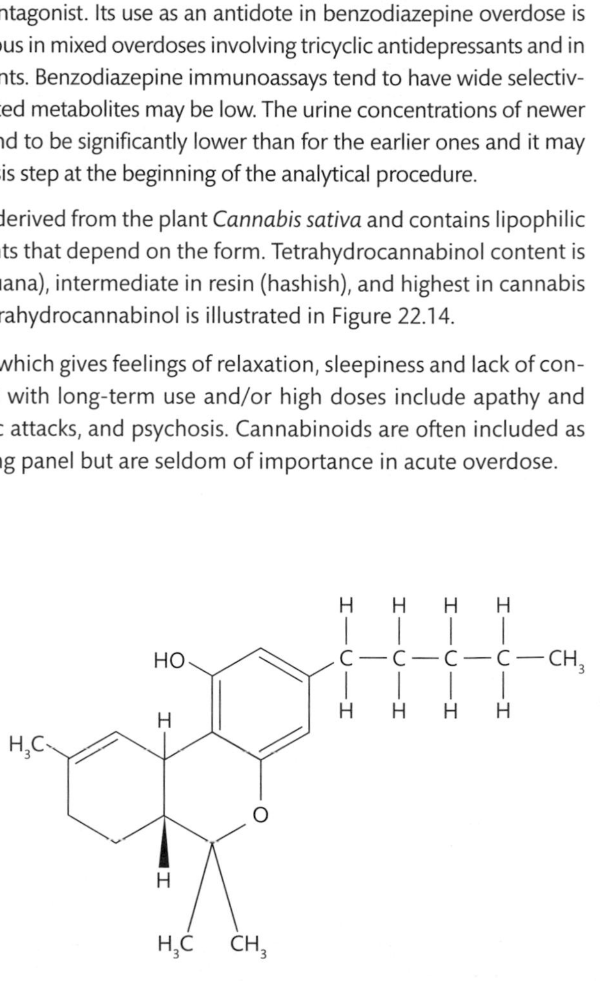

FIGURE 22.14
Tetrahydrocannabinol.

SUMMARY

- It is important to identify and measure poisons so that patients get the best treatment.

- Ensure that all samples are properly identified and keep meticulous records, as toxicology results may have legal implications.

- Samples must be collected into the correct containers and care must be taken to avoid contamination throughout collection and analysis.

- Identification of a group of similar poisons may be more important than quantitative measurement of a specific poison.

- A good clinical history combined with a knowledge of the clinical and biochemical features of poisoning is needed for the effective selection of toxicology investigations.

FURTHER READING

- **Moffat AC, Osselton MD, and Widdop B (2004)** *Clarke's Analysis of Drugs and Poisons*. **London: Pharmaceutical Press.**

  Initial chapters cover the practical issues of analytical toxicology, including detailed methods of analysis and their scientific basis. An essential reference book for the monographs, which give physico-chemical data, analytical procedures, data on doses and concentrations in therapeutic and toxic situations, and pharmacokinetic parameters, for hundreds of substances.

- **NPIS and ACB (2002)** Guidelines—laboratory analyses for poisoned patients; joint position paper. *Annals of Clinical Biochemistry* **39**, 328–39.

- **Shaw LM, Kwong TC, Rosano TG, Orsulak PJ, Wolf BA, and Magnani B (2001)** *The Clinical Toxicology Laboratory: Contemporary Practice in Clinical Toxicology*. **Washington, DC: AACC Press.**

  A course text in the diagnosis and treatment of poisoning, with self-assessment questions.

- **Watson ID and Proudfoot A (2002)** *Poisoning and Laboratory Medicine*. **Kent: ACB Venture Publications.**

  A very readable text with monographs on 32 common poisons and illustrative cases.

- **Widdop B (2002)** Analysis of carbon monoxide. *Annals of Clinical Biochemistry* **39**, 378–91.

QUESTIONS

22.1 State whether TRUE or FALSE for each statement:

(a) Drugs cause nearly three times the number of non-drug poisoning deaths

(b) Gastric lavage is always useful to decrease absorption of ingested poisons

(c) Thin-layer chromatography is widely used to measure drug concentrations

(d) Paracetamol measurement is appropriate in a patient who has ingested a large number of unspecified tablets more than four hours ago

(e) Spectrophotometric measurement of carboxyhaemoglobin is used in the diagnosis of carbon monoxide poisoning

22.2 Which ONE of the following is an antidote for ethylene glycol poisoning?

(a) Methanol

(b) Oxygen

(c) Fomepizole

(d) Naloxone

(e) N-acetylcysteine

22.3 A 26-year-old male is brought to hospital with injuries sustained in a fight outside a nightclub. On examination he is agitated, has a rapid pulse, increased respiratory rate, and dilated pupils. Which of the following urine tests are likely to be positive?

(a) Opiates

(b) Barbiturates

(c) Amphetamines/methylamphetamines

(d) Cannabinoids

(e) Benzodiazepines

22.4 Match the following poisonings with the technique commonly used for its detection/measurement:

(a)	paraquat	(1)	gas chromatography
(b)	methanol	(2)	colour spot test
(c)	lead	(3)	atomic absorption spectrometry
(d)	heroin	(4)	HPLC
(e)	amitriptyline and metabolite	(5)	immunoassay

22.5 List four routes by which poisons may enter the body.

Answers to self-check questions, case study questions, and end-of-chapter questions are available in the Online Resource Centre accompanying this book.

 Go to www.oxfordtextbooks.co.uk/orc/ahmed/

Glossary

ABC transporters members of the ATP binding cassette superfamily of transmembrane pumps.

Abetalipoproteinaemia an inherited disorder characterized by severe deficiency of beta-lipoproteins, abnormally low cholesterol and the presence of abnormal red cells (acanthocytes) in the blood.

Acanthocytes abnormal red blood cells that have thorny projections of protoplasm.

Acarbose an oral drug that lowers blood glucose because it inhibits the enzymes that digest starches in food and so there is a lower rise in blood glucose particularly after meals. It is classified as an alpha-glucosidase inhibitor.

Accuracy describes the ability to produce the true value for a test on a sample.

Achlorhydria absence of acid in the stomach.

Acid any compound that donates H^+ in water, i.e. proton donors.

Acidaemia increased concentration of H^+ ions in the blood.

Acidosis a clinical condition of acidity in the blood and body tissues.

Acrodermatitis enteropathica characterized by an inherited inability to absorb dietary zinc. Patients present with poor growth, skin rash, chronic diarrhoea, and sparse hair.

Acromegaly is a disease characterized by enlargement of the bones and soft tissue of the face, hands, and feet due to excessive secretion of growth hormone from the anterior pituitary gland, due to a pituitary tumour.

Action limits are test results set locally for each analyte. Values obtained outside this range mean a course of action must be taken.

Acute coronary syndrome refers to a group of clinical signs and symptoms of acute myocardial ischaemia (chest pain due to an inadequate blood supply to the heart muscle). Acute coronary syndrome includes unstable angina, STEMI, and NSTEMI.

Acute hypoglycaemia a rapid fall in blood glucose below 2.5 mmol/L.

Acute interstitial nephritis a type of nephritis which occurs suddenly and affects the interstitium of the kidneys around the tubules.

Acute kidney injury a rapid, often reversible, deterioration in kidney function.

Acute liver failure rapid appearance of severe complications (e.g. hepatic encephalopathy and reduced protein synthesis) after first signs of liver disease such as jaundice indicating that the liver has lost 80–90% of hepatocytes.

Acute myocardial infarction is the sudden blockage of an artery supplying blood to the heart.

Acute pancreatitis a sudden inflammation of the pancreas often presenting clinically as severe abdominal pain.

Acute phase response a response to injury or infection characterized by a change (increase or decrease) in the concentration of several proteins in the plasma.

Acute tubular necrosis a condition where there is death of tubular cells in the kidney tubules.

Addison's disease arises due to underactivity of the adrenal cortex where patients present with weakness, weight loss, hypotension, skin pigmentation, and GIT problems.

Adenocarcinoma a cancer of epithelia that originates in glandular tissue.

Adenoma a benign tumour of glandular origin.

Adenosine triphosphate when hydrolysed to adenosine diphosphate energy is liberated for use in cellular reactions.

Adherence correct following of medical (dietary/treatment) advice by patients. Often referred to as compliance.

Adipocytes are fat storing cells found in adipose tissue.

Adipose tissue tissue that primarily stores fat.

Adrenal cortex beneath the capsule of the adrenal gland as three distinct zones of glomerulosa, fasciculata, and reticularis cells. It produces aldosterone, cortisol, and DHEAS. Its activity is controlled by angiotensin and ACTH.

Adrenal failure cortisol is not produced by the adrenal cortex due to Addison's disease, failure of the adrenal to respond to ACTH, or failure of the pituitary to secrete ACTH. In some cases aldosterone production also fails.

Adrenal medulla innermost zone of the adrenal cortex producing adrenaline and noradrenaline.

Adrenaline also called epinephrine. It is secreted by the adrenal medulla in response to stress and it activates the autonomic nervous system.

Adrenarche a period before puberty when the adrenal cortex develops a reticularis zone that secretes DHEAS. The adrenarche is associated with appearance of pubic hair at 5–8 years of age.

Adrenocorticotrophic hormone (adrenocorticotrophin) a peptide hormone secreted by the anterior pituitary. It is the main regulator of cortisol production by the adrenal cortex.

Adrenoleukodystrophy an inherited condition characterized by progressive brain damage, failure of the adrenal glands, and eventually death.

Advanced glycation endproducts toxic glycated protein products that cause a proinflammatory state and the build up of reactive oxygen species in cells.

Aerobic requiring oxygen.

Albumin is the predominant protein in the circulation and is responsible for maintaining the oncotic pressure of plasma.

Aldosterone steroid hormone produced in the outer zona glomerulosa of the adrenal cortex. It acts on the kidney and other sites to retain sodium ions and secrete potassium.

Aldosterone to renin ratio this test is useful to make a diagnosis of Conn's syndrome (primary hyperaldosteronism).

Aldosterone 18-glucuronide metabolite of aldosterone excreted in the urine and a useful marker of aldosterone production.

Alkalaemia reduced concentration of H^+ ions in the blood.

Alkaline phosphatases are a group of enzymes that hydrolyse phosphate esters in alkaline solutions.

Alkalosis a clinical condition of high alkalinity in the blood and body tissues.

Alpha cell an islet cell which secretes glucagon.

Alpha glucosidase inhibitor an anti-diabetes drug that lowers post-prandial glucose levels by suppressing the absorption of carbohydrate in the gut. Acarbose is an example of an α glucosidase inhibitor.

Alport's syndrome an inherited condition where patients have kidney failure, hearing loss, and visual problems. Affected patients almost always present with presence of blood in the urine.

Ambiguous genitalia appear as a large clitoris or small penis that may or may not require surgery. Genetic disorders of enzymes of the adrenal cortex can lead to excess male hormones in a genetic female or lack of male hormone production in the adrenals and gonads in a male that does not develop a penis.

Amenorrhoea absence of menstrual periods.

Aminotransferases are a family of enzymes that catalyse the transfer of amino acid groups. The members of the family relevant to assessment of liver damage are aspartate aminotransferase and alanine aminotransferase.

Amiodarone an anti-arrhythmic drug that is used to treat patients with an irregular heart beat.

Amniotic fluid the watery liquid within the amniotic sac that cushions and protects the developing foetus in the womb.

Amphipathic refers to a molecule that has a polar end which likes water and a non-polar end that does not like water.

Ampholytes molecules containing both acidic and basic groups and mostly existing as zwitterions within a certain pH range.

Amylin a peptide hormone co-secreted with insulin by β cells which acts to suppress glucagon release and slow gastric emptying. Pramlintide is a stable, injectable amylin analogue.

Amyloid A a protein that belongs to a family of apolipoproteins associated with high-density lipoprotein in the plasma.

Anabolic the synthetic phase of metabolism, i.e. the building of complex molecules from simpler ones.

Anaerobic without oxygen.

Analgesics medications used to relieve pain.

Analyser maintenance logs are records to confirm that an analyser has been maintained to the required standard.

Analyte the substance which is measured during an analysis.

Analytical sensitivity is the ability of an assay to detect the smallest amount or concentration of an analyte.

Analytical specificity is the ability of an assay to measure only the analyte in question.

Analytical step the analytical stage of analysis—this is usually undertaken within an instrument. An example would be the testing undertaken within an automated instrument to derive a glucose result.

Anaphylactic severe allergic reaction to food, medicines, and insect stings.

Anastomosis an artificial connection between two hollow organs, e.g. two separate parts of the intestine.

Androgen a steroid related to testosterone, with activity as male hormone affecting secondary sexual characteristics of the male (penis, beard, musculature).

Android male-like.

Androstenedione a 19-carbon steroid hormone made in the adrenal glands and in the gonads as an intermediate step in the pathway that synthesizes testosterone and oestradiol.

Angina cardiac (heart) pain which can occur in stable or unstable forms.

Angioplasty a surgical technique used to treat patients with a blocked artery that reduces blood flow to the heart.

Angiotensin converting enzyme a circulating enzyme that catalyses the conversion of angiotensin I to angiotensin II. Angiotensin converting enzyme is a vasoconstrictor and is part of the renin-angiotensin system that controls aldosterone synthesis.

Angiotensin I a peptide hormone formed by the action of renin on angiotensinogen. It is converted by ACE to angiotensin II, which stimulates aldosterone synthesis in the adrenal cortex.

Angiotensin II a peptide hormone formed by the action of ACE on angiotensin I. Angiotensin II stimulates aldosterone synthesis in the adrenal cortex.

Angiotensin II receptor antagonists also called angiotensin receptor blockers (ARBs). They block the action of angiotensin II.

Angiotensinogen the protein substrate for renin that leads to release of angiotensin I.

Anion gap the calculated difference in total concentrations (in mmol/L) between the two major plasma cations (sodium and potassium) minus the two major anions (chloride and bicarbonate).

Anorexia nervosa an eating disorder where affected individuals eat very little or starve themselves due to a fear of gaining weight.

Anovulation the absence of ovulation in the ovaries.

Antenatal diagnosis determining the nature of a disease in the embryo or foetus prior to birth.

Anthropometric measurements of the human body and the capacities of various body systems.

Antiatherogenic protects against formation of atheromas.

Antibiotics drugs that kill or suppress the growth of bacteria.

Antibody gamma globulin protein found in the body that can recognize and neutralize foreign particles, e.g. bacteria and viruses.

Anticoagulant a substance that prevents clotting of blood.

Anticonvulsant a drug that stops or prevents convulsions.

Antidepressant drug used to treat depression.

Antidotes are compounds that counteract poisons and are used to treat poisoned patients.

Antigen a molecule that is recognized by the immune system and can stimulate an antibody response.

Antimalarial a drug used to prevent or treat malaria.

Antineoplastic a drug that prevents the growth of cancerous cells.

Antioxidant a substance capable of protecting biomolecules from the damaging effects (oxidation) of reactive molecules called free radicals.

Antipsychotic drugs used to treat psychosis.

Antispasmodic a drug used to prevent spasms of muscles.

Anti-thrombotic a drug which prevents thrombus formation.

Anuria absence of urine.

Apical at the apex or tip (at the luminal surface of the epithelial cell).

Apoferritin intracellular iron storage protein which binds with iron preventing cell toxicity.

Apolipoproteins proteins that bind to fats to form lipoproteins.

Apoproteins the protein components of lipoproteins required for transport of lipids.

Apparent volume of distribution the volume in which the drug would have to be uniformly distributed to give the observed concentration in the plasma.

Appetite the desire to eat food and expressed as hunger.

Arginine an amino acid which amplifies glucose-induced insulin secretion and suppresses glucagon secretion.

Arrhythmias an abnormal rate of heart contractions.

Artefactual refers to something that is made/caused by humans.

Ascites is the term given to an accumulation of fluid in the peritoneal cavity.

Ascitic fluid the serous fluid that collects in the peritoneal cavity when there is an imbalance of plasma flow into and

out of the lymphatic system and is often found in cirrhotic liver and malignant diseases.

Ataxia lack of coordination of muscle movements.

Atherogenesis is the formation of atheromas in the arteries.

Atheroma a deposit of mostly lipid, cell debris and calcium found on the inner walls of blood vessels which can cause a blockage.

Atherosclerosis a process of thickening and hardening of the inside walls of blood vessels due to the laying down of atheromata. It can lead to cardiovascular disease.

Atorvastatin a member of the statin class of drugs (trade name Lipitor) which is effective in lowering levels of tri-acylglycerol and LDL cholesterol in the blood, thus improving the lipid profile and lowering cardiovascular risk.

ATPase a membrane bound enzyme catalysing the decomposition of ATP with release of energy.

Atrophic wasted away.

Audit a quality improvement process that aims to improve patient care by reviewing care against defined criteria and the implementation of change.

Autoimmune immune response against own cells and tissues.

Autosomal is inherited via non-sex chromosomes.

Autosome a chromosome, which is not a sex chromosome, occurring in pairs in diploid cells.

Azoospermia complete absence of sperm in semen.

Bariatric surgery is surgery of the stomach and/or intestines reserved for very obese patients in order to help them to lose weight.

Base any compound that accepts H^+, i.e. proton acceptors.

Base deficit the amount of base that must be added to each L of blood to return the pH to 7.40 at a temperature of 37°C and a PCO_2 of 5.3 kPa (40 mmHg).

Base excess the number of mmoles of acid required to restore 1 L of blood, *in vitro*, maintained at a PCO_2 of 5.3 kPa (40 mmHg), back to pH 7.4.

Basement membrane a layer of extracellular material on which the basal surfaces of epithelial cells sit.

Basolateral part of the epithelial cell membrane which forms its basal and lateral surfaces.

Bence-Jones protein immunoglobulin light chains (paraproteins) produced by plasma cells and excreted in the urine where they are measured for diagnosis of multiple myeloma.

Benign a tumour that does not invade and destroy the tissue in which it originates or spread to distant sites in the body.

Beri-Beri a nutritional disorder that arises due to vitamin B_1 deficiency and characterized by nervous degeneration and often death from heart failure.

Beta blockers drugs that antagonize action of adrenaline (a beta adrenergic substance) thus reducing stress to the heart muscle. They are often used to lower heart rate and blood pressure.

Beta cell an islet cell which secretes insulin.

Beta-2 adrenergic agonists drugs that mimic the action of noradrenaline, a neurotransmitter that causes stimulation of the sympathetic nervous system.

Bicarbonate-carbonic acid buffer system major buffering system in blood. The bicarbonate reacts with H^+ ions forming carbonic acid that can dissociate to give CO_2 and H_2O). The latter are removed by the respiratory and renal systems respectively.

Biguanides a class of anti-diabetes drugs which improve insulin sensitivity at target tissues and decrease glucose production by the liver. Metformin is most commonly used (trade name Glucophage).

Bile is a complex chemical mixture produced in the liver, stored in the gallbladder and secreted into the gastrointestinal tract. It is essential for the absorption of many compounds from ingested food, especially fats in the small intestine.

Biliary atresia a rare condition where the bile ducts between the liver and small intestine are blocked or absent. If untreated it can result in liver failure.

Biliary cirrhosis is an autoimmune disease of the liver characterized by progressive destruction of the small bile ducts in the liver resulting in build up of bile within the liver.

Bilirubin is the yellow-green pigment which when present in the bloodstream in excess is responsible for jaundice.

Bioavailability is the fraction of the given dose of a drug that reaches the systemic circulation.

Bioelectrical impedance a method for measuring the proportion of body fat by examining the difference in the resistance to flow of electrical current between fatty and lean tissue.

Biological therapies where molecules in the disease process have been identified and can be targeted by specific monoclonal antibodies.

Biomarker a metabolic intermediary molecule, signal molecule, enzyme, or protein that significantly changes in response to a disease state and can be used to monitor onset or progress of disease, or to predict the outcome in response to treatment.

Birefringence or double refraction where a ray of light is split into two parallel rays polarized perpendicularly.

Bisphosphonates drugs that inhibit osteoclast-mediated bone resorption.

Blind loop syndrome medical condition caused by intestinal obstruction, slowing or stopping movement of digested food and allowing bacterial growth which can reduce nutrient absorption.

Blood-brain barrier specialized capillaries which act as a protective membrane that separates brain cells from circulating blood.

Body mass index is an expression of body weight in relation to height. It is calculated by dividing weight (in kilos) by height (in metres) squared and is then categorized according to set intervals as normal, overweight, obese, etc.

Bowman's capsule a thin double membrane that surrounds the glomerulus of a nephron.

Bradycardia a reduction in heart rate below normal.

Bronchodilators drugs that relax and widen the bronchial airways, thereby improving passage of air into the lungs.

Bruton's disease, also known as X-linked agammaglobulinaemia where affected males are unable to produce antibodies and therefore protect themselves from infection.

Buccal mucosa mucous membrane of the inside of the cheek; is continuous with the mucosa of the soft palate the under surface of the tongue, and the floor of the mouth.

Buffer is a solution of a weak acid with its conjugate base and resists change in pH.

Buffering capacity is the ability of a buffer to resist change in pH.

Bulimia nervosa a disease where there are uncontrolled episodes of overeating that are usually followed by purging or self-induced vomiting. Often these individuals are involved in misuse of laxatives or other medications and have a strong desire to lose weight.

Calcification the hardening of tissue as a result of calcium deposition.

Calcitriol a form of vitamin D_3 which increases calcium absorption and its production is stimulated by hypocalcaemia.

Calibration logs are records which confirm that the analyte to be measured has been calibrated according to manufacturer's instructions within the correct timescale.

Calprotectin a calcium and zinc binding protein that can be measured in the faeces.

Captopril tests inhibitor of ACE used to treat hypertension. Normally, captopril decreases aldosterone production. The activity of an aldosterone secreting tumour is sustained in the captopril test.

Carbohydrates a group of organic compounds used as a source of energy and include sugars and starch.

Carboxyhaemoglobin the substance formed after binding of carbon monoxide to haemoglobin.

Carcinoid tumours rare types of neuroendocrine tumours affecting the GIT.

Cardiac arrest when heart beat and cardiac function stops suddenly, thus causing loss of blood circulation.

Cardiac output volume of blood pumped by the heart ventricles per minute.

Cardiovascular disease is arterial disease affecting the brain and other parts of the circulation as well as the coronary arteries.

Carpal tunnel syndrome is a painful condition caused by compression of a nerve in the carpal tunnel resulting in weakness, numbness, and tingling in the hands and fingers. It is more common in diabetes.

Carry over the transfer of sample or reagent from one sample to another usually by reagent probes.

Cascade tests are generated to aid diagnosis in response to results from a preceding test.

Catabolic destructive phase of metabolism, i.e. the breakdown of complex molecules to simpler ones, usually with the release of energy.

Cataract eye disease that causes cloudiness or opacification of the lens and impairs vision.

Catecholamine a hormone produced by the adrenal medulla in response to fright. Adrenaline, noradrenaline, and dopamine are catecholamines.

Catechol-O-methyl transferase one of the enzymes that inactivate catecholamines.

Cellular necrosis permanent damage to a cell that leads to its death.

Central obesity excessive fat accumulation around the abdomen and waist.

Cerebral oedema is a life-threatening build up of water in brain tissues causing swelling and increased pressure.

Cerebrospinal fluid is the serum-like clear fluid that occupies the cavities of the brain and spinal cord, and the subarachnoid space.

Cerebrovascular accident is a stroke caused by a lack of oxygen to brain tissues after a rupture or occlusion in a cerebral blood vessel.

Charcot foot/joint is a severe progressive degeneration of joints, especially in the foot, seen frequently in patients with diabetes.

Chelating agents compounds that can draw metallic ions into their chemical structure and are often used to absorb ions from solutions.

Chemoreceptors receptors that are sensitive to and activated by chemical stimuli.

Chemotherapy treatment with drugs that kill cancerous cells.

Chloride shift exchange of chloride for bicarbonate across the erythrocyte membrane as part of the buffering mechanism in blood.

Chlorpropamide a drug in the sulphonylurea class used to treat type 2 diabetes mellitus.

Cholangiocarcinoma a cancer affecting the bile ducts through which bile drains from the liver into the small intestines.

Cholestasis is the term given to the failure to excrete bile and is derived from the Greek word 'stopped bile'.

Cholesterol is an abundant sterol produced by the liver. It is a precursor to steroids and is present in cell membranes.

Chromaffin cell neuroendocrine cells of the adrenal medulla. They are visualized by staining with chromium salts that turn to a brown colour.

Chromogranin A a secretory protein of neuroendorine cells. Plasma concentrations are elevated in carcinoid tumours.

Chromatography a technique for separating compounds in a mixture that is based on their differences in distribution between the mobile and stationary phases utilized in the procedure.

Chronic kidney disease progressive loss of kidney function over a period of months or years.

Chronic obstructive airways disease refers to chronic bronchitis and emphysema which are often co-existing diseases of the lungs where the airways become narrow causing a reduction in flow of air into and out of the lungs presenting as shortness of breath.

Chronic pancreatitis long-term inflammation of the pancreas that can alter its structure and function.

Chvostek's sign an abnormal contraction of facial muscles after the facial nerve is stimulated.

Chylomicrons are triacylglycerol rich particles that transport dietary fats from GIT to liver and adipose tissues.

Circadian rhythm a 24-hour cycle in a biochemical or other process. For example, plasma concentrations of cortisol are highest around 8 am and lowest around midnight in normal humans. Also called a diurnal rhythm.

Cirrhosis occurs when deposits of fibrous tissue occur in chronic liver disease and is an irreversible stage of liver damage.

Clathrin is a protein, which is recruited to segments of the plasma membrane of cells destined to become vesicles.

Clearance the volume of plasma from which drug is removed per unit time, usually expressed as mL/min or L/h.

Clinical sensitivity the ability of an assay to detect only *patients* with a particular disease.

Clinical specificity the ability of an assay to detect only that *disease process.*

Clone genetically identical cells derived from a single cell.

Clonidine test Catecholamine secretion is normally suppressed by clonidine whereas secretion is not suppressed in an adrenal medullary tumour.

Co-dominant inheritance occurs when an abnormal gene interferes with the function of the normal gene.

Coefficient of variation is a measure of precision, and describes the distribution of measurements for a repeated series of tests on the same sample. It is expressed as a percentage and the lower the value the more precise the method.

Coeliac disease autoimmune disease of the small intestine due to sensitivity of the intestinal lining to the protein gliadin, causing atrophy of cells in the intestine.

Colonoscopy the endoscopic examination of the colon with a fibre optic camera on a flexible tube passed through the anus.

Colostrum a type of milk produced by mammary glands in late pregnancy and directly after birth.

Common variable immunodeficiency a group of primary immunodeficiencies which have a common set of symptoms including low levels of gamma globulins but different causes.

Compensation physiological response to change in acid-base balance with the aim of limiting change in pH.

Competitive immunoassays techniques that makes use of the competition between an antigen and the same labelled antigen for binding to the homologous antibody in order to identify and quantify the specific antigen in a sample.

Compound E abbreviation for cortisone 17α, 21dihydroxy-4-pregnen-3,11,20 trione.

Compound F abbreviation for cortisol (hydrocortisone) 11β,17α,21-trihydroxy-4-pregnen-3,20 dione.

Compound S abbreviation for 11-deoxycortisol 17α, 21-dihydroxy-4-pregnen-3,20 dione.

Congenital adrenal hyperplasia a condition caused by a genetic defect in an enzyme required for cortisol synthesis. In the absence of cortisol production, the pituitary gland secretes ACTH in an attempt to stimulate cortisol production. Prolonged ACTH stimulation leads to growth (hyperplasia) of the adrenal cortex.

Congenital hyperinsulinism a variety of congenital disorders resulting in hypoglycaemia caused by excessive insulin secretion.

Conn's syndrome a condition caused by the overproduction of aldosterone from the adrenal cortex or a tumour, resulting in low potassium concentrations and high blood pressure.

Continuous flow analysis an automation technique that utilizes a continuous stream of reagent, pumped through a set flow path and into which samples are introduced at set intervals. The moving reaction zones, associated with the sequential samples are separated by air bubbles and subjected to a defined incubation period, determined by the length of tubing used. The reaction zone is monitored when the flowing reagent stream passes through the detection system, usually a flow cell of a spectrophotometer.

Core laboratories are laboratory areas where over 90% of tests are performed.

Coronary artery is an artery arising from the base of the aortic arch supplying the heart muscle.

Coronary heart disease is a narrowing of the blood vessels supplying blood and oxygen to the heart muscle.

Cortex outer layer of adrenal gland that can synthesize glucocorticoids and mineralocorticoids.

Cortical refers to the outer hard layer of bone.

Corticosteroids steroids secreted by the adrenal cortex.

Corticosterone a major glucocorticoid in the rat and other animals, and a precursor of aldosterone in humans.

Corticotrophin releasing hormone a peptide secreted from the hypothalamus in response to low circulating cortisol concentrations and acts to stimulate ACTH secretion from the anterior pituitary.

Cortisol binding globulin transport protein for cortisol in the blood. It becomes saturated when cortisol concentrations in plasma exceed 500 nmol/L

Cortisol secretion rate a measure of rate of cortisol secretion used to assess adrenal cortisol secretion.

Cortisol the major glucocorticoid in humans. It is a hormone synthesized in the zona fasciculata of the adrenal cortex in response to stress.

Counter-regulatory hormones oppose the action of insulin by increasing blood glucose levels and include glucagon, adrenaline, cortisol, and growth hormone.

Cows' milk allergy an abnormal immune response to milk proteins resulting in the rapid appearance of symptoms following consumption of cows' milk.

C-peptide connecting peptide cleaved from proinsulin in the synthesis of insulin.

C-reactive protein an acute phase protein whose concentration in the blood rises in response to inflammation.

Creatinine a break-down product of creatine phosphate in muscles. Its serum concentration is used to assess renal disease.

Cretinism a syndrome that arises due to congenital deficiency of thyroid hormones; children present with dwarfism, mental retardation, and are often deaf and mute.

Crohn's disease inflammatory disease of the digestive system which may affect any part of the GIT from mouth to anus but most often the terminal ileum.

Cushing's disease a tumour of the pituitary gland secreting ACTH described by Harvey Cushing. Associated with high cortisol secretion from the adrenal cortex

Cushing's syndrome the clinical picture of cortisol excess seen in Cushing's disease. Can be due to an adrenal tumour secreting cortisol, an ACTH secreting tumour outside the pituitary (ectopic), or intake of high doses of glucocorticoids.

Cystic fibrosis an inherited condition characterized by deposition and clogging of lungs, intestines, and pancreas with thick mucus.

Cystic fibrosis transmembrane conductance regulator a transporter protein and ion channel that transports chloride ions across epithelial membranes.

Cystinosis a defect in metabolism of cystine causing it to accumulate in the blood.

Cystinuria loss of amino acid cystine in the urine due to a defect in renal reabsorption by the kidney tubules.

Cytochrome P450 enzyme that most commonly inserts an atom of oxygen into a substrate. Usually part of an electron transport chain and a family of enzymes that have a key role in drug metabolism.

Cytogenetics the study of the structure of chromosomes.

Cytotoxic substance that is toxic to cells.

DAX-1 a nuclear receptor protein important in the development of the adrenal cortex.

De novo **synthesis** the formation of complex molecules from simple molecules such as sugars or amino acids.

Deconjugation for example separation of amino acid from bile acid to form free bile acid.

Defaecation passage of excreta through the anus.

Dehydroepiandrosterone a steroid androgen hormone secreted by the adrenal cortex.

Dehydroxylation removal of a hydroxyl group.

Delta cell an islet cell which secretes somatostatin.

Delta check examination of any value obtained for an analyte against the previous result.

Dermatitis herpetiformis a skin condition associated with coeliac disease; lesions like those found in herpes.

11-deoxycorticosterone an intermediate in the synthesis of aldosterone and a potent mineralocorticoid.

11-deoxycortisol intermediate in cortisol synthesis. Its plasma concentrations are raised when the enzyme 11-hydroxylase is defective or blocked by drug action for example metyrapone.

21-deoxycortisol intermediate in cortisol synthesis. Its plasma concentrations are raised when the enzyme 21-hydroxylase is defective.

Desmolase cytochrome P450 enzyme in steroidogenesis that cuts a carbon-carbon bond. Cholesterol 20,22 desmolase is a side chain cleavage enzyme (gene is CYP11A1). Androgens are formed by the action of 17,20 desmolase or lyase (gene is abbreviated CYP17A1). Hydroxylations precede the carbon-carbon bond cleavage.

Detoxification removal of toxic effects of drugs from the body (usually done by the liver).

Dexamethasone test a synthetic analogue of cortisol and a potent glucocorticoid. If 0.5 milligrams are taken at night, it will suppress cortisol release so that plasma cortisol concentrations at 8 am will be less than 100 nmol/L in a normal subject. The low dose test involves taking 0.5 mg of the drug six-hourly. To distinguish Cushing's syndrome 2 mg every six hours will not suppress cortisol in the case of an ectopic ACTH secreting tumour but will suppress a pituitary secreting tumour.

Diabetes insipidus is a disorder where a large amount of dilute urine is produced due to a deficiency of ADH.

Diabetes mellitus a condition brought about by a total lack of or a functional defect in the action of insulin, resulting in an inability to get glucose from the blood into target tissues.

Diabetes UK the largest organization in the UK working for people with diabetes. It funds research and campaigns to help people live with this condition.

Diabetic ketoacidosis a lack of control, mainly in type 1 diabetes, characterized by high blood glucose concentrations and a lack of insulin which results in the formation of ketones in an acidic bloodstream. Potentially life threatening.

Diagnosis the decision reached by a clinician after examining and investigating the patient regarding his or her clinical condition.

Diarrhoea production of large amounts of loose and watery stools.

Dietary fibre indigestible parts of plant foods that move with food through the digestive system, retaining water and facilitating defaecation.

Differential diagnosis identifies which of the two or more diseases are responsible for the clinical features shown by the patient.

Digoxin a drug that helps an injured or weakened heart to work more efficiently by increasing its contraction, thereby sending more blood through the body.

Dipeptidyl peptidase IV an enzyme found in the intestine and kidney which catalyses the hydrolysis of GLP-1 from its active form to its inactive form.

Dipeptidyl peptidase IV inhibitors drugs which inhibit the action of DPP IV, thereby increasing the activity of GLP 1. Sitagliptin and vildagliptin are DPP IV inhibitors.

Disaccharidase deficiency deficiency of enzymes capable of digesting disaccharides.

Discrete analyser an analyser that undertakes each sample analysis in an individual vessel physically separated from the others as opposed to continuous flow analysis.

Discretionary analyser an analyser capable of undertaking only those tests selected by the operator, which is different from the standard continuous flow multi-channel systems that would undertake all of the tests on the analytical system.

Diuretic a drug that promotes the loss of urine from the body.

Diurnal rhythm a 24-hour cycle in a biochemical or other process. Plasma concentrations of cortisol are highest around 8 am and lowest around midnight in normal humans. Also called a circadian rhythm.

Diverticulae out pouching of a hollow structure; in jejunum can become colonized, in colon can become infected and can perforate.

Diverticular disease a disease characterized by the presence of small bulges in the large intestines that are often inflamed.

Dopamine a member of the catecholamine family, dopamine is a precursor to noradrenaline and then adrenaline in the biosynthetic pathways for these neurotransmitters.

Dopamine hydroxylase catalyses conversion of dopamine to noradrenaline.

Drug any substance that after absorption into the body alters normal bodily function. Drugs are often components of medication and used in the treatment or prevention of a disease.

Dumping syndrome condition due to food moving too rapidly from the stomach into the small intestine.

Dupuytren's contracture a condition in which the fingers and the palm of the hand thicken and shorten, causing the palm to curve inwards. It is more prevalent in patients with diabetes.

Dyslipidaemia abnormal amounts of lipids or lipoproteins in the blood.

Dysphagia difficulty in swallowing.

Dysrhythmia an abnormal heart rhythm.

Ectopic hormone a hormone formed by a tissue outside its normal endocrine site of production.

Electrocardiogram a graph showing the electrical activity of the heart muscle.

Elemental diets food is given in its simplest formulation, i.e. carbohydrate as glucose, protein as amino acids, fat as medium chain triacylglycerols, which are water-soluble.

Elimination body process of removing a drug.

Embolization the insertion of a material into a blood vessel in order to obstruct blood flow.

Emphysema chronic disease of the lung characterized by reduced respiratory function.

Encephalopathy diseases affecting the brain.

Endocrine referring to a hormone which is secreted in one location but exerts its action, via the bloodstream, in another location.

Endocrinology is the study of the endocrine glands and the hormones they produce.

Endocytosis uptake of material by a cell by invagination of its plasma membrane.

Endometriosis a condition where pieces of the womb lining or endometrium are found outside the womb in the ovaries, Fallopian tubes, bladder, and vagina.

Endomysial of endomysium.

Endomysium layer of connective tissue that ensheathes a muscle fibre, mainly reticular fibres.

Endopeptidases proteolytic enzymes that hydrolyse peptide bonds of non-terminal amino acids.

Endoscopic retrograde cholangiopancreatography a procedure that combines use of endoscopy with X-rays for investigation of disorders of the bile and pancreatic ducts.

Endoscopy looking inside the body for medical reasons.

Endosome an intracellular organelle enclosed by a membrane derived from the plasma membrane. They may fuse with other endosomes or lysosomes, or become acidified by a proton pump.

Endothelial cell flat, nucleated lining cell found in body cavities or blood vessels.

Endothelium the inner lining of blood vessels.

End-stage renal disease a condition where renal function is inadequate to support life.

Enteric coating a coating on the outside of a tablet that helps to prevent it disintegrating in the acid in the stomach.

Enterocytes cells lining the intestines and colon and involved in secretion of digestive enzymes and absorption of nutrients.

Enterohepatic circulation between liver and small intestine.

Enteropathy a disease of the intestinal tract.

Enzyme a protein that functions as a biological catalyst.

Epitope also known as an antigenic determinant that is the site on a antigen molecule to which an antibody will attach.

Erythropoietin a glycoprotein hormone that stimulates synthesis of erythrocytes.

Ethanol the alcohol found in alcoholic drinks. Also known as ethyl alcohol.

Euthyroid is normal thyroid function.

Excretion the process of removing waste products of metabolism.

Exenatide (trade name Byetta) is an injectable analogue of GLP-1, isolated from the saliva of the hela monster.

Exocrine glands which secrete their products into ducts.

Exocytosis is a process by which membrane-enclosed vesicles inside the cell fuse with the plasma membrane thereby releasing their contents to the outside.

Exopeptidases proteolytic enzymes that hydrolyse peptide bonds to release terminal amino acids.

Extrahepatic outside the liver.

Extrahepatic biliary atresia a rare extrahepatic condition that can cause liver failure due to a blockage or absence of the bile duct between the liver and the small intestine.

Faecal elastase refers to the enzyme pancreatic elastase found in the faeces and its concentration indicates exocrine pancreatic function.

Faecal osmotic gap used to distinguish between an osmotic and secretory diarrhoea.

Fanconi syndrome a disorder in which the proximal tubules of the kidneys are diseased causing reduced reabsorption of electrolytes and nutrients back into the blood such as glucose, amino acids, phosphate, and bicarbonate.

Fasciculata middle region of adrenal cortex that produces glucocorticoids.

Fatty acid β oxidation defects a deficiency in the process of burning stored or dietary fat to produce NADH, FADH, and acetyl co-enzyme A.

Fenestra window-like opening.

Ferritin iron storage protein.

First-order elimination is the rate of elimination of a drug that is directly proportional to its plasma concentration.

First-pass metabolism refers to the removal of a drug from the plasma after absorption and before it reaches the systemic circulation.

Fistulae abnormal connections between two epithelium-lined organs or vessels that normally do not connect.

Fludrocortisone a synthetic steroid that acts like aldosterone as a mineralocorticoid.

Foam cells lipid-rich macrophages that are one of the first cells found at the site of fatty streaks in blood vessels and represent the start of atherosclerotic plaque formation.

Foetal adrenal cortex major zone of the adrenal cortex that grows in size during foetal life and disappears over the first six months after birth. It also secretes DHEAS.

Free radical atom or molecule with an unpaired electron. Free radicals are highly reactive and capable of damaging biomolecules in the human body.

Frozen shoulder pain and stiffness around the shoulder joint due to changes in muscles and soft tissues around this area.

Fructosamine glycation of serum proteins used to indicate short-term glycaemic control.

G cells gastrin secreting cells found in the gastric glands of the pyloric antrum region of the stomach.

Galactorrhoea spontaneous flow of milk from mammary glands that is not associated with childbirth or feeding.

Galactosaemia an inherited metabolic disorder where the enzyme needed to metabolize galactose found in milk is lacking. The resulting build-up of galactose can cause liver problems and can be life threatening.

Gallstones small, hard, stone-like objects that form and obstruct the gallbladder or bile ducts. They are composed of cholesterol, bile pigments, and calcium salts.

Gamma camera a device used to image gamma radiation emitting radioisotopes.

Gamma-glutamyl transpeptidase is an enzyme responsible for the transfer of glutamyl groups from gamma-glutamyl peptides to other peptides or amino acids.

Ganglioneuroma tumour of the nervous system that can be in the eye or adrenal glands.

Gas chromatography a chromatographic method used to separate out volatile molecules using a gaseous mobile phase.

Gastric antrum the distal part of the stomach.

Gastric bypass surgery a procedure for weight loss in the obese which has shown dramatic improvement and even reversal of diabetes in patients undergoing treatment. The procedure is thought to bring about a more favourable gut hormone profile.

Gastric fluid the fluid produced in the stomach.

Gastric lavage is a process of washing out the stomach using water to remove stomach contents (often poisons).

Gastrin peptide hormone that stimulates release of gastric juice in the stomach.

Gastrinoma a gastrin secreting tumour.

General chemistry analyser an automated instrument used in clinical biochemistry laboratories. The automated reactions may be chemical, enzymatic, or based on immunoassay and all give rise to a change in absorbance that is monitored to generate quantitative results. In most general chemistry analysers there will also be ISEs that will test for sodium, potassium, and chloride.

Genetic counselling procedure by which families are given advice about the nature and consequences of an inherited disorder affecting their children and the various options available to them.

Gestational diabetes mellitus biochemically defined diabetes which occurs or is first diagnosed during pregnancy.

Ghrelin a hormone involved in appetite control.

Gilbert's syndrome an inherited condition with mild jaundice that is harmless and does not require treatment.

GK gene that encodes enzyme glucokinase.

Gliadin a glycoprotein constituent of gluten present in wheat, barley, and rye; α, β, and δ, gliadin sensitivity results in gluten sensitive enteropathy or coeliac disease.

Gliclazide a drug classified as a sulphonylurea and is an oral hypoglycaemic agent.

Glimepiride a medium to long-acting sulphonylurea anti-diabetic drug.

Glipizide an oral medium to long-acting anti-diabetic drug from the sulphonylurea class.

Glomerular filtration rate the rate in mL per minute that small substances, such as creatinine and urea, are filtered through the kidney's glomeruli. It is a measure of the number of functioning nephrons.

Glomerulonephritis nephritis accompanied by inflammation of the capillary loops of the glomeruli of the kidney. It occurs in acute, subacute, and chronic forms.

Glomerulosa outer region of adrenal cortex that produces mineralocorticoids.

Glomerulosclerosis hardening of the glomerulus in the kidneys.

Glomerulus a tuft of blood vessels found in each nephron of the kidney that are involved in the filtration of the blood to form urine.

GLP-1 (7-36) amide the active form of glucagon-like peptide 1.

GLP-1 (9-36) amide the inactive form of glucagon-like peptide 1.

Glucagon a hormone secreted by the pancreatic α cell that is involved in the release of energy substrates such as glucose from the tissues into the bloodstream.

Glucagon-like peptide 1 is an incretin hormone produced in the intestine which amplifies glucose-induced insulin secretion, suppresses glucagon secretion, slows gastric emptying, and suppresses appetite. The active form is GLP-1 (7-36) amide and the inactive form is GLP-1 (9-36) amide.

Glucagon-like peptide 1 receptor an incretin hormone receptor that stimulates release of insulin from pancreatic beta cells in conjunction with sugars absorbed from the gut.

Glucagonoma a glucagon secreting tumour.

Glucocorticoid a steroid hormone produced by the adrenal cortex that affects protein, glucose and fat metabolism.

Glucokinase a hexokinase enzyme which phosphorylates glucose, especially in pancreatic beta cells.

Gluconeogenesis synthesis of glycogen from non-carbohydrate sources (such as amino acids).

Glucose 6-phosphate a metabolic intermediate in the glycolytic pathway.

Glucose-dependent insulinotropic peptide formerly known as gastric inhibitory peptide is an incretin hormone secreted by the intestine which amplifies glucose-induced insulin secretion and may also be involved in the storage of fat into adipocytes.

Glucose transporters a family of molecular structures which, when inserted into the cell membrane, facilitate the movement of glucose into the cell. Glut 2 transporters are found on β cell membranes, glut 4 transporters are found on target tissue cells.

Glucose-induced insulin secretion is insulin secretion stimulated by glucose.

Glucuronide steroids when conjugated with glucuronic acid increases their water solubility and assists renal excretion.

GLUT 2 referred to as glucose transporter 2, is a transmembrane carrier protein that allows passive glucose movement across cell membranes.

GLUT 4 transporters are insulin-regulated glucose transporters found in adipose tissues and striated muscle that are responsible for glucose translocation into the cell.

Glutamic acid decarboxylase an enzyme which converts glutamate to GABA. Autoantibodies to GAD are common in patients with type 1 diabetes.

Gluten a composite of the proteins gluten and glutenin found in wheat, rye, and barley.

Glycated haemoglobin haemoglobin that has undergone non-enzymatic glycosylation by glucose. Used to indicate long-term glycaemic control in diabetic subjects.

Glycation non-enzymatic modification of proteins by sugars to give glycated proteins.

Glycogen storage disease 1 an inborn error of metabolism caused by a deficiency of glucose 6 phosphatase resulting in hypoglycaemia, a build up of lactic acid, liver problems, and growth defects.

Glycogenesis synthesis of glycogen in the liver.

Glycogenolysis breakdown of glycogen in the liver to release free glucose.

Glycolipids are lipids covalently linked to sugar residues.

Glycolysis is the metabolic breakdown of glucose to pyruvate, generating ATP in the process.

Glycolytic pathway the series of phosphorylative reactions during pyruvic acid production.

Glycosylation addition of sugar residues to molecules (usually proteins).

Goblet cells glandular, single columnar epithelial cells which secrete mucin.

Goitre is characterized by a swelling in the neck due to enlargement of the thyroid gland.

Goodpasture's disease is a condition characterized by glomerulonephritis and haemorrhage of the lungs caused by antibodies to the glomerular basement membrane.

Gout clinical condition characterized by the deposition of crystals of monosodium urate in synovial fluid.

G-protein cell membrane proteins that act as intermediaries between hormone receptors and effector enzymes allowing cells to regulate their metabolism in response to hormonal stimulation.

Granuloma a mass of tissue that occurs at the site of inflammation, injury, or infection and is part of the healing process.

Graves' disease is an autoimmune condition characterized by overactivity of the thyroid gland.

Growth hormone a hormone released by the pituitary gland (also called somatotropin) which promotes growth and cell reproduction.

Gut associated lymphoid tissue diffuse system of small areas of lymphoid tissue found in the gastrointestinal tract and other sites in the body.

Gynecomastia abnormal enlargement of breasts in males.

Gynoid female-like.

H_2 - receptor antagonists drugs which block the action of histamine on parietal cells in the stomach.

Haem a prosthetic group which consists of an iron atom at the centre of a porphyrin group found in haemoglobin and myoglobin.

Haematuria blood in the urine.

Haemochromatosis characterized by excessive iron absorption and its subsequent deposition in the liver, pancreas, and endocrine glands.

Haemodialysis dialysis of blood to remove toxic substances.

Haemoglobin A_{1c} a glycated form of haemoglobin measured to establish average blood glucose levels in patients with diabetes.

Haemolysis the rupture of erythrocytes and release of their contents into the plasma, which becomes pink or red because of the presence of haemoglobin.

Half-life the time taken for a drug to fall to half its original concentration.

Haplotypes are alleles that are transmitted together.

Hartnup disease an autosomal recessive condition that affects absorption of certain amino acids, in particular tryptophan.

Hashimoto's thyroiditis is characterized by chronic inflammation of the thyroid gland due to autoantibodies directed against the thyroid gland causing reduced secretion of thyroid hormones.

Heart failure state in which the heart is unable to deliver blood at a rate that meets the needs of body tissues to function normally.

Hemizygote an individual is a hemizygote or is hemizygous when he or she has only one allele of a gene rather than the usual two.

Hepatectomy surgical removal of the liver. This may be complete or partial.

Hepatic encephalopathy brain damage from toxic substances that are not detoxified due to liver disease.

Hepatitis inflammation of the liver.

Hepatocellular carcinoma a malignant tumour of the liver.

Hepatocyte nuclear factor a group of proteins that act as transcription factors in liver cells and are involved in gene regulation.

Hepatocytes account for 80% of the cells in the liver and carry out most of the functions except excretion. When hepatocytes are inflamed and subsequently damaged a person has hepatitis.

Hepatoma a malignant tumour of the liver.

Hepatomegaly is enlargement of the liver.

Hepatotoxicity toxic damage to the liver.

Hereditary fructose intolerance an inherited disorder of fructose metabolism seen in young babies. Affected individuals lack an aldolase enzyme so are unable to process fructose. A characteristic symptom of the disorder is hypoglycaemia, especially after the ingestion of fructose or sucrose.

Hereditary fructose intolerance an inherited disorder where fructose cannot be metabolized due to an enzymatic deficiency.

Hereditary haemochromatosis an inherited condition characterized by excessive iron absorption and its subsequent deposition in the liver, pancreas, and endocrine glands. Affected patients have a bronze coloured skin and often suffer from liver failure and diabetes.

Hereditary spherocytosis a form of inherited anaemia characterized by spherical shaped fragile erythrocytes that can often lead to haemolytic anaemia.

Heteroplasmy is the presence of a mixture of more than one type of mitochondrial DNA in a cell. Since cells contain hundreds of mitochondria with hundreds of copies of mtDNA, it is possible for mutations to affect only some of these copies, whilst the rest are unaffected.

Heterozygote an individual is a heterozygote or is heterozygous for a gene when they have different alleles for this gene in each of the homologous chromosomes.

Heterozygous an individual with a pair of genes for a particular characteristic that are not identical.

High density lipoprotein a class of lipoprotein that transports cholesterol to the liver from other tissues, it has a high proportion of protein and is cardioprotective.

High Km hexokinase a weak binding enzyme that phosphorylates six carbon sugars like glucose.

High risk samples are samples from patients carrying hepatitis B or C, or HIV viruses, or any other category III pathogen.

Hirsutism the presence of hair in females found on face, chest, back, and abdomen often due to excessive androgen production.

HLA DQ2 a serotype group within HLA DQ serotyping system with a strong association with autoimmune disease.

HLA DR3 an antigen encoded by the D locus on chromosome 6 and found on lymphoid cells. Has a strong association with juvenile diabetes, coeliac disease, and Graves' disease.

HLA DR4 an antigen encoded by the D locus chromosome 6 and found on lymphoid cells. Has a strong association with juvenile diabetes and rheumatoid arthritis.

Homeostasis model assessment an algorithm designed in Oxford, UK, used to model pancreatic β cell function (Homa-B) with tissue insulin sensitivity (Homa-S) under steady state (i.e. fasting) conditions. It is especially useful for comparing different groups.

Homocysteine an amino acid produced by the body. High levels of blood homocysteine have been associated with an increased risk of coronary heart disease.

Homocystinuria an inherited defect in metabolism of homocystine causes it to accumulate in the blood with its subsequent loss in the urine.

Homozygote an individual is a homozygote or is homozygous for a gene when it has the same alleles for this gene in each of the homologous chromosomes.

Homozygous an individual with a pair of genes for a particular characteristic that are identical.

Hormone a chemical released from an endocrine gland into the blood which circulates in the blood and acts on a target cell.

Hormone sensitive lipase an enzyme that liberates NEFA from fats for use as energy. It is suppressed by high levels of insulin and stimulated by glucagon and catecholamines.

Hospital information system the central patient database for the hospital and its patients. It is used to generate requests for diagnostic procedures and is also a repository for diagnostic test results.

Human leucocyte antigen complex a group of antigens found on the surface of nucleated cells in the body, coded for by the major histocompatibility complex of humans, therefore allowing the immune system to distinguish between self and non-self.

Humoral found in fluids (not cellular).

Hungry bone syndrome causes hypocalcaemia following surgical treatment of hyperparathyroidism in patients who have had prolonged secondary or tertiary hyperparathyroidism. In this condition, calcium from the plasma is rapidly deposited in the bone causing hypocalcaemia.

Hydrocortisone pharmaceutical name for cortisol.

Hydrophilic attracts water.

Hydrophobic repels water.

17-Hydroxyprogesterone a steroid hormone produced during the synthesis of glucocorticoids and sex hormones.

Hydroxylase enzyme that adds one or more hydroxyl groups to a steroid or other molecule.

3β-Hydroxysteroid dehydrogenase an alcohol oxidoreductase that catalyses removal of two hydrogen atoms from a steroid.

Hyperaldosteronism condition characterized by increased concentrations of aldosterone.

Hyperammonaemia a metabolic disorder characterized by high concentrations of ammonia in the blood.

Hyperamylasaemia high concentrations of amylase in the blood.

Hyperandrogenism excessive secretion of androgens in females.

Hyperbaric oxygen use of pressurized oxygen to treat oxygen-deprived tissues following injury.

Hyperbilirubinaemia a higher than normal concentration of bilirubin in the blood.

Hypercalcaemia abnormally high plasma calcium concentrations.

Hypercapnia high concentration of carbon dioxide in blood.

Hypercholesterolaemia abnormally high levels of blood cholesterol.

Hyperchylomicronaemia abnormally high levels of blood chylomicrons.

Hypercortisolism excessive secretion of cortisol.

Hyperemesis gravidarum is severe vomiting during pregnancy.

Hyperglycaemia a higher than normal level of glucose in the bloodstream.

Hyperkalaemia an electrolyte disorder characterized by concentrations of potassium in the serum or plasma samples above the reference range.

Hypermagnesaemia abnormally high concentration of magnesium in the blood.

Hypernatraemia an electrolyte disorder characterized by concentrations of sodium in serum or plasma samples above the reference range.

Hyperosmolar hyperglycaemic syndrome a type of diabetic emergency seen mainly in patients with type 2 diabetes associated with severe hyperglycaemia and high serum osmolality. Slow and careful replacement of fluids is essential to avoid life-threatening brain injury.

Hyperosmolar non-ketotic state often occurs in type 2 diabetes and is characterized by high blood glucose but without increased concentrations of blood ketones.

Hyperparathyroidism excessive secretion of PTH from the parathyroid glands.

Hyperphosphataemia abnormally high phosphate concentration in the blood.

Hyperphosphatasaemia an abnormally high concentration of blood alkaline phosphatase.

Hyperplasia abnormal proliferation of cells and tissues.

Hyperprolactinaemia presence of high levels of plasma prolactin.

Hyperpyrexia very high body temperature.

Hypertension high blood pressure.

Hyperthyroidism is a condition characterized by an overactive thyroid gland.

Hypertriglyceridaemia abnormally high levels of blood triacylglycerols.

Hyperuricaemia concentrations of uric acid above the reference range.

Hyperventilation overbreathing, causing a decline in PCO_2 in blood.

Hypervitaminosis D the condition resulting from excessive consumption of vitamin D.

Hypoalbuminaemia low plasma concentration of albumin.

Hypoaldosteronism deficiency of aldosterone often accompanied by decreased renin production.

Hypoalphalipoproteinaemia inherited condition characterized by deficiency of blood high-density lipoproteins.

Hypobetalipoproteinaemia a hereditary disorder where there are abnormally low levels of blood beta-lipoproteins, lipids, and cholesterol.

Hypocalcaemia abnormally low blood calcium concentrations.

Hypocalciuria reduced loss of calcium in the urine.

Hypocapnia low concentration of carbon dioxide in the blood.

Hypochlorhydria low production of acid in the stomach.

Hypocholesterolaemia low levels of blood cholesterol.

Hypoglycaemia plasma glucose below its reference range.

Hypokalaemia plasma potassium below its reference range.

Hypomagnesaemia abnormally low concentrations of blood magnesium.

Hyponatraemia an electrolyte disorder characterized by concentrations of sodium in plasma samples below the reference range.

Hypoparathyroidism inadequate secretion of PTH from the parathyroid glands.

Hypophosphataemia abnormally low phosphate concentration in the blood.

Hypopituitarism reduced activity of the pituitary gland.

Hypotension low blood pressure.

Hypothermia a fall in body temperature below that required for normal metabolism.

Hypothyroidism is a condition characterized by underactivity of the thyroid gland.

Hypouricaemia concentrations of uric acid below the reference range.

Hypoventilation reduced breathing.

Hypovolaemia a decline in the volume of blood.

Hypoxia lack of oxygen affecting the whole body or a region of the body.

Icterus also known as jaundice, refers to the high concentrations of the yellow coloured pigment, bilirubin that appears in plasma and in the body tissues most frequently in liver disease.

Idiopathic arising from an unknown cause.

Idiopathic ketotic hypoglycaemia hypoglycaemia of unknown cause characterized by a fall in insulin release and presence of ketosis. Usually occurs in thin, small-for-date children.

Immunoassay a biochemical test that enables measurement of the concentration of a substance in a biological fluid. Often immunoassays rely on competition between a hormone and a labelled hormone for binding sites on antibodies.

Immunochemistry analyser an automated instrument that uses immunoassay technology for testing.

Immunoradiometric assay an assay based on binding of an antigen by a specific radioactively labelled antibody.

Immunosuppressant a drug than can suppress or lower the immune response of the body.

Impaired fasting glycaemia defined by the WHO as a fasting plasma glucose level, measured on a laboratory analyser, of 6.1 mmol/L to 6.9 mmol/L and a two-hour plasma glucose (after 75 g oral glucose) of less than 7.8 mmol/L.

Impaired glucose tolerance defined by the WHO as a two-hour post 75 g glucose load plasma glucose level, measured on a laboratory analyser, greater than 7.8 mmol/L and less than 11.1 mmol/L, with a non-diabetic (i.e. less than 7.0 mmol/L) fasting glucose level.

Implantation this refers to attachment of the embryo to the uterus lining.

Incretin hormones secreted in the intestine, which amplify glucose induced insulin secretion from the pancreas.

Infarction cell necrosis following prolonged ischaemia that has caused irreversible damage to the cell.

Infertility inability to conceive in females or to induce conception in males.

Inherited tubulopathies are heterogeneous disorders often characterized by electrolyte disturbances and include defects of the proximal tubule, transport channel defects of the loop of Henle, and distal tubular defects.

Inorganic macronutrients are those inorganic nutrients required in amounts of a gram or more per day and include calcium, sodium, potassium, chloride, phosphorus, and magnesium.

Insulin a hormone secreted by pancreatic β cells involved in growth and particularly in the correct delivery of glucose from the bloodstream to the tissues.

Insulin-like growth factor a peptide hormone that shows similarity to insulin and mediates growth.

Insulin induced hypoglycaemia test hypoglycaemia stimulates ACTH release from the pituitary, which in turn increases cortisol release from the adrenal cortex.

Insulin resistance failure of the body to respond to insulin.

Interleukin 6 a protein that can improve the body's response to infection and disease.

Internal standard a chemical that is added in a constant amount to blanks, standards, and samples during an assay and corrects for differences in the amounts of solution applied to the instrument, e.g. a HPLC column.

International normalized ratio is a measure of how long it takes for blood to clot and is the most dynamic indicator of liver function or dysfunction.

Intramuscular within a muscle.

Intravascular haemolysis breakdown of red blood cells in blood vessels.

Intravenous within a vein.

Intrinsic factor a substance produced in the stomach mucosa that binds to and is required for absorption of vitamin B_{12}.

Ion-selective electrode an analytical device that measures the concentration of the ion of interest.

Irritable bowel syndrome affects motility of the intestine causing abdominal pain, bloating, constipation, and/or diarrhoea.

Ischaemia state that arises when a tissue's demand for oxygen exceeds its supply and an insufficient amount of oxygen reaches its cells.

Ischaemic heart disease is disease affecting the coronary arteries and depriving the heart muscle of blood and oxygen.

Islets of Langerhans are small clusters of cells found throughout the pancreas which make up the endocrine part of the gland, named after Paul Langerhans, who first described them.

Isoelectric point the pH at which a particular molecule carries no net electrical charge.

Isoenzymes different forms of the same enzyme. Each isoenzyme is composed of different combinations of individual subunits or polypeptides. Each polypeptide is encoded by a different gene and thus has a different structure.

Jaffe reaction the reaction of creatinine with alkaline picrate to form a coloured compound. Used by most laboratories to measure creatinine.

Jaundice yellow discolouration of skin due to high concentrations of bilirubin in the blood.

Juxtaglomerular apparatus the structure close to the glomerulus that regulates function of each nephron.

K_{ATP} **channel** a potassium ion channel found in pancreatic beta cells that is sensitive to the level of ATP.

Kernicterus damage to brain and nervous tissue due to deposition of unconjugated bilirubin in neonates.

Keshan disease a congestive cardiomyopathy caused by deficiency of selenium in the diet.

Ketoacidosis a metabolic acidosis caused by the presence of ketone bodies usually in uncontrolled diabetes mellitus type 1.

Ketogenesis the process of ketone production.

Ketoisocaproate a keto acid of leucine and an insulin secretagogue.

Ketones/ketone bodies are organic carbonyl containing molecules which can build up in the bloodstream when fats are broken down for energy in the absence of carbohydrate fuel and insulin.

Ketosteroid a steroid with a carbonyl (ketone) group; usually refers to 17-ketosteroid where the ketone group is at C17.

Ketosteroid reductase an enzyme that removes two hydrogen atoms from a steroid.

Kinetic continuous monitoring or rate reaction an assay in which measurements are determined at a series of defined time intervals.

Kinetic fixed time or end point reaction an assay in which the reaction is stopped and the amount of product estimated after a set length of time.

Kringle the name given to a structural motif in proteins, so called because its shape evokes that of a Danish cake of the same name.

Kussmaul respiration a deep rapid respiration typically occurs in uncontrolled type 1 diabetes.

Kyphosis is abnormal curvature of the spine causing a hunchback type of appearance.

Labelling the attachment of a tag to a molecule in order to detect its presence in a reaction. It was widely used in immunoassays where the radioactive element was Iodine^{125} but a wide range of other labels are now used including enzymes, such as alkaline phosphatase and chemiluminescent molecules such as acridinium esters.

Laboratory information system a database that holds all the laboratory diagnostic information, which is interfaced to the laboratory analysers and to the hospital information system.

Lactase deficiency inability to digest lactose due to deficiency of the enzyme lactase.

Lactation secretion of milk from mammary glands after childbirth.

Lacteals lymphatic vessels that absorb dietary fats in the villi of the jejunum.

Lactic acidosis a condition caused by accumulation of lactic acid in the blood thereby lowering its pH (acidosis).

Lamina propria a thin layer of loose connective tissue which lies beneath the epithelium, and together they constitute the mucosal layer.

Laparoscopic adjustable gastric banding a surgical treatment for obesity which can improve blood glucose control.

Laxative a substance that causes emptying of the bowels.

Leucine an amino acid which can stimulate insulin secretion.

Ligand a molecule or molecular group capable of recognizing and binding to another molecule or group to form a larger complex.

Limited joint mobility a form of arthritis found in type 1 diabetics affecting the hand. It is characterized by limited movements of fingers and tightening and thickening of skin on palms.

Lipaemia the presence of lipids and in particular triacylglycerols in the plasma producing a cloudy effect.

Lipase an enzyme that hydrolyses a fatty acid ester bond.

Lipids a group of organic molecules that are insoluble in water but soluble in non-polar solvents such as ether. Lipids are key components of cell membranes.

Lipodystrophy fat deposition in the muscle.

Lipolysis the metabolic breakdown of lipids.

Lipoprotein lipase an enzyme which breaks down lipids in lipoproteins. It is stimulated by insulin and may play a role in insulin resistance.

Lipoproteins a particle composed of lipids and proteins that transports fats in the body.

Lipoprotein-X an abnormal lipoprotein found in serum of patients with obstructive jaundice, and is an indicator of cholestasis. In patients with familial LCAT deficiency, there is an inverse relationship between plasma Lp-X levels and LCAT activity.

Liquid chromatography a chromatographic method used to separate out molecules using a liquid mobile phase.

Low density lipoprotein the lipoprotein that transports most cholesterol in the blood. High levels cause atherosclerosis and increase cardiovascular risk.

Lyase enzyme that catalyses the breaking of a carbon to carbon bond.

Lymphangiectasia dilation of lacteals of villi with reduced lymph flow.

Lymphoma type of cancer that originates in lymphocytes of the immune system.

Lyonization is the process which causes all X chromosomes of the cells in excess of one to be inactivated on a random basis.

Macroamylasemia high plasma amylase activity due to reduced clearance by the kidneys.

Macrocytic anaemia anaemia characterized by larger than normal sized erythrocytes.

Macronutrients nutrients required in large amounts of a gram or more per day for normal body function.

Major histocompatability complex the most gene-dense part of the genome found on chromosome 6. It codes for proteins involved in the immune and autoimmune system. This area also codes for the expression of a large number of cell surface proteins.

Malabsorption abnormal, often inadequate absorption of nutrients from the GIT.

Malignant a tumour that can invade and destroy the tissue in which it originates and can spread to distant sites in the body.

Malnutrition poor nutrition arising from either inadequate or excessive intake of one or more nutrients.

Malpighian body capsule containing the Bowman's capsule and glomerulus at the end of the nephrons.

Maltase a disaccharidase enzyme that converts maltose to glucose.

Manhattan median is the intersection for the median lines parallel to the x and y axis in the Youden plot.

Mass spectrometry a technique that measures the mass/charge ratio of the ions produced when a molecule or atom is ionized, vaporized, and introduced into a vacuum. It is used to determine chemical structures of molecules.

Mastopathy hardening of the breast tissue, more common in patients with diabetes.

Maturity onset diabetes in the young several forms of relatively mild diabetes caused by mutations in different hepatocyte nuclear factor genes.

Medulla the inner region of the adrenal gland that secretes adrenaline.

Megaloblastic anaemia arises due to a deficiency of vitamin B_{12} or folic acid and characterized by the presence of numerous large immature and dysfunctional cells or megaloblasts in the bone marrow.

Melanocyte stimulating hormone peptide cleaved from POMC that stimulates pigment (melanin) synthesis in skin cells.

Menkes disease also called the kinky hair disease is a genetic disorder due to copper deficiency. Patients suffer from mental retardation, brittle fragile hair, poor vision, and seizures.

Menopause this is a stage in a woman's life when the ovaries can no longer produce eggs.

Mesangial cells these are connective tissue cells of the glomerulus.

Metabolic relating to metabolism or metabolic reactions.

Metabolic acids are derived from sources other than carbon dioxide and are not excreted by the lungs. Usually produced from incomplete metabolism of carbohydrates, fats, and proteins.

Metabolic bone disease refers to a number of diseases characterized by defective bones due to abnormalities of minerals such as calcium, phosphate, and magnesium.

Metabolic syndrome a group of conditions that occur together with patients presenting with obesity, insulin resistance (or diabetes), hypertension, and hyperlipidaemia.

Metabolism the chemical reactions that occur in cells to maintain life.

Metanephrine also called metadrenaline, a metabolite of adrenaline.

Metastatic tumour that has spread to a site in the body distant from its origin.

Metformin a biguanide drug used to improve insulin sensitivity at target tissues and to reduce glucose production by the liver.

Methyl succinate a mitochondrial fuel substrate and an insulin secretagogue.

Metyrapone inhibitor of 11-hydroxylase.

Micelles are formed when bile salts surround digested fats so they can be absorbed by cells lining the small intestine.

Microalbuminuria an increase in the urinary excretion of albumin, which in patients with diabetes is an early sign of kidney disease.

Microcephaly a congenital condition where affected individuals have an abnormally small head in relation to the rest of the body.

Microheterogeneity heterogeneous on a very small scale.

Micronutrients nutrients required in only small amounts of less than one gram per day for normal body function.

Microsomal triacylglycerol transfer protein a protein which plays a central role in lipoprotein assembly.

Microvilli tiny projections found on the surface of cells of the villi in the small intestine. Their presence increases the surface area for absorption of nutrients.

Milk-alkali syndrome a rare condition characterized by hypercalcaemia and alkalosis that arises due to excess consumption of milk and alkaline antacids.

Mineralocorticoid a steroid hormone that acts to retain sodium ions and excrete potassium and hydrogen ions at target sites.

Miosis excessive contractions of the pupils of the eye.

Mixed dyslipidaemia a combination of high blood LDL cholesterol and triacylglycerols, but low HDL cholesterol that is often found in patients suffering from diabetes mellitus or the metabolic syndrome.

Mixed function oxidases enzymes that catalyse oxidation-reduction reactions where one atom of oxygen is incorporated into the substrate and the other atom of oxygen undergoes reduction and combines with hydrogen to form water.

Mobilferrin protein carrier for iron in the gut epithelium.

Molar absorptivity also called the molar extinction coefficient, is a measure of how strongly a substance absorbs light at a specified wavelength. A relatively small concentration change in a substance with a high molar absorptivity, such as NADH and p-nitrophenol, will result in a large change in absorbance that can easily be monitored.

Monitoring the use of diagnostic testing to track changes in a disease condition.

Monoacylglycerols glycerides where one fatty acid chain is covalently bound to a glycerol molecule through an ester linkage.

Monoamine oxidase catalyses oxidative deamination of monoamines where oxygen is used to remove an amine group. Monoamine oxidases are used in breakdown of catecholamines.

Monoclonal gammopathies disorders characterized by abnormal synthesis of a single antibody.

Monoclonal usually refers to antibodies produced from a single clone of cells and consisting of identical antibody molecules.

Monogenic disease caused by a mutation in a single gene.

Mucins high molecular weight, highly glycosylated proteins secreted onto mucosal surfaces.

Multidrug resistance ability of cells, especially microorganisms, to evolve so that they can resist many different drugs used against a particular disease.

Multiple endocrine neoplasia pituitary tumours also involving parathyroid gland and pancreas.

Multiple myeloma cancer of the bone marrow where there is uncontrolled proliferation of plasma cells.

Multiple sclerosis autoimmune disease where the immune system attacks its own central nervous system causing demyelination of nerves.

Munchausen's syndrome a mental condition where sufferers repeatedly try to obtain treatment for an illness they don't have.

Mutation is a change in the genetic material of a cell.

Mydriasis excessive dilation of pupils of the eye due to drugs or disease.

Myocardial infarction death of heart tissue due to blockage of arteries supplying blood to the heart.

Myxoedema is the clinical condition characterized by underactivity of the thyroid gland and reduced release of thyroid hormones.

Nateglinide a meglitinide drug that lowers blood glucose by stimulating the release of insulin from the pancreas.

National Service Framework for Diabetes a commitment by the NHS to improve and harmonize the access to and quality of care for patients with diabetes.

Natriuretic peptides are molecules that promote increased sodium and water loss by the kidneys.

Necrobiosis lipoidica a patchy degeneration of the elastic and connective tissue of the skin, most often on the lower shins and usually associated with diabetes.

Necrosis cell death due to injury or disease.

Neonatal apnoea temporary cessation of breathing in neonates.

Neonatal jaundice yellow discolouration of the skin and other tissues in a newborn infant.

Nephritis inflammation of the kidney with focal or diffuse proliferation or destructive processes that may involve the glomerulus, tubule, or interstitial renal tissue.

Nephrocalcinosis deposition of calcium salts in the kidneys that often result in their reduced function.

Nephrolithiasis the formation of kidney stones.

Nephron the anatomical and functional unit of the kidney, consisting of the renal corpuscle, the proximal convoluted tubule, the descending and ascending limbs of Henle's loop, the distal convoluted tubule, and the collecting tubule.

Nephropathy damage to or disease of the kidney, a chronic complication of diabetes.

Nephrosclerosis a disease characterized by hardening of the arteries and arterioles in the kidneys.

Nephrotic syndrome general name for a group of diseases involving defective kidney glomeruli, characterized by massive proteinuria and lipiduria with varying degrees of oedema, hypoalbuminaemia, and hyperlipidaemia.

Nephrotoxicity toxicity to the kidneys.

Neuroblastoma neuroendocrine tumour in children.

Neurocrine relates to molecules released from nerve cells.

Neuroglycopenia inadequate glucose supply to the brain.

Neuropathy a disease or abnormality of the nervous system, pathological change in structure and function of peripheral neurones, a chronic complication of diabetes.

Neurotransmitter chemical that transfers a signal from a neuron to another cell, e.g. acetylcholine.

Neutral endopeptidase an enzyme found mainly in the kidney which can break down active GLP-1.

Neutropenia a decrease in the number of neutrophils in the blood.

Nicotinamide adenine dinucleotide, reduced (NADH) is used alternately with NAD as a reducing or oxidizing agent, respectively, in metabolic reactions.

Night blindness an inability to see clearly in low light conditions due to a deficiency of vitamin A.

Nitric oxide synthesized by nitric oxide synthase it relaxes smooth muscle especially within vascular walls.

Non-esterified fatty acid a free fatty acid as opposed to one esterified with glycerol.

Non-islet cell tumour hypoglycaemia an uncommon complication of malignancy that arises due to overproduction of insulin-like growth factor-2 (IGF-2), which results in stimulation of insulin receptors and increased utilization of glucose causing hypoglycaemia.

Non-steroidal anti-inflammatory drug one that is not steroid in nature and used to treat inflammation.

Noradrenaline a neurotransmitter and hormone released in response to low blood pressure and stress.

Normal fasting glycaemia normal blood glucose concentrations during fasting.

Normetanephrine metabolite of metanephrine.

Nourishment provision of substances that support or maintain life.

Nucleoside compounds consisting of a sugar covalently joined to a nitrogenous base.

Nucleotides phosphorylated nucleosides.

Nutrient a substance that is metabolized by the body to provide energy and for building tissues.

Nutrition the study of food in relation to its absorption and usage by the body.

Nystagmus rapid involuntary movements of eyeballs.

Obesity excess body fat accumulation with a BMI greater than 30 kg/m^2.

Obstructive jaundice increased serum bilirubin concentration caused by the obstruction of the flow of bile into the duodenum.

Occupational exposure is exposure to substances, e.g. chemicals or radiation, etc. while at work.

Oedema swelling in tissues due to accumulation of fluid.

Oligospermia less than 20 million sperms per mL of semen often with poor motility.

Oliguria reduced urine production.

Omental pertaining to the omentum, a large fatty structure that drapes over the intestines like an apron.

Oncotic pressure osmotic pressure due to plasma proteins (mainly albumin) that is required to keep fluid within blood vessels.

Oral glucose tolerance test a test to measure the body's ability to break down and use carbohydrate. It is used to diagnose diabetes and prediabetic conditions such as impaired fasting glycaemia or impaired glucose tolerance.

Osmolality a measure of the osmotic concentration, an indicator of hydration and fluid balance, usually expressed as moles of solute per kilogram of solvent or mOsm/kg.

Osmolar gap the difference between the measured and the calculated osmolality.

Osmometer an instrument used to measure osmolality.

Osmoreceptors a group of cells in the hypothalamus that monitor changes in blood osmolality.

Osmotic diuresis loss of fluids, electrolytes, and glucose in the urine of patients with uncontrolled diabetes.

Osteoblasts cells that make bone tissue.

Osteoclasts cells that break down bone releasing minerals such as calcium.

Osteoid organic matrix consisting of proteins and polysaccharides and secreted by osteoblasts; converted to bone after mineralisation.

Osteolytic lesions areas of bone loss that can be seen as holes on X-rays and usually caused by infiltration of cancer cells into the bone.

Osteomalacia softening of the bones in adults due to vitamin D deficiency.

Osteoporosis a condition characterized by a fall in the bone mass and density, causing the bones to become fragile and increases their susceptibility to fracture.

Ovulation refers to release of an ovum from a mature Graafian follicle.

Oxidation the addition of oxygen to molecules.

Paget's disease a metabolic bone disease where there is bone destruction followed by abnormal regrowth resulting in bone deformity.

Pancreatectomy surgical removal of the pancreas.

Pancreatic polypeptide an islet hormone secreted by the PP cell. Its function remains to be fully elucidated.

Pancreatitis inflammation of the pancreas, which may be life threatening.

Paraprotein immunoglobulin produced by proliferation of plasma cells in bone marrow of patients with multiple myeloma. They form a discrete band following electrophoresis as they are all exactly the same immunoglobulins.

Parathyroidectomy surgical excision of the parathyroid gland.

Parenteral feeding a procedure where nutrients are taken into the body and enter the blood directly without having to go through the digestive system, e.g. intravenously or entry of nutrients via a vein.

Parietal cell gastric epithelium cells that secrete gastric acid and intrinsic factor.

Parotitis inflammation of one or both parotid glands.

Pathology network describes a group of laboratories working together to share best practice and provide cost-effective patient care.

Patient identification demographics information that is unique to that individual and is stored on the HIS system. The details held include full name, date of birth, sex, address, and hospital number of the patient.

Pellagra a disease that occurs due to a deficiency of niacin where the patients present with dermatitis, dementia, and diarrhoea.

Peri-anal around the anus.

Perilipin a protein that coats lipid droplets stored in adipocyte cells in adipose tissue. It protects these lipids from lipases capable of digesting the triacylglycerols.

Peri-oral around the mouth.

Peristalsis rhythmic contraction of smooth muscles to propel contents through the digestive tract.

Peritoneal dialysis a process for removal of waste from the body using a peritoneal membrane in the abdomen as a filter.

Pernicious anaemia a form of megaloblastic anaemia due to vitamin B_{12} deficiency, the latter often caused by a lack of intrinsic factor required for its absorption.

Peroxisome proliferator-activated receptor γ a type 2 nuclear receptor which regulates fatty acid storage and glucose metabolism and improves insulin sensitivity. PPARγ is a target for several anti-diabetic drugs.

Peroxynitrite an oxidizing and nitrating agent that can damage biomolecules, such as DNA and proteins in cells.

Persistent hyperinsulinaemic hypoglycaemia of infancy a persistent form of congenital hyperinsulinism, normally caused by a mutation in the gene encoding the K_{ATP} channel.

Pessaries drugs that are delivered by insertion into the vagina.

Pesticides chemicals used to kill pests such as insects and rodents.

Petechiae small red spots on the skin that occur due to leakage of blood from blood vessels.

Phaeochromocytoma a rare tumour, usually of the adrenal gland, causing too much release of adrenaline and noradrenaline resulting in increased heart rate and high blood pressure.

Pharmacokinetics study of absorption, distribution, degradation, and elimination of drugs in the body.

Phenylketonuria an inherited disorder where there is an inability to metabolize the amino acid phenylalanine due to complete or partial deficiency of the enzyme phenylalanine hydroxylase.

Phospholipids are lipids made up of glycerol and fatty acids, with at least one phosphate group attached.

Phosphorylated reacted or combined with a phosphate group.

Phytates also known as inositol hexakisphosphates and are the main storage form of phosphorus in plant tissues.

Pinocytosis is a process whereby soluble molecules are internalized from outside the cell through the formation of vesicles.

Pioglitazone a drug with antihyperglycaemic effects belonging to the thiazolidinedione class of drugs.

Plasmacytoma cancer of plasma cells in the bone marrow.

Plasmapheresis is a procedure in which plasma is removed from donated blood, whereas the remaining components, i.e. largely red blood cells, are returned to the donor.

Plasminogen is the inactive precursor of plasmin, a fibrin-degrading enzyme important in the control of clot formation.

Pneumatic tube system transports samples in protective pods, through a network of plastic pipes to various destinations throughout the hospital.

Podocytes epithelial cells that are attached to the outer surface of the glomerular capillary basement membrane forming foot processes.

Point of care testing a term employed to describe analyses that are undertaken outside the laboratory setting usually at a patient's bedside or in a clinic or ward side room.

Poison a substance that can cause injury, illness, or death of a living organism.

Polyarthritis any type of arthritis where five or more joints are affected.

Polyclonal usually refers to a mixture of antibodies against an antigen, where each antibody recognizes a different epitope of the antigen.

Polycystic kidney disease a disease which causes renal failure characterized by large kidneys containing multiple cysts.

Polycystic ovary syndrome a condition that occurs in females, characterized by follicles that fail to ovulate and remain as multiple cysts in the ovaries.

Polydipsia excessive drinking, a symptom of untreated diabetes.

Polygenic a disease attributable to the effects of several genes and their interaction with external factors, such as lifestyle.

Polyunsaturated refers to lipids with multiple double bonds in their structure.

Polyuria excessive urination, a symptom of untreated diabetes.

Porphyrias a group of rare disorders that arise due to deficiencies in the enzymes involved in the biosynthesis of haem.

Post-analytical step a procedure that is required after the actual analytical process is finished. Examples include the physical process of storing or archiving the specimen, alternatively it may be knowledge based and related to interpretation of results.

Potentiators often refer to drugs that enhance the activity of other drugs.

PP cell an islet cell that secretes pancreatic polypeptide.

Pramlintide an analogue of amylin which is a small peptide hormone released into the blood by beta cells of the pancreas along with insulin after a meal.

Prandial glucose regulators a group of short-acting drugs that stimulate the β cell to secrete insulin and are taken at mealtimes. Repaglinide is an example.

Pre-analytical step a procedure that is required prior to the actual analytical process. Examples include centrifugation of whole blood to obtain the plasma or serum when a process requires a cell-free solution.

Precision describes the ability to reproduce the same result consistently.

Precocious puberty early signs of puberty such as breast development in girls; increased size of penis and testes in boys.

Pre-eclampsia development of hypertension in a female during pregnancy.

Pregnane steroid with 21 carbon atoms.

Pregnanetriol the main metabolite of 17-hydroxyprogesterone.

Pregnenolone intermediate in steroid synthesis formed from cholesterol by the action of the side chain cleavage enzyme.

Premature adrenarche pubic hair before the age of six years in girls and seven in boys.

Preproinsulin the biosynthetic precursor of proinsulin.

Primary aldosteronism aldosterone excess state. Renin activity will be suppressed.

Primary biliary cirrhosis an autoimmune liver disease characterized by progressive destruction of the bile canaliculi within the liver.

Primary immunodeficiencies disorders where part of the immune system is absent or not functioning properly.

Primary sample the tube utilized for the collection of the specimen from the patient that may be taken through the analytical process without a requirement to transfer into a secondary storage vessel.

Primary sclerosing cholangitis a chronic inflammatory disease affecting the liver bile ducts and obstructing flow of bile into the gut.

Primary tumour tumour that is at the site from which it first developed.

Prodrugs inactive drugs that only convert to their active form once within the body.

Progesterone a steroid hormone whose function is to prepare the endometrium for pregnancy.

Prognathism is enlargement of one jaw (usually the bottom jaw).

Prognosis the predicted clinical outcome of a disease.

Proinsulin the biosynthetic precursor of insulin.

Prolactinoma benign tumour of the pituitary that secretes large amounts of prolactin.

Proopiomelanocortin a neuropeptide that can be processed in the pituitary to release active peptides such as ACTH and MSH.

Prophylactic drug that is used to protect or defend against disease, as opposed to its treatment.

Protein-energy malnutrition a form of malnutrition due to inadequate intake of proteins and calories.

Proteins organic compounds made from amino acids linked together by peptide bonds. Proteins are required for growth and development.

Proteinuria excess protein in the urine.

Prothrombotic promotes formation of thrombin during blood clotting.

Proton pump inhibitors drugs which block the the H^+/K^+ ATPase of the gastric parietal cells.

Pseudogout clinical condition characterized by the deposition of crystals of calcium pyrophosphate in synovial fluid.

Pseudohyponatraemia low serum sodium due to an artefactual cause.

Pseudohypoparathyroidism a condition in which body cells are resistant to PTH. Patients therefore have hypocalcaemia or hyperphosphataemia, but the PTH is still high due to the hypocalcaemia.

Psoriasis is a skin disease characterized by thick patches of inflamed red skin covered with thick silvery scales that flake away from the skin.

Purines a group of chemical compounds having a double ring structure consisting of a six-membered pyrimidine ring fused to a five-membered imidazole ring.

Pyelonephritis an inflammation of the kidney and its pelvis as a result of infection.

Pyloric sphincter a ring of smooth muscle (valve) at the end of the pyloric canal leading from the stomach to the duodenum.

Pyloric stenosis is the narrowing (stenosis) of the outlet of the stomach and means that food cannot readily enter the duodenum. Often causes vomiting.

Pyruvate the product of glycolysis which can then undergo further metabolism to produce energy.

Pyuria presence of white blood cells in the urine.

Radioimmunoassay immunoassay where a radioactive label is used.

Random access analyser an analyser able to undertake analysis out of the defined sequence.

Random error usually due to an operator's inability to reproduce a process or part of it consistently, causing measurement to vary either side of the mean and producing scattered results.

Range check examination of an analyte value to determine whether the result is physiologically possible.

Reactive oxygen species are highly reactive forms of oxygen that are capable of damaging biomolecules. They include oxygen-derived free radicals and some compounds capable of forming free radicals.

Readily releasable pool a pool of primed β cell insulin granules which are secreted immediately on influx of calcium.

Reagent logs are records confirming that reagents in use are in date and stored at the correct temperature.

Receptor for advanced glycation end product binding to this receptor induces a proinflammatory state in target cells, especially the vascular endothelium.

Reductase an enzyme that catalyses a reduction such as addition of two hydrogen atoms to a steroid.

Refeeding syndrome a syndrome characterized by metabolic disturbances that arise following feeding or nutritional intervention in individuals who have been starved or have suffered from severe malnutrition.

Reference methods these are methods against which new methods are compared and validated.

Reference range a range of values for an analyte from 95% of healthy individuals that are used to interpret results for patients in which that analyte has been measured.

Renal colic pain often caused by kidney stones.

Renal insufficiency reduced ability of the kidney to perform its functions such as excretion of waste products.

Renal osteodystrophy decreased mineralization of bones caused by hormonal and electrolyte abnormalities of kidney disease.

Renal tubular acidosis a rare condition characterized by accumulation of acid in the body due to failure of the kidneys to lose acid in the urine.

Renin activity rate of production of angiotensin I from angiotensinogen.

Renin concentration concentration of the protein renin.

Renin protein secreted from the kidney into the circulation in response to low sodium or low blood pressure in the kidney. Renin acts on angiotensinogen in the blood to produce angiotensin I. This is the first step in a cascade called the renin-angiotensin-aldosterone axis. An increase in aldosterone production leads to retention of sodium in the kidney tubules.

Repaglinide a blood glucose lowering drug that belongs to the meglitinide class.

Request form a document that accompanies a diagnostic sample providing patient identification, sample details, the tests requested, clinical details, and source of the request.

Resorption refers to the process in bones when osteoclasts break down bone to release calcium that can subsequently enter blood.

Respiratory acids are acids that are exhaled from the lungs.

Respiratory related to the process of breathing or the respiratory system.

Reticularis the inner region of the adrenal cortex that produces androgens.

Reticuloendothelial system part of the immune system and includes phagocytic cells such as macrophages that protect against incoming bacteria and foreign particles.

Retinitis pigmentosa are a group of inherited disorders characterized by abnormalities of the photoreceptors (primarily the rods) in the retina and which often lead to blindness.

Retinopathy disease or damage to the retina of the eye, a chronic complication of diabetes.

Rhabdomyolysis breakdown of skeletal muscle due to injury. There is a release of myoglobin, which may cause renal failure.

Rheumatoid arthritis a systemic autoimmune inflammatory disease characteristically affecting the joints.

Rickets insufficient hardening of the bones, usually of the limbs, due to inadequate intake of vitamin D.

Rosiglitazone a drug that lowers blood glucose and belongs to the thiazolidinedione class of drugs.

Roux-en-Y bypass a procedure whereby the size of the stomach is reduced and a part of the small intestine is bypassed.

Saint Vincent Declaration a formal commitment, signed in Italy in 1989 by European nations, to improve the quality of care of patients with diabetes.

Satiety feeling full after eating food.

Saturated refers to lipids that have no double bonds in their structure.

Scurvy a disease that arises due to vitamin C deficiency where patients present with bleeding gums and a rash of tiny bleeding spots around their hair follicles.

Secondary tumour tumour that has developed as a result of metastasis from the original tumour site.

Selectivity extent to which a particular method can be used to determine an analyte under given conditions in the presence of other similar analytes.

Shift is when six or more results occur on one side of the mean, rather than scattered about it as in a trend.

Sitagliptin an oral antihyperglycaemic drug that belongs to the dipeptidyl peptidase-4 (DPP-4) inhibitor class.

Sitosterolaemia is an inherited condition where sitosterols or plant sterols cannot be broken down and build up in the blood causing high cholesterol levels and increase the likelihood of xanthomas.

Sitosterols plant derived sterols with a similar structure to cholesterol.

Somatostatin an islet hormone, secreted by the δ cell, which exerts inhibitory effects on other islet cells.

Somatostatinoma a tumour secreting somatostatin.

Specimen dead volume the minimum volume of specimen required on an analyser to allow a probe or other sampling mechanism to remove the specimen required for the analysis of the test.

Specimen matrix a term used to describe the general characteristics of a specimen which may influence the behaviour of the material during its analysis.

Spermatogenesis the process of sperm production in the testes.

Sphincter of Oddi a muscular valve controlling flow of bile/pancreatic juice from the common bile duct into the duodenum.

Spironolactone inhibitor of the action of aldosterone.

Standard bicarbonate is the calculated concentration of bicarbonate in a blood sample corrected to a PCO_2 of 5.3 kPa (40 mmHg). That is, it is the expected bicarbonate concentration if the PCO_2 is 'normal'.

Standard operating procedures documents that describe in detail, exactly how a measurement is performed to give a valid result.

Starvation a state of severe hunger due to prolonged and severe reduction in intake of nutrients.

Steady-state when the rate of administration of a drug matches its rate of elimination.

Steatorrhoea presence of excess fat in faeces; stools may also float, and have an oily appearance.

Steatosis abnormal retention of fats within cells.

Steroidogenic acute regulatory protein a protein in the mitochondrial wall that regulates uptake of cholesterol.

Steroidogenic factor 1 is a regulator of adrenal development.

Strong acid this ionizes entirely to form H^+ ions in solution.

Strong ion difference is the difference in mmol/L between the fully dissociated cations and anions in the plasma.

Subcutaneous below the skin.

Sublingual below the tongue.

Sub-optimal less than optimum.

Substrates the substances upon which an enzyme acts. Substrates can be natural or artificial.

Sub-therapeutic refers to drug doses given at concentrations lower than those used in treatment of disease.

Sucrase-isomaltase glucosidase enzyme found in brush border and whose deficiency causes sucrose intolerance.

Sulphatase enzyme that hydrolyses the sulphate group from a sulphated steroid.

Sulphate steroids when conjugated with sulphuric acid which increases their water solubility and aids their renal excretion.

Sulphonylurea receptor 1 a receptor on β cell membranes that is activated by sulphonylurea drugs such as tolbutamide and is intimately linked to the activity of the K_{ATP} channel such that binding of sulphonylurea to this receptor closes the K_{ATP} channel.

Sulphonylureas stimulate insulin release from β cells by their action on the SUR1 receptor and closure of K_{ATP} channels. Tolbutamide, gliclazide, glipizide, and glimepiride are sulphonylureas.

Sulphotransferase an enzyme that adds a sulphate group to a steroid hydroxyl group to raise its water solubility and increase renal excretion.

Suppositories drugs that are delivered by insertion into the rectum.

Synacthen a synthetic peptide of the amino acids 1-24 of ACTH.

Syndactyly where two or more digits are fused together, e.g. fingers.

Synovial fluid the liquid lubricant secreted into the synovial cavities of joints.

Systemic clearance overall clearance of a drug from the body, often represented as the sum of all organ clearances.

Systematic error produces measurements differing from the true value by a constant amount in the same direction.

Systemic lupus erythematosus is a multisystem, inflammatory disease of autoimmune origin that can affect the skin, joints, and kidneys.

Systemic sclerosis an autoimmune disease causing scarring and fibrosis in the skin and other organs causing thickness or firmness. In some patients it affects only face and fingers but in others can affect tissues throughout the body.

T cell mediated immune response from T cells (does not involve antibodies).

Tachycardia an increase in the heart rate above normal.

Tamm-Horsfall glycoprotein a monomeric glycoprotein (68 KDa) not derived from plasma but produced by the loop of Henle and present in the urine in large aggregates of several million Da.

Tenosynovitis inflammation of the fluid filled sheath (synovium) surrounding a tendon.

Test:request ratio the average number of tests undertaken on each sample within a laboratory over a defined period of time.

Testosterone active male hormone.

Tetany involuntary muscle contractions due to calcium deficiency.

Tetrahydro-steroid a steroid metabolite where four hydrogen atoms have been added to the structure of the steroid hormone.

Therapeutic drug monitoring measurement of plasma drug concentrations in order to provide effective dosages with minimal toxicity.

Therapeutic range is the plasma concentration range for a drug over which it shows clinical benefit with minimal toxicity for the vast majority of patients.

Therapeutic window the concentration range of a drug over which it has a therapeutic effect but minimal toxicity.

Thiazide diuretics a group of compounds that decrease reabsorption of sodium by the kidneys, promoting loss of sodium in the urine and therefore water. They enhance excretion of sodium and chloride equally.

Thiazolidinediones also called glitazones, bind to PPARγ to increase insulin sensitivity. Rosiglitazone and pioglitazone are thiazolidinedione drugs.

Thrombolytic therapy use of medication that will break up blood clots.

Thrombus a blood clot within a blood vessel that often obstructs blood flow.

Thyroglobulin is a globular glycoprotein found in the thyroid gland that acts as a precursor for thyroid hormones. It contains tyrosine residues that become iodinated when synthesizing thyroid hormones.

Thyroidectomy partial or complete surgery (removal) of the thyroid gland.

Thyrotoxicosis the clinical condition characterized by overactivity of the thyroid gland and excessive release of thyroid hormones.

Thyrotrophs are cells in the anterior pituitary that can release TSH.

Thyroxine binding globulin is a globular protein that binds and transports thyroid hormones in the circulation.

Tolbutamide a sulphonylurea drug used in the treatment of diabetes.

Total laboratory automation describes a system that automates most of the common laboratory processes including sample centrifugation, analysis, storage, and disposal.

Toxicity the degree to which a substance is able to cause damage to a living organism.

Toxicology the study of poisons.

Toxins poisonous substances produced by living cells or organisms.

Trabecular the inner spongy part of a bone.

Trace elements these are minerals required in very small amounts, often mg quantities, in the diet for normal body function.

Training log books are records that are kept to confirm an operator has been trained in the use of an instrument and/or in performing a test.

Transcobalamin carrier protein for vitamin B_{12} in blood and required for transport to tissues.

Transcription factor a protein that binds to a specific DNA sequence to influence generation of an RNA for protein synthesis.

Transferrin a globular protein in blood that can transport iron.

Treatment the application of medicines and therapies to a disease condition.

Trend this is when six or more results show a consecutive move in the same direction, whether up or down from the mean.

Triacylglycerols are glycerides in which the glycerols are esterified with three fatty acids. They are found in vegetable and animal fats. Also known as triglycerides.

Tropical sprue malabsorption disease common in tropical regions and characterized by abnormal flattening of the villi. Has an unknown aetiology and patients present with acute diarrhoea, fever, and malaise.

Trousseau's sign is a spasm of the hand and forearm that occurs on pressing the upper arm and is an indication of hypocalcaemia.

Tumour abnormal mass of tissue that results from uncontrolled cell growth.

Tumour necrosis factor a cytokine produced by monocytes and macrophages that can cause inflammation but also has antineoplastic effects.

Turnaround time this represents the time taken from sample collection to the clinician receiving the result.

Type 1 diabetes a severe form of diabetes characterized by the autoimmune destruction of the β cells of the pancreatic islet, usually occurring in childhood/teenage years.

Type 2 diabetes a form of diabetes characterized by severe insulin resistance and a refractoriness of the endocrine pancreas to secrete insulin. Usually brought about in middle age due to lack of exercise and a diet too rich in carbohydrate and fat.

Tyrosinaemia a metabolic disorder characterized by high blood tyrosine concentrations.

Tyrosine hydroxylase the precursor molecule for the catecholamine neurotransmitters dopamine and noradrenaline, and the hormone adrenaline.

Ulcerative colitis a disease characterized by chronic inflammation of the colon.

Ulcers breaks in the lining of the GIT that fail to heal and are often accompanied by inflammation.

Ultrafiltration a method of separating macromolecules such as proteins from smaller molecules by filtration through a membrane, often under pressure.

Unsaturated refers to lipids that contain at least one double bond within their fatty acid chain.

Uraemia an excess in the blood of urea, creatinine, and other nitrogenous end products of protein and amino acid metabolism; often referred to as azotaemia.

Urate the salt derived from uric acid that accumulates in body tissues and may crystallize within joints as monosodium urate.

Urea is the water-soluble end product of the metabolic pathway in the liver of the breakdown of proteins and amino acids.

Ureagenesis formation of urea from the metabolism of amino acids.

Uric acid the secretory product of purine metabolism in humans.

Uricolysis the breakdown of uric acid to carbon dioxide and ammonia.

Uricosuric drug one that lowers the plasma concentration of uric acid.

Urinary steroid profile chromatographic separation of steroids. From the pattern of steroid elution with time (profile) the components can be identified. Profiles are characteristic of adrenal disorders. There can be low production of enzyme products with high levels of substrates.

Urine the sterile, clear fluid excreted by the kidneys, passed through the ureters, stored in the bladder, and discharged through the urethra.

Urolithiasis formation of renal stones composed of urates in patients suffering from prolonged hyperuricaemia.

Vagus nerve one of the cranial nerves that originates in the brain stem and innervates the viscera.

Vanilla a flavour chemical with a structure similar to catecholamines.

Vanillylmandelic acid a product of the metabolism of catecholamines.

Ventilation rate at which gases enter or leave the lungs.

Very low density lipoprotein large lipoproteins rich in triacylglycerols which circulate through the blood giving up their triacylglycerols to muscle and adipose tissue. The particle remnants are then converted to LDL.

Vildagliptin an oral anti-hyperglycemic agent that belongs to the dipeptidyl peptidase-4 (DPP-4) inhibitor class of drugs.

Villi tiny, finger-like projections that come out from the wall of the small intestine.

Virilism the development of male characteristics in a female such as body hair, muscular bulk, and deepening of the voice.

Vitamins organic compounds used as nutrients in small amounts for normal cellular function.

Voltage-sensitive calcium channel a channel involved in glucose induced insulin secretion from the pancreatic β cell.

Volume of distribution fluid volume in which drugs are distributed in the body.

Volume-regulated anion channel a channel thought to be involved in glucose-induced insulin secretion from the pancreatic β cell.

Waist hip ratio a ratio of the circumference of the waist (measured around the navel) to the circumference of the hips (measured at the iliac crest). It gives a measure of fat accumulated around the abdomen and is thought to be a more sensitive measure of risk associated obesity.

Waldenström's macroglobulinaemia a rare cancer affecting white blood cells, called B lymphocytes, causing them to proliferate principally in lymph nodes and spleen and produce large amounts of IgM.

Weak acid these ionize only partially to form H^+ ions in solution.

Wernicke-Korsakoff syndrome a disorder characterized by a degenerative brain and paralysis of the eye muscles. It occurs due to lack of vitamin B_1.

Whipple's triad a collection of three criteria that suggest a patient has hypoglycaemia. (1) the symptoms are known or likely to be caused by hypoglycaemia; (2) a low plasma glucose at time of symptoms; and (3) relief of symptoms when glucose is given.

Wilson's disease an inherited disorder of copper metabolism where free copper deposits in the liver causing cirrhosis or in the brain causing mental retardation.

Within laboratory turnaround time is the time between the sample being received in the laboratory to result production.

Workcell is a logical and productive grouping of analysers, tools, and personnel to maximize the efficiency and effectiveness of the analytical processes.

Xanthomas cholesterol deposits in the tendons.

Xenobiotics are exogenous compounds that may be toxic to the body and are rendered safe by metabolism in the hepatocytes.

Zellweger syndrome a genetic disorder characterized by reduction or absence of peroxisomes in the cells of the brain, liver, and kidneys.

Zero-order kinetics when the *rate* of elimination of drug is constant, i.e. not proportional to the plasma concentration.

Zieve's syndrome an acute metabolic syndrome that occurs following withdrawal of alcohol after prolonged use. Affected patients have haemolytic anaemia, hyperlipoproteinaemia, jaundice, and abdominal pain.

Zinc an element that is co-secreted with insulin from the β cell and is thought to activate α cell K_{ATP} channels which reduce glucagon secretion.

Zollinger Ellison syndrome a condition characterized by severe and multiple ulceration in the GIT due to excessive production of gastric juice in the stomach. Caused by a tumour secreting large amounts of gastrin.

Index